Sleep and Sleep Disorders:
A Neuropsychopharmacological Approach

Sleep and Sleep Disorders:
A Neuropsychopharmacological Approach

Malcolm Lader
King's College London
University of London
Institute of Psychiatry at the Maudsley
London, U.K.

Daniel P. Cardinali
Departamento de Fisiología
Facultad de Medicina
Universidad de Buenos Aires
Buenos Aires, Argentina

S.R. Pandi-Perumal
Comprehensive Center for Sleep Medicine
Department of Pulmonary, Critical Care and Sleep Medicine
Mount Sinai School of Medicine
New York, New York, U.S.A.

Landes Bioscience / Eurekah.com
Georgetown, Texas
U.S.A.

Springer Science+Business Media
New York, New York
U.S.A.

Sleep and Sleep Disorders:

A Neuropsychopharmacological Approach

LANDES BIOSCIENCE / EUREKAH.COM
SPRINGER SCIENCE+BUSINESS MEDIA, INC.

ISBN: 0-387-27681-5 Printed on acid-free paper.

Springer Science+Business Media, Inc., 233 Spring Street, New York, New York 10013, U.S.A.
http://www.springeronline.com

Please address all inquiries to the Publishers:
Landes Bioscience / Eurekah.com, 810 South Church Street, Georgetown, Texas 78626, U.S.A.
Phone: 512/ 863 7762; FAX: 512/ 863 0081
http://www.eurekah.com
http://www.landesbioscience.com

Printed in the United States of America.

9 8 7 6 5 4 3 2 1

COVER image: The publishers wish to thank Dr. Steve Frucht and Lorin Bernsohn for the use of their artwork within our cover image.
Reprinted with permission from Frucht SJ, Bernsohn L. Visual Hallucination in PD, Neurology 2002: 59,1965.

Library of Congress Cataloging-in-Publication Data

Sleep and sleep disorders : a neuropsychopharmacological approach
/ [edited by] Malcolm Lader, Daniel P. Cardinali, S.R. Pandi-Perumal.
p. ; cm.
Includes index.
ISBN 0-387-27681-5 (alk. paper)
1. Sleep--Physiological aspects. 2. Sleep disorders. 3. Neuropharmacology. 4. Chronopharmacology. I. Lader, Malcolm Harold. II. Cardinali, Daniel P. III. Pandi-Perumal, S. R.
[DNLM: 1. Sleep--physiology. 2. Sleep Disorders--drug therapy. 3. Sleep Disorders--psychology. 4. Sleep--drug effects. 5. Sleep Disorders--physiopathology. WL 108 S63033 2005]
QP425.S668 2005
612.8'21--dc22

2005028061

CONTENTS

EDITORS

Malcolm Lader
King's College London
University of London
Institute of Psychiatry at the Maudsley
London, U.K.
Email: spklmhl@iop.kcl.ac.uk

Daniel P. Cardinali
Departamento de Fisiología
Facultad de Medicina
Universidad de Buenos Aires
Buenos Aires, Argentina
Email: cardinal@mail.retina.ar
Chapter 11

S.R. Pandi-Perumal
Comprehensive Center for Sleep Medicine
Department of Pulmonary, Critical Care and Sleep Medicine
Mount Sinai School of Medicine
New York, New York, U.S.A.
Email: pandiperumal@gmail.com
Chapters 7, 33

CONTRIBUTORS

Mehmet Yucel Agargun
Department of Psychiatry
and Neuroscience Research Unit
Yuzuncu Yil University School of Medicine
Van, Turkey
Email: myagargun@kure.com.tr
Chapter 29

Jean-Jacques Askenasy
Department of Neurology
Sackler School of Medicine
Tel-Aviv University
Ramat Aviv, Israel
Email: ajean@post.tau.ac.il/jeanj@bezeqint.net
Chapter 26

Chun Bai
Center for Sleep Disorders
Section of Sleep Medicine
Division of Epilepsy and Clinical Neurophysiology
Department of Neurology
SUNY Downstate Medical Center
Brooklyn, New York, U.S.A.
Email: chun.bai@downstate.edu
Chapter 33

Michel Billiard
Department of Neurology
School of Medicine
Gui de Chauliac Hospital
Montpellier, Cedex, France
Email: mbilliard@wanadoo.fr
Chapter 17

Grethe Støa Birketvedt
Department of Gastromedicine
University Hospital of Tromsø
Tromsø, Norway
Email: gsb42nor@aol.com
Chapter 28

Luis I. Brusco
Departamento de Fisiología
Facultad de Medicina
Universidad de Buenos Aires
Buenos Aires, Argentina
Chapter 11

Germaine Cornélissen
Halberg Chronobiology Center
University of Minnesota
Minneapolis, Minnesota, U.S.A.
Email: corne001@tc.umn.edu
Chapter 7

António Martins da Silva
Department of Neurological Disorders
and Senses
Hospital Santo António
ICBAS/IBMC
University of Porto
Porto, Portugal
Email: martinsdasilva@hgsa.min-saude.pt
Chapter 33

Karl Doghramji
Psychiatry and Human Behavior
Thomas Jefferson University
Philadelphia, Pennsylvania, U.S.A.
Email: karl.doghramji@jefferson.edu
Chapter 20

Jon R. Florholmen
Department of Gastromedicine
University Hospital of Tromsø
Tromsø, Norway
Chapter 28

Nicholas P. Franks
Merck Sharp and Dohme
Harlow, Essex, U.K.
Chapter 6

Analía M. Furio
Departamento de Fisiología
Facultad de Medicina
Universidad de Buenos Aire
Buenos Aires, Argentina
Chapter 11

Franz Halberg
Halberg Chronobiology Center
University of Minnesota
Minneapolis, Minnesota, U.S.A.
Chapter 7

Raed Hawa
Sleep Research Laboratory
University Health Network
Toronto Western Hospital
Toronto, Ontario, Canada
Email: raed.hawa@uhn.on.ca
Chapter 24

James J. Herdegen
Sleep Center
University of Illinois Medical Center of Chicago
Chicago, Illinois, U.S.A.
Email: jherdege@uic.edu
Chapter 18

Babita Kara
Acquired Brain Injury Clinic
Toronto Western Hospital
University Health Network
University of Toronto
Toronto, Ontario, Canada
Chapter 23

Satvinder Kaur
School of Life Sciences
Jawaharlal Nehru University
New Delhi, India
Chapter 5

Leonid Kayumov
Sleep Research Laboratory
University Health Network
Toronto Western Hospital
Toronto, Ontario, Canada
Chapters 16, 24

Esa R. Korpi
Institute of Biomedicine Pharmacology
Biomedicum Helsinki
University of Helsinki
Helsinki, Finland
Email: esa.korpi@helsinki.fi
Chapter 1

Milton Kramer
Maimonides Medical Center
and
School of Medicine
New York University
and
Albert Einstein College of Medicine
New York, New York, U.S.A.
Email: milton1929@yahoo.com
Chapter 14

Daniel F. Kripke
University of Southern California Medical School
La Jolla, California, U.S.A.
Email: dkripke@ucsd.edu
Chapter 15

Velayudhan Mohan Kumar
Department of Physiology
All India Institute of Medical Sciences
Ansari Nagar, New Delhi, India
Chapter 3

Dieter Kunz
Psychiatrische Universitätsklinik der Charité
St. Hedwig Krankenhaus
Humboldt-Universität zu Berlin
Berlin, Germany
Email: dieter.kunz@charite.de
Chapter 9

Moshe Laudon
Neurim Pharmaceuticals Ltd.
Tel-Aviv, Israel
Email: MosheL@Neurim.com
Chapter 25

Cynthia Liberczuk
Departamento de Fisiología
Facultad de Medicina
Universidad de Buenos Aires
Buenos Aires, Argentina
Chapter 11

Alan Lowe
Sleep Research Laboratory
University Health Network
Toronto Western Hospital
Toronto, Ontario, Canada
Chapter 24

Richard Mahlberg
Psychiatrische Universitätsklinik der Charité
St. Hedwig Krankenhaus
Humboldt-Universität zu Berlin
Berlin, Germany
Chapter 9

Birendra N. Mallick
Department of Neurobiology
School of Life Sciences
Jawaharlal Nehru University
New Delhi, India
Email: remsbnm@yahoo.com
Chapter 5

Mervyn Maze
Merck Sharp and Dohme
Harlow, Essex, U.K.
Chapter 6

Melissa Mendez
Department of Neurology
New York University
Comprehensive Epilepsy Center
New York, New York, U.S.A.
Email: MendezMel@aol.com
Chapter 33

Henry J. Moller
Sleep Research Unit
Department of Psychiatry
Toronto Western Hospital
Toronto, Ontario, Canada
Chapter 16

Daniel Monti
Western Psychiatric Institute and Clinic
University of Pittsburgh School of Medicine
Pittsburgh, Pennsylvania, U.S.A.
Chapter 32

Jaime M. Monti
Pharmacology and Therapeutics
Clinics Hospital
Montevideo, Uruguay
Email: jmonti@mednet.org.uy
Chapter 32

Haruo Nagayama
Department of Neuropsychiatry
Faculty of Medicine
Oita University
Hasama-machi, Oita-gun, Japan
Email: nagayama@med.oita-u.ac.jp
Chapter 7

Laura E. Nelson
Merck Sharp and Dohme
Harlow, Essex, U.K.
Email: laura_nelson2@merck.com
Chapter 6

Eugene Ng
Department of Newborn
and Developmental Paediatrics
Health Sciences Centre
Sunnybrook and Women's College
Toronto, Ontario, Canada
Email: eugene.ng@sw.ca
Chapter 19

Lennard P. Niles
Department of Psychiatry and Behavioural
Neurosciences
McMaster University
Hamilton, Ontario, Canada
Email: niles@mcmaster.ca
Chapter 8

David J. Nutt
Psychopharmacology Unit
School of Medical Sciences
University of Bristol
Bristol, U.K.
Chapter 31

Sebastiaan Overeem
Department of Neurology
Radboud University Nijmegen Medical Center
Nijmegen, The Netherlands
and
Department of Neurology
Leiden University Medical Center
Leiden, The Netherlands
Email: s.overeem@neuro.umcn.nl
Chapter 17

Hanefi Ozbek
Department of Pharmacology
and Neuroscience Research Unit
Yuzuncu Yil University School of Medicine
Van, Turkey
Chapter 29

J.F. Pagel
Rocky Mountain Sleep Disorders Center
Pueblo, Colorado, U.S.A.
Email: pueo34@juno.com
Chapter 27

Dinesh Pal
School of Life Sciences
Jawaharlal Nehru University
New Delhi, India
Chapter 5

Hyung Park
Department of Psychiatry
Dartmouth Medical School
Hanover, New Hampshire, U.S.A.
Email: Hyung.Park@dartmouth.edu
Chapter 22

Timo Partonen
Department of Mental Health and Alcohol Research
National Public Health Institute
Helsinki, Finland
Email: Timo.partonen@ktl.fi
Chapter 12

Marisa Pedemonte
Neurofisiologia, Departamento de Fisiologia
Facultad de Medicina
Universidad de la República
Montevideo, Uruguay
Chapter 2

Mirko Petrovic
Department of Gerontology and Geriatrics
Heymans Institute for Clinical Pharmacology
and Pharmacotherapy
Ghent University School of Medicine
Ghent University Hospital
Ghent, Belgium
Email: mirko.petrovic@ugent.be
Chapter 13

Wilfred R. Pigeon
Sleep and Neurophysiology Research Laboratory
University of Rochester
Rochester, New York, U.S.A.
Email: Wilfred_Pigeon@URMC.Rochester.edu
Chapter 22

Vijay Ramesh
Department of Psychiatry
Harvard Medical School and Veterans Affairs
Medical Center
West Roxbury, Massachussettes
Email: ramesh_vijay@hms.harvard.edu
Chapter 3

Vadim S. Rotenberg
Abarbanel Mental Health Center
Bat-Yam, Israel
Email: vadir@post.tau.ac.il
Chapter 4

Michael J. Sateia
Sleep Disorders Center
Dartmouth-Hitchcock Medical Center
Lebanon, New Hampshire, U.S.A.
Email: michael.j.sateia@dartmouth.edu
Chapter 22

Chanth Seyone
Department of Psychiatry
Acquired Brain Injury Clinic
Toronto Western Hospital
University Health Network
University of Toronto
Toronto, Ontario, Canada
Email: chanth.seyone@uhn.on.ca
Chapter 23

Vibhuti Shah
Department of Paediatrics
Mount Sinai Hospital
University of Toronto
Toronto, Ontario, Canada
Chapter 19

Colin M. Shapiro
Sleep Research Unit
Department of Psychiatry
Toronto Western Hospital
University Health Network
Toronto, Ontario, Canada
Email: colin.shapiro@uhn.on.ca
Chapters 16, 24

Marcello Spinella
Division of Social and Behavioral Sciences
Richard Stockton College of New Jersey
Pomona, New Jersey, U.S.A.
Email: marcello.spinella@stockton.edu
Chapter 34

Stephen M. Stahl
Department of Psychiatry
University of California San Diego
San Diego, California, U.S.A.
Email: smstahl@neiglobal.com
Foreword

Ricardo Tarrasch
Department of Psychology
Faculty of Social Sciences
Tel-Aviv University
Tel-Aviv, Israel
Email: ricardo@post.tau.ac.il
Chapter 25

Stephen Thankachan
School of Life Sciences
Jawaharlal Nehru University
New Delhi, India
Chapter 5

Ricardo A. Velluti
Neurofisiologia, Departamento de Fisiologia
Facultad de Medicina
Universidad de la República
Montevideo, Uruguay
Email: rvelluti@fmed.edu.uy
Chapter 2

Marinus N. Verbaten
Department of Psychopharmacology
Utrecht Institute for Pharmaceutical Sciences
Utrecht University
Utrecht, The Netherlands
Chapter 21

Joris C. Verster
Department of Psychopharmacology
Utrecht Institute for Pharmaceutical Sciences
Utrecht University
Utrecht, The Netherlands
Email: j.c.verster@pharm.uu.nl
Chapter 21

Edmund R. Volkerts
Department of Psychopharmacology
Utrecht Institute for Pharmaceutical Sciences
Utrecht University
Utrecht, The Netherlands
Chapter 21

Alan G. Wade
CPS Research
Glasgow, U.K.
Email: alangwade@ntlworld.com
Chapter 30

Sue J. Wilson
Psychopharmacology Unit
University of Bristol
Bristol, U.K.
Email: sue.wilson@bristol.ac.uk
Chapter 31

Irina V. Zhdanova
Department of Anatomy and Neurobiology
Boston University School of Medicine
Boston, Massachusetts, U.S.A.
Email: zhdanova@bu.edu
Chapter 10

Nava Zisapel
Department of Neurobiochemistry
Tel-Aviv University
Tel-Aviv, Israel
Email: navazis@post.tau.ac.il
Chapter 25

FOREWORD

The neuropsychopharmacology of sleep is now awakening after a long slumber. Once a niche subspecialty where most of the breakthroughs of the last decade in the past century were scientific rather than therapeutic, sleep medicine has now arrived as one of the most exciting areas of new therapeutics in neuropsychopharmacology. As is common when several new treatments arrive in a field, numerous paradigm shifts are afoot for the clinical conceptualization of the most common and important sleep disorders, including chronic insomnia, shift work and other chronobiological sleep disorders and several others. A book to update not only sleep experts but also general neuropsychopharmacologists about these new developments is now much needed.

Thus, this volume has been assembled by three notable opinion leaders whose expertise spans the breadth of general neuropsychopharmacology and the depth of sleep medicine. They have wisely chosen authors for some of the most compelling and timely of the "hot topics" in this field, and have captured the highlights of a field on the move, organizing contributions into three sections. The first section addresses some of the most recent advances in *basic pharmacology* that relate to the subsequent two sections on *chronopharmacology* and on *clinical pharmacology*.

Notably, important advances in the pharmacology of the $GABA_A$ receptor subtype are first reviewed and linked to the mechanism of action of many of the new hypnotics now entering the market for the treatment of insomnia. Other neurotransmitters and neurophysiologic mechanisms linked to sleep are also covered in the basic pharmacology section. Next, several chapters deal with the rapidly emerging field of chronopharmacology, with an extensive chapter on time dependent drug effects, a chapter on possible applications of this idea to seasonal affective disorder and several chapters on melatonin, not only a regulator of chronobiology, but a model for several of the potential new therapeutics on the horizon for various sleep disorders.

Finally the most extensive section of the book is the collection of contributions on clinical pharmacology. Several chapters deal with the current controversy over chronic insomnia, and whether this is an important or serious clinical problem whose treatments enjoy risk-benefit considerations that weigh in favor of chronic treatment. The controversy is a very important one to address and very timely since the case is not yet settled by the field. Discussions in the chapters included here range on the one hand from "Pharmacologic Calvinism," with the idea being to avoid cosmetic psychopharmacology for a lifestyle problem, to the dawning recognition on the other hand that patients can be disabled by chronic insomnia and may need chronic treatment for persisting symptoms, especially with agents that may have little or no potential for dependency with long term use.

A number of other clinical topics included in this book speak to the interdisciplinary nature of treating sleep disorders today, and the important comorbidities that sleep disorders have with numerous other conditions, from depression, to pain, Parkinsonism, and many more. Special clinical topics address the effects that antidepressants, antipsychotics and even herbal medicines have upon sleep.

In summary, it is a great time for this text and a great text for this time. Many of us need to become updated and better informed about the neuropsychopharmacology of sleep as this field enters mainstream psychiatry, neurology and medicine, and this book is a useful tool for the specialist and generalist alike.

Stephen M. Stahl

PREFACE

Many recent discoveries in both laboratory and clinical settings have greatly increased our understanding of sleep medicine and the relevant psychopharmacology. These advances are reported in the clinical neuroscience literature, as well as in specialized sleep publications and other journals. Sleep medicine is thus becoming more interdisciplinary while researchers in other areas of neuroscience and psychopharmacology are beginning to become interested in the subject of sleep. A parallel development is that these fields are now exploiting techniques of increasing complexity. The consequence of these trends is that it is increasingly challenging for the sleep physiology researcher and for the clinician to assimilate, let alone master, the relevant findings in each of these topics.

To address this challenge, the editors of the present volume have assembled articles that summarize and review a carefully chosen selection of the latest discoveries concerning sleep medicine, sleep physiology and sleep pharmacology. To this end, a limited number of outstanding contributions have been sought from acknowledged experts in their respective fields. The goal of the volume has therefore been to present the more recent developments and advances in the fields of sleep and neuropsychopharmacology, and to provide a context for considering them both in depth and from multidisciplinary perspectives. This volume thus brings together the collective expertise of clinicians and basic researchers who represent a range of interests in neuroscience, neuropharmacology, sleep physiology, and biological rhythms.

The editors were well aware that exhaustive reviews of the field could have run the risk of producing an unwieldy tome that would jeopardize the ultimate objective of creating a practical and useful resource. It is our hope that our effort has succeeded in presenting a thoughtful balance of basic experimental and clinical facts and viewpoints, and that it will further serve as a foundation for understanding and ultimately treating sleep disorders.

Inasmuch as we envision continuing updates of this volume, readers are encouraged to contact us with any thoughts and suggestions for topics to be included in future editions. Needless to say in the process of assembling the significant amount of material for the present compendium, there may have been occasional inaccuracies or omissions. We take full responsibility for those, and we would also appreciate having them drawn to our attention.

Malcolm Lader
Daniel P. Cardinali
S.R. Pandi-Perumal

Section I:
Basic Pharmacology

CHAPTER 1

GABA$_A$ Receptor Subtypes in Sedation and Hypnosis

Esa R. Korpi

Abstract

Drugs, such as sedative-hypnotics and anesthetics, are able to strongly regulate the vigilance state by affecting the main fast inhibitory neurotransmitter receptor system in the brain, the GABA$_A$ receptor system. Agonists, such as classical benzodiazepines, are today the most widely used hypnotics. This review summarizes the recent molecular and behavioral pharmacological advance to pinpoint the actions of various drugs on selected amino acid residues of the receptor subunits. As possible targets for future drug development, the properties of several GABA$_A$ receptor subtypes having critical localization in sleep-related brain nuclei/regions are reviewed on the basis of novel mouse models. It is predicted that the molecular heterogeneity of the GABA$_A$ receptor system holds promise for better and more selective drugs for sleep disorders.

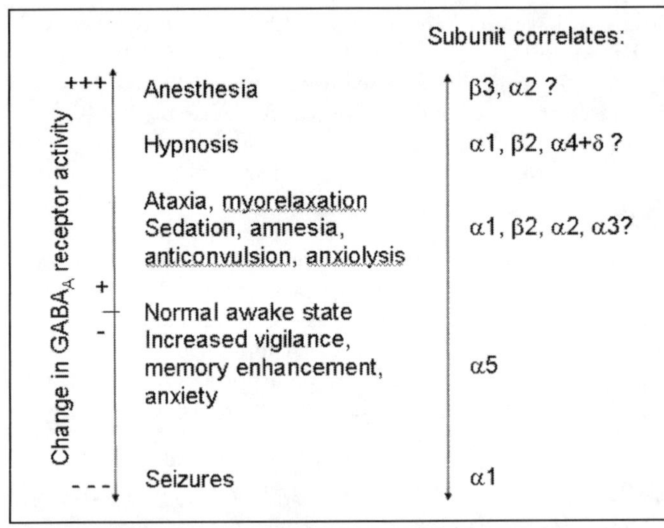

Figure 1. Behavioral correlates of altered GABA$_A$ receptor function. The whole scale of behavioral effects can be provoked by agonists and inverse agonists of the GABA$_A$ receptor benzodiazepine site, and all these actions are practically abolished by receptor antagonists, such as flumazenil. Estimates of the receptor subunit correlates are based on this review and on that of Korpi et al[9] and the references therein.

Introduction

There is a strong effort for improving sedative, hypnotic and anesthetic drugs to be able to better control the drug effects and to better avoid wide-ranging effects on those neuronal systems not actually needed for therapeutic effects. In the literature, sedation, hypnosis and even anesthesia are sometimes defined loosely and used interchangeably. In this review I will use sedation to describe the activity-depressing effect, hypnosis the sleep-inducing or -promoting effect and anestesia the loss of consciousness. Sleep is defined by natural sleep often with classifications by EEG power and frequency into various sleep phases.

Presently, more and more molecular pharmacological/biological data on the main mammalian fast-acting inhibitory neurotransmitter receptor system, γ-aminobutyric acid type A receptors (GABA$_A$), is being used to assess the roles and significances of various subtypes of the receptor in sedation, hypnosis, sleep and anestesia. GABA$_A$ receptors are known for mediating a range of behaviors (Fig. 1), all also being produced by the allosteric effects at their benzodiazepine binding sites. The purpose of this review is to first describe the basics of GABA$_A$ receptor molecular pharmacology and then evaluate the roles of subtypes on the basis of recent studies on novel animal models. Also the latest developments on the mechanisms of action of current sleeping medications are being reviewed as related to the GABA$_A$ receptors.

Molecular Heterogeneity of Brain GABA$_A$ Receptors

Since the cloning of the first GABA$_A$ receptor subunit, the α1 subunit, almost 20 years ago,[1] there has been a rapid progress in finding novel subunits with interesting features. There are several recent reviews on the topic [2-9] and here only basic facts are summarized.

Presently, there are at least 16 different GABA$_A$ receptor subunits, all encoded by different genes. The molecular heterogeneity of the receptor subtypes stems mainly from sequence differences between the subunits and on this basis the subunits can be classified into α (1-6), β (1-3), γ (1-3), δ, ε, π, and θ subunit classes. The function will then depend on the subunit combinations of the pentameric receptor complex, i.e., on receptor subtypes. Various subunits have either widespread or very restricted expression profiles in various brain cells, most neurons are anyway responding to the transmitter, γ-aminobutyric acid, GABA. Therefore, the receptor subtypes are also based on differential expression profiles of the subunits in different brain cells. The main problem still in relating receptor subtypes to drug actions is the lack exact knowledge on the native subunit combinations, since theoretically there can be tens of thousands of different combinations. It is evident from studies on native brain receptors that this number is much smaller (see ref. 3), but we still cannot

Sleep and Sleep Disorders: A Neuropsychopharmacological Approach, edited by Malcolm Lader, Daniel P. Cardinali and S.R. Pandi-Perumal.

exclude the possibility that a minor subtype located in a neuronal circuit critical for a behavior, such sleep, might be more important than a subtype that is abundant and widespread. In this review, we will focus on certain receptor subtypes, but at the same time we have to keep in mind that more detailed information will be on its way in the near future through the application of transgenic mouse models.

The main subunit combination is the α1βXγ2 receptor (X = 2 and/or 3) that accounts for about 40% of all $GABA_A$ receptors. These subunits are widespread and often expressed in the same brain regions/neurons. When α1 and β2 subunits are expressed in heterologous expression systems, they produce functional receptors that have high sensitivity to GABA (for a review see ref. 7). The receptor channel conductance, however, is rather small. When the γ2 subunit is expressed together with α1 and β2 subunits, the receptors will have a lower GABA sensitivity but higher conductance, which corresponds to the properties of main native $GABA_A$ receptor populations. The smaller conductance of the α1β channels might also explain why the knockout mice without γ2 subunits are not viable.[10] However, the γ2 subunits have been shown to regulate the targeting of $GABA_A$ receptors to synapses via interaction with gephyrin, a protein needed also for clustering of inhibitory glycine receptors.[11] Without the γ2 subunits, receptors are more localized on extrasynaptic membranes, where they may also mediate important regulatory function by sensoring low concentrations of GABA spilled over from neighbouring synapses.[12]

$GABA_A$ receptor forms an integral anion-selective channel (Fig. 2), which normally mediates fast chloride ion influx producing hyperpolarization of the neuronal plasma membrane and thereby reducing any firing activity of the postsynaptic neuron. This action can simply explain the sedative and hypnotic action of $GABA_A$ receptor agonists, such as benzodiazepines, neurosteroids, barbiturates, alcohol, anesthetics and GABA-site agonists such as gaboxadol. All these compounds produce sedation and most of them also modulate sleep in animals and humans. $GABA_A$ receptor function requires low intracellular Cl^- concentrations, which are produced by chloride pumping action of various ion transporters. An interesting example is KCC2 transporter, whose expression is developmentally regulated in many neurons.[13] Early in the development, KCC2 transporter is not expressed and GABA responses of the $GABA_A$ receptors are depolarizing due to lack of proper electrochemical Cl^- gradients. Although various $GABA_A$ receptor subtypes differ in kinetics of the channel function,[7] it is not known how these differential properties would make a contribution in physiology. The subcellular location of the receptor subtypes in the neurons and their activation by presynaptic neurons may be more important for physiology than any channel property.

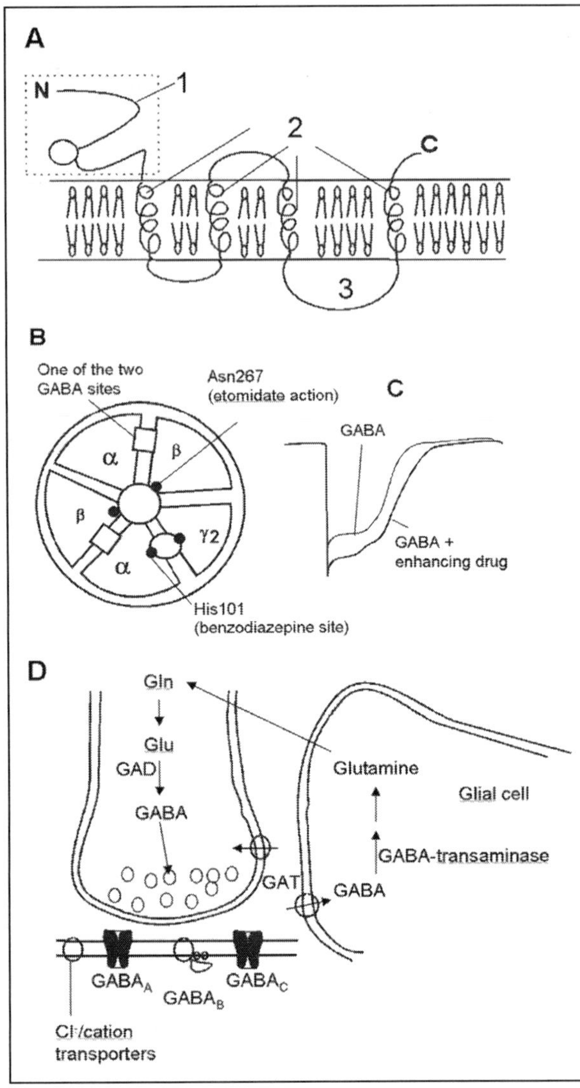

Figure 2. Schematics of $GABA_A$ receptor structure and function and the basic pharmacology of GABAergic synapse. A) Topography of a $GABA_A$ receptor subunit partially embedded in the lipid bilayer. 1) N-terminal extracellular domain responsible for transmitter and ligand linding and coupling of the binding sites with ion channel. This part is also important for the allosteric effects and for the assembly of various receptor subunits into functional receptors. 2) Four transmembrane regions forming the anion channel are responsible for binding of hydrophobic ligands, ion selectivity and channel binding sites. 3) Intracellular loop between TM3 and TM4 forms the motif for regulatory phosphorylation sites and for the intracellular factors anchoring the receptors in appropriate locations e.g., on the postsynaptic thickening. B) Hypothetical binding sites for GABA and allosteric modulators, such as benzodiazepines and etomidate in a pentameric receptor complex. C) Allosteric activation of $GABA_A$ receptor may increase the peak height of the response and/or prolong the response as compared to the response by GABA alone. D) GABAergic synapse with synthetic and catabolic enzymes, receptors and transporters. Examples of prototypical or clinically used drugs for various target sites are shown. GAD, glutamate decarboxylase, a GABA-synthesizing enzyme; GABA transaminase, a GABA metabolising enzyme; $GABA_B$ receptor is G-protein coupled heptahelix receptor and $GABA_C$ a ligand-gated GABA-activated receptor composed of ρ subunits with largely unknown effects in the brain and retina.

Pharmacological Subtypes of $GABA_A$ Receptors

A simple classification of $GABA_A$/benzodiazepine receptors can be made on the basis of benzodiazepine affinities. Figure 3 describes the main features of this classification. In terms of clinical pharmacology, it is useful to make the classification on the basis of diazepam and zolpidem with the help of one experimental ligand, Ro 15-4513. The high-affinity binding of flumazenil-sensitive tritium-labelled Ro 15-4513 is very little affected by natural molecular heterogeneity of αβγ2 receptors, and, therefore, it has been used as the ligand for receptor subtyping.[14]

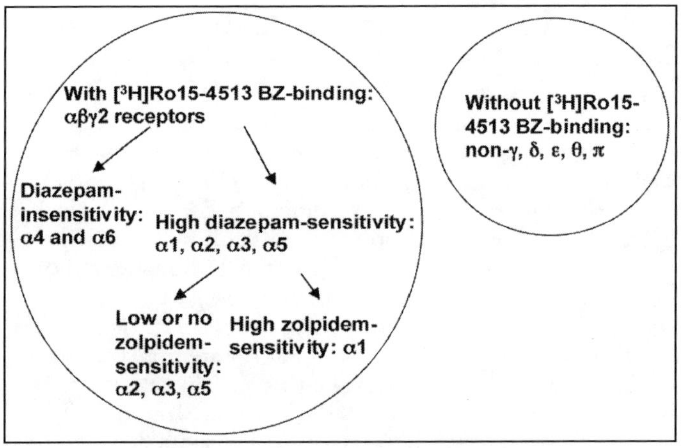

Figure 3. Classification of $GABA_A$ receptor populations on the basis of the ligand affinities to the benzodiazepine (BZ) sites labelled by [^{3}H]Ro 15-4513. See refs. 9, 14 for more detailed discussion.

The classical benzodiazepine diazepam acts as an agonist (increases allosterically the affinity of GABA) on $\alpha 1/2/3/5$ subunit-containing $\alpha\beta\gamma 2$ receptors, but fails to act on $\alpha 4$ and $\alpha 6$ subunit-containing receptors. Therefore, we can exclude the $\alpha 4$ and $\alpha 6$ subunit-containing receptors as mediators of diazepam-induced sedation and hypnosis.

The $\alpha 1$ subunit has a histidine residue at the position of about 100 in the extracellular domain and this residue is obligatory for classical benzodiazepine actions.[15] Homologous histidine is also found in other benzodiazepine-sensitive receptor subtypes having α 2, α 3 or $\alpha 5$ subunits, but the insensitive subunits $\alpha 4$ and $\alpha 6$ harbor arginine in the same position. This residue is critical for the benzodiazepine action as a domain of the binding site and/or a domain for regulating allosteric interactions.[16]

Recent modeling studies have indicated that this histidine/arginine residue is close to the interface between α and $\gamma 2$ subunits.[8] Indeed, for the action of benzodiazepines, also the $\gamma 2$ subunit is obligatory. Figure 2 summarizes the schematic structure and function of the $GABA_A$ receptor at subunit and pentameric levels, and indicates the most important residues for GABA and benzodiazepine ligand binding.

Also many other drugs produce sedation and hypnosis by facilitating the action of $GABA_A$ receptor. Barbiturates and neurosteroids seem to have rather little subtype specificity and their actions are not dependent on the $\gamma 2$ subunits. A general intravenous anesthetic etomidate requires the $\beta 2$ or $\beta 3$ subunits for its efficacy and it has been pointed out that the residues in the 2nd transmembrane region have a critical role either for the binding or for the allosteric interactions of this drug.[17] Homologous or closely located amino acid residues of various other $GABA_A$ receptor subunits have also shown to be important for the high-concentration actions of ethanol and volatile anesthetics.[18] Some of these residues may form hydrophobic binding pockets for the drugs in the transmembrane domains.[19]

The GABA binding site has been also resolved in the middle of the extracellular domain of the $GABA_A$ receptor subunits. Like the benzodiazepine binding site, it is apparently also located in the interface between two subunits, now between the α and β subunits. There are many different residues that have been found to affect the efficacy and binding of GABA and antagonists in various α and β subunit variants (see for a review ref. 9). There seems to be some brain regional variation in the potency and efficacy of various GABA site agonists and antagonists,[20] suggesting that different receptor subtypes would be differentially affected by the compounds. In terms of sleep research, perhaps the most interesting agonist is gaboxadol (THIP, 4,5,6,7-tetrahydroisoxazolo[5,4-*c*]pyridin-3-ol) that was initially considered as a partial agonist of $GABA_A$ receptors.[21] Muscimol is a prototypic full agonist selective for $GABA_A$ receptors. In autoradiography with rat brain sections, both of these compounds affect the receptor channels more like full agonists at the main receptor subtypes,[20] but this assay fails to see the effects on less abundant receptor subtypes that might still have important regulatory functions for subsets of neurons. The regional distribution of [^{3}H]muscimol binding suggests that high-affinity interaction is limited to $\alpha 6$, $\alpha 4$ and δ subunit-containing receptors especially in the thalamic and cortical regions and in the cerebellar granule cell layer.[22] It remains to be tested whether gaboxadol has a similar high-affinity binding distribution as muscimol in native receptors. In recombinant receptors, gaboxadol has shown variable efficacy depending on the subunit combination: It acts efficiently on $\alpha 2$, $\alpha 5$ and $\alpha 6$ subunit-containing receptors, and less efficiently in $\alpha 1$ and $\alpha 3$ receptors.[23] More recently, Brown et al[24] have found a higher affinity and efficacy of gaboxadol than GABA at $\alpha 4\beta 3\delta$ receptor subtype, which in the absence of $\gamma 2$ subunits should be largely targeted to extrasynaptic membranes. The δ subunit-containing receptors may thus form the major extrasynaptic receptor subtype and their properties such as very high sensitivity and very slow desensitisation[25,26] fit well to monitor low extracellular agonist concentrations and to regulate neuronal excitability on that basis.[12] The δ subunit does not support benzodiazepine sites. The expression of $\alpha 4$ and δ subunits is high, e.g., in the thalamus, where $\alpha 1\beta X\gamma 2$ receptor is not as abundant as in many other regions.[27,28] Therefore, it is possible that some direct GABA agonists might offer profiles of receptor subtype stimulations clearly different from those of benzodiazepines.

Molecular nature of drug interactions is thus becoming clearer, but the significance of the interactions has not been fully resolved in whole animals. However, in recent years we have witnessed several important findings with novel mouse models that help us further to understand the molecular basis of sedation and sleep. I want to stress here that we are making a long leap from simple molecular interaction between a drug and receptor subtype to behavioral effects without yet knowing all mechanisms that are operative between these processes, e.g., in neuronal circuits. At the moment, it is not possible to analyse all receptor subtype information at the functional level of neuronal pathways, making it difficult to form any universal hypotheses on the actions of a given drug on sleep parameters.

Receptor Subtype(s) Responsible for Sedative and Hypnotic Actions of Benzodiazepine Ligands

Minute molecular modifications in protein structure can affect profoundly either the function or drug sensitivities. This principle was recognized for the first time in vivo with alcohol- and benzodiazepine-sensitive ANT rat line that harbors an arginine to glutamine mutation at the 100th position of the $GABA_A$ receptor $\alpha 6$ subunits.[29] The mutation is due to a single nucleotide change, resulting in dramatic change in pharmacology: The cerebellar granule cell-specific $\alpha 6$ subunit is normally insensitive to benzodiazepine full agonists, but the mutation makes these cells abnormally sensitive to diazepam. Behaviorally, these ANT animals show abnormally high diazepam sensitivity of motor performance in a test requiring rapid adaptation to altered posture.[30] Importantly, these animals seem to get as sedated as control animals after administration of zolpidem,[31] which indicates that the α6 subunit-containing $GABA_A$ receptors are not involved in sedation or hypnosis induced by benzodiazepine ligands. This same principle, but in opposite manner, has been now used to produce knockin mouse lines to study the roles of other α subunits.

The first mouse line was the $GABA_A$ receptor α1(His101Arg) knockin line that has the major $GABA_A$ receptor subtype altered in a way that benzodiazepine agonists lose their affinity and allosteric modulation.[32,33] These mice have only α2βXγ2, α3βXγ2 and α5βXγ2 subunit-containing receptors that are remaining for diazepam actions, which was also verified by ligand autoradiography in brain sections. With measurements of locomotor activity and other drug effects, it became evident that in α1(His101Arg) knockin mice diazepam no more induces sedation, amnesia nor has a full anticonvulsant action. Similarly, zolpidem fails to induce sedation in the mutants.[34] These experiments add α2, α3 and α5 subunit-containing receptors to the list of $GABA_A$ receptors that are not critically needed for benzodiazepine-induced sedation. Since the α4 subunit also has arginine as the "critical" amino acid,[35] it is insensitive to benzodiazepine agonists and can, therefore, also eliminated from the list. In conclusion, sedation induced by benzodiazepines requires $GABA_A$ receptors of α1βXγ2 subunit combination. These receptor subtypes are widely expressed in the brain and, therefore, this information per se fails to reveal any neuronal pathway or circuit where increased GABAergic inhibition would lead to sedative effects.

A major surprise was the observation that diazepam-promoted sleep induction or sleep EEG alterations were not altered in mice with α1(His101Arg) knockin mutation,[36] suggesting that sedation and hypnotic effects differ in receptor mechanisms. The mutant mice had the effects of diazepam at the onset of light period on sleep latency, amount of sleep, sleep EEG (increase in higher frequencies and suppression of slow-wave activity) were similar as or even more pronounced as the wild-type animals. These results indicate that in mouse, the hypnotic effects of benzodiazepine ligands are dependent on other than α1 subunit-containing receptor subtypes.

Of the remaining possibilities of α2, α3 and α5 subunit-containing receptors, the α3 subunit mutants [α3(His126Arg)] have been studied and found not to contribute to diazepam-induced sleep behavior or sleep EEG,[37] which leaves the α2 and α5 subunit-containing receptors as the main candidates. Since zolpidem does not act on α5 subunit-containing receptors, the α2 receptors remain the main possibility for the primary target of benzodiazepine-induced sleep alterations. The exclusions of the α1 and α3 receptor subtypes are surprising, since these receptors are abundantly expressed in the thalamic and cortical regions involved in cortical synchronization.[27,28] The α2 subunit knockin mutation [α2(His101Arg)] has abolished the anxiolytic action of diazepam but not its sedative action.[38] The α2 subunit is enriched in the axon initial segment of cortical principal neurons. Therefore, any potentiation of its actions can have strong effect on descending activity from the cortex. The role of α2βγ2 receptors remains to be established in sleep induction and EEG.

Possible Roles of Other Major $GABA_A$ Receptor Subunits/Subtypes

Although the α subunits determine much of the pharmacology and γ2 subunits are obligatory for benzodiazepine efficacy, also other subunits may have important functional and pharmacological significance. The β subunits are fairly similar in terms of pharmacology and function as members of the pentameric receptor complex. However, there is one particular amino acid residue in these subunits that differentiates the β1 from β2 and β3. Several drugs act only in the presence of β2 or β3 subunit. These drugs include a wide-spectrum antiepileptic loreclezole, an α6 subunit-selective antagonist furosemide and an intravenous anesthetic etomidate,[17,39,40] suggesting that this domain is important for allosteric effects within the receptor. Interestingly, a knockin mouse line β3(Asn265Met) has been created and found to have lost the anesthetic actions of etomidate and propofol, but not clearly those of neurosteroids and inhalation anesthetics.[41] If anesthetic actions have common mechanisms with slow-wave sleep mechanisms, then the β3 subunit-containing receptors might be a part of the molecular machinery needed for these functions. Interestingly, β2(Asn265Ser) knockin mice have normal anesthetic responses to etomidate, but have reduced sedation at low doses and during recovery from anesthesia,[42] suggesting that even if the β2 subunit-containing receptors make about half of the brain $GABA_A$ receptors their stimulation causes more sedation than anesthesia, while a smaller population of β3 subunit-containing receptors are obligatory for normal anesthetic efficacy. This hypothesis might be consistent with α1 assembling with β2 and α2 with β3 subunits (see above for benzodiazepine actions via α subunits).

The β3 subunit appears to be important for the brain function and development, and deficiency of this subunit causes high neonatal mortality and severe neurological phenotype in about 5% of the mice that survive into adulthood.[43] Deletions of this subunit may cause at least some of the symptoms found in human neurodevelopmental disorder Angelman syndrome, which is characterized by hyperactivity, sleep disorders, epilepsy, poor learning and frequent, easily provoked laughter/smiling.[44] Inactivation of the β3 subunit causes dramatic changes in overall $GABA_A$ receptor numbers and other subunits in many brain regions,[45,46] making it difficult to draw any clear conclusions on functional receptor subtypes. However, the β3 knockout mice have slowing of the EEG rhythm with high amplitudes and epileptic spikes.[47] Detailed examination of sleep characteristics and EEG of these mice, revealed no differences in the amount of NREM sleep, but reduced REM sleep.[48] The transient increase in EEG power in the 12-16 Hz frequency during transition to REM sleep was blunted in the mutants and not much affected by midazolam as compared to the wild-type animals. Midazolam failed to affect REM sleep time in the mutants, while it reduced it in the controls. This benzodiazepine agonist reduced the slow-wave EEG power similarly in the knockouts and wild-types. These data suggest that the profound changes, such as reduced recurrent inhibition within the reticular thalamic nuclei leading to increased oscillatory syncrony, in the thalamocortical circuitry (β3 subunit is the major β subunit in the reticular thalamic nuclei)[27] found in the β3 knockout mice[49] do not much affect the slow-wave sleep, but may reduce the susceptibility to benzodiazepine modulation and the amount of the REM sleep. Keeping in mind the likely compensatory changes in receptors and pathways in the knockout mice and the unaltered diazepam-induced sleep modulation in α3(His126Arg) mice (see above), further studies are needed and other receptor subtypes and neuronal pathways should be considered.

In clinical practice, the Angelman patients are often treated for sleep disorders with melatonin that has own 7-transmembrane domain G-protein coupled receptors.[50] Preclinical work has suggested that melatonin might function as a benzodiazepine-like compound with effects on sleep and anxiety being at least partially blocked by flumazenil, the selective benzodiazepine site antagonist.[51] However, the mechanisms of action of melatonin should still be further studied in humans (see ref. 52), but its apparent efficacy in Angelman patients indicates that the β3 subunit expression is not obligatory for its actions.

Possible Roles of Minor GABA$_A$ Receptor Subunits/Subtypes

The expression of ε and θ subunits is very restricted in the brain: There is a strong enrichment of mRNAs for both of these subunits in the locus coeruleus and less so in some other brainstem nuclei.[53,54] Their genes are located in the X-chromosome as is that of the α3 subunit,[9] which might suggest similar expression profiles for the members of this GABA$_A$ receptor subunit gene cluster. However, it is likely that while α3 is expressed everywhere where ε and θ subunits are expressed, it is also expressed in ε and θ subunit-deficient brain regions. Since neither of these subunits forms benzodiazepine sites nor replace the γ2 subunit, the α3βε and α3βθ receptors are most likely insensitive to benzodiazepines and located outside the synapses. Their pharmacology has not been reported yet, but at least the ε subunits expressed with α1 subunits produce benzodiazepine- and anesthetic-insensitive receptors[55] that might even have constitutively open anion channels in the absence of agonists.[56] The GABA$_A$ receptors in the locus coeruleus have been implicated in the regulation of REM sleep that is increased by agonists and decreased by antagonists.[57] This apparently differs from the effects of systemic benzodiazepines, which reduce the REM sleep. It is thus possible that some effects of nonbenzodiazepine drugs on sleep are mediated by distinct GABA$_A$ receptor populations insensitive to benzodiazepines. These receptors could also be extrasynaptic.

Another subunit that promotes extrasynaptic localisation of GABA$_A$ receptors especially with α4 (forebrain, thalamus) and α6 (cerebellum) subunits is the δ subunit.[58] It may actually be present in many extrasynaptic receptors that are poorly desensitized during agonist application.[12] The effects of direct GABA$_A$ receptor agonists, such as muscimol and gaboxadol, and those of the benzodiazepines on sleep are often qualitatively different,[59-61] and there may be only partial cross-tolerance to their behavioral actions in general. One possible explanation is that GABA$_A$ agonists that are not quickly removed from the extracellular space, as they are not substrates for GABA transporters, act at low concentrations on tonically active, high-sensitive extrasynaptic receptors, while benzodiazepines require the presence of higher concentrations of synaptically released endogenous agonist that is quickly removed from synaptic clefts. This is supported by the findings showing recombinant α4βδ receptors being exceptionally sensitive to gaboxadol.[24] It remains to be seen whether these preclinical data can be applied to treatment of human sleep disorders. At least, gaboxadol has been found to be efficient in humans.[62-64]

Concluding Remarks

Figure 1 summarizes the possible correlations of GABA$_A$ receptor subunits and behavioral consequences of altered receptor activity. These correlations are partially preliminary (marked in the Figure with the question mark), and more detailed data will be produced by several research groups in the near future. This will give both good opportunities and challenges for drug discovery and development. However, as mentioned above, I have made a long leap from receptor molecules to behaving animals and not really filled the necessary intermediate mechanisms. In recent years, also strong advance has been made to localise "new" sleep-associated nuclei in the rostral hypothalamic area.[65-67] And indeed, the neurons from sleep-promoting ventrolateral preoptic area are GABAergic with galanin as a neuropeptide modulator. These neurons inhibit the activity of several nuclei, including the locus coeruleus, during phases of either nonREM or REM sleep. These findings make it important to put research effort on the characterization of functional GABA$_A$ receptor subtypes of the hypothalamic preoptic neurons and their target regions for developing novel subtype-selective drugs. These drugs might then offer the selectivity and low adverse effect profiles that we would like to have from sleeping pills!

Acknowledgements

This work was supported by the Academy of Finland and by the Sigrid Juselius Foundation. The comments of Dag Stenberg on the manuscript are gratefully acknowledged.

References

1. Schofield PR, Darlison MG, Fujita N et al. Sequence and functional expression of the GABA$_A$ receptor shows a ligand-gated receptor super-family. Nature 1987; 328:221-227.
2. Wisden W, Seeburg PH. GABA$_A$ receptor channels: From subunits to functional entities. Curr Opin Neurobiol 1992; 2:263-269.
3. McKernan RM, Whiting PJ. Which GABA$_A$ receptor subtypes really occur in the brain? Trends Neurosci 1996; 19:139-143.
4. Sigel E, Buhr A. The benzodiazepine binding site of GABA$_A$ receptors. Trends Pharmacol Sci 1997; 18:425-429.
5. Moss SJ, Smart TG. Constructing inhibitory synapses. Nat Rev Neurosci 2001; 2:240-250.
6. Rudolph U, Crestani F, Mohler H. GABA$_A$ receptor subtypes: Dissecting their pharmacological functions. Trends Pharmacol Sci 2001; 22:188-194.
7. Hevers W, Lüddens H. The diversity of GABA$_A$ receptors. Pharmacological and electrophysiological properties of GABA$_A$ channel subtypes. Mol Neurobiol 1998; 18:35-86.
8. Ernst M, Brauchart D, Boresch S. et al. Comparative modeling of GABA$_A$ receptors: Limits, insights, future developments. Neuroscience 2003; 119:933-943.
9. Korpi ER, Gründer G, Lüddens H. Drug interactions at GABA$_A$ receptors. Prog Neurobiol 2002; 67:113-159.
10. Günther U, Benson J, Benke D et al. Benzodiazepine-insensitive mice generated by targeted disruption of the γ2 subunit gene of γ-aminobutyric acid type A receptors. Proc Natl Acad Sci USA 1995; 92:7749-7753.
11. Kneussel M, Betz H. Receptors, gephyrin and gephyrin-associated proteins: Novel insights into the assembly of inhibitory postsynaptic membrane specializations. J Physiol 2000; 525:1-9.
12. Mody I. Distinguishing between GABA$_A$ receptors responsible for tonic and phasic conductances. Neurochem Res 2001; 26:907-913.
13. Rivera C, Voipio J, Payne JA et al. The K^+/Cl^- cotransporter KCC2 renders GABA hyperpolarizing during neuronal maturation. Nature 1999; 397:251-255.
14. Lüddens H, Korpi ER, Seeburg PH. GABA$_A$/benzodiazepine receptor heterogeneity: Neurophysiological implications. Neuropharmacology 1995; 34:245-254.

15. Wieland HA, Lüddens H, Seeburg PH. A single histidine in $GABA_A$ receptors is essential for benzodiazepine agonist binding. J Biol Chem 1992; 267:1426-1429.
16. Dunn SM, Davies M, Muntoni AL et al. Mutagenesis of the rat $\alpha 1$ subunit of the γ-aminobutyric $acid_A$ receptor reveals the importance of residue 101 in determining the allosteric effects of benzodiazepine site ligands. Mol Pharmacol 1999; 56:768-774.
17. Belelli D, Lambert JJ, Peters JA et al. The interaction of the general anesthetic etomidate with the γ-aminobutyric acid type A receptor is influenced by a single amino acid. Proc Natl Acad Sci USA 1997; 94:11031-11036.
18. Mihic SJ, Ye Q, Wick MJ et al. Sites of alcohol and volatile anaesthetic action on $GABA_A$ and glycine receptors. Nature 1997; 389:385-389.
19. Jenkins A, Greenblatt EP, Faulkner HJ et al. Evidence for a common binding cavity for three general anesthetics within the $GABA_A$ receptor. J Neurosci 2001; 21:RC136.
20. Rabe H, Picard R, Uusi-Oukari M et al. Coupling between agonist and chloride ionophore sites of the $GABA_A$ receptor: A gonist/ antagonist efficacy of 4-PIOL. Eur J Pharmacol 2000; 409:233-242.
21. Braestrup C, Nielsen M, Krogsgaard-Larsen P et al. Partial agonists for brain GABA/benzodiazepine receptor complex. Nature 1979; 280:331-333.
22. Korpi ER, Mihalek RM, Sinkkonen ST et al. Altered receptor subtypes in the forebrain of $GABA_A$ receptor δ subunit-deficient mice: Recruitment of $\gamma 2$ subunits. Neuroscience 2002; 109:733-743.
23. Ebert B, Thompson SA, Saounatsou K et al. Differences in agonist/ antagonist binding affinity and receptor transduction using recombinant human γ-aminobutyric acid type A receptors. Mol Pharmacol 1997; 52:1150-1156.
24. Brown N, Kerby J, Bonnert TP et al. Pharmacological characterization of a novel cell line expressing human $\alpha 4\beta 3\delta$ $GABA_A$ receptors. Br J Pharmacol 2002; 136:965-974.
25. Saxena NC, Macdonald RL. Assembly of $GABA_A$ receptor subnits: Role of the δ subunit. J Neurosci 1994; 14:7077-7086.
26. Hevers W, Korpi ER, Lüddens H. Assembly of functional $\alpha 6\beta 3\gamma 2\delta$ $GABA_A$ receptors in vitro. Neuroreport 2000; 11:4103-4106.
27. Wisden W, Laurie DJ, Monyer H et al. The distribution of 13 $GABA_A$ receptor subunit mRNAs in the rat brain. I. Telencephalon, diencephalon, mesencephalon. J Neurosci 1992; 12:1040-1062.
28. Pirker S, Schwarzer C, Wieselthaler A et al. $GABA_A$ receptors: Immunocytochemical distribution of 13 subunits in the adult rat brain. Neuroscience 2000; 101:815-850.
29. Korpi ER, Kleingoor C, Kettenmann H et al. Benzodiazepine-induced motor impairment linked to point mutation in cerebellar $GABA_A$ receptor. Nature 1993; 361:356-359.
30. Hellevuo K, Kiianmaa K, Korpi ER. Effect of GABAergic drugs on motor impairment from ethanol, barbital and lorazepam in rat lines selected for differential sensitivity to ethanol. Pharmacol Biochem Behav 1989; 34:399-404.
31. Wong G, Sarviharju M, Toropainen M et al. Pharmacologic actions of subtype-selective and novel GABAergic ligands in rat lines with differential sensitivity to ethanol. Pharmacol Biochem Behav 1996; 53:723-730.
32. Rudolph U, Crestani F, Benke D et al. Benzodiazepine actions mediated by specific γ-aminobutyric $acid_A$ receptor subtypes. Nature 1999; 401:796-800.
33. McKernan RM, Rosahl TW, Reynolds DS et al. Sedative but not anxiolytic properties of benzodiazepines are mediated by the $GABA_A$ receptor $\alpha 1$ subtype. Nat Neurosci 2000; 3:587-592.
34. Crestani F, Low K, Keist R et al. Molecular targets for the myorelaxant action of diazepam. Mol Pharmacol 2001; 59:442-445.
35. Benke D, Michel C, Möhler H. $GABA_A$ receptors containing the $\alpha 4$-subunit: Prevalence, distribution, pharmacology, and subunit architecture in situ. J Neurochem 1997; 69:806-814.
36. Tobler I, Kopp C, Deboer T et al. Diazepam-induced changes in sleep: Role of the $\alpha 1$ $GABA_A$ receptor subtype. Proc Natl Acad Sci USA 2001; 98:6464-6469.
37. Kopp C, Rudolph U, Keist R et al. Diazepam-induced changes on sleep and the EEG spectrum in mice: Role of the $\alpha 3$-$GABA_A$ receptor subtype. Eur J Neurosci 2003; 17:2226-2230.
38. Low K, Crestani F, Keist R et al. Molecular and neuronal substrate for the selective attenuation of anxiety. Science 2000; 290:131-134.
39. Wingrove PB, Wafford KA, Bain C et al. The modulatory action of loreclezole at the γ-aminobutyric acid type A receptor is determined by a single amino acid in the $\beta 2$ and $\beta 3$ subunit. Proc Natl Acad Sci USA 1994; 91:4569-4573.
40. Thompson SA, Arden SA, Marshall G et al. Residues in transmembrane domains I and II determine γ-aminobutyric acid type. A receptor subtype-selective antagonism by furosemide. Mol Pharmacol 1999; 55:993-999.
41. Jurd R, Arras M, Lambert S et al. General anesthetic actions in vivo strongly attenuated by a point mutation in the $GABA_A$ receptor $\beta 3$ subunit. Faseb J 2003; 17:250-252.
42. Reynolds DS, Rosahl TW, Cirone J et al. Sedation and anesthesia mediated by distinct $GABA_A$ receptor isoforms. J Neurosci 2003; 23:8608-8617.
43. Homanics GE, DeLorey TM, Firestone LL et al. Mice devoid of γ-aminobutyrate type A receptor $\beta 3$ subunit have epilepsy, cleft palate, and hypersensitive behavior. Proc Natl Acad Sci USA 1997; 94:4143-4148.
44. Clayton-Smith J, Laan L. Angelman syndrome: A review of the clinical and genetic aspects. J Med Genet 2003; 40:87-95.
45. Ramadan E, Fu Z, Losi G et al. $GABA_A$ receptor $\beta 3$ subunit deletion decreases $\alpha 2/3$ subunits and IPSC duration. J Neurophysiol 2003; 89:128-134.
46. Sinkkonen ST, Homanics GE, Korpi ER. Mouse models of Angelman syndrome, a neurodevelopmental disorder, display different brain regional $GABA_A$ receptor alterations. Neurosci Lett 2003; 340:205-208.
47. DeLorey TM, Handforth A, Anagnostaras SG et al. Mice lacking the $\beta 3$ subunit of the $GABA_A$ receptor have the epilepsy phenotype and many of the behavioral characteristics of Angelman syndrome. J Neurosci 1998; 18:8505-8514.
48. Wisor JP, DeLorey TM, Homanics GE et al. Sleep states and sleep electroencephalographic spectral power in mice lacking the $\beta 3$ subunit of the $GABA_A$ receptor. Brain Res 2002; 955:221-228.
49. Huntsman MM, Porcello DM, Homanics GE et al. Reciprocal inhibitory connections and network synchrony in the mammalian thalamus. Science 1999; 283:541-543.
50. Kokkola T, Laitinen JT. Melatonin receptor genes. Ann Med 1998; 30:88-94.
51. Wang F, Li J, Wu C et al. The $GABA_A$ receptor mediates the hypnotic activity of melatonin in rats. Pharmacol Biochem Behav 2003; 74:573-578.
52. Nave R, Herer P, Haimov I et al. Hypnotic and hypothermic effects of melatonin on daytime sleep in humans: Lack of antagonism by flumazenil. Neurosci Lett 1996; 214:123-126.
53. Sinkkonen ST, Hanna MC, Kirkness EF et al. $GABA_A$ receptor ε and θ subunits display unusual structural variation between species and are enriched in the rat locus ceruleus. J Neurosci 2000; 20:3588-3595.
54. Moragues N, Ciofi P, Tramu G et al. Localisation of $GABA_A$ receptor ε-subunit in cholinergic and aminergic neurones and evidence for codistribution with the θ-subunit in rat brain. Neuroscience 2002; 111:657-669.
55. Davies PA, Hanna MC, Hales TG et al. Insensitivity to anaesthetic agents conferred by a class of $GABA_A$ receptor subunit. Nature 1997; 385:820-823.
56. Davies PA, Kirkness EF, Hales TG. Evidence for the formation of functionally distinct $\alpha\beta\gamma\varepsilon$ $GABA_A$ receptors. J Physiol 2001; 537:101-113.
57. Mallick BN, Kaur S, Saxena RN. Interactions between cholinergic and GABAergic neurotransmitters in and around the locus coeruleus for the induction and maintenance of rapid eye movement sleep in rats. Neuroscience 2001; 104:467-485.
58. Nusser Z, Sieghart W, Somogyi P. Segregation of different $GABA_A$ receptors to synaptic and extrasynaptic membranes of cerebellar granule cells. J Neurosci 1998; 18:1693-1703.
59. Lancel M, Faulhaber J. The $GABA_A$ agonist THIP (gaboxadol) increases nonREM sleep and enhances delta activity in the rat. Neuroreport 1996; 7:2241-2245.

60. Lancel M, Cronlein TA, Faulhaber J. Role of GABA$_A$ receptors in sleep regulation. Differential effects of muscimol and midazolam on sleep in rats. Neuropsychopharmacology 1996; 15:63-74.
61. Frölund B, Ebert B, Kristiansen U et al. GABA$_A$ receptor ligands and their therapeutic potentials. Curr Top Med Chem 2002; 2:817-832.
62. Faulhaber J, Steiger A, Lancel M. The GABA$_A$ agonist THIP produces slow wave sleep and reduces spindling activity in NREM sleep in humans. Psychopharmacology (Berl) 1997; 130:285-291.
63. Lancel M, Wetter TC, Steiger A et al. Effect of the GABA$_A$ agonist gaboxadol on nocturnal sleep and hormone secretion in healthy elderly subjects. Am J Physiol Endocrinol Metab 2001; 281:E130-137.
64. Mathias S, Steiger A, Lancel M. The GABA$_A$ agonist gaboxadol improves the quality of post-nap sleep. Psychopharmacology (Berl) 2001; 157:299-304.
65. Lu J, Bjorkum AA, Xu M et al. Selective activation of the extended ventrolateral preoptic nucleus during rapid eye movement sleep. J Neurosci 2002; 22:4568-4576.
66. McGinty D, Szymusiak R. Hypothalamic regulation of sleep and arousal. Front Biosci 2003; 8:s1074-1083.
67. Mignot E, Taheri S, Nishino S. Sleeping with the hypothalamus: Emerging therapeutic targets for sleep disorders. Nat Neurosci 2002; 5(Suppl):1071-1075.

CHAPTER 2

Sleep Hippocampal Theta Rhythm and Sensory Processing

Marisa Pedemonte and Ricardo A. Velluti

Introduction

Ancient human cultures have developed diverse forms of a device that, based on sensory stimulation, is used to put babies to sleep: the rocking cradle. Vestibular and somatosensory stimulation produced by the rocking movements, complemented by eye closure and other stimulation such as constant temperature and the mother's voice/song (lullaby) activating the auditory system, are able to induce sleep. On the other hand, it is a common experience that reducing the sensory afferent volleys to the brain can facilitate sleep.

A series of experimental data will be presented demonstrating the sensory input relevance in the organization of the sleep and wakefulness cycle. Firing rate shifts in auditory and visual neurones, changes in the pattern of discharge, and, most important, the temporal correlation of the spike timing with the hippocampal theta rhythm, will be set forth.

Sleep, a huge change in the brain physiology, depends on both, a series of active processes and passive mechanisms, e.g., functional sensory deafferentation[3,19] and neural networks changing organization. Although many signs of active processes have been shown, there are not enough experimental data to support a final decision about the relative contribution of passive processes.[16,27,43,46] However, both approaches may be partially reconciled conceding that the deafferentation may be provoked by an inhibitory influence acting, e.g., upon the ascending activating reticular system.

Our main purpose is to provide an experimental aspect of sensory data analysis, its relation to sleep and the hippocampal theta rhythm as an internal *zeitgeber* (time giver) for auditory and visual information processing.

The Hippocampal Theta Rhythm

The hippocampal theta rhythm is a well-known feature of the hippocampal electrogram in humans and other mammals although its functions remain partially unknown.[15,29,39,49]

Since the beginning, attention processes have been associated to the theta rhythm. Figure 1 shows a classical example on this matter. When a cat observed himself in a mirror exhibited a theta rhythm burst in the hippocampus.[14]

Although more prominent in active wakefulness and paradoxical sleep, the hippocampal theta can also be observed during slow wave sleep.[12,21] It has been related with phasic phenomena during paradoxical sleep,[11,22] with movements[5] and with autonomic control of the heart rate.[32,34]

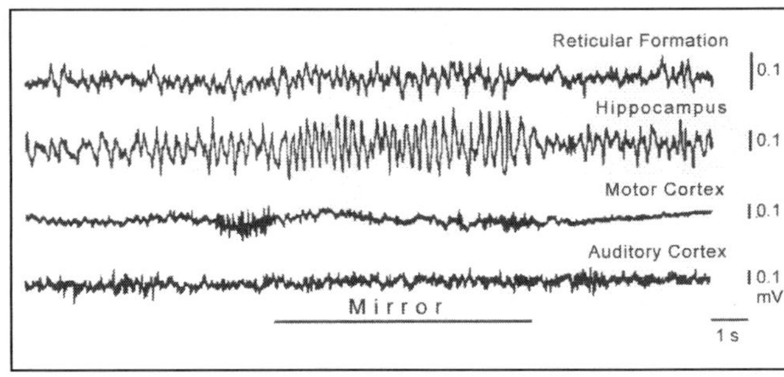

Figure 1. Recording in an awake cat showing enhancement of theta rhythm in the hippocampus when the animal see itself at a mirror (black bar). (Modified from Grastyan et al, 1959.)

Since the pioneering work of Scoville and Milner,[40] almost five decades of research resulted in the recognition of the hippocampus as a brain region implicated in learning and memory processes[4,42] in several species including humans.[1,45] Besides, theta blocking by septal lesion provokes memory impairment.[13]

Moreover, the hippocampus is involved in the neural coding of spatial position[2,28,51] necessarily associated with the sensory input and its processing. As an experimental animal traverses space, the hippocampal place neurones firing progressively changes to an earlier phase of the ongoing theta rhythm.[23,44] This may be relevant to long-term potentiation which is sensitive to the theta phase, i.e., potentiation increases at the theta peak while depression is associated with the throughs.

The theta wave may affect spatially distant neurons by inducing fluctuations in cellular excitability due to membrane potential oscillations.[11,20] Moreover, intracranial recording from human cortices have revealed theta oscillations in several brain regions including the neocortex, suggesting that theta waves may not reflect volume conduction from the hippocampus but the existence of theta generators in the brain surface.[17,18]

Our approach was to study the hippocampal theta influences on the unitary activity of the sensory systems in the context of wakefulness and sleep. The interactions were present in both, sleep and waking behaviour.

By studying the unitary activity of several auditory nuclei, the lateral geniculate visual thalamic neurones and their relationship to the hippocampal theta, we have found that this rhythm may play a role as an internal clock. We postulate it constitutes a low

Sleep and Sleep Disorders: A Neuropsychopharmacological Approach, edited by Malcolm Lader, Daniel P. Cardinali and S.R. Pandi-Perumal.

frequency *zeitgeber* associating a temporal dimension to the processing of auditory signals in the lateral superior olive,[47] inferior colliculus,[31,47] primary auditory cortex,[33,48] and to the visual processing in the thalamus.[10]

Auditory and Visual Neuronal Activity in Sleep and Wakefulness

Auditory Neurones

Changes in the Discharge Rate

Since sleep is a special physiological condition, it is possible to maintain that the incoming auditory information- always present- is differently processed in different states. Accordingly, the brain will select what input to focus on and determine to what neuronal network sensory neurones are engaged.

Approximately half of the auditory cortex neurones studied showed changes in the response to tone bursts in sleep compared to quiet wakefulness (Fig. 2). Those neurons that changed can be related to sleep processes still unknown, participating in a different neuronal assembly. There has been no auditory unit that stopped firing as the guinea pig enters sleep, thus, the auditory system is continuously monitoring the environment.[7,26,30,35,36,47,48]

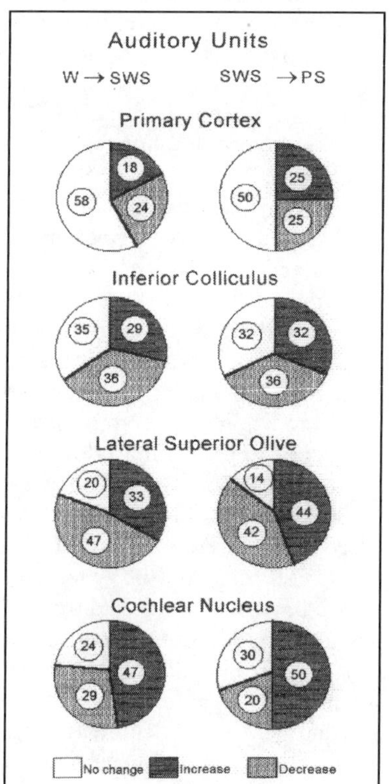

Figure 2. Neuronal sound evoked activity along the auditory pathway during different behavioural states in guinea pigs. Pie charts show percentages of neuronal firing changes on passing from wakefulness (W) to slow wave sleep (SWS), and from SWS to paradoxical sleep (PS) at four auditory loci: auditory cortex (A1), central nucleus of the inferior colliculus, lateral superior olive and ventral cochlear nucleus. No neurons were recorded that became silent on passing to sleep in the auditory regions studied (modified from Velluti and Pedemonte, 2002).

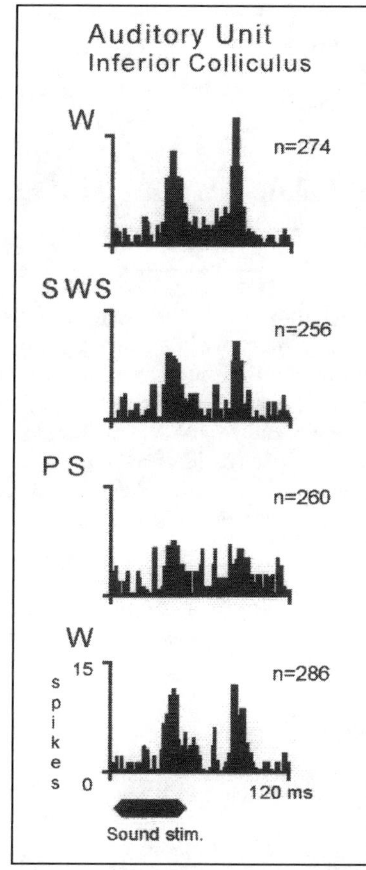

Figure 3. Tone-evoked response of an inferior colliculus neuron during waking (W), slow wave (SWS) and paradoxical sleep (PS) in guinea pig. Post-stimulus time histograms showed no significant changes in the firing rate (Mann-Whitney U-test) on passing from W to sleep phases, meanwhile the pattern of discharge during 120 ms exhibited differences. Two peaks present during W decrease in SWS and almost disappear during PS to recover, in the following W control, the same pattern and firing rate as the initial W. Prereceptorial effects were eliminated by removing the middle ear ossicles and delivering the sound directly to the ear. Sound stimulation: tone-burst at the characteristic frequency; intensity of 10 dB above the threshold; 50 ms duration and 5 ms rise-fall Modified with permission from Morales-Cobas et al. J Sleep Res 1995; 4:242-251. ©1995 Blackwell Publishing.

Changes in Discharge Pattern

Besides discharge rate, the pattern in which the firing develops may be relevant for the processing. In addition to changes in the firing rate throughout sleep and waking, a set of neurons exhibited shifts in the pattern of discharge (Fig. 3). Although some neurones showed no significant changes in discharge rate, the temporal distribution of spikes was different when the animal entered slow wave or paradoxical sleep. The discharge pattern was recovered in the following waking period, used as control.[26]

Response to Natural Guinea Pig's Calls

The study of neuronal response to natural stimuli may introduce to the analysis of their processing during sleep. It is known that a significant auditory stimulus may awake a person more easily than a non significant one. Besides, we have carried out stimulation with non significant natural call by just inverting it in time. Cortical auditory neurons (A1) exhibited firing shifts on passing from

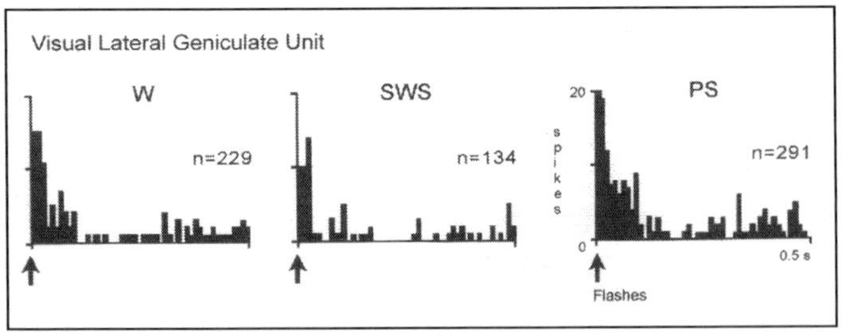

Figure 4. Visual evoked unitary firing in the lateral geniculate nucleus during wakefulness (W), slow wave (SWS) and paradoxical sleep (PS) in guinea pig. Post-stimulus time histograms showed statistically significant changes in the firing rate (Mann-Whitney U-test), both on passing from W to SWS and from SWS to PS. Light stimulation: flashes, 2/s, high intensity, 30 cm distance of the animal head, pupils pharmacologically dilated. The flashes' intensity and the pupils' dilation allowed a constant stimulation of the retina even during periods of eye movements.

wakefulness to slow wave sleep when stimulated with a natural call. Moreover, the pattern of discharge exhibited differences when the stimulus was presented direct or inverted in time.[37]

Visual Neurones

Unlike the auditory system, which is always relatively "open" to the environment, the activity in the visual system is expected to decrease during sleep due to closing of the eyelids. However, the discharge rate of visual neurones also showed changes during sleep. Figure 4 exhibits an increase in firing rate during paradoxical sleep. Visual neurons keep responding to a flash during sleep.

Not only the discharge rate but also the firing pattern may vary on passing from waking to sleep phases (Fig. 4). The incoming visual information reaches the visual centres differently depending on the current behavioural state. The central nervous system, acting through its sensory efferent system, may control its own input (see review ref. 46).

Role of the Theta Rhythm

The theta rhythm may be activated by several physiological variables, such as attention, movements, etc. Thus, the resulting cross-correlation may be dependent on the most relevant signal at a particular time. The changes associated with attention may appear when the input varies or when an unknown internal factor becomes relevant. There is a relationship between the theta power and the presence of phase locking with a sensory neuron, although other factors could condition such temporal correlation.

Auditory neurones from the lateral superior olive and central nucleus of the inferior colliculus exhibited phase locking to the hippocampal theta rhythm.[47] Although being nonrhythmic, the spontaneous activity of inferior colliculus neurones analysed during wakefulness exhibited phase locking to the hippocampal theta (Fig. 5). When the same unit was stimulated with a continuous pure tone at the neurone's characteristic frequency—adding a specific evoked activity—the neurone become more synchronized with the theta rhythm's frequency. Since there was no significant increase in the firing number (n=287 vs. n=289), it means that the theta rhythm input was, in this case, a relevant influence resulting in a spike autocorrelation increased rhythmicity.[31] This led us to conclude that the temporal correlation between both the rhythm and the unit is functionally significant.

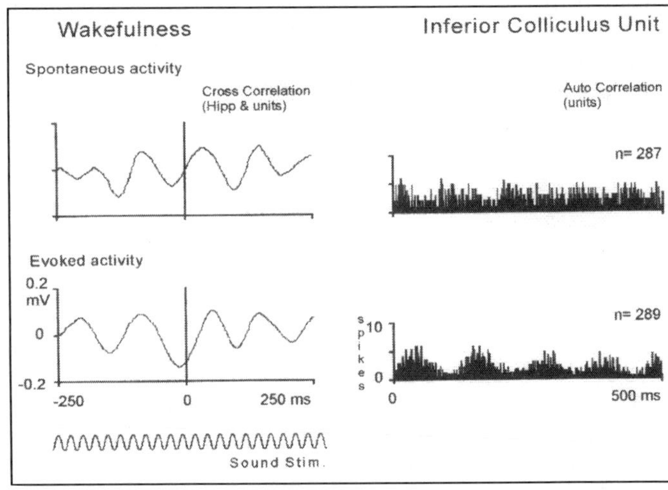

Figure 5. Temporal correlation between hippocampal theta rhythm (Hipp) and the spontaneous activity of an inferior colliculus neurone.in an awake guinea pig. The correlation of the spikes with the theta rhythm was studied with spike-triggered averaging of the hippocampal electrogram. Both, spontaneous and evoked activity showed temporal correlation with hippocampal theta rhythm (bipolarly recorded). The cross-correlation was considered positive when it became flat after "shuffling" the spikes series included in the data.[8] The autocorrelation histograms showed that the almost nonrhythmic discharge pattern during spontaneous activity (upper recording) becomes rhythmic at theta frequency during the evoked activity (lower recording), thus stressing the relevance of theta input onto the auditory cell. Sound stimulation: continuous tone at the unit´s characteristic frequency (1.1 kHz), 10 dB above threshold (modified from Pedemonte et al, 1996).

Since visual information includes temporal cues, our analysis was centred on the correlation between hippocampal theta rhythm and lateral geniculate activity.[10] Phase-relationships between hippocampal theta and unitary firing were found with both spontaneous and light evoked activity during wakefulness, slow wave and paradoxical sleep. This temporal correlation was dynamic, exhibiting changes related to the sleep-waking cycle and perhaps to attention shifts, e.g., in Figure 6 (right) the cross-correlation appeared when the flash stimulation ceased during a slow wave sleep epoch.

Auditory and visual units exhibited temporal correlation with the hippocampal theta rhythm during wakefulness, slow wave and paradoxical sleep.[33] An interesting finding is that the phase locking with hippocampal theta may be provoked by changes in the sensory input. We have found that auditory as well as visual neurones, may change from a nonphase-locked condition into a phase locked one after a change in its sensory input.

Figure 6 shows an example of an auditory cortex (A1) neurone (left) and a visual geniculate neuron (right), that became phase locked to theta when the stimulus changed during a slow wave sleep period.

The following experimental approach included guinea pig´s natural call ("whistle", 700 ms duration). During rhythmic and random presentations of natural calls, most A1 neurones exhibited phase locking to the hippocampal theta waves in wakefulness, slow wave and paradoxical sleep. The theta phase locking was also observed when the sound (natural call) was presented in reverse, i.e., inverted in time.[37]

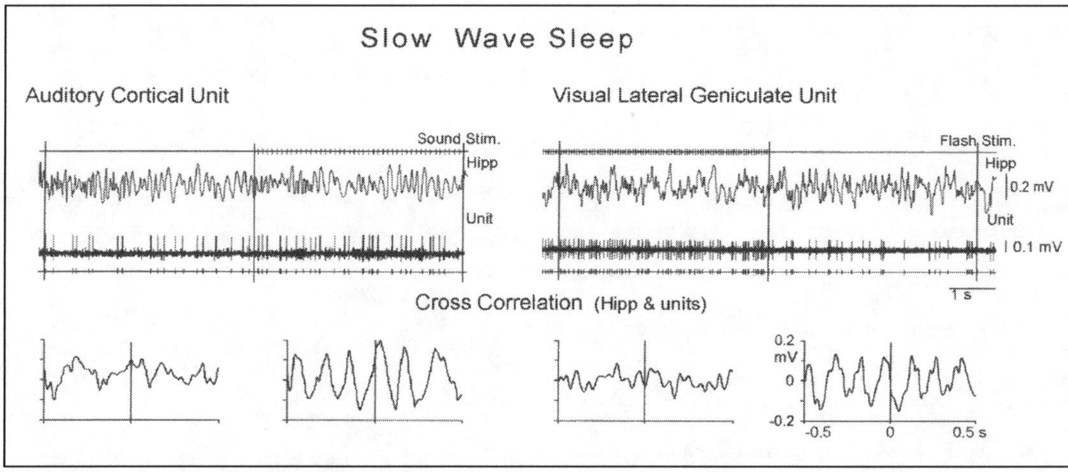

Figure 6. Firing and cross-correlation of two sensory unit- auditory and visual- when their input is changed during slow wave sleep. Top, single recordings of the hippocampus (Hipp), the unitary discharge and the digitised signals. Sound and flash stimulation (Stim) represent the synchronizing pulses. Bottom, four cross-correlations performed on the data sets limited by vertical lines in the above traces. Both are showing a change from the no phase locked condition to a temporal correlation (phase locking) with the hippocampal theta rhythm, when a shift in the sensory input occurred. Whereas the auditory cortical unit (A1) phase locked to the hippocampal theta only when the sound stimulation began, the visual neurone showed the opposite phenomenon. The lateral geniculate neuron exhibited no phase locking to the theta when the flashes of light were present. With the flashes off, the correlation with hippocampal theta appeared (modified from Velluti and Pedemonte, 2002).

Conclusions

Our experimental approaches have assessed different aspects of the neuronal activity that give us an insight on how sensory information processing and sleep mechanisms reciprocally affect each other, participating in the processing and/or in sleep promoting functions. The neuronal discharge level and pattern of discharge changes in response to constant stimuli indicate that the CNS is modulating (selecting?) and perhaps distributing the incoming auditory information according to its current state. Auditory,[46] somatosensory,[38,41] and visual neurones studied[9,23,25] exhibited changes in their firing rates in correlation with stages of sleep and wakefulness. This is consistent with the hypothesis of a general shift in the neuronal networks involved in sensory processing during sleep that may participate in the switch to a sleeping mode of cell assemblies. A number of neurones at different auditory *loci,* from brain stem to the cortex itself, presented significant quantitative and qualitative changes in their evoked firing rates and pattern. On the other hand, another group of neurones, recorded in every nucleus of the pathway, did not show behavioural related changes in firing rates on passing to sleep. Moreover, no neuron belonging to any pathway level or cortex was observed to stop firing on passing to sleep.

A close temporal correlation between hippocampal theta rhythm and unitary firing was also demonstrated at several stages in the auditory pathway,[31,33,47,48] and in visual neurons at the thalamic lateral geniculate nucleus.[10] At a neural population scale, this phase-locking may result in a composite final signal that could be used in processes like attention, movements, and, in particular auditory/visual sensory input processing. Furthermore, we hypothesize that the phase locking to the hippocampal theta adds a temporal dimension to the sensory processing, perhaps necessary for time related perception also during sleep. Given that every stimulus develops in time, the CNS must have a way to encode this parameter. Hippocampal theta, being one of the most regular brain-generated low frequency rhythms, may participate in this internal clock.

The temporal relationship between the sensory neuronal firing and the hippocampal theta field activity is a changing phenomenon whose variation may depend on the interaction of a set of signals: (a) the hippocampal theta rhythm amplitude and frequency, (b) the current state of the brain, awake or asleep, and (c) the incoming sensory information. A neuronal assembly may shift its discharge pattern by changing the interaction between these three input signals, e.g., facing a novel stimulus may change the brain condition evoking a new phase locking with a higher power of hippocampal theta waves.

The parallel recording of hippocampal theta field activity and cortical auditory multiunit firing revealed a precise temporal organization of population events during wakefulness, slow wave and paradoxical sleep. The notion of a discontinuous exchange of information between hippocampus and cortical areas is supported by the data. The phase locking of cortical auditory units and hippocampal theta mainly occurs when a novel stimulus or an on/off condition of the same one are applied during wakefulness, slow wave or paradoxical sleep, indicative of a "top down" theta action.[50]

Corollaries

a. The auditory units that did not change firing in sleep may be related to the environmental monitoring during sleep.

b. Those units that shift their firing may be related to unknown processes during sleep, or perhaps, act as active signals in sleep related neuronal networks.

c. The auditory neurons that keep responding during sleep, as well as those exhibiting theta phase-locking, convey information that could be the first step in the complex auditory learning function, which is consistent with a recent report of learning during sleep in human newborns.[6]

d. The activity-dependent development of the brain during early life may not only occur during wakefulness.[24] We suggest that it also occurs associated to auditory and visual incoming information during the long periods of sleep in newborns and infants.

During early ontogenetic development, and maybe in adults, the sensory information reaches the CNS not only during wakefulness but also during sleep. This continuous sensory input may "sculpt" the brain and participate in the adaptation to novel conditions.

References

1. Basar E, Schurmann M, Sakowitz O. The selectively distributed theta system: Functions. International J Psychophysiol 2001; 39:197-212.
2. Best PJ, White AM, Minai A. Spatial processing in the brain: The activity of hippocampal place cells. Annu Rev Neurosci 2001; 24:459-486.
3. Bremer F. Cerveau "isole" et physiologie du sommeil. C R Soc Biol 1935; 118:1235-1241.
4. Brown MW, Aggleton JP. Recognition memory: What are roles of the perirhinal cortex and hippocampus? Nat Rev Neurosci 2001; 2:51-61.
5. Buño W, Velluti JC. Relationship of hippocampal theta cycle with bar pressing during self-stimulation. Physiol Behav 1977; 19:615-621.
6. Cheour M, Martynova O, Naatanen R et al. Speech sounds learned by sleeping newborns. Nature 2002; 415:599-600.
7. Edeline JM, Dutrieux G, Manunta G et al. Diversity of receptive field changes in auditory cortex during natural sleep. Eur J Neurosci 2001; 14:1865-1880.
8. Fuentes J, Buño W, García-Austt E. Simulation of post-synaptic activities in hippocampal cells during theta rhythm. Brain Res Bull 1981; 7:157-162.
9. Gambini JP, Pedemonte M, Velluti RA. Sleep-wakefulness modulation of lateral geniculate visual information: Unitary study and hippocampal theta phase-locking. Sleep Res Online 1999; (Suppl 1) 2:99.
10. Gambini JP, Velluti RA, Pedemonte M. Hippocampal theta rhythm synchronized visual neurons in sleep and waking. Brain Res 2002; 926:137-141.
11. García-Austt E. Hippocampal level of neural integration. In: Ajmone-Marsan E, Reinoso-Suárez F, eds. Cortical integration. Basic archicortical and cortical association levels of neuronal integrations. New York: Raven Press, 1984:91-104.
12. Gaztelu JM, Romero-Vives M, Abraira V et al. Hippocampal EEG theta power density is similar during slow-wave sleep and paradoxical sleep. A long-term study in rats. Neurosci Lett 1994; 172:31-34.
13. Givens BS, Olton DS. Cholinergic and GABAergic modulation of medial septal area: Effect on working memory. Behav Neurosci 1990; 104:849-855.
14. Grastyán E, Lissák K, Madarász I et al. Hippocampal electrical activity during the development of conditioned reflexes. Electroenceph Clin Neurophysiol 1959; 11:409-430
15. Green JD, Arduini AA. Hippocampal electrical activity in arousal. J Neurophysiol 1954; 17:403-420.
16. Hess WR. Le sommeil comme une fonction physiologique. J Physiol 1949; 41:61A-67A Paris.
17. Kahana MJ, Sekuler R, Caplan JB et al. Human theta oscillations exhibit task dependence during virtual maze navigation. Nature 1999; 399:781-784.
18. Kahana MJ, Seelig D, Madsen JR. Theta returns. Curr Opinion Neurobiol 2001; 11:739-744.
19. Kleitman N. Sleep and Wakefulness. Chicago, London: The University of Chicago Press, 1963.
20. Kocsis B, Vertes RP. Dorsal raphe neurons: Synchronous discharge with theta rhythm of the hippocampus in the freely behaving rat. J Neurophysiol 1992; 68:1463–1467.
21. Komisariuk B. Synchrony between limbic system theta activity and rhythmical behaviour in rats. J Comp Physiol Psychol 1970; 10:482-492.
22. Lerma J, García-Austt E. Hippocampal theta rhythm during paradoxical sleep, Effects of afferent stimuli and phase-relationships with phasic events. Electroenceph Clin Neurophysiol 1985; 60:46-54.
23. Livingstone MS, Hubel DH. Effects of sleep and arousal on the processing of visual information in the cat. Nature 1981; 291:554-561.
24. Marks GA, Shaffery JP, Oksenberg A et al. A functional role for REM sleep in brain maturation. Behavioural Brain Res 1995; 69:1-11.
25. McCarley R, Benoit O, Barrionuevo G. Lateral geniculate nucleus unitary discharge in sleep and waking: State- and rate- specific aspects. J Neurophysiol 1983; 50:798-817.
26. Morales-Cobas G, Ferreira MI, Velluti RA. Sleep and waking firing of inferior colliculus neurons in response to low frequency sound stimulation. J Sleep Res 1995; 4:242-251.
27. Moruzzi G. The sleep-waking cycle. Ergebnisse der Physiologie 1972; 64:1-165.
28. O'Keefe J, Recce ML. Phase relationship between hippocampal place units and EEG theta rhythm. Hippocampus 1993; 3:317-330.
29. O'Keefe J, Burgess N. Theta activity, virtual navigation and the human hippocampus. Trends Cognit Sci 1999; 3:403-406.
30. Pedemonte M, Peña JL, Morales-Cobas G et al. Effects of sleep on the responses of single cells in the lateral superior olive. Arch Ital Biol 1994; 132:165-178.
31. Pedemonte M, Peña JL, Velluti RA. Firing of inferior colliculus auditory neuron is phase-locked to the hippocampus theta rhythm during paradoxical sleep and waking. Exp Brain Res 1996; 112:41-46.
32. Pedemonte M, Rodríguez A, Velluti RA. Hippocampal theta waves as an electrocardiogram rhythm timer in paradoxical sleep. Neurosci Lett 1999; 276:5-8.
33. Pedemonte M, Pérez-Perera L, Peña JL et al. Sleep and wakefulness auditory processing: Cortical units vs. hippocampal theta rhythm. Sleep Res Online 2001; 4:51-57.
34. Pedemonte M, Goldstein-Daruech N, Velluti RA. Temporal correlation between heart rate, medullary units and hippocampal theta rhythm in anesthetized, sleeping and awake guinea pigs. Autonomic Neurosci 2003; 467:107: 99-104.
35. Peña JL, Pedemonte M, Ribeiro MF et al. Single unit activity in the guinea-pig cochlear nucleus during sleep and wakefulness. Arch Ital Biol 1992; 130:179-189.
36. Peña JL, Pérez-Perera L, Bouvier M et al. Sleep and wakefulness modulation of the neuronal firing in the auditory cortex of the guinea-pig. Brain Res 1999; 816:463-470.
37. Pérez-Perera L. Actividad unitaria de la corteza auditiva: Ritmo theta del hipocampo y respuesta a vocalizaciones en el ciclo vigilia-sueño. Tesis de Maestría. Programa de Desarrollo de Ciencias Básicas-Facultad de Ciencias. Montevideo Uruguay 2002.
38. Pompeiano O. Mechanisms of sensorimotor integration during sleep. In: Stellar E, Sprague JM, eds. Progress in Physiological Psychology. New York, London: Academic Press, 1970:1-179.
39. Raghavachari S, Rizzuto D, Caplan J et al. Gating of human theta oscillations by a working memory task. J Neurosci 2001; 21:3175-3183.
40. Scoville WB, Milner B. Loss of recent memory after bilateral hippocampal lesions. J Neurol Neurosurg Psych 1957; 20:11-21.
41. Soja PJ, Cairns BE, Kristensen MP. Transmission through ascending trigeminal and lumbar sensory pathways: Dependence on behavioral state. In: Lydic R, Baghdoyan HA, eds. Handbook of Behavioral State Control. Boca Raton, London, New York, Washington: CRC Press, 1998:521-544.
42. Sutherland GR, McNaughton B. Memory trace reactivation in hippocampal and neocortical neuronal ensembles. Curr Opin Neurobiol 2000; 10:180-186.
43. Szymusiak R. Magnocellular nuclei of the basal forebrain: Substrates of sleep and arousal regulation. Sleep 1995; 18:478-500.
44. Skaggs WE, McNaughton BL, Wilson MA et al. Theta phase precession in hippocampal neuronal populations and the compression of temporal sequences. Hippocampus 1996; 6:149-172.
45. Tesche CD, Karhu J. Theta oscillations index human hippocampal activation during a working memory task. Proc Natl Acad Sci USA 2000; 97:919-924.
46. Velluti RA. Interactions between sleep and sensory physiology. J Sleep Res 1997; 6:61-77.
47. Velluti RA, Peña JL, Pedemonte M. Reciprocal actions between sensory signals and sleep. Biol Signals Recept 2000; 9:297-308.
48. Velluti RA, Pedemonte M. In vivo approach to the cellular mechanisms for sensory processing in sleep and wakefulness. Cell Mol Neurobiol 2002; 22:501-516.
49. Vertes RP, Kocsis B. Brainstem-diencephalo-septo-hippocampal systems controlling the theta rhythm of the hippocampus. Neurosci 1997; 81:893-926.
50. von Stein A, Sarnthein J. Different frequencies for different scales of cortical integration: From local gamma to long range alpha/theta synchronization. Int J Psychophysiol 2000; 38:301-313.
51. Wallenstein GV, Eichenbaum H, Hasselmo ME. The hippocampus as an associator of discontiguous events. Trends Neurosci 1998; 21:317-323.

CHAPTER 3

Monoaminergic Mechanisms in the Regulation of Sleep-Wakefulness:

Special Emphasis on Preoptic Noradrenergic System

Vijay Ramesh and Velayudhan Mohan Kumar

Introduction

The catecolaminergic, serotonergic and histaminergic projections arises from different nucleus groups in different regions of the brain (Fig. 1). They have an important role in the control of behavioral state, body temperature, reproduction and other vital functions. Among these, catecolaminergic projection system, by the action of noradrenaline (NA) released by it at different brain areas, especially on the preoptic area (POA), has a major influence on modulating sleep-wakefulness (S-W). In this chapter, we will discuss in detail, the role of POA in S-W. The roles of these neurotransmitters on other brain areas are beyond the scope of this chapter.

von Economo's finding in 1929 on sleeping sickness was the first milestone in the understanding of the involvement of the hypothalamus in the regulation on S-W. While studying the post-mortem material of an epidemic *encephalitis lethargica*, he described two symptomatic patterns of the disease associated with two different localizations of inflammatory lesions in the nervous system. In those cases in which somnolence and opthalmoplegy were distinguishing symptoms, the lesions were regularly found in the posterior wall of the third ventricle, extending caudally to the level of the occulomotor nucleus. In contrast to this, there were other cases in which insomnia was observed in addition to chorea. In these patients, inflammations were associated with the rostral hypothalamus, the tuberal region and the adjacent portion of the striatum. From these observations, von Economo concluded that the rostral hypothalamic zone was a part of a "sleep regulating center" which, when appropriately excited, actively inhibited the thalamus and cerebral cortex and caused "brain sleep". Therefore, he concluded that the rostral hypothalamus is a "*schlafsteuerungszentrum*" or "sleep center". The importance of the POA in the induction of sleep, in animals, was later experimentally substantiated by many studies.[6,20,31,40-42,53,56-58,73,82,84,96,97,107] The POA has noradrenergic terminals.[22,71] It also has tyrosine hydroxylase, and its rate-limiting enzymes.[70,112] There are α and β adrenergic receptors in the POA, though β receptors are relatively fewer in this area.[13]

Morphology and Anatomy of the Preoptic Area

The POA has been extensively studied in a variety of mammals,[1,4,8,12,15,17,33,37,62,78,100] The medial part of the POA, the medial preoptic area (mPOA) is rostrally bound by the nuclei of the diagonal band of Broca and accumbens. Caudally its border is not well defined and is continuous with the anterior hypothalamic region. Medially, it is delimited by the third ventricle and laterally by the anterior amygdaloid area. Dorsally, it is related to the anterior commissure and ventrally to the optic chiasma. It is considered to be derived embryologically from unevaginated telencephalon medium.[14,100] Morphologically, the rostral hypothalamus and the preoptic region are inseparable which has given birth to the belief amongst some workers that the POA may be a hypothalamic derivative, arising caudal and ventral to the lamina terminalis and the commissural plate.[49,89]

In rats, the POA is a small area in the basal forebrain, measuring less than 1.5 mm in the rostro-caudal extent. The POA can be divided cytoarchitectonically into three zones: (a) lateral (b) medial and (c) periventricular.[8,33,101] The lateral and medial POA are clearly identified in monkeys, but not so in cats.[93,100]

Lateral Zone

The lateral POA is characterized by medium sized neurons, scattered among the fibers of a conspicuous dorsomedial division of the medial forebrain bundle.[8,33,100] The central part of this area is called pars preoptica of medial forebrain bundle. The medial part contains fibers and many neurons, and is called the lateral preoptic nucleus.[8,33,39,100] Lateral preoptic zone merges medially with the mPOA and caudally with the lateral hypothalamus without any clearcut demarcation. Rostrally, it is delimited by the diagonal band of Broca and nucleus accumbens septi. Dorsally, it is separated from the bed nucleus of stria terminalis by a thin layer of relatively cell free zone.[100]

Medial Zone

The medial zone consists of loose, heterogeneous, medium sized and relatively densely packed neurons and appears to be free of fibers. It is limited dorsally by the anterior commissure and ventrally, by the optic chiasma. Laterally, it merges with lateral POA without any clearcut demarcation. Caudally, it merges with the less cell dense anterior hypothalamic area. Rostrally, it is confluent with the septum, and the basomedial aspect, with the diagonal band and its nucleus. The nucleus of the stria terminals is wedged between the lateral and medial POA dorsally. Densely

Sleep and Sleep Disorders: A Neuropsychopharmacological Approach, edited by Malcolm Lader, Daniel P. Cardinali and S.R. Pandi-Perumal.

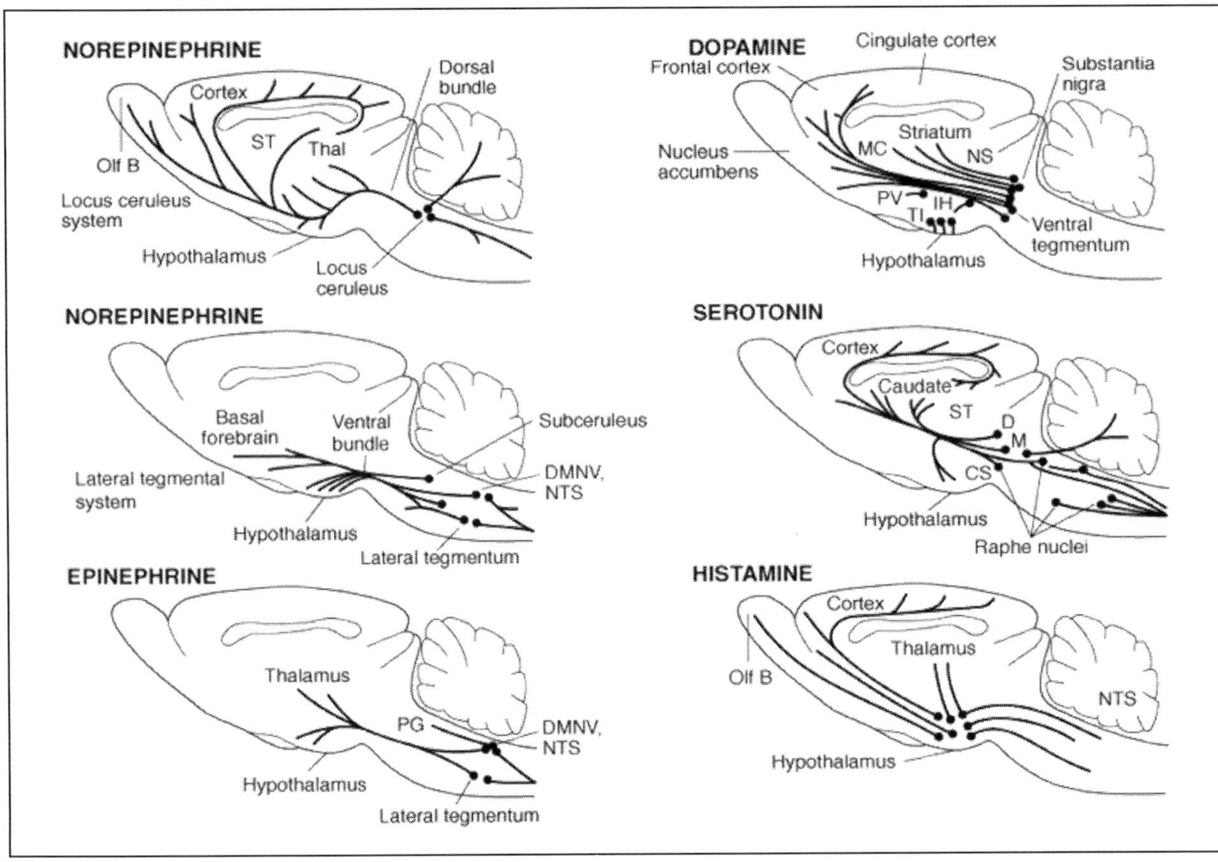

Figure 1. Monoaminergic pathways in rat brain. The pathways in human appear to be similar. The 2 principal noeadrenergic systems (locus ceruleus and lateral tegmental) are shown separately. Olf B= olfactory bulb; Thal= thalamus; ST= stria terminalis; DMNV= dorsal motor nucleus of vagus; NTS= nucleus of tractus solitarius; PG= periaqueductal gray; NS= niagrostriatal system; MC= mesocortical system; PV= periventricular system; IH= incertohypothalamic system; TI= tuberoinfundibular system; D,M, and CS= dorsal, medial, and central superior raphe nuclei. (Reproduced with permission from Review of Medical Physiology by William F. Ganong; Appleton and Lange, Norwalk, Connecticut.)

grouped column of small round cells which run along the ventral part of the third ventricle, immediately, caudal to the lamina terminalis and preoptic periventricular nucleus is called the medial preoptic nucleus.[8]

Periventricular Zone

This is a thin region of cells lining which are oriented parallel to the third ventricle. Preoptic periventricular nucleus is a collection of small densely grouped cells, which is most prominent rostral to the third ventricle. As the ventricle and the lamina terminalis appear, the periventricular nucleus turns laterally and a pair of prominent bar like collection of cells is seen as part of its ventral extremity. Caudal to this level, the preoptic periventricular nucleus cells are loosely grouped and finally succeeded by the periventricular cells of the anterior hypothalamic area.[8,100]

The description, so far attributed to the POA, is as seen by gross examination and light microscopy. Further anatomical information regarding the functional importance, has been gained from histochemical and biochemical studies of neurotransmitters present in this area. In this chapter we limit our discussions mainly to the noradrenergic system, although discussion on other monoaminergic neurotransmitters will be briefly touched upon.

Noradrenergic System in Rat Brain

Noradrenergic System

Histofluroscence and spectrophoto-fluoremetric techniques have shown that noradrenergic fibers are present in the mPOA.[2,7,43,65,74,110] This has been further substantiated by immunofluroscence studies.[47,102] Moderate amounts of NA have been shown to be present in the mPOA by radioisotope studies.[36,48,77,113] 6-hydroxydopamine (6-OHDA) was used to map central noradrenergic pathways.[38,90] Using the above mentioned techniques, several ascending and descending noradrenergic pathways have been identified in the brain stem of rat.[7,60,110] These pathways arise from catecholamine containing cell bodies, in the brain stem. They are classified into twelve groups, designated as A1 to A14.[18] These cell groups are present in the medulla oblongata, pons, mesencephalon, caudal thalamus, mediobasal hypothalamus, zona incerta, caudal hypothalamus, periventricular grey of the anterior hypothalamus and POA[7,26,61] (Tables 1 and 2).

There are two major ascending pathways originating from the noradrenergic cell groups and supplying the whole of diencephalon and cortex which are classified as the dorsal noradrenergic bundle (DNA) and the ventral noradrenergic bundle (VNA).

Table 1 . Noradrenergic system (origin of cell groups)

Lateral Tegmental Medullary System	
A1	Lateral tegmental medullary cell group (LTMC) in the ventrolateral part of the anterior medullary tegmentum of medulla oblongata.
A2	Dorsal medullary cell group in the dorso-medial part of the medulla oblongata in the NTS, the commissural nucleus and the dorsal motor nucleus of the vagus (X).
A3	LTMC in the ventro-lateral part of the posterior medullary tegmentum.
A4	Cell group in the roof of the 4th ventricle, laterally and dorsally along the medial aspect of the superior cerebellar peduncle.
Lateral Tegmental Pontine System	
A5	Lateral tegmental pontine cell group in the rostral part of the facial nucleus upto parabrachial nucleus just rostal to the LC.
A7	Cell group in the parabrachial region primarily in the area ventro-lateral to the superior cerebellar peduncle, rosatral to cell group A5 in the lateral pontine reticular formation.
Locus Coeruleus (LC) System	
A6	Cell bodies densely packed within the LC proper, found in the gray matter of pontine brainstem at the floor of the 4th ventricle where they intermingle with the cells of the subcoeruleus.

Table 2. Noradrenergic system (projection system)

Dorsal Noradrenergic Bundle (DNA)
This tract ascends through the mesencephalon just ventro-lateral to the central gray, to enter the MFB. Before entering the MFB some fibers leave the DNA to innervate habenula and thalamus (via the internal and external thalamic laminae). Many NA fibers through MFB innervates dorsal thalamus, anterior thalamus (via the mammillothalamic tract), ventral amygdala, ventral hippocampus, and ventral cortical areas (via the ventral amygdalofugal pathway), striatum and subthalamic extra-pyramidal areas (via ansa lenticularis). Near the rostral pole of the MFB at the level of the septum, DNA breaks up into 5 final groups which travel the rostral habenula and dorsal thalamus (via stria medullaris), amygdala (via stria terminalis), olfactory regions (via MFB) and the cerebral cortex (via the cingulum) respectively. Also gives afferents to dorsal hypothalamus.
Ventral Noradrenergic Bundle (VNA)
A portion of the LC fibres more ventrally from the LC join the CTT, which traverses the VTA, until it reaches the caudal hypothalamic zones including paraventricular nuclei.
Dorsal Periventricular Tract (DPT)
Fibers from LC enter the PAG and ascend as components of the dorsal longitudinal fasiculus. Axons can be followed rostrally as far as the middle of the hypothalamus where they join DNA that are destined to innervate the caudal and dorsal hypothalamus. Fibers from the LC join the superior cerebellar peduncle to innervate the cerebellum (Purkinje cells). LC also projects to the medulla oblongata and spinal cord via fibers which descend in the CTT and few of which descend in the dorsal periventricular system. LC also innervate all 6 layers of neocortex.

Dorsal Noradrenergic Bundle

The cell bodies of the DNA arise in the nucleus of locus coeruleus. The DNA bundle occupies the dorsal part of the ascending group of noradrenergic axons. At the caudal level of nucleus pontis, these axons turn dorso-medially to form a completely separate, dorsal bundle of axons. At the rostral level of nucleus mamillaris they turn ventro-laterally to join the ascending noradrenergic and dopaminergic axons which then ascend in the medial forebrain bundle and in the septum and then caudally into the cingulum. It seems that the pathway radiate branches to the corpora geniculata and the thalamic nuclei. Lesion studies have shown that this pathway also innervates the cortex and the hippocampus.

Ventral Noradrenergic Bundle

The cell groups, which give rise to the VNA bundle, comprise A1, A2, A5, and A7 cell groups. These cell groups are seen in the medulla oblongata and the pons. The axons from these cell bodies ascend in the mid reticular formation, turn ventromedially along the medial forebrain bundle. These systems give rise to the noradrenergic nerve terminals in the lower brain stem, the mesencephalon and diencephalon. In the medulla oblongata and the pons, the VNA and the DNA system overlap and contribute together to the terminal areas. However, in the mesencephalon and the diencephalon, the majority of the noradrenergic fibers derive from the ventral system. The VNA bundle innervates the whole of the hypothalamus, and most notably, the nucleus dorso-medialis, hypothalamus, nucleus periventricularis, the area ventral to the fornix, the nucleus paraventricularis, nucleus supraopticus and the POA. Further rostrally, the ventral pathway supplies the terminals of the densely innervated nucleus interstialis stria and terminates in the ventral part.[110]

Role of the Preoptic Area in Sleep-Wakefulness

Lesion and Brain Section Studies

In 1946, Nauta, employing the knife cut lesion technique showed that the rats became insomniac, restless and irritable after lesion of the POA. They reacted vigorously even to minor stimuli. Those rats, which survived up to13 days, did not show a return to what he called the "capacity of sleeping". He thought that the POA was an important region for sleeping and he termed this area as "sleep center". Nagel and Satinoff in 1980 reported hyperactivity in rats after bilateral electrolytic region of the mPOA. Sterman et al, in 1964 confirmed Nauta's findings in cats. This was based on behavioral observations as well as EEG records. McGinty and Sterman[67] reported that large bilateral preoptic lesions produced complete sleeplessness in cats. Smaller lesions resulted in significant reduction in slow wave sleep (SWS) as well as paradoxical sleep (PS). The severity of sleep suppression was related to the size and localization of lesion placed specifically within the POA.[63] These lesions shortened the mean periodicity of the sleep awake cycle with decrease in SWS and no alteration in PS.[63] Cell specific neurotoxic lesion studies counter the argument that the effects produced by other means also destroy the fibers of passage. Asala et al., in 1990, reported decrease in day time SWS in rats as a result of radio frequency lesion of mPOA. This also disturbed the circadian distribution of sleep. Lesions produced by neurotoxins such as kainic acid and NMDA which spares the fibres of passage in the POA, reduce both SWS and PS.[40,42,103] Studies on the POA warming and lesion also show that POA regulates delta EEG activity within NREM sleep, an index of the depth of sleep.[68]

Stimulation Studies

Electrical, chemical and thermal stimulation of POA have also shown involvement of the POA in the regulation of S-W. Sterman and Clement[97] on the basis of behavioral and electrophysiological observations reported that bilateral stimulation of the POA in unanesthetized, freely moving cats, produced sleep. Low frequency stimulation was effective in inducing sleep. The effect of low frequency stimulation (5-25 cycles/sec) on induction of sleep was confirmed by Hernandez-Peon (1962) and Yamaguchi et al.[117] These studies were also carried out in conscious cats. In addition they showed that high frequency stimulation (200-300 Hz) induced cortical EEG desynchronization and some signs indicative of behavioral arousal. Thus, it can be concluded that both sleep and arousal responses can be obtained from electrical stimulation of POA, depending upon the rate and site of stimulation.

Warming the POA produces sleep.[5,79,88,115] Radio frequency diathermic warming of the POA in cats and opossum induced sleep[88] while cooling the POA produced huddled posture.[25,87] Roberts and Robinson[88] have suggested that the POA thermoreceptors may provide input to sleep mechanisms situated in this area. Stimulation of central receptors by changing blood temperature is likely to be an important source of impulses driving the sleep inducing structures of basal forebrain.[72]

In 1965, Hernandez-Peon showed that application of ACh at the POA elicited EEG synchronization and sleep, whereas application of NA at same site elicited EEG desynchronization and arousal in cats. The induction of arousal after application of NA crystals in the preoptic area was confirmed in cats by Yamaguchi et al.[117] Garcio-Arraras and Pappenheimer[28] demonstrated that microinjection of muramyl peptide in the same area promotes sleep. Application of serotonin (5-HT) crystals in the POA produced drowsiness and SWS in freely moving rats.[117] But, Datta et al[19] showed that 5-HT application at the same site did not have any change in SW.

Unit Recording Studies

Unit recording studies also support the notion that an active site for the induction of SWS may reside in the POA.[75] Mallick et al[64] showed that a majority (55%) of neurons of POA showed alterations in their firing rate during transient changes in EEG. Among these 62.5% showed increased firing during synchronization and the remaining 37.5% showed increased firing during desynchronization of the EEG.[24] Findlay and Hayward[24] showed in freely moving rabbits that the majority of the neurons in the hypothalamus, including the POA showed higher firing rates during sleep than during wakeful state. In cats, the maximal discharge rate was obtained during SWS and REM sleep.[45] A small percentage of neurons (10%) showed a maximal discharge rate during wakeful period. These observations thus support the earlier stimulation and lesion studies that showed that thePOA is involved in both sleep and wakefulness functions.

Techniques of Studying Monoaminergic Mechanisms in the Regulation of Sleep-Wakefulness in Experimental Animals

Many studies have been conducted on many animal species, including man, in understanding the basic mechanisms of sleep and wakefulness. Mainly, the animals include rats, mice and cats, wherein the researchers have employed various techniques like, intracerebral microinjections, lesions, unit recordings and molecular biology. Basically, to study the effect of drugs on S-W, we need to implant chronically indwelling electrodes for recording EEG, EMG and EOG and a cannula for perfusing the brain tissue with the drug in question or to withdraw the CSF samples for further analysis (Fig. 2). These electrographic recordings will enable the investigator to determine the behavioral state of the animal. The two main states of behavior, the wakeful state and the sleep state can be further classified into active wake (W1), quiet wake (W2), light SWS (S1), deep SWS (S2) and PS (Fig. 3). EEG during W1 stage had low amplitude, low frequency (desynchronized) waves. EMG and EOG record gross body movements and eyeball movement artifacts, respectively. The animal showed grooming, scratching and orienting activities during this

Figure 2. Photograph showing a rat in a recording cage with chronic head implants and cable for recording sleep-wake state and indwelling cannula for microinjection.

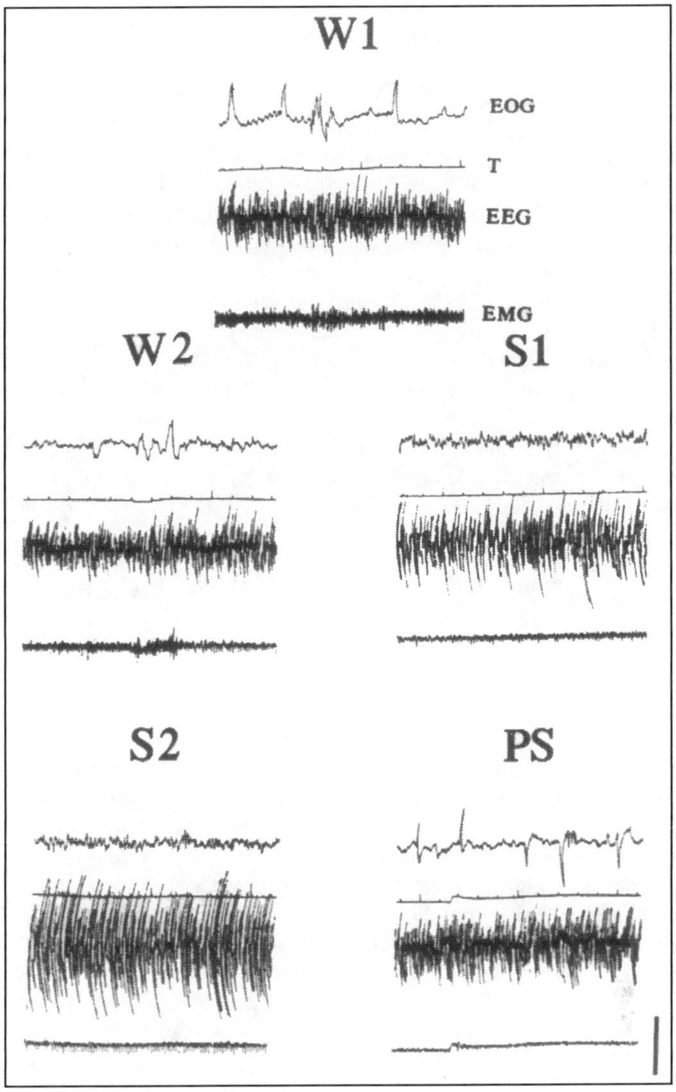

Figure 3. Polygraphic recording showing high speed (10 mm/sec) of electrooculogram (EOG), electroencephalogram (EEG) and electromyogram (EMG) of rat during different sleep-wake stages. W1, active wake; W2, quiet wake; S1, light slow wave sleep; S2, deep slow wave sleep; PS, paradoxical sleep/ rapid eye movment sleep; T, time shows 1/ sec signal. Cal bar: 600 μv (EOG), 300 μv (EEG) and 200 μv (EMG).

period. During W2, the EEG remained desynchronized. EMG though high did not show any movement artifacts. EOG activity were practically absent though some slow rolling movement could be seen. The animals were mostly sitting quietly during this period. S1 stage was characterized by low frequency, high amplitude (synchronized) EEG spindle waves. There was considerable reduction in EMG activity. The rats assumed a sleeping posture during this period. There was no EOG activity. S2 was characterized by continuous electrocortical slow wave activity, where synchronized waves were not seen as separate spindles. There was further reduction in EMG and no EOG activity was noticed. The PS was characterized by desynchronized EEG, drastic reduction in EMG amounting to muscle atonia and spiky waves in the EOG.

Though important studies were conducted by several scientists on different species to find out the role of the monoaminergic system on S-W, in this chapter, we will be discussing the role of noradrenergic fibers of the POA in regulating sleep. The main contents of this chapter are based on the studies conducted in the All India Institute of Medical Sciences on Wistar rats weighing between 225-250 g unless otherwise mentioned. They were housed in separate cages in an animal room having controlled temperature (26 ± 2°C) and light-on period from 05.00 h to 19.00 h. Food and water were provided ad libitum. A brief description of the methodology of each section is given below. For more information readers are advised to refer original articles.

Changes in Sleep-Wakefulness after Destruction of NE Fibers in the POA

In this group of studies, afferent fibers were selectively destroyed by intracerebral injection of neurotoxins. S-W was assessed on the basis of EEG, EMG and EOG recordings, through electrodes chronically implanted, under pentabarbitone sodium anesthesia (40 mg/kg b.w.) as described elsewhere.[50,58,82] The behavior of the animals was continuously monitored. After the rats recovered from surgery, they were adapted for three days to move around freely in the recording cage with the attached recording cables. After obtaining the control record of S-W for 24 h, catecholaminergic fibers in the POA were destroyed by injection of 8 μg of 6-OHDA into the POA through an injector cannula, under pentobarbital anesthesia. S-W was recorded for 24 h on different days after the catecholaminergic fibers were destroyed. The rats were treated with 25 mg/kg, i.p. desmethylimipramine (DMI), 30 min before injection of the same dose of 6-OHDA at the POA, for those groups of rats in which dopaminergic fibers were to be selectively destroyed.

The thermal preference of the rats was examined by keeping them individually in a specially designed environmental chamber.[76,86] The chamber had three interconnected compartments maintained at 24°C, 27°C and 30°C. The animals moved around freely from one chamber to the other. One rat, at a time, was kept in the environmental chamber. The duration of the stay in each compartment by the rats, during five prelesion days was compared with the post-lesion readings. The injection sites and the degree of fiber degeneration were histologically confirmed after the study by staining the fibers through the glyoxylic acid method.[108]

The lesion technique is one of the most important tools for studying the function of any brain area. Most of the studies after transactions, electrolytic or radiofrequency lesions of the POA in rats resulted in insomia.[3,73] But the major problem of these techniques was that the lesion area was not confined and could not be controlled. The observed effects could have been either due to the destruction of the neurons or fibers of passage, including the afferent terminals. The discovery that 6-OHDA could selectively destroy catecholaminergic neurons, provided a very useful tool for further investigations in this field.[34,66,105] More recently with immunolesioning techniques, we can selectively destroy specific neuronal population; for e.g., delivery of 192 IgG-saporin complex into the field of cholinergic field, will destroy only cholinergic neurons sparing the rest of the neuronal population and fibers.

The rats, whose catecholaminergic fibers in the POA had been destroyed by 6-OHDA, showed an increase in wakefulness.[50,58] Though the increase was small, the finding was reproducible and long lasting.[82] Pretreatment of animals with DMI, before the injection of 6-OHDA, prevented the destruction of noradrenergic fibers, destroying only the dopaminergic fibers. In these rats there was no change in S-W in these rats. The results suggest that the noradrenergic have a hypnogenic influence at the level of the POA.

The lesion of catecholaminergic terminals of the POA, produced hyperthermia, in addition to sleep changes. During this period of increased body temperature, which was more prominent during the first week, the rats preferred to stay in a lower ambient temperature.[76] Selection of a lower ambient temperature could be a behavioral correction to bring down the elevated body temperature. But this altered thermal preference obviously discards the possible impact of lower ambient temperature on sleep (Fig. 4). Low ambient temperature produces a decrease in sleep in normal rats.[106] This suggests that the catecholaminergic terminals of the POA help to interlink sleep with thermoregulation.

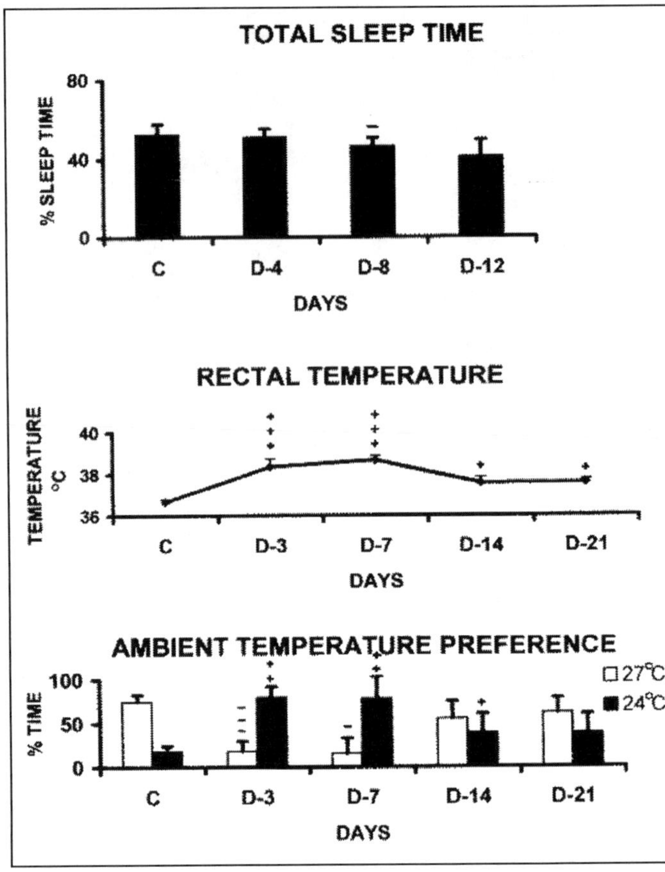

Figure 4. The Figure shows the total sleep time, rectal temperature and ambient temperature preference of rats before (C) and after destruction of catecholaminergic fibers in the preoptic area, by local injection of 6-hydroxydopamine. The bar diagram on top shows the percentage of total sleep (slow wave sleep plus paradoxical sleep) time that was calculated from 24 h recordings done before lesion (C) and on 4th (D-4), 8th (D-8) and 12th (D-12) day after catecholaminergic fibers lesion of the preoptic area. The rectal temperatures of the rats before (C) and on 3rd (D-3), 7th (D-7), 14th (D-14) and 21st (D-21) day after the lesion, are shown in the middle line drawing. The lowest bar diagram shows the time spent by the rats in the compartments maintained at 24 and 27°C, before (C) the lesion of catecholaminergic fibers, and also on 3rd (D-3), 7th (D-7), 14th (D-14) and 21st day (D-12) after the lesion. The ambient temperature preferences of rats were studied in a specialized multi-compartment chamber. The different compartments were maintained at different temperatures. Rats moved freely from one interconnected compartment to the other. Y-axis shows the percentage of time spent by the rats in different temperatures. Data are mean ± SD. + $p<0.05$, ++ $p<0.01$, +++ $p<0.001$ significant increase compared to prelesion (C), and - $p<0.05$, --- $p<0.001$ significant decrease compared to prelesion value.

Alterations in Norepinephrine Release and Dendritic Spine Densities at the Preoptic Area after Acute Total Sleep Deprivation

In this study, the levels of NA, monoamine metabolites and the morphological changes like spine density in the mPOA were studied after total sleep deprivation, and they were compared with their level in the cortex.[85] The rats were sleep-deprived for 48 h, by placing them in a rotating drum.[11] These rats along with those from the control groups were sacrificed, and the mPOA was dissected out. The motor cortex was also dissected out and studied, for comparison.[85] The POA and motor cortex from one control and one experimental group were weighed and prepared for the estimation of monoamines and their metabolites.[59] On the other hand, the brain tissues from two other groups were processed for the assessment of dendritic morphology in 100-120 μm thick Golgi stained sections.[32]

Sleep deprivation is one among the powerful tools to assess the quality of sleep. Recovery of sleep after prolonged wakefulness, most certainly will bring about NREM and REM sleep rebound and increase in delta power in normal animals. These changes may not be seen in pathological conditions. The probable involvement of the neurotransmitters at the level of the POA in the regulation of sleep could be studied by assessing their levels in this area, after sleep deprivation. If there is an alteration in the release of some transmitters during sleep deprivation, it is likely to produce an alteration in the spine density. Excessive synaptic activity has been shown to increase the spine density in the neocortex and the mPOA.[30,91,111]

Acute total sleep deprivation for 48 h led to a significant decrease in noradrenaline in the mPOA, though there was no significant change in motor cortex (Fig. 5). The level of metabolites of monoamines significantly increased only at the mPOA. Total sleep deprivation might have caused excess release of noradrenaline at the mPOA. Its subsequent breakdown would have resulted in the reduction of its level, and an increase in the products of the metabolism of monoamine transmitters. This may indicate that

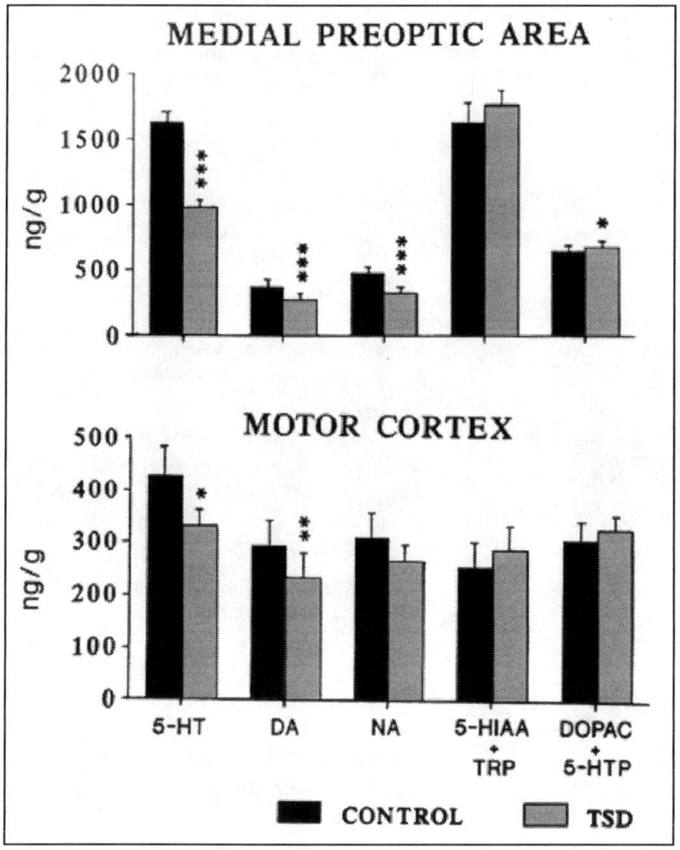

Figure 5. Changes in the level of NA, DA, 5-HT, 5-HIAA and TRP, DOPAC and 5-HTP in control and total sleep deprivation groups of the mPOA and motor cortex neurons. Data are mean ± SD. * $p< 0.05$, ** $p< 0.01$, *** $p< 0.001$. (Reproduced with permission from Ramesh et al, Sleep Research Online, 1999.)

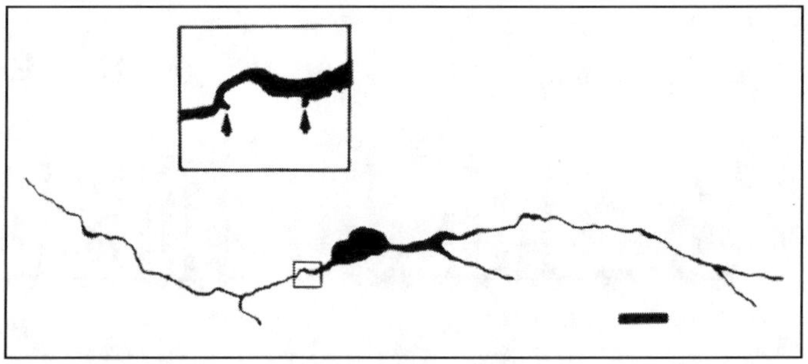

Figure 6. Representative example of camera lucida tracings of medial preoptic area neuron. Inset: Enlarged tracing of a small portion to show the dendritic spines. Scale bar: 16.12 µm; inset: 4.06 µm. (Reproduced with permission from Ramesh et al, Sleep Research Online, 1999.)

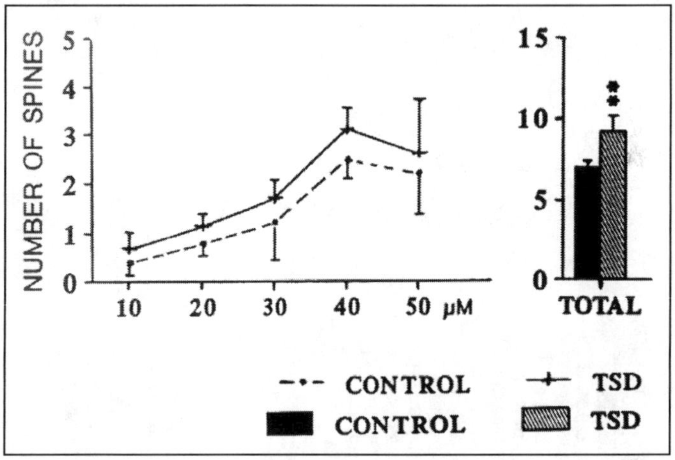

Figure 7. Graph showing the number of spines (control and total sleep deprived groups) on successive segments of 10 µm up to 50 µm, and a total length (50 µm), of the mPOA neurons. Data are mean ± SD. ** $p< 0.01$. (Reproduced with permission from Ramesh et al, Sleep Research Online, 1999.)

noradrenaline was released in excess because of increase in sleep pressure, resulting from sleep deprivation. There was an increase in the number of dendritic spines at the mPOA in rats that were sleep-deprived (Figs. 6 and 7). This could be due to enhanced afferent input signals, caused by sustained sleep pressure.[85]

Changes in Sleep Produced by Local Administration of Adrenergic Agonists and Antagonists at the mPOA

The changes in sleep were studied in free moving animals, after the injection of adrenergic agonists and antagonists at the mPOA, through chronically implanted cannulae. Experiments were conducted on different groups of rats in which S-W was assessed as described above. The cannulae were chronically implanted stereotaxically, under anesthesia, for intracerebral injection into the mPOA. After they recovered from operative trauma, the recordings were done before and after bilateral injection of adrenergic agonists, antagonists and their vehicles.[53,54] Not more than one injection was given to any animal to avoid the effect of injury. The site and spread of injection were confirmed by the histology of the brain after infusing ferric chloride (2%) into the same site of drug injection (Fig. 8).

Changes elicited by local injection of adrenergic agonists and antagonists at the POA have given some insight into the role of the noradrenergic system in S-W. Injection of NA at the mPOA produced arousal in rats (Fig. 9).[53] Microinjection of β agonist,

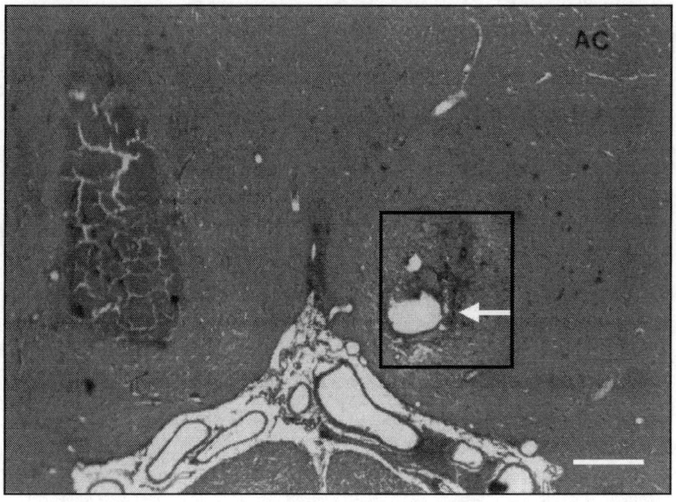

Figure 8. Photomicrograph of the rat brain section showing cannula tracts and the site of injection. AC, anterior commisure. Arrow shows the tip of the cannula. Scale bar: 600 µm.

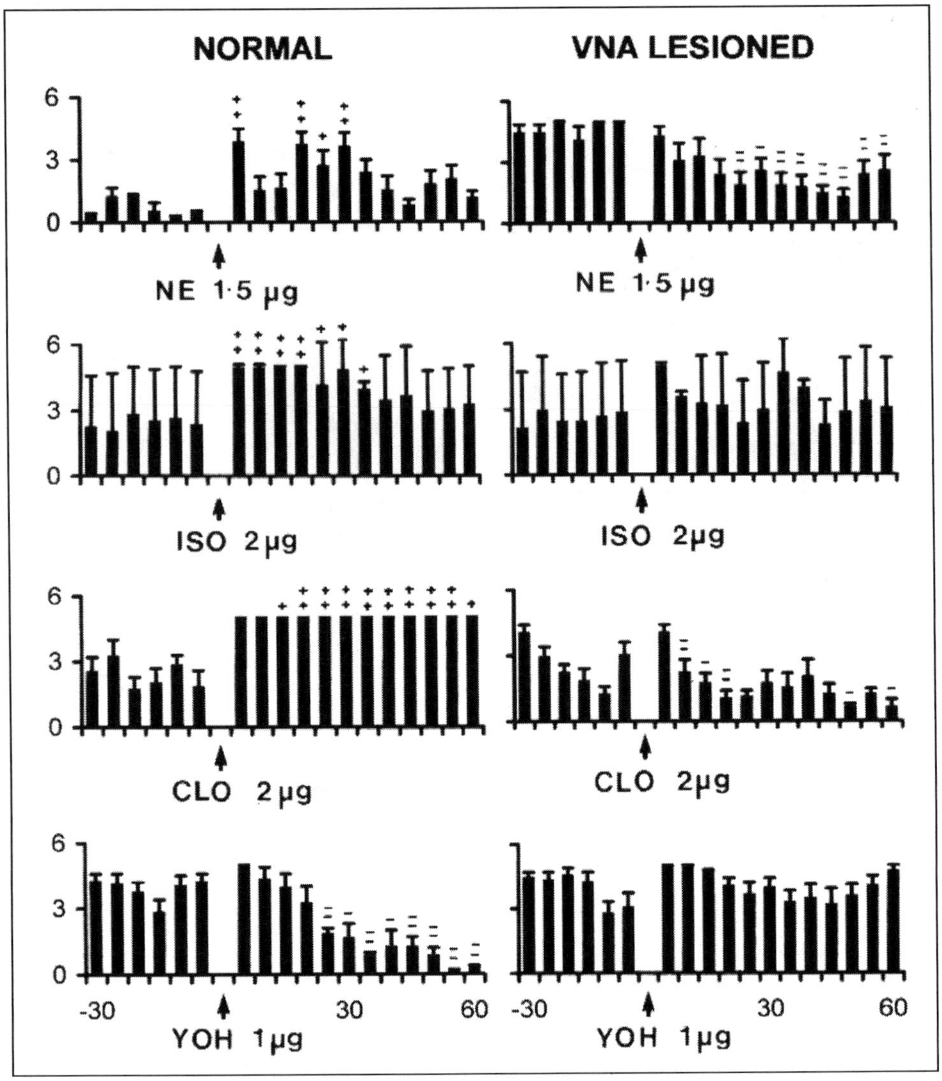

Figure 9. Effects of injection of norepinephrine (NE), isoproterenol (ISO), clonidine (CLO) and yohimbine (YOH) at the medial preoptic area on sleep-wakefulness were studied in different groups of rats with chronically implanted cannulae. Figure shows the differences in the effects produced by the drugs in normal rats, and those in which catecholaminergic fibers (including NE fibers) in the POA were destroyed by injecting 6-OHDA at the VNA. Y-axis shows the period of wakefulness, in minutes, assessed from every 5-min bin of record. X-axis shows the time in minutes before and after drug injection. Arrow indicates injection point. Drug injected (along with amount administered) is shown below the arrow. Studies were done during day and night, and some of drugs were tried in different dozes also. Values for half an hour before the drug injection (shown as negative values), and one hour after the drug injection, from selected records, are only given in this Figure. Author's earlier publications may be consulted for details. Data are mean ± SD. + $p < 0.05$, ++ $p < 0.01$ significant increase compared to control (vehicle) injection, and - $p < 0.05$, --- $p < 0.01$ significant decrease compared to control injection.

isoproterenol, into the mPOA also produced arousal.[94] (Fig. 9) Still, the α-adrenergic receptors may be considered more important, as only α-adrenergic antagonists produced the opposite effect, i.e., injection-bound sleep, and the β-receptor blocker did not produce any alteration in S-W. [54]

Increased wakefulness observed after injection of adrenergic agonists at the mPOA, may appear contrary to expectations, as the destruction of the noradrenergic fibers also produced increased wakefulness. Studies using α_2 adrenergic agents helped to explain this paradox. NA injected at the POA can act on α_1 or α_2 adrenergic receptors, apart from β-receptors. α_2 agonist, clonidine produced arousal when injected at the mPOA (Fig. 9). Clonidine would have activated presynaptic α2 autoreceptors, causing a decrease in endogenous NA. Yohimbine, an α_2 antagonist, induced sleep, when applied at the mPOA (Fig. 9). These results indicate that NA, when injected at the mPOA, produced arousal by acting on α_2 receptors.[84]

Local application of NA in mPOA produced hypothermia (Fig. 10), whereas its blocker, phenoxybenzamine induced increase in body temperature (Fig. 11). On the other hand, propranolol did not produce any change in body temperature[21] (Fig. 12). Clonidine and yohimbine did not produce any change in body temperature in normal rats, however clonidine in VNA lesioned animals induced hypothermia.[83]

Local injection studies, including those cited here, may be criticized for application of neurotransmitter agonists and antagonists in unphysiological amounts, and it may not mimic normal physiological action.[9,29,69,80,116] So, any interpretation of the local injection studies should take into account all the evidences accumulated using other techniques. Though the studies quoted here are based on drug injections in the mPOA, FOS immunocytochemistry studies showed that the ventrolateral preoptic (VLPO) neurons are specifically activated during sleep.[92] The VLPO neurons are inhibited by NA.[27] It was proposed that reciprocal inhibitory interaction of VLPO neurons with noradrenergic system is a key factor for promoting sleep.

Action of Adrenergic Agents on the Presynaptic and the Postsynaptic Receptors

Changes in sleep elicited by injection of adrenergic agents at the mPOA, after the NE terminals were destroyed, gave an idea about the involvement of the presynaptic and postsynaptic receptors. The electrodes for recording EEG, EMG and EOG, and bilateral guide cannulae for injection of drugs at the mPOA were chronically implanted in these rats, as described in the previous section. The VNA, which contained the noradrenergic fibers projecting to the mPOA, were bilaterally lesioned, by injecting 8 μg of 6-OHDA, in the brain stem. When the rats had completely

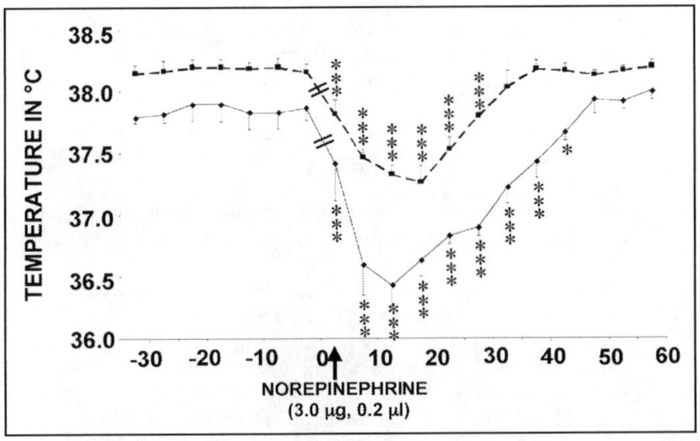

Figure 10. The graph shows the effect of injection of NA (0.2 µl, 3 µg) into mPOA on the rectal temperature (mean ±SD) of rats. Dotted line for night recording and continuous line for day recording. * $p < 0.05$, *** $p < 0.001$ level of significance of change compared to its own preinjection values.

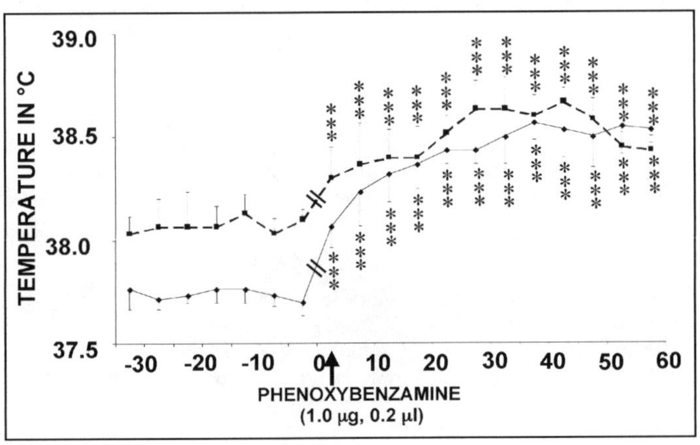

Figure 11. The graph shows the effect of injection of phenoxybenzamine (0.2 µl, 1 µg) into mPOA on the rectal temperature (mean ±SD) of rats. Dotted line for day temperature. *** $p < 0.001$ level of significance of change compared to its own preinjection values.

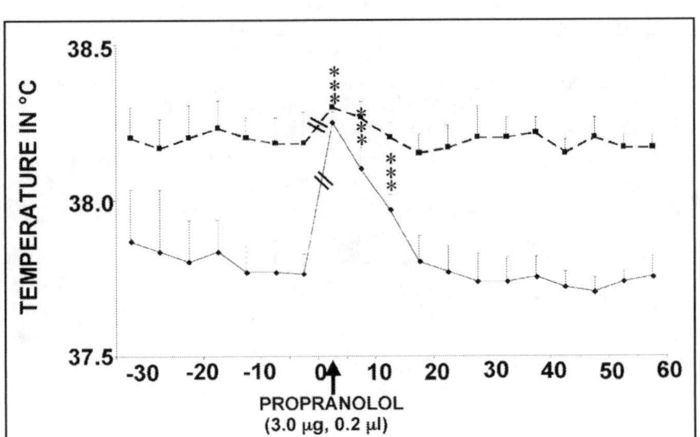

Figure 12. The graph shows the effect of injection of propranolol (0.2 µl, 1 µg) into mPOA on the rectal temperature (mean ± SD) of rats. Dotted line for night temperature and continuous line for day temperature. *** $p < 0.001$ level of significance of change compared to its own preinjection values. None of these readings was significantly different from readings obtained at identical periods after saline injection.

recovered from the postoperative trauma, noradrenergic agonists, antagonists and their vehicles were injected at the mPOA. As the noradrenergic terminals were destroyed in these animals, the injected drugs could not have acted on the presynaptic noradrenergic receptors.[58,83,94]

In the previous section it was suggested that NA produced arousal by acting on presynaptic α_2 receptors at the POA. Locally applied NA could act on both postsynaptic and presynaptic receptors[95] was suggested.[10] In order to test the proposition that, hypothalamically-injected NA acted on presynaptic receptors, NA was locally administered at the POA in rats whose catecholaminergic terminals were degenerated.

NA injection at the mPOA induced sleep in the VNA lesioned animals (Fig. 9). As the presynaptic adrenergic receptors were lesioned at the VNA, the response elicited must have been due to the action of NA on the postsynaptic receptors.[58] Clonidine administration at the mPOA in these animals, induced sleep (Fig. 9). So, it can be assumed that the increased wakefulness was produced by the action of this drug on the presynaptic receptors in normal rats. Similarly, the effect after yohimbine application was significantly attenuated after the VNA lesion.[83]

Local application of isoproterenol, into the mPOA, in the VNA lesioned animals, did not produce any significant change in S-W, though it produced arousal in normal rats (Fig. 9). Thus, the increase in wakefulness obtained on isoproterenol administration was probably the result of its action on the presynaptic adrenergic terminals.[94] The possible involvement of sexual arousal in the isoproterenol-induced increase in wakefulness is discussed elsewhere.[94]

Based on these findings, it can be concluded that the changes in S-W produced by local application of adrenergic agonists and antagonists, in normal rats, were due to their action on the presynaptic terminals.

Influence of Body Temperature and Sex Behavior Changes on the Observed Alterations in Sleep-Wakefulness

In order to find out the possible changes in sleep as a result of alterations in body temperature, the rectal temperature of the animals were monitored along with S-W. All animals selected for the study of rectal temperature were well trained to accept rectal probes for monitoring the rectal temperature. Details of operative procedure for the drug administration in the mPOA were described in earlier sections. A thermistor probe system was used to monitor rectal temperature every 5-min throughout the experiment. A thin coat of local anesthetic gel was applied to the rectum and the probe was gently inserted into the rectum. The probe was then firmly fastened to the base of the tail, before introducing the animal into the recording cage.

Many sleep researchers tend to ignore the important role played by the POA in many other functions like thermoregulation and reproduction.[23,31,55-57] Any stimuli which brings about changes in these functions would certainly affect S-W. It is possible that the changes in S-W, resulting from the injection of drugs, may be dependent on the body temperature changes. It has been shown that systemic injection of phentolamine produces reduction in sleep and fall in body temperature. Whenever the fall in body temperature was prevented, there was no reduction in sleep.[46] It was suggested that the arousal induced by injection of the drug could have resulted from a fall in body temperature, rather than from a direct action of the drug on the arousal inducing system. Stimulation of the noradrenergic system at the POA brings about a fall in body temperature, by acting on α_1 receptors.[21,83] So, it is possible that the arousal induced by injection of NA at the POA could have also resulted from a fall in body temperature, rather than from a direct action of the drug at the POA.

It may even be suggested that the changes in sleep and temperature might have been brought about through stimulation of the same neuronal circuits in the POA. However, this suggestion is not probable as neurotransmitter agonists and antagonists, injected at the POA, did not always produce simultaneous alterations in sleep and body temperature.[19,20,104] It was proposed that there are two separate sets of NE terminals at the POA for regulation of sleep and body temperature.[83]

Role of DA, 5-HT and Histamine in Sleep-Wakefulness and Thermoregulation

There are not many studies on the action of DA on S-W at the level of the POA. However some studies have indicated that blockade of D_1 receptors promote REM sleep in rats.[109] The activity of the serotoninergic system varies in phase with the sleep-wake cycle, which in turn is also associated with changes in several physiological function, including EEG activity, brain temperature and locomotion. Microdialysis studies by Python et al (2001) have shown that 5-HT increased just before the rats fell asleep and then decreased during sleep in POA. Certain group of cells, possibly GABAergic, are inhibited by 5-HT but are unaffected by histamine.[27] They also propose that the reciprocal inhibitory interaction of such VLPO neurons with the noradrenergic, serotoninergic and cholinergic waking system to which they project may have a key role in promoting sleep. Recent studies on genomic events in the POA suggests that the release of 5-HT during wakfulness leads to a homeostatic regulation of SWS (for review see ref. 44). Local application of 5-HT in the POA did not have any effect on S-W, although it induced hyperthermia.[19] Strecker et al[99] have shown in cats that the extracellular histamine levels in the preoptic/anterior hypothalamic area are the highest in wakefulness followed by SWS and lowest during REM sleep. Other studies have linked the orexinergic system in modulating histamine release. Microdialysis application of orexin-A to the TMN increased histamine release from the mPOA. It is also suggested that the onset of sleep may require an inhibition of histaminergic neurons by the GABAergic system.

References

1. Ambach G, Kivovics P, Palkovits M. The arterial and venous blood supply of the preoptic region in the rat. Acta Morphol Acad Sci Hung 1978; 26:21-41.
2. Anden NE, Dahlstrom A, Fuxe K et al. Ascending noradrenaline neurons from the pons and the medulla oblongata. Experientia 1966; 22:44-45.
3. Asala SA, Okano Y, Honda K et al. Effects of medial preoptic area lesion on sleep and wakefulness in unrestrained rats. Neurosci Lett 1990; 114:300-304.
4. Atlas D, Ingram IB. Topography of the brain stem of the rhesus monkey with special reference to the diencephalon. J Comp Neurol 1937; 66:263-289.
5. Benedec G, Obal Jr F, Szekeres L et al. Cortical synchronization induced by thermal stimulation of the preoptic area in immobilized rats. Acta Physiol Acad Sci Hung 1976; 48:65-72.
6. Benedec G, Obal Jr F, Szekeres L et al. Two separate synchronizing mechanisms in the basal forebrain: Study of the synchronizing effects of the rostral hypothalamus, preoptic region and olfactory tubercle. Arch Ital Biol 1979; 117:164-185.
7. Bjorklund A, Nobin A. Fluorescence histochemical and microspectroflurometric mapping of dopamine and noradrenaline cell groups in the rat diencephalon. Brain Res 1973; 51:193-205.
8. Bleier R, Cohn P, Siggelkow IR. A cytoarchitectonic atlas of the hypothalamus and hypothalamic third ventricle of the rat. In: Morgane PJ, Panksepp J, eds. Handbook of the Hypothalamus, New York: Marcel Dekker Inc., 1979; 137-220.
9. Bloom FE, Hoffer BJ, Siggins GR et al. Effects of serotonin on central neurons: Microiontophoretic administration. Fed Proc 1972; 31:97-106.
10. Booth DA. Mechanism of action of norepinephrine in eliciting an eating response on injection into the rat hypothalamus. J Pharmacol Exp Ther 1968; 160:336-348.
11. Borbely AA, Neuhaus HU. Sleep-deprivation: Effects on sleep and EEG in the rat. J Comp Physiol 1979; 133:71-87.
12. Broadwell RD, Bleier R. A cytoarchitectonic atlas of the mouse hypothalamus. J Comp Neurol 1976; 167:315-339.
13. Clark AJM, Butcher SP, Winn P. Evidence for functional separation of alpha-1 and alpha-2 noradrenaline receptors by presynaptic terminal reuptake mechanisms. Psychopharmacol 1991; 102:366-374.
14. Clark WEL. Morphological aspects of the hypothalamus. In: Clark WEL, Beatte J, Riddoch G et al, eds. The Hypothalamus. Edinburg: Oliver and Boyd, 1938.
15. Conrad LA, Pfaff DW. Autoradiographic tracing of projections from preoptic area and anterior hypothalamus in rats. Proc Soc Neurosci 1974; 176.
16. Conard LC, Pfaff DW. Efferents from medial basal forebrain and hypothalamus in the rat. I. An autoradiographic study of the medial preoptic area. J Comp Neurol 1976a; 169:185-220.
17. Conrad LA, Pfaff DW. Efferents from medial basal forebrain and hypothalamus in the rat. II. An autoradiographic study of the anterior hypothalamus. J Comp Neurol 1976; 169:221-262.
18. Dahlstrom A, Fuxe K. Evidence for the existance of monoamine containing neurons in the central nervous system. Demonstration of monoamines in the cell bodies of brain stem neurons. Acta Physiol Scand 1964; 62:1-55.
19. Datta S, Mohan Kumar V, Chhina GS et al. Effect of application of serotonin in medial preoptic area on body temperature and sleep-wakefulness. Ind J Exp Biol 1987; 25:681-685.
20. Datta S, Mohan Kumar V, Chhina GS et al. Interrelationship of thermal and sleep-wakefulness changes elicited from the medial preoptic area in rats. Exp Neurol 1988; 100:40-50.
21. Datta S, KumarVM, Chhina GS et al. Tonic activity of medial preoptic noradrenaline mechanism for body temperature maintenance in sleeping and awake rats. Brain Res Bull 1985; 15:447-451.
22. Day TA, Blessing W, Willoughby JO. Noradrenergic and dopaminergic projections to the medial preoptic area of the rat. A combined horseradish peroxidase/catecholamine fluorescence study. Brain Res 1980; 193:543-548.
23. Dhawan JK, Kumar VM, Govindaraju V et al. Changes in MRI and sex behavior after 6-OHDA injection at the medial preoptic area. Brain Res Bull 1998; 45:333-339.
24. Findlay AL, Hayward JN. Spontaneous activity of single neurons in the hypothalamus of rabbits during sleep and waking. J Physiol (Lond) 1969; 201:237-258.
25. Freeman WJ, Davis DD. Effect on cats of conductive hypothalamic cooling. Am J Physiol 1959; 197: 145-148.
26. Fuxe K, Hanson CF. Central catecholamine neurons and conditioned avoidance behaviour. Psychopharmacologia 1967; 11:439-447.
27. Gallopin T, Fort P, Eggermann E et al. Identification of sleep-promoting neurons in vitro. Nature 2000; 404:992-995.
28. Garcia-Arraras LE, Pappenheimer JR. Site of action of sleep inducing muramyl peptide isolated from human urine microinjection studies in rabbit brains. J Neurophysiol 1983; 49:528-533.
29. Gardner D, Kandel ER. Diphasic postsynaptic potential: a chemical synapse capable of mediating conjoint excitation and inhibition. Science 1972; 176:675-678.
30. Globus A, Scheibel AB. Loss of dendritic spines, as an index of presynaptic terminal patterns. Nature 1996; 212:463-465.
31. Gulia KK, Mallick HN, Kumar VM. Orexin A (hypocretin-1) application at the medial preoptic area potentiates male sexual behavior in rats. Neuroscience 2003; 116:921-923.

32. Gundappa G, Desiraju T. Deviations in brain development of F2 generation on caloric undernutrition and scope of their prevention by rehabilitation: Alterations in dendritic spine production and pruning of pyramidal neurons of lower laminae of motor cortex and visual cortex. Brain Res 1988; 456:205-223.
33. Gurdjian ES. The diencephalon of the albino rat. Studies on the brain of the rat No. 2. J Comp Neurol 1927; 43:1-114.
34. Hartmann E, Chung R, Draskoczy P et al. Effects of 6-hydroxydopamine on sleep in the rat. Nature 1971; 233:425-426.
35. Hernandez-Peon R. A cholinergic hypnogenic limbic forebrain-hindbrain circuit. In: Jovet M, eds. Aspects anatomofonctionnels de la Physiologie du Sommeil, Paris Centre Natl Rech Sci 1965:63-88.
36. Hohn KG, Wuttke WP. Cahnges in the catecholamine turn over in the anterior part of the medio-basal hypothalamus and the medial preoptic area in response to hyperprolactinemia in ovariectomised rats. Brain Res 1978; 156:241-252.
37. Humphrey T. The telencephalon of the bat I. the noncortical nuclear masses and certain pertinent fiber connections. J Comp Neurol 1936; 65:603-711.
38. Jacobowitz D, Kostrzewa R. Selective action of 6-hydroxydopa on noradrenergic terminals: Mapping of preterminal axons of the brain. Life Sci 1971; 10: 1321-1342.
39. Jacobowitz DM, Palkovits M. Topographic atlas of the catecholamine and acetylcholinesterase containing neurons in the rat brain. I. Forebrain (Telencephalon, diencephalon). J Comp Neurol 1974; 157:13-28.
40. John J, Kumar VM. Effect of NMDA lesion of medial preoptic neurons on sleep and other functions. Sleep 1998; 21:585-597.
41. John J, Kumar VM, Gopinath G. Recovery of sleep after fetal preoptic transplantation in the medial preoptic area lesioned rats. Sleep 1998; 21:598-603.
42. John J, Kumar VM, Gopinath G et al. Changes in sleep-wakefulness after kainic acid lesion of the preoptic area in rats. Jap J Physiol 1994; 44:231-242.
43. Jonsson G, Sachs C. Degenerative and nondegenerative effects of 6-hydroxydopamine on adrenergic nerves. J Pharmacol Exp Ther 1972; 180:625-635.
44. Jouvet M. Sleep and serotonin: An unfinished story. Neuropsychopharmacology 1999; 21(2 suppl):24S-27S.
45. Kaitin KI. Preoptic area unit activity during sleep and wakefulness in the cat. Exp Neurol 1984; 83:347-357.
46. Kent S, Satinoff E. Influence of ambient temperature on sleep and body temperature after phentolamine in rats. Brain Res 1990; 511:227-33.
47. Kizer JS, Muth E, Jacobowitz DM. The effects of bilateral lesions of the ventral noradrenergic bundle on endocrine induced changes of tyrosine hydroxylase in the rat median eminence. Endocrinology 1976; 98:886-893.
48. Kobayashi RM, Palkovits M, Kopin IJ et al. Biochemical mapping of noradrenergic nerves arising from the rat locus coeruleus. Brain Res 1974; 77:269-279.
49. Kuhlenbeck H. A summary of development, structure, function and pathology. Kerger, Basel: 1954.
50. Kumar VM. Noradrenaline mechanism in the regulation of sleep-wakefulness: A special role at the preoptic area. In: Kumar VM, Mallick HN, Nayar U, eds. Sleep-wakefulness. Wiley-Eastern, New Delhi: 1993:25-34.
51. Kumar VM. Role of hypothalamus in sleep. Biomedicine 2000; 20:55-66.
52. Kumar VM. Role of noradrenergic fibres of preoptic area in regulating sleep. J Chem Neuroanat 2003; 26(2):87-93.
53. Kumar VM, Datta S, Chhina GS et al. Sleep-wake responses elicited from medial preoptic area on application of norepinephrine and phenoxybenzamine in free moving rats. Brain Res 1984; 322:322-325.
54. Kumar VM, Datta S, Chhina GS et al. Alpha adrenergic system in medial preoptic area involved in sleep-wakefulness in rats. Brain Res Bull 1986; 16:463-468.
55. Kumar VM, John J, Govindaraju V et al. Magnetic resonance imaging of NMDA induced lesion of the medial preoptic area and changes in sleep, temperature and sex behaviour. Neurosci Res 1996a; 24:207-214.
56. Kumar VM, Khan NA, John J. Male sexual behaviour not abolished after medial preoptic lesion in adult rats. NeuroReport 1996b; 7:1481-1484.
57. Kumar VM, Khan NA. Role of the preoptic neurons in thermoregulation in rats. Arch Clin Exp Med 1998; 7:24-27.
58. Kumar VM, Sharma R, Wadhwa S et al. Sleep-inducing function of noradrenergic fibres in the medial preoptic area. Brain Res Bull 1993; 32:153-158.
59. Lakshmana MK, Raju TR. An isocratic assay for norepinephrine, dopamine and 5- Hydroxytryptamine using their native fluorescence by high-performance liquid chromatography with fluorescence detection in discrete brain areas of rat. Analytical Biochem 1997; 246:166-170.
60. Lindvall O, Bjorklund A. The organization of the ascending catecholamine neuron system in the rat brain as revealed by the glyoxilic acid fluorescence metrhod. Acta Physiol Scand 1974; 412(Suppl):1-48.
61. Lindvall O, Bjorklund A, Falck B et al. Recent developments in aldehyde-induced monoamine fluorescence: The aluminum-formaldehyde (ALFA) method applied to immature and adult central nervous tissue. Histochem J 1981; 13:583-598.
62. Loo YT. The forebrain of the opossum, Didelphis virginiana. J Comp Neurol 1931; 52:1-48.
63. Lucas EA, Sterman MB. Effect of forebrain lesion on a polycyclic sleep-wake cycle and sleep-wake patterns in cats. Exp Neurol 1975; 46:368-388.
64. Mallick BN, Chinna GS, Sundaram KR et al. Activity of preoptic neurons during synchronization and desynchronization. Exp Neurol 1983; 81:586-597.
65. Martinovic JV, McCann SM. Effects of lesions in the ventral noradrenergic tract produced by microinjection of 5-hydroxy-dopamine on gonadotropin release in the rat. Endocrinology 1977; 100: 1206-1213.
66. Matsuyama S, Coindet J, Mouret J. 6-Hydroxydopamine intracisternale et sommeil chez le rat. Brain Res 1973; 57:85-95.
67. McGinty D, Sterman MB. Sleep suppression after basal forebrain lesions in the cat. Science 1968; 160:1253-1255.
68. McGinty D, Szymusiak R. Hypothalamic regulation of sleep and arousal. Front Biosci 2003; 8:s1074-1083.
69. Miller JJ, Mogenson GJ. Effect of septal stimulation on lateral hypothalamic unit activity in the rat. Brain Res 1971; 32:125-142.
70. Mohankumar PS, Thyagarajan S, Quadri SK. Tyrosine hydroxylase and DOPA decarboxylase activities in the medical preoptic area and arcuate nucleus during the estrous cycle: Effects of aging. Brain Res Bull 1997; 42:265-271.
71. Moore RY, Bloom FE. Central catecholamine neurons systems anatomy and physiology of the norepinephrine and epinephrine systems. Ann Rev Neuronal 1979; 2:113-168.
72. Moruzzi G. The sleep-waking cycle. Ergb Physiol 1972; 64: 1- 67.
73. Nauta WJH. Hypothalamic regulation of sleep in rats. An experimental study. J Neurophysiol 1946; 9:285-316.
74. Nicholson G, Greeley G, Humm J et al. Lack of effect of noradrenergic denervation of the hypothalamus and medial preoptic area on the feedback regulation of gonadotropin secretion and the estrous cycle of the rat. Endocrinology 1978; 103:559-566.
75. Ogawa Y, Kawamura H. Increase of multiple unit activity during slow wave sleep in cat preoptic area. Brain Res Bull 1988; 20:897-902.
76. Pal R, Mallick HN, Kumar VM. Role of catecholaminergic terminals in the preoptic area in behavioural thermoregulation in rats. Indian J Physiol Pharmacol 2002; 46:434-440.
77. Palkovits M, Brownstein M, Saavedra JM et al. Norepinephrine and dopamine content of hypothalamic nuclei of the rat. Brain Res 1974; 77:13--149.
78. Papez JW, Arenson LR. Thalamic nuclei of the Pithecus (Macacus) rhesus; ventral thalamus. Arch Neurol Psychiat 1934; 32:1-26.
79. Parmeggiani PL, Zamboni G, Cianci T et al. Influence of anterior hypothalamic heating on the duration of fast wave sleep episodes. Electroencephalogr Clin Neurophysiol 1974; 36:465-470.
80. Paton WDM, Perry WLM. The relationship between depolarization and block in the cat's superior cervical ganglion. J Physiol 1953; 119:43-57.

81. Python A, Steimer T, de Saint Hilaire Z et al. Extracellular serotonin variations during vigilance states in the preoptic area of rats: A microdialysis study. Brain Res 2001; 910:49-54.
82. Ramesh V, Kumar VM. Changes in sleep-wakefulness after 6-hydroxydopamine lesion of the preoptic area. Neuroscience 2000; 98:549-553.
83. Ramesh V, Kumar VM. The role of alpha-2 receptors in the medial preoptic area in the regulation of sleep-wakefulness and body temperature. Neuroscience 1998; 85:807-818.
84. Ramesh V, Kumar VM, John J et al. Medial preoptic alpha-2 adrenoceptors in the regulation of sleep-wakefulness. Physiol Behav 1994; 57:171-175.
85. Ramesh V, Lakshmana MK, Rao BSS et al. Alterations in monoamine neurotransmitters and dendritic spine densities at the medial preoptic area after sleep deprivation. Sleep Res Online 1999; 2:49-55.
86. Ray B, Mallick H, Kumar VM. Role of the medial preoptic area in thermal preference of rats. Indian J Physiol Pharmacol 2001; 45:445-450.
87. Roberts WW, Berquist EH, Robinson TCL. Thermoregulatory grooming and sleep-like relaxation induced by local warming of preoptic area and anterior hypothalamus in opossum. J Comp Physiol Psychol 1969; 67:182-188.
88. Roberts WW, Robinson TCL. Relaxation and sleep induced by warming of the preoptic region and anterior hypothalamus in cats. Exp Neurol 1969; 25:284-294.
89. Rose JE. The ontogenic development of the rabbits diencephalon. J Comp Neurol 1942; 77:61-130.
90. Sachs C, Jonsson G. Degeneration of central and peripheral noradrenaline neurons produced by 6-hydroxy-DOPA. J Neurochem 1972; 19:1561-1575.
91. Sanchez-Toscano F, Sanchez M, Garzon J. Changes in the number of dendritic spines in the medial preoptic area during a premature long-term social isolation in rats. Neurosci Lett 1991; 122:1-3.
92. Sherin JE, Shiromani PJ, McCarley RW et al. Activation of ventrolateral preoptic neurons during sleep. Science 1996; 271:216-219.
93. Simerly RB, Swanson LW. The organization of neural inputs to the medial preoptic nucleus of the rat. J Comp Neurol 1986; 246:312-342.
94. Sood S, Dhawan JK, Ramesh V et al. Role of medial preoptic area beta adrenoceptors in the regulation of sleep-wakefulness. Pharmacol Biochem Behav 1997; 57:1-5.
95. Starke K. Presynaptic α-autoreceptors. Rev Physiol Biochem Pharmacol 1987; 107:73-145.
96. Sterman MB, Clemente CD. Forebrain inhibitory mechanisms: Cortical synchronization induced by basal forebrain stimulation. Exp Neurol 1962a; 6: 91-102.
97. Sterman MB, Clemente CD. Forebrain inhibitory mechanisms: Sleep patterns induced by basal forebrain stimulation in the behaving cat. Exp Neurol 1962b; 6:103-117.
98. Sterman MB, Knauss TK, Lehmann D et al. Alteration of sleep patterns following basal forebrain lesions. Fed Proc 1964; 23:209.
99. Strecker RE, Nalwalk J, Dauphin LJ et al. Extracellular histamine levels in the feline preoptic/anterior hypothalamic area during natural sleep-wakefulness and prolonged wakefulness: An in vivo microdialysis study. Neuroscience 2002; 113:663-670.
100. Swanson LW. An autoradiographic study of the efferent connections of the preoptic regions in rats. J Comp Neurol 1976a; 167:227-256.
101. Swanson LW. The locus coeruleus: A cytoarchitectonic, golgi and immunohistochemical study in the albino rat. Brain Res 1976b; 110:39-56.
102. Swanson LW, Hartman BK. The central adrenergic system. An immunofluorescence study of the localization of cell bodies and their efferent connections in the rat utilizing dopamine-beta-hydroxylase as a marker. J Comp Neurol 1975; 163: 467-506.
103. Szymusiak R, McGinty D. Sleep suppression following kainic acid induced lesions of the basal forebrain. Exp Neurol 1986; 94:598-614.
104. Talwar A, Mohan Kumar V. Effect of carbacol injection in the medial preoptic area on sleep-wakefulness and body temperature in free moving rats. Ind J Physiol Pharmacol 1994; 38:11-16.
105. Thoenen H, Tranzer JP. The pharmacology of 6-hydroxydopamine. Ann Rev Pharmacol 1973; 13:169-180.
106. Thomas TC, Kumar VM. Effect of ambient temperature on sleep-wakefulness in normal and medial preoptic area lesioned rats. Sleep Res Online 2000; 3:141-145.
107. Thomas TC, Kumar VM. Effect of ambient temperature on brain temperature and sleep-wakefulness in medial preoptic area lesioned rats. Indian J Physiol Pharmacol 2002; 46:287-297.
108. Torre de la JC, Surgeon JW. A methodological approach to rapid and sensitive monoamine histofluorescence using a modified glyoxylic acid technique: The SPG method. Histochemistry 1976; 49:81-93.
109. Trampus M, Ongini E. The D1 dopamine receptor antagonist SCH 23390 enhances REM sleep in the rat. Neuropharmacology 1990; 29:889-893.
110. Ungerstedt U. Stereotaxic mapping of the monoamine pathways in the rat brain. Acta Physiol Scand 1971; 82:367:1-48.
111. Valverde F. Apical dendritic spines of the visual cortex and light deprivation in the mouse. Exp Brain Res 1967; 3:337-352.
112. Vathy I, Rimanoczy A, Eaton RC et al. Modulation of catecholamine turnover rate in brain regions of rats exposed prenatally to morphine. Brain Res 1994; 662:209-215.
113. Versteeg DH, Van Der Gugten J, De Jong W et al. Regional concentrations of noradrenaline and dopamine in rat brain. Brain Res 1976; 113:563-574.
114. von Economo C. Schlabheorie. Ergb Physiol 1929; 28:312-339.
115. Von Euler C. Prog Brain Res 1964; 5:12--131.
116. Wachtel H, Kandel ER. Conversion of synaptic excitation to inhibition at a dual chemical synapse. J Neurophysiol 1971; 34:56-58.
117. Yamaguchi N, Merczynski TJ, Ling GM. The effects of electrical and chemical stimulation of the preoptic region and some nonspecific thalamic nuclei in unrestrained, waking animals. Electroencephlogr Clin Neurophysiol 1963; 15:154.

CHAPTER 4

REM Sleep Function and Brain Monoamine Regulation:

An Application of the Search Activity Concept

Vadim S. Rotenberg

The present chapter contains the discussion of the very complicated and controversial topic of brain monoamines activity in REM sleep in relationships with the main REM sleep functions. The author is going to present many contradictory experimental data in this area. He will make an attempt to overcome at least some of these contradictions by using the search activity concept that prescribes to REM (paradoxical – PS) sleep a function of the restoration of search activity that determines subject's resistance to stress and various noxious factors. It is for this reason one of the main features of the behavior of highly developed species.

Activity of the Monoamines Containing Neurons in REM Sleep

The sleep-wakefulness cycle is characterized by a very definite dynamic of the discharge of cell groups of the central nervous system from wakefulness through nonREM (NREM) sleep to REM sleep. Active waking and REM sleep are characterized by very similar (high) discharge rates of most cell groups in neocortex and different subcortical and brainstem areas in contrast to quiet waking and NREM sleep where the same cell groups display a low activity (see also review of ref. 90).[89,97] While this similarity between active waking and REM display the main tendency in brain neuronal activity, some cell groups in the brainstem are active only during REM sleep (REM-on cells[35,80,89]). According to the data of lesion studies[83,91] it is possible to suggest that REM-on cells are strongly related to the generation of REM sleep with all its significant physiological features. However, whether these REM-on cells are responsible not only for the phenomenology of REM sleep but also for REM sleep functions remains an open question. There is some evidence that it is the frontal lobe that is related to the production of REM sleep dreams and consequently may relate to the behavioral and psychological REM sleep functions.[24,95]

Some other cell groups are active during all types of waking behavior and, on the relatively lower level, also during NREM sleep. However they are almost totally inactive during REM sleep (REM-off cells).[18,30,42,59] All these REM-off cell groups are either noradrenergic or serotonergic,[88] and that is especially important for the topic of the present chapter. Moreover, these REM-off cells are concentrated in the main brain sources of the noradrenergic and serotonergic activity – in locus coeruleus (LC) and in the midline raphe. It means that noradrenergic and serotonergic activity is present in all main functional states (in active and quiet waking and in NREM sleep) except REM sleep and is especially high in active wakefulness the latter being similar to REM sleep according to the discharge rate of most brain neurons that are acethylcholinergic in their nature.[23] REM-off noradrenergic and serotoninergic cells do not seem to be responsible for the REM sleep phenomenology because the depletion of norepinephrine (NE) and serotonin (5-HT), by the electrolytic lesion of the LC and raphe nuclei does not prevent REM sleep as a physiological phenomenon.[14,28,33,93] However, it does not mean that such lesions and depletion has no effect on the REM sleep functions. For instance, ponto-geniculo-occipital (PGO) spikes that are normally tied to REM state, correspond in wakefulness to orienting activity[39] and consequently may relate to REM sleep psychological functions (see ref. 68) became released into all states after lesion of 5-HT containing neurons.[93] There are data (see ref. 85) that serotoninergic neurons are related to the regulation of saccadic eye movement.

Thus, REM sleep differs from wakefulness according to the low activity of noradrenergic and serotoninergic neurons. REM sleep represents the functioning of the cortex without the influence of norepinephrine[27] and it is reasonable to suggest that such low activity may in some way relate to the peculiarity of REM sleep functions or to the regulation of REM sleep in different functional conditions.[90] At the same time, the mesencephalic dopamine containing neurons discharge on the equal rate in wakefulness and in all sleep stages.[23] According to Siegel and Rogawski[90] it means that dopamine containing cells have no essential role in sleep generation or in sleep functions. However, if from all monoaminergic neurons only dopaminergic neurons are continuously active in REM sleep, this activity has probably a special meaning and relates to the peculiarity of REM sleep functions.[23]

Monoamines Containing Neurons in Wakefulness: Towards the Functional Meaning

Gottesmann[23] has reviewed the role of different neurotransmitters in waking activity. According to this review, making the long story short, acetylcholine (Ach) is responsible for the general activation of cortical neurons initiated by the reticular activating system of the brainstem.[37] This general nondifferentiated activation of the cortex is important for maintaining stable tonic

Sleep and Sleep Disorders: A Neuropsychopharmacological Approach, edited by Malcolm Lader, Daniel P. Cardinali and S.R. Pandi-Perumal.

vigilance and preventing coma. Atropine, an antagonist of acetylcholine, produces EEG slow waves – a state opposite to vigilance.[105] However, a general cortical activation promoted by acetylcholine is only a nonspecific predisposition to the goal-oriented selective activity that requires differentiation (discrimination) between meaningful and meaningless information elicited by the environment. Such discrimination is based on the partial flexible inhibition of cortical neuronal activity and as a result – on the increase of the signal-to-noise ratio that makes neuronal activity task-relevant. NE and 5-HT in wakefulness are responsible for this partial cortical inhibition.[27,108] Thus, mental functioning during the waking state depends upon two types of neurotransmitters:[23] activators which support the general mobilization of cortical functions, and inhibitors controlling and modulating this activation in order to make mental functions flexible and relevant to the task.

Dopamine in the normal waking brain plays an important role in motivational processes providing "reward" and "reinforcement", and in novelty seeking that includes exploratory behavior, attention, exhilaration and excitement in response to novel stimuli.[8,10,15,23,36] According to Wise et al[107] and Wise and Colle[106] dopamine mediates naturally rewarding experiences (like pleasure from food, sex, drugs). However it is involved not only in appetitive events and in an approach behavior but also in aversive ones.[81,82] Paradoxically, such aversive behavior in some of its aspects seems to be attractive for the subject and does not contradict to the general concept of "reinforcement". In this context it is possible to ascribe rewarding experiences also to the dopamine dependent psychotic symptoms like hallucinations and delusions that are very resistant to any treatment except for the antagonists of DA receptors.[32] According to some recent investigations[96] in the cortex, and especially in the frontal cortex DA transporters are under the strong modulating influence of NE nerve terminals while in basal ganglia NE has a little regulatory role for DA. NE reuptake blocker increases not only NE but even in a more prominent way DA concentration in the cortex and only NE in subcortical area.

Rem Sleep: Resensitization of the Postsynaptic Noradrenegic Receptors?

REM sleep state is unique according to the complete cessation of the noradrenergic LC cells activity. It is reasonable to believe that such cessation has a special physiological meaning, and Siegel and Rogawski[90] were the first who proposed a coherent and comprehensive theory of this topic. These authors hypothesized that the inhibition of activity of the NE containing cells in REM sleep is required to maintain the sensitivity of NE postsynaptic receptors, with consequent benefits for all types of behavior in wakefulness that utilize these receptors. During wakefulness all adaptive forms of behavior have to be flexible and require constant activity of the noradrenergic system. Such almost nonstop activity is unavoidably leading to the desensitization of the NE postsynaptic receptors, and this negative feedback finally causes the decrease of the NE system efficiency. REM sleep that appears with regular intervals provides this system with an opportunity to restore its functional activity without interference with the ongoing waking behavior. Siegel and Rogawski hypothesized that NE release sets in motion two processes having opposite effect on REM sleep duration. In the first process, NE release or its functional enhancement suppresses or "substitute" for REM sleep by increasing the activity of negative feedback circuits monitoring the efficiency of NE receptor action. In this step, the noradrenergic system is active and does not require REM sleep for its restoration. In the second process, the release or potentiation of NE action is hypothesized to downregulate/desensitize NE receptors, this downregulation producing the increased REM sleep pressure. Thus, according to this concept, the cessation of NE cells activity in REM sleep actually contributes to the activity of the noradrenergic system. The evidence confirming this point of view can be found in Siegel and Rogawski.[90] Much more important and relevant is to consider and to discuss numerous data that seem to contradict this theory.

1. Depression in humans and learned helplessness in animals[88] are characterized by the stable reduction of monoamine (NE and 5-HT) transmission in brain synapses. Thus, according to the theory it would be reasonable to expect the decrease of REM sleep requirement in these states as an outcome of the already established chronic sensitization of the postsynaptic noradrenergic receptors. However, according to Adrien et al[1] there is a positive correlation between experimentally induced learned helplessness and percentage of paradoxical (REM) sleep. The increased REM sleep pressure in depression is shown by the reduction of REM sleep latency, a relative increase of REM sleep in the first cycle, an increased number of the short sleep cycles, and by the absence of the first night effect.[11,40,61,78,79] Healthy long sleepers are characterized by the relatively increased REM sleep[27] and at the same time by the inclination toward subdepressive reactions.[103] A disposition to depressive reactions is also characteristic of narcoleptic patients who show a constantly high REM sleep requirement.[9] When the level of depression is moderate, an increased REM sleep requirement realizes itself in the increased REM sleep. The relationship between the severity of depression and REM sleep is nonlinear:[66] when the MMPI scale D (depression dominate) and does not exceed 75 T points REM sleep grows longer (compared to the magnitude of this scale of up to 65 T points). When the scale gets higher, REM sleep becomes reduced. Thus, before depression starts to destroy sleep structure, it determines a tendency towards REM sleep increase. From my point of view, all these data do not correspond to the Siegel and Ragowski theory.

2. Reserpine treatment causes a depletion of NE[87] but produces elevation of REM sleep. Siegel and Ragowski[90] explain this REM sleep elevation as an attempt to upregulate NE receptors in response to NE depletion. However, this explanation looks circular: REM sleep is characterized by the marked reduction of noradrenalin cells activity and such reduction has to overcompensate NE depletion.

3. It was shown in many investigations, that antidepressants—monoamine oxidase inhibitors (MAOI) that enhance the noradrenergic transmission in synapses—suppress REM sleep for the all period of prescription in animals, in healthy subjects and in depressed patients.[17,31,41,64] This period of prescription can take a few weeks. According to the theory we are discussing, it was possible to expect not a decrease but rather an increase of REM sleep because of a long-lasting and intense stimulation of the noradrenergic system. Siegel and Ragowski are aware of this contradiction. They suggest that the process of NE receptors downregulation caused by enhanced NE transmission can take from few minutes to few weeks. However, from our point of view such time course is too broad for this micro-physiological process and instead of making such proposition it is more reasonable to search for another explanation.

4. In the frame of the discussed theory, it seems difficult to explain some data of partial REM deprivation by using awakenings in REM sleep. If after momentary awakening, animals were maintained in a condition of active and emotional wakefulness (i.e., wakefulness based on the enhanced noradrenergic activity) neither the accumulation of REM need nor the postdeprivation REM rebound appear.[13,51] It is necessary to take into consideration, that REM sleep was already reduced before awakenings and nevertheless fragments of active wakefulness were able to satisfy the accumulated REM need.

5. Siegel and Rogawski predicted that the sensitivity of all LC innervated postsynaptic NE receptors should be reduced by prolonged sleep/REM sleep loss. However, Tsai et al[98] have shown that density and affinity of adrenergic binding sites did not decrease after 10 days of total sleep deprivation. Thus sleep deprivation made no expected changes in central NE receptor regulation.

In spite of all these contradictions, we do not conclude that the theory of Siegel and Rogawski is not relevant at all. The cessation of the activity of noradrenergic cells in REM is a fundamental fact that needs explanation, and the resensitization of the postsynaptic noradrenergic receptors may be a real task of such cessation. However, the abovementioned contradictions show that this theory has limitations and is probably relevant only in some particular conditions, and, secondly, that it is not an exhaustive one and has to be supplemented by additional suggestions of REM sleep functions related to the monoamine activity that may be helpful in solving contradictions.

Search Activity Concept, REM Sleep Functions and Brain Monoamines

I suggest search activity (SA) concept to represent such a supplementary theory.[65,68,74]

By search activity is understood activity designed to change the situation or the subject's attitude to it in the absence of a definite forecast of the results of such activity (i.e., in the case of pragmatic indefiniteness), but with constant monitoring of the results at all stages of activity. This definition makes it clear that certain behavioral categories cannot be classed with search behavior. This primarily applies to all forms of stereotyped behavior having a quite definite forecast of results. Panicky behavior at first sight may seem to imitate search behavior but differ from it by the disturbance of the feedback between the activity and its regulation. During a panic the results of the activity are not considered at any stage and cannot be used for the correction of behavior. No line of activity can be traced to its conclusion and panicky behavior easily becomes imitative, approaching stereotyped behavior. Finally, the opposite of search behavior is the state of renunciation of search, which in animals may assume the form of freezing or learned helplessness and in humans corresponds to depression and maladaptive (neurotic) anxiety.[76]

Search activity is a component of many different forms of behavior: self-stimulation in animals, creative behavior in humans, as well as exploratory and active defense (fight/flight) behavior in all species. In all these forms of activity the probability forecast of the outcome is indefinite, but there is a feedback from the behavior and its outcome enabling the subject to correct his behavior in accordance with the outcome. One of the best indications of search activity in animals is a high-amplitude and well-organized hippocampal theta-rhythm (for details see refs. 68, 76).

The need for a new classification of behavior based on the presence or absence of search activity is determined by its important biological meaning. In research conducted together with V. Arshavsky, we found that all forms of behavior which include search activity increase body resistance to the different forms of artificial pathology (artificial epilepsy, artificial extrapyramidal disturbances caused by neuroleptics, anaphylactoid edema, artificial arrythmia of cardiac contractions, etc.), while renunciation of search decreases body resistance, suppresses immune system and predisposes subject to somatic disorders.[65,74,77] We concluded that the process of search activity by itself independently of whether it is successful or not (according to the pragmatic results of the behavior) protects the subject from somatic disorders.

However, if search activity is so important for survival and if renunciation of search is so destructive and harmful, it would be reasonable to assume a special brain mechanism able to restore search activity after temporary and occasional renunciation of search. According to the search activity concept, PS fulfils this function. A covert search activity in PS during dreams compensates for the lack of search activity in the previous wakefulness and ensures the resumption of search activity in the subsequent wakefulness. This claim is based on the following findings:

1. Renunciation of search evoked by the direct stimulation of ventro-medial hypothalamus causes an increase of PS in the subsequent sleep, while after search behavior evoked by the brain stimulation PS decreases.[74]

2. Depression in humans and learned helplessness in animals are accompanied by increased PS requirement (decreased PS latency and increase of PS in the first sleep circle). A correlation is detected between learned helplessness and PS percentage.[1]

3. Both PS and search activity in wakefulness are characterized by regular and synchronized hippocampal theta-rhythm. Moreover, the more pronounced the theta-rhythm in wakefulness, the less pronounced it is in the subsequent PS.[51] PS in animals regularly contains ponto-geniculo-occipital (PGO) waves, which in wakefulness correspond to orienting activity.[39] The presence of the PGO spikes in PS means that the subject is predisposed to react to novel stimuli, including spontaneous change of dream content.

4. If nucleus coeruleus in the brain stem is artificially destroyed and as a result muscle tone does not drop during PS, animals demonstrate complicated behavior that can be generally described as orienting activity[48] or search behavior.

If behavior in stressful situation contains search activity (aggression or active avoidance), PS decreases without subsequent rebound because such behavior in wakefulness does not require the restoration of search activity in PS. This approach can explain also data of Oniani and his coworkers.[13,51] These investigators performed awakenings of animals on every PS onset during sleep. When they have produced just short fragments (2-3 seconds) of nonemotional wakefulness, a typical effect of PS deprivation appeared: PS onset frequency increased in comparison to the baseline level and it was also PS rebound in the post-deprivation period. However, if after momentary awakening animals were maintained in the condition of active and emotional wakefulness equal in length to PS mean duration, neither the accumulation of PS need nor the post-deprivation PS rebound appeared. Darchia et al[13] stressed that fragments of active wakefulness are able to satisfy even the accumulated PS need, and from our point of view this effect can be explained by the dominance of search activity in the evoked wakefulness. Short total sleep deprivation (4-12 hours) performed by awakenings decreases sleep latency and increases SWS and delta power in the subsequent sleep.However PS is not increased after such deprivation.[26]

In contrast, immobilization stress makes the manifestation of search behavior in wakefulness inavailable, and as a result the need in the subsequent compensatory PS increases.

Very similar conditions are created during the sleep deprivation on the wooden platform.[16,60,73] Of course, it is not a real immobilization, however animals free behavior in this condition is restricted and search activity is almost blocked. In addition, animals are regularly frustrated in their attempts to satisfy their natural need in sleep, or in PS. Such regular frustration is a condition for learned helplessness as a concrete manifestation of renunciation of search.[71] As a result, the need in PS increases, however PS is suppressed together with the total sleep. Such a combination of the increased requirement in search activity with PS deprivation can explain the main outcomes of the total sleep deprivation.

On the one hand, in surviving animals recovery sleep is marked by a dramatic rebound of PS after immobilization stress.[16] NREM sleep rebound was not observed although most of the lost sleep was of the NREM sleep type. It means that the requirement in PS caused by the combination of the PS deprivation and the frustration of behavioral search activity is more important for the organism than the requirement in NREM sleep which in this particular condition is less obligatory. Moreover, after the PS rebound it is a quick reversal of the somatic outcomes of the prolonged sleep deprivation.

Dreams in REM sleep represent a very specific kind of search activity, which, however, is compatible with the above-mentioned notion of search activity: the healthy subject is usually active in his dreams[67] and the more active the dream characters and the dreamer himself the more prominent is the improvement of subject's mood;[38] at the same time the dreamer is unable to make a definite probability forecast according to dream events. Search activity in dreams is more flexible, less organized and less goal-directed than in wakefulness, and even if dreamer is moderately self-reflective in dream[54] it is obvious that he/she is less self-reflective than in wakefulness.

It is worth stressing that dreams provide a good opportunity for the compensatory search activity after giving up in waking behavior.[72] First, the subject is separated from the reality while sleeping, including those aspects of reality that caused renunciation of search. Thus, the subject is free to start from the beginning. Second, within his dream, the subject is very free in his decisions: he can try to solve the his actual problem in a metaphoric manner, or he can start solving another problem, one that displace the actual problem[25] since the search process itself is the main restorative factor. Polysemantic image thinking that is active in dreams is more flexible than logical thinking and is free from the probability forecast.[75] Since I assume that the final aim of dream work is not the real solution of the actual problem but only the restoration of search activity, all the above features contribute substantially to this restoration.

Concerning the relationships of the brain monoamine system to search behavior, the following hypothesis has been developed.[65] Search activity can start in the presence of a certain critical level of the brain monoamines (in particular, norepinephrine) which are utilized as "oil" in the course of search behavior. Search activity itself, once it starts, further stimulates the synthesis of the brain monoamines and ensures their availability. There are some reasons to belief that search activity in wakefulness decreases the sensitivity of the inhibitory presynaptic alpha2-adrenoreceptors thus preventing the inhibition of monoamines neurons. For instances, it was suggested that the sensitivity of these receptors is decreased in REM sleep deprivation[3,45] and we have suggested (see below) that symptoms of the relatively short REM deprivation correspond to the notion of search activity. Thus, the more pronounced the search activity, the sooner the turnover and synthesis of monoamines will be, in turn maintaining search behavior (positive feedback system). For search activity to begin, the brain monoamine concentration must exceed a critical level. If it drops below its level, search activity is canceled.

In a state of renunciation of search, the above-mentioned positive feedback system does not function. Furthermore, in this state, which manifests itself particularly in depression, monoamines display a tendency to drop. This may be explained by the fact that renunciation of search is usually combined with distress, which causes intense monoamine expenditure without subsequent restoration due to the absence of search activity. Thus, according to this hypothesis, monoamine functioning complete a vicious circle: renunciation of search leads to a drop in the brain monoamines level, which in turn leads to the renunciation of search's becoming more prominent.

This theoretical approach has some important practical outcomes. For instance, conceptualizing depression as a renunciation of search leads to the revised approach to the mechanisms of clinical treatment.[69]

To overcome depression characterized by the exhaustive "vicious" circle (renunciation—decreased brain monoamine turnover—renunciation), it is necessary not only to restore brain monoamines level but also to "switch on" the opposite positive feedback (increased brain monoamines – search activity – further increase of brain monoamines). Only when renunciation of search is replaced by search activity does brain monoamines stabilize on an appropriate level. As a result, the number and/or sensitivity of the postsynaptic receptors in the brain are diminished, which probably correlates with the clinical efficacy of antidepressant treatment. Thus, the therapeutic tactic has to be directed to the behavioral and intellectual activation of patients in the course of drug treatment. This hypothesis can explain paradoxical data of the reduction of the depressive symptoms in unexpected stressful conditions.

According to the initial hypothesis[65] the relationships between monoamines and REM sleep have been presented as following: in the state of renunciation of search the restoration of brain monoamines requires search activity in REM sleep dreams; its start requires, like in wakefulness, an above critical level of brain monoamines however this level in REM sleep is lower as for the start of search activity in wakefulness. On the other hand, a high monoamines turnover that corresponds to the prominent search activity in wakefulness reduces REM sleep without the subsequent REM sleep rebound, it means reduces the REM sleep requirement. This hypothesis explained REM sleep increase along with a moderate reduction of norepinephrine system activity and REM sleep suppression following the pronounced inhibition of this system.[19,34] However, this initial hypothesis does not fit the above data of the total cessation of NE cells activity in REM sleep because according to this initial hypothesis this activity had to restore the course of REM sleep in parallel with search activity in this state.

By taking into consideration these and many other data from recent investigations, in the present chapter I am going to revise and modify the initial hypothesis. This modification partly includes the hypothesis of Siegel and Ragowski.[90] However, the corner-stone of the modified hypothesis is the proposition of Gottesmann[23] according to the role of different monoamines in mental activity, particularly in REM sleep.

According to this modified hypothesis, search activity in wakefulness is based on the combination of activating (Ach and DA dependent) and inhibitory (NE and 5-HT dependent) influences on cortical neurons. This combination determines the regulation of search behavior, its goal direction, its relative restriction according to the actual tasks and its relevance to the objective reality. Due to this regulative inhibitory influences search activity in normal wakefulness although relatively flexible, is neither infinite nor omnipotent: it has limits.

In REM sleep, due to the cessation of the inhibitory NE and 5-HT neurons and the absence of its modulating activity, search activity being based exclusively on the DA system became free, unrestricted, labile and almost chaotic. It displays itself in dreams. According to Solms,[95] dreaming itself occurs only if and when the initial activation stage engages the dopaminergic circuits of the ventromedial forebrain. Dopaminergic agents increase the frequency, vivacity and duration of dreaming without similarly affecting the frequency, intensity and duration of REM sleep.[29] It is not an occasion that many prominent authors have underlined the similarity between dreams and psychosis (like positive symptoms in schizophrenia) the latter being also related to the hyperactivity of DA system. This topic was discussed in detail by Gottesmann.[23] Positive symptoms in schizophrenia have been already considered as a form of misdirected and maladaptive search activity.[70] However, the main difference between them and dreams is that hallucinations and delusions appear during wakefulness, interfere with the reality perception and disturb the adaptive behavior while dreams appear in REM sleep when subject is naturally separated from the reality and predisposed to such extravagant compensatory search activity in the virtual world. Another difference is that dreams are using the rich potential of the right-hemispheric polysemantic image thinking and are acting mostly in the domain of visual system while delusions and hallucinations are mostly in the domain of the left-hemispheric verbal system, and moreover—they are the outcome of the functional disability of image thinking.[70]

If search activity in REM sleep (in dreams) is based predominantly on the nonmodulated activity of DA system, it has a lot of advantages. First of all, as it was already stressed, it makes search activity in dreams unrestricted and almost omnipotent. Secondly, the temporal cessation of NE activity in REM sleep may help to restore the sensitivity of the postsynaptic NE receptors, as Siegel and Ragowski proposed, and the restored sensitivity of the NE system is very important for the well-regulated and goal-directed search activity (and any mental activity) in the subsequent wakefulness.

The application of the search activity concept to the REM sleep-brain monoamines interrelationships provides an opportunity to reconsider some theoretical assumptions avoiding contradictions.

1. If the main task of REM sleep (PS) is the restoration of search activity in the subsequent wakefulness and the restoration of physiological mechanisms that provide search activity then all conditions that enhance search activity in waking behavior abolish the demand (requirement) in REM sleep. It is a reason why different antidepressants and amphetamine suppress REM sleep without rebound effect. At the same time, it allows us to make a very important assumption that even intense and long lasting search activity (in chronic stress that is not replaced by distress, in short sleepers etc.), in opposite to the routine, stereotyped activity, does not cause the downregulation (desensitization) of the postsynaptic noradrenergic receptors. Perhaps it can be explained by the very intense turnover of brain monoamines – they are released, used for search behavior and immediately replaced by the new portion.

2. On the other hand, search activity concept explains the increased REM sleep pressure as a response on the depletion of brain monoamines caused by reserpine with its depression-like effect on behavior.

3. It was found in some investigations[49,102,104] that not all antidepressant agents suppress REM sleep and increase REM sleep latency (decrease REM sleep requirement). Nefazodone increased or at least does not decrease REM sleep and shifted it to earlier in the night. Bupropion reduced REM latency and increased REM sleep percent and REM time. It looks quite opposite to the outcome of other antidepressant agents on sleep structure. However, if we accept the proposition that the natural REM sleep function is the restoration of search activity and the hypothesis, partly confirmed in our previous investigations, that REM sleep in depression is functionally inefficient[65,66,68,73] then it is possible to speculate that some antidepressant agents may help to abolish depression by the restoration of the functional efficacy of REM sleep. In such cases REM sleep may increase like in long sleepers who are using sleep for mood restoration without antidepressive treatment.

4. The revised search activity concept helps to explain the alteration of sleep structure on different doses of neuroleptic treatment: small and moderate doses of neuroleptics increase the total REM sleep time, whereas large doses suppress it.[43] It is possible to suggest that small and moderate doses of neuroleptics decrease search activity in wakefulness (see refs. 43, 70) thus increasing the REM sleep requirement, while high doses suppress search activity in REM sleep based on DA activity and as a result abolish the need in this state.

5. According to Siegel and Ragowski,[90] the sensitivity of all LC innervated postsynaptic NE receptors should be downregulated by prolonged sleep and REM sleep deprivation. Such desensitization was predicted as an outcome of the stable and long lasting NE cells activity in wakefulness. This proposition was not confirmed after 10 days of total sleep deprivation (TSD) in rats on the rotating platform surrounded by water:[98] density and affinity of adrenergic binding sites did not decrease, although it was a typical effect of sleep deprivation on body weight and energy expenditure and a massive PS rebound after even 5 days of sleep deprivation. However, TSD in this condition may not maintain a high NE discharge rate typical for the normal wakefulness because this condition gives no place for search behavior, frustrate animal and finally causes renunciation of search[73] presumably accompanied by brain monoamines depletion.

By discussing data of sleep and PS deprivation by the water-tank technique it is necessary to bear in mind that the behavioral and physiological reaction on such deprivation has two opposite stages (see review of ref. 65). In the first stage animals exposed to such deprivation after turning back to normal conditions exhibit increased activity that may combine search and stereotyped behavior: hypersexuality, hyperphagia, increased motor activity in the open field, decreased latency for the object approach, increased object exploration, diminished anxiety, intensified self-stimulation.[46,50] It is like a rebound effect after frustration and this rebound effect confirms that the compensatory sources of the organism are still not lost. (It is interesting that a short-lasting REM sleep deprivation increases the explorative (search) behavior and reduces the latency to the object approach even in animals with a damaged locus coeruleus and damaged NE system.[46] This enhanced behavior activity after REM deprivation might be at least partly based on the activity of DA system, activity that cannot be realized in REM sleep due to its deprivation. This proposition is in agreement with the assumption of the role of DA system in search activity and was experimentally confirmed by Asakura et al[3,4] who have shown the involvement of dopamine D2 receptor mechanism in the REM deprivation induced increase in swimming activity).

However, if sleep/PS deprivation in these stressful conditions lasts a sufficiently long time (for PS deprivation – more than 96 hours) the brain and body's reserves deplete and renunciation of search will prevail even after the cessation of the condition of deprivation. Animals after such prolonged deprivation remained passive and "depressive" for a long time. Mollenhour et al[47] assumed that the weakening of active (aggressive) behavior in the case of prolonged PS deprivation is connected with the exhaustion of brain monoamines (NE). We cannot exclude the exhaustion of DA also. As it was already mentioned, according to Rechtschaffen et al[60] a prolonged sleep/PS deprivation inevitably causes death. Thus while discussing the outcome of PS deprivation it is necessary to take into consideration these two stages. Brain monoamines sources have to be high enough to allow an animal to display an active behavior after sleep/PS deprivation.

The investigation of Asakura et al[3] seems to confirm this assumption. Clonidine increases swimming activity in the forced swimming test, and a short-lasting REM sleep deprivation intensifies this clonidine response while monoamine depletion contradicts this effect of REM sleep deprivation.

6. Another abovementioned contradiction is related to the role of brain NE in REM sleep preservation and functional flexibility. On the one hand, a total destruction of LC with consequent depletion of NE in most brain areas does not prevent REM sleep. On the other hand, PS rebound after 10 hours of sleep deprivation on the small platform was significantly decreased after a single injection of a neurotoxic substance which induces long-term degeneration of NE fibers coming from LC[21] and the same substance decreases PS augmentation after the immobilization stress[22] Search activity concept presents a following explanation of these contradictory data. REM-on cells localized in medulla and responsible for the generation of REM sleep as a physiological phenomenon are independent of NE system and of the whole brain and are continuing their activity even being totally separated from the higher parts of the brain. However after such separation they are working in a very stereotyped automatic way. But being responsible for the restoration of search activity REM sleep is flexible and changes in its duration only when this function is required and is available. When monoamines ("oil" of search activity) became depleted, or monoamines systems are blocked by other reasons, search activity in wakefulness in any case cannot be restored by the mean of REM sleep and the latter has no functional reasons to change in duration. Actually, a systemic administration of serotonergic or noradrenergic antagonists reduces REM sleep expression and increases the intervals between REM sleep episodes, perhaps reducing the rate of accumulation of REM sleep propensity.[7] Another and also very relevant reason for this lack of REM flexibility was presented by Gonzales et al.[22] They suggested that the degeneration of NE system prevents the development of stress (distress) and it is the reason why REM sleep does not increase. It is very possible: renunciation of search (whether depression or helplessness or frustration) that requires REM sleep for its compensation is always accompanied by distress and an absence of search behavior without distress may not elicit REM sleep.

7. Mirmiran et al[44] Vogel et al[101] Vogel and Hagler[100] and Feng and Ma[17] have found that if active sleep in postnatal species that resembles REM sleep of adults is suppressed by mean of antidepressants without a corresponding increase of wakefulness, it causes subsequently depressive-like disorders in adults. On the first glance it looks like a paradox, however if an active sleep in postnatal period is a state that lay a basis for the development of search activity in adulthood than these results are understandable because in this case the suppression of active sleep leaves subject without predispositions to an adaptive behavior.

8. While discussing the outcome of different psychotropic drugs on brain monoamines and REM sleep it is necessary to take into consideration that such outcome may differ in patients and in normal subjects.

In animals and healthy subjects clonidine, an effective alpha2-adrenergic receptor agonist suppress REM sleep and this effect is blocked by alpha2-adrenoreceptors antagonist yohimbine,[57] while depressed patients display a blunted effect of clonidine on REM sleep.[86] Depressed patients demonstrated also a blunted growth hormone response to clonidine.[84] According to these data, Schittecatte et al[86] hypothesized a subsensitivity of central alpha2-adrenoreceptors in depression. However, this hypothesis is in strong contradiction to the hypothesis that endogenous depression is characterized by supersensitivity of alpha2-adrenoreceptors, in particular inhibitory presynaptic alpha2-adrenoreceptors[53] and that the delayed positive effect of antidepressant medications is related to the desensitization of these receptors that takes time.[92]

From our point of view, by discussing these contradictions it is necessary to take into consideration that clonidine and yohimbine are only imitating the natural conditions in which presynaptic and postsynaptic alpha2-adrenoreceptors are activated or inhibited. In natural conditions, alpha2-adrenoreceptors are activated by the NE transmission as a consequence of the high activity of the NE neurons. This high activity of the NE system provides conditions for active interrelations with the environment and the requirement of REM sleep in these conditions became decreased. Thus, if clonidine stimulates postsynaptic adrenoreceptors, it is very understandable that in normal subjects this stimulation causes the suppression of REM sleep and an increase of swimming activity in the forced swimming test in experimental animals. This explanation is confirmed by data that the response on clonidine treatment is dose-dependent, because clonidine stimulates postsynaptic adrenoreceptors in doses higher than those required for the stimulation of the presynaptic inhibitory alpha2-adrenoreceptors. In animals after a relatively short REM sleep deprivation (48-72 hours) the lower dose of clonidine has a stimulating effect on swimming, and it is in agreement with our assumption that a short REM sleep deprivation stimulates search activity and has an activating influence on the brain monoamines systems.

This proposition is confirmed also by data that clonidine is working in the same direction as imipramine (antidepressant that increases the concentration of NE in synapses) and both, directly or indirectly, stimulate alpha2-adrenoreceptors and increase swimming activity in forced swimming test,[2,4] and it is very natural that increased activity is accompanied by reduced REM sleep.

In depressed patients, in contrast to healthy subjects, clonidine does not suppress REM sleep, and when it is used after treatment with serotonin reuptake blocker REM sleep even display a tendency to increase.[86] Schittecate et al, explained the blunted response of REM sleep on clonidine as a sign of the down-regulation of alpha2-adrenoreceptors in depression. However, from our point of view it is also another possibility: the response on clonidine may be low if adrenoreceptors are already high activated, up-regulated, high sensitive due to the stable low level of monoamines in the synaptic clefts, and its activation by mean of clonidine do not add to much to this initial and unhelpful activation. This approach seems to fit with data that clonidine starts to suppress REM sleep in depressed patients 48 hours after treatment with mirtazapine (alpha2-adrenoreceptor blocker).

Conclusion

Our general conclusion is that the main function of REM sleep is the restoration of search activity in the subsequent wakefulness. In wakefulness search activity in a normal state is relevant to the reality, goal directed and task oriented and sustained by the interrelationships between brain activating and activity modulating brain monoamines. Renunciation of search in wakefulness is accompanied by the decreased activity of most brain monoamine systems, particularly of brain norepinephrine. In the functionally sufficient REM search, activity is based on the nonmodulated brain dopamine activity (that makes search activity in dreams extremely flexible and available for restoration). At the same time REM sleep provides the condition for the resensitization of the norepinephrine postsynaptic receptors that is important for the continuation of search activity in the subsequent wakefulness. The present model helps to explain many controversial data in REM sleep-brain monoamines relationships.

References

1. Adrien J, Dugovic CH, Martin P. Sleep wakefulness patterns in helpless rats. Physiol Behav 1991; 49:257-262.
2. Asakura W, Matsumoto K, Ohta H et al. Effect of alpha2-adrenergic drugs on REM sleep deprivation–induced increase in swimming activity. Pharmacol Biochem Behav 1993; 46:111-115.
3. Asakura W, Matsumoto K, Ohta H et al. Monoamine depletion attenuates the REM sleep deprivation-induced increase in clonidine response in the forced swimming test. Pharmacol Biochem Behav 1994; 49:79-84.
4. Asakura W, Matsumoto K, Jhta H et al. Involvement of dopamine D2 receptor mechanism in the REM sleep deprivation-induced increase in swimming activity in the forced swimming test. Pharmacol Biochem Behav 1994; 48:43-46.
5. Asakura W, Matsumoto K, Watanabe H. REM sleep deprivation treatment enhances the effect of clozapine in the forced swimming test. Gen Pharmac 1995; 26:1225-1228.
6. Baldessarini RJ. Chemotherapy in psychiatry: Principles and practice. 2nd ed. Cambridge: Harvard University Press, 1985.
7. Benington JH, Heller HC. Monoaminergic and cholinergic modulation of REM-sleep timing in rats. Brain Res 1995; 681:141-146.
8. Benjamin J, Patterson Ch, Greenberg BD et al. Population and familial association between the D4 dopamine receptor gene and measures of novelty seeking. Nat Genet 1996; 12:81-84.
9. Beutler L, Ware J, Karacan I et al. Differentiating psychological characteristics of patients with sleep apnea and narcolepsy. Sleep 1981; 4:39-47.
10. Cloninger CB, Adolfsson R, Svrakic NM. Mapping genes for human personality. Nat Genet 1996; 12:3-4.
11. Coble PA, Kupfer DJ, Shaw DH. Distribution of REM latency in depression. Biol Psychiatry 1981; 16:453-466.
12. Cohen RM, Pickar D, Garnett D et al. REM sleep suppression induced by selective monoamine oxidase inhibitors. Psychopharmacology (Berl) 1982; 78:137-140.
13. Darchia N, Oniani T, Gvilia I et al. Analysis of competitive interrelationship of wakefulness and paradoxical sleep using two different methods of parado[ical sleep deprivation. Abstracts of the 3rd International Congress of WFSRS. Dresden, 1999; 521.
14. Dement WC, Mitler MM, Henricksen SJ. Sleep changes during chronic administration of parachlorphenylalanine. Rev Can Biol 1972; 31:69-75.
15. Ebstein RP, Novick O, Umansky R et al. Dopamine D4 receptor (D4DR) exon III polymorphism associated with the human personality trait of novelty seeking. Nat Genet 1996; 12:78-80.
16. Everson CA. Functional consequences of sustained sleep deprivation in the rat. Behav Brain Res 1995; 69:43-54.
17. Feng P, Ma Y. Clomipramine suppresses postnatal REM sleep without increasing wakefulness: Implications for the production of depressive behaviors. Sleep 2002; 25:177-184.
18. Fornal C, Auerbach S, Jacobs BL. Activity of serotonin-containing neurons in nucleus raphe magnus in freely moving cats. Exp Neurol 1985; 88:590-608.
19. Gaillard JM. Brain catecholaminergic activity in relation to sleep. In: Priest RG, Pletscher A, Ward J, eds. Sleep Research: Proceedings of the Northern European Symposium on Sleep Research. Basle: 1979:35-41.
20. Gillin JC, Wyatt RJ, Fram D et al. The relationship between changes in REM sleep and clinical improvement in depressed patients treated with amitriptyline. Psychopharmacol (Berlin) 1978; 59:267-272.
21. Gonzales MM, del C, Valatx JL et al. Role of the locus coeruleus on the mechanism of the sleep rebound. J Sleep Res 1994; 3(Suppl 1):92.
22. Gonzalez MM, del C, Debilly G et al. Sleep increase after immobilization stress: Role of the noradrenergic locus coeruleus system in rat. Neuroscience Letters 1995; 202:5-8.
23. Gottesmann C. The neurochemistry of waking and sleeping mental activity: The disinhibition-dopamine hypothesis. Psychiatry and Clinical Neurosciences 2002; 56:345-354.
24. Greenberg R. Where is the forest? Where is the dream? Behavioral and Brain Sciences 2000; 23:943-945.
25. Greenberg R, Pearlman CH. The private language of the dream. In: J Natterson, ed. The dream in clinical practice. New York: Aronson, 1980; 85-96.
26. Gvilia I, Oniani T, Darchia N et al. Analysis of the effect of total sleep deprivation in rats. Abstracts of the 3rd International Congress of WFSRS. Dresden, 1999:533.
27. Hartmann E. The Functions of Sleep. New Hawen, CT: Yale University Press, 1973.
28. Hartmann E, Chung R, Draskoczy PR et al. Effects of 6-hydroxydopamine on sleep in the rat. Nature (Lond.) 1971; 233:425-427.
29. Hartmann E, Russ D, Oldfield M et al. Dream content: Effects of L-DOPA. Sleep Research 1980; 9:153.
30. Hobson JA, McCarley RW, Nelson JP. Location and spike-train characteristics of cells in anterodorsal pons having selective decreases in firing in rat during desynchronized sleep. J Neurophysiol 1983; 50:770-783.
31. Jobert M, Jahnig P, Schulz H. Effect of two antidepressant drugs on REM sleep and EMG activity during sleep. Neuropsychobiology 1999; 39:101-109.

32. Jones HM, Pilowsky LS. Dopamine and antipsychotic drug action revisited. British Journal of Psychiatry 2002; 181:271-275.
33. Jones BE, Harper ST, Halaris AE. Effects of locus coeruleus lesions upon cerebral monoamine content, sleep wakefulness states and the response to amphetamine in the cat. Brain Res 1977; 124:473-496.
34. Kafi S, Bouras C, Constantinidis J et al. Paradoxical sleep and brain catecholamines in the rat after single and repeated administration of alpha-methyl-parathyrosine. Brain Res 1977; 135:123-134.
35. Kamamori N, Sakai K, Jouvet M. Neuronal activity specific to paradoxical sleep in the ventromedial medullary reticular formation of unrestrained cats. Brain Res 1980; 189:251-255.
36. Kapur Sh. Psychosis as a state of aberrant salience: A framework linking biology, phenomenology, and pharmacology in schizophrenia. American J Psychiatry 2003; 160:13-23.
37. Kinai T, Scerb JC. Mesencephalic reticular activating system and cortical acetylcholine output. Nature 1965; 205:80-82.
38. Kramer M. The selective mood regulatory function of dreaming: An update and revision. In: Moffitt A, Kramer M, Hoffmann R, eds. The function of dreaming. New York, Albany: State University of New York Press, 1993:139-196
39. Kuiken D, Sikora S. The impact of dreams on waiking thoughts and feelings. In: Moffitt A, Kramer M, Hoffmann R, eds. The function of dreaming. New York: States University of New York Press, 1993:419-476.
40. Kupfer DJ, Ulrich RF, Coble PA et al. The application of automated REM and slow wave sleep analysis (normal and depressives). Psychiatr Res 1984; 13:325-334.
41. Landolt HP, de Boer LP. Effect of chronic phenelzine treatment on REM sleep: Report of three patients. Neuropsychopharmacology 2001; 25:S63-67.
42. McGinty DJ, Harper KM. Dorsal raphe neurons depression of firing during sleep in cats. Brain Res 1976; 101:569-575.
43. Mendelson WB, Gillin JCh, Wyatt RJ. Human sleep and its disorders. New York: Plenum Press, 1977.
44. Mirmiran M, Van de Poll NE, Corner MA. Suppression of active sleep by chronic treatment with chlorimipramine during early postnatal development: Effect upon adult sleep and behavior in the rat. Brain Res 1981; 204:129-146.
45. Mogilnicka E, Pilc A. Rapid eye movement sleep deprivation inhibits clonidine-induced sedation in the rat. Europ J Pharmacol 1981; 71:123-126.
46. Mogilnicka E, Boissard CG, Hunn C. Suppresant effect of REM sleep deprivation on neophobia in normal rats and in rats with selective DSP-4 induced damage of locus coeruleus neurons. Pharmacol Biochem Behav 1985; 23:93-97.
47. Mollenhour MN, Voorhees JW, Davis SF. Sleepy and hostile: The effects of REM sleep deprivation on shock-elicited aggression. Anim Learn Behav 1977; 5:148-152.
48. Morrison A. Central active states overview. In: Beckman AL, ed. The neural basis of behavior. New York: Spectrum, 1982:3-18
49. Nofzinger EA, Reynolds III CF, Thase ME et al. American J Psychiatry 1995; 152:274-276.
50. Ogilvie RD, Broughton RJ. Sleep deprivation and measures of emotionality in rat. Psychophysiology 1976; 13:249-260.
51. Oniani T, Lortkipanidze L. Effect of paradoxical sleep deprivation on the learning and memory. In: Oniani TN, ed. Neurophysiology of motivation, memory and sleep-wakefulness cycle Tbilisi. Metzniereba: 1985:214-234
52. Perez NM, Benedito MAC. Activities of monoamine oxidase (MAO) A and B in discrete regions of rat brain after rapid eye movement (REM) sleep deprivation. Pharmacol Biochem Behav 1997; 58:605-608.
53. Piletz JE, Halaris A, Ernsberger PR. Psychopharmacology of imidazoline and alpha-2 adrenergic receptors: Implications for depression. Crit Rev Neurobiol 1994; 9:29-66.
54. Purcell S, Moffitt A, Hoffmann R. Waking, dreaming and self-regulation. In: Moffitt MA, Kramer R, Hoffmann, eds. The functions of dreaming. New York, Albany: State University of New York Press, 1993:197-260
55. Putkonen PTS. Alpha- and- beta-adrenergic mechanisms in the control of sleep stages. In: Priest RG, Pletscher A Ward J, eds. Sleep research: Proceedings of the Northern European Symposium on Sleep Research Basle. 1979:19-34.
56. Putkonen P, Putkonen A. Suppression of paradoxical sleep following hypothalamic defense reactions in cats during normal conditions and recovery from PS deprivation. Brain Res 1971; 26:334-347.
57. Putkonen PTS, Leppvuori A, Stenberg D. Paradoxical sleep inhibition by central alpha-2 adrenoreceptor stimulant, clonidine, antagonized by alpha-2 receptor blocker, yohimbine. Life Sciences 1977; 21:1059-1066.
58. Radulovacki M, Micovic N. Effects of REM sleep deprivation and desipramine on beta-adrenergic binding sites in rat brain. Brain Res 1982; 235:393-396.
59. Rasmussen K, Morilak DA, Jacobs BL. Single unit activity of locus coeruleus neurons in the freely moving cat. I. During naturalistic behavior and in response to simple and complex stimuli. Brain Res 1986; 371:324-334.
60. Rechtschaffen A, Gilliland M, Bergmann B et al. Physiological correlates of prolonged sleep deprivation in rats. Science 1983; 221:182-184.
61. Reynolds CF III, Kupfer D. Sleep in depression. In: Williams RZ, Karacan I, Moore CA, eds. Sleep disorders, diagnosis and treatment. New York: John Wiley, 1988:147-164.
62. Riemann D, Velthaus S, Laubenthal S et al. REM-suppressing effects of amitriptyline and amitriptyline N-oxide after acute medication in healthy volunteers: Results of two uncontrolled pilot trials. Pharmacopsychiatry 1990; 23:253-258.
63. Ross RJ, Ball WA, Gresh PJ et al. REM sleep suppression by monoamine reuptake blockade: Development of tolerance with repeated drug administration. Biol Psychiatry 1990; 28:231-239.
64. Ross RJ, Gresh PJ, Bull WA et al. REM sleep inhibition by desipramine: Evidence for an alpha-1- adrenergic mechanism. Brain Res 1995; 701:129-34.
65. Rotenberg VS. Search activity in the context of psychosomatic disturbances, of brain monoamines and REM sleep function. Pavlovian J Biolog Sci 1984; 19:1-15.
66. Rotenberg VS. The nature of nonlinear relationship between the individual's present state and his sleep structure. In: Koella W, Obal F, Schulz H, Visser P, eds. Sleep. Stuttgart and New York: Gustav Fischer Verlag, 1988a:86:134-137.
67. Rotenberg VS. Functional deficiency of REM sleep and its role in the pathogenesis of neurotic and psychosomatic disturbances. Pavlovian J0 of Biological Science 1988b; 23:1-3.
68. Rotenberg V. REM sleep and dreams as mechanism of search activity recovery. In: Moffitt A, Kramer M, Hoffmann R, eds. Function of Dreaming. New York: State University of New York press, 1993:261-292.
69. Rotenberg VS. The revised monoamine hypothesis: Mechanism of antidepressant treatment in the context of behavior. Integr Physiol Behav Sci 1994; 29:182-188.
70. Rotenberg VS. An integrative psychophysiological approach to brain hemisphere functions in schizophrenia. Neuroscience and Biobehavioral Reviews 1994; 18:487-495
71. Rotenberg VS. The psychobiological dream functions: A new solution for old contradiction. In: Rozenberg JJ, ed. Sense and Nonsense. Philosophical, clinical and ethical perspectives. The Magnes Press, the Hebrew University Jerusalem, 1996:187-197
72. Rotenberg VS. Learned helplessness and sleep: Discussion of contradictions. Homeostasis 1996; 37:89-92.
73. Rotenberg VS. Sleep after immobilization stress and sleep deprivation: Common features and theoretical integration. Critical Reviews in Neurobiology 2000; 14:225-231.
74. Rotenberg VS, Arshavsky VV. Search activity and its impact on experimental and clinical pathology. Activitas Nervosa Superior (Praha) 1979; 21:105-115.
75. Rotenberg VS, Arshavsky VV. Psychophysiology of hemispheric asymmetry: The "entropy" of right hemisphere activity. Integr Physiol Behav Sci 1991; 26:183-188.
76. Rotenberg VS, Boucsein W. Adaptive versus maladaptive emotional tension. Genet Soc Gen Psychol Monographs 1993; 119:207-232.
77. Rotenberg VS, Sirota P, Elizur A. Psychoneuroimmunology: Searching for the main deteriorating psychobehavioral factor. Genet Soc Gen Psychol Monogr 1996; 122:329-346.
78. Rotenberg VS, Kayumov L, Indursky P et al. REM sleep in depressed patients: Different attempts to achieve adaptation. J Psychosomati Res 1997; 42:565-575.

79. Rotenberg VS, Shamir E, Barak Y et al. REM sleep latency and wakefulness in the first sleep cycle as markers of major depression. A controlled study vs. schizophrenia and normal controls. Progress in Neuro-Psychopharmacology & Biological Psychiatry 2002; 26:1211-1215.
80. Sakai K. Some anatomical and physiological properties of pontomesencephalic tegmental neurons with special reference to the PGO waves and postural atonia during paradoxical sleep in the cat. In: Hobson JA, Brazier MA, eds. The Reticular Formation Revisited Raven. New York: 1980:427-447.
81. Salamone JD. The involvement of nucleus accumbens dopamine in appetitive and aversive motivation. Behav Brain Res 1994; ,61:117-133.
82. Salamone JD, Cousins MS, Snyder BJ. Behavioral functions of nucleus accumbens dopamine: Empirical and conceptual problems with the anhedonia hypothesis. Neurosci Biobehav Rev 1997; 21:341-359.
83. Sastre JP, Sakai K, Jouvet M. Persistance du sommeil paradoxal chez le chat après destruction de l'aire gigantocellulaire du tegmentum pontique par l'acide kainique. CR Acad Sci 1979; 289D:959-964.
84. Schatzberg AF, Schildkraut JJ. Recent studies on norepinephrine systems in mood disorders. In: Bloom FE, Kupfer DJ, eds. Psychopharmacology: The Fourth Generation of Progress. New York: Raven Press, 1995:911-920.
85. Schenk CH, Mahowald MW, KimSW et al. Prominent eye movements during NREM sleep and REM sleep behavior disorder associated with fluoxetine treatment of depression and obsessive-compulsive disorder. Sleep 1992; 15:226-235.
86. Schittecatte M, Dumont F, Machowski R et al. Mirtazapine, but not fluvoxamine, normalizes the blunted REM sleep response to clonidine in depressed patients: Implications for subsensitivity of alpha2- adrenergic receptors in depression. Psychiatry Research 2002; 109:1-8.
87. Schwartz JC, Costentin J, Martres MP et al. Modulation of receptor mechanisms in the CNS: Hyper-and hyposensitivity to catecholamines. Neuropharmacology 1978; 17:665-685.
88. Seligman MEP. Helplessness. On depression, development and death. San Fransisco: Greeman WH, 1975.
89. Siegel JM, Wheeler RL, McGinty DJ. Activity of medullar reticular formation neurons in the unrestrained cat during waking and sleep. Brain Res 1979; ,179:49-60.
90. Siegel JM, Rogawski MA. A function for REM sleep: Regulation of noradrenergic receptor sensitivity. Brain Res Rev 1988; 13:213-233.
91. Siegel JM, Tomaszewski KS, Nienhuis R. Behavioral organization of reticular formation: Studies in the unrestrained cat II. Cells related to facial movements. J Neurophysiol 1986; 50:17-723.
92. Siever LJ, Davis KL. Overview: Towards a dysregulation hypothesis of depression. American Journal of Psychiatry 1985; 142:1017-1031.
93. Simon RP, Gershon MD, Brooks DC. The role of the raphe nuclei in the regulation of ponto-geniculo-occipital wave activity. Brain Res 1973; 58:313-330.
94. Soldatos CR, Stefanis CN, Bergiannaki JD et al. An experimental antidepressant increases REM sleep. Progr. Neuropsychopharmacol. Biol Psychiatry 1988; 12:899-907.
95. Solms M. Dreaming and REM sleep are controlled by different brain mechanisms. Behavioral and Brain Sciences 2000; 23:843-850.
96. Stahl SM. Neurotransmission of cognition, pt I: Dopamine is a hitchhiker in frontal cortex: Norepinephrine transporters regulate dopamine (Brainstorms) J Clin Psychiatry 2003; 64:4-5.
97. Steriade M, Hobson JA. Neuronal activity during the sleep-waking cycle. Progr Neurobiol 1976; 6:155-376.
98. Tsai L-L, Bergmann BM, Perry BD. Effects of chronic total sleep deprivation on central noradrenergic receptors in rat brain. Brain Res 1993; 602:221-227.
99. Vogel GW. Evidence for REM sleep deprivation as the mechanism of action of antidepressant drugs. Prog Neuropsychopharmacol Biol Psychiatry 1983; 7:343-349.
100. Vogel G, Hagler M. Effects of neonatally administered iprindole on adult behaviors of rats. Pharmacol Biochem Behav 1996; 55:157-161.
101. Vogel G, Neill D, Koris D et al. REM sleep abnormalities in a new animal model of endogenous depression. Neurosci Biobehav Rev 1990; 14:77-83.
102. Vogel G, Cohen J, Mullis D et al. Nefazodone and REM sleep: How do antidepressant drugs decrease REM sleep? Sleep 1998; 15:795-796.
103. Wagner M, Mooney D. Personality characteristics of long and short sleepers. Journal of Clinical Psychology 1975; 31:434-436.
104. Ware JC, McBrayer RH. REM sleep and nefazodone. Sleep 1998; 21:795-796.
105. Wikler A. Pharmacological dissociation of behavior and EEG sleep patterns in dogs: Morphine, N-allylmorphine and atropine. Proc Soc Exp Biol Medical 1952; 79:261-265.
106. Wise RA, Colle LM. Pimozide attenuates free feeding: Best scores analysis reveals a motivational deficit. Psychopharmacology (Berl) 1984; 84:445-451.
107. Wise RA, Spindler J, deWitt H et al. Neuroleptic induced "anhedonia" in rats: Pimozide blocks reward quality of food. Science 1978; 201:262-264.
108. Woodward DJ, Moises HC, Waterhouse BD et al. Modulatory action of norepinephrine in the central nervous system. Federation Proceeding 1979; 38:2109-2116.

CHAPTER 5

Role of Wakefulness Area in the Brainstem Reticular Formation in Regulating Rapid Eye Movement Sleep

Birendra N. Mallick, Satvinder Kaur, Stephen Thankachan and Dinesh Pal

Abstract

Rapid eye movement sleep is a unique paradoxical state within sleep period. Normally it follows deep sleep, is maintained for varying duration and may terminate in either sleep or wake state. During REM sleep some neurons increase firing, the REM-ON neurons, while some others cease firing, the REM-OFF neurons. Although the mechanism is not completely known, these REM sleep -related neurons are likely to play a significant role in the initiation and maintenance of REM sleep. It was proposed that GABA may be involved in the cessation of REM-OFF neurons and the classical sleep and waking areas in the brain stem are likely to modulate the REM-ON and REM-OFF neurons for the regulation of REM sleep.

Results from our single neuronal activity experiments in freely behaving animals confirmed that the brain stem area, which induce wakefulness inhibit the REM-ON neurons but stimulate the REM-OFF neurons. Microinjection studies revealed that the increase in REM sleep by the cholinergic input (possibly from REM-ON neurons) to the locus coeruleus (where REM-OFF neurons are located) is mediated through GABA. Thus, it is proposed that during wakefulness the REM-ON neurons are inhibited while the REM-OFF neurons are active. During sleep gradually the awake-related neurons slow down withdrawing their effects on REM sleep related neurons. This causes an increase in the REM-ON neuronal activity inducing release of acetylcholine on the GABA-ergic neurons in the locus coeruleus. GABA then inhibits the REM-OFF neurons resulting in the initiation of REM sleep. The presence of GABA in optimum concentration maintains the duration of REM sleep episode.

Introduction

Rest and activity are basic instinct behavioral phenomena of living systems present across species. It is believed that in higher species these phenomena further evolved into complex processes like sleep and wakefulness. Sleep-wakefulness is a subjective behavioral phenomenon. Therefore, for its identification and quantification in an objective manner some of the electrophysiological signals are taken into consideration. The three primary electrophysiological signals that are commonly considered for the purpose are that from the brain, the electroencephalogram (EEG), from the muscles, the electromyogram (EMG) and due to the movements of the eyes, the electrooculogram (EOG). Based on characteristic signals from these records, wakefulness has been divided into two stages viz. active wakefulness (AW) and quiet wakefulness (QW) while sleep has been divided into three stages viz. slow wave sleep (SWS), deep sleep (DS) and rapid eye movement (REM) sleep (Fig. 1). REM sleep is a unique phenomenon that has been identified in its present form in the mid-twentieth century.[1] The characteristic features of REM sleep are presence of low voltage fast electrical waves in the cortical EEG, rapid bursts of eye movements in the EOG and atonia in the EMG recorded from antigravity muscles (usually recorded from the neck muscles). REM sleep episodes repeat several times within sleep period and its frequency of occurrence as well as duration per bout increases with the progress in depth of sleep.[2]

The mechanism of generation of REM sleep is not completely understood. However, it is known that there are at least two types of neurons related to REM sleep. During REM sleep one type of neurons ceases firing—the REM-OFF neurons—while the second type increases firing—the REM-ON neurons. Some of the characterising features of REM sleep viz. desynchronization of EEG and eye movements, resemble signs associated to wakefulness, while some other signs viz. muscle atonia, resemble signs associated to sleep. It may also be noted that normally REM sleep appears at certain depth of sleep and it does not follow wakefulness although it may terminate either into sleep or waking state (Figs. 2A and B). However, some characteristic sign(s) associated to REM sleep might be triggered and expressed during wakefulness in some disorders e.g., narcolepsy.[3] Therefore, it is most likely that for proper regulation of REM sleep, the neurons in the waking area(s) in the brain maintain a communication with the neurons related to REM sleep. In order to understand these issues, interactions between the REM-ON and the REM-OFF neurons and responses of these neurons to input from waking area(s) in the brain will be discussed here.

Possible Interaction between REM-ON and REM-OFF Neurons for the Regulation of REM Sleep

REM-ON neurons were initially identified to be cholinergic and are located in the latero-dorsal and pedunculo-pontine tegmental (LDT/PPT) areas of the brain stem.[4,5] However, a later report suggest that some noncholinergic neurons could also behave as REM-ON neurones.[6] REM-OFF neurons are reported to be aminergic and are located in locus coeruleus (nor-epinephrinergic neurons)[7,8] raphe region (serotonergic neurones)[9] and in the posterior hypothalamus (histaminergic

Sleep and Sleep Disorders: A Neuropsychopharmacological Approach, edited by Malcolm Lader, Daniel P. Cardinali and S.R. Pandi-Perumal.

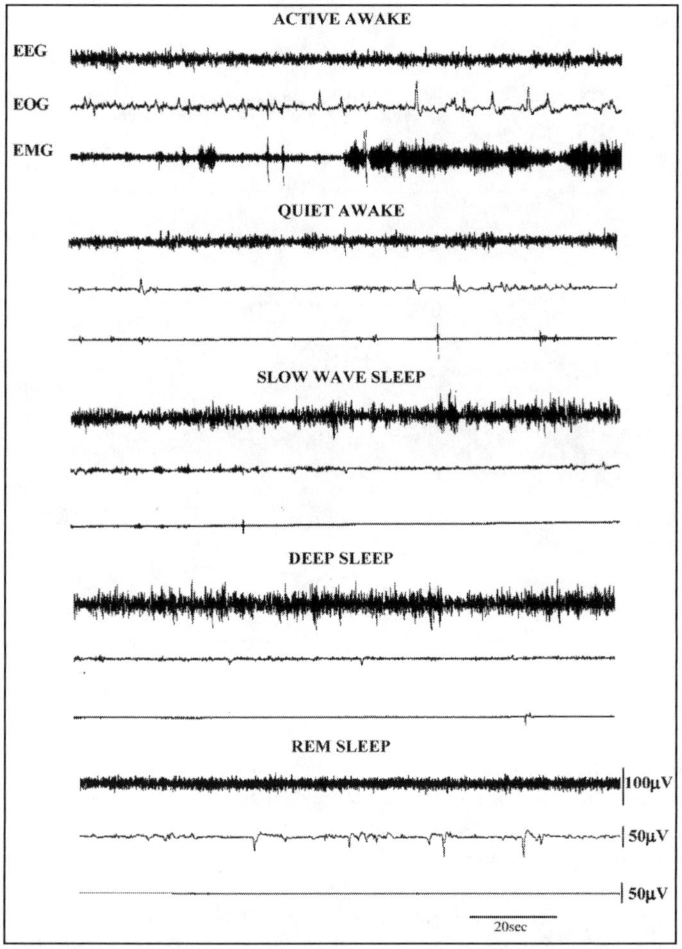

Figure 1. Polygraphic traces showing simultaneous recording of EEG, EOG and EMG in freely moving rat classifying five stages of sleep-wakefulness.

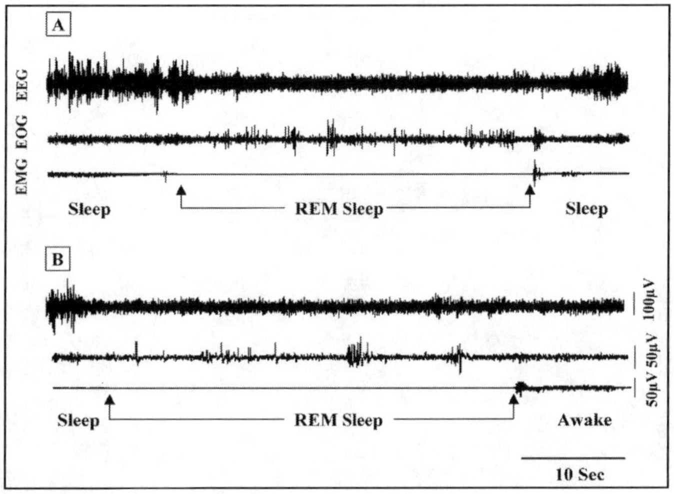

Figure 2. Polygraphic traces showing transition from sleep to REM sleep and then again to either (A) sleep or (B) wakefulness.

neurons).[10,11] In addition to the presence of REM-OFF neurons in locus coeruleus (LC) there are other evidence that support involvement of LC neurons in REM sleep. It was found that the firing rate of these neurons decrease during REM sleep deprivation—possibly a compensatory phenomenon[12]—continuous activation (noncessation) of these neurons by mild electrical stimulation reduced REM sleep[13] and cholinergic stimulation of dorsolateral pons, including LC, increased REM sleep.[14-17] Both temporary as well as permanent inactivation of LC by local cooling[18] and lesioning [19,20] respectively, increased REM sleep.

The cholinergic REM-ON neurons in LDT/PPT increase firing, while the nor-epinephrinergic REM-OFF neurons in LC cease firing during spontaneous[5,8,12,21] as well as induced[22] REM sleep. It was proposed that a reciprocal connection exists between these two types of neurons for the regulation of REM sleep.[21,23] Such interaction may be supported by the facts that LC receives ACh-ergic inputs, [24] ACh-ergic receptors have been identified on the LC neurons[25] and levels of ACh increased in and around LC during REM sleep.[26] However, a closer and detailed study revealed that acetylcholine or its agonist did not inhibit the nor-epinephrinergic neurons in LC that are supposed to be REM-OFF type.[27] Therefore, it was proposed that an inhibitory neurotransmitter, may be GABA, could be inhibiting the REM-OFF neurons in LC and that the ACh released from the cholinergic REM-ON neurons could be mediating its effects through GABA-ergic neurons for the regulation of REM sleep.[28,29] The possible role of GABA in LC region for the regulation of REM sleep will be further discussed.

The involvement of GABA in LC for the regulation of REM sleep may be supported by the fact that GABA-ergic interneurons and terminals are present in LC,[30-33] GABA receptors are present on the neurons in LC[34,35] and GABA levels increase in LC during REM sleep.[36] The hypothesis was further supported by our study that blocking of GABA-A receptors by microinjection of its antagonist, picrotoxin, into LC significantly reduced REM sleep.[37] However, it was not known how the cholinergic inputs from REM-ON neurons to the LC got translated into a GABA-ergic input for the regulation of REM sleep. Besides, it was also not known whether the cholinergic and the GABA-ergic inputs to the LC had different roles to play for initiation and maintenance of REM sleep. The investigation demanded a study of micro-anatomical connections between the proposed neurons and the role of respective neurotransmitters released from those neurons in regulating bio-behavioral response, REM sleep in this case. The complexity of the problem was such that just micro-anatomical (histological) investigation could not have resolved the complex issue and provided an answer to such micro chemo-anatomicobehavioral question. The study needed to be conducted in freely behaving animals combined with microinjection. It was proposed that in LC the input from the cholinergic REM-ON neurons possibly acted on the GABA-ergic neurons for the regulation of REM sleep. The working model of the hypothesis was that if the cholinergic input was mediated through the GABA-ergic neurons, the agonist of the former would be ineffective when the latter was blocked by antagonist, while the agonist of the latter would be effective even in presence of antagonist of the former. Therefore, experiments were conducted to locally microinject in LC the agonist or the antagonist of these neurotransmitters, either alone or in selected combinations. The experiments were carried out in chronically prepared freely moving rats and the effects on total REM sleep, rate of generation of REM sleep and duration of REM sleep per episode were studied.

Experiments were conducted on male Wistar rats (250-300 g), maintained in 12:12 L: D cycle with food and water ad libitum. Under surgical anaesthesia (Nembutal, pentobarbital sodium, 35mg/kg, ip) rats were implanted with bilateral EEG, EOG and EMG electrodes for chronic sleep-wakefulness recording.[38] A pair

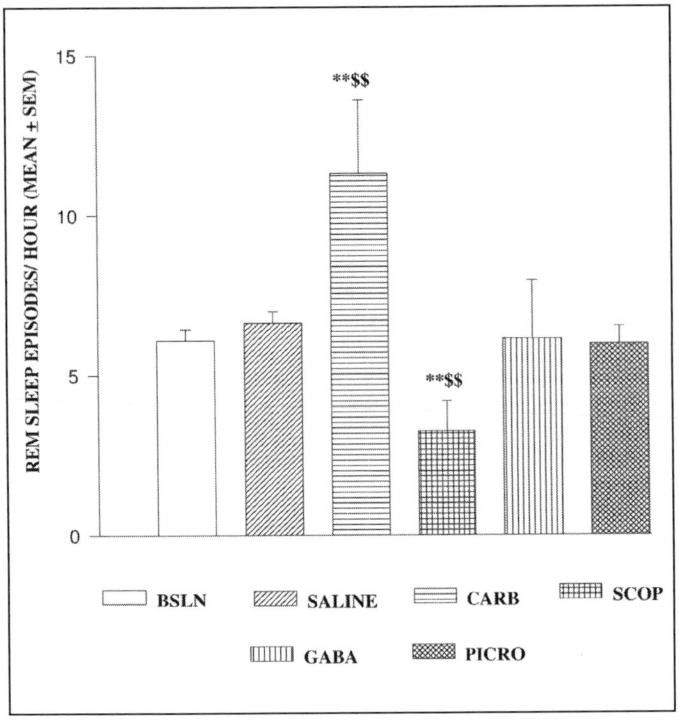

Figure 3. The frequency (Mean ± SEM) of REM sleep per hour for the entire recording period in normal (BSLN) and after bilateral microinjection of saline, carbachol (CARB), scopolamine (SCOP), GABA and picrotoxin (PICRO) into the locus coeruleus are shown in this figure. ** Significant as compared to baseline, $ significant as compared to saline; **, $, p< 0.01.

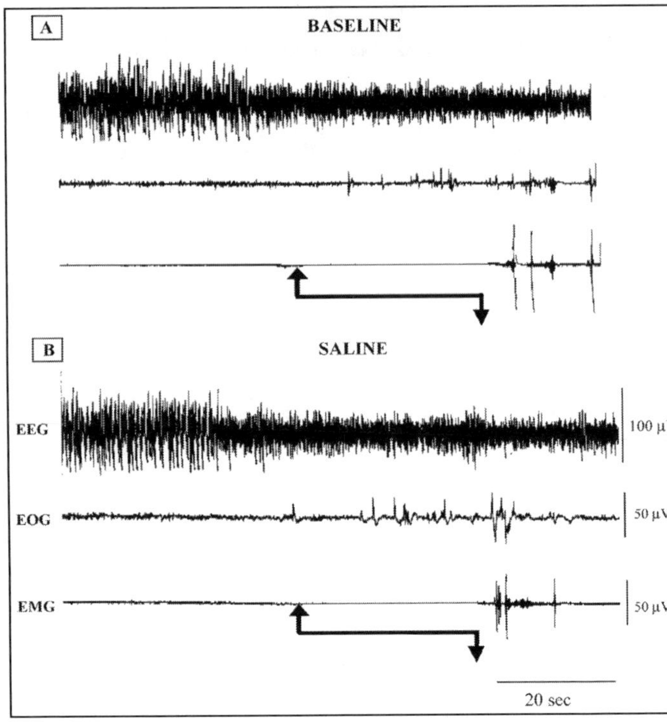

Figure 4. Representative polygraphic traces of EEG, EOG and EMG associated to REM sleep episode (marked by arrows), recorded before (baseline) (A) and after bilateral microinjection of saline into the locus coeruleus in a freely moving rat (B).

of guide cannulae with blockers were implanted bilaterally in LC. After recovery from surgical trauma and acclimatisation to the recording environment, rats were connected to a polygraph to record the electrophysiological correlates of sleep-wakefulness. Baseline sleep-wakefulness recording was done for 8 hours (between 9am-6pm). In the experimental group, 250nl of either saline or 1% (0.25 mg in 250 nl) carbachol (cholinergic agonist) or scopolamine (cholinergic antagonist) or 0.1% (250 ng in 250 nl) GABA or picrotoxin (GABA A-antagonist) was injected bilaterally into the LC either alone or in a sequential combination. The injection was performed using an injector cannula connected to a 2 µl Hamilton syringe by polyethylene tubing. The combination injections involving cholinergic and GABA-ergic agonists/antagonists were done such that injection of antagonist always preceded the agonist. After the experiment, 2% pontamine sky blue was injected through the same guide cannulae using the same injector. Thereafter, under deep Nembutal anaesthesia (45 mg/kg, i.p.) the brain was perfused intracardially and site as well as spread of injection were histologically identified. The polygraphic records were analysed into active wakefulness (AW), quiet wakefulness (QW), slow wave sleep (SWS), deep sleep (DS) and REM sleep as per the criteria followed earlier.[13,39] Total REM sleep, frequency of generation of REM sleep per hour and mean duration of REM sleep per episode were calculated. The effects of injection on frequency of REM sleep as well as duration of REM sleep per episode were statistically compared with that of baseline and saline values.

Percent change in REM sleep during baseline (BSNL) and after microinjection of saline and other agonists and antagonists are shown in Figure 3. It was observed that cholinergic agonist/antagonist affected the frequency of REM sleep initiation; the agonist increased while the antagonist decreased the frequency of generation of REM sleep. On the other hand, GABA and picrotoxin affected the mean duration of REM sleep per episode; the former increased while the latter decreased the duration of REM sleep. Representative polygraphic tracings through REM sleep episode under baseline and after microinjection of saline are shown in Figures 4A and B; while that after microinjection of carbachol and GABA are shown in Figures 5A and B. These results suggest that GABA-ergic input to the LC modulated the duration of REM sleep per episode, while that of the cholinergic modulated the frequency of generation of REM sleep. These results also indicated that GABA acted after the cholinergic system had initiated the action, because maintenance of any process would be required only after the process has been initiated. Thus, individual injection studies brought forward the possibility that cholinergic inputs, presumably from the REM-ON neurons, acted on GABA-ergic neurons for initiation of REM sleep and that GABA acted on a system that has been primed by ACh to maintain the duration of REM sleep. Therefore, to confirm, LC was microinfused with any of the following three combinations (a) GABA-ergic antagonist followed by cholinergic agonist; or (b) cholinergic antagonist followed by GABA; or (c) cholinergic antagonist followed by GABA-ergic antagonist. These were done with the assumption that if cholinergic input acted on the GABA-ergic neurons, carbachol in presence of picrotoxin would induce an effect similar to that of picrotoxin alone, while GABA in presence of scopolamine would show an effect similar to that of the GABA alone.

The results from the combination injection studies showed that picrotoxin followed by carbachol in LC significantly decreased REM sleep due to a reduction in REM sleep duration per

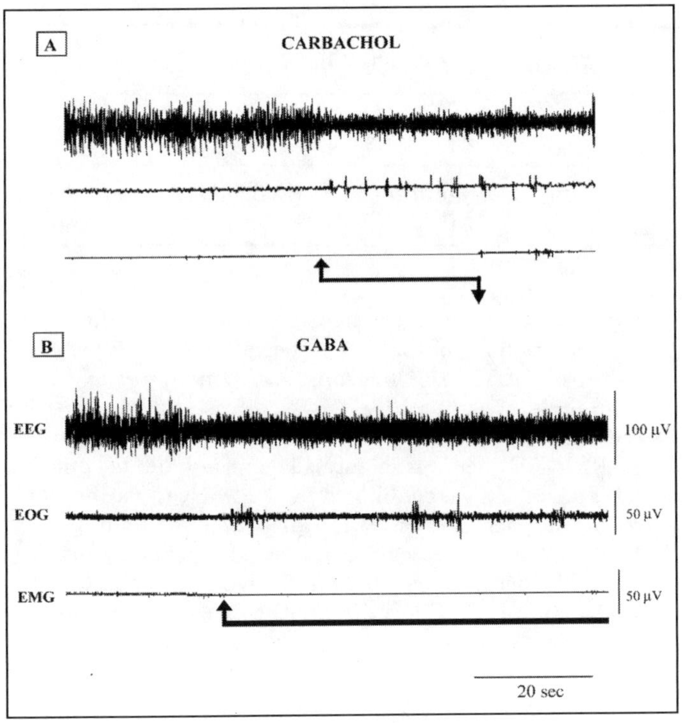

Figure 5. Representative polygraphic traces of EEG, EOG and EMG showing REM sleep episode (marked by arrows) after bilateral microinjection of carbachol (A) and GABA (B) in the locus coeruleus of freely moving rat.

episode, while scopolamine followed by GABA microinjection significantly increased REM sleep due to an increase in mean duration per episode of REM sleep. Thus, when cholinergic and GABA-ergic agonist or antagonist (as the case may be) were injected into the LC in any combination, the effect of the GABAergic agonist/antagonist prevailed over that of the cholinergic. The most likely explanation for getting such a result is that the cholinergic input in the LC was acting on the GABA-ergic neurons.[40] Thus, the observation supported our contention that cholinergic influence in LC was mediated through GABA. The cholinergic-sensitive GABA-ergic neurons are present in and immediately around LC.[6] Although a total loss of REM sleep was not observed after blocking either the cholinergic receptor or GABA-ergic receptor, there was an almost complete suppression of REM sleep when scopolamine and picrotoxin were injected together. This supports mutually permissive and cooperative role between cholinergic and GABA-ergic systems in LC. However, possibility of additional GABA-ergic input to LC from any other source could not be ruled out. Subsequently, it has been shown that GABA-ergic input from prepositus hypoglossus to LC may also modulate REM sleep.[41] It is known that the PrH receives input from PPT,[42] the site of REM-ON neurons. Thus REM-ON neurons may excite the GABAergic neurons in PrH, which in turn inhibit the REM-OFF neurons in LC and generate REM sleep. Based on these results the connection between the neurons are shown in Figure 6.

Does Waking Area in the Brain Stem Have a Role to Play in REM Sleep Regulation?

It is well known that normally REM sleep appears at certain depth of sleep period i.e., it does not normally follow wakefulness, although it may terminate into either sleep or wakefulness

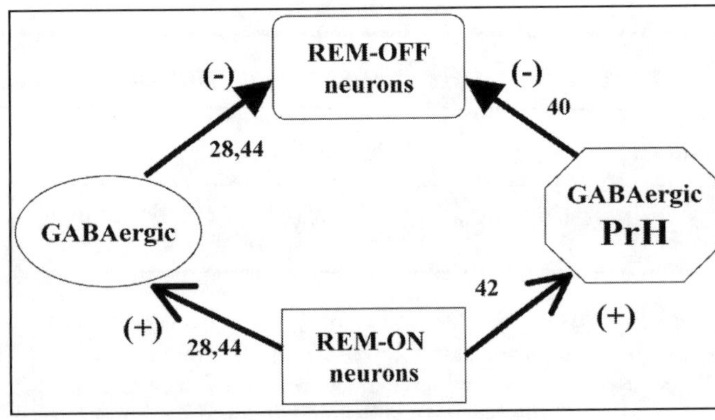

Figure 6. Neural connections between REM-ON and REM-OFF neurons for the regulation of REM sleep are shown. The numbers show the Reference number and abbreviations are as in the text.

(Fig. 2A and B). Some of the electrophysiological signals associated to REM sleep viz. EEG desynchronization and rapid bursts of eye movements are apparently similar to those observed during wakefulness.[43,44] However, wakefulness and REM sleep are separated by a distinctly different behavioral state viz., the nonREM sleep. None of the models proposed for the generation of REM sleep[21,23,29] could explain that why does not REM sleep or its signs appear during wakefulness. Recently, we have suggested that there are likely to be separate groups of neurons in the brain stem responsible for EEG desynchronization during wakefulness and REM sleep states.[45] Hence, it is likely that the neurons responsible for expression of REM sleep associated signs remain suppressed during wakefulness and that as and when they get triggered at certain depth of sleep, REM sleep is expressed. Therefore, it was proposed that REM sleep related neurons i.e., REM-ON and REM-OFF neurons in the brain stem may be differentially modulated by the awake inducing area in the brainstem. It was hypothesised that the wakefulness inducing area in the brain stem reticular formation i.e., the midbrain reticular formation (MRF) would inhibit REM-ON neurons and excite REM-OFF neurons during wakefulness.

Experiments were conducted on freely moving normally behaving cats (2.5-3.4 kg). Under surgical anaesthesia cats were prepared for chronic recording of sleep-wakefulness (S-W) and single neuronal activity.[12,45] In brief, under surgical anaesthesia bilateral EEG, EOG, EMG and PGO electrodes and a stainless steel tripolar epoxy coated stimulating electrode in the MRF were implanted. Single neuronal activity was recorded with the help of microwires introduced into the brain stem through guide cannula mounted on a mechanical microdrive fixed to the skull. After recovery from surgical trauma, the cats were habituated to the recording chamber and were attached via shielded cable to a Grass polygraph to record electrophysiological signals viz., EEG, EOG, EMG and PGO under unrestrained and freely moving conditions. The physiological wakefulness-inducing region in the MRF was confirmed by delivering for a few seconds high frequency stimulation (100Hz, 200-300 µA, 300 µsec) that resulted in induction of desynchronization of the EEG that outlasted the period of stimulation.

Thereafter, once a well isolated signal (S:N 3:1) from single neuron was encountered it was simultaneously recorded along with EEG, EOG, EMG and PGO in separate channels of a polygraph. At least three spontaneous sleep-wake-REM sleep cycles were recorded along with single neuronal activities. In order to ascertain the behavior of each of these neurons their mean firing

Table 1. Effect of 1 Hz stimulation of MRF on the activity of REM-ON and REM-OFF neurons

Stimulation	REM - ON			REM - OFF		
	Excitation	Inhibition	No Change	Excitation	Inhibition	No Change
MRF	0	9	1	7	0	0

rates during QW were statistically compared with the mean firing rates during other states by applying analysis of variance (ANOVA) coupled with Newman-keul's test. Accordingly, the neurons were classified into REM sleep related or nonrelated types. Among REM sleep related neurons there were REM- ON and REM- OFF neurons. The former increased firing only during REM sleep, while the latter ceased firing only during REM sleep. Thereafter, to study the influence of wakefulness inducing area on each of those neurons, the MRF was stimulated with 1Hz rectangular pulses of 500-600 μA, 300 μsec. The effect of stimulation was recorded by overlapping 10-15 stimulus bound responses on a digitizing oscilloscope. Ten such responses were observed and a consistent response (7 out of 10) was noted. The difference in the concentration of neuronal spikes (activities) before and after the stimulus artefact on the overlapped figure was compared. An increase in the cluster of spikes was taken as excitation and a decrease, as inhibition while a comparable spike concentration as no change. The time delay between the artefact and the onset of response i.e., the excitation or the inhibition was taken as latency of response of that particular neuron. The duration of the response (excitation or inhibition) was also estimated. After completion of the recording sessions, under deep anaesthesia (sodium pentobarbital 45 mg/Kg), the recording and stimulation sites were marked by electrolytic lesion (50-100μA for 10sec) by passing anodal current (D.C. lesion maker, Grass, USA). Later, under overdose of anaesthesia the cats were euthanasized by intracardial perfusion. The stimulating and the recording sites were histochemically identified in 40μm coronal sections stained with cresyl violet or haematoxylin and eosin. The sites of the recorded neurons were also reconstructed.

The effect of MRF stimulation was studied on a total of 63 neurons including 10 REM-ON and 7 REM-OFF neurons (Table 1). Most of the REM-ON neurons (9 out of 10) were inhibited (Fig. 7A) while one neuron remained unaffected. All the 7 REM-OFF neurons were excited (Fig. 7B) by MRF stimulation. Thus, the MRF excited the REM-OFF while inhibited the REM-ON neurons. Among the rest of the 46 neurons studied, the MRF excited about 50% of the neurons whose firing rate increased during spontaneous waking period. Thus, the results showed that a majority of the neurons whose firing rate increased during spontaneous wakefulness, including the REM-OFF neurons, were excited by the MRF while the REM-ON neurons were inhibited.

Physiological Significance

It may be noted that wake-inducing area, the MRF, exerts an opposite influence on REM-OFF and REM-ON neurons - the former were excited while the latter inhibited. Based on these results, we hypothesise and propose that during wakefulness when the wake related neurons in MRF are active,[46-48] they excite the norepinephrinergic REM-OFF neurons in LC, which then remain active through waking period. This view may be supported by the fact that activation of REM-OFF neurons is reported to prevent REM sleep[13] and is likely to increase the level of norepinephrine in the brain causing cortical activation and desynchronization of the EEG[49-51] Therefore, it is possible that activation of REM-OFF neurons may contribute to EEG desynchronization associated with wakefulness, but not to that of REM sleep.

The REM-ON neurons that usually remain inhibited during normal wakefulness were inhibited by the MRF. It may be supported by the fact that activation of the site containing the REM-ON neurons increased REM sleep.[52] It is reported that the neurons in the peri-LC that contains the REM-ON neurons are responsible for muscle atonia during REM sleep.[53-56] All the results considered together provide possible explanation for neural mechanism as to why does not muscle atonia, associated with REM sleep, appear during wakefulness although the EEG becomes desynchronized. Further, the information may be extended

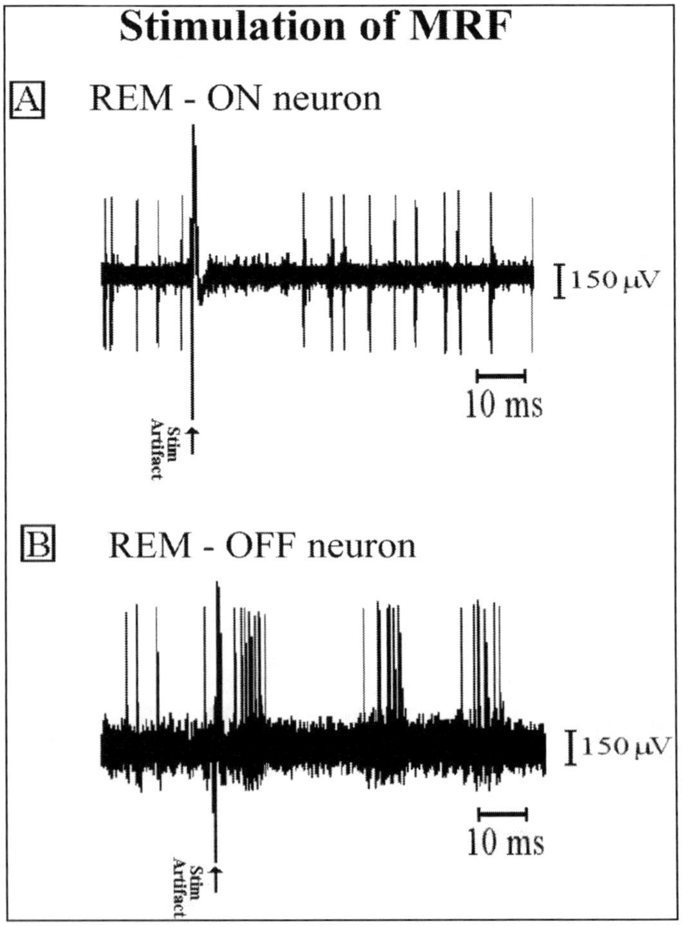

Figure 7. Ten 1 Hz stimulus bound overlapped responses of a REM-ON neuron (A) and a REM-OFF neuron (B) to stimulation of midbrain reticular formation (MRF) are shown here. The REM-ON neuron was inhibited, while the REM-OFF neuron was excited.

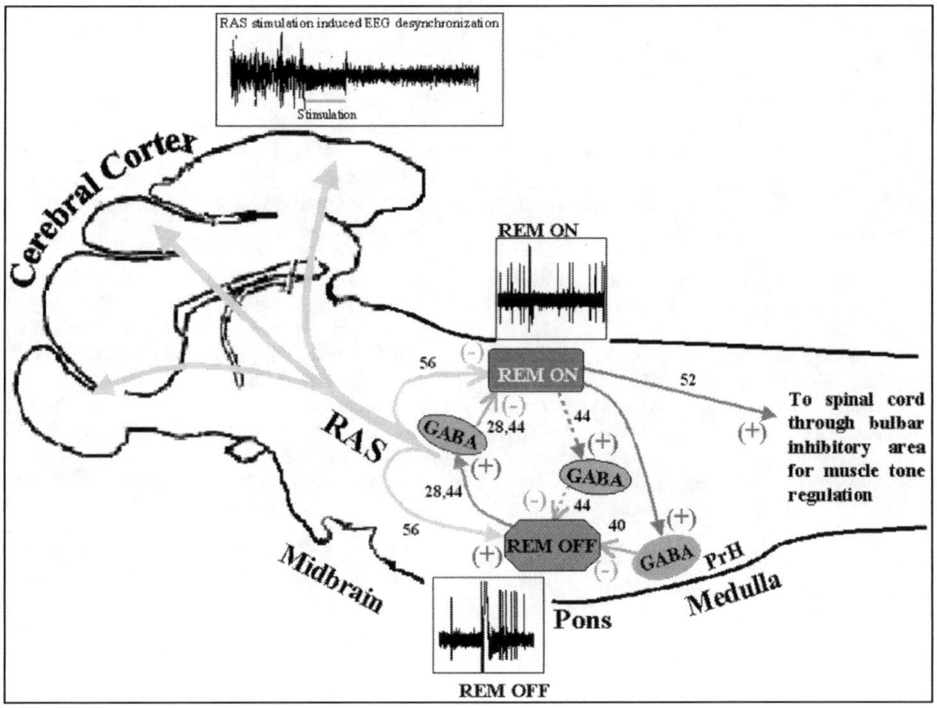

Figure 8. Composite neuronal connections showing modulation of REM-OFF neurons by wakefulness inducing area for the regulation of REM sleep. The numbers in the figure represent the reference numbers on which the model is based on. RAS - reticular activating system.

that in case of narcolepsy possibly there occurs an error in this pathway resulting in muscle atonia during wakefulness.

As a mechanism of action at the neuronal level it may be said that during wakefulness the MRF neurons are active causing stimulation of REM-OFF neurons and inhibition of REM-ON neurons resulting in absence of REM sleep. This may be supported by the fact that the REM-OFF neurons are normally active during all the stages except during REM sleep while the REM-ON neurons behave in an opposite manner. As a mechanism of action one or more or all three of the following possibilities may exist. (1) that MRF neurons exert an independent inhibitory and excitatory effects on the REM-ON and the REM-OFF neurons, respectively; (2) that the MRF exerts an excitatory effect on the REM-OFF neurons that in turn (through GABAergic neurons) then inhibit the REM-ON neurons;[57] and three, that the MRF exerts an inhibitory effect on the REM-ON neurons and that in turn (through GABAergic neurons) exert an excitatory effect on the REM-OFF neurons.[57] At the onset of sleep the activity of the MRF neurons is reduced[47] that gradually withdraws the excitatory and the inhibitory effects from the REM-OFF and the REM-ON neurons, respectively. Gradually sleep is induced when the sleep inducing neurons further increase firing and the wake active neurons further reduce or cease firing. The reduction or cessation of the wake active neurons withdraws the inhibition from REM-ON neurons, which in turn also withdraws the GABA mediated inhibition from REM-OFF neurones.[40] Finally, the activation of REM-ON and the inhibition of REM-OFF neurons initiate REM sleep. The neural connections as proposed above have been shown in Figure 8.

Acknowledgements

Financial supports from CSIR, DBT, ICMR and UGC, India are acknowledged.

References

1. Aserinsky E, Kleitman N. Regularly occurring periods of eye motility and concomitant phenomena during sleep. Science 1953; 118:273-274.
2. Kleitman N. Sleep and wakefulness. Chicago: University of Chicago Press, 1939.
3. Guilleminault C. Narcolepsy syndrome. In: Kryger MH, Roth T, Dement WC, eds. Principles and Practice of Sleep Medicine. WB Saunders Co 1989:338–346.
4. El Mansari M, Sakai M, Jouvet M. Unitary characteristics of presumptive cholinergic tegmental neurons during sleep-waking cycle in freely moving cats. Exp Brain Res 1989; 76:519-529.
5. Kayama Y, Ohta M, Jodo E. Firing of possibly cholinergic neurons in the rat laterodorsal tegmental nucleus during sleep wakefulness. Brain Res 1992; 569:20-21.
6. Sakai K, Kayama Y. Are there cholinergic and noncholinergic paradoxical sleep-on neurones in the pons? NeuroReport 1996; 7:2449-2453.
7. Chu NS, Bloom FE. Activity patterns of catecholamine-containing pontine neurons in the dorsolateral tegmentum of unrestrained cats. J Neurobiol 1974; 5:527-544.
8. Aston-Jones G, Bloom FE. Activity of norepinephrine containing locus coeruleus neurons in behaving rats anticipates fluctuations in the sleep-waking cycle. J Neurosci 1981; 1:876-886.
9. McGinty DJ, Harper RW. Dorsal raphe neurons: Depression of firing during sleep in cats. Brain Res 1976; 101:569-575.
10. Vanni-Mercier G, Sakai K, Jouvet M. Waking state specific neurons in the caudal hypothalamus of the cat. C R Acad Sci 1984; 298:195-220.
11. Lin JS, Luppi PH, Salvert D et al. Histamine containing neurons in the cat hypothalamus. C R Acad Sci 1986; 303:371-376.
12. Mallick BN, Siegel JM, Fahringer H. Changes in pontine unit activity with REM sleep deprivation. Brain Res 1989; 515:94-98.
13. Singh S, Mallick BN. Mild electrical stimulation of pontine tegmentum around locus coeruleus reduces rapid eye movement sleep. Neurosci Res 1996; 24:227-235.
14. Baghdoyan HA, Rodrigo-Anglio ML, McCarley RW. Site -specific enhancement and suppression of desynchronised sleep signs following cholinergic stimulation of three brainstem areas. Brain Res 1984; 306:39-52.

15. Vanni-Mercier G, Sakai K, Lin JS et al. Mapping of cholinoceptive brainstem structures responsible for the generation of paradoxical sleep in the cat. Arch Ital Biol 1989; 127:133-164.
16. Yamamoto K, Mamelak AN, Quattrochi JJ et al. A cholinoceptive desynchronized sleep induction zone in the anterodorsal pontine tegmentum: Locus of the sensitive region. Neuroscience 1990; 39:279-293.
17. Datta S, Calvo JM, Quattrochi JJ et al. Long term enhancement of REM sleep following cholinergic stimulation. Neuroreport 1991; 2:619-222.
18. Cespuglio R, Gomez ME, Faradji H et al. Alterations in the sleep-waking cycle induced by cooling of the locus coeruleus area. EEG Clin Neurophysiol 1982; 54:570-578.
19. Braun CM, Pivik RT. Effects of locus coeruleus lesions upon sleeping and waking in the rabbit. Brain Res 1981; 230:133-151.
20. Caballero A, De Andres I. Unilateral lesions in locus coeruleus area enhance paradoxical sleep. EEG Clin Neurophysiol 1986; 64:339-346.
21. Hobson JA, McCarley RW, Wyzinski PW. Sleep cycle oscillation: Reciprocal discharge by two brain stem neuronal groups. Science 1975; 189:55-58.
22. Thankachan S, Islam F, Mallick BN. Behavior of brainstem neurons to spontaneous and induced EEG desynchronization during wakefulness and rapid eye movement sleep in freely moving cats. Sleep Res 1997; 26:54.
23. Sakai K. Executive mechanisms of paradoxical sleep. Arch Ital Biol 1988; 126:239-257.
24. Jones B. Immunohistochemical study of choline acetyltransferase-immunoreactive processes and cells innervating the pontomedullary reticular formation in the rat. J Comp Neurol 1990; 295:485-514.
25. Baghdoyan HA, Mallios VJ, Duckrow RB et al. Localization of muscarinic receptor subtypes in brain stem areas regulating sleep. NeuroReport 1994; 5:1631-134.
26. Kodama T, Takahashi Y, Honda Y. Enhancement of acetylcholine release during paradoxical sleep in the laterodorsal tegmental field of the cat brain stem. Neurosci Lett 1990; 114:277-282.
27. Egan TM, North RA. Acetylcholine acts on m_2-muscarinic receptors to excite rat locus coeruleus neurons. Eur J Pharmacol 1985; 85:733-735.
28. Alam MN, Kumari S, Malick BN. Role of GABA in acetylcholine induced locus coeruleus mediated increase in REM sleep, Sleep Res 1993; 22:541.
29. Mallick BN, Kaur S, Jha SK et al. Possible role of GABA in the regulation of REM sleep with special reference to REM-OFF neurons. In: Mallick BN, Inoue S, eds. Rapid Eye Movement Sleep. Marcel and Dekker Inc 1999:153-166.
30. Iijima K, Ohtomo K. Immunohistochemical study using GABA antiserum for demonstration of inhibitory neurons in the rat locus coeruleus. Am J Anat 1988; 181:183-194.
31. Jones BE. Paradoxical sleep and its chemical/ structural substrates in the brain. Neuroscience 1991; 40:637-656.
32. Ford B, Holmes CJ, Mainville L et al. GABA-ergic neurons in the rat pontomesencephalic tegmentum: Codistribution with cholinergic and other tegmental neurons projecting to the posterior lateral hypothalamus. J Comp Neurol 1995; 363:177-196.
33. Peyron C, Luppi PH, Rampon C et al. Location of the GABA-ergic neurons projecting to the dorsal raphe nucleus and the locus coeruleus of the rat. Soc Neurosci Abstr 1995; 21:373.
34. Olpe HR, Steinmann MW, Hall RG et al. GABA-A and GABA-B receptors in locus coeruleus: Effects of blockers. Eur J Pharmacol 1988; 149:183-185.
35. Luque JM, Malherbe P, Richards JG. Localization of GABA-A receptor subunit mRNAs in the rat locus coeruleus. Mol Brain Res 1994; 24:219-226.
36. Nitz D, Siegel JM. GABA release in the locus coeruleus as a function of the sleep/ wake state. Neuroscience 1997; 78:795-801.
37. Kaur S, Saxena RN, Mallick BN. GABA in locus coeruleus regulates spontaneous rapid eye movement sleep by acting on GABA-A receptors in freely moving rats. Neurosci Lett 1997; 223:105-108.
38. Alam MN, Mallick BN. Differential acute influence of medial and lateral preoptic areas on sleep wakefulness in freely moving animals. Brain Res1990; 525:242-248.
39. Timo-Iaria C, Negrao N, Schmidek WR et al. Phases and states of sleep in the rat. Physiol Behav 1970; 5:1057-1062.
40. Mallick BN, Kaur S, Saxena RN. Interactions between cholinergic and GABAergic neurotransmitters in and around the locus coeruleus for the induction and maintenance of rapid eye movement sleep in rats. Neuroscience 2001; 104:467-485.
41. Kaur S, Saxena RN, Mallick BN. GABAergic neurones in prepositus hypoglossi regulate REM sleep by its action on locus coeruleus in freely moving rats. Synapse 2001; 42:141-150.
42. Higo S, Ito K, Fuchs D et al. Anatomical interconnections of the pedunculopontine tegmental nucleus and the nucleus prepositus hypoglossi in the cat. Brain Res 1990; 536:79-85.
43. Dement WC. The occurrence of low voltage, fast electrophysiologic patterns during behavioral sleep in the cat. EEG Clin Neurophysiol 1958; 10:291-295.
44. Jouvet M. Around the discovery of REM sleep in cats. In: Mallick BN, Inoue S, eds. Rapid Eye Movement sleep. V-IX. New York: Marcel and Dekker, 1999.
45. Mallick BN, Thankachan S, Islam F. Differential responses of Brain stem neurons during spontaneous and stimulation induced desynchronization of the cortical EEG in freely moving cats. Sleep Res Online 1998; 14:132-146.
46. Huttenlocher PR. Evoked and spontaneous activity in single units of medial brainstem during natural sleep and waking. J Neurophysiol 1961; 24:451-468.
47. Kasamatsu T. Maintained and evoked unit activity in the mesencephalic reticular formation of the freely behaving cat. Exp Neurol 1970; 28:450-470.
48. Manohar S, Noda H, Adey WR. Behavior of mesencephalic reticular neurons in sleep and wakefulness. Exp Neurol 1972; 34:140-157.
49. Tanaka DJ. Labeled NA release from rat cerebral cortex following electrical stimulation of LC. Brain Res 1976; 106:384-389.
50. Vanderwolf CH, Baker GB. The role of brain noradrenaline in cortical activation and behavior: A study of lesions of the locus coeruleus, medial thalamus and hippocampus-neocortex and of muscarinic blockade in the rat. Behav Brain Res 1996; 78:225-234.
51. Mallick BN, Adya HVA, Thankachan S. REM sleep deprivation alters factors affecting neuronal excitability: Role of norepinephrine and its possible mechanism of action. In: Mallick BN, Inoue S, eds. Rapid Eye Movement Sleep. Marcel and Dekker Inc., 1999:338-354.
52. Thakkar M, Portas C, McCarley RW. Chronic low-amplitude electrical stimulation of the laterodorsal tegmental nucleus of freely moving cats increases REM sleep. Brain Res 1996; 723:223-227.
53. Sakai K, Sastre JP, Kanamori N et al. State-specific neurons in the ponto-medullary reticular formation with special reference to the postural atonia during paradoxical sleep in the cat. In: Pompeiano O, Ajmone Marsan C, eds. Brain mechanisms and perceptual awareness. New York: Raven Press,1981:405-429.
54. Chase MH, Morales FR, Boxer PA et al. Effect of stimulation of the nucleus reticularis gigantocellularis on the membrane potential of cat lumbar motor neurons during sleep and wake-fulness. Brain Res 1986; 386:237-244.
55. Morales FR, Engelhardt JK, Soja PJ et al. Motoneuron properties during motor inhibition produced by microinjection of carbachol into pontine reticular formation of the decerebrate cat. J Neurophysiol 1987; 57:1118-1129.
56. Lai YY, Siegel JM. Muscle tone suppression and stepping produced by stimulation of midbrain and rostral pontine reticular formation. J Neurosci 1990; 10:2727-2734.
57. Thankachan S, Islam F, Mallick BN. Role of wake inducing brain stem area on rapid eye movement sleep regulation in freely moving cats. Brain Res Bulletin 2001; 55:43-49.

CHAPTER 6

The Mechanistic Relationship between NREM Sleep and Anesthesia

Laura E. Nelson, Nicholas P. Franks and Mervyn Maze

Abstract

The mechanisms by which both natural sleep and anesthesia generate and maintain a loss of consciousness are currently the focus of much investigation. The neuronal networks of substrates mediating endogenous regulation of consciousness level are complex. Much research has focused on understanding the neural correlates of wakefulness, NREM sleep, REM sleep and transitions between sleep and wake states, but a molecular basis for these behavioral alterations is only beginning to emerge. Mechanisms governing anesthesia, the pharmacological modulation of consciousness level, which shares the key common feature of loss of response to external stimuli with endogenous sleep, are even less clear. Several qualitative similarities between sleep and anesthesia suggest that pharmacological "sleep" may be transduced via activation of existing neurological pathways involved in promoting natural sleep. This chapter reviews how these mechanisms may intersect and outlines the neurochemical, pharmacological, and anatomical evidence that two classes of anesthetic drugs exert their hypnotic effects, at least in part, by duplicating activities of specific brain regions important for initiating and maintaining endogenous NREM sleep. Experiments demonstrate that anesthetic agents that are proven, or postulated, to act on α_2-adrenoceptors (e.g., dexmedetomidine, clonidine) and $GABA_A$ receptors (e.g., muscimol, propofol, and pentobarbital, isoflurane) induce a loss of consciousness, at least in part, via activation of endogenous nonrapid eye movement (NREM) sleep-promoting pathways at different junctions. One critical difference relates to the fact that the noradrenergic neurons within the locus coeruleus maintain their "awake" activity during hypnosis produced by compounds putatively mediated via the $GABA_A$ receptor while this collection of neurons is inactive during hypnosis produced by α_2-adrenoceptor agonists. This crucial difference may represent the reasons for the qualitative differences in the sedative/hypnotic response produced by these classes of anesthetic agents.

The Relationship between NREM Sleep and Anesthesia

Although there are significant physiological differences between the two states, there are also many similarities between sleep and anesthesia. Key components of general anesthesia include sedation/hypnosis, analgesia, amnesia, and muscle relaxation, while sleep can be characterized by a loss of responsiveness to external, nonpainful stimuli that are normally arousing, and a fulfilment of an essential biological need. Their most obvious shared characteristic is a reduced responsiveness to external stimuli. Despite the fact that unlike general anesthesia (which is often portrayed metaphorically as "going to sleep"[1]), natural sleep occurs spontaneously and in response to an essential biological need, is associated with dreaming and a characteristic ultradian pattern of EEG activity, and is immediately fully reversible with sufficient external stimuli. Significant commonalities that suggest a potential mechanistic relationship or overlap are discussed below.

K-complexes (the occurrence of single, episodic, large amplitude waves), sleep spindles (0.5-3.0 second short runs of 12-14 Hz), and a progressive degrees of anesthesia are associated with an increasing predominance of slow waves (high amplitude lower frequency waves (delta 1-4 Hz, theta 4-7 Hz), three characteristic electroencephalographic features of nonrapid eye movement (NREM) sleep, are also associated with anesthesia at lower (sedative/hypnotic, but not surgical) doses. For example, although the precise neural mechanisms mediating K-complexes remain unknown, they have been observed in cats under NMDA receptor antagonist ketamine and α_2-adrenoceptor agonist xylazine anesthesia (and it has been reported that K-complexes in cats and humans are similar).[2]

The bispectral index (BIS) monitor of anesthetic depth employs a method of EEG signal analysis to identify a coherence of the sine wave frequencies of the Fourier-transformed raw EEG. It is an algorithm that identifies interfrequency correlation of component frequencies in the EEG and the presence of slow waves in the algorithm it generates.[3] It has been further observed that because the degree of synchronisation of cortical EEG activity that occurs with natural sleep resembles that of lighter anesthesia, the BIS index reflects this degree of synchronisation. Consequently, the BIS anesthesia monitor can be used to accurately assess the assess the transition from consciousness to natural sleep state (i.e., the onset and "depth" of sleep) including identifying periods of REM as reflected by desynchronized cortical activity.[4,5] Although some variability in the BIS value marking human sleep onset is observed, a threshold BIS value might be defined to monitor and detect sleep onset in the clinical setting.[5]

Labelled positron emission tomography (PET) scans of human brains during anesthesia have demonstrated regional changes similar to those seen during sleep.[6] Further, PET/metabolic scanning[6] and microelectrode recordings of thalamic relay neuronal activity[7] show distinctive reductions in thalamic activity during

Sleep and Sleep Disorders: A Neuropsychopharmacological Approach, edited by Malcolm Lader, Daniel P. Cardinali and S.R. Pandi-Perumal.

anesthesia, which are also known to stimulate natural sleep-like changes in thalamocortical electrical activity.

Changes in the propensity to sleep (sleep deprivation) can modify responses to inhaled and intravenous anesthetic agents. Twenty four hours of sleep deprivation (induced by the disk-over-water paradigm, in which animals are placed on a 45 cm elevated disk that rotates when sleep is detected by computerized EEG/EMG monitoring causing the rat to wake up to avoid falling in a water hazard[8]) decreases the latency to loss of righting reflex (LORR) by 40% for propofol and 55% for isoflurane, and prolongs the time to recovery.[9]

Microinjection of intravenous anesthetics (propofol),[10] benzodiazepines (triazolam),[11] or barbiturates (pentobarbital)[12] into the medial preoptic area (MPA) of rats dose-dependently increases the subsequent time spent in natural sleep[10] (a different effect than inducing LORR, and anesthetic endpoint, in their own right), as assessed by traditional sleep study analysis of percentage of time spent in NREM slow wave EEG. The MPA, along with the entire preoptic area, has long been associated with natural sleep regulation via lesion and microinjection studies.[11-13] In addition, microinjection of the $GABA_A$ receptor antagonist picrotoxin into the MPA reduces sleep in rats.[14] The MPA contains serotonergic, norepinephrinergic, and GABAergic neurons, and propofol has been shown to act on GABAergic,[15] serotonergic,[16] and cholinergic[17,18] neurotransmitter systems receptors. In addition, sleep deprivation potentiates the ability of inhaled and intravenous anesthetic agents to induce LORR[9] and prolonged sedation with propofol does not result in a sleep deprivation-like state.[19]

Several investigators have looked at the effects of anesthesia on acetylcholine release in the medial pontine reticular formation, a key REM sleep-promoting nucleus. Acetylcholine release in the pontine reticular formation can be decreased by the fentalyl, morphine, and halothane (but not remifentanyl) and recovery from halothane anesthesia increases acetylcholine release here.[20,21] In humans, the acetylcholinesterase inhibitor physostigmine can reverse unconsciousness induced by propofol.[22]

Of course there also are many significant differences between anesthetic and sleep-induced "unresponsiveness", such as the ability of sleep to fulfil an essential biological need. Physiological processes take on a repeating cyclical variability while anesthesia attempts to achieve a physiological steady state. At deeper levels of anesthesia neural and electrophysiological function cease to resemble sleep. One electroencephalographic feature associated with anesthesia (as well as coma and trauma associated with cerebral anoxia and tumour infiltration), but not seen during natural sleep, is burst suppression (intermittent sequences of high-voltage slow waves and sharp waves, alternating with periods of depressed background activity or complete EEG flatness). With deepening degrees of burst suppression (when silent EEG periods exceed 30 seconds), thalamic cells cease firing and there is a virtual disconnect of all brain circuits implicated in the generation of the EEG.

What about REM sleep? There is no neurophysiologic correlate of REM sleep during anesthesia, though a relationship has been speculated upon. Selective activation of brainstem systems can be used to both create a REM sleep model and to counteract the spindle sleep pattern of anesthesia in animals. Further, the microinjection of narcotic agents onto brainstem neurons responsible for REM sleep can induce a naloxone-reversible and μ opioid receptor specific decrement in REM sleep.[23-26] It is hypothesized that the very REM-like abnormal electrooculograph and increased cortical activity that are observed during the "excitement phase" of anesthetic induction and emergence may be modulated by mechanisms that converge on REM sleep-promoting pathways.

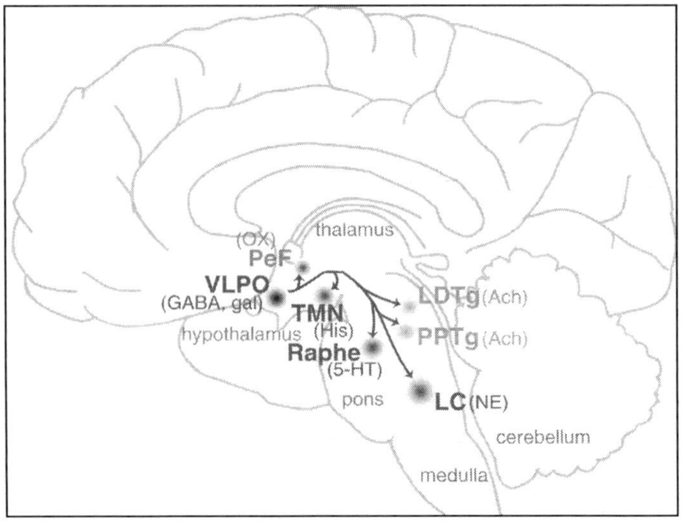

Figure 1. Efferent projections from VLPO. This cartoon illustrates the neural substrates of nonrapid eye movement (NREM) sleep; the efferent sleep-promoting (purple) projections from the ventrolateral preoptic nucleus (VLPO). VLPO is uniquely active during sleep, and believed to release inhibitory GABA and galanin onto the cell bodies and proximal dendrites of the monoaminergic (red) and cholinergic (orange) nuclei of the ascending arousal system. The ensuing dwindling of release of the neurotransmitters of arousal into the forebrain and cortex is thought responsible for the induction of loss of consciousness. Abbreviations: ACh= acetylcholine; GABA= γ-aminubutyric acid; gal= galanin; His= histamine; 5-HT= 5-hydroxytryptophan (serotonin); LC= locus coeruleus; LDTg= laterodorsal tegmental nucleus; NE= norepinephrine; NREM= nonrapid eye movement; PeF= perifornical area of the lateral hypothalamus; OX= orexin (hypocretin); PPTg= pedunculopontine tegmental nucleus; TMN= tuberomammillary nucleus. Reproduced with permission from Nelson LE, Maze M. Neural substrates for behavior; consciousness. In: Evers AS, Maze M. Anesthetic pharmacology: physiologic principles and clinical practive. A companion to Miller's anesthesia. Churchill Livingstone 2004, 15:227-243. ©2004 Elsevier.

Neural Mechanisms of NREM Sleep

NREM sleep is an actively generated state involving anatomically discrete supraspinal pathways in the hypothalamus and brainstem. The CNS pathways driving it are currently being characterized. Current understanding of the neural mechanisms of NREM sleep is briefly summarized below (see Fig. 1) and more extensive review elsewhere in this book. It has also been long known that both electrical and chemical stimulation of the basal forebrain produces NREM sleep, and that the anterior hypothalamus is the most active area in the induction of NREM sleep. Recent investigations have described "sleep-promoting"[27] neurons in this region and the adjacent basal forebrain, specifically in the ventrolateral preoptic nucleus (VLPO), which are under inhibitory control by norepinephrine and serotonin[28,29] (see Fig. 2) and discharge maximally during sleep while remaining relatively inactive during wakefulness. VLPO neurons, which form a dense cluster just lateral to the optic chiasm (the VLPO cluster; important for NREM sleep) and a diffuse population of cells

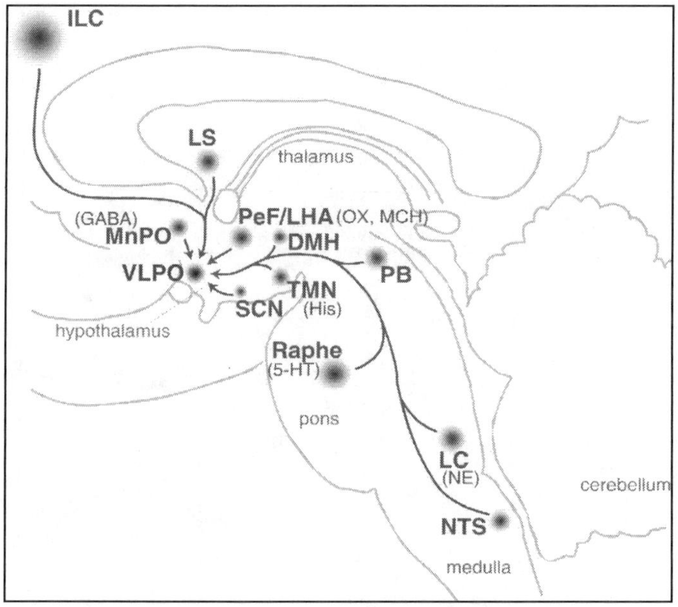

Figure 2. Afferent projections to VLPO. Chou et al described the efferent projections to the ventrolateral preoptic nucleus (VLPO; purple) after conducting extensive anterograde and retrograde tracing studies in the rat,[28] and sleep-promoting pathways are believed to be similar in humans.[31] The VLPO is innervated by the monoaminergic arousal promoting tuberomammillary nucleus (TMN), raphé nuclei, and locus coeruleus (LC), the median preoptic nucleus (MnPO; also believed to be sleep-promoting), the lateral hypothalamus including orexinergic neurons in the perifornical area (PeF). Other hypothalamic projections from the dorsomedial hypothalamic nucleus (DMH) and suprachiasmatic nucleus (SCN) also reach VLPO. In addition, VLPO receives projections from autonomic areas such as the infralimbic cortex (ILC) and parabrachial nucleus (PB) and limbic areas such as the lateral septum (LS). Abbreviations: DMH= dorsomedial hypothalamic nucleus; GABA= γ-aminobutyric acid; His= histamine; 5-HT= 5-hydroxy tryptophan (serotonin); ILC= infralimbic cortex; LC= locus coeruleus; LHA= lateral hypothalamic area; LS= lateral septum; MCH= melanin concentrating hormone; MnPO= median preoptic nucleus; NE= norepinephrine; NTS= nucleus of the solitary tract; PB= parabrachial nucleus; SCN= suprachiasmatic nucleus. Reproduced with permission from Nelson LE, Maze M. Neural substrates for behavior; consciousness. In: Evers AS, Maze M. Anesthetic pharmacology: physiologic principles and clinical practive. A companion to Miller's anesthesia. Churchill Livingstone 2004, 15:227-243. ©2004 Elsevier.

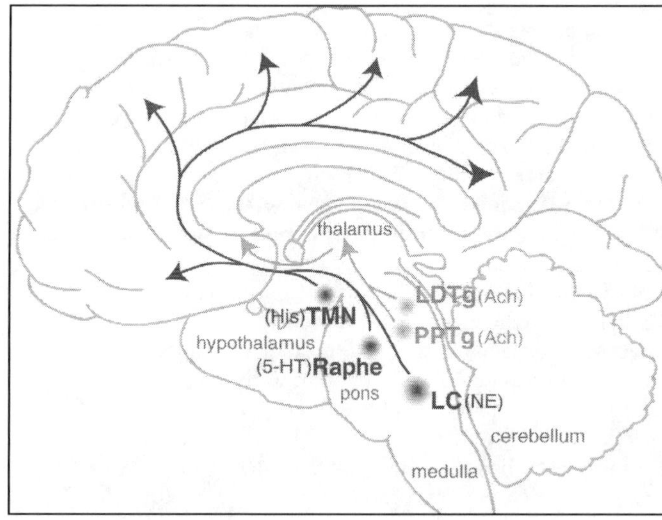

Figure 3. Neural substrates of arousal. This cartoon illustrates that projections from the brainstem through the hypothalamus (red) or thalamus (orange) into the cortex and forebrain form the two branches of the ascending arousal system (termed the ascending arousal system). Monoaminergic projections (red) from the pontine norepinephrinergic locus coeruleus (LC), midbrain serotoninergic raphé nuclei, and hypothalamic histaminergic tuberomammillary nucleus (TMN) diffusely innervate the forebrain areas and regulate cortical and hypothalamic activity to promote arousal. Similarly, cholinergic (orange) projections from the pedunculopontine and laterodorsal tegmental nuclei (PPTg and LDTg) innervate the thalamus and forebrain to modulate wakefulness. Abbreviations: ACh= acetylcholine; His= histamine; 5-HT= 5-hydroxytryptophan (serotonin); LC= locus coeruleus; LDTg= laterodorsal tegmental nucleus; NE= norepinephrine; PPTg= pedunculopontine tegmental nucleus; TMN= tuberomammillary nucleus. Reproduced with permission from Nelson LE, Maze M. Neural substrates for behavior; consciousness. In: Evers AS, Maze M. Anesthetic pharmacology: physiologic principles and clinical practive. A companion to Miller's anesthesia. Churchill Livingstone 2004, 15:227-243. ©2004 Elsevier.

extending medially and dorsally from this cluster (the extended VLPO; more important for REM sleep[30]), are defined by three, characteristics: they are (1) uniquely sleep-active (show c-Fos expression during sleep),[27] (2) contain the colocalized (80%) inhibitory neurotransmitters GABA and galanin across species, while surrounding neurons contain GABA only,[31,32] and (3) project to the arousal-promoting[33] TMN.[32] VLPO neurons also extend inhibitory projections to all of the other wake-active ascending monoaminergic, cholinergic, and orexinergic arousal-promoting sites in the brain; the noradrenergic LC, serotonergic dorsal raphé (DR), cholinergic pedunculopontine tegmental nucleus (PPTg) and laterodorsal tegmental nucleus (LDTg), and the orexinergic perifornical area (PeF) in the lateral hypothalamic area (LHA);[32] and inhibit their release of arousal-promoting neurotransmitters into the cortex, forebrain and subcortical areas (see Figs. 1, 3).

A decrease in firing of the noradrenergic neurons in the LC of the pons releases the LC's tonic inhibition of GABAergic VLPO neurons, which are then activated and release GABA and galanin into LC, TMN and other arousal promoting centers.[32] At the level of the LC, this has an inhibitory effect, further decreasing firing in the LC, and therefore further decreasing norepinephrine's tonic inhibition of the VLPO neurons. At the level of the TMN, descending projections from the VLPO release GABA and galanin.[32] This inhibition of the TMN by the VLPO is believed to play a key role in causing sleep.

The inhibition of the TMN (as well as the LC, DR, PeF, and LDTg/PPTg) by GABA and galanin released by VLPO neurons[34] is believed to play a key role in causing NREM sleep. Galanin and GABA are observed in the TMN, and $GABA_A$ IPSPs in the TMN region are observed when the VLPO is stimulated.[35] Discrete bilateral lesions induced by microinjection of the nonspecific excitotoxin ibotenic acid into the VLPO induce persistent insomnia in rats.[36] This may be explained by the fact that GABA-mediated anesthetic drugs are believed to act by inhibiting the wake-active (c-Fos-immunoreactive during wakefulness) TMN. Discrete injections of $GABA_A$ receptor agonist muscimol cause sedation and at higher doses hypnosis, and potentiate the hypnotic effects of anesthetic agents, direct injections of anesthetic agents induce sedation, and of $GABA_A$ receptor antagonist gabazine attenuate the hypnotic effect of anesthetics.[37]

Other pathways may also be involved in the generation of NREM sleep. Earlier it was suggested that there are two main hypnogenic centers involved in the generation of NREM sleep: the preoptic area (VLPO) and the nucleus of the solitary tract (NTS) in the medulla.[38] Subsequent research has demonstrated that low frequency stimulation of the vago-aortic nerve or of the NTS produces a slow wave EEG, distension of the carotid sinus (an NTS stimulus) induces behavioral sleep, discharge of certain NTS neurons, which are hypothesized to be reciprocally interconnected with cells in the midbrain EEG arousal region, increases during NREM sleep, and inactivation of the lower brainstem, particularly the NTS, induces a marked arousal. However, a causal involvement of NTS in NREM sleep has not been definitively established, and available evidence suggests that it is a far weaker NREM sleep-promoting center than the VLPO.

Switching between Sleep and Wake States

Switching between these sleep and wake states is thought to be controlled by a reciprocal relationship of mutual inhibition between the activities of neurons in the VLPO and the major monoamine areas (TMN, LC, and midbrain raphé nuclei) of the ascending reticular arousal system.[39] The VLPO innervates the arousal system and inhibits activity during sleep, and the nuclei of the ascending arousal system inhibit the VLPO during wakefulness (see Fig. 3). Therefore, when the VLPO's firing rate is high during sleep it inhibits the monoaminergic arousal nuclei (see Fig. 1) and thereby further disinhibits, or activates, its own firing. When the monoaminergic neurons fire rapidly during wakefulness, they act to inhibit the VLPO, and disinhibit, or further activate, their own firing.

Saper et al reported[39] that this reciprocal relationship tends toward two stable firing patterns (sleep or wake) and away from intermediate transition states (and drew comparisons to the bistable circuit known to electrical engineers as a 'flip-flop'; i.e., when one side is firing rapidly, there is a "resistance" to "switching" such that it occurs infrequently but rapidly), so that only large scale influences such as circadian drive to sleep or a sufficient degree of sleep drive or deprivation exert pressure on the circuit until it reaches a critical threshold and the firing patterns reverse rapidly.[39] When the firing of the sleep-promoting VLPO side of the circuit is weakened by ablation of the VLPO, the circuit is destabilized and animals experience insomnia plus an increased drive to sleep which may bring the circuit balance closer to the transition state.[36] A destabilising deficit on the waking side of this sleep switch induces the inappropriate abrupt transitions from wakefulness to sleep (particularly REM or fragments of REM sleep, such as cataplexy or the loss of muscle tone while awake) seen in the sleep disorder narcolepsy (see Fig. 4) and elsewhere in this book for review of REM sleep mechanisms; not covered in this chapter).

Saper et al further hypothesized that the recently discovered orexin- (also known as hypocretin) containing neurons in the perifornical area of the lateral hypothalamus act as a stabilising "finger" helping to hold the sleep switch pointing toward wakefulness, and prevent switching to sleep. Dysfunction of this "finger" and the consequent susceptibility to sudden and inappropriate transitions is seen in narcolepsy, which is characterized by an orexinergic deficiency.[40] The TMN, LC, and raphé nuclei all receive input from orexinergic neurons, contain orexin receptors, and inhibit REM sleep,[41] so that narcoleptics' inappropriate

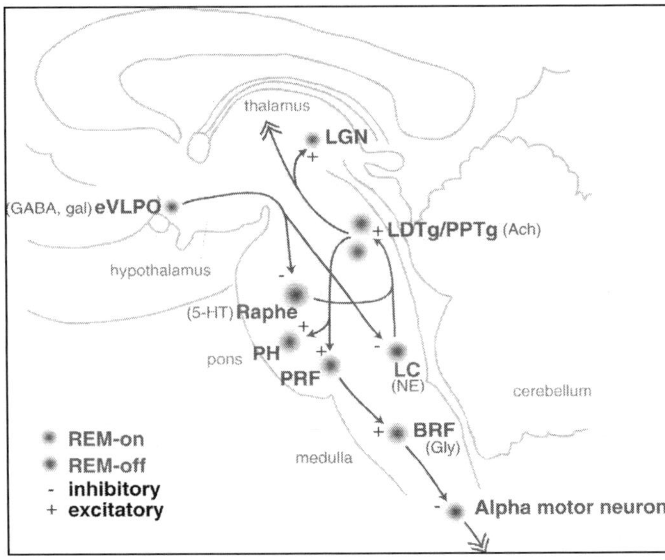

Figure 4. Neural substrates of rapid eye movement (REM) sleep. The cholinergic neurons of the pedunculopontine and laterodorsal tegmental nuclei (PPTg and LDTg) are very active during REM sleep and make three key projections. First they innervate the thalamus and cortex to modulate the electroencephalogram (EEG) synchronisation seen during REM sleep. Second they innervate the pontine reticular formation (PRF), thalamic lateral geniculate nucleus (LGN), and medulary prepositus hypoglossi nucleus (PH), which are generate the pontine-geniculo-occipital (PGO) waves that characteristically denote the onset of REM sleep. Third, the PPTg and LDTg neurons innervate the glycinergic brainstem reticular formation (BRF), which in turn mediates REM atonia by inhibiting alpha motor. Abbreviations: Ach= acetylcholine; BRF= brainstem reticular formation; EEG= electroencephalogram; eVLPO= extended VLPO; GABA= γ-aminubutyric acid; gal= galanin; 5-HT= 5-hydroxy tryptophan (serotonin); LC= locus coeruleus; LGN= lateral geniculate nucleus; LDTg= laterodorsal tegmental nucleus; NE= norepinephrine; PH= medullary prepositus hypoglossi nucleus; PPTg= pedunculopontine tegmental nucleus; PRF= pontine reticular formation; REM= rapid eye movement. Reproduced with permission from Nelson LE, Maze M. Neural substrates for behavior; consciousness. In: Evers AS, Maze M. Anesthetic pharmacology: physiologic principles and clinical practive. A companion to Miller's anesthesia. Churchill Livingstone 2004, 15:227-243. ©2004 Elsevier.

transitions into REM or fragments of REM sleep may be modulated by the weakening of the sleep switch reinforcement "finger". Specifically, the absence of excitatory orexin input is hypothesized to weaken the arousal system's inhibition of the extended VLPO (which is thought to modulate REM while the VLPO cluster modulates NREM sleep[30]), allowing more frequent transitions into REM.[39]

Sites of Anesthetic Action

The state of anesthesia is a behavioural syndrome comprised of a constellation of behavioural components; sedation/hypnosis, amnesia, analgesia, and muscle relaxation. Although millions of people worldwide have been anesthetized every year, the mechanisms by which anesthetic agents produce their central nervous system (CNS) effects remain incompletely understood. The invention of anesthesia is not certain either. In 4000 B.C. the euphoric effects of the poppy were first reported in Sumerian medical texts. Inca and preInca shamans performing trepanations (holes drilled in patient's skulls to free evil humours) from around 1000 A.D. chewed coca leaves and spat into the wound, affecting local anesthesia. General anesthesia by inhalation developed in the 1840s involving two agents. The soporific effects of diethyl ether ("sweet vitriol") had been known since the 14th century, and nitrous oxide ("laughing gas") was synthesized by Joseph Priestly in 1772. In 1842, Crawford Long, a Georgia physician who had experienced the then-fashionable "ether frolics" successfully administered diethyl ether to James W. Venable for surgical removal of a neck tumour but did not publish his accomplishment immediately. In 1844, the dentist Horace Wells publicly demonstrated the use of inhaled nitrous oxide for tooth extraction at Massachusetts General Hospital in Boston, but removed the N_2O-containing bag too soon and the patient screamed in pain so was denounced as a fake. Two years later the dentist William Thomas Green Morton publicly extracted a tooth from patient Gilbert Abbot at Massachusetts General Hospital in Boston using a specially designed glass apparatus used containing ether (which he termed "letheon", and was only later convinced to use the term "anesthesia") and is today credited as the father of anesthesia. For nearly a century, scientists have been trying to elucidate exactly how anesthesia works.

Until recently anesthesia was considered to be an emergent phenomenon caused by a generalized suppression of neuronal electrical activity. The mechanism of this nonspecific inhibition was thought related to a century old-observation made independently in 1897 by H.H. Meyer in Germany and C.E. Overton in England that anesthetic potency is proportional to lipophilicity (the so-called "Meyer-Overton" correlation[42]); thus was born the theory that anesthetics act on the lipid bilayer to interfere with membrane fluidity and affect the functioning of cellular processes.[43] This once prevalent unitary theory of anesthetic mechanisms postulated that anesthesia arises by way of a common effect on all neurons,[44,45] but has been displaced by an understanding that the state of general anesthesia is actually comprised of a constellation of behavioral components (including sedation/hypnosis, amnesia, analgesia, and muscle relaxation or immobility) mediated at discrete and independent sites in the CNS.

Early lipid theories were challenged in the 1980s by Franks and Lieb who demonstrated that anesthetics target and affect specific protein function, even when no membrane is present, and concluded that anesthetics exert their effect by binding to hydrophobic (lipid-like) pockets (normally less than 2% of the size of the total protein volume on proteins and forming weak van der Waals bonds.[46] Coincident with this was the understanding of the structure and chemistry of individual amino acid residues which generated a new appreciation of the potential lipophilicity of proteins based on amino acid sequence. It is now believed that anesthetic agents selectively target a relatively small number of specific membrane post-synaptic receptors (e.g., $GABA_A$, glycine, 5-HT_3, and nicotinic acetylcholine receptors), as well as two-pore domain K^+ channels[47,48] and other proteins (e.g., voltage-gated channels, cytoskeletal actin, microtubules, G-protein coupled receptors, and gap junction proteins), which are less sensitive to anesthetics but more abundant and/or directly involved in activities related to consciousness.[48] Several different ion channels controlling neuronal excitability have been demonstrated to be targets of a host of anesthetic agents at clinically relevant concentrations, including $GABA_A$ receptors, nicotinic acetylcholine receptors and NMDA receptors.[48-53] Anesthetics are believed to act by either reducing the efficacy of excitatory neurotransmission and/or increasing the efficacy of inhibitory neurotransmission. Two key targets of anesthetic agents, the $GABA_A$ receptor and the α_2-adrenoceptor, are discussed below.

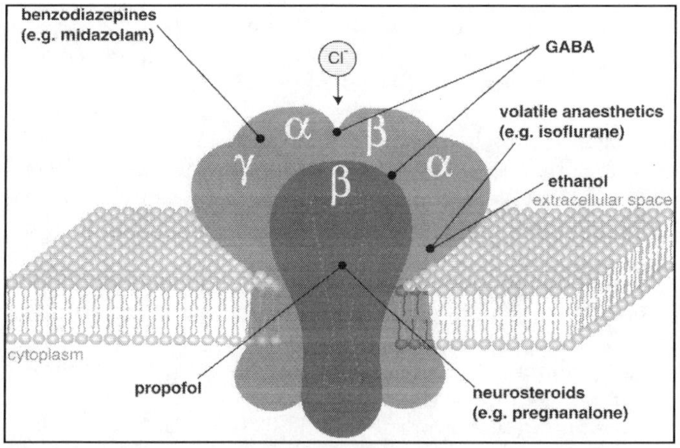

Figure 5. $GABA_A$ receptor and postulated sites of drug action. This cartoon depicts the stoichiometry, subunit configuration, and agonist and modulator binding sites of the $GABA_A$ receptor, a member of a superfamily of ligand-gated ion channels, including the nicotinic acetylcholine, glycine, and 5-HT_3 receptors.[93] Activation of the $GABA_A$ receptor by GABA leads to an increased influx of chloride ions, resulting in membrane hyperpolarization and neuronal inhibition. $GABA_A$ receptors are hetero-oligomeric (pentameric) protein complex in their native state, composed of five subunits with molecular weights of approximately 40-60 kDa. At least 18 different receptor subunits genes have been separated into the sequence homology subfamilies alpha (α_1 - α_6), beta (β_1 - β_3), gamma (γ_1 - γ_3), rho (σ_1 - σ_3), delta (δ), pi (π), epsilon (ε), theta (θ), building a ligand gated chloride channel.[94,95] Two additional subunits (β_4 and γ_4) have been identified in chick brain and five isoforms of the σ-subunit in the retina of the white perch. Several splice variants of the subunits also exist. The γ_2-subunit is expressed as short and a long form variant, as are α_6- is human β_3-subunit. Most functional subtypes of mammalian $GABA_A$ receptor contain two α, two β and one γ subunit (as shown here), but there is a large number of other combinations possible, and each is predicted to have its own characteristics and pharmacology.[96]

$GABA_A$ Receptor

Inhibitory $GABA_A$ receptors, which are comprised of multivariant pentameric combinations of two α (α_1-α_6), two β (β_1-β_3), and one γ (γ_1-γ_3), δ, or ε subunit assembled to form a membrane-bound GABA gated chloride channel (see Fig. 5), play a key role in sleep and are believed to be one of the primary targets of general anesthetic agents. GABA-containing neurons are located throughout the brain, including the brainstem, thalamus, hypothalamus, basal forebrain and cortex, and GABA is released into the cerebral cortex in the highest concentration during NREM sleep. Clinically relevant concentrations of anesthetics (e.g., barbiturates, etomidate, propofol, neuroactive steroids, benzodiazepines, and volatile anesthetics) markedly enhance $GABA_A$ receptor mediated chloride current or GABAergic neurotransmission in neurons in vivo as well as in a variety of recombinant expression systems in vitro[54] to ultimately enhance chloride ion flow, resulting in hyperpolarization and decreased cell firing.

"GABAergic" general anesthetics are far more effective at enhancing the actions of GABA at its $GABA_A$ receptors than they are at directly activating the receptors in the absence of GABA (e.g., propofol is enhancing the action of endogenous GABA at the $GABA_A$ receptors than it is as a direct activator[55]). The direct connection between the activation of $GABA_A$ receptors and the hypnotic response is supported by the fact that the $GABA_A$ receptor agonist muscimol and the GABAergic anesthetic agents propofol and pentobarbital transduce their hypnotic effects via acting on $GABA_A$ receptors in regionally discrete anatomical sites with in an endogenous NREM sleep-promoting pathway; the VLPO and TMN[37] (see three sections below).

Benzodiazepines interact with all the benzodiazepine-sensitive α subunits (α_1, α_2, α_3, and α_5), with the classical high affinity stereospecific benzodiazepine binding site located between the α- and γ_2 subunits. It is known that a knock-in point mutation (histidine to arginine at position 101) introduced into the murine α_1-subunit (which contributes to 60% of $GABA_A$ receptors in the brain) gene renders homomeric α_1 $GABA_A$-receptors insensitive to benzodiazepine agents in vitro, and in vivo these α_1(H101R) mice are insensitive to the sedative and amnestic but not anxiolytic actions of benzodiazepines. This suggests that the multiple actions of benzodiazepines can be molecularly distinguished; the hypnotic action appears to be mediated by the $GABA_A$ receptors α_3 subunit,[56] the sedative action by α_1 subunit[57-59] and the anxiolytic action by α_2, α_4, and α_5 subunits.[60]

One clinical application of this observation is zalepon (CL284,846), the pyrazolopyrimidine hypnotic agent developed for the treatment of insomnia which binds preferentially to $GABA_A$ receptors containing the α_1-subunit 8-20 fold stronger than to α_2-, α_3-, and α_5-subunit containing receptors, as well as to those containing the γ_3-subunit (though the contribution of γ_3-subunit mediated action is unclear).[61] Other agents have been developed for their anxiolytic effects with a distinct lack of effect on α_1-subunits, including L-838417, SL651498, pagoclone, pazinaclone, and PNY-101017. Pharmacological analysis animal models in which particular $GABA_A$ receptor subunits are either inactivated (knock-out strategy) or point-mutated (knock-in strategy), as described above, is expected to lead further $GABA_A$ receptor subtype-targeted drugs with selective therapeutic action in the near future.

α_2-Adrenoceptor

The α_2-adrenoceptor is a seven-transmembrane domain G-protein coupled receptor which is largely (but not exclusively) presynaptic on the CNS and post-synaptic in the PNS. There are three known subtypes (A, B, and C) (see Fig. 6). α_2-adrenoceptor agonists (e.g., dexmedetomidine, clonidine, or xylazine) are anesthetic agents used widely in clinical and veterinary settings for their sedative and analgesic effects. Dexmedetomidine (Precedex™, Abbott Labs) is the pharmacologically active dextroisomer of medetomidine and displays specific and selective α_2-adrenoceptor agonism. It was approved by the U.S. Food and Drug Administration in 1999 as a short-term medication (< 24 hour) for sedation in the intensive care unit and has also been shown to act as a general anesthetic.[62] Its specificity for the α_2-adrenoceptor, and relative selectivity for the α_{2A}-adrenoceptor subtype (which is responsible for its sedative and sedative properties[63]), renders it a more effective sedative and analgesic agent than clonidine and makes it an ideal probe with which to investigate anesthetic mechanisms.

One of the highest densities α_2-adrenoceptor has been detected in the LC,[64] the predominant norepinephrine-containing nucleus in the brain and an important modulator of vigilance[65] and sedative effects of α_2-adrenoceptor activation have been

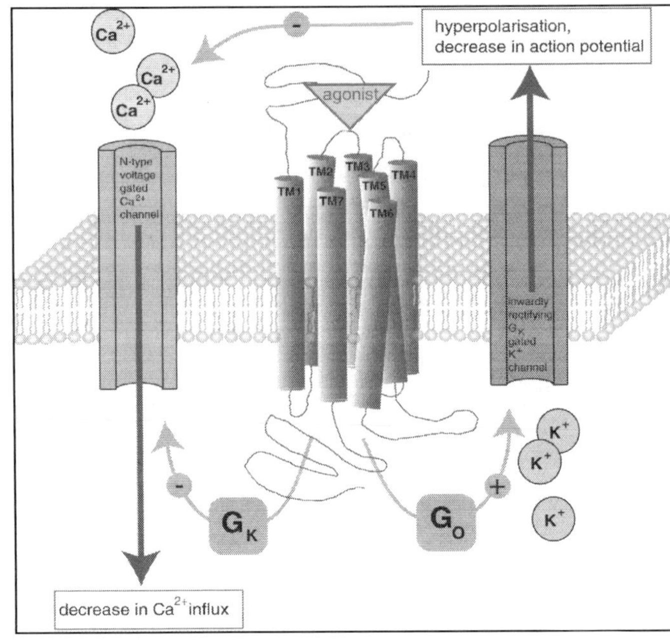

Figure 6. α_2-adrenoceptor. The seven-transmembrane (TM) domain α_2-adrenoceptor coupled by pertussis toxin-sensitive G-proteins to various effectors, including adenylyl cylcase and ion channels. When an α_2-adrenoceptor agonist, such as dexmedetomidine, binds to the receptor binds, two pathways are activated; a G_O/G_I mediated direct inhibition of Ca^{2+} influx through N-type voltage-gated Ca^{2+}channels resulting in a hyperpolarization and an activation of inwardly rectifying G_K gated K^+ channels resulting in an efflux of K^+ ions and consequent further membrane hyperpolarization and reduction of firing.

attributed to the LC,[66,67] which densely innervates and tonically inhibits the sleep-active VLPO during wakefulness.

After a selective α_2-adrenoceptor agonist binds to α_2-adrenoceptors in the LC, transmembrane signalling results in activation of an inwardly rectifying potassium channel facilitating a K^+ efflux and inhibition of voltage-gated Ca^{2+} channels. The resulting hyperpolarization decreases the firing rate of LC neurons and allows presynaptic inhibition of their terminals.[68-70] From an anesthetic viewpoint, hyperpolarization of norepinephrinergic neurons in the LC appears to be a key factor in the mechanism of action of dexmedetomidine and other α_2-adrenoceptor agonist.[66,67,71] Relative to its effects on the α_1-adrenoceptor, dexmedetomidine is eight times more specific for the α_2-adrenoceptor than clonidine, the next most selective α_2-adrenoceptor agonist. This is important because activation of α_1-adrenoceptors counteracts the sedative action mediated at the α_2-adrenoceptor.[72]

How Does a Decrease in Noradrenergic Firing of Noradrenergic Firing in the Locus Coeruleus (the Known Site of Initiation of the Sedative Response to α_2-Adrenoceptor Agonists) Lead to a Loss of Consciousness?

Given that the LC is known to play a key role in modulating arousal, vigilance, and sleep-wake states, it is logical to ask whether dexmedetomidine converges on endogenous sleep-promoting pathways downstream of the LC to exert its sedative effects.

Animal experiments indicate that mechanism of sedation induced by the α_2-adrenoceptor agonist dexmedetomidine does indeed activate an endogenous NREM sleep pathway at neural substrates within the brainstem and hypothalamic regions. First pretreatment with systemic $GABA_A$ receptor agonist gabazine was shown to decrease the hypnotic potency of systemically-administered dexmedetomidine in rats, suggesting that $GABA_A$ receptors may be involved downstream of the LC, the key site for initiation of dexmedetomidine hypnosis and a crucial nucleus for modulating sleep, arousal, and vigilance. Next, the changes c-Fos protein expression (an immediate early gene product commonly used as a surrogate marker of neuronal activation) induced by 90 minutes of dexmedetomidine-induced hypnosis were demonstrated to be qualitatively similar to those induced during NREM sleep (a decrease at the LC, an increase at the VLPO, and a decrease at the TMN).[37,73] This suggested a likelihood that these observed changes are being initiated at α_2-adrenoceptors, an idea that was subsequently supported by the observation that the above described changes in c-Fos expression can be prevented by systemic administration of the α_2-adrenoceptor antagonist atipamezole at a dose sufficient to block dexmedetomidine-induced hypnosis.

That these changes in neuronal activity (c-Fos expression) are causally important for the sedative response to dexmedetomidine was then suggested by the observation mice genetically modified to have mutated α_{2A}-adrenoceptor subtype receptors (in which dexmedetomidine cannot induce hypnosis) do not exhibit these regionally discrete changes in c-Fos expression in the sleep promoting pathways. Gene targeting was used to point mutate a conserved aspartate at position 79 in transmembrane domain two of α_{2A}-adrenoceptor subtype (the α_2-adrenoceptor is a 7-transmembrane domain G-protein coupled receptor) to an asparagine in the C57B6 strain mouse genome, rendering the receptor dysfunctional in the eponymous D79N mouse. The α_{2A} subtype is known to be responsible for the sedative component of dexmedetomidine's anesthetic action,[63,74] while α_{2B} and α_{2C} subtypes are relatively uninvolved in sedation but may play a larger role in analgesia.[75] In D79N animals, dexmedetomidine cannot induce LORR at any dose,[63] an observation used in this study to further confirm genotype.

Finally the hierarchical nature of this pathway and its relevance was suggested by the demonstration that discrete bilateral lesions of the VLPO significantly reduced the sedative potency of dexmedetomidine (but not muscimol).[37,73]

These results were a novel demonstration of a discrete neuronal pathway causally connected with the hypnotic response to a general anesthetic, which is significant for two reasons. Firstly, while the actions of anesthetic agents at the level of the individual receptor are relatively well defined, whether or not these drugs act at nonspecific or specific anatomical sites within the brain has not been clearly demonstrated. Secondly, because many anesthetic and hypnotic/sedative agents are known to have disruptive effects on endogenous sleep cycles.[76-80]

Future refinement of drugs targeting the α_2-adrenoceptor may yield a pharmacological sedation with more of the restorative features of endogenous sedation. It is already known that in the intensive care setting patients receiving dexmedetomidine experience a clinically effective sedation yet are still easily and uniquely arousable (as is natural sleep), an effect not observed with any other clinically available sedatives.[81] Other recent data from an fMRI study in human volunteers seemed to confirm that the Blood Oxygen Level Dependent (BOLD) signal, which positively correlates with local brain activity, changes during dexmedetomidine-induced sedation in a manner similar to that during natural sleep, while changes induced by the benzodiazepine midazolam were markedly different.[82]

Are Endogenous Sleep-Promoting Pathways Also Important for Mediating the Sedation Induced by Intravenous Anesthetic Agents Putatively Mediated by the $GABA_A$ Receptor?

Do anesthetics putatively mediated by the $GABA_A$ receptor also converge on endogenous NREM sleep-promoting pathways to exert their sedative effects? Behavioral studies similar to those described above verified the involvement of the $GABA_A$ receptor in the transduction of the sedative response to muscimol, pentobarbital, and propofol. Second, immunohistochemical studies of c-Fos suggested that these "GABAergic" agents converge upon sleep pathways at a different junction than dexmedetomidine (at the LC, as described above), and established the TMN as a possible site at which $GABA_A$ receptors could transduce the sedative response.[37] Next, experiments were performed demonstrating that a GABAA receptor agonist, muscimol, administered directly into the TMN produced a dose-dependent sedation, and that propofol and pentobarbital microinjected into this site induced a lightly sedates state (sedation with maintenance of LORR). Importantly, injection of muscimol into other sites (such as the LC and hypothalamic sites near the TMN) had no effect, indicating the anatomically selective nature of the response. Finally, it was demonstrated that discrete administration $GABA_A$ receptor antagonist gabazine was discretely microinjected into the TMN of rats attenuated the hypnotic response induced by systemically administered GABAergic anesthetic agents.[37] Figure 7 summarizes the differing convergences of propofol, pentobarbital, and dexmedetomidine on endogenous sleep-promoting pathways.

Clinical Relevance

Could a better understanding of the interactions of anesthetic agents with endogenous sleep promoting pathways facilitate the creation of a pharmacological sedation that could confer the restorative effects of natural sleep? As discussed above, prolonged sleep deprivation alters electrocortical (rebound increase in REM and NREM EEG), respiratory (impairs respiratory muscle strength[83] as well as CO_2 and O_2 homeostasis[84]), psychiatric (irritability, disorientation, agitation, psychosis[85]), and immune (increased susceptibility to infection[86]) functions, and can ultimately results in death,[87-89] all of which are reversible with physiological sleep.[90]

Owing to intensive individualized care and monitoring, chronic pain, discomfort, and anxiety, severe sleep deprivation is a common problem in the ICU,[91] even though it is known that adequate sleep is critically necessary for physiologic, cognitive, and behavioral homeostasis in humans.[90] Despite marked physiological and behavioral differences between pharmacological sedation and natural sleep, ICU patients are often continuously sedated for days or weeks to facilitate care, allay anxiety and pain, and possibly dispel the effects sleep deprivation. The pharmacologically sedated state may not effectively or efficiently reverse the negative consequences of sleep deprivation (and indeed may potentiate these harmful effects in some cases), but the finding that prolonged propofol sedation does not result in a sleep deprived state[19] suggests that pharmacological sedation which can confer the restorative properties of natural sleep may be possible.

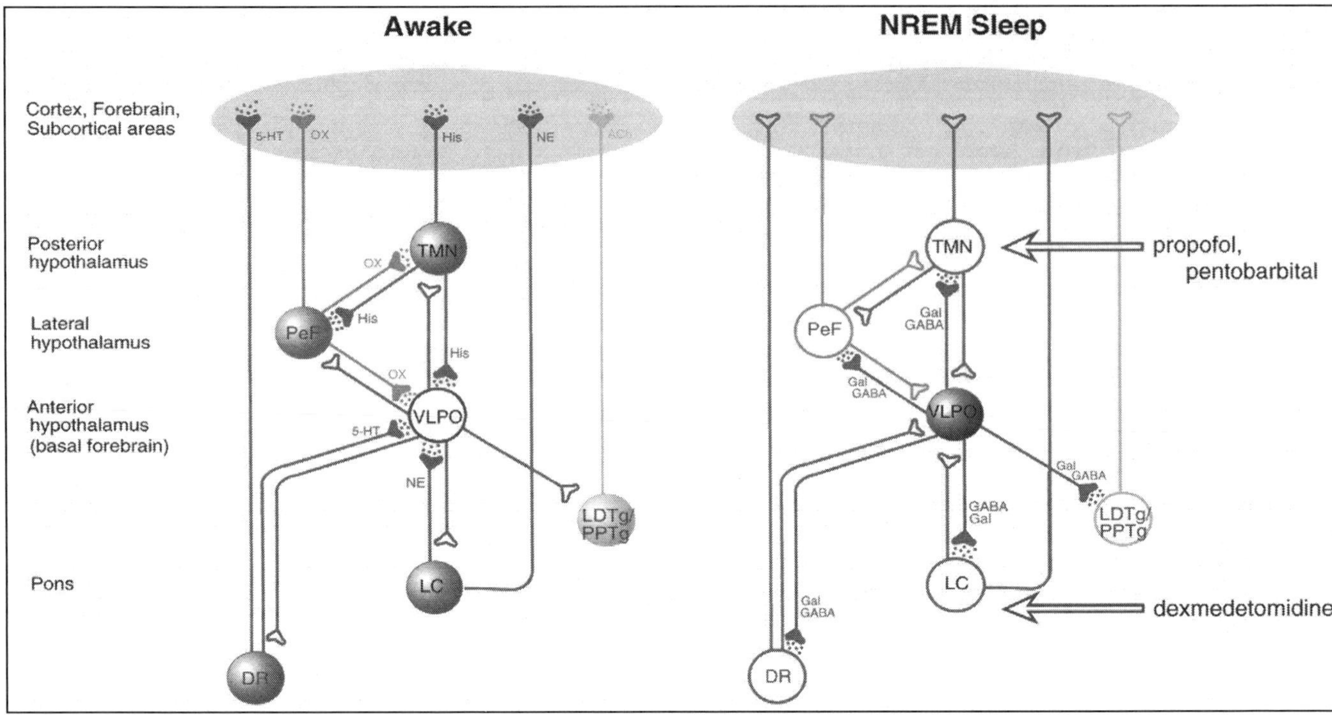

Figure 7. Summary diagram: anesthetic convergences on NREM sleep-promoting pathways. This diagram summarizes data indicating that different classes of anesthetics (as characterized by their molecular mechanisms) "hijack" different portions of the endogenous sleep-promoting circuitry to exert their sedative effects. Dexmedetomidine appears to converge on these sleep pathways at the level of the LC, while anesthetics putatively mediated at least partly via the $GABA_A$ receptor (e.g., propofol, pentobarbital) converge further downstream at the level of the TMN. Abbreviations: ACh= acetylcholine; DR= dorsal raphé nuclei; GABA= γ-aminobutyric acid; GBZ= gabazine; His= histamine; 5-HT= 5-hydroxy tryptophan (serotonin); LC= locus coeruleus; LDTg= laterodorsal tegmental nuclei; NE= norepinephrine; NREM= nonrapid eye movement; OX= orexin (hypocretin); PeF= perifornical area; PPTg= pedunculopontine tegmental nuclei; TMN= tubero-mammillary nucleus; VLPO= ventrolateral preoptic nucleus. Adapted with permission from Nelson et al.[37,73]

For example, the rapid-acting anesthetic propofol, originally developed for use as an intravenous anesthetic for out patient use, was recently introduced as an intensive care unit (ICU) sedative. Its rapid onset and offset allows physicians to sedate patients to near unresponsiveness for extended periods while retaining the ability to wake them up rapidly,[92] and have consequently led to the advocacy of its use to promote sleep in the ICU, although little evidence supports such a strategy. As discussed above, propofol's molecular mechanism of action is believed to be binding to the $GABA_A$ receptor at a site distinct from the benzodiazepine binding site and act to allosterically potentiate of the activity of GABA.[15] However, the question that arises is does propofol-induced sedation promotes or mimics physiological sleep? Unlike endogenous sleep, propofol sedation does not demonstrate an orderly progression of EEG states and is not entirely reversible with external stimuli. In addition, little evidence exists suggesting that propofol-induced sedation can satisfy the biological drive need to for natural sleep. Might prolonged periods of continuous sedation overlapping with naturally occurring sleep periods result in sleep deprivation? One rat study reported that upon emergence from prolonged propofol-induced sedation, no EEG (rebound increases in REM or NREM) or behavioral signs of sleep deprivation are observed.[19]

Summary

Different classes of anesthetic agents converge on pathways known to modulate natural NREM sleep at different junctions to exert their sedative effects. The $GABA_A$ receptors in these sleep-promoting pathways, particularly in the TMN, appear to play a key role in the sedative/hypnotic component of anesthesia.

References

1. Shafer A. Metaphor and anesthesia. Anesthesiology 1995; 83(6):1331-42.
2. Amzica F, Steriade M. The K-complex: Its slow (<1-Hz) rhythmicity and relation to delta waves. Neurology 1997; 49(4):952-9.
3. Rampil IJ. A primer for EEG signal processing in anesthesia. Anesthesiology 1998; 89(4):980-1002.
4. Sleigh JW, Andrzejowski J, Steyn-Ross A et al. The bispectral index: A measure of depth of sleep? Anesth Analg 1999; 88(3):659-61.
5. Tung A, Lynch JP, Roizen MF. Use of the bis monitor to detect onset of naturally occurring sleep. J Clin Monit Comput 2002; 17(1):37-42.
6. Alkire MT, Pomfrett CJ, Haier RJ et al. Functional brain imaging during anesthesia in humans: Effects of halothane on global and regional cerebral glucose metabolism. Anesthesiology 1999; 90(3):701-9.
7. Vahle-Hinz C, Detsch O, Siemers M et al. Local GABA(A) receptor blockade reverses isoflurane's suppressive effects on thalamic neurons in vivo. Anesth Analg 2001; 92(6):1578-84.
8. Bergmann BM, Kushida CA, Everson CA et al. Sleep deprivation in the rat: II. Methodology. Sleep 1989; 12(1):5-12.
9. Tung A, Szafran MJ, Bluhm B et al. Sleep deprivation potentiates the onset and duration of loss of righting reflex induced by propofol and isoflurane. Anesthesiology 2002; 97(4):906-11.
10. Tung A, Bluhm B, Mendelson WB. The hypnotic effect of propofol in the medial preoptic area of the rat. Life Sci 2001; 69(7):855-62.
11. Mendelson WB, Martin JV. Characterization of the hypnotic effects of triazolam microinjections into the medial preoptic area. Life Sci 1992; 50(15):1117-28.

12. Mendelson WB. Sleep induction by microinjection of pentobarbital into the medial preoptic area in rats. Life Sci 1996; 59(22):1821-8.
13. McGinty DJ, Sterman MB. Sleep suppression after basal forebrain lesions in the cat. Science 1968; 160(833):1253-5.
14. Ali M, Jha SK, Kaur S et al. Role of GABA-A receptor in the preoptic area in the regulation of sleep-wakefulness and rapid eye movement sleep. Neurosci Res 1999; 33(3):245-50.
15. Hales TG, Lambert JJ. The actions of propofol on inhibitory amino acid receptors of bovine adrenomedullary chromaffin cells and rodent central neurones. Br J Pharmacol 1991; 104(3):619-28.
16. Shyr MH, Tsai TH, Yang CH et al. Propofol anesthesia increases dopamine and serotonin activities at the somatosensory cortex in rats: A microdialysis study. Anesth Analg 1997; 84(6):1344-8.
17. Violet JM, Downie DL, Nakisa RC et al. Differential sensitivities of mammalian neuronal and muscle nicotinic acetylcholine receptors to general anesthetics. Anesthesiology 1997; 86(4):866-74.
18. Flood P, Ramirez-Latorre J, Role L. Alpha 4 beta 2 neuronal nicotinic acetylcholine receptors in the central nervous system are inhibited by isoflurane and propofol, but alpha 7-type nicotinic acetylcholine receptors are unaffected. Anesthesiology 1997; 86(4):859-65.
19 Tung A, Lynch JP, Mendelson WB. Prolonged sedation with propofol in the rat does not result in sleep deprivation. Anesth Analg 2001; 92(5):1232-6.
20. Keifer JC, Baghdoyan HA, Becker L et al. Halothane decreases pontine acetylcholine release and increases EEG spindles. Neuroreport 1994; 5(5):577-80.
21. Mortazavi S, Thompson J, Baghdoyan HA et al. Fentanyl and morphine, but not remifentanil, inhibit acetylcholine release in pontine regions modulating arousal. Anesthesiology 1999; 90(4):1070-7.
22. Meuret P, Backman SB, Bonhomme V et al. Physostigmine reverses propofol-induced unconsciousness and attenuation of the auditory steady state response and bispectral index in human volunteers. Anesthesiology 2000; 93(3):708-17.
23. Keifer JC, Baghdoyan HA, Lydic R. Sleep disruption and increased apneas after pontine microinjection of morphine. Anesthesiology 1992; 77(5):973-82.
24. Lydic R, Keifer JC, Baghdoyan HA et al. Microdialysis of the pontine reticular formation reveals inhibition of acetylcholine release by morphine. Anesthesiology 1993; 79(5):1003-12.
25. Capece ML, Baghdoyan HA, Lydic R. Opioids activate G proteins in REM sleep-related brain stem nuclei of rat. Neuroreport 1998; 9(13):3025-8.
26. Kshatri AM, Baghdoyan HA, Lydic R. Cholinomimetics, but not morphine, increase antinociceptive behavior from pontine reticular regions regulating rapid-eye-movement sleep. Sleep 1998; 21(7):677-85.
27. Sherin JE, Shiromani PJ, McCarley RW et al. Activation of ventrolateral preoptic neurons during sleep. Science 1996; 271(5246):216-9.
28. Chou TC, Bjorkum AA, Gaus SE et al. Afferents to the ventrolateral preoptic nucleus. J Neurosci 2002; 22(3):977-90.
29. Gallopin T, Fort P, Eggermann E et al. Identification of sleep-promoting neurons in vitro. Nature 2000; 404(6781):992-5.
30. Lu J, Bjorkum AA, Xu M et al. Selective activation of the extended ventrolateral preoptic nucleus during rapid eye movement sleep. J Neurosci 2002; 22(11):4568-76.
31. Gaus SE, Strecker RE, Tate BA et al. Ventrolateral preoptic nucleus contains sleep-active, galaninergic neurons in multiple mammalian species. Neuroscience 2002; 115(1):285-94.
32. Sherin JE, Elmquist JK, Torrealba F et al. Innervation of histaminergic tuberomammillary neurons by GABAergic and galaninergic neurons in the ventrolateral preoptic nucleus of the rat. J Neurosci 1998; 18(12):4705-21.
33. Lin JS, Sakai K, Jouvet M. Evidence for histaminergic arousal mechanisms in the hypothalamus of cat. Neuropharmacology 1988; 27(2):111-22.
34. Steininger TL, Alam MN, Gong H et al. Sleep-waking discharge of neurons in the posterior lateral hypothalamus of the albino rat. Brain Res 1999; 840(1-2):138-47.
35. Yang QZ, Hatton GI. Electrophysiology of excitatory and inhibitory afferents to rat histaminergic tuberomammillary nucleus neurons from hypothalamic and forebrain sites. Brain Res 1997; 773(1-2):162-72.
36. Lu J, Greco MA, Shiromani P et al. Effect of lesions of the ventrolateral preoptic nucleus on NREM and REM sleep. J Neurosci 2000; 20(10):3830-42.
37. Nelson LE, Guo TZ, Lu J et al. The sedative component of anesthesia is mediated by $GABA_A$ receptors in an endogenous sleep pathway. Nat Neurosci 2002; 5(10):979-984.
38. Bremer F. Cerebral hypnogenic centers. Ann Neurol 1977; 2(1):1-6.
39. Saper CB, Chou TC, Scammell TE. The sleep switch: Hypothalamic control of sleep and wakefulness. Trends Neurosci 2001; 24(12):726-31.
40. Chemelli RM, Willie JT, Sinton CM et al. Narcolepsy in orexin knockout mice: Molecular genetics of sleep regulation. Cell 1999; 98(4):437-51.
41. Marcus JN, Aschkenasi CJ, Lee CE et al. Differential expression of orexin receptors 1 and 2 in the rat brain. J Comp Neurol 2001; 435(1):6-25.
42. Meyer HH. Zur theorie de alkoholnarkose. I. Mitt. Welche eigenschaft der anasthetika bedingt ihre narkotische wirkung? Arch Exp Path Pharmakol 1899; 42:109.
43. Kaufman RD. Biophysical mechanisms of anesthetic action: Historical perspective and review of current concepts. Anesthesiology 1977; 46(1):49-62.
44. Roth SH. Membrane and cellular actions of anesthetic agents. Fed Proc 1980; 39(5):1595-9.
45. Eger EI2, Koblin DD. Unitary versus multiple mechanisms of anesthesia. Anesthesiology 1995; 83(6):1368.
46. Franks NP, Lieb WR. Do general anaesthetics act by competitive binding to specific receptors? Nature 1984; 310(5978):599-601.
47. Patel AJ, Honore E, Lesage F et al. Inhalational anesthetics activate two-poredomain background K+ channels. Nat Neurosci 1999; 2(5):422-6.
48. Franks NP, Lieb WR. Molecular and cellular mechanisms of general anaesthesia. Nature 1994; 367(6464):607-14.
49. Anis NA, Berry SC, Burton NR et al. The dissociative anaesthetics, ketamine and phencyclidine, selectively reduce excitation of central mammalian neurones by N-methyl-aspartate. Br J Pharmacol 1983; 79(2):565-75.
50. Jevtovic-Todorovic V, Todorovic SM, Mennerick S et al. Nitrous oxide (laughing gas) is an NMDA antagonist, neuroprotectant and neurotoxin. Nat Med 1998; 4(4):460-3.
51. Franks NP, Dickinson R, de Sousa SL et al. How does xenon produce anaesthesia? Nature 1998; 396(6709):324.
52. Krasowski MD, Harrison NL. General anaesthetic actions on ligand-gated ion channels. Cell Mol Life Sci 1999; 55(10):1278-303.
53. Thompson SA, Wafford K. Mechanism of action of general anaesthetics—new information from molecular pharmacology. Curr Opin Pharmacol 2001; 1(1):78-83.
54. Mihic SJ, Ye Q, Wick MJ et al. Sites of alcohol and volatile anaesthetic action on GABA(A) and glycine receptors. Nature 1997; 389(6649):385-9.
55. Adodra S, Hales TG. Potentiation, activation and blockade of $GABA_A$ receptors of clonal murine hypothalamic GT1-7 neurones by propofol. Br J Pharmacol 1995; 115(6):953-60.
56. Reynolds DS, Rosahl TW, Cirone J et al. Sedation and anesthesia mediated by distinct GABA(A) receptor isoforms. J Neurosci 2003; 23(24):8608-8617.
57. McKernan RM, Rosahl TW, Reynolds DS et al. Sedative but not anxiolytic properties of benzodiazepines are mediated by the GABA(A) receptor alpha1 subtype. Nat Neurosci 2000; 3(6):587-592.
58. Reynolds DS, McKernan RM, Dawson GR. Anxiolytic-like action of diazepam: Which GABA(A) receptor subtype is involved? Trends Pharmacol Sci 2001; 22(8):402-403.
59. Rudolph U, Crestani F, Benke D et al. Benzodiazepine actions mediated by specific gamma-aminobutyric acid(A) receptor subtypes. Nature 1999; 401(6755):796-800.
60. Low K, Crestani F, Keist R et al. Molecular and neuronal substrate for the selective attenuation of anxiety. Science 2000; 290(5489):131-4.
61. Sanger DJ, Morel E, Perrault G. Comparison of the pharmacological profiles of the hypnotic drugs, zaleplon and zolpidem. Eur J Pharmacol 1996; 313(1-2):35-42.

62. Ebert TJ, Hall JE, Barney JA et al. The effects of increasing plasma concentrations of dexmedetomidine in humans. Anesthesiology 2000; 93(2):382-94.
63. Lakhlani PP, MacMillan LB, Guo TZ et al. Substitution of a mutant alpha2a-adrenergic receptor via "hit and run" gene targeting reveals the role of this subtype in sedative, analgesic, and anesthetic-sparing responses in vivo. Proc Natl Acad Sci USA 1997; 94(18):9950-5.
64. MacDonald E, Scheinin M. Distribution and pharmacology of alpha 2-adrenoceptors in the central nervous system. J Physiol Pharmacol 1995; 46(3):241-58.
65. Aston-Jones G, Rajkowski J, Kubiak P et al. Locus coeruleus neurons in monkey are selectively activated by attended cues in a vigilance task. J Neurosci 1994; 14(7):4467-80.
66. Correa-Sales C, Rabin BC, Maze M. A hypnotic response to dexmedetomidine, an alpha 2 agonist, is mediated in the locus coeruleus in rats. Anesthesiology 1992; 76(6):948-52.
67. Correa-Sales C, Nacif-Coelho C, Reid K et al. Inhibition of adenylate cyclase in the locus coeruleus mediates the hypnotic response to an alpha 2 agonist in the rat. J Pharmacol Exp Ther 1992; 263(3):1046-9.
68. Williams JT, North RA. Catecholamine inhibition of calcium action potentials in rat locus coeruleus neurones. Neuroscience 1985; 14(1):103-9.
69. Williams JT, Henderson G, North RA. Characterization of alpha 2-adrenoceptors which increase potassium conductance in rat locus coeruleus neurones. Neuroscience 1985; 14(1):95-101.
70. Birnbaumer L, Abramowitz J, Brown AM. Receptor-effector coupling by G proteins. Biochim Biophys Acta 1990; 1031(2):163-224.
71. Nacif-Coelho C, Correa-Sales C, Chang LL et al. Perturbation of ion channel conductance alters the hypnotic response to the alpha 2-adrenergic agonist dexmedetomidine in the locus coeruleus of the rat. Anesthesiology 1994; 81(6):1527-34.
72. Guo TZ, Tinklenberg J, Oliker R et al. Central alpha 1-adrenoceptor stimulation functionally antagonizes the hypnotic response to dexmedetomidine, an alpha 2-adrenoceptor agonist. Anesthesiology 1991; 75(2):252-6.
73. Nelson LE, Lu J, Guo TZ et al. The \alpha\$_2$-adrenoceptor agonist dexmedetomidine converges on an endogenous sleep pathway to produce its hypnotic response. Anesthesiology 2003; 98(2):428-436.
74. Hunter JC, Fontana DJ, Hedley LR et al. Assessment of the role of alpha2-adrenoceptor subtypes in the antinociceptive, sedative and hypothermic action of dexmedetomidine in transgenic mice. Br J Pharmacol 1997; 122(7):1339-44.
75. Guo TZ, Davies MF, Kingery WS et al. Nitrous oxide produces antinociceptive response via alpha2B and/or alpha2C adrenoceptor subtypes in mice. Anesthesiology 1999; 90(2):470-6.
76. Aurell J, Elmqvist D. Sleep in the surgical intensive care unit: Continuous polygraphic recording of sleep in nine patients receiving postoperative care. Br Med J (Clin Res Ed) 1985; 290(6474):1029-32.
77. Cronin AJ, Keifer JC, Davies MF et al. Postoperative sleep disturbance: Influences of opioids and pain in humans. Sleep 2001; 24(1):39-44.
78 Krska J, MacLeod TN. Sleep quality and the use of benzodiazepine hypnotics in general practice. J Clin Pharm Ther 1995; 20(2):91-6.
79 Shiihara Y, Nogami T, Chigira M et al. Sleep-wake rhythm during stay in an intensive care unit: A week's long-term recording of skin potentials. Psychiatry Clin Neurosci 2001; 55(3):279-80.
80. Sveinsson IS. Postoperative psychosis after heart surgery. J Thorac Cardiovasc Surg 1975; 70(4):717-26.
81. Venn RM, Bradshaw CJ, Spencer R et al. Preliminary UK experience of dexmedetomidine, a novel agent for postoperative sedation in the intensive care unit. Anaesthesia 1999; 54(12):1136-42.
82. Jones MEP, Coull JT, Egan TD et al. The effects of dexmedetomidine and midazolam on functional brain activity during rest and tasking. Anesthesiol 2003; in press.
83. Chen HI, Tang YR. Sleep loss impairs inspiratory muscle endurance. Am Rev Respir Dis 1989; 140(4):907-9.
84. White DP, Douglas NJ, Pickett CK et al. Sleep deprivation and the control of ventilation. Am Rev Respir Dis 1983; 128(6):984-6.
85. Kornfeld DS. Psychiatric view of the intensive care unit. Br Med J 1969; 1(636):108-10.
86. Brown R, Pang G, Husband AJ et al. Suppression of immunity to influenza virus infection in the respiratory tract following sleep disturbance. Reg Immunol 1989; 2(5):321-5.
87. Everson CA, Bergmann BM, Rechtschaffen A. Sleep deprivation in the rat: III. Total sleep deprivation. Sleep 1989; 12(1):13-21.
88. Everson CA. Sustained sleep deprivation impairs host defense. Am J Physiol 1993; 265(5Pt 2):1148-54.
89. Everson CA, Toth LA. Systemic bacterial invasion induced by sleep deprivation. Am J Physiol Regul Integr Comp Physiol 2000; 278(4):905-16.
90. Rechtschaffen A, Bergmann BM, Gilliland MA et al. Effects of method, duration, and sleep stage on rebounds from sleep deprivation in the rat. Sleep 1999; 22(1):11-31.
91. Krachman SL, D'Alonzo GE, Criner GJ. Sleep in the intensive care unit. Chest 1995; 107(6):1713-20.
92. Beller JP, Pottecher T, Lugnier A et al. Prolonged sedation with propofol in ICU patients: Recovery and blood concentration changes during periodic interruptions in infusion. Br J Anaesth 1988; 61(5):583-8.
93. Ortells MO, Lunt GG. Evolutionary history of the ligand-gated ion-channel superfamily of receptors. Trends Neurosci 1995; 18(3):121-127.
94. Bonnert TP, McKernan RM, Farrar S et al. Theta, a novel gamma-aminobutyric acid type A receptor subunit. Proc Natl Acad Sci USA 1999; 96(17):9891-9896.
95. Mehta AK, Ticku MK. An update on $GABA_A$ receptors. Brain Res Brain Res Rev 1999; 29(2-3):196-217.
96. Barnard EA, Skolnick P, Olsen RW et al. International Union of Pharmacology. XV. Subtypes of gamma-aminobutyric acidA receptors: Classification on the basis of subunit structure and receptor function. Pharmacol Rev 1998; 50(2):291-313.

Section II: Chronopharmacology

CHAPTER 7

Time-Dependent Psychotropic Drug Effects:

Hints of Pharmacochronomics, Broader than Circadian Time Structures

Haruo Nagayama, Germaine Cornélissen, S.R. Pandi-Perumal and Franz Halberg

Abstract

The importance of timing medications is noted in the context of the effects of psychotropic drugs. The information here assembled as yet is examined mostly by inspection with the unaided eye and conventional (rather than time series-related) statistics. An effect of time, shown by an analysis of variance, however, awaits an inferential statistical estimation of the cycles' parameters and of their uncertainties. In summarizing drug effects, only peak times may be tabulated—time-macroscopically (*tma*)—as clock-hours and times in relation to the synchronizing 12-hourly alternation of light and darkness, a proxy for a marker rhythm. A large body of such carefully collected information here included, however, awaits further time-microscopic (*tmi*) computer-implemented time series analyses that rely on all available data. Among many other procedures, curve-fitting assesses the uncertainties involved in detecting a reproducible rhythm and/or provides interval as well as point estimates of parameters, such as amplitudes, A, and acrophases, ϕ, when a single component is fitted. The magnitude and orthophase are the predictable extent of change within a cycle and peak of the fitted model when two or more components are considered. The period involved should be estimated as soon as the length of the time series permits. The A and ϕ values here computed from mean values taken off a graph should be only an incentive to *tmi* analyze the original data, so that charts can be mapped that are based on all of the data, rather than depending on the vagaries of peak locations.

Putative mechanisms underlying variations, if not rhythms, in drug efficacy are noted. Some *tmi* considerations are added as concepts and tools for further work that takes more than synchronization with the lighting regimen and/or an obvious living routine into account. Chronomes—time structures, consisting of deterministic and other chaos, trends and a broader-than-circadian spectral element—are pertinent to pharmacology. Chronomes in us resonate with chronomes in our environment, far beyond the photic day and calendar year. Whether the transyears, e.g., of ~1.3 and/or 1.6 years, among other biological (evolutionary?) near-matches of non-photic environmental cycles are also pertinent to the time-structure-based pharmacochronomics of psychotropic drugs prompts the suggestion to collect data replicated along the scales of months and years as well as along that of a day. Estimations of the characteristics of circadian, circaseptan or circannual rhythms based on just one cycle in a day, a week or a year are comparable to taking the pulse for only a heartbeat, i.e., one second! Treatment timed by marker rhythm, rather than by clock-hour, can not only save the amount of needed drug and reduce side effects; it also aims at optimizing the conventional desired effect and to pursue the goal of detecting new effects by focusing on elevated disease risk and the timely development of countermeasures. In the field of drugs affecting sleep and broader brain function, the importance of assessing time structure remains a worthwhile challenge, based on evidence that constitutes a complementary system to the partial system of timing psychotropic drugs.

A Broad Rhythm Spectrum

Most physiologic functions have a rhythm with one cycle in approximately 24 hours,[1] among many other rhythms with frequencies covering over 10 orders of magnitude.[2,3] There are also trends with age or disease risk elevation and changes resolvable as probabilistic or other chaos.[2-17] The endpoints of rhythms, trends and chaos constitute chronomes (time structures).[8] Mechanisms underlying chronomes are the topic of chronobiology, the study of diversity in time, complementing genetics, the study of diversity in space. Genetics led after 1950 to chronobiology.[1,18] and eventually to genomics, the mapping of genomes. Chronobiology led to chronomics, the mapping of chronomes in us that resonate with, or are synchronized by, chronomes around us. Chronomics are concerned with mapping the characteristics and uncertainties of changes with time of variables in us and in our environment near and far. Like many other phenomena, drug efficacy varies markedly as a function of chronomes, among others along the scales of 24 hours (e.g., refs. 19-31) and the seasons.[22,31]

To be investigated are the roles played by rhythms of non-photic origin, such as the week,[31] about half-weekly,[32-35] about 8-hourly,[32,35-37] about-half-yearly[38,39] about 1.3 to about 1.6-yearly (transyearly)[39-41] periods, as well as circadecadals, circadidecadals and circaquindecadals[2,3,8,9] that may contribute to unwarranted "substitution" therapy, implemented in ignorance of circadecadal cycles.[42] Infradian rhythms in still broader chronomes remain a challenge to endeavors in diagnosis or treatment, including the use of psychotropic drugs, such as

Sleep and Sleep Disorders: A Neuropsychopharmacological Approach, edited by Malcolm Lader, Daniel P. Cardinali and S.R. Pandi-Perumal. ©2006 Landes Bioscience/Eurekah.com and Springer Science+Business Media.

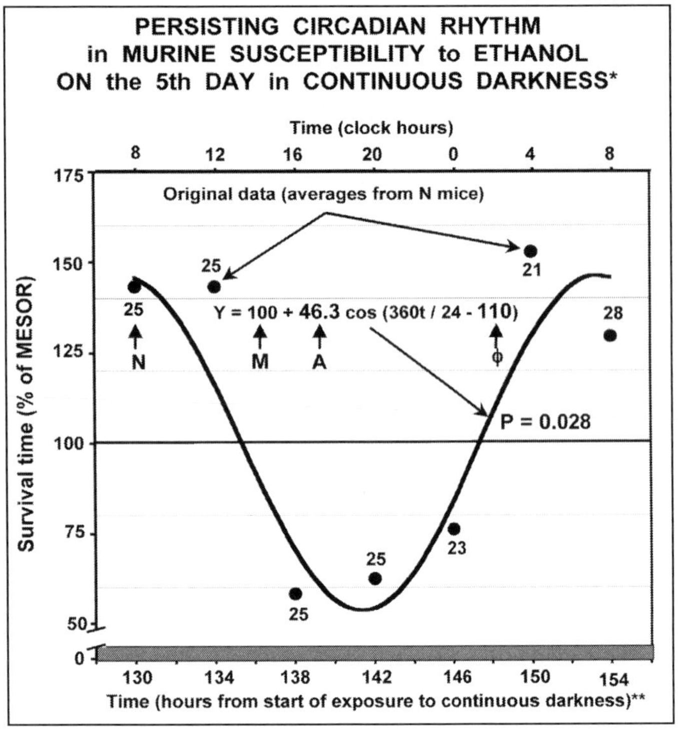

Figure 1. The statistical significance of an anticipated (!) circadian rhythmicity[1,18,20] can be validated by the rejection of the zero-amplitude (no circadian rhythm) assumption with the cosinor method. The procedure of standardization of the experimental animal laboratory by rendering the lighting regimen as constant as possible (e.g., by instituting continuous darkness, as in the case of the study presented herein, or by continuous light, and by providing a controlled temperature, reducing noise and other obvious stimuli, notably cyclic ones) has been used for half a century to demonstrate an independence from the lighting regimen, no more. The endogenicity of circadians is supported in many cases by the results of molecular biology. Results in continuous darkness, however, do not suffice to prove it, although they support this hypothesis. © Halberg.

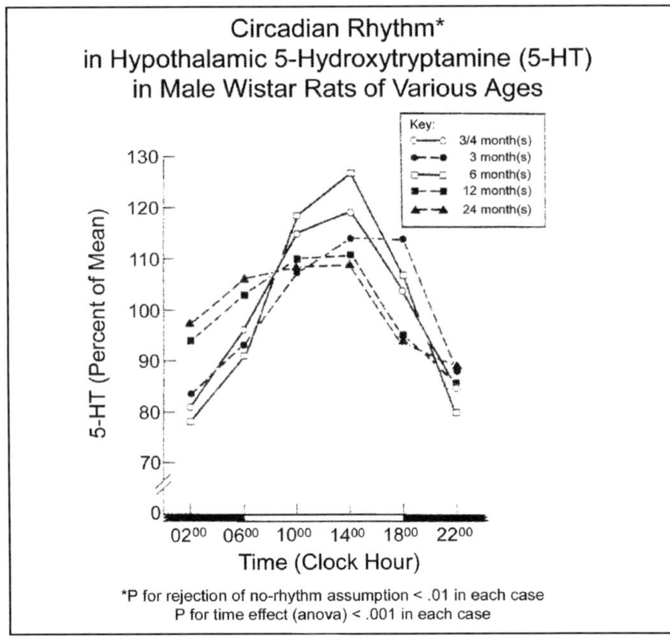

Figure 2A. *Tma* time course along the 24-hour scale of murine hypothalamic 5-hydroxytryptamine (5-HT); expressed as % of series mean. *Tmi* quantification required to estimate any age-dependent amplitudes shown elsewhere.[53] Original data of Bhaskaran and Radha.[173] © Halberg.

antipsychotics, antidepressants, mood stabilizers, benzodiazepines, barbiturates, psychostimulants, and a variety of neurotransmitter agonists,[19-30] in view of infradian cycles documented by Derer, Gjessing, Reimann and Richter, among others.[1-3,43-50]

The variations in drug effects may not be due exclusively and probably not primarily to the stage of rhythms documented for the pharmacokinetics of many drugs,[51] but may stem substantially from a partly built-in, lighting- and temperature-regimen-independent rhythm in susceptibility-resistance, documented for ethanol effects in alternating light and darkness[52] and shown by 1960, Figure 1,[20] to persist in the absence of this lighting regimen. The task of aligning this effect upon the brain and many others with circadian rhythms in the intracerebral neurotransmission system, shown in the *tma* Figures 2A and B and the *tmi* Figures 2C-E[53] is a halting first step toward focus upon mechanisms.

counts of certain circulating blood cells using (as controls) a 24-hour synchronized group of mice were compared with counts from another group on the same lighting regimen;[54] results from comparisons of the same two groups at different clock-hours led to different, even opposite results, since one of the groups was phase-shifted by a diet restricted in calories offered in the morning, usually a time of rest for nocturnally feeding mice, as realized before publication (details in refs. 1 and 55). The same results cannot be expected when comparing the same two phase-shifted groups at different times of day.

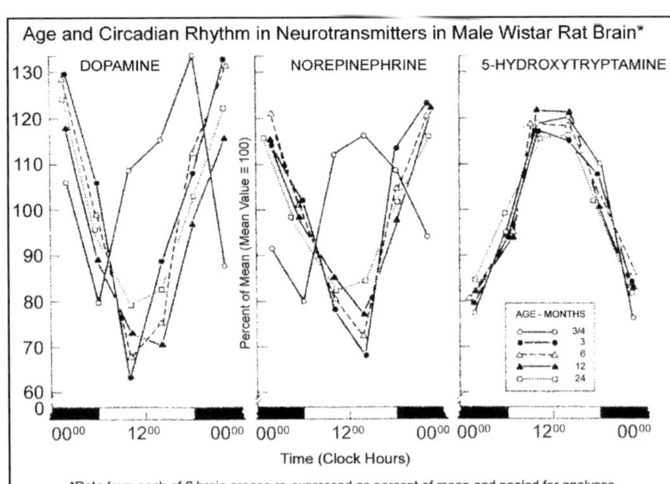

Figure 2B. *Tma* approach reveals age-dependent differences in the circadian timing of dopamine and norepinephrine but not of 5-hydroxytryptamine (5-HT) in data pooled from 6 brain areas; original data from each area expressed as % of series mean and pooled to construct the curves here shown, analyzed elsewhere[53] and in Figures 2C-E. Original data of Bhaskaran and Radha.[173] © Halberg.

Puzzles

Among biologists in 1949, fixing the time of day was deemed a sufficient precaution to "eliminate the effect of rhythms" and (for far too many, as now) repeated sampling at different clock-hours was regarded as sufficient for an approach to assess the effect of drugs on what became circadian rhythms.[1] In 1949,

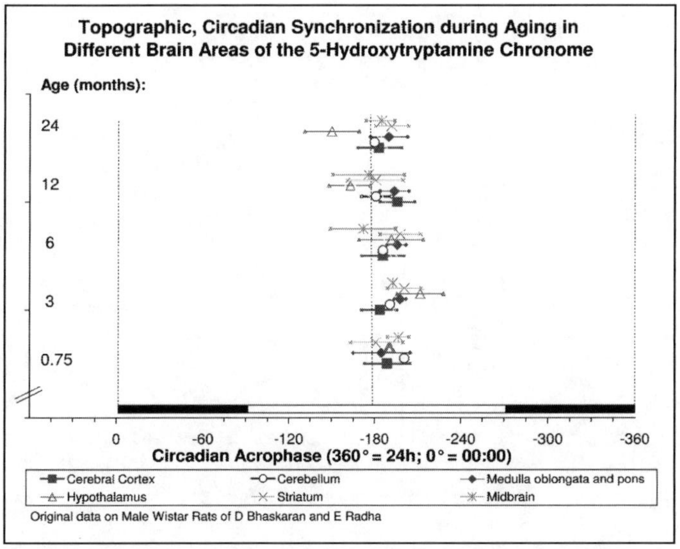

Figure 2C. The *tmi* approach allows, among others, the specification of times of peak with their uncertainty shown as 95% confidence intervals for 5-hydroxytryptamine (5-HT) complementing the tma approach, which refers solely to a peak in a portion of the daily light and/or dark span. Since the *tma* approach can be subjective, notably when there are multiple peaks of different shapes, it is complemented or replaced by objective hypothesis testing for the occurrence of a rhythm and parameter estimations.[61,62] Rhythm characteristics can then be compared by parameter tests in inferential statistical terms.[62] Original data of Bhaskaran and Radha.[173] © Halberg.

Solving this first puzzle led to the study of blinded mice. As anticipated, conflicting results from a comparison of a 24-hour synchronized group with one that was phase-drifting after blinding [56] constituted a new puzzle: now at the same times of day, on different calendar dates, i.e., on different days after operation, different results were obtained. The group of mice without eyes was desynchronized with a period shorter than 24 hours, it was drifting out of phase and back into phase with the 24-hour synchronized group showing and again not showing differences in group comparisons. This can be expected whenever one compares a 24-hour synchronized group with one that phase-drifts with another slightly different frequency (the solution of the second puzzle[1,56]). The adrenal and then pituitary and hypothalamic as well as cellular mechanisms and eventually galactic and/or helio-geomagnetic[2,3] contributions underlying rhythmic behavior had to be explored and ascertained before any data could be interpreted.[1]

Pursuit of the answers to those puzzles, albeit resolved as phase-shifts and -drifts, was perhaps regarded as experimental overkill and actually labeled "paranoia" half a century ago,[1] their obvious immediate pertinence to any problem in biomedicine using rhythmic controls notwithstanding. At one clock-hour eosinopenia, at another clock-hour eosinophilia, was found in ovariectomized calorie-restricted vs. ad-libitum-feeding mice, i.e., for groups with drastically different incidences of breast cancer. The interpretation was and remains exciting that an adrenocortical activation, gauged by eosinopenia, inhibited and in another comparison of the same two groups enhanced mammary carcinogenesis. At another clock-hour, however, in a comparison of the same two groups, the animals with the lowered cancer incidence had eosinophilia. (This happened in the era of adrenalectomies on women with breast cancer!) The degree of generality of such

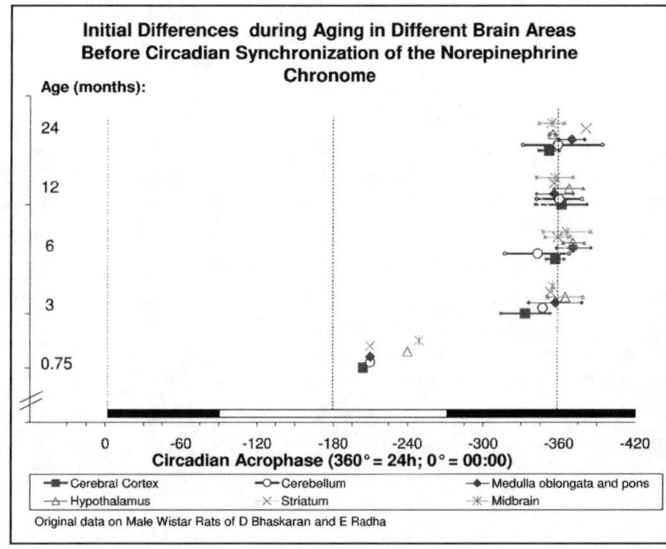

Figure 2D. *Tmi* quantification of age difference in the timing of a circadian rhythm in cerebral norepinephrine. Original data of Bhaskaran and Radha.[173] © Halberg.

confusing results due to a phase shift in one of two rhythms being compared conveyed the fact that an assessment of rhythms (not only of circadians but also of rhythmic changes with any other frequency) while an important aspect of a built-in biological time measurement (pursued by others and certainly by the original investigator himself [FH][1]), are also very much more. It became essential to replace imaginary baselines with chronomes, time structures that constitute the control of everyday photically and non-photically influenced physiology (by contrast to conventional physiology that draws a curtain of ignorance, i.e., of time-unqualified "normal" ranges, over everyday rhythmic function). Eventually, everyday pharmacology came about[7,10-12,19-30,57-60] and even everyday cosmology, for which living matter (from prokaryotes to humans) constitutes specialized sensitive radiation detectors.[2,3]

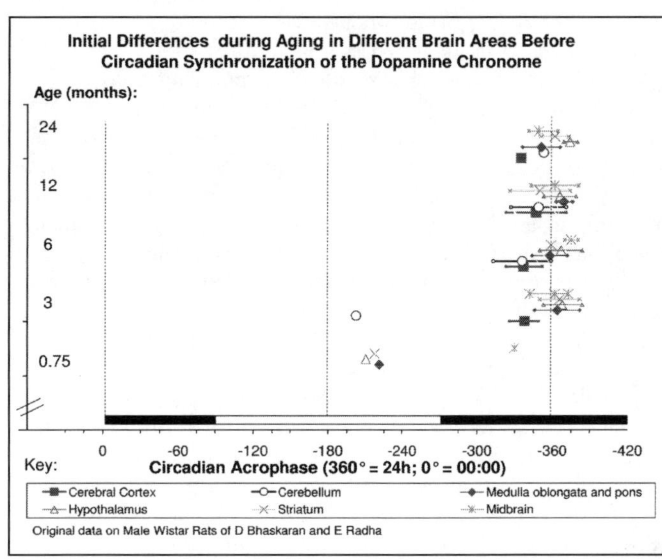

Figure 2E. *Tmi* quantification of dopamine timing as a function of age and brain area. Original data of Bhaskaran and Radha.[173] © Halberg.

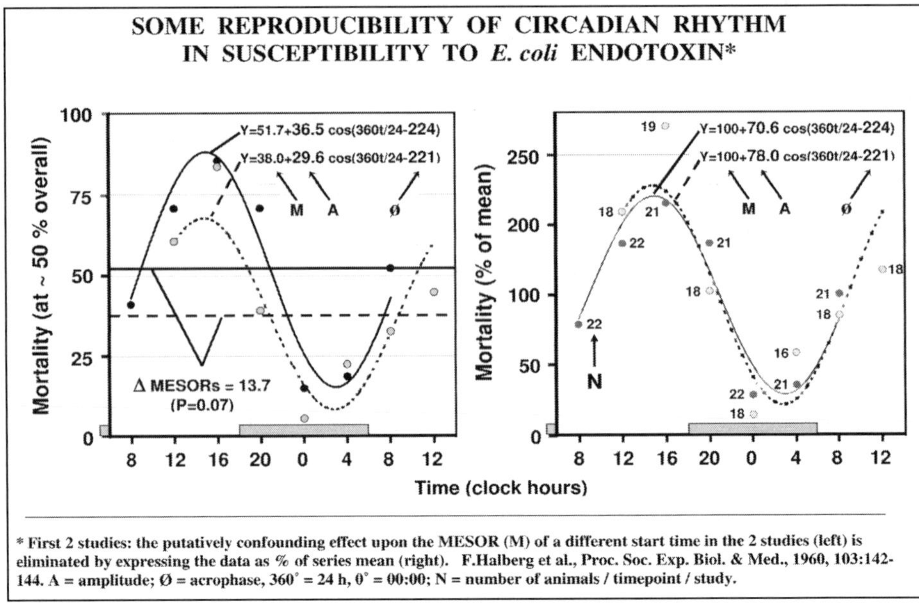

Figure 3. The response to endotoxin along the scale of a 12-hourly alternation of light and darkness in two experiments is surprisingly close but is to be qualified, e.g., by putative effects of only a 4-hour difference in the starting time. The importance of the starting time was found earlier for certain blood cell counts and hormones. In any event, note differences in each study between over 80% vs. a very low mortality from the same dose of endotoxin in groups of mice of similar genetic background, sex and age (left) or changes of more than 100% above the series mean to more than 50% below that mean (right). Parameter tests suggest a possible difference in MESOR (possibly associated with a different starting time, among many other possibilities). Note that the tests do not detect a difference in A or ϕ.[67,68] © Halberg.

Hours of Changing Susceptibility

In several cases still in the 1950s, the rhythms themselves (used as "controls" of what happens "spontaneously" as a function of time under the seemingly standardized conditions of a laboratory) contributed the difference between life and death - whether the stimulus was physical, like noise[63,64] or whole-body irradiation;[65,66] a bacterial endotoxin,[67,68] Figure 3, or a drug, such as ouabain,[69,70] an adrenocortical inhibitor, Figure 4,[71] or ACTH,[72] Figure 5.[15,30,72] The time of maximal susceptibility could be phase-shifted to any desired clock-hour, as shown in Figure 6 for the case of susceptibility to noise,[64,73,74] yet the rhythm persisted in continuous darkness, as shown in Figure 1 for the case of susceptibility to ethanol.[20]

As a minimum, it becomes essential to specify, among others, the times of any intervention by drugs or otherwise, preferably in relation to a marker rhythm, such as that of core temperature. In a study of a protective effect against endotoxin by chlorpromazine, on mice kept on a specified lighting regimen, the times of interventions, such as surgery, and the times of challenge were thus standardized.[225] Temporal specification by whole body marker rhythms was suggested as a laboratory and clinical routine by 1960,[1] including a study of circulating 5-hydroxytryptamine in humans,[226] strengthened "by the periodicity approach which rules out possible differences brought about by sampling in different stages of a circadian rhythm" (ref. 226; cf. ref. 227). Marker rhythm use differed from the earlier recognition of important differences in outcome as a function of time of day without further specification.[120,228-237] Eventually, the telemetry of core temperature and motor activity measured at 10-minute intervals on rats kept on 8 different staggered lighting regimens, were recommended, and hundreds of animals used in any one study could thus each be studied to obtain new information.[206] The individualized automatically monitored longitudinal study, eventually by telemetry, remains the answer for clinical trials. By such monitoring, the sample size for the study of a desired drug effect[238,239] can be drastically reduced.[240]

More and more variables sampled with sufficient density for several 24-hour cycles, on the same individual and/or on different comparable groups of individuals under comparable conditions, happened to exhibit some individually reproducibly changing structures in time. These patterns could be altered or distorted

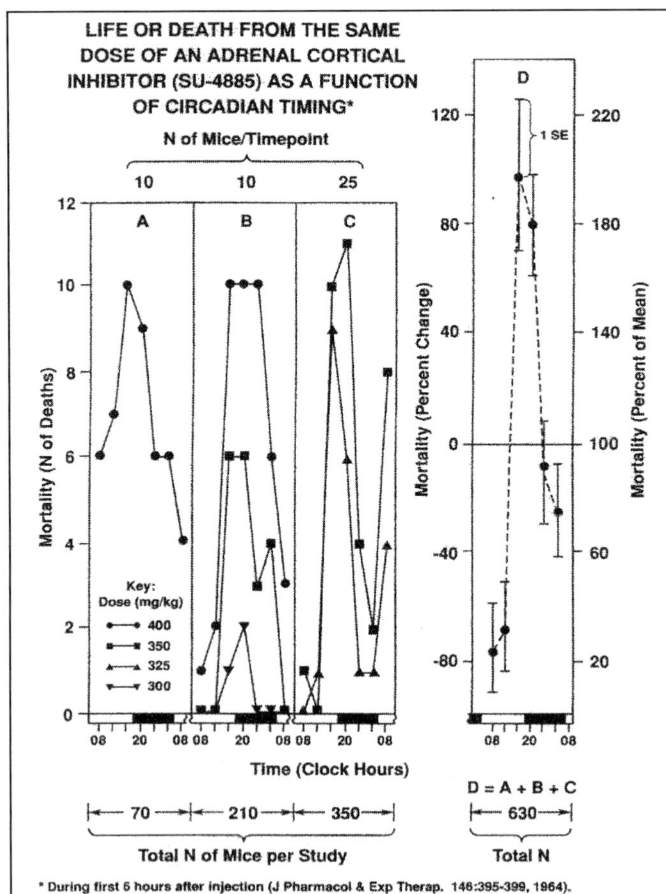

Figure 4A. A very drastic and highly reproducible *tma*-revealed dependence on the light-dark synchronized circadian stage characterizes mortality from an agent affecting (in lower doses the adrenal cortex and) in the several high but graded doses used, the brain to the point of inducing anesthesia of 80 to 200-minute duration in survival if not death.[71] *Tmi* or *tma* P < 0.01. © Halberg.

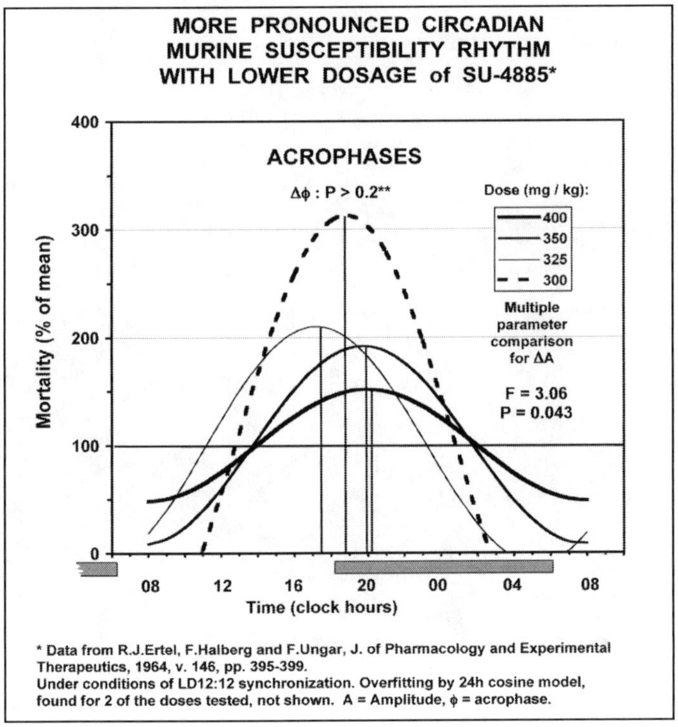

Figure 4B. Dose affects circadian amplitude of neuron susceptibility resistance cycle to SU-4885, assessed by single component model. © Halberg.

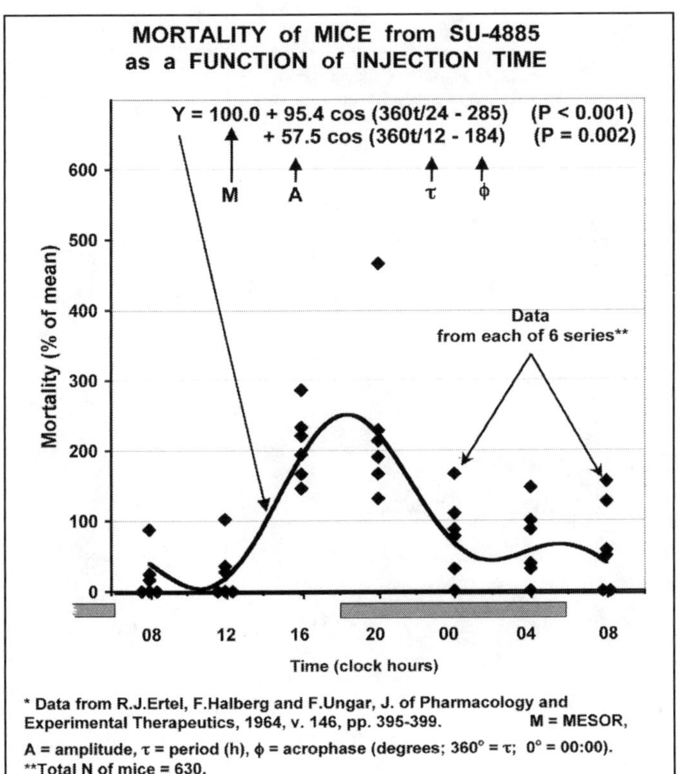

Figure 4C. Two-component model approximates, more closely than 24-h fit, the circadian susceptibility of mice to SU-4885. © Halberg.

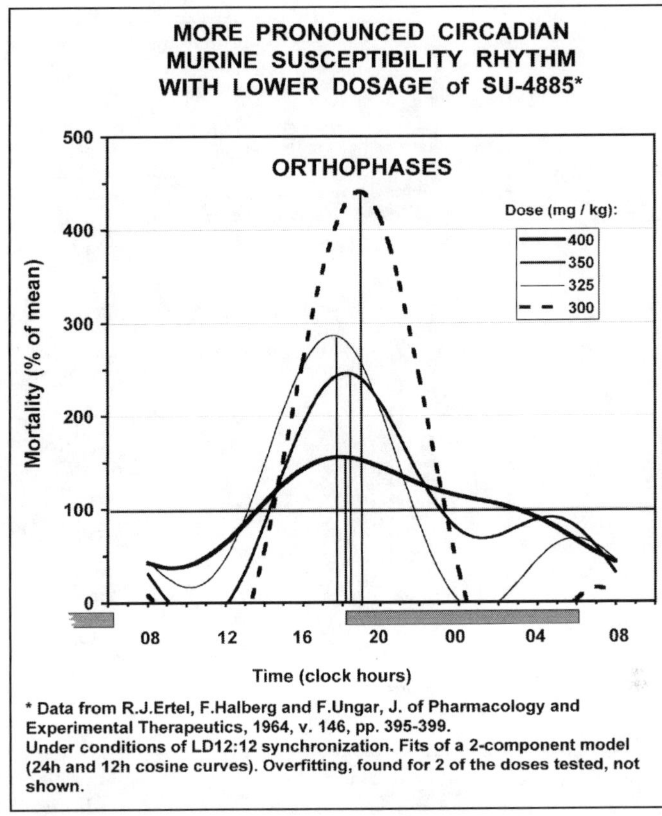

Figure 4D. Two-component model derived orthophases are tighter than acrophases (Fig. 4B) of murine susceptibility cycle to SU-4885. © Halberg.

by circumstances other than the lighting regimen, the availability of food and other obvious factors, e.g., by magnetic storms.[75,76] Figure 7A depicts changes in magnetic disturbance recorded globally during two studies with the same approach, 4 years apart. During a day of sampling in September 1980 and during the preceding 10 days, geomagnetics were relatively quiet. During 3 additional days of sampling and during the 2 preceding days in September 1984, there was increased geomagnetic disturbance. Figures 7B and 7C show intracardiac pressure during these days in September 1980 and 1984. All rhythm characteristics differ with statistical significance between the two studies. Figure 7D shows a near-antiphase between the profiles during relatively quiet times and stormy geomagnetic activity. Three other studies in 1984-1985, in the absence of magnetic storms, showed circadian acrophases similar to those observed in 1980. Similar studies on cerebral variables are overdue, as is pharmacological investigation during spans with different geomagnetic disturbance.

When sampled long enough, time structures can change as a function of developmental and/or other trends and/or can be buried in noise, i.e., in patterns that remain unresolved, unless continued sampling may perhaps reveal them as rhythms with a cycle longer than the length of the original data span examined. Eventually deterministic chaos also emerged in sufficiently dense data.[14-17] Chronobiology, a science of the interplay of make-ups in time around and in us, came about, and in turn spawned chronomics, a critical step beyond biological clocks and calendars, Figure 8.

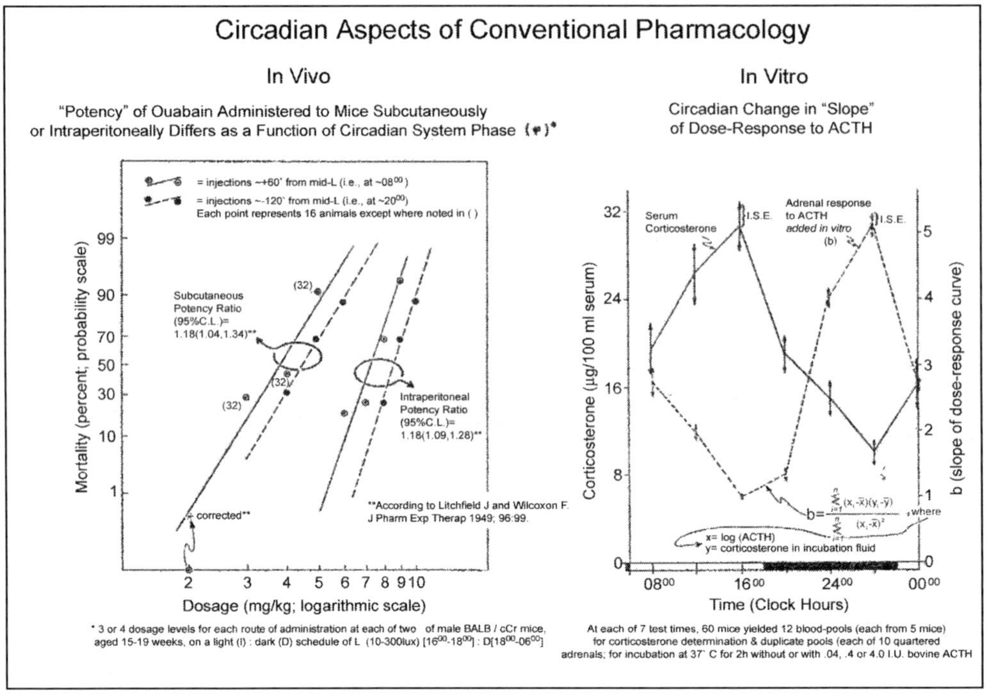

Figure 5. Conventional pharmacology usually presents a slope for dose responses without temporal qualification. The slope of a dose response relationship varies under different conditions including different host rhythm stages. Moreover, different results can be obtained for the dose-response, as in the case of mortality from ouabain in vivo on the left, vs. drastic change in the slope, found on the right for the corticosterone response to ACTH (the dotted line), in vitro, in the absence of interacting neural and humoral agents, other than those present in the quartered adrenal gland at the time of its removal for incubation.[72] © Halberg.

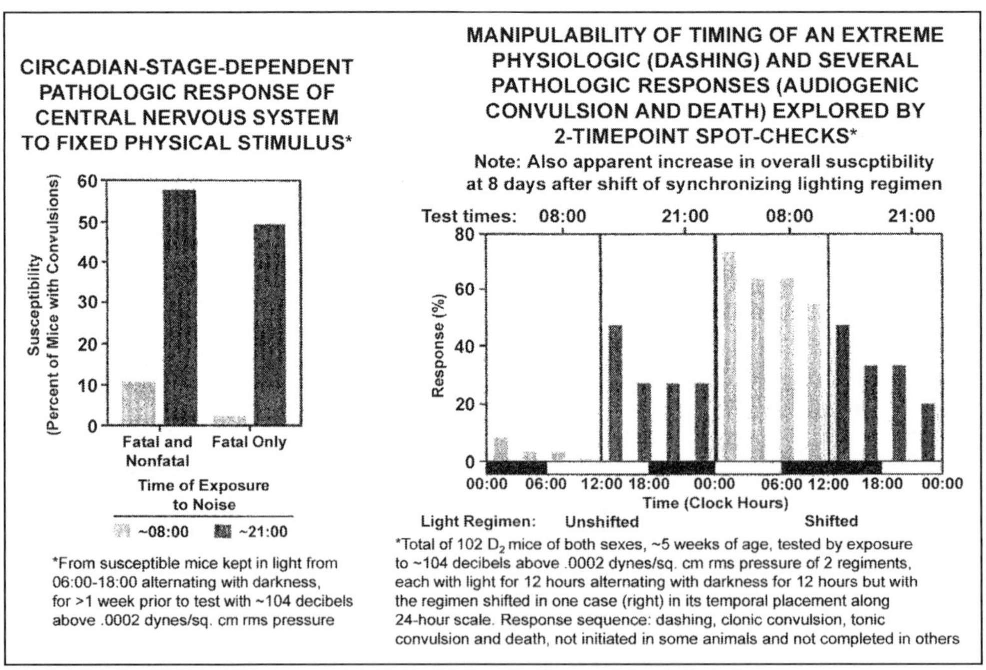

Figure 6. A two-timepoint approach is valid in the presence of prior information to demonstrate a circadian rhythm in the response of mice susceptible to audiogenic stimulation, as revealed by dashing, clonic or tonic convulsions and death.[64] © Halberg.

Ubiquity of Pertinence or Pretention?

By resolving characteristics such as periods and for each fundamental period, corresponding amplitudes, phases and waveforms, always with the uncertainties of these estimated parameters in ever longer biological and environmental time series, up to several decades,[2,3] a biospheric-environmental spectral reciprocity[3,39] and interdigitated feedsidewards[4-6] were uncovered among the rhythms of chronomes. These characteristics, in their turn, became dynamic reference standards of everyday (within the normal range) science, among others, of everyday toxicology and

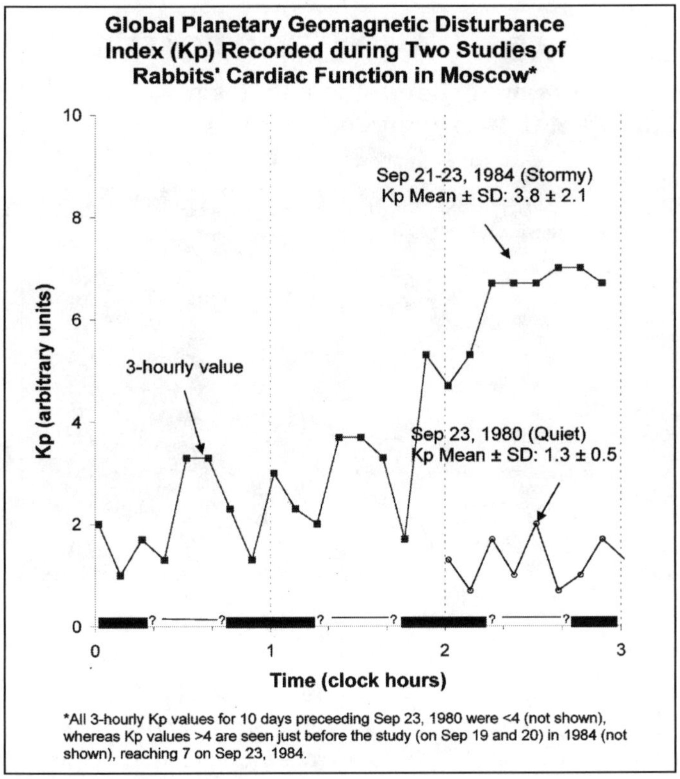

Figure 7A. Time course of geomagnetic disturbance during two spans bracketing cardiac pressure measurements in September 1980 and 1984. © Halberg.

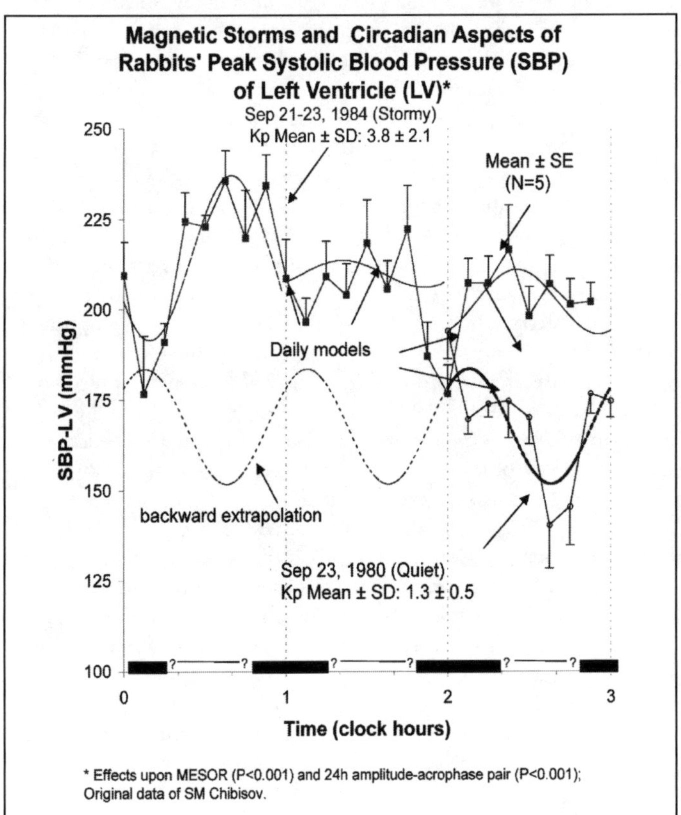

Figure 7B. Time course of pressure measurements in the left ventricle of rabbits during a geomagnetic storm vs. a quiet span, respectively. Original data of Sergei M. Chibisov and Tamara Breus.[75] © Halberg.

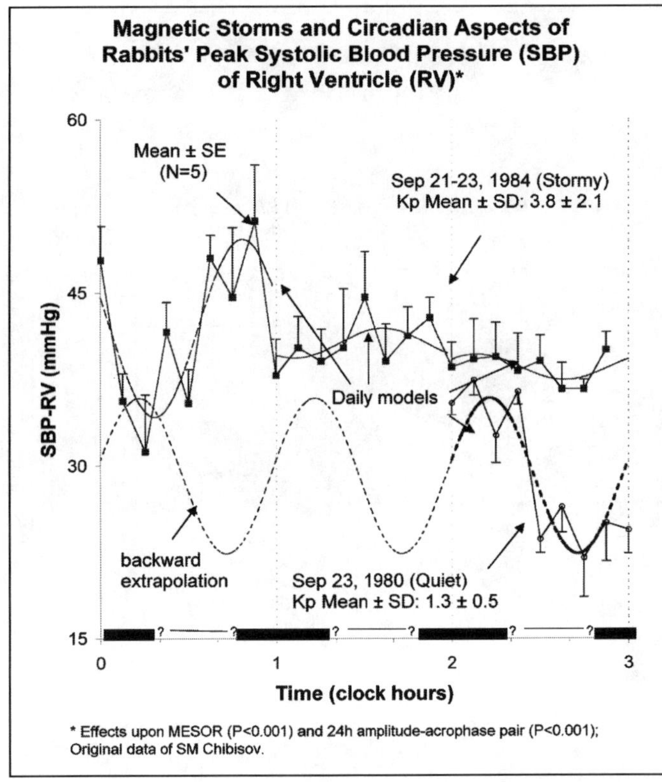

Figure 7C. Time course of pressure measurements in the right ventricle of rabbits during a geomagnetic storm vs. a quiet span, respectively. Original data of Sergei M. Chibisov and Tamara Breus.[75] © Halberg.

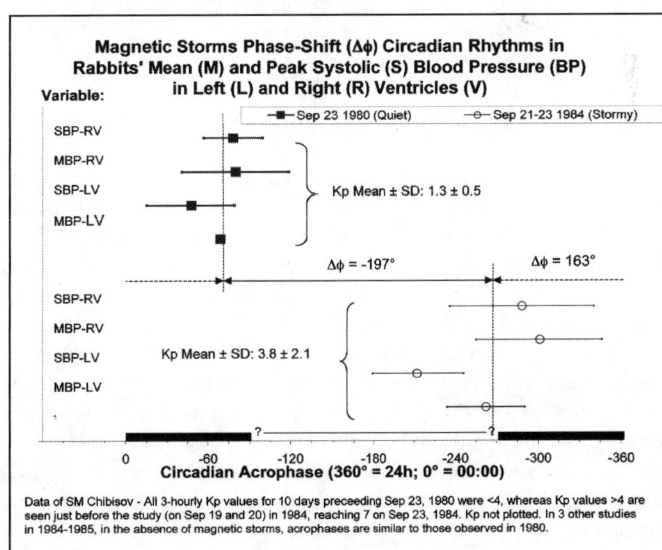

Figure 7D. Near-antiphase in timing of circadian rhythms of pressure in rabbit hearts during magnetically quiet and stormy time spans. Three relatively quiet spans in 1984 analyzed subsequently are in keeping with results shown herein. For differences in MESOR and circadian amplitude see Figures 7B and 7C. Original data of Sergei M. Chibisov and Tamara Breus.[75] © Halberg.

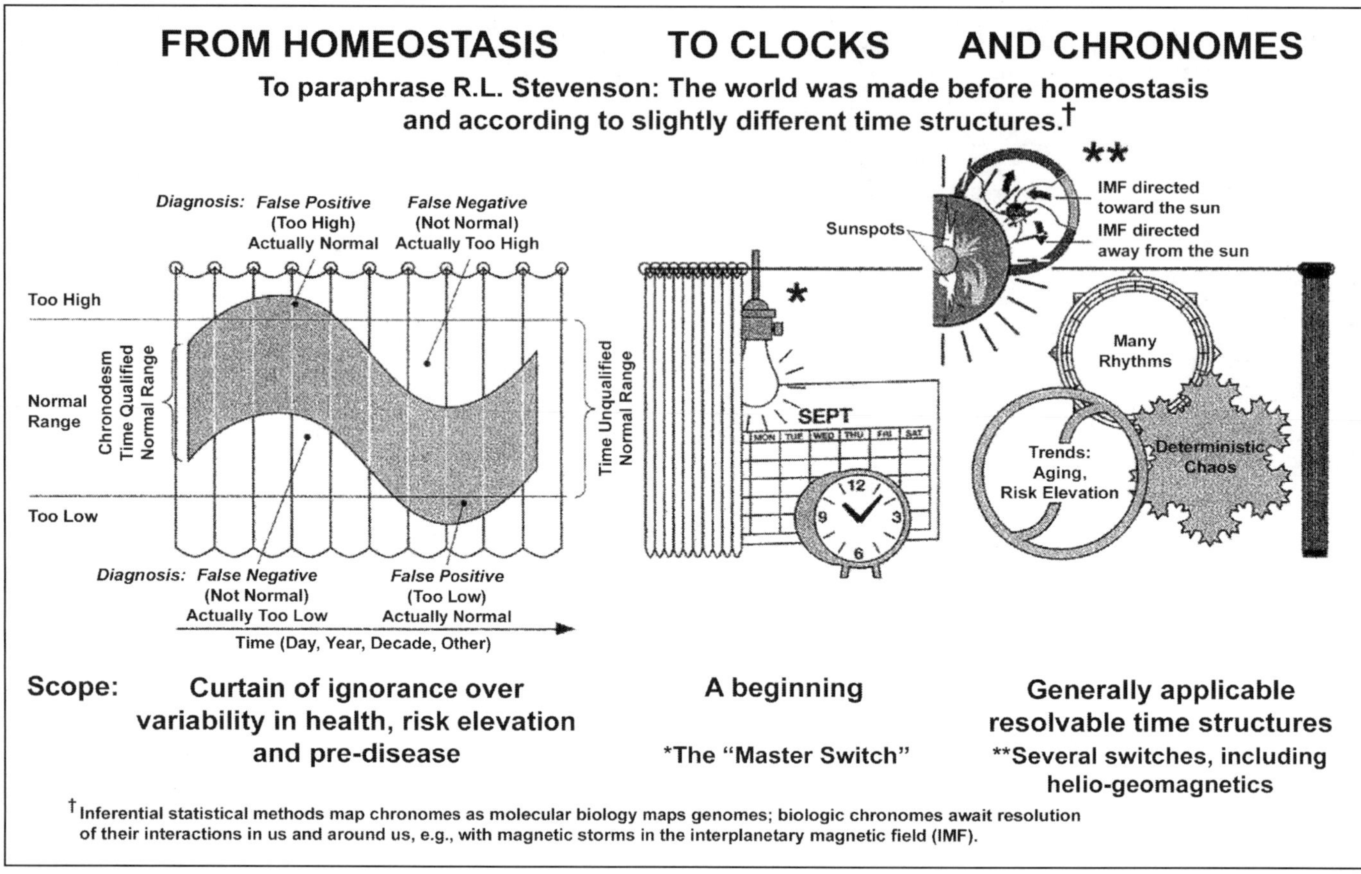

Figure 8. Concepts of pharmacologists about rhythms may turn from a master clock and calendar (middle) serving regulation for constancy (left) to an integrative internal-external collateral hierarchy for physiological coordination. While physiological processes remain largely within certain ranges in health, homeostasis seeks only departures from such "normal values" to diagnose overt disease. Variability within the normal range is then often dealt with as if it were narrow, random and trivial, the body striving for at least a relative "constancy"; differences such as those in the foregoing figures, notably in Figure 7, are then ignored. The alternative, not complement, to this status quo, on the left of this figure, is learning about the rules of rhythmic, chaotic and trendy variations, many of which take place within the "usual value" ranges. Rhythms led to the postulation of a "biological clock" that would enable the body to keep track of time. By removal and replacement experiments, a "clock" was located first in the adrenal, shown to be responsible for some but not for other circadians that persisted after brain ablation.[7] The fact that single cells and even bacteria are genetically coded for a spectrum of rhythmic variation indicates that the concept of a master "clock" needs extension within an organism and in the context of the environment. Beyond biological clocks (circadian) and calendars (circannual), we recognize, among others, biological near-matches of an environmental near-week, a month, a half-year, an about 1.3- (trans) year, and a decade, all non-photic cycles. When the giant alga Acetabularia, a prominent model of a "clock", is placed into continuous light, its spectrum of electrical activity reveals the largest amplitude for a component of about 1 week rather than of 1 day. When over a decade of studies on this alga are pooled to constitute a time series, an about 10-year cycle emerges in the data set as a whole. Long-term longitudinal, but not yet entire lifetime monitoring of critical variables complements current linked cross-sectional (hybrid) reference values required for preventive health and environmental care. Changes occurring within the usual value range as longer and still longer cycles are resolvable as chronomes (time structures), with a (predictable multifrequency) rhythmic element, that allows us to measure the dynamics of everyday life, in order to obtain, e.g., warnings before the fait accompli of disease, so that prophylactic measures can be instituted in a timely way, to detect heretofore largely undetected or unquantified environmental effects which are silent not only to the person involved at high risk but also to conventional health care and science. This is a challenge for individualized chronotherapy. On the top right, above the horizontal line, an abstract circle with three arrows in opposite directions oversimplifies the factually more complex sector structure of the interplanetary magnetic field. The fourth arrow is covered by another circle straddling the horizontal line, showing solar flares. The parameters of the solar wind are much more variable than originally visualized when first described, as sketched by irregular solar flares. The three circles completely below the horizontal line represent the three elements of a chronome. The right-hand section as a whole symbolizes associations of helio- and geomagnetic variability, with the biosphere, e.g., with myocardial infarctions and strokes that are accumulating and are just the tip of the iceberg. A profound effect of magnetism (recognized by Gilbert in 1600) is apparent in the human ECG, notably in humans living in the auroral region. In external-internal interactions, a broad spectrum of rhythms (in both the environment and in living matter) organizes deterministic and other chaos and trends. Trends pursued long enough may become low-frequency cycles, e.g., for the detection of any risk elevation and for the taking of timely action. © Halberg.

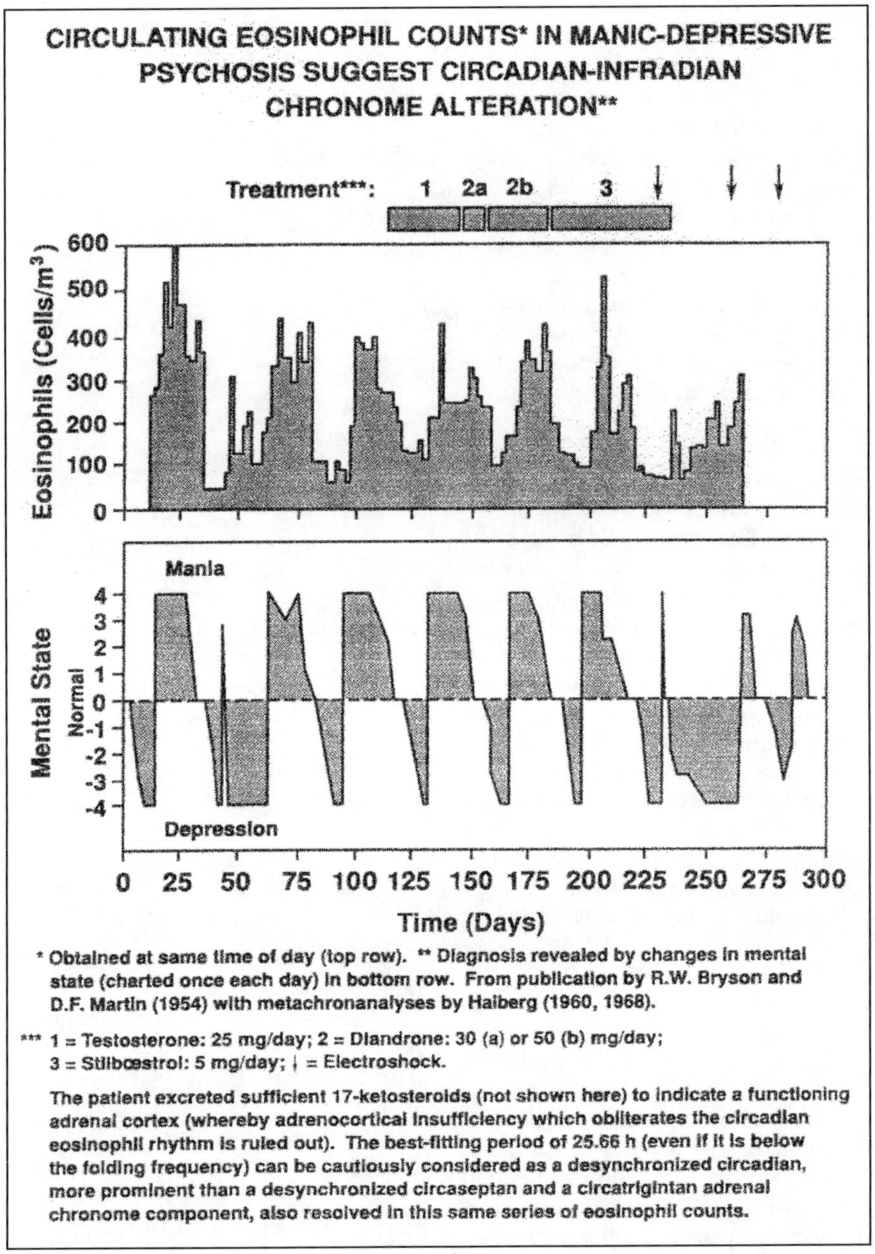

Figure 9. Counts obtained at the same time of day every second day are compatible with the desynchronization of the circadian rhythm in circulating blood eosinophil counts from the 24-hour routine, Table 1. These counts may gauge the adrenal cycle's behavior and may underlie a possible association with also-shown drastic changes in mood in a manic-depressive psychosis studied by R.W. Bryson and D.F. Martin.[20] © Halberg.

pharmacology. It was tempting to put the prefix "chrono-" in front of the names of these disciplines. Doing so served to distinguish from the silent majority those who took time structure into account, at least by a *tma* and preferably by a *tmi* approach, until a time in the still-distant future when the prefix "chrono" can be skipped, since the need for temporal (controls as) maps of schedules with their uncertainties (for chronomics) is then generally recognized by the majority, including timing as a function of specified marker rhythms on the package insert of drugs.

Semantics aside, attention to temporal reference standards turned out to be critical for the recognition of harbingers of a highly elevated disease risk in the usual "normal" range, and thus for the preventive treatment of earliest alterations: prehabilitation[77] to save on the cost of rehabilitation. To those who had not seen the data of Figures 1-6, i.e., had not seen firsthand that rhythms tipped the scale between life and death in the laboratory, chronobiology as a science at first appeared to be "unnecessary, presumptuous and inaccurate" – to cite a dean's letter almost certainly not meant for publication.[1] The issue became the disciplinary stake. Eventually, the afore-cited critic, for a publication summarizing his life's pursuits, requested, received and published a chronobiological *tmi* phase chart, Figure 9, which included susceptibility-resistance cycles to drug effects.[1] In this sense, the dean, a time-macroscopist *par excellence*, set an example for unity and complementarity among those who do time-microscopy and time-macroscopy, who all contribute to transdisciplinary science based on the assessment of rhythms and/or broader time structures. This unity of *tma* and *tmi* approaches, with *tmi*, of course, including *tma*, and eventually vice versa, prompted this historical introduction to the present review, in keeping with Shakespeare that the past (see Tables 3-7 and Figs. in ref. 30) is prologue.

Methodology

With respect to chronobiology and chronomics and thus for chronopharmacology, there is the need to incorporate, better sooner than later, what is now accepted for most published research; there is a need for inferential statistical hypothesis testing and, in any event, whenever possible for an interval as well as point estimation of characteristics in the context of methods for time series analyses.[7,61,62] In addition to computer programming and the underlying mathematics, chronomics builds upon the accumulating prior knowledge in the form of maps for documented rhythms that after further specification of species, gender, age and topographic region become invaluable reference values. For health care, in the presence of a multifactorial statistical causality underlying risk elevation as well as disease and great inter-individual differences, there is the added need for an individualized inferential statistical approach in an evidence-based medicine aiming to diagnose risk elevation and to treat it before (not only after) the fact of disease.[77] The current gold standard in medical research consists of controlled clinical trials that ignore inter-individual differences, assuming homogeneous groups. Sooner or later, sole reliance upon this approach will have to be replaced by a consideration of the person involved. There is no substitute for individualized self-experimentation with inferential statistical N-of-1 studies. Everybody can test whether to pass on the salt (if one responds to sodium intake with an increase in blood pressure) or use the salt shaker to lower one's blood pressure (there are also such persons, albeit rarely);[8] also to find out whether one's risk of severe vascular disease is low, perhaps around 4%, or very high, nearing 100%. In the latter case the risk should be lowered, and this lowering, e.g., by a drug should again be ascertained in inferential statistical terms before a severe disease occurs, so that, e.g, a massive stroke is prevented.[39,40,77]

In concerns for both a controlled transdisciplinary science and literacy in preventive health maintenance, chronobiologists can benefit from unity. They will accept the biological week sooner if they actually see, in a gliding spectrum, a towering week in a human baby's circulation, and a dominance of the circadian, e.g., in heart rate or blood pressure, only toward the end of the first month of life.[1,78] They will agree more rapidly on the importance of the biological week if they are interested in cancer prevention and find that sinusoidally increasing and then decreasing daily doses of an immunomodulator can actually inhibit the subsequently implanted malignant growth, whereas in equal daily doses the same total weekly amount enhances the immunocytoma's growth.[8,31] They will also find the built-in week in sleep research.[79-84] As instrumentation for sleep studies at home complements activities in sleep centers, infradian studies will become more readily possible. Basic scientists will note that time-microscopy resolves what the naked eye may miss, as, for instance, possibly the trans-year, a rhythm longer than the calendar year.[39-41] Everybody may wish for a MESOR rather than an arithmetic average, when the former, because of a smaller standard deviation, but not the latter, allows the validation of a therapeutic effect for the given, e.g., MESOR-hypertensive person involved.[85,86]

Circadian Desynchronization

Important for sleep, as well as psychiatric research, is the now generally accepted fact of an (external) desynchronization (from

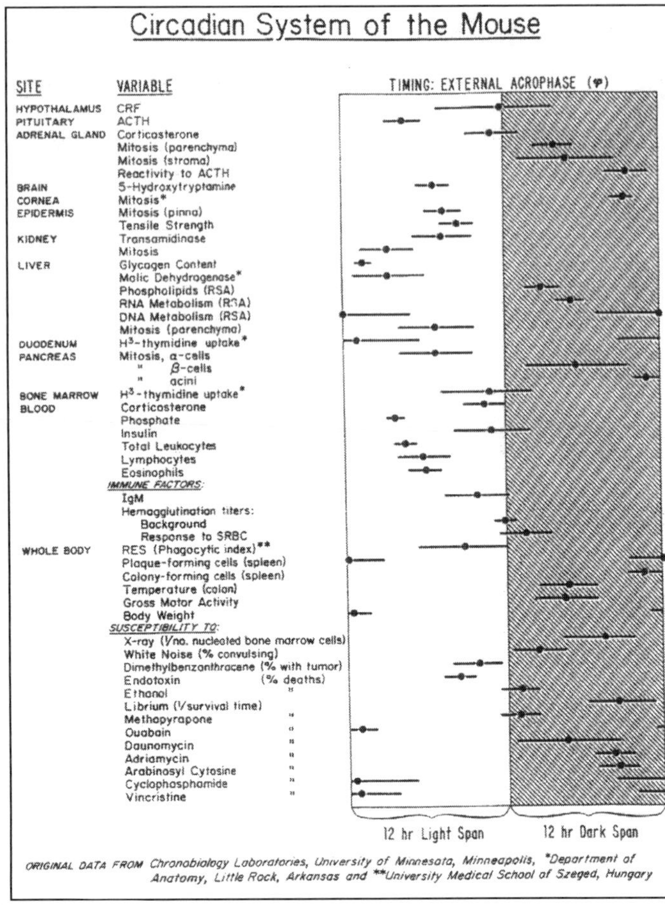

Figure 10. Circadian phase chart as part of broader chronomics awaiting further extension to psychopharmacological topics. Note statistically significant difference indicated by nonoverlapping 95% confidence intervals between two agents affecting the nervous system, librium and, in the high doses used, methopyrapone. © Halberg.

a 24-hour routine) of the period of many circadian rhythms under conditions of constant darkness or continuous light as well as after blinding[1] in the experimental laboratory. Circadian desynchronization is documented for humans in isolation from society, studied long enough *tma* in bunkers,[87] special facilities,[88-90] or an improvised isolation room in a hospital.[91] Longer time series collected in caves[92-98] (reviewed in ref. 99) were analyzed time-microscopically. Desynchronization has been extended to circaseptans for spelunkers in caves[100,101] and even to ordinary conditions.[102] *Tma* studies on desynchronization, their importance notwithstanding, remain beyond our scope herein. Within our scope is the *tma* lesson from Figure 10[20] summarized by *tmi* extrapolations in Table 1. The circadian adrenocortical cycle (then) gauged indirectly by counts of circulating blood eosinophil cells may desynchronize under conditions of ordinary life in a patient with extreme mood swings. Whenever eventually it can be validated in appropriately densely sampled series, and should it have a bearing on the etiology of sleep or other disorders, a circadian desynchronization could be the basis of a causal chronotherapy consisting of resynchronization by light[73,74,103,104] and/or other means.[87]

Table 1. Adrenocortical or medullary desynchronization in manic-depressive psychosis?*

Indirect internal bioassay bearing on whether the adrenal cortex desynchronizes from the daily routine with a period differing from 24 hours, and whether this periodicity bears at least a numerical relation to swings in mood that in this case were bipolar

			Tentative Extrapolations			
Period (Hours)	(Days)	A ± SE (Cells/mm^3)	Eosinophil Rhythm Loses: Hours per Day	24 Hours in *x* Days	Mental State Best-Fitting Period, τ, in Days (Amplitude ± SE)	|x - τ| (Days)
Total series: Mean ± SD=221 ± 125 cells/mm^3 (N=128)						
24.66	1.03	86 ± 13	0.66	36.36	35.29	1.07
622	25.92	54 ± 15			(2.10 ± 0.35)	
406	16.92	35 ± 15				
First untreated span: Mean ± SD=261 ± 53 cells/mm^3 (N=45)						
24.56	1.02	158 ± 21	0.56	42.86	40.96	1.90
649	27.04	123 ± 26			(3.44 ± 0.56)	
462	19.75	54 ± 31				

* Periods and corresponding amplitudes from a least-squares fit of some functions to a series of blood eosinophil cell counts of blood drawn at 48-h intervals (~10 am) for 256 days, with concomitant semi-quantitative evaluation of mental state, by Dr. Ronald W. Bryson. Data here analyzed are approximate, taken from a graph by two individuals, thus producing two similar time series. Treatments consisted of the administration of testosterone (25 mg, daily), diandrone (30 or 50 mg, daily), stilbestrol (5 mg, daily) or electroconvulsive therapy, in this order. Bryson RW. Psycho-endocrinology. Thesis, #1561, Glasgow University, 1958. "A document freely available for consultation on the premises of Glasgow University Library"; Bryson RW, Martin DF. 17-ketosteroid excretion in a case of manic-depressive psychosis. Lancet 1954; ii:365-367; Halberg F. Symposium on "Some current research methods and results with special reference to the central nervous system." Physiopathologic approach. Amer J Ment Defic 1960; 65:156-171; Halberg F. Physiologic considerations underlying rhythmometry, with special reference to emotional illness. Symposium on Biological Cycles and Psychiatry. In: Symposium Bel-Air III. Cycles biologiques et psychiatrie/publié sous la direction du professeur J. de Ajuriaguerra. Geneva: Georg/Paris: Masson et Cie; 1968:73-126.

Temporal Changes in the Effect, Notably Efficacy, of Psychotropic Drugs

Antipsychotics

A 6-timepoint approach, developed much earlier for many other studies, Figures 2-4, eventually applied to a long series of drugs reviewed herein, including antipsychotics, was available for psychopharmacology by 1961.[19]

Mortality and/or survival time from i.p. injection of Librium (7-chloro-2-methylamino-5-phenyl-3H-1, 4-benzodiapezine-4-oxide HCl) was tested by indirect periodicity analysis on inbred D_8 and B_1 mice. In each of several experiments, separate groups, each of 21-65 mice, were given a fixed dose of Librium, at 4-hr intervals, the first group injected at 08 of one day, the last at 04 or 08 of the next. Number of survivors was recorded at 4-hr intervals for the first 48 hrs post-injection and then at 12-hr intervals up to one week. Mortality was higher in mice injected during daily dark period (18 to 06) than during light period, with a peak usually at 24, but occasionally at 20 or 04. Significance of susceptibility rhythm below 1% level was ascertained by a procedure for locating peak of a time series. Circadian changes in susceptibility interact with age effects, susceptibility increasing significantly with age. The circadian peak in susceptibility to this psychotherapeutic drug has a timing similar to other susceptibility rhythms involving the central nervous system, e.g., to ethanol and to audiogenic abnormality; it falls into a period of increased electrocerebral and gross motor activity.

By 1957, a marker rhythm, namely circulating blood eosinophil cell counts, was used in studying the effect of reserpine (ref. 105, c.f. refs. 106,107) Studies on drug efficacy, dependent on the time of administration mostly in relation to the alternation of light (L) and darkness (D) in the laboratory (LD12:12)[105] as well as in humans,[105] have been reported for reserpine,[106,107] Figure 11, librium,[19] chlorpromazine (ref. 108; cf. ref. 29), and thereafter for haloperidol,[23] tetrabenazine, spiperone, and pimozide (see Table 2; cf. ref. 21-29). The *tma* waveform, including the time of peak, can vary with the kind of drug, the dose, and the measured variable. In rats, the optimal time of administration for maximal antiapomorphine effects (expressed as the "peak time") was *tma* in the early daily dark span for chlorpromazine and in the early light span for haloperidol. The maximal effect for spiperone was in the early daily light or early daily dark span (HN, unpublished data) and for pimozide, in the middle of the dark span (HN, unpublished data). The timing of drug efficacy along the circadian scale differed among drugs, even when the same endpoints were being compared.

The peak time for the sedative effect of chlorpromazine[24] at a dose of 2.5 to 5 mg/kg was the middle of the light span, whereas for a dose of 10 mg/kg, the peak time shifted to the middle of the dark span. A similar tendency was observed for haloperidol; i.e., at 0.5 to 1 mg/kg, the peak time was in the early light span, and

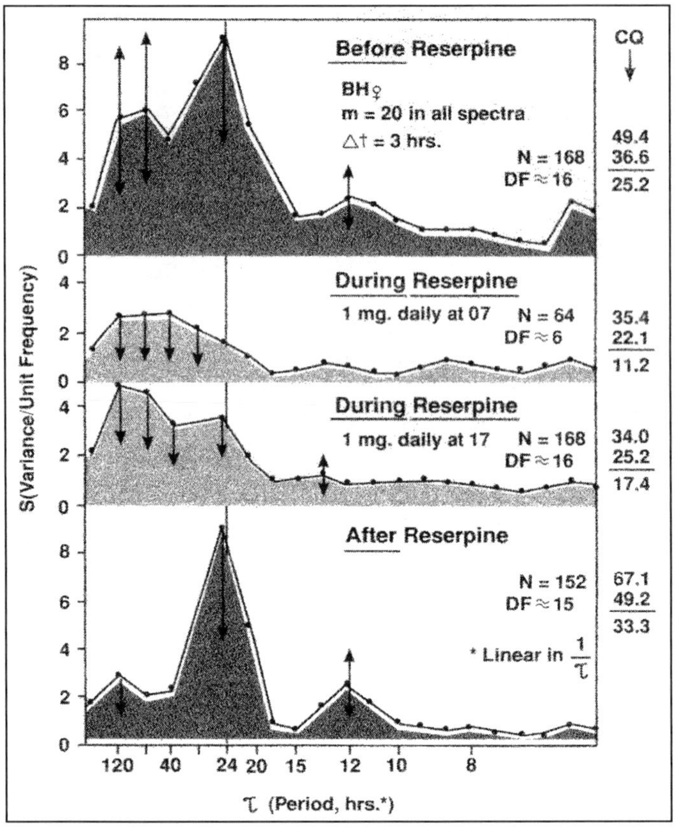

Figure 11. Effect of a psychotropic drug, reserpine, assessed *tmi* by variance spectra in the core temperature of a girl measured for several months 6-hourly around-the-clock (not shown). A circadian component is prominent in spectra on top and bottom (before and after reserpine administration); the variance is spread out into adjacent frequency domains during reserpine administration (middle). The alteration of the thermovariance spectrum during two spans of reserpine administration (hatched) occurs in the absence of fever or hypothermia. N: number of observations; DF= degrees of freedom; CQ= circadian quotient; m= determines resolving power by spectral analysis.[108] © Halberg.

at 2 to 8 mg/kg, it was in the middle dark span, indicating that the peak varied with the dose, for the same drug. Generally, the peak time shifted forward as a function of increased dose. This tendency in the rhythm of drug efficacy, however, is dependent on the variable measured. A dose-dependent shift of the peak time for the antiapomorphine effects was not observed for either chlorpromazine or haloperidol.

The peak time for the sedative effects of chlorpromazine was the middle light or middle dark span; for antiapomorphine effects, the early dark span; and for hypothermic effects, the middle light span.[109] The peak time for the sedative effects of haloperidol was either the early light span or the middle of the dark span, whereas it was the early light span for the antiapomorphine effects. These results suggested that the peak time varied with the variable measured, even for the same drug.

Antidepressants and Mood Stabilizers

The variations in drug effects have been clinically applied using antidepressants (Table 3). Lofepramine was shown to have a greater antidepressant effect during a 3-week course of therapy when administered at 24:00 than when administered at 08:00 or 16:00.[110] Furthermore, the antidepressant effects of clomipramine during 4 weeks of therapy varied depending on the time of administration.[25] In this case, administration at noon was more effective than administration in the morning or evening, suggesting that the variation in the drug's efficacy could be different from that of lofepramine, a possibility awaiting systematic concomitant testing at comparable test times.

In laboratory animal studies, fluoxetine suppressed the intake of carbohydrates only when administered in the early dark span but not at the other two times examined.[111] The effects on feeding behavior have been investigated with the 5-HTIA receptor agonist, 8-0H-DPAT,[112] and L-norepinephrine,[113] neither of which is an antidepressant. Feeding behavior decreased after administration of 8-0H-DPAT in the early dark span but increased after administration in the middle dark span or the latter half of the dark span. Also, feeding behavior was decreased by L-norepinephrine administered in the early dark span but was increased after administration in the latter half of the light span. The fact that there are not only quantitative differences in drug effects but also opposite (qualitatively different) effects depending on the time of administration further emphasizes the importance of timing. The timing of food, notably on a diet restricted in calories, can play a critical role in this context and must be taken into consideration in the laboratory[1,33,114-116] and in the field.[117]

Benzodiazepines and Barbiturates

The results of rotarod tests in mice conducted after administration of lorazepam are different depending on the time of administration (Table 2).[118]

Many animal studies of the effects on sleep duration of pentobarbital.[119-122] and hexobarbital.[123-125] have been conducted. Most of these studies were interpreted to yield maximal effects after administration in the latter half of the light span or the early dark span.[126] If replicated, the study summarized in Figure 12A would suggest in turn that results can be dramatically dose-dependent.[122] A *tmi* analysis establishes differences in acrophase and in amplitude as well as age and dose effects, but no differences in serum or brain concentrations at awakening that could account for the pharmacodynamics seen (Table 4). Intracerebral dopamine concentrations after administration of phenobarbital[127] vary within the regions of the brain. For the caudate nucleus and midbrain, the peak time is the early light span and latter half of the light span, respectively, and for the cerebellum and cortex, the peak time is in the late light or middle dark span (Table 2).

Mortality from the barbiturate, methohexital, is also a function of circadian stage at administration time.[128] Again, this is not a direct effect of differences in the blood concentration of the drug. It seemed important to write "... that in no case was the time of highest mortality associated with the time of highest blood levels".[128]

Psychostimulants

Most of the animal studies of amphetamines[109,129,130] have concluded that the peak time is the latter half of the dark span, while one report.[109] indicated the presence of another peak time in the latter half of the light span (see Tables 2 and 3).

Agonists and Related Compounds

For the subcutaneous injection of 8-0H-DPAT in rats, the peak time is the middle dark span for the effect on the 5-HTIA receptor[131] or hypothermia;[132] the peak time for intraventricular injection is the same[26] (see Table 2). These results, among others, also suggest that the peak efficacy times of the subcutaneous

Table 2. Circadian variation in laboratory animals in the effects of psychotropic drugs

Drug	Animal, L:D, Timepoints and Mode of Administration*[a]
Antipsychotics	
Chlorpromazine	SD rat, 12:12, 6, i.p., 7.5 mg/kg
Chlorpromazine	SD rat, 12:12, 4, i.p., 2.5-10 mg/kg
Chlorpromazine	SD rat, 12:12, 4, i.p., 1.2-10 mg/kg
Haloperidol	SD rat, 12:12, 4, i.p., 0.5-8 mg/kg
Haloperidol	SD rat, 12:12, 4, i.p., 0.12-0.5 mg/kg
Haloperidol	SD rat, 12:12, 2, i.p., 0.25 mg/kg, 3 wk
Haloperidol	SD rat, DD (7 days), 4, i.p., 0.5 mg/kg
Haloperidol	SD rat, DD (60 hrs), 4, i.p., 0.5 mg/kg
Tetrabenazine	SD rat, 12:12, 8, i.p., 50 mg/kg
Reserpine	SD rat, 14:10, 6, i.p., 1.2 mg/kg
Spiperone	SD rat, 12:12, 4, i.p., 0.006-0.12 mg/kg
Pimozide	SD rat, 12:12, 4, i.p., 0.12-1.0 mg/kg
Antidepressants and mood stabilizers	
Fluoxetine	SD rat, 12:12, 3, (A) to hypothalamus, 3.2-100nmol, (B) i.p., 0.6-10 mg/kg
Valproic acid	ICR mouse, 12:12, 6, p.o., 600 mg/kg
Benzodiazepines and barbiturates	
Lorazepam	Swiss-Webster mouse, 12:12, 4, p.o., 3 mg/kg
Pentobarbital	Swiss albino mouse, 12:12, 4, i.p., 60 mg/kg
Pentobarbital	SD rat, 12:12, 24, i.p., 35 mg/kg
Pentobarbital	SD rat, 12:12, 4, i.p., 50 mg/kg
Pentobarbital	BALB/cCr mouse, 12:12, 6, i.p., 65-87 mg/kg
Hexobarbital	NIH mouse, 12:12, 6, i.p., 125 mg/kg
Hexobarbital	SD rat, 12:12, 6, i.p., 150 mg/kg
Hexobarbital	Wistar rat, 12:12, 8, i.p., 150 mg/kg
Phenobarbital	SD rat, 12:12, 8, i.p., 50 mg/kg
Psychostimulants	
d-amphetamine	SD rat, 12:12, 6, i.p., 1.5 mg/kg
Amphetamine	Wistar rat, 9:15, 20, i.p., 5 mg/kg
Methamphetamine	dd mouse, 12:12, 6, s.c., 1.0, 2.0 mg/kg
Agonists and related compounds	
Serotonin	Wistar rat, 12:12, 12, ionophoresis
5-hydroxytryptophan	Wistar rat, 12:12, 8, i.p., 75 mg/kg
5-MeODMT	CFLP mouse, 12:12, 16, i.p., 5 mg/kg
5-MeODMT	BK.TO mouse, 12:12, 4, i.v., 5 mg/kg
p-chloroamphetamine	BK.TO mouse, 12:12, 4, i.v., 10 mg/kg
RU24969	CFLP mouse, 12:12, 8, i.p., 0.625 mg/kg
8-OH-DPAT	CFLP mouse, 12:12, 8, i.p., 5 mg/kg
8-OH-DPAT	SD rat, 12:12, 3, into raphe, 0.4-1.6 nmol
8-OH-DPAT	Wistar rat, 12:12, 6, s.c., 0.25 mg/kg
8-OH-DPAT	Wistar rat, DD (6 days), 6, s.c., 0.25 mg/kg
8-OH-DPAT	Wistar rat, 12:12, 6, s.c., 0.16 mg/kg
8-OH-DPAT	Wistar rat, 12:12, 6, i.c.v., 0.05 mg/kg
DOI Wistar rat, 12:12, (A) 6, s.c., 0.5, (B) 2,	i.c.v., 0.5 mg/kg
Apomorphine	SD rat, 12:12, 4, i.p., 0.6-80 mg/kg
Apomorphine	Wistar rat, 12:12, 6, i.p., 0.6-80 mg/kg
Apomorphine	dd mouse, 12:12, 6, s.c., 0.5, 1.0 mg/kg
Apomorphine	SAF rat, 12:12, 6, i.p., 15 mg/kg/day, 2 days
l-norepinephrine	Charles River rat, 12:12, 6, to hypothalamus, 25 g

Table continued on next page

Table 2. Continued

Variables Investigated and Peak Time*[b]	Reference
Hypothermia: L6	Wolfe et al, 1977[109]
Sedation: L6 (2.5, 5.0 mg/kg), D6 (10 mg/kg)	Nagayama et al, 1978[24]
Anti-apomorphine effect: D0	Nagayama et al, 1982[21]
Sedation: L0 (0.5, 1 mg/kg), D6 (2, 4, 8 mg/kg)	Nagayama et al, 1979[22]
Anti-apomorphine effect: L0	Nagayama et al, 1979[22]
Anti-apomorphine effect: L0>D6	Nagayama et al, 1982[29]
Sedation	Nagayama et al, 1987[140]
Anti-apomorphine effect	Nagayama et al, 1987[140]
Sedation: D3	Nagayama et al, 1977[28]
Brain norepinephrine release: D3-5	Black et al, 1969[208]
Anti-apomorphine effect: L0, D0	Nagayama, unpublished
Anti-apomorphine effect: D6	Nagayama, unpublished
(A) and (B) carbohydrate intake: suppressed only during D0	Weiss et al, 1991[111]
ES seizure threshold: L6, D10	Ohdo et al, 1996[150]
Rotarod test: D9-L3>D3	Henauer et al, 1984[118]
Sleep duration: L8	Davis, 1962[120]
Sleep duration: D1	Scheving et al, 1968[119]
Sleep duration: L10	Friedman & Walker, 1969[121]
Sleep duration: L1 (65, 71), L0 (79), L6 (87 mg/kg)	Nelson & Halberg, 1973[122]
Sleep duration: L7	Vesell, 1968[123]
Sleep duration: L8	Nair, 1974[125]
Sleep duration: D0	Mueller, 1974[124]
Dopamine level: L1 (caudate), L10 (midbrain), L10-D7 (cerebellum, cortex)	Owasoyo et al, 1979[127]
Locomotion L10, D10	Wolfe et al, 1977[109]
Head shake:D9	Urba-Holmgren et al, 1977[129]
Locomotion: D9	Kuribara & Tadokoro, 1982[130]
Neuronal activity in SCN: D4	Mason, 1986[134]
Head twitch: L6-9	Moser & Redfern, 1986[135]
Head twitch: L7, 5-HT: no rhythm	Moser & Redfern, 1985[136]
Head twitch: 0600	Singleton & Marsden, 1981[137]
Head twitch: 14:00	Singleton & Marsden, 1981[137]
Locomotion: no rhythm	Moser & Redfern, 1985[133]
Hypothermia: no rhythm	Moser & Redfern, 1985[133]
Feeding: increase (D6, D10.5), decrease (D0)	Currie & Coscina, 1993[112]
5-HT syndrome: D6	Lu & Nagayama, 1996[131]
5-HT syndrome	Lu & Nagayama, 1997[141]
Hypothermia: D6	Lu & Nagayama, 1997[132]
5-HT syndrome: D6	Nagayama & Lu, 1997[26]
Wet-dog shake: (A) L10, (B) L10>D10	Nagayama & Lu, 1996[138]
Stereotypy: L0	Nagayama et al, 1978[152]
Stereotypy: L6 (1.0), L10 (3.0)	Nakano et al, 1980[27]
Locomotion: D9.5	Kuribara & Tadokoro, 1982[130]
Hypothermia, effect and tolerance: L7	Williams et al, 1993[139]
Feeding: increase (L10.5), decrease (D3)	Margules et al, 1972[113]

*[a]: L:D: light (L)-dark (D) cycle (hr) in the housing room. DD: continuous dark. Time points represent number of time points of drug administration. *[b]: Peak time represents time of administration for maximum effects. L, D, and figures represent hours after the beginning of light and dark spans, respectively.

injection are not due to pharmacokinetics (e.g., drug absorption, excretion, metabolism, and distribution) but rather to a variation in the susceptibility of the postsynaptic 5-HTIA receptor, which is the site of action for 8-0H-DPAT. In one study on mice, however, a rhythm in the hypothermic effects of the drug was not detected.[133] Presumably, 8-0H-DPAT acts postsynaptically on the 5-HTIA receptor in the rat but on other sites in the mouse and the inter-species difference has to be considered in studies on rats[132-135] as compared to investigations on mice.[136]

The effects of 5-HT,[134] 5-hydroxytryptophan,[135] 5-MeODMT (a nonselective 5-HT receptor agonist),[136,137] p-chloroamphetamine,[137] and DOI (a selective 5-HT2A/2c agonist)[138] on the 5-HT system undergo changes depending on the time of administration, although there is some controversy. These

Table 3. Circadian variation in humans of the effects of psychotropic drugs

Drug	Open/Blind	Subject, N of Subjects, Time Points and Mode of Administration*[a]	Variables Investigated and Major Findings	Reference
Healthy persons				
Amitryptiline	Open	Healthy, 10, 2, p.o., 50 mg, single	Decrease in salivary flow: 0900 > 2100	Nakano & Hollister, 1983[213]
Triazolam	Open	Healthy, 10, 2, p.o., 0.5 mg/day, single	Performance decrement: 2200 > 0700	Smith et al, 1986[214]
Dextroamphetamine	Open	(A) post-menopausal, 5, 2, i.v., 0.15 mg/kg, single (B) young man, 6, 2, i.v., 0.15 mg/kg, single	Growth hormone response: (A) 1830 > 0900 (B) 1830 = 0900	Halbreich et al, 1980[215]
Dextroamphetamine	Open	Healthy, 10, 2, i.v., 0.15 mg/kg, single	Cortisol response: 1830 > 0900	Halbreich et al, 1980[215]
Dextromethamphetamine	Open	Healthy, 10, 2, p.o., 30 mg/70 kg, single	Improvement in cognition and fatigue: 0840 > 2040	Shappell et al, 1996[217]
d-fenfluramine	Open	Healthy, 7, 2, p.o., 30 mg, single	Cortisol response: 1400 > 0800	Monteleone et al, 1997[218]
Patients				
Lofepramine	Open	Endogenous depression, 30, 3, p.o, 210 mg/dy, once/day, 3 wks	Antidepressive effect: 2400 > 0800, 1600	Philipp & Marsneros, 1978[110]
Fluoxetine	Double blind	Major depression, 120, 2, p.o., flexible dose, once/day, 5 wks	Antidepressive effect, side effect: AM = PM	Usher et al, 1991[219]
Clomipramine	Double blind	Major depression, 40, 3, p.o., 150 mg/day, once/day, 4 wks	Antidepressive effect: 1220 > 0820, 2030	Nagayama et al, 1991[25]
Lithium	Open	Cluster headache, 15, 2, p.o., 900 mg/day	Therapeutic effect: 0930 = 2130	Ferrara et al, 1984[220]
Levodopa	Open	Parkinson's disease, 5, 2, p.o., 250 mg, single	Motor response: 0800 = 1200	Frankel et al, 1990[221]

*[a]: Time points represent the number of time points of drug administration.

discrepancies may be related to the receptor specificity of individual drugs. The efficacy of apomorphine, a dopamine agonist, peaks in the light span for both stereotypy[27,130] and hypothermia[139] and in the dark span for locomotion.[130]

Several additional studies have used only two administration times. When a statistically significant difference is observed, as was anticipated at two specific time points picked as peak and trough based on prior denser sampling, this result is in keeping with the presence of a circadian variation, albeit no parameters can be estimated. The presence or absence of a temporal variation cannot be judged when no difference is observed between the results at two arbitrarily chosen time points since these can happen to be at midline crossings. In the absence of prior information, it is much more efficient to use 6 time points, even if the total of comparable animals is as small as 6, i.e., there is only one individual per timepoint. In such a case, the cosinor approach can test the statistical significance of an anticipated rhythm. A *tmi* test for the absence of a rhythm by curve-fitting has allowed the rejection of the null hypothesis of a zero circadian amplitude with only 1 subject at each of only 5 equidistant 4.8-hourly timepoints. In this review, studies using only two time points are included in Table 3 for reference only; studies at 4 or more time points in the laboratory or studies at three or more time points are considered, but are not recommended in the absence of prior information and when completed await a *tmi* analysis, as do denser studies.

Endogenous Aspects of Circadian Rhythmicity in the Effect of Psychotropic Drugs

In general, external and internal factors contribute to changes with a 24-hour cycle. Surveys using constant light or constant darkness have been used in an attempt to discriminate between the extent of an exogenous contribution by lighting and the other exogenous cycles to the circadian rhythmicity. This approach, however, can never rule out the contribution of external cycles other than the lighting acting, for instance, during exposure to continuous darkness, e.g., in Figure 1. With this qualifying restriction, the rhythmicity of the host responding to the effects of psychotropic drugs has been demonstrated under conditions of constant darkness,[140,141] suggesting that it is a circadian rhythm in this narrower sense of independence from the lighting regimen.[20]

Chronopharmacokinetics and Chronopharmacodynamics

There are temporal aspects of pharmacokinetics, including rhythmic (e.g., circadian) variations of drug bioavailability, even

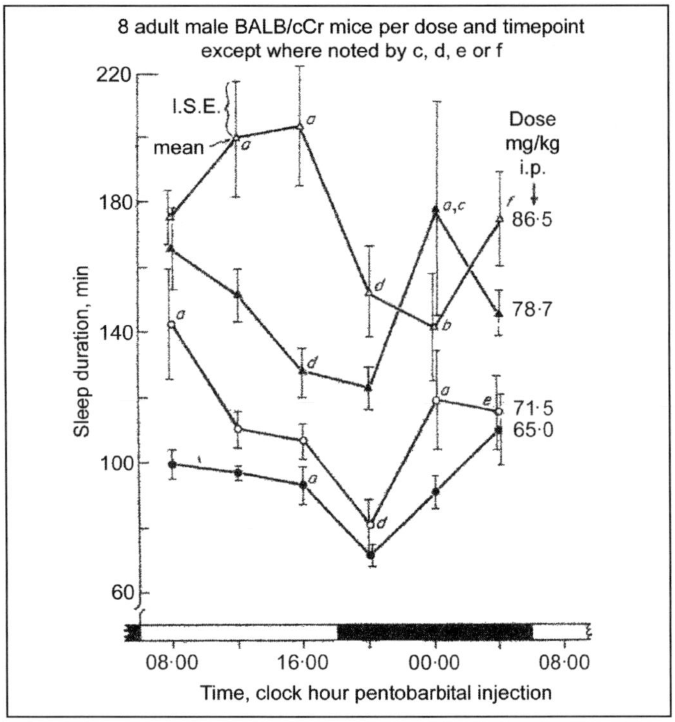

Figure 12A. Sleep duration as a function of pentobarbital dose and circadian phase at the time of injection. A variation with dose-dependent timing. a= includes single extremely long sleeping time (> 3 S.D. above mean of remainder); b= includes extremely short sleeping time (> 3 S.D. below mean of remainder); c= 3 mice did not sleep; d= 1 mouse did not sleep; e= 1 mouse died; f= 1 mouse deleted because of incorrect injection.[122] © Halberg.

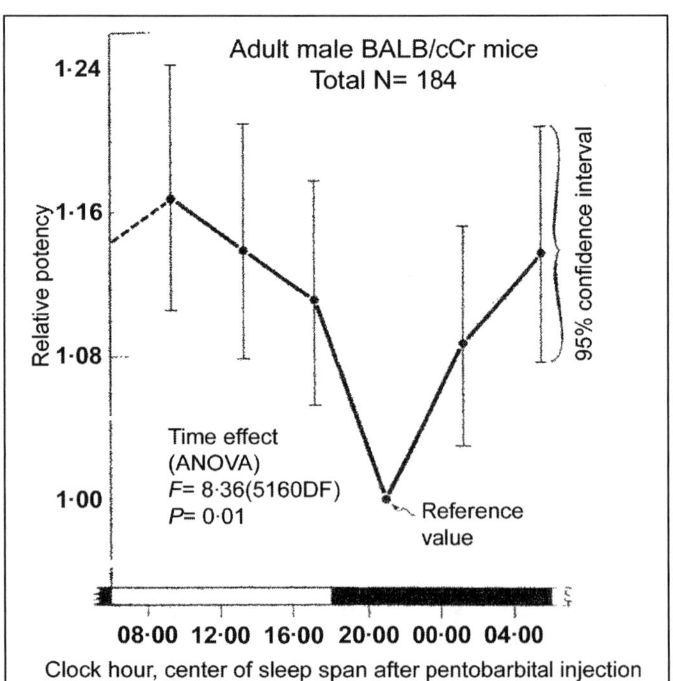

Figure 12B. Circadian variation in susceptibility of mice to pentobarbital injection. Expressed as a variation in relative potency (from parallel-line bioassay), with sleep duration as a measure of response.[122] © Halberg.

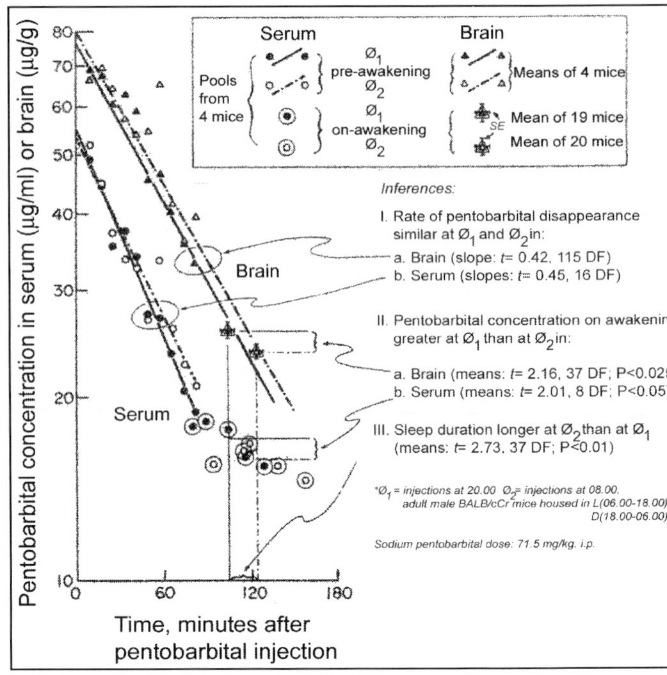

Figure 12C. Pentobarbital concentration in brain and serum of mice as a function of time after injection at circadian phases of short sleep duration (ϕ_1) and long sleep duration (ϕ_2). ϕ_1: injections at 20:00; ϕ_2: injections at 08:00. Adult male BALB/c Cr mice housed in L(06:00-18:00)D(18:00-06:00). Sodium pentobarbital dose: 71.5 mg/kg i.p.[122] © Halberg.

in the case of a continuous infusion at a constant rate, Figure 13A;[142] drug or metabolite excretion in urine, Figure 13B,[143] feces, sweat or saliva is also periodic. Statistically significant circadian rhythms have been demonstrated for the parameters used to characterize classical pharmacokinetics, Table 5.[144] The more critical aim of a chronopharmacology, however, is the systematic mapping of the variations in the undesired and desired drug effects, Figure 13C.[145,146] Cancer chronotherapy, guided by marker rhythms, remains a model for transfer of information from the laboratory to the clinic.[146]

Rhythms in the Pharmacokinetics of Psychotropic Drugs

There are many reports on the changes in the pharmacokinetics of psychotropic drugs depending on the time of administration. In human studies, however, the effects of administration time have been compared at only two time points within 24 hours, not enough information to determine whether a circadian rhythm is present in these particular psychopharmacokinetics.

In animal experiments (Table 6), rhythms are observed in the activities of enzymes involved in drug metabolism in the liver for imipramine[147] and hexobarbital.[123-125,147] The rhythms of those enzyme activities might help to elucidate the underlying mechanism of the rhythms in the drug effect. Rhythms in enzyme activity must be reflected, however, by the blood concentrations of a drug in order for the reflection of the rhythms in enzyme activity to be considered one of the drug's effects. Studies comparing drug concentrations among four or more different administration times are as yet inconclusive.[109,148-151]

To find out whether a *tma* variation in drug efficacy is the result only of a *tma* variation in pharmacokinetics, including

Table 4. ***No detectable differences in pentobarbital concentration of waking brain (above) and serum (below) in mice receiving different doses of pentobarbital at circadian phases of high (08:00) or low (20:00) susceptibility****

Injection Time	Dose (mg/kg) 65.0	71.5	78.7	86.5	
	Mean Pentobarbital Concentration (μg/g) ± SE, Following Dose:				
08:00	19.7 ± 0.6	17.7 ± 1.3	—	20.9 ± 0.6	*F* = 32 (1,40 D.F.)
20:00	23.0 ± 0.5	25.2 ± 2.0	—	24.7 ± 0.7	P < 0.01
	F = 1.23 (2,40 D.F.); P > 0.05				
	Pentobarbital Concentration (μg/g) ± SE, Following Dose:				
08:00	16.6	13.7	17.1	15.3	F = 40
	16.0	15.7	14.6	15.4	(1.8 D.F.)
20:00	19.8	21.8	19.5	18.4	P < 0.01
	18.9	20.2	18.0	22.0	
	F = 0.15 (3,8 D.F.): P > 0.05				

*Animals housed in L(06:00-18:00):D(18:00-06:00). Each value above represents 8 brains except for the 71.5 and 86.5 dose at 20:00 in which case 7 brains were available for analysis; values below represent a blood pool from 4 mice. Above, injection time-dose interaction F=2.44 (2,40 D.F.); P > 0.05; below, injection time-dose interaction F = 1.42 (3,8 D.F.); P > 0.05.

absorption, metabolism, excretion, and distribution of the drug, it is desirable to demonstrate rhythmic changes in drug concentrations in both blood and brain. In addition, it is desirable to demonstrate a reasonable relation between the waveforms of the variations of the drug concentrations and of the effects. Many authors have investigated a variation in drug efficacy in relation to the drug's concentration in the brain; all of the studies denied a correlation between the changes in drug efficacy and the intracerebral drug concentration[22-24,28,29,109,118,122,152] (Table 6). Statistically significant differences in the intracerebral drug concentration among different administration times that were associated with a different effect were not found. The hypothesis that any rhythm in the effect of psychotropic drugs might depend critically upon a circadian rhythm of the drug concentration in the blood and in the brain is not supported.[122,128]

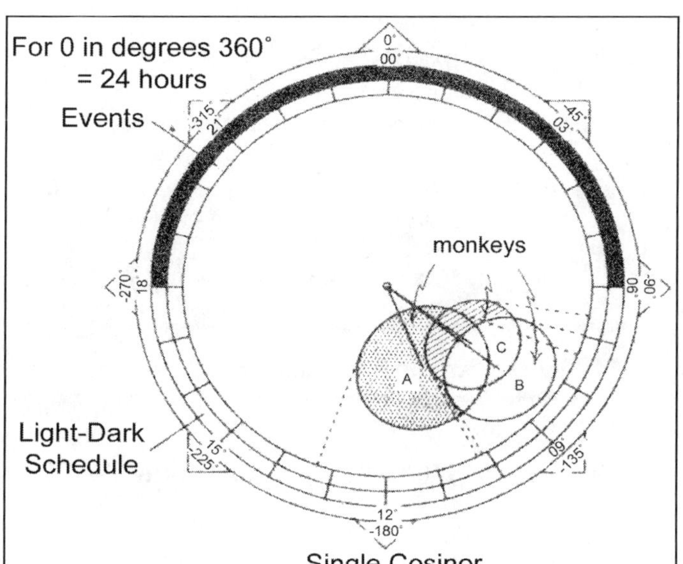

Figure 13A. Plasma ethosuximide concentration (μg/ml) in rhesus monkeys during four months of approximately constant drug infusion. Samples collected during five weeks, for approximately 1 day/week, at 2-h intervals. Drug concentration in plasma varies with circadian rhythm despite constant rate of infusion.[51] Original data of Patel, Levy and Lockard.[142] © Halberg.

Circannual Variation in the Efficacy of Psychotropic Drugs?

The circadian variation in the efficacy of 8-0H-DPAT exhibits a variation along the scale of 1 year in rats.[22] This variation was observed first in 5 different surveys 3 months apart, in which the phase of the rhythm shifted along the 1-year scale. The fifth survey, carried out about a year after the first one in January, reproduced the results of the earlier January study quite well to the unaided eye. Furthermore, a difference of as much as 9 hours was found in the peak time among surveys in different months. The variation was observed in animals raised in a laboratory isolated from the natural environment; hence it seemed possible that it was not solely due to variations caused by seasonal changes, such as those in the environmental temperature and/or the daily photofraction, from which the animals were shielded. Some endogenous factors were postulated as underlying differences in Figures 2 and 4 of reference 22, some displayed and analyzed in Figures 14A and B. These figures are in keeping with the assumption that differences (among studies replicated in different months of the same year, with the same drug) in the phase of circadian rhythms in drug efficacy may be partly explained by the presence of a circannual rhythm.

An about 1.3- to 1.6- (trans)yearly variation[39-41,153] was recently discovered, a possibility which cannot be rigorously tested on the data taken off the published Figures 2 and 4 of reference 22. These data, obtained at intervals of several months, cover only 1 year, i.e., are much too sparse and short, but are nonetheless the first of their kind in neuropharmacology, Figure 14. An earlier

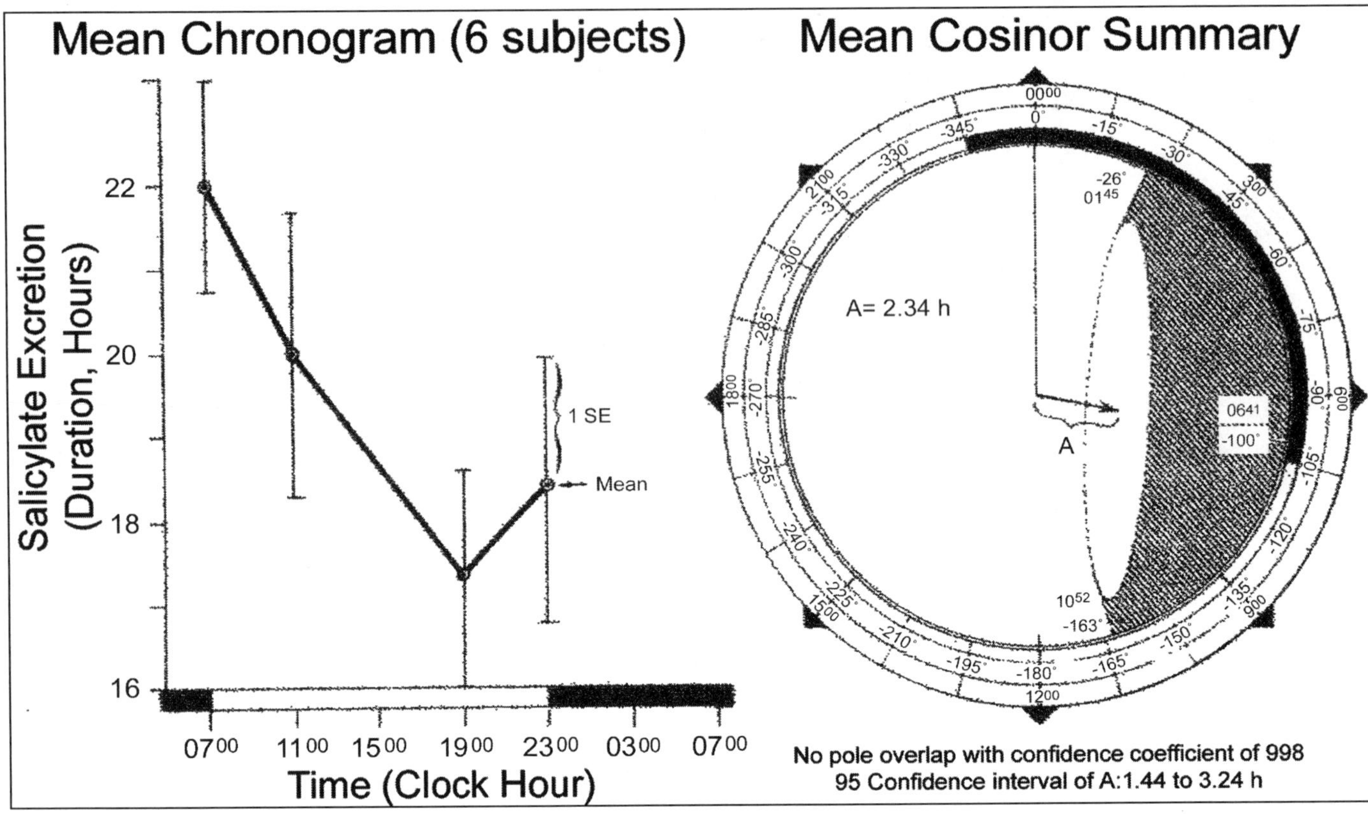

Figure 13B. Salicylate excretion by healthy human adults receiving a standardized dose of aspirin at several different times along the 24-h scale.[51] Chronopharmacokinetics revealed indirectly by *tma* differences in the excretion of salicylate in studies with the sodium salt of acetylsalicylic acid by Reinberg et al in Paris, showing mean duration of excretion at different clock-hours (left) and *tmi* summary (right).[143] © Halberg.

immunological *tmi* study of different circadian and circaseptan patterns (ref. 31; cf. ref. 50) yielded different results in 2 different seasons, but this is a completely different approach.[237] With the qualification that our purpose herein is only hypothesis formulation, the very scarce data in the figures of reference 22 were fitted with linear-nonlinear rhythmometry. This *tmi* approach allowed the rejection of the no-1.3-year cycle assumption with a probability of borderline statistical significance (P=0.07) by contrast to the fit of a precise year, which was not statistically significant, as shown in Table 7. As implied by the invaluable data of Nagayama and Lu,[22] the results warrant urgent focus upon both the scales of days and years, in a chronomial design that should eventually include decades[3] and has already detected the interaction of about-daily and about-weekly[31,50] as well as about-yearly changes in the effect of an immunomodulator.[31,50,237]

Efficacy of Psychotropic Drugs and the Neurotransmission System

Variations in the Intracerebral Neurotransmission System

Since the effects of a drug, depending on the time of administration, vary in no detectable relation to the amount of drug that enters the brain, the presence of a rhythm in the susceptibility to drugs of the brain itself can be postulated. The results of direct

Table 5. Chronoavailability indices for oral erythromycin in 24 healthy human volunteers*

Index Summarized	Percent Rhythm	MESOR ± SE	Amplitude (95% CI)	Acrophase (95% CI)
Area (μg/ml · hr)	8	5.24 ± 0.14	1.41 (0.92, 1.91)	-180° (-159, -200)
Minimum (μg/ml)	22	0.44 ± 0.01	0.31 (0.25, 0.37)	-186° (-178, -195)
Maximum (μg/ml)	8	1.65 ± 0.04	0.42 (0.27, 0.57)	-173° (-142, -193)
Time to maximum	11	3.49 ± 0.06	0.68 (0.48, 0.89)	-144° (-127, -161)
Terminus (μg/ml)	10	0.90 ± 0.03	0.36 (0.28, 0.44)	-163° (-150, -176)

* P from (cosinor) F-test < 0.001 in each case. Acrophase in angular degrees; 360° ≡ 24 h; 15° = 1h; reference 00:00. Meta-analysis[51] of original data from DiSanto et al.[144]

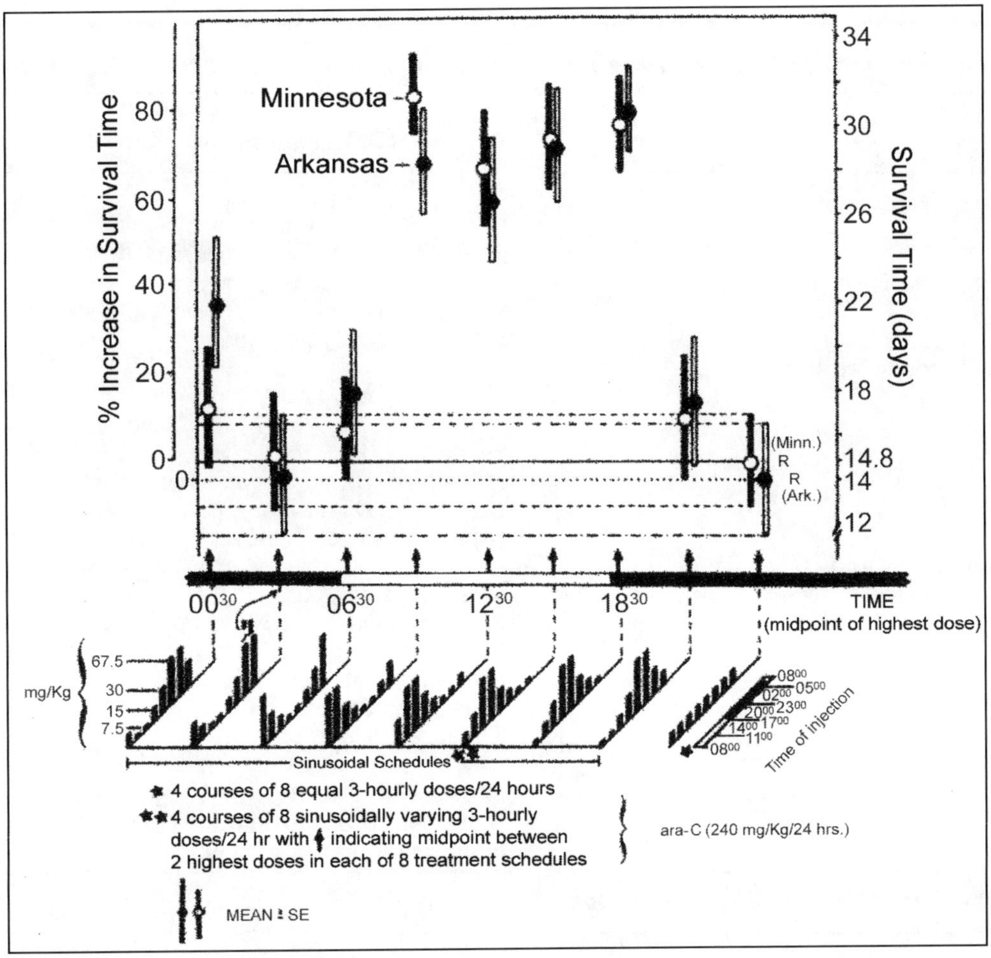

Figure 13C. Treatment of leukemic mice with ara-C on sinusoidal schedules to which non-leukemic mice are more tolerant also yields higher cure rate. As compared to nonsinusoidal (equal-dose) schedules, some sinusoidal schedules are much better, in terms of cure as well as tolerance, whereas others may be worse, a demonstration of the importance of timing that in this case improves cures.[146] © Halberg.

drug administration into the brain[26,113] strongly support this hypothesis. This assumption is also supported by differences as a function of the dose, the variable measured, and the brain region involved, which could all show a different time dependence, even for the same drug. The variables measured may contribute variations in efficacy with different phases depending upon the phase of the drug susceptibility rhythm that may vary further from region to region in the brain.

Variations have been reported for various neurotransmitters, receptors, enzymes, and the second messenger system in the brain, as documented by determinations in various regions other than the pineal body, made at least at six different timepoints. Because the presence of a variation under constant conditions of illumination[20,154-156] and its *tma*-reported disappearance after ablation of the suprachiasmatic nucleus[157] in the rat, the rhythm in this species was considered to be endogenous. By a *tmi* approach, when averages taken off Figure 2B in reference 157 are meta-analyzed, the best-fitting period is mostly at 12 hours, and allows the rejection of the no-circasemidian-rhythm assumption with a P of 0.022 for controls and a P of 0.072 for the SCN-ablated rats.[157] But this finding awaits further testing, since a circasemidian rhythm was not predicted. An analysis of the original data is also desirable before inferences are drawn. In a study on a different species, the hamster, a dependence of the *tma* variation in brain 5-HT is reported to depend upon the lighting regimen.[158]

The presence of a circadian rhythm has been reported for neurotransmitters and related substances, such as 5-HT,[53,159-161,163,170,175,181] norepinephrine,[53,161,162,168,171-174,177,182-185] dopamine,[53,127,165,173,177,183,186-188] 5-HIAA,[164,166,169,176-180,188-190] tryptophan, tyrosine, 5-hydroxytryptophan, 3-methoxy-4-hydroxyethyleneglycol, homovanillic acid, epinephrine, 3,4-dihydroxyphenylglycol, 3,4-dihydroxyphenylacetic acid, 3-methoxy-4-hydroxy-phenylethyleneglycol, acetylcholine, GABA, and cholecystokinin. The *tma* phases were different among these substances. Also, the *tma* phases were different among various animal species,[167,168] strains,[189] brain regions,[171,177,180] and ages,[53,173] even within a given study. The variations also differ by discrete regions obtained by more vs. less detailed dissection, even within the same cortex.[154] Most of these rhythms have a single *tma* peak. A bimodal variation is observed, however, for some substances.

A variation with a *tma* 24-hour period is observed for receptors of 5-HT, alpha-adrenergic,[156,191-194] beta-adrenergic,[155,191-195] dopamine,[192,196,197] GABA, benzodiazepine, imipramine, muscarinic, adenosine, and naloxone. Discrepancies in the *tma* phases observed in these receptors are again found, as in the case of neurotransmitters. These *tma* properties vary in complicated ways according to animal species,[167,168] brain region,[53,154] age,[198] and season.[190,191,196,199]

Table 6. Circadian variations in blood and brain concentrations, hepatic metabolism or clearance

Drug	Animal, L:D, Timepoints and Mode of Administration*[a]
Antipsychotics	
Chlorpromazine	SD rat, 12:12, 4, i.p., 7.5 mg/kg
Chlorpromazine	SD rat, 12:12, 2, i.p., 10 mg/kg
Haloperidol	SD rat, 12:12, 2, i.p., 2 mg/kg
Haloperidol	SD rat, 12:12, 2, i.p., 0.25 mg/kg, 3 wks
Tetrabenazine	SD rat, 12:12, 2, i.p., 50 mg/kg
Antidepressants and mood stabilizers	
Imipramine	Long-Evans rat, 12:12, 4, in vitro
Amitriptyline	Wistar rat, 12:12, 6, (A) i.v., 3 mg/kg, (B) p.o., 64 mg/kg
Lithium	Wistar rat, 12:12, 4, p.o. (in food), 70nmol/kg/day, 4-6 wks
Lithium	ICR mouse, 12:12, 6, i.p., 3 mmol/kg
Valproic acid	ICR mouse, 12:12, 6, (A) p.o., 600 mg/kg, (B) s.c. continuously, 1072.5 g/hr, 3-4 days, (C) i.v., 50 mg/kg
Benzodiazepines and barbiturates	
Lorazepam	Swiss-Webster mouse, 12:12, 2, p.o., 3 mg/kg
Pentobarbital	BALB/cCr mouse, 12:12, 2, i.p., 65-86.5 mg/kg
Hexobarbital	NIH mouse, 12:12, 6, in vitro
Hexobarbital	Long-Evans rat, 12:12, 4, in vitro
Hexobarbital	SD rat, 12:12, 6, in vitro
Hexobarbital	Wistar rat, 12:12, 8, in vitro
Psychostimulants and dopamine antagonist	
d-amphetamine	SD rat, 12:12, 4, i.p., 1.5 mg/kg
Femcamfamine	Wistar rat, 12:12, 4, i.p., 10 mg/kg
Apomorphine	SD rat, 12:12, 2, i.p., 10 mg/kg

Variables Investigated and Major Findings (Peak Time)*[b]	Reference
Plasma and brain level (30-180 min): no variation (cf. hypothermia: L6>L2, D0, D10)	Wolfe et al, 1977[109]
Plasma and brain level (1-24 hr): D6=D0 (cf. sedation D6>L0)	Nagayama et al, 1978[24]
Plasma and brain level (2-15 hr): L0=D0 (cf. sedation L0>D0)	Nagayama et al, 1979[23]
Plasma and brain level: L0=D6 (cf. anti-apomorphine effect L0>D6)	Nagayama et al, 1982[29]
Plasma and brain level (10 min - 8 hr): L6=D6 (cf. sedationD6>L6)	Nagayama et al, 1977[28]
Enzymatic activity: peak D3.5	Jori et al, 1971[147]
(A) AUC, MRT, t1/2, Vdss: peak L2-4; CI: peak D5 (B) MRT:L11, t1/2; L10; ka: D1; CI: D4.5; AUC: no variation	Rutkowska et al, 1992[149]
Serum level: L3-7	Olesen & Thomsen, 1985[221]
CL: D2	Shito et al, 1992[148]
(A) plasma level: L10, (B) AUC: L10>D6; CI, V: D6>L10; K, t1/2: L10=D6	Ohdo et al, 1991[223]
(A) plasma level (30 min): L6, D10, (B) plasma level: L17, (C) plasma level: L10>D6 (15, 30 min), L10=D6 (45-75 min)	Ohdo et al, 1996[150]
Brain level (15-30 min): L3=D3 (cf. rotarod test L3>D3)	Henauer et al, 1984[118]
Serum & brain level: L2=D2; awakening level: D2>L2 (cf. sleep duration L2>D2)	Nelson & Halberg, 1973[122]
Enzymatic activity: D7 (cf. sleep duration: L7)	Vesell, 1968[123]
Enzymatic activity: D3.5	Jori et al, 1971[147]
Enzymatic activity: D4 (cf. sleep duration: L8)	Nair, 1974[125]
Enzymatic activity: L0 (cf. sleep duration: D0)	Mueller, 1974[124]
Plasma and brain level (30-120 min): no variation (cf. hyperactivity: two peaks)	Wolfe et al, 1977[109]
Plasma level: D2 (30 min), no variation (60 min)	Planeta et al, 1994[151]
Brain level (20-180 min): L0=D6 (cf. stereotypy L0>D6)	Nagayama et al, 1978[152]

*[a]: L:D: light (L)-dark (D) cycle (hr) in the housing room. DD: continuous dark. Time points represent number of time points of drug administration. *[b]: Peak time represents time of administration for maximum effects. L, D, and figures represent hours after the beginning of light and dark spans, respectively.

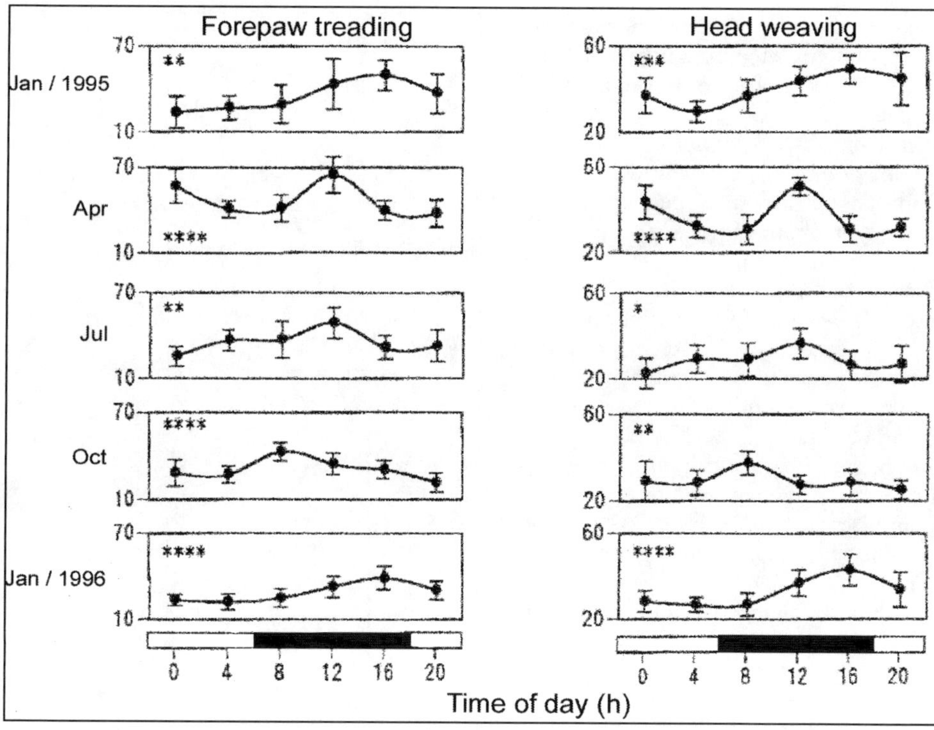

Figure 14A. Circadian rhythms in the 8-OH-DPAT-induced forepaw treading and head weaving (counts), observed in five separate surveys performed at 3-monthly intervals. Values represent the means ± SD of eight animals. The dark bar corresponds to the dark period. 1: P<0.050; 2: P<0.01; 3: P<0.001; 4: P<0.0001 by ANOVA (a) or LSF (b). From Nagayama and Lu.[22] Reproduced with permission from Springer Verlag.

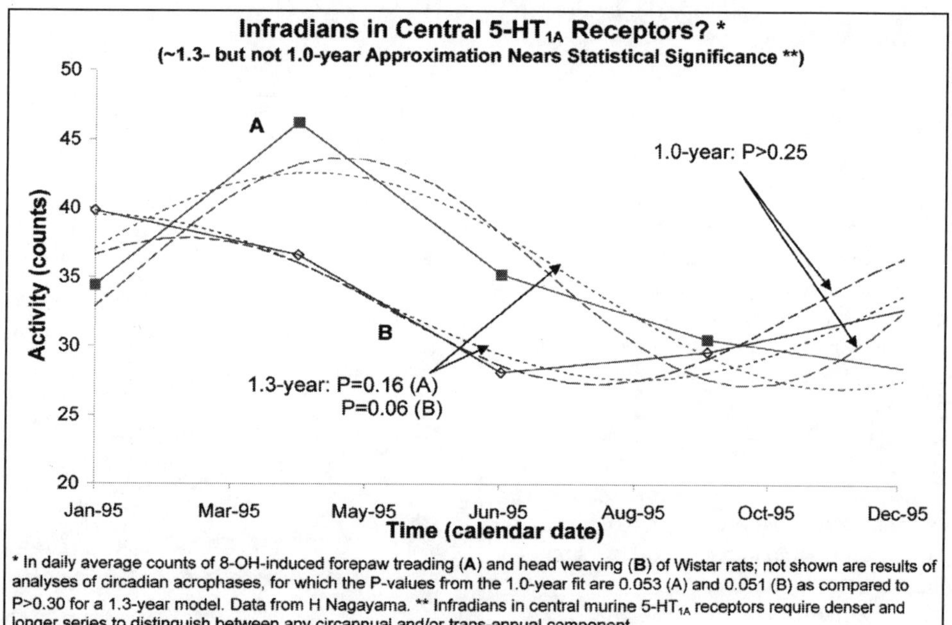

Figure 14B. Analyses carried out only to suggest the need for differentiating between any about-yearly, near-yearly, about 1.3-yearly and about 1.6-yearly components in important data heretofore regarded as solely seasonal or rather circannual variations.[22] © Halberg.

The enzyme activities related to neurotransmitters also have a circadian variation. Rhythm has been reported for the activities of tryptophan hydroxylase,[180,200-203] tyrosine hydroxylase, monoamine oxidase, tyrosine aminotransferase, dopamine-beta-hydroxylase, phenylethanolamine N-methyltransferase, choline acetyltransferase, acetylcholinesterase, and glutamic acid decarboxylase. Although there are only a few of these reports, the results indicate that variations differ among animal strains,[201,204] ages,[173] and brain regions.[53,173,180]

The second messenger system has a variation in cyclic adenosine monophosphate and cyclic guanidine monophosphate and in adenylate cyclase and phosphodiesterase activities. The presence of the variation and the phase are affected by animal strain[205] and brain region.[154]

Individual neurotransmitter systems have partly inherent variations. Reasons for the inter-study differences can be that their phase depends on the species, strain, and age of the animals and on brain regions. Even if these factors are taken into account, however, a discrepancy still remains. Seasonal or transyearly changes in phase[190,191,196] may be a reason for this discrepancy, but season-controlled studies as yet are rarely performed. Systematic sampling along the scale of infradian rhythms could greatly reduce the discrepancies. Another important issue is the method used to measure variations. Usually, a variable is not continuously measured over time in a single animal; instead, a group mean obtained from several animals is used to characterize the rhythm, which is measured only once in each animal at a given time point. Thus, it is desirable to establish a method by which the content or concentration of intracerebral substances in individual cerebral regions can be measured continuously for a long time in a single animal under freely moving conditions, and/or a marker can be longitudinally assessed, e.g., by telemetry[206] of core temperature motor activity or the electroencephalogram for pertinent features of its chronome.

It is unlikely that these rhythms change without mutual dependency. It seems that the signals emitted from the suprachiasmatic nuclei are transmitted to each neurotransmitter system to produce a rhythm network that allows integral coordination of the brain as a whole through interactions among the respective rhythms. Some of these rhythms have amplitudes as large as several hundred percent. These results suggest that these rhythms must be regarded as one of the most important factors for the investigation of various brain functions. In addition, studies of the mechanisms of the rhythm in drug efficacy can be used to investigate the causal mechanisms of drug action and brain function in general.

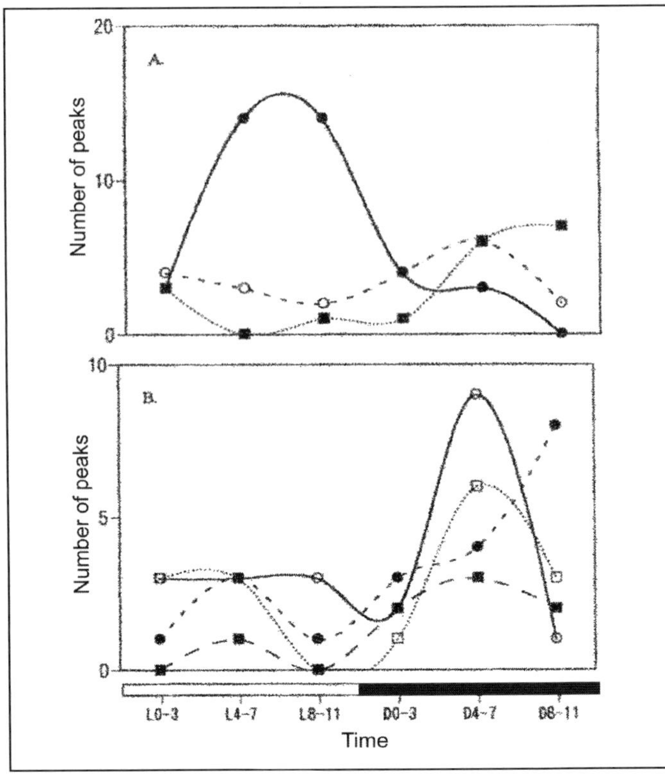

Figure 15A. Time-macroscopic circadian variations in the 5-hydroxytryptamine (5-HT) and norepinephrine-dopamine systems reproduced from H. Nagayama,[126] awaiting time-microscopic analyses beyond Figures 2C-E.

Rhythms of the Cerebral 5-HT and Norepinephrine-Dopamine Systems

The rhythms of the neurotransmission system in the brain might be affected by a large number of factors. To gain a general understanding of their operation, the tendencies of the 5-HT and norepinephrine-dopamine systems in the rat brain, which are closely related to the rhythm in the efficacy of psychotropic drugs, were analyzed. The numbers (frequencies) of peak times in the rhythms of substances in the above-mentioned studies are presented in Figures 15A and B.

Figure 2 in ref. 125 (top), and Figure 15 herein, show the peak times of 5-HT; of its metabolite, 5-HIAA, and of the activity of tryptophan hydroxylase. Tryptophan hydroxylase activity increases in the latter half of the dark span, resulting in enhanced 5-HT synthesis. Thus, the accumulation of 5-HT reaches a maximum between the middle and the latter half of the light span, and the utilization of 5-HT is enhanced in the middle dark span, leading to an increase of 5-HIAA. An increase in the extracellular 5-HIAA during the dark span in rats has been observed using in vivo voltametry[190] and in vivo microdialysis.[188] These findings are consistent with the findings of Hery et al.[207]

That the 5-HT releasers, tetrabenazine[28] and reserpine,[208] are more effective during a dark span may reflect the activated release of 5-HT during this span. The effect of the 5-HT uptake blocker, clomipramine,[25] is stronger at noon in humans. This may also be due to an increase in 5-HT release at noon, in the middle of the activity span. Drugs that have presynaptic effects seem to be more susceptible to the direct effects of their transmitter's rhythm.

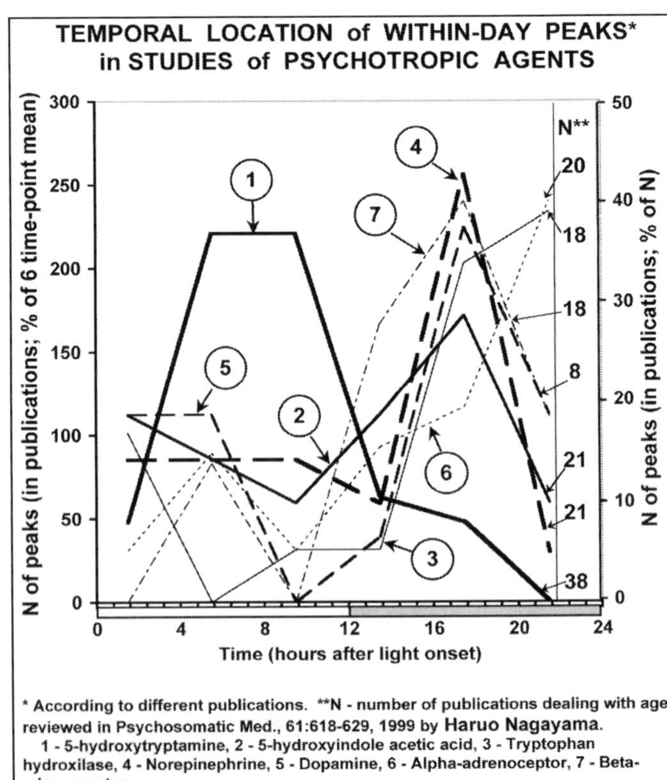

Figure 15B. Pooled data from Figure 15A expressed as a percentage of each series mean value. © Halberg.

*Table 7. Fits of a calendar year (left) and of an ~1.3 (trans)-year (right) to daily average counts of 8-OH-DPAT-associated forepaw treading and head weaving**

	8766.0 Hours (1 Year) (= 360°)					11395.8 Hours (1.3 Year) (= 360°)				
Variable	PR	P	MESOR ± SE	Double Amplitude ± SE	Acrophase ± SE	PR	P ± SE	MESOR SE	Double Amplitude ± SE	Acrophase
Forepaw treading	75	0.254	35.41 ± 2.27	16.52 ± 6.86	-108° ± 21	84	0.164	34.76 ± 1.77	15.66 ± 5.06	-73° ± 18
Head weaving	73	0.270	32.55 ± 1.65	10.56 ± 4.60	-40° ± 26	94	0.059	33.53 ± 0.75	11.90 ± 2.14	-3° ± 10

*Five mean values taken off graph in reference 22 do not allow demonstration of a rhythm, but all original data may allow rejection of the zero-amplitude assumption.

Table 8. Circadian rhythm in serum 5-hydroxytryptamine of male human subjects

Group Investigated (N of Subjects) (Daily Span of Rest or Sleep & Darkness)	Mean of All Series* ± SE μg/100 ml	Circadian Amplitude (A) (95% Confidence Interval of A) μg/100 ml	Percent of Mean	Circadian Acrophase (ϕ) (95% Confidence Arc) ϕ (Local Time) Ref: 00^{00}	ϕ (Degrees) Ref: Mid-Dark Span
Healthy mature men (5) [21^{00}-06^{00}]	19.0 ± 1.2	1.7 (0.8, 2.5)	8.9 (4.2, 13.2)	09^{12} (04^{48}, 13^{08})	-116° (-50, -175)
Mentally retarded males (11)⁺ [21^{00}-06^{00}]	21.8 ± 1.0	2.2 (0.1, 4.2)	10.1 (0.4, 19.3)	09^{30} (05^{24}, 14^{44})	-117° (-59, 199)

*Other mean values in μg/100 ml (± SE) obtained in this study were: a) for 8 male patients with Down syndrome, 17.0 ± 1.0; b) for 4 healthy mature women, 25.6 ± 2.6; c) for 9 female patients with mental retardation, 7 of them with Down syndrome, 19.4 ± 1.01. ⁺Of the 11 mentally retarded subjects, 8 had Down syndrome and one a coexisting convulsive disorder.

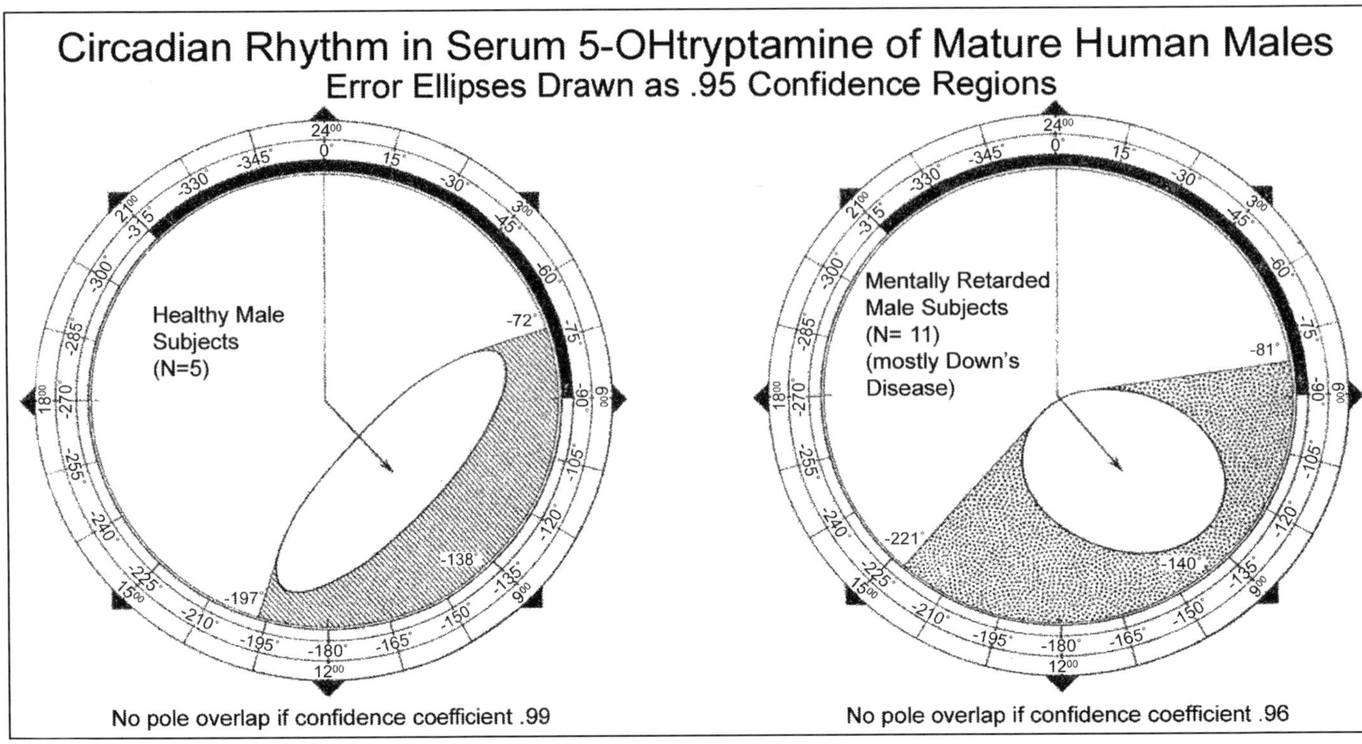

Figure 16. Cosinor summary of two samples of short time series on serum 5-HT. Results of fitting a 24-hour cosine summarized by directed line representing amplitude-weighted sample phase—acrophase—and by an error ellipse representing the corresponding 95% confidence region. Error ellipses do not overlap the pole (in keeping with the assumption of a statistically significant circadian rhythm). Directed lines point to circadian acrophase; the length of these lines indicates the sample amplitude of the circadian rhythm. Note agreement of two sample acrophase values. From Halberg et al.[227] © Halberg.

The waveform of the rhythm in the efficacy of 8-0H-DPAT is the mirror image of the waveform of the rhythm in the wet-dog-shake response to DOI. DOI is a selective agonist for the 5-HT2A and 5-HT2C receptors, and the wet-dog-shake response is a 5-HT2A receptor-mediated behavior that is not related to the 5-HT2C receptors. Because a rhythm in the effect of agonists is thought to be most greatly affected by the susceptibility rhythm of the relevant receptors, the 5-HT1A and 5-HT2A receptors are considered to have mirror-image waveforms in their susceptibility rhythm. The rhythm of a neurotransmitter may not directly mediate the rhythm of its corresponding receptors.

Norepinephrine and dopamine reach a peak during the middle dark span, followed by an increase in the numbers of alpha- and beta-adrenergic receptors in the middle to terminal dark span (Fig. 15 herein, bottom in ref.125). Many studies have demonstrated that the effects of amphetamine, an indirect dopamine agonist,[23,109,129] reach a peak in the latter half of the dark span. In this span, when the level of the dopamine metabolite, homovanillic acid (not presented in the figure), reaches a peak, dopamine release might easily occur. That amphetamine enhances dopamine release might be the reason that the rhythm in the efficacy of amphetamine reaches a peak in the latter half of the dark span.

In the norepinephrine-dopamine system, presynaptically acting drugs may be affected by the rhythm of the relevant transmitters and their substrates. This is not the case for postsynaptically acting drugs. The dopamine agonist, apomorphine, has a peak of the rhythm in efficacy in the light span in rats.[27,28,139,152] Different antipsychotics can have different peak times. Moreover, in *tma* studies of rhythms of the dopamine receptor, the reported times differ from study to study. This discrepancy is thought to be due mainly to the fact that agonists that are highly specific for individual receptor subtypes have not been used. It may be necessary to clarify the waveform of rhythm for each receptor subtype and for each brain region.

Need for Replication

The curve for 5-HT on top of Figure 15A, peaking in the middle of the light span,[125] corresponds to the curve in Figure 2A on the same variable, also analyzed *tmi* in Figure 2C. Thus, there is a *tma* replication. The *tmi* approach to the same data of Figure 2D on 5-HT, with a single cosinor, yields a large amplitude circadian 5-HT rhythm that has an acrophase within the 95% confidence interval reported elsewhere,[53,161] whether only a 24-hour cosine curve is fitted, or whether a second (12-hour) harmonic is fitted, which approximates the deviation from sinusoidality much better and yields an orthophase (a phase based on a 2-component fit of cosine curves with periods of 24 and 12 hours) also within the mapped 95% confidence interval of Figure 2D.

Figure 16 shows the *tmi*-assessed circadian rhythm of circulating 5-HT in healthy and mentally retarded persons. Table 8 also shows a numerical difference in mean circadian amplitude and MESOR, along with a surprising reproducibility in acrophase. Tentatively, one can consider the agreement in a characteristic (and each characteristic is to be tested separately) as a replication. One must realize, however, that even with 11 replications of a 6-timepoint study at 4-hour intervals one can find two studies (in February and March) that differ greatly in phases, until proof is offered to the contrary as a function of circannual variation (Fig. 12.8 in ref. 51).

Table 9. Demonstration by the lack of a statistically significant difference by parameter tests of the replicability of an anticipated circadian rhythm in murine susceptibility to* E. coli *endotoxin*

			Original Values		% of MESOR	
Series N	**N**	**MESOR**	**Amplitude**	**Acrophase**	**Amplitude**	**Acrophase**
Series 1	7	51.75	36.51	-224°	70.56	-224°
Series 2	7	38.02	29.65	-221°	77.98	-221°
Test of equality of parameters						
Parameter(s)	DF	F	P	DF	F	P
MESOR	(1, 7)	4.25	0.078	(1, 6)	—	—
Amplitude	(1, 7)	0.52	0.494	(1, 6)	0.10	0.759
Acrophase	(1, 6)	0.02	0.890	(1, 6)	0.02	0.890
(A, ϕ)	(2, 8)	0.29	0.758	(2, 8)	0.06	0.940
(M, A, ϕ)	(3, 8)	1.65	0.255	(3, 8)	0.04	0.988

*The lack of statistical significance in parameter tests is a necessary but not sufficient condition; the borderline statistical significance of a difference in MESOR (bold) in the original data may be associated with a different start time.

Need for Checking Replications

Table 9 shows how one can test for the reproducibility of a rhythm by the results of parameter tests[1] on the data of Figure 3 on the susceptibility rhythm to endotoxin. Disparities among studies will be reduced if the mapping combines *tmi* and *tma*. Under these conditions, discrepancies can lead to novel findings by reducing the contribution of differences in procedures, such as differences in start time from those due entirely to unknown factors or chance.

Discussion: History Is Prologue

Under the title "When to treat", it was noted in 1975 that:

Rhythms by implication are too often regarded as immediate consequences of cycles in our environment, in our activities, meals and other social schedules. It is not generally realized that a number of rhythms have been shown to persist during starvation, dehydration, in bedrest and in the absence of some of those environmental cycles with which they have been previously synchronized. Moreover, they persist with periods differing from precisely 24 hours or from precisely 1 year...

Free-running circadian rhythms... may require treatment at different clock hours on different days, until, or unless, the rhythms can be therapeutically resynchronized with a routine. In health, standardized living routines can synchronize a network of interacting neural, endocrine and cellular rhythms. Drugs have been used to manipulate timing in patients...

Circadian rhythms in rodents can be synchronized by periodic environmental factors, such as recurrent within-day changes in lighting. Under conditions of synchronization by a lighting regimen and/or meal-timing, it can be shown that a rhythmic organization determines important differences in the effectiveness or toxicity of various drugs when the same dose is given at various circadian times. ...The scheduling of therapy and even that of food consumption thus seems long overdue. ...One then prescribes "à la carte" and must not advocate a "table d'hôte menu", i.e., a rigidly fixed prescription of timing for a population as a whole.[209]

Caution in Considering Panchronotherapy

(The recommendation for timing of a given treatment) may differ for particular "morning-active" ("larks") and "evening-active" individuals ("owls") or for those who by choice or circumstance work on odd routines - night watchmen, policemen, nurses, pilots on transmeridian routes, to cite but a few examples. Again, whenever possible, clinical investigation should identify temporal reference variables or markers and should map these body functions under defined conditions in a way similar to the completed mapping (of variables in experimental animals, e.g., in Fig. 10). Nonetheless the identification and generalization of the body's time dimension involves at least initially much more than the simple specification of one or several clock hours of drug administration.

This (need for time-microscopy) can be made even clearer by a historical analogy. The microscopic examination of tissues became a routine part of diagnostic procedure only a (relatively) short time ago. For all too long medicine had used the traditional criteria of " tumor, rubor, calor and dolor " (to diagnose inflammation vs. cancer). But eventually the focus on swelling, reddishness, heat and pain was replaced by microscopic examination of tissues, which gave access to phenomena previously beyond the reach of observation (such as a malignant growth). Now a similarly revolutionary aid to diagnosis has become available, namely the computer, which allows the investigator to resolve components of biological time series and to identify rhythmic patterns otherwise inaccessible.

Among the preliminary findings in research that may be exploited by using the lately developed yet now rather readily accessible tool for calculations is the recognition that the timed administration of a drug can tip the scale between life and death. We must try to identify the times when a given patient is best able to resist the toxicity of the drug. Failure to take advantage of chronobiologic methods, once they are available, may one day come to seem as negligent as it would seem now to make a diagnosis of a tumor without microscopy.[209]

Updates are available in general terms (ref. 146; cf. refs. 209-223, 240-243).

Clinical Application of Rhythms in Drug Efficacy

What, then, can be expected when the results of animal experiments are applied to humans?[126]

We attempted to assess the clinical relevance in humans of our findings of the rhythm phase observed in rodents by administering a comparable clinical dose, taking into account that antiapomorphine effects correlate well with antipsychotic effects and that the pattern of daily activity in humans is opposite of that in rodents. As a result, it was speculated that chlorpromazine would be most effective in producing sedative and antipsychotic effects when administered at midnight and immediately after rising, respectively. For haloperidol, administration in the evening would be best for obtaining either a sedative or antipsychotic effect. That the phase of rhythm differs depending on the parameters measured suggests that the time of administration can be selected to optimally treat a target symptom or to minimize an adverse reaction. Furthermore, because different drugs have different phases of rhythm in efficacy even for the same symptom, this may further enhance the applicability of time of drug administration.

These suggestions, however, do not mean that once-a-day administration is always optimal. Also, in the case of divided administration, an uneven division corresponding with rhythms in efficacy and adverse reactions can be used instead of the commonly used even division. Whether administration should be made once daily at the time when drug susceptibility is maximal, or whether a large dose should be administered at the time of low susceptibility along with unevenly divided smaller doses at the time points of high susceptibility, should be studied for individual drugs.

Utilizing this phenomenon, we may be able to reduce the daily drug doses that are currently prescribed. Considering that psychotropic drugs are usually administered in large doses and over long periods of time and that the drugs show a variety of adverse reactions, the clinical significance of this phenomenon is potentially quite significant. The rhythmicity in efficacy was also observed in chronically treated animals,[29] suggesting further clinical application.

Even if a rhythm in drug efficacy is present in humans, not only similarities but also differences in phase rhythm are likely to exist between animals and humans. Therefore, it is necessary to reconfirm these matters in humans, referring to results from animal experiments. Because it has been confirmed by animal experiments that individual drugs have different phases, the waveform of rhythm for individual drugs based on their clinical doses must be confirmed. The biological rhythms of individual patients, particularly shift workers, must also be taken into account. In addition, the time of administration is generally an important factor in clinical investigations.

Some adverse effects of psychotropic drugs have different consequences depending on the time of onset in relation to a patient's particular lifestyle. For example, drowsiness due to psychotropic drugs is an adverse reaction disturbing daily life when it occurs during the day but may be considered beneficial when it occurs at night. This is certainly one important matter we confront with administration of antipsychotic drugs to patients. This matter, however, is independent of the importance of rhythms in drug efficacy and adverse reactions. For example, the same sedative effect described above may be a required drug effect all day long for some patients.[126]

As Simply as Possible but not Simpler

There is a circadian variation in the efficacy of psychotropic drugs. This phenomenon has been studied in many animal species and confirmed in humans. Although there are various discrepancies among reports, such discrepancies will be considerably decreased by season-controlled administration studies (also controlled for broader chronomes). Some discrepancies may still remain unexplained but may be due to other methodological difficulties. In the research of rhythms, experimental animals should be raised under strict conditions (and the consideration of ever broader chronomes). Room temperature and humidity must be kept constant, and only artificial illumination should be used to accurately control the time of illumination and complete darkness. In addition, noise from the environment must be muted, and the animal room should be cleaned at random time intervals under a dim red light. Only animals raised under such conditions for at least 3 to 4 weeks should be used for experiments.[126]

Synchronization with an LD12:12 or other regimen can be validated by telemetry at least of core temperature and preferably by the telemetry of pertinent variables, providing new timing endpoints, to study long-term effects by using chronomic endpoints as gauges of drug effects. As a number of geophysical variations become known for their ability to shift or otherwise to affect rhythm characteristics, the need arises to record the changes and, if need be, to assure constancy by shielding or by compensating for these changes such as those in geomagnetic disturbances.

Illuminating conditions on drug administration during a dark span as well as conditions for observing the responses of animals after drug administration should also be controlled strictly. In some previous reports, some of these conditions were not described distinctly.

For wider clinical application of this phenomenon, it is necessary to identify the rhythm phases of each drug effect and the adverse reaction to a number of psychotropic drugs. Moreover, there are many other problems to overcome, such as whether once-a-day administration is desirable, whether unevenly divided administration is desirable, and, in the case of unevenly divided administration, the most desirable division (ref. 126; cf. ref. 224 for the case of a drug with such a division). It is strongly suggested, however, that more effective treatment can be obtained by utilizing this phenomenon.

These rhythms seem to be endogenous circadian rhythms resulting from the rhythmicity in drug susceptibility of the brain, which is not dependent on drug pharmacokinetics. The rhythm is attributed to the rhythms in the neurotransmission system, such as neurotransmitters, receptors, and second messengers (that, nevertheless, remain open systems to their environment).

The causal mechanism of the rhythms in the efficacy of presynaptically acting drugs may be explained to some extent by the rhythm of the neurotransmitters, whereas that of postsynaptically acting drugs is difficult to explain on the same basis. It is necessary to elucidate the waveform of rhythm for each receptor subtype and for each brain region. It is also necessary to develop a technique capable of continuously measuring the rhythm of the neurotransmitter systems from individual animals over a long period of time.[126]

Eventually, with the monitoring of some indirect markers of body time, such as motor activity, core temperature, heart rate and blood pressure available, we may visualize prescriptions not as to clock-hour as an indicator of timing but by reference to the timing of a marker rhythm. In cancer chronotherapy, this approach has doubled the two-year survival rate by timing.[145] A literature search for Medline-indexed publications about chronobiology over the past four decades shows an increase from very few per year in 1963 to thousands in 2002. There is also a steady increase in publications on "chronotherapy", with the majority of these being time-macroscopic. Time-microscopy, however, may resolve some of the discrepancies resulting from trusting only the unaided eye and sole reliance on analyses that cannot separate the year from the transyear. *Tma* procedures on limited data cannot examine the possible beating already demonstrated by *tmi* for human pulse and blood pressure in series of 6- to 36-year length.[40,41]

The longer the series, the easier it is to distinguish between a calendar year-synchronized component and a transyear with a period of about 1.3 years. In many of the long series, both components can be demonstrated concomitantly, suggesting that the transyear is a separate component in its own right. The shortest span allowing at least a separation of the two components is about 4 years, the common denominator between the 1.0 and 1.3-year periods. Because the two periods are close to each other, they can beat notably when their amplitudes are similar, as observed for some blood pressure and pulse series. Analyses over spans shorter than about 4 years will not be able to distinguish between the two components, revealing only a single spectral peak, which may be prominent when both components are in phase and reinforce each other. The same spectral peak may, however, be undetectable when both components are out of phase and cancel each other out. Such methodological considerations prompt the recommendation to proceed, whenever possible, with longitudinal monitoring over several years, and to record the date as well as the time of each experiment, so as to allow the pooling of results from several investigators.

Caveat

One can neither assess the pulse in a second nor circadians or circannuals in a day or a year, respectively. But the data in hand suffice to suggest that this task has to be systematically shouldered unless "chronotherapy" is to remain an often discrepant endeavor. When rhythms with multiple frequencies interact, such interactions have to be resolved. Just as we monitor parking garages and supermarkets to cope with rape and theft, we have to monitor the pertinent physiology and pharmacology for a timely and timed administration of treatment, guided by marker rhythms for diagnosing the problems on hand and for their timely and timed treatment.

Desynchronization, or Rather Asynchronization and Beating

By comparison to the set of multiple new biological "years", in the perspective of the 1950s,[1] the biological day (a circadian rhythm) seemed to be simple to deal with. More specifically, circadian rhythms in laboratory animals could be synchronized by an alternation of light and darkness on a 24-hour schedule, would be phase-shifted by the temporal displacement of this regimen along the 24-hour scale, or could be desynchronized from this lighting regimen, e.g., after blinding.[1,7,93] Likewise, in humans, circadians could be desynchronized from the daily routine with which they had been synchronized, e.g., in caves[101] but not necessarily after (e.g., surgical, in the case of cancer) removal of the eyes.[245] A fixed prominent thermic and photic environmental counterpart of the biological rhythms was as obvious as night and day. It may seem that the same applies to the biological counterpart of equally obvious changes from winter to summer at middle and high latitudes. But this need not be the case. We have to consider more than a synchronization and desynchronization of a rhythm from the calendar year, when we find divergent phases in seasonal studies. "Discrepancies" can be due to different cycles characterizing the spectrum in the neighborhood of a year, i.e., there can be asynchronization – an a priori lack of synchronization.

The need for caution in interpreting the wobbly near-year emerges from analyses of time series of different lengths of possibly influencing environmental variables, such as sunspots, the sun's coronal index and geomagnetic indices, among many others. Comparisons of a few environmental time series covering the same span as the biological data yield different results, when they are analyzed as a whole vs. after decimations, Table 10, for the several components with small amplitude and a wobbly period in the spectral region of a calendar year and of neighboring cycle lengths. Moreover, in Wolf numbers for a 19.7- (rather than a 15-) year span and in yet another set of various environmental time series each covering 29 years, different environmental cycle lengths were found in keeping with prior analyses of an also very wobbly cycle of 6.75-day length, that came to the fore consistently in the entire now-available series of the geomagnetic indices K_p and aa. This average cycle length of 6.75 days can coexist with a cycle length of precisely 7 days.[246] No component with a period shorter than 4 years was apparent when sunspot data from 1749 to 2003 were examined.

With such qualifications, we can describe biological "years" other than the calendar year, only for a given span without generalization beyond it. Only thus, if, in a given span, the 95% CI of a given component does not cover the precise year, this finding should document that it is a distinctly different entity, but not necessarily a consistent one. Should point estimates lie between the precise 1 and the precise 2 calendar year(s)' length, and their 95% CIs cover neither of these lengths, this finding would allow us to regard all of these components with non-zero amplitudes, with 95% CIs not overlapping each other, as biological "transyears", without assuming that they are stationary. As to a discussion of external desynchronization vs. asynchronization of structures in and around us, we focus only on analyses of environmental (and biological) data with gaps for spans corresponding to those in the physiological data. All other extensive analyses of environmental data, not shown, are mentioned only to visualize problems arising from gaps and/or from varying time series lengths and to serve as a warning for planning future studies without gaps and for as long as possible.

Against this background, we present the analysis of a thus-far unique endocrine time series (with gaps) of daily 17-ketosteroid (17KS) excretions and urine volumes excreted during 15 consecutive years by a man (CH) in his mid-forties at the start of study, who was clinically healthy insofar as he died in his sleep in his mid-eighties without any known pathology in the interim. Turning back to Table 10, there is more than one periodic component resolved in CH's urine volumes and in the 17KS (determined in these urines). All components in the 17KS spectrum also differ in period from the precise year by non-overlapping 95% CIs just as a single about-yearly peak differed in Figure 3 of the original publication on the same data in 1965,[102] Figure 17. The 95% CIs of components in each of the two variables do not overlap each other. In the region of the near-year, a comparison of the 95% CIs of cycles in 17KS vs. those in urine volume, are also non-overlapping, i.e., there may be an internal asynchronization (Fig. 17). The 95% CIs in Table 10 reveal each component as an independent entity, asynchronized, i.e., bearing a fixed phase relation neither with each other nor with respect to the calendar year.

The time course of phases in chronobiologic serial sections carried out with different interval lengths chosen for analysis (not shown) is in keeping with the periods listed in Table 10. The term "asynchronized" is used to say that these components are different entities, and that in relatively short series the other-than-circannual cycles can be confused with a circannual rhythm. In sufficiently long series they appear to free-run from

Table 10. ***Overdue transdisciplinary imaging in time modeled by a series of daily (and denser) determinations of human 17-ketosteroid (17KS) excretion in the transyear and decadal charts***

	Transyear (TY)		Near-Year (NY)		Year (Y)	Amplitude Ratio (%)	
Data Series (Units)	**Period (Years) (95% CI)**	**Amplitude (95% CI)**	**Period (Years) (95% CI)**	**Amplitude (95% CI)**	**Amplitude (95% CI)**	A_{TY}/A_{NY}	A_{TY}/A_{Y}
17-KS (daily) (mg/24 h)	1.313 (1.289, 1.338)	0.23 (0.16, 0.31)	1.036 (1.023, 1.049)	0.26 (0.18, 0.34)	0.19 (0.12, 0.26)	88.5	121.1
Urine volume (daily) (ml)	1.392 (1.352, 1.432)	36.82 (18.20, 55.43)	1.000 (0.987, 1.013)	60.14 (41.73, 78.55)	60.14 (43.44, 76.83)	61.2	61.2
K_p (daily) (AU)	1.301 (1.235, 1.367)	0.07 (0.00, 0.14)	1.056 (1.036, 1.077)	0.15 (0.08, 0.22)	0.05 (—, —)	46.7	140.0
Wolf N (daily) (AU)	1.262 (1.233, 1.291)	8.53 (4.38, 12.67)	1.049 (1.017, 1.080)	5.40 (1.21, 9.58)	2.97 (—, —)	158.0	287.2
K_p (matching dates) (AU)	1.224 (1.163, 1.284)	0.09 (0.00, 0.18)	1.052 (1.018, 1.085)	0.12 (0.04, 0.21)	0.07 (—, —)	75.0	128.6
Wolf N (matching dates) (AU)	1.388 (1.316, 1.461)	5.62 (0.46, 10.78)	1.000 (0.965, 1.035)	6.05 (0.90, 11.20)	6.05 (1.38, 10.72)	92.9	92.9

N data = 3719; N days = 5476; AU: arbitrary units

Time structure of solar activity (gauged by Wolf numbers) and geomagnetic disturbance (gauged by global index K_p) with focus on any transyear, near-year and/or calendar year during 1932-2003 (monthly values)

	Transyear (TY)		Near-Year (NY)		Year (Y)	
Variable	**Period (Years) (95% CI)**	**Amplitude (AU) (95% CI)**	**Period (Years) (95% CI)**	**Amplitude (AU) (95% CI)**	**Period (Years) (95% CI)**	**Amplitude (AU)**
K_p (1)	1.294 (1.28, 1.31)	0.08 (0.00, 0.16)	1.056 (1.05, 1.06)	0.09 (0.01, 0.16)	—	—
Wolf N (2)	—	—	—	—	—	—

1. After removal of model consisting of cosine curves with trial periods of 35 (estimated as 32.2), 10.5 (estimated as 10.526) and (fixed) 0.5 y.

2. None of these components can be resolved in longer 255-year span (1749-2003) either.

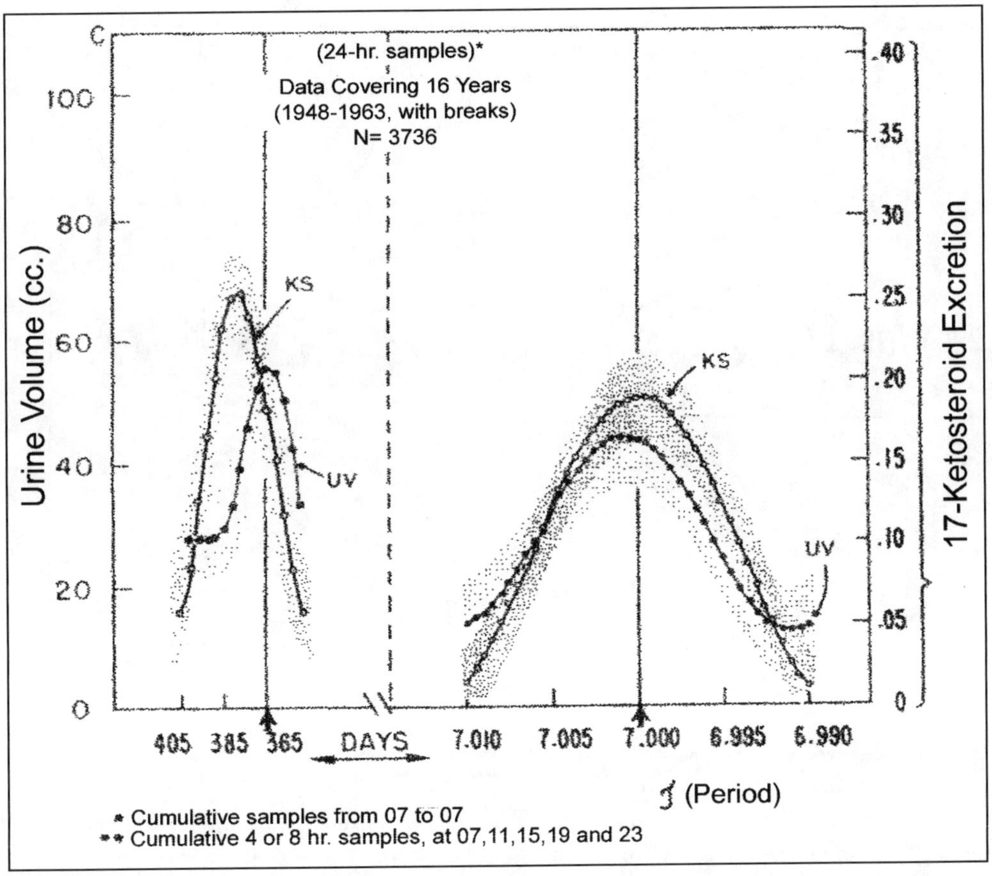

Figure 17A. Apparent circannual (left) and circaseptan (right) internal desynchronization in a clinically healthy man, CH, of urine volume and urinary 17-ketosteroid excretion from each other and from the calendar year; uncertainties were not resolved and not interpreted at the time of publication in 1965.[102] © Halberg.

the year and may be misinterpreted as being desynchronized from the calendar year. Any drift in the phase domain, however, can indicate the a priori presence of cycles differing from 1-year length, as a feature of possibly built-in self-cycling, with a very wobbly period changing with modulations by decadal, multidecadal and other cycles with lower frequencies. Conceivably, the frequency (the period) rather than the phase may be built into the organism. The seemingly free-running insofar as asynchronized periodicity may respond to random stimuli, that result in a "single stimulus manifestation", the phenomenon underlying a superficial "single stimulus induction" by stimuli that in themselves carry no information on cycle length and may elicit the cyclic response at any time, with the zero phase being the time of their application, a fact demonstrated for circadians, known for circaseptans since antiquity, and here applicable perhaps for circannuals and transannuals around and possibly in us.

The untangling of the relative time-varying contributions of these components, as demonstrated in Figure 18, is the more interesting since these periodicities may be beating. A minimal cycle length is needed to demonstrate a beat, e.g., of an about-yearly and/or near-yearly and/or trans-yearly component, Figure 18. Here lies a challenge to everyday physiology and of a chronopharmacology based on follow-ups on the important data in Figures 14A and 15A, originally presented as a putative circannual rhythm. A near-year may be closer to terrestrial magnetism as such and a transyear more the contribution of the solar wind acting upon (but not solely accounting for) terrestrial magnetism. In our organism we may have acquired the capacity of self-cycling with several of these (beating) frequencies.

Clarifying the solar or terrestrial mechanisms underlying different, distinct biological spectral components coded in the course of evolution and/or resonating in organisms with near-matching environmental periodicities requires long time series. Most likely information documenting both of these possible conditions may lead to a better understanding of any differences in desired or undesired effects of drugs along scales longer than circadian. Such information from follow-ups on the data in Figures 14 and 15 in turn may lead to the design of new molecules for use on a daily basis, apart from treatment patterned on a calendar-yearly vs. a terrestrial magnetism involving (and/or responsive) near-yearly and/or probably solar wind magnetics involving (and/or responsive) trans-yearly schedule(s).

Conclusion

A future chronopharmacology and chronotherapy may be based (preferably on automatic continuous) physiological monitoring and inferential statistical sequential as-one-goes analyses, guiding drug administration (preferably by a pump). This can be done automatically today for sensing, e.g., glucose, blood pressure or heart rate, among other variables. In a figurative tomorrow there may be an automatically closed loop of coordinated devices for sensing, analyzing and treating disease risk elevations rather than only overt illnesses. All the software and hardware for all elements necessary for such a system are

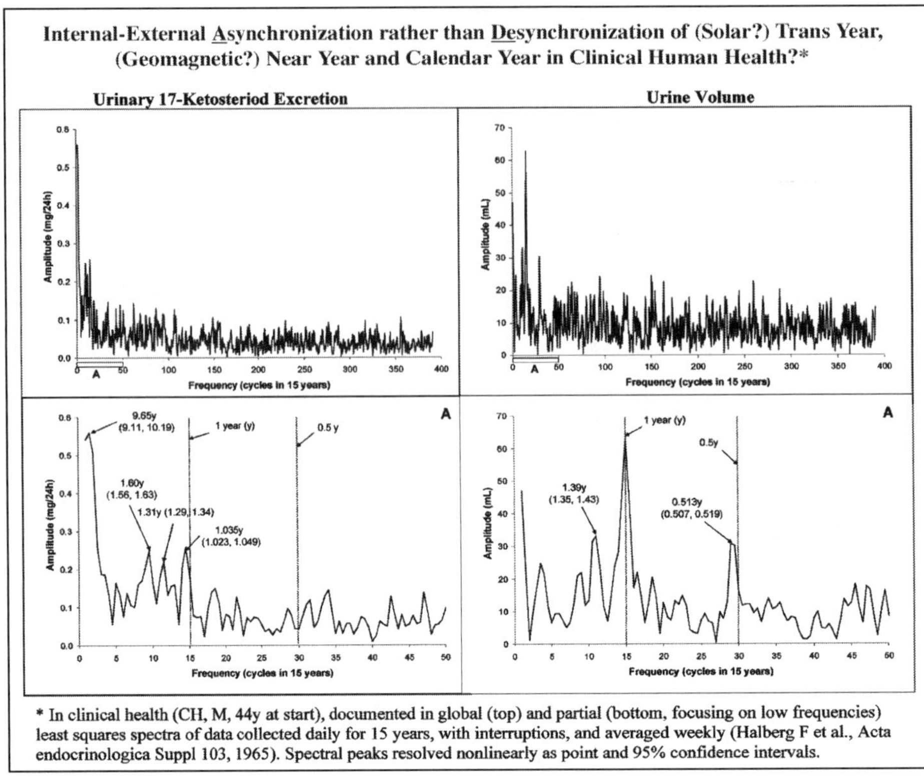

Figure 17B. Meta-analyses of the original data underlying Figure 17A.[102] Several peaks stand out clearly in the low-frequency region underscored by a black bar, area A, on top, but cannot be easily separated by inspection. Hence, this area is reproduced in more detail in the bottom half of this figure. For 17-ketosteroid (17KS) excretion (left), we find a most prominent circadecadal component on the extreme left. This is a peak in its own right, since the first spectral estimate on its left is lower. By comparison to the circadecadal peak, several other components are relatively small but have each a statistically significant amplitude in the spectral region of biological years; for 17KS (left) but not for urine volume (right), all components differ from the precise calendar year (all with 95% confidence intervals (CI), not overlapping the precise year length). By contrast, the most prominent component in the spectrum of urine volume on the right is 1-year synchronized. For urine volume, there is a transyear and a trans-half-year. The phases of any one of the transyearly 17KS components, when they are analyzed in chronobiologic serial sections, are drifting with respect to the phases of other components, as if they were desynchronized from each other and from the calendar year. In this case of a clinically healthy man (CH), however, this is because these are distinct components with different cycle lengths. These cycles may never have been synchronized with each other, a finding in Table 10 and in this figure that introduces the concept of an about-yearly a priori asynchronization of several components that may be beating, as shown for the blood pressure of another man, GSK, and by simulations in Figure 18. © Halberg.

cost-effectively available today for blood pressure and heart rate that can be immediately tested for use as markers for timing, among others, psychotropic drug administration. The greatest benefit, however—a task for the future—is anticipated from the elucidation of mechanisms acting in the normal range to increase disease risk. These could be circadian desynchronizations, and/or circaseptan de- or, rather, asynchronizations and/or largely circannual and transannual self-cycling. Interactions as intermodulations (feedsidewards) among these components, if causally related to risk elevation and amenable to optimization with desired effects, may lead to a preventive causal therapy, correcting earliest rhythm alteration. Such a timely and timed chronotherapy is already on the horizon for the case of blood pressure.[40] In this case, a drug—long-acting carteolol—is available to dampen not only a circadian and a circaseptan[247] but also a circannual and transannual amplitude[248] (Fig. 19). Certainly in the case of an excessive circadian blood pressure amplitude, the circadian damping constitutes a great documented benefit.[249] The question whether broader spectral damping and/or other manipulations of variability are desirable remains the task of promising research.

Chronopharmacology and chronotoxicology, in vitro and in vivo, started as circadian endeavors, first time-macroscopically, then also time-microscopically. Both approaches reveal that time-dependent responses can be critical as such, that they complement dose responses and that, in some variables, the dose-response itself undergoes dramatic changes with time, as does the toxic-therapeutic ratio.[145] As a function of time, a dose response, as a steep, straight line, can move along the dose scale, remaining more or less parallel at different administration times, as in the cases of phenobarbital or ouabain, or conversely the slope of a dose response can actually disappear, as in the case of the corticosterone response to ACTH. When time-macroscopic focus upon circadians has been systematically extended by replications along the scale of a year, discrepancies or differences, respectively, do occur. These can remain vexing sources of variation or phenomena that represent new endpoints for optimizing drug effects.

As physiological monitoring and automatic drug administration systems become practical, many superficial discrepancies may be recognized as features of a broad spectrum of rhythms in an even broader chronome. Some components of the spectral chronome element, because of their low prominence can be detected only time-microscopically; all these components await

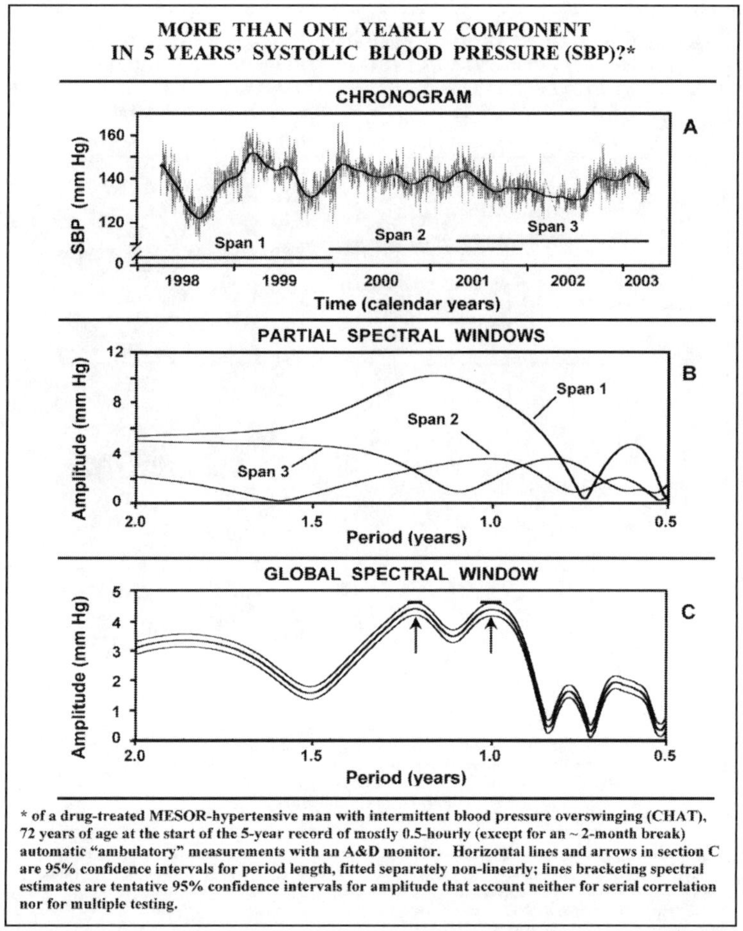

Figure 18A. Multiple spectral components in the spectral region of a year in the blood pressure of GSK, a MESOR-hypertensive man. These cannot be resolved time-macroscopically (top) nor reliably time-microscopically in series of only 2-year length (middle), but stand out *tmi* in 5 years of data (bottom). © Halberg.

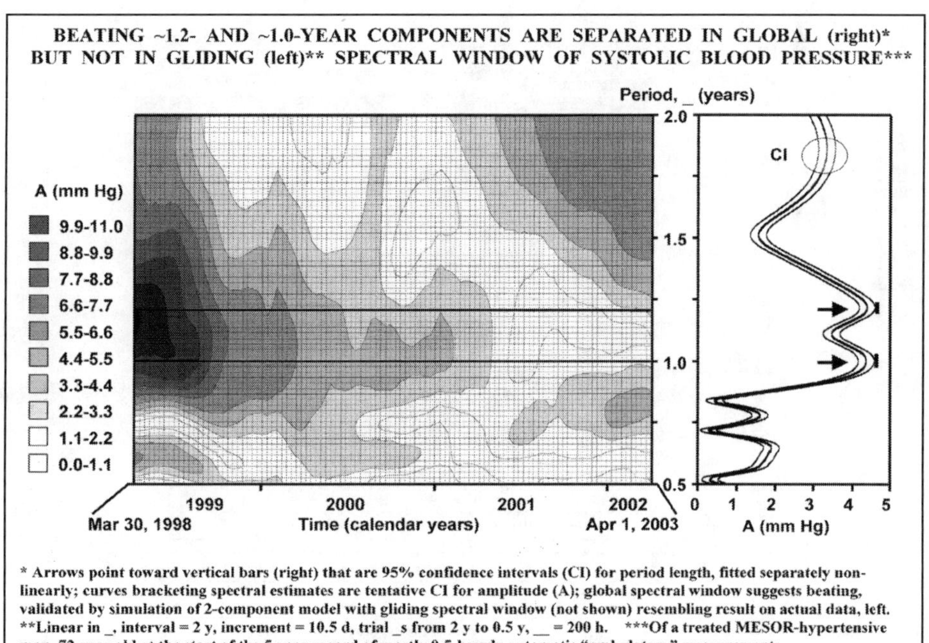

Figure 18B. The two peaks of Figure 18A (reproduced again on the right) are aligned with a gliding spectral window of GSK, suggesting a beat. Note heavy shading in the left when the two components are likely in phase, reinforcing each other, and blank space, when they cancel each other. © Halberg.

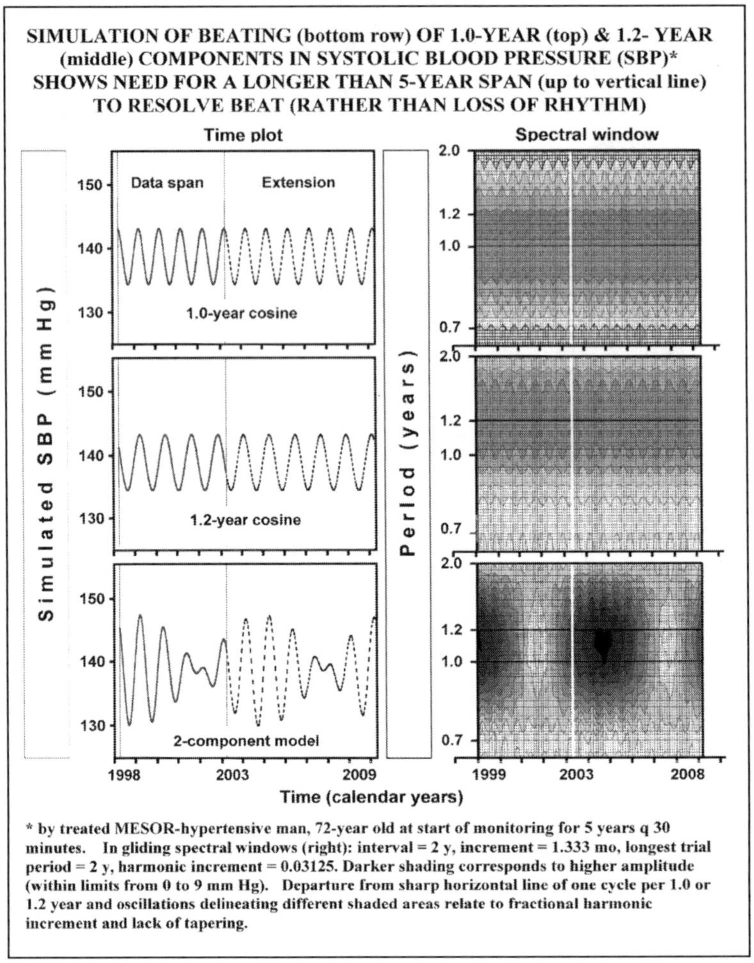

Figure 18C. Simulating the beat of Figure 18B underlines the need for long series in dealing with beating of multiple circannual components (cf. Figure 18D). © Halberg.

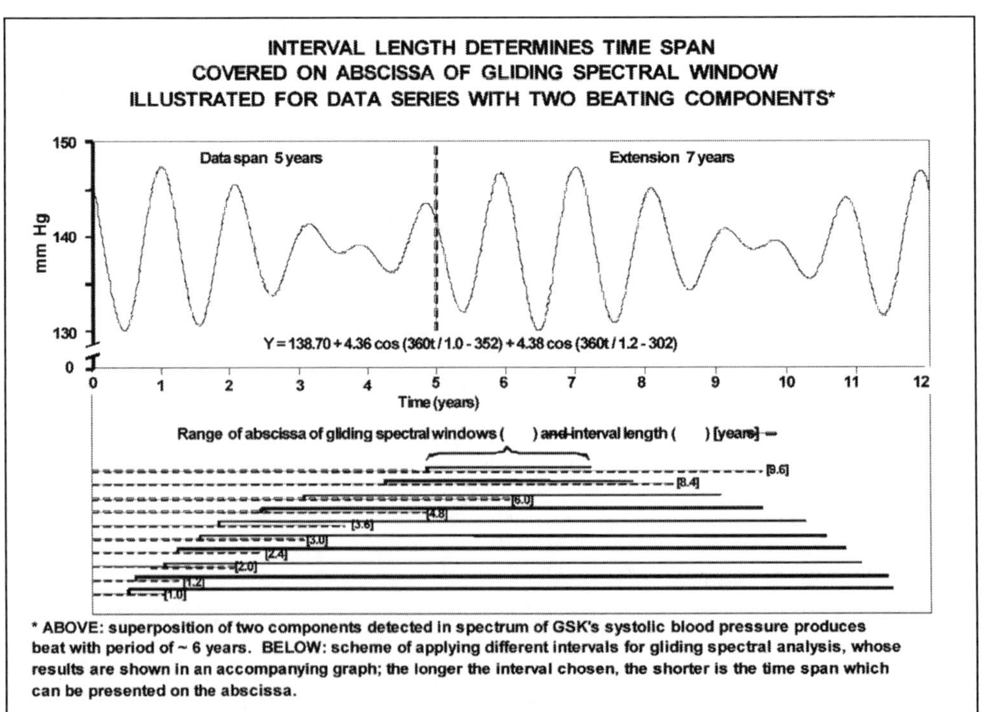

Figure 18D. Further simulation, complementing Figure 18C, shows need for a long interval for analysis to interpret the data. © Halberg.

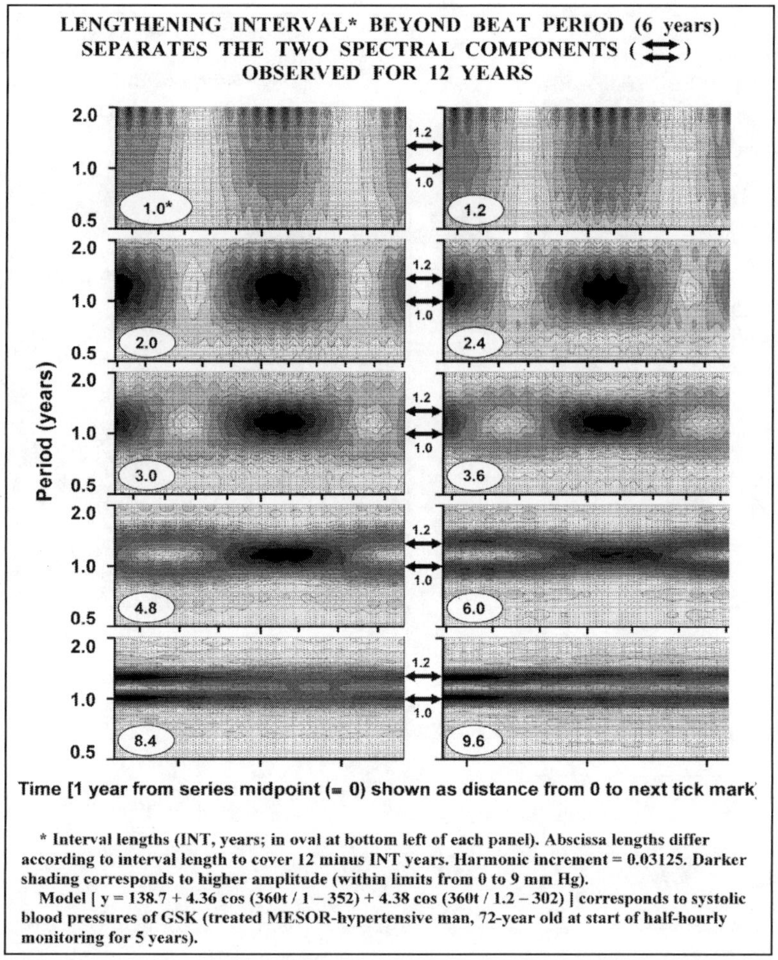

Figure 18E. Separation achieved with lengthening interval in the abstract model. © Halberg.

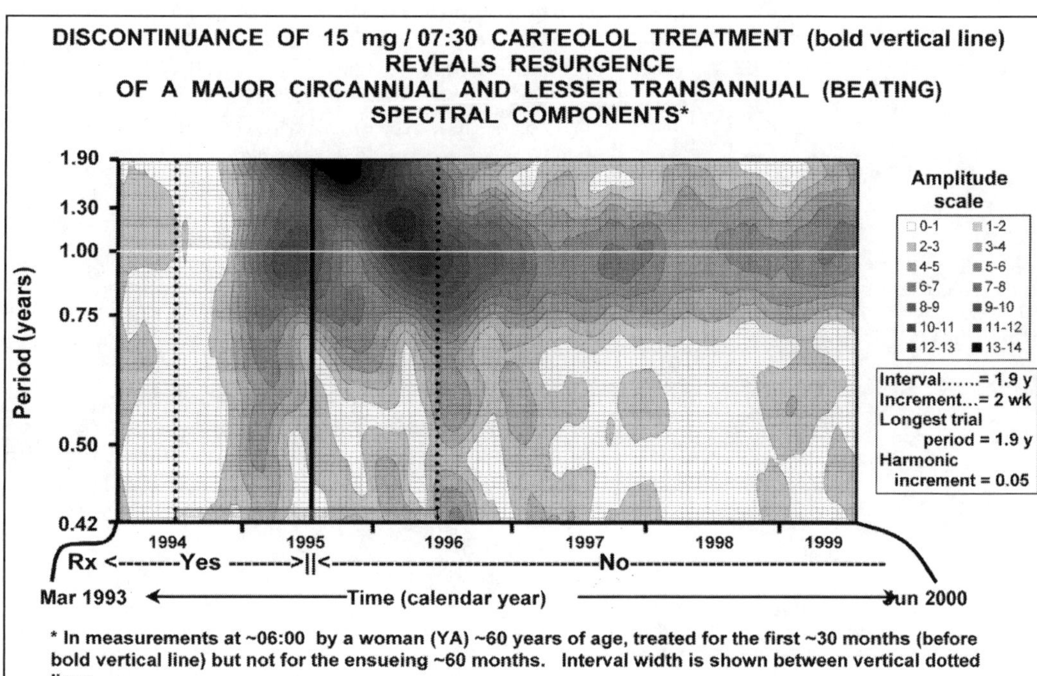

Figure 19A. Initial damping of circannual and transannual spectral components in the presence of treatment for 30 months in an adult woman, YA. Original data of Yoshihiko Watanabe. A new pharmacodynamic effect, emerging upon discontinuation of treatment, with apparent beating. © Halberg.

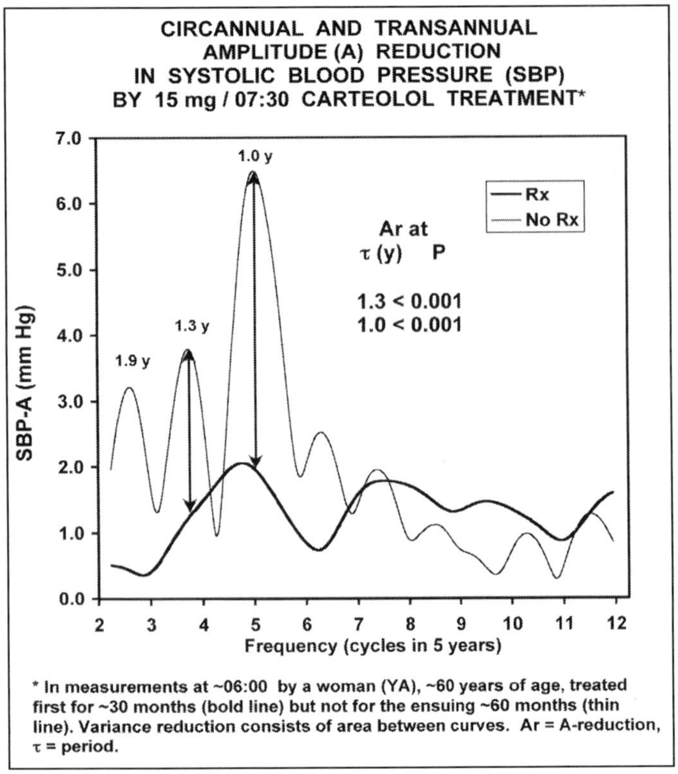

Figure 19B. Damping of circannual amplitude of systolic blood pressure by carteolol is statistically significant. Original data of Yoshihiko Watanabe. © Halberg.

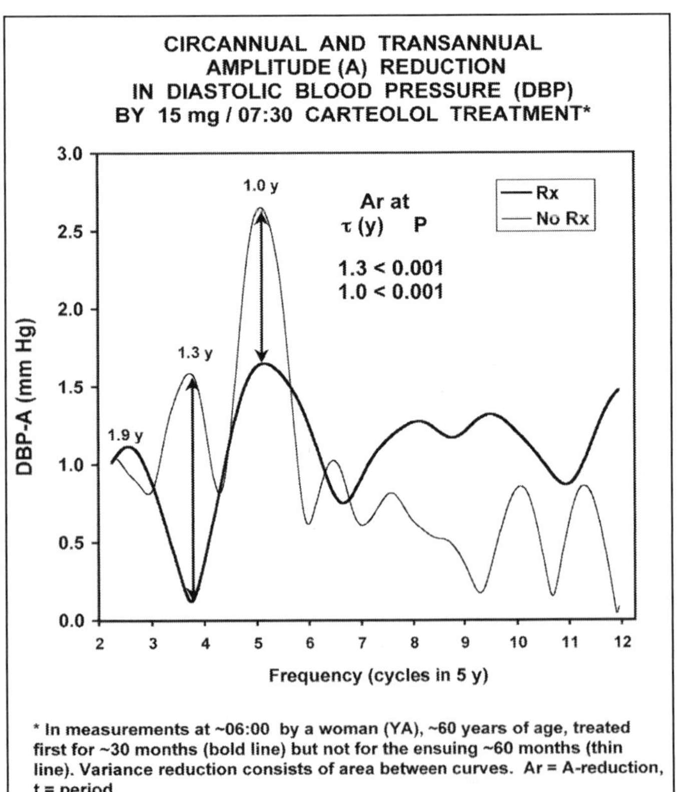

Figure 19C. Damping of circannual amplitude of diastolic blood pressure by carteolol is statistically significant. Original data of Yoshihiko Watanabe. © Halberg.

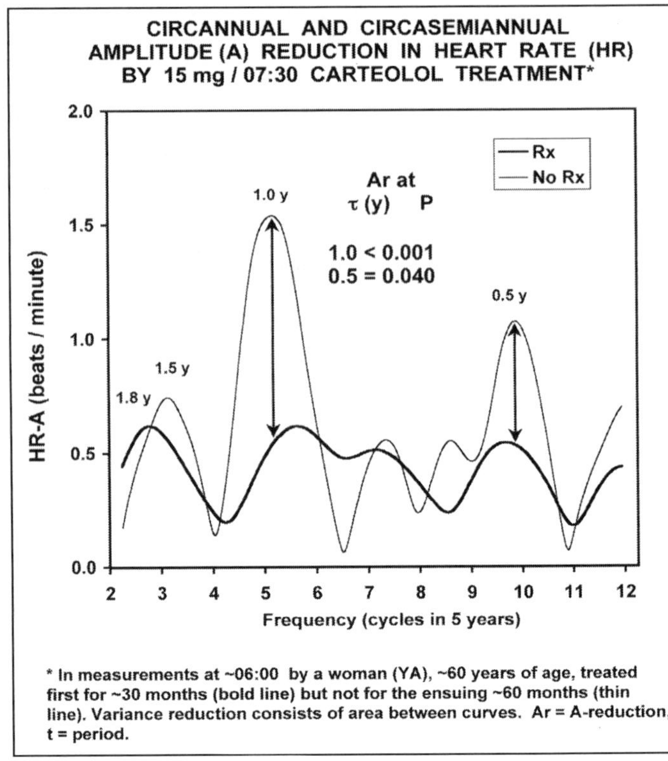

Figure 19D. Damping of circannual and circasemiannual amplitude of heart rate by carteolol is statistically significant. Original data of Yoshihiko Watanabe. © Halberg.

time-microscopic quantification carried out with a chronobiologic design as a phase-zero endeavor preceding the current three stages of drug testing, a topic for another review.[244] Even when time-macroscopy (the naked eye) suffices to establish the phenomenon, it is useful to have parameters such as one or several validated periods; and, at each period, an amplitude and phase, and, when indicated, parameters such as magnitude and orthophase, the latter two characteristics derived from multiple cosine fits), determined with their uncertainties. Treatment should be made as simple as possible but not simpler.

Abbreviations:

tma = time-macroscopic;
tmi = time-microscopic;
DOI = 2,5-dimethoxy-4-iodophenyl-2-aminopropane hydrochloride;
GABA = Gamma-aminobutyric acid;
5-HIAA = 5-hydroxyindole acetic acid;
5-HT = serotonin;
5-MeODMT = 5-methoxy-N.N-dimethyotryptamine;
8-0H-DPAT = 8-hydroxy-2-(di-N-propylamino)tetralin.

References

1. Halberg F, Cornélissen G, Katinas G et al. Transdciplinary unifying implications of circadian findings in the 1950s. J Circadian Rhythms 2003; 1:2. http://www.jcircadianrhythms.com/content/1/1/2
2. Cornélissen G, Halberg F, Breus T et al. Non-photic solar associations of heart rate variability and myocardial infarction. J Atmos Solar-Terr Phys 2002; 64:707-720.
3. Halberg F, Cornélissen G, Otsuka K et al. Cross-spectrally coherent ~10.5- and 21-year biological and physical cycles, magnetic storms and myocardial infarctions. Neuroendocrinol Lett 2000; 21:233-258.

4. Halberg F, Cornélissen G, Katinas GS et al. Feedsidewards: intermodulation (strictly) among time structures, chronomes, in and around us, and cosmo-vasculo-neuroimmunity. About ten-yearly changes: what Galileo missed and Schwabe found. In: Conti A, Maestroni GJM, McCann SM et al, eds. Neuroimmunomodulation (Proc. 4th Int. Cong. International Society for Neuroimmunomodulation, Lugano, Switzerland, September 29-October 2, 1999). Ann NY Acad Sci 2000; 917:348-376.
5. Halberg F. Quo vadis basic and clinical chronobiology: promise for health maintenance. Am J Anat 1983; 168:543-594.
6. Sanchez de la Pena S. The feedsideward of cephalo-adrenal immune interactions. Chronobiologia 1993; 20:1-52.
7. Halberg F. Chronobiology. Annu Rev Physiol 1969; 31:675-725.
8. Halberg F, Cornélissen G, Otsuka K et al. Chronomics. Biomedicine and Pharmacotherapy 2001; 55(Suppl 1):153-190.
9. Halberg F, Cornélissen G, Schack B et al. Blood pressure self-surveillance for health also reflects 1.3-year Richardson solar wind variation: spin-off from chronomics. Biomed Pharmacother 2003; 57(Suppl 1):58s-76s.
10. Otsuka K, ed. Proceedings, 1st International Symposium Workshop on Circadian Rhythms and Clinical Chronobiology, 11 Nov 2000, Tokyo, Japan. Biomed Pharmacother 2001; 55(Suppl 1):7s-190s.
11. Otsuka K, ed. Proceedings, 2nd International Symposium Workshop on Circadian Rhythms and Clinical Chronobiology, 17 Nov 2001, Tokyo, Japan. Biomed Pharmacother 2002; 56(Suppl 2):231s-382s.
12. Otsuka K, ed. Proceedings, 3rd International Symposium Workshop on Circadian Rhythms and Clinical Chronobiology, 9 Nov 2002, Tokyo, Japan. Biomed Pharmacother 2003; 57(Suppl 1):1s-198s.
13. Cornélissen G, ed, Schwartzkopff O, Niemeyer-Hellbrügge P et al. Time structures - chronomes - in child development. International Interdisciplinary Conference, Nov. 29-30, 2002, Munich, Germany. Neuroendocrinol Lett 2003; 24 (Suppl 1). 256 pp.
14. Otsuka K, Cornélissen G, Halberg F. Circadian rhythmic fractal scaling of heart rate variability in health and coronary artery disease. Clinical Cardiology 1997; 20:631-638.
15. Burioka N, Cornélissen G, Halberg F et al. Relationship between correlation dimension and indices of linear analysis in both respiratory movement and electroencephalogram. Clin Neurophysiol 2001; 112:1147-1153.
16. Burioka N, Cornélissen G, Halberg F et al. Approximate entropy of human respiratory movement during eye-closed waking and different sleep stages. Chest 2003; 123:80-86.
17. Burioka N, Cornélissen G, Otsuka K et al. Linear and nonlinear indices of variability in respiratory movement, the electroencephalogram and the electrocardiogram. Neuroendocrinol Lett 2003; 24 (Suppl 1):223-230.
18. Halberg F, Visscher MB. Regular diurnal physiological variation in eosinophil levels in five stocks of mice. Proc Soc Exp Biol (N.Y.) 1950; 75:846-847.
19. Marte E, Halberg F. Circadian susceptibility rhythm of mice to librium. Fed Proc 1961; 20:305.
20. Halberg F. Symposium on "Some current research methods and results with special reference to the central nervous system." Physiopathologic approach. Amer J ment Defic 1960; 65:156-171.
21. Nagayama H, Takagi A, Takahashi R. Toward chronotherapy in psychiatry. In: Takahashi R, Halberg F, Walker CA, eds. Toward chronopharmacology. Oxford: Pergamon Press, 1982:43-50.
22. Nagayama H, Lu JQ. Circadian and circannual rhythms in the function of central 5-HT_{1A} receptors in laboratory rats. Psychopharmacology 1998; 135:279-283.
23. Nagayama H, Takagi A, Sakurai Y et al. Chronopharmacological study of neuroleptics. III. Circadian rhythm of brain susceptibility to haloperidol. Psychopharmacology 1979; 63:131-135.
24. Nagayama H, Takagi A, Sakurai Y et al. Chronopharmacological study of neuroleptics. II. Circadian susceptibility rhythm to chlorpromazine. Psychopharmacology 1978; 58:49-53.
25. Nagayama H, Nagano K, Ikezaki A et al. Double-blind study of the chronopharmacotherapy of depression. Chronobiol Int 1991; 8:203-209.
26. Nagayama H, Lu JQ. Circadian rhythm in the response to intracerebroventricular administration of 6-OH-DPAT. Brain Res 1997; 758:92-96.
27. Nakano S, Hara C, Ogawa N. Circadian rhythm of apomorphine-induced stereotypy in rats. Clin Pharmacol Ther 1980; 12:459-461.
28. Nagayama H, Takagi A, Tateishi T et al. Circadian susceptibility rhythm to neuroleptics: tetrabenazine. Psychopharmacology 1977; 53:61-66.
29. Nagayama H, Takagi A, Yoshimoto S et al. Circadian rhythm of brain susceptibility to haloperidol during chronic administration. Pharmacol Biochem Behav 1982; 18:311-314.
30. Reinberg A, Halberg F. Circadian chronopharmacology. Ann Rev Pharmacol 1971; 2:455-492.
31. Halberg E, Halberg F. Chronobiologic study design in everyday life, clinic and laboratory. Chronobiologia 1980; 7:95-120.
32. Tarquini B, Perfetto F, Tarquini R et al. Endothelin-1's chronome indicates diabetic and vascular disease chronorisk. Peptides 1997; 18:119-132.
33. Tarquini B, Cornélissen G, Perfetto F et al. About-half-weekly (circasemiseptan) component of the endothelin-1 (ET-1) chronome and vascular disease risk. Peptides 1997; 18:1237-1241.
34. Halberg F, Breus TK, Cornélissen G et al. International Womb-to-Tomb Chronome Initiative Group: Chronobiology in space. Keynote, 37th Ann. Mtg. Japan Soc. for Aerospace and Environmental Medicine, Nagoya, Japan, November 8-9, 1991. University of Minnesota/Medtronic Chronobiology Seminar Series, #1, December 1991, 21 pp. of text, 70 figures.
35. Katinas GS, Halberg F, Cornélissen G et al. About 8- and ~84-hour rhythms in endotheliocytes as in endothelin-1 and effect of trauma. Peptides 2001; 22:647-659.
36. Herold M, Cornélissen G, Löckinger A et al. About 8-hourly variation of circulating human endothelin-1 (ET-1) in clinical health. Peptides 1998; 19:821-825.
37. Löckinger A, Köberle D, Koenig PS et al. Neuropeptide chronomics in clinically healthy young adults: circaoctohoran and circadian patterns. Peptides 2004; 25:533-542.
38. Cornélissen G, Halberg F, Pöllmann L et al. Circasemiannual chronomics: half-yearly biospheric changes in their own right and as a circannual waveform. Biomed Pharmacother 2003; 57(Suppl 1):45s-54s.
39. Halberg F, Cornélissen G, Stoynev A et al. Season's Appreciations 2002 and 2003. Imaging in time: The transyear (longer-than-the-calendar year) and the half-year. Neuroendocrinol Lett 2003; 24:421-440.
40. Halberg F, Cornélissen G, Schack B. Self-experimentation chronomics for health surveillance and science: also transdisciplinary civic duty? Behav Brain Sci 2004; 27:267-269.
41. Cornélissen G, Masalov A, Halberg F et al. Multiple resonances among time structures, chronomes, around and in us. Is an about 1.3-year periodicity in solar wind built into the human cardiovascular chronome? Human Physiology 2004; 30:86-92.
42. Halberg F, Cornélissen G, Watanabe Y et al. Near 10-year and longer periods modulate circadians: intersecting anti-aging and chronoastrobiological research. J Gerontol A Biol Sci Med Sci 2001; 56:M304-M324.
43. Dérer L. Rhythm and proliferation with special reference to the six-day rhythm of blood leukocyte count. Neoplasma 1960; 7:117-134.
44. Richter CP. Biological Clocks in Medicine and Psychiatry. Springfield, Illinois: Charles C. Thomas; 1965. 109 pp.
45. Reimann H. Periodic diseases. Philadelphia: F.A. Davis; 1963. 189 pp.
46. Gjessing R. Beiträge zur Kenntnis der Pathophysiologie des katatonen Stupors. Mitteilung I: Über periodisch rezidivierenden katatonen Stupor, mit kritischem Beginn und Abschluss. Arch Psychiatr 1932; 96:319-392.
47. Gjessing R. Beiträge zur Kenntnis der Pathophysiologie des katatonen Stupors. Mitteilung II: Über aperiodisch rezidivierend verlaufenden katatonen Stupor, mit lytischem Beginn und Abschluss. Arch Psychiatr 1932; 96:393-413.
48. Gjessing R. Beiträge zur Kenntnis der Pathophysiologie des katatonen Stupors. Mitteilung III: Über periodisch rezidivierende katatone Erregung, mit kritischem Beginn und Abschluss. Arch Psychiatr 1936; 104:355-416.

49. Simpson HW, Gjessing L, Fleck A et al. Phase analysis of the somatic and mental variables in Gjessing's case 2484 or intermittent catatonia. In: Scheving LE, Halberg F, Pauly JE, eds. Chronobiology, Proc. Int. Soc. for the Study of Biological Rhythms, Little Rock, Ark. Stuttgart: Georg Thieme Publishers/Tokyo: Igaku Shoin Ltd, 1974:535-539.
50. Halberg F. The week in phylogeny and ontogeny: opportunities for oncology. In vivo 1995; 9:269-278.
51. Halberg F. Chronopharmacology and chronotherapy. In: Carpenter DO, editor. Cellular Pacemakers. New York: John Wiley and Sons Inc., 1982:261-297.
52. Haus E, Halberg F. 24-hour rhythm in susceptibility of C mice to a toxic dose of ethanol. J Appl Physiol 1959; 14:878-880.
53. Radha E, Shankaraiah K, Halberg F et al. Developmental, circadian and aging aspects of dopamine, norepinephrine and 5-HT in rat brain regions. In: Redfern PH, Campbell I, Xavier JA, Martin KF, eds. Circadian Rhythms in the Central Nervous System, Proc. IX Conf. IUPHAR, Sat. Symp., Bath, England, August 4-5, 1984. London: Macmillan, 1985:199-209.
54. Halberg F, Visscher MB. A difference between the effects of dietary calorie restriction on the estrous cycle and on the 24-hour adrenal cortical cycle in rodents. Endocrinology 1952; 51:329-335.
55. Halberg F, Visscher MB, Bittner JJ. Eosinophil rhythm in mice: Range of occurrence; effects of illumination, feeding and adrenalectomy. Amer J Physiol 1953; 174:109-122.
56. Halberg F, Visscher MB, Bittner JJ. Relation of visual factors to eosinophil rhythm in mice. Amer J Physiol 1954; 179:229-235.
57. Cavallini M, Halberg F, Sutherland DER et al. Optimization by timing of oral cyclosporine to prevent acute kidney allograft rejection in dogs. Transplantation 1986; 41:654-657.
58. Lemmer B, editor. Chronopharmacology: cellular and biochemical interactions. New York: Marcel Dekker; 1989.
59. Reinberg A, Smolensky MH. Biological rhythms and medicine. Cellular, metabolic, physiopathologic, and pharmacologic aspects. New York: Springer: 1983:305.
60. Touitou Y, Haus E, eds. Biological Rhythms in Clinical and Laboratory Medicine. Berlin: Springer-Verlag, 1992:730.
61. Halberg F. Chronobiology: methodological problems. Acta med rom 1980; 18:399-440.
62. Cornélissen G, Halberg F. Chronomedicine. In: Armitage P, Colton T, eds. Encyclopedia of Biostatistics, v. 1. Chichester, UK: John Wiley & Sons Ltd., 1998:642-649.
63. Halberg F, Bittner JJ, Gully RJ et al. 24-hour periodicity and audiogenic convulsions in I mice of various ages. Proc Soc Exp Biol (NY) 1955; 88:169-173.
64. Halberg F, Jacobson E, Wadsworth G et al. Audiogenic abnormality spectra, 24-hour periodicity and lighting. Science 1958; 128:657-658.
65. Halberg F. Temporal coordination of physiologic function. Cold Spr Harb Symp Quant Biol 1960; 25:289-310. See discussion of LD_{50}, actually documented with data on susceptibility to whole-body irradiation.
66. Haus E, Halberg F, Loken MK et al. Circadian rhythmometry of mammalian radiosensitivity. In: Tobias A, Todd P, eds. Space Radiation Biology. New York: Academic Press; 1973:435-474.
67. Halberg F, Spink WW, Albrecht PG et al. Resistance of mice to brucella somatic antigen, 24-hour periodicity and the adrenals. J Clin Endocrinol 1955; 15:887.
68. Halberg F, Johnson EA, Brown BW et al. Susceptibility rhythm to E. coli endotoxin and bioassay. Proc Soc Exp Biol (NY) 1960; 103:142-144.
69. Halberg F, Haus E, Stephens A. Susceptibility to ouabain and physiologic 24-hour periodicity. Fed Proc 1959; 18:63.
70. Halberg F, Stephens AN. Susceptibility to ouabain and physiologic circadian periodicity. Proc Minn Acad Sci 1959; 27:139-143.
71. Ertel RJ, Halberg F, Ungar F. Circadian system phase-dependent toxicity and other effects of methopyrapone (SU-4885) in mice. J Pharmacol Exp Ther 1964; 146:395-399.
72. Ungar F, Halberg F. Circadian rhythm in the in vitro response of mouse adrenal to adrenocorticotropic hormone. Science 1962; 137:1058-1060.
73. Halberg F, Barnum CP, Silber RH et al. 24-hour rhythms at several levels of integration in mice on different lighting regimens. Proc Soc exp Biol (NY) 1958; 97:897-900.
74. Halberg F, Halberg E, Barnum CP et al. Physiologic 24-hour periodicity in human beings and mice, the lighting regimen and daily routine. In: Withrow RB, ed. Photoperiodism and Related Phenomena in Plants and Animals. Ed. Publ. No. 55. Washington DC: AAAS, 1959:803-878.
75. Breus TK, Pimenov KYu, Cornélissen G et al. The biological effects of solar activity. Biomed Pharmacother 2002; 56(Suppl.2):273s-283s.
76. Masalov AV, Syutkina EV. Magnetic storms and neonatal blood pressure and heart rate chronomes. Neuroendocrinol Lett 2003; 24(Suppl 4):113-118.
77. Cornélissen G, Halberg F, Schwartzkopff O et al. Chronomes, time structures, for chronobioengineering for "a full life". Biomed Instrum Technol 1999; 33:152-187.
78. Watanabe Y, Nintcheu-Fata S, Katinas G et al. Methodology: partial moving spectra of postnatal heart rate chronome. Neuroendocrinol Lett 2003; 24(Suppl1):139-144.
79. Hecht K, Cornélissen G, Fietze I et al. Circaseptan aspects of self-assessed sleep protocols covering 70 nights on 33 clinically healthy persons. Perceptual and Motor Skills 2001; 95:258-266.
80. Balzer H-U, Hecht K. Chronobiologische Aspekte des Schlafverhaltens. In: Hecht K (Hrsg.); Engfer A, Peter JH, Poppei M. Schlaf, Gesundheit, Leistungsfähigkeit. Berlin: Springer Verlag, 1993:49-50.
81. Hecht K. Schlaf und die Gesundheits-Krankheits-Beziehung unter dem Aspekt des Regulationsbegriffes von Virchow. In: Hecht K (Hrsg.); Engfer A, Peter JH, Poppei M. Schlaf, Gesundheit, Leistungsfähigkeit. Berlin: Springer Verlag, 1993:3-12.
82. Walter S, Balzer H-U, Hecht K. Computergestützte Analyse des Schlafprotokolls zur Verifizierung von zirkaseptanen Rhythmen und zum Nachweis von stabilen und unstabilen Zuständen des Schlafverhaltens. Wiss. Ztschr. der Humboldt-Universität zu Berlin. Reihe Medizin 1989; 38/4; 446-450.
83. von Broen B. Computergesteuerte Pilotstudie zur Bedeutung des zirkaseptanen Biorythmus de Schlafverhaltens in der medizinischen Grundbetreuung. Ein Vergleich von Gesunden, Schlafgestörten und Neurotikern. Dissertation Med. Fak. der Humboldt-Universität zu Berlin; 1988.
84. Balzer H-U, Hecht K, Siems R et al. Zirkaseptaner Rhythmus des Schlafverhaltens. In: Schuh J, Gattermann R, Romanow JA. Chronobiologie - Chronomedizin. Wissenschaftliche Beiträge 36(P30). Wittenberg: Martin-Luther-Universität Halle/Saale, 1987:211-214
85. Cornélissen G, Syutkina EV, Halberg F et al. Chronobiologic blood pressure self-monitoring from the beginning until the cure of sickness or death. Human Physiology 1998; 24: 118-125. [In Russian.] Also Fisiologiya Cheloveka 1998; 24 (5): 92-99 [in Russian]; also Human Physiology 1998; 24(5):601-607 (in English).
86. Halberg F, Cornélissen G, International Womb-to-Tomb Chronome Initiative Group: Resolution from a meeting of the International Society for Research on Civilization Diseases and the Environment (New SIRMCE Confederation), Brussels, Belgium, March 17-18, 1995: Fairy tale or reality? Medtronic Chronobiology Seminar #8, April 1995, 12 pp. text, 18 figures. URL: http://www.msi.umn.edu/~halberg/
87. Wever RA. The Circadian System of Man: Results of Experiments under Temporal Isolation. New York: Springer-Verlag, 1979:276.
88. Mills JN. Circadian rhythms during and after three months in solitude underground. J Physiol (Lond) 1964; 174:217-231.
89. Mills JN, Minors DS, Waterhouse JM. Periods of different components of human circadian rhythms in free-running experiments. Int J Chronobiol 1973; 1:344.
90. Mills JN, Minors DS, Waterhouse JM. The circadian rhythms of human subjects without timepieces or indication of the alternation of day and night. J Physiol (Lond) 1974; 240:567-594.
91. Haus E. Biologic aspects of a chronopathology. Ph.D Thesis, University of Minnesota, June 1970:362.
92. Siffre M, Reinberg A, Halberg F et al. L'isolement souterrain prolongé. Étude de deux sujets adultes sains avant, pendant et apres cet isolement. Presse méd 1966; 74:915-919.

93. Ghata J, Halberg F, Reinberg A et al. Rythmes circadiens désynchronisés (17-hydroxycorticostéroïdes, température rectale, veille-sommeil) chez deux sujets adultes sains. Ann Endocrinol (Paris) 1968; 29:269-270.
94. Ghata J, Halberg F, Reinberg A et al. Rythmes circadiens désynchronisés du cycle social (17-hydroxycorticostéroïdes, température rectale, veille-sommeil) chez deux sujets adultes sains. Ann Endocrinol (Paris) 1969; 30:245-260.
95. Halberg F, Reinberg A, Haus E et al. Human biological rhythms during and after several months of isolation underground in natural caves. Nat Speleol Soc Bull 1970; 32:89-115.
96. Hillman DC, Siffre M, Milano G et al. Free-running psycho-physiologic circadians and three-month pattern in a woman isolated in a cave. New Trends in Experimental and Clinical Psychiatry 1994; 10:127-133.
97. Hillman DC, Siffre M, Milano G et al. Urinary about-84-hour (circasemiseptan) variations of a woman isolated in a cave and cosmic ray effects. New Trends in Experimental and Clinical Psychiatry 1994; 10:173-178.
98. Siffre M, Hillman DC, Halberg F. About 3.5-day (circasemiseptan) aspects of urinary temperature in isolation from society. Chronobiologia 1994; 21:157-158.
99. Halberg F. Physiologic considerations underlying rhythmometry, with special reference to emotional illness. Symposium on Biological Cycles and Psychiatry. In: Cycles biologiques et psychiatrie. Symposium Bel-Air III. Geneva: Georg/Paris: Masson et Cie; 1968. p. 73-126.
100. Sanchez de la Peña S, Halberg F, Galvagno A et al. Circadian and circaseptan (about-7-day) free-running physiologic rhythms of a woman in social isolation. Proc. 2nd Ann. IEEE Symp. on Computer-Based Medical Systems, Minneapolis, June 26-27, 1989. Washington DC: Computer Society Press; 1989. p. 273-278.
101. Halberg F, Cornélissen G, Sonkowsky RP et al. Chrononursing (chronutrics), psychiatry and language. New Trends in Experimental and Clinical Psychiatry 1998; 14:15-26.
102. Halberg F, Engeli M, Hamburger C et al. Spectral resolution of low-frequency, small-amplitude rhythms in excreted 17-ketosteroid; probable androgen induced circaseptan desychronization. Acta endocrinol (Kbh) 1965; 50(Suppl103):5-54.
103. Wetterberg L. Light and biological rhythms (Frontiers in medicine). J Internal Medicine 1994; 235:5-19.
104. Halberg F, Visscher MB. Some physiologic effects of lighting. Proceedings of the First International Photobiological Congress (4th International Light Congress) Amsterdam (August) 1954:396-398.
105. Albrecht P, Halberg F, Bittner JJ. Reserpine effects in the mouse and the adrenal. Physiologist 1957; 1:6.
106. Halberg F. Circadian temporal organization and experimental pathology. VII Conferenza Internazionale della Società per lo Studio dei Ritmi Biologici. Siena: Edizioni Panminerva Medica, September, 1960:1-20.
107. Halberg F, Adkins G, Marte E et al. Reserpine effect upon the variance spectrum of human rectal temperature. Fed Proc 1962; 21:347.
108. Halberg F. Circadian (about 24-hour) rhythms in experimental medicine. Proc roy Soc Med 1963; 56:253-257.
109. Wolfe GO, Bousquet WF, Schnell RC. Circadian variations in response to amphetamine and chlorpromazine in the rat. Commun Psychopharmacol 1977; 1:29-37.
110. Philipp M, Marneros A. Chronobiology and its implications for pharmacotherapy of endogenous depression. Pharmacopsychiatry 1978; 11:235-40.
111. Weiss GO, Rogacki N, Gueg A et al. Effect of hypothalamic and peripheral fluoxetine injection on natural patterns of macronutrient intake in the rat. Psychopharmacology 1991; 105:467-76.
112. Currie PJ, Coscina DV. Diurnal variations in the feeding response to 8-0H-DPAT injected into the dorsal or median raphe. Neuroreport 1993; 4:1105-7.
113. Margules DL, Lewis MJ, Dragovich JA et al. Hypothalamic norepinephrine: circadian rhythms and the control of feeding behavior. Science 1972; 178:640-643.
114. Nelson W, Cadotte L, Halberg F. Circadian timing of single daily "meal" affects survival of mice. Proc Soc Exp Biol (NY) 1973; 144:766-769.
115. Stokkan K-A, Yamazaki S, Tei H et al. Entrainment of the circadian clock in the liver by feeding. Science 2001; 291:490-492 (Jan 19).
116. Damiola F, Minh NL, Preitner N et al. Restricted feeding uncouples circadian oscillators in peripheral tissues from the central pacemaker in the suprachiasmatic nucleus. Genes & Development 2000; 14:2950-2961.
117. Halberg F, Haus E, Cornélissen G. From biologic rhythms to chronomes relevant for nutrition. In: Marriott BM, ed. Not Eating Enough: Overcoming Underconsumption of Military Operational Rations. Washington DC: National Academy Press; 1995:361-372.
118. Henauer S, Lombrozo L, Hollister LE. Circadian variations of lorazepam-induced neurologic deficits. Life Sci 1984; 35:2193-2197.
119. Scheving LE, Vedral DF, Pauly JE. A circadian susceptibility rhythm in rats to pentobarbital sodium. Anat Rec 1968; 160:741-749.
120. Davis WM. Day-night periodicity in pentobarbital response of mice and the influence of socio-psychological condition. Experientia 1962; 18:235-237.
121. Friedman AH, Walker CA. Amines, blood histamine and glucose levels in relationship to circadian changes in sleep induced by pentobarbitone sodium. J Physiol (Lond) 1969; 202:133-146.
122. Nelson W, Halberg F. An evaluation of time-dependent changes in susceptibility of mice to pentobarbital injection. Neuropharmacology 1973; 12:509-24.
123. Vesell ES. Genetic and environmental factors affecting hexobarbital metabolism in mice. Ann NY Acad Sci 1968; 151:900-912.
124. Mueller O. Circadian rhythmicity in response to barbiturates. In: Scheving LE, ed. Chronobiology. Stuttgart: Georg Thieme Publishers/ Tokyo: Igaku Shoin Ltd.; 1974:187-190.
125. Nair V. Circadian rhythm in drug action: a pharmacologic, biochemical, and electron microscopic study. In: Scheving LE, ed. Chronobiology. Stuttgart: Georg Thieme Publishers/Tokyo: Igaku Shoin Ltd.; 1974:182-186.
126. Nagayama H. Influences of biological rhythms on the effects of psychotropic drugs. Psychosomatic Medicine 1999; 61:618-629.
127. Owasoyo JO, Walker CA, Whitworth UG. Diurnal variation in the dopamine level of rat brain areas: effect of sodium phenobarbital. Life Sci 1979; 25:119-122.
128. Marte E, Nelson DO, Halberg F et al. Circadian rhythms in murine susceptibility to the anesthetics halothane and methohexital. In: Walker CA, Winget CM, Soliman KFA, eds. Chronopharmacology and Chronotherapeutics. Tallahassee, Florida: Florida A & M University Foundation; 1981:89-94.
129. Urba-Holmgren R, Holmgren B, Aegir M. Circadian variation in an amphetamine induced motor response. Pharmacol Biochem Behav 1977; 7:571-2.
130. Kuribara H, Tadokoro S. Circadian variation in methamphetamine- and apomorphine-induced increase in ambulatory activity in mice. Pharmacol Biochem Behav 1982; 17:1251-6.
131. Lu J-Q, Nagayama H. Circadian rhythm in the response of central 5-HTIA receptor to 8-0H-DPAT in rats. Psychopharmacology 1996; 123:42-45.
132. Lu J-Q, Nagayama H. Circadian rhythm in the hypothermic response to serotonin 1A receptor agonist 8-0H-DPAT in rats. Chronobiol Int 1997; 14:267-73.
133. Moser PC, Redfern PH. Lack of variation over 24 hours in responses to stimulation of 5-HTI receptors in the mouse brain. Chronobiol Int 1985; 2:235-8.
134. Mason R. Circadian variation in sensitivity of suprachiasmatic and lateral geniculate neurons to 5-hydroxytryptamine in the rat. J Physiol (Lond) 1986; 377:1-13.
135. Moser PC, Redfern PH. Circadian rhythms in behaviours mediated by 5-HT receptor stimulation in the rat. In: Reinberg A, Smolensky M, Labrecque G, eds. Annual review of chronopharmacology. Vol. 3. Oxford: Pergamon Press, 1986: 13-6.
136. Moser PC, Redfern PH. Circadian variation in behavioural responses to central 5-HT receptor stimulation in the mouse. Psychopharmacology 1985; 86:223-7.
137. Singleton C, Marsden CA. Circadian variation in the head twitch response produced by 5-methoxy-N1, N1-dimethyl-tryptamine and p-chloramphetamine in the mouse. Psychopharmacology (Berl) 1981; 74:173-6.

138. Nagayama H, Lu J-Q. Circadian rhythm in the responsiveness of central 5-HT2A receptors to DOI in rats. Psychopharmacology 1996; 127:113-6.
139. Williams RLL, Soliman KA, Mizinga KM. Circadian variation in tolerance to the hypothermic action of CNS drugs, Pharmacol Biochem Behav 1993; 46:283-8.
140. Nagayama H, Takagi A, Takahashi R. Circadian susceptibility rhythm to haloperidol under constant conditions. Experientia 1987; 43:625-6.
141. Lu J-Q, Nagayama H. Circadian rhythm in the function of central 5-HTIA receptors is endogenous in nature. Cell Mol Life Sci 1997; 53:224-6.
142. Patel IH, Levy RH, Lockard JS. Time-dependent kinetics II: Diurnal oscillations in steady-state plasma ethosuximide levels in rhesus monkeys. J Pharm Sci 1977; 66:650-653.
143. Reinberg A., Zagula-Mally ZW, Ghata J, Halberg F. Circadian rhythm in duration of salicylate excretion referred to phase of excretory rhythm and routine. Proc Soc exp Biol (NY) 1967; 124:826-832.
144. Di Santo A, Chodos D, Halberg F. Chronobioavailability of three erythromycin test preparations assessed by each of four indices: time to peak, peak, nadir and area. Chronobiologia 1975; 2 (Suppl. 1):17.
145. Halberg F, Haus E, Cardoso SS et al. Toward a chronotherapy of neoplasia: Tolerance of treatment depends upon host rhythms. Experientia (Basel) 1973; 29:909-934.
146. Halberg F, Cornélissen G, Wang ZR et al. Chronomics: circadian and circaseptan timing of radiotherapy, drugs, calories, perhaps nutriceuticals and beyond. J Exp Therapeutics Oncol 2003; 3:223-260.
147. Jori A, Di Salle E, Santini V. Daily rhythmic variation and liver drug metabolism in rats. Biochem Pharmacol 1971; 20:2965-9.
148. Shito T. Ando T, Okamoto T et al. Chronopharmacokinetics and chronotoxicity of lithium in mice eating normal and low-sodium diets. Chronobiol Int 1992; 9:114-23.
149. Rutkowska A, Piekoszewski W, Brandys J. Circadian changes in the elimination of amitriptyline in rats. Pol J Pharmacol Pharm 1992; 44:671-7.
150. Ohdo S, Ogawa N, Nakano S et al. Influence of feeding schedule on the chronopharmacological aspects of sodium valproate in mice. J Pharmacol Exp Ther 1996; 278:74-81.
151. Planeta CS, DeLucia R, Aizenstein ML et al. Daily variation in plasma concentration of fencamfamine and striatal dopamine receptors in rats. Braz J Med Biol Res 1994; 27:737-41.
152. Nagayama H, Takagi A, Nakamura E et al. Circadian susceptibility rhythm to apomorphine in the brain. Commun Psychopharmacol 1978; 2:301-10.
153. Cornélissen G, Halberg F, Richardson JD et al.About 1.3-year Richardson component of solar wind speed detected in human circulation and myocardial infarctions (MI). Abstract, 5th Japanese Annual Conference on Chronocardiology and Hypertension, Sept. 12, 2003, Sapporo, Japan. Session (3) Ambulatory BP, 27.
154. Kafka MS, Benedito MA, Roth RH et al. Circadian rhythms in catecholamine metabolites and cyclic nucleotide production. Chronobiol Int 1986; 3:101-15.
155. Wirz-Justice A, Kraeuchi K, Campbell IC et al. Adrenoceptor changes in spontaneous hypertensive rats. Brain Res 1983; 262:233-42.
156. Cagampang FAR, Rattrap M, Powell IF et al. Circadian changes of glutamate decarboxylase 65 and 67mRNA in the rat suprachiasmatic nuclei. Neuroreport 1996; 7:1925-8.
157. Kafka M, Marangos PJ, Moore BY. Suprachiasmatic nucleus ablation abolishes circadian rhythms in rat brain neurotransmitter receptors. Brain Res 1985; 327:344-7.
158. Ferraro JS, Stager RW. Diurnal variations in brain serotonin are driven by the photic cycle and are not circadian in nature. Brain Res 1990; 512:121-4.
159. Quay WB. Regional and circadian differences in cerebral cortical serotonin concentrations. Life Sci 1965; 4:379-84.
160. Scheving LE, Harrison WH, Gordon P et al. Daily fluctuation (circadian and ultradian) in biogenic amines of the rat brain. Am J Physiol 1968; 214:166-73.
161. Bhaskaran D, Halberg F, Shankaraiah K et al. Circadian acrophase synchronization of 5-hydroxytryptamine (5-HT) in Wistar rat brain areas from 21 days to 24 months of age. Chronobiologia 1982; 9:343.
162. Asano Y. The maturation of the circadian rhythm of brain norepinephrine and serotonin in the rat. Life Sci 1971; 10:883-94.
163. Scapagnini U, Moberg GP, Van Loon GR et al. Relation of brain 5-hydroxytryptamine content to the diurnal variation in plasma corticosterone in the rat. Neuroendocrinology 1971; 7:90-6.
164. Scott C, Laverty R. Diurnal rhythms in brain regional 5-hydroxytryptamine metabolism: effects of age and 6-hydroxydopamine treatment. Proc Univ Otago Med Sch 1974; 52:52-4.
165. Simon M, George R. Diurnal variations in plasma corticosterone and growth hormone as correlated with regional variations in norepinephrine, dopamine and serotonin content of rat brain. Neuroendocrinology 1975; 17:125-38.
166. Hillier JG, Redfern PH. Twenty-four rhythms in serum and brain indoleamine concentrations: tryptophan-5-hydroxylase and monoamine oxidase activity in the rat. Int J Chronobiol 1976; 4:197-210.
167. Morgan WW, Pfeil KA, Reiter RJ et al. Comparison of changes in tryptophan and serotonin in regions of the hamster and the rat brain over a twenty-four hour period. Brain Res 1976; 117:77-84.
168. Philo R, Rudeen PK, Reiter RJ. A comparison of the circadian rhythms and concentrations of serotonin and norepinephrine in the telencephalon of four rodent species. Comp Biochem Physiol C 1977; 57:127-30.
169. Hery F, Chouvet G, Kan JP et al. Daily variations of various parameters of serotonin metabolism in the rat brain. II. Circadian variations in serum and cerebral tryptophan levels: lack of correlation with 5-HT turnover. Brain Res 1977; 123:137-45.
170. Banky Z. Data on the ontogenesis of the diurnal serotonin rhythm in the mesencephalon and hypothalamus, in relation to the plasma corticosterone rhythm in the rat. Acta Physiol Acad Sci Hung 1981; 58:181-7.
171. Owasoyo JO, Whitworth UGJ, Walker CA. Circadian rhythm in rat brain norepinephrine and serotonin: effects of picrotoxin and pentylenetetrazol. In: Walker CA, Winget CM, Soliman KA, eds. Chronopharmacology and chronotherapeutics. Tallahassee (FL): Florida A & M University Foundation; 1981:95-101.
172. Semba J, Toru M, Mataga N. Twenty-four hour rhythms of norepinephrine and serotonin in nucleus suprachiasmatics, raphe nuclei, and locus ceruleus in the rat. Sleep 1984; 7:211-8.
173. Bhaskaran D, Radha E. Circadian variations in the monoamine levels and monoamine oxidase activity in different regions of the rat brain as a function of age. Exp Gerontol 1984; 19:153-70.
174. Agren H, Koulu M, Saavedra JM et al. Circadian covariation of norepinephrine and serotonin in the locus coeruleus and dorsal raphe nucleus in the rat. Brain Res 1986; 397:353-8.
175. Wesemann W, Rotsch M, Schulz E et al. Circadian rhythm of serotonin binding in rat. I. Effect of the light-dark cycle. Chronobiol Int 1986; 3:135-9.
176. Greco AM, Gambardella P, Sticchi R et al. Tricyclic imipramine modification of the circadian rhythms of hypothalamic serotonin, its precursors and acid catabolite in individually housed rats. Chronobiol Int 1988; 5:217-25.
177. Koulu M, Bjelogrlic N, Agren H et al. Diurnal variation in the concentrations of catecholamines and indolamines in the median eminence and in the intermediate and posterior lobes of the pituitary gland of the male rat. Brain Res 1989; 503:246-52.
178. Gamvardella P, Greco AM, Sticchi R et al. Circadian rhythm variations in the adult rat induced by low and high protein diets administered at various stages of development. Chronobiol Int 1990; 7:43-50.
179. Voog L, Erickson T. Diurnal rhythms in rat brain large neutral amino acids (LNAAS), monoaminers and monoamine metabolites. J Neural Transm 1992; 87:215-24.
180. Poncet L, Denoroy L, Jouvet M. Daily variations in vivo tryptophan hydroxylation and in the contents of serotonin and 5-hydroxyindole acetic acid in discrete brain areas of the rat. J Neural Transm Gen Sect 1993; 92:137-50.

181. Cagampang FAR, Inouye S-IT. Diurnal and circadian changes of serotonin in the suprachiasmatic nuclei: regulation by light and an endogenous pacemaker. Brain Res 1994; 639:175-9.
182. Kohno Y, Tanaka M, Nakagawara R et al. Study on diurnal variation of noradrenaline release in three brain regions of rats. Kurume Med J 1980; 27:227-32 .
183. Barden N, Chevillard C, Saavedra JM. Diurnal variations in rat posterior pituitary catecholamine levels. Neuroendocrinology 1982; 34:148-50.
184. Schweiger U, Warnhoff M, Pirke K-M. Norepinephrine turnover in the hypothalamus of adult male rats: alteration of circadian patterns by semistarvation. J Neurochem 1985; 45:705-9.
185. Stanley BG, Schwartz DH, Hernandez L et al. Patterns of extracellular norepinephrine in the paraventricular hypothalamus: relationship to circadian rhythm and deprivation-induced eating behavior. Life Sci 1989; 45:275-82.
186. Dluzen DE, Ramirez VD. Daily changes in in vitro spontaneous dopamine efflux from the corpus striatum of male rats. Chronobiol Int 1987; 4:477-82.
187. Schade R, Vick K, Sohr R et al. Correlative circadian rhythms of cholecystokinin and dopamine content in nucleus accumbens and striatum of rat brain. Behav Brain Res 1993; 59:211-4.
188. Paulson PE, Robinson TE. Relationship between circadian changes in spontaneous motor activity and dorsal versus ventral striatal dopamine neurotransmission assessed with on-line microdialysis. Behav Neurosci 1994; 108:624-35.
189. Morgan WW, Yndo CA, McFadin LS. Daily rhythmic changes in the content of serotonin and 5-hydroxyindole acetic acid in the cerebral cortex of mice. Life Sci 1974; 14:329-38.
190. Weiner N, Clement H-W, Gemsa D et al. Circadian and seasonal rhythms of 5-HT receptor subtypes, membrane anisotropy and 5-HT release in hippocampus and cortex of the rat. Neurochem Int 1992; 21:7-14.
191. Kafka MS, Wirz-Justice A, Naber D. Circadian and seasonal rhythms in alpha- and beta-adrenergic receptors in the rat brain. Brain Res 1981; 207:409-19.
192. Wirz-Justice A, Tobler I, Kafka MS et al. Sleep deprivation: effects on circadian rhythms of rat brain neurotransmitter receptors. Psychiatry Res 1981; 5:67-76.
193. Kraeuchi K, Wirz-Justice A, Morimasa T et al. Hypothalamic alpha 2- and beta-adrenoceptor rhythms are correlated with circadian feeding: evidence from chronic methamphetamine treatment and withdrawal. Brain Res 1984; 321:83-90.
194. Kafka MS, Benedito MA, Blendy JA et al. Circadian rhythms in neurotransmitter receptors in discrete rat brain regions. Chronobiol Int 1986; 3:91-100.
195. Weiland NG, Wise PM. Diurnal rhythmicity of beta-1- and bcta-2-adrenergic receptors in ovariectomized, ovariectomized estradiol-treated and proestrous rats. Neuroendocrinology 1989; 50:655-62.
196. Naber D, Wirz-Justice A, Kafka MS et al. Seasonal variations in the endogenous rhythm of dopamine receptor binding in- rat striatum. Biol Psychiatry 1981; 16:831-5.
197. Jenni-Eiermann S, von Hahn HP, Honegger CG. Diurnal rhythms in neurotransmitter receptor binding and choline acetyltransferase activity: different patterns in two rat lines of Wistar origin. Brain Res 1986; 370:54-60.
198. Jenni-Eiermann S, von Hahn HP, Honegger CG. Circadian variations of neurotransmitter binding in three age groups of rats. Gerontology 1985; 31:138-49.
199. Kafka M, Wirz-Justice A, Naber D et al. Circadian acetylcholine receptor rhythm in rat brain and its modification by imipramine. Neuropharmacology 1981; 20:421-5.
200. Kan JP, Chouvet G, Hery F et al. Daily variations of various parameters of serotonin metabolism in the rat brain. I. Circadian variations of tryptophan-5-hydroxylase in the raphe nuclei and the striatum. Brain Res 1977; 123:125-36.
201. Natali JP, McRae-Degueurce A, Chouvet G et al. Genetic studies of daily variations of first-step enzymes of monoamines metabolism in the brain of inbred strains of mice and hybrids. I. Daily variation of tryptophan hydroxylase activity in the nuclei raphe dorsalis, raphe centralis and in the striatum. Brain Res 1980; 191:191-203.
202. Cahill A, Ehret CF. Circadian variations in the activity of tyrosine hydroxylase, tyrosine amino transferase, and tryptophan hydroxylase: relationship to catecholamine metabolism. J Neurochem 1981; 37:1109-15.
203. Redfern PH, Sinei K. 24-Hour variation in synaptosomal tryptophan-5-hydroxylase activity in the rat brain. In: Redfern PH, Campbell I C, Davies JA, Martin KF, editors. Circadian rhythms in the central nervous system. Weinheim, Germany: Verlagsgesllschaft; 1985:193-198.
204. Natali JP, McRae-Degueurce A, Keane P et al. Genetic studies of daily variation of first-step enzymes of monoamines metabolism in the brain of inbred strains of mice and hybrid. II. Daily variation of tyrosine hydroxylase activity in the locus coeruleus. Brain Res 1980; 191:205-13.
205. Joanny P, Chouvet G, Giannellini F et al. Brain diurnal levels of adenosine 3', 5'-cyclic monophosphate in C57 BL/6 and BALB/C mice. Chronobiol Int 1984; 1:37-40.
206. Halberg F, Nelson W, Runge WJ et al. Plans for orbital study of rat biorhythms. Results of interest beyond the Biosatellite program. Space Life Sci 1971; 2:437-471.
207. Hery F, Rouer E, Glowinski J. Daily variations of serotonin metabolism in the rat brain. Brain Res 1972; 43:445-65.
208. Black I, Parker L, Axelrod J. A daily rhythm in the rate of depletion of brain norepinephrine by reserpine. Biochem Pharmacol 1969; 18:2688-91.
209. Halberg F. Quando trattare / When to treat. Hæmatologica (Pavia) 1975; 60:1-30.
210. Reinberg AE. Chronopharmacology of H_1-receptor antagonists: experimental and clinical aspects (allergic diseases). In: Redfern PH, Lemmer B, eds. Handbook of Experiental Pharmacology, Vol. 125: Physiology and Pharmacology of Biological Rhythms. Berlin: Springer-Verlag; 1997:589-606.
211. Reinberg AE, Ashkenazi IE. Interindividual differences in chronopharmacologic effects of drugs: a background for individualization of chronotherapy. Chronobiology int 1993; 10:449-460.
212. Reinberg AE. Concepts in chronopharmacology. Annu Rev Pharmacol Toxicol 1992; 52:51-66.
213. Nakano S, Hollister LE. Chronopharmacology of amitriptyline. Clin Pharmacol Ther 1983; 33:453-9.
214. Smith RB, Kroboth PD, Phillips JP. Temporal variation in triazolam pharmacokinetics and pharmacodynamics after oral administration. J Clin Pharmacol 1986; 26:120-4.
215. Halbreich U, Asnis GM, Halpern F et al. Diurnal growth hormone responses to dextroamphetamine in normal young men and post-menopausal women. Psychoneuroendocrinology 1980; 5:339-44.
216. Halbreich U. Sachar EJ, Asnis GM et al. Diurnal cortisol responses to dextroamphetamine in normal subjects. Psychoneuroendocrinology 1981; 6:223-9.
217. Shappell SA, Kearns GL, Valentine JL et al. Chronopharmacokinetics and chronopharmacodynamics of dextromethamphetamine in man. J Clin Pharmacol 1996; 36:1051-63.
218. Monteleone P, Bortolotti F, Oreazzo C et al. Differences between morning and afternoon hormonal responses to D-fenfluramine in healthy humans. Psychoneuroendocrinology 1997; 22:79-87.
219. Usher RW, Beasley CMJ, Bosomworth JC. Efficacy and safety of morning versus evening fluoxetine administration. J Clin Psychiatry 1991; 52:134-6.
220. Ferrari E, Bossolo PA, Bono G et al. Chronobiological approach to the treatment of cluster headache by lithium carbonate. In: Reinberg A, Smolensky M, Labrecque G, editors. Annual review of chronopharmacology. Vol. 1. Oxford: Pergamon Press; 1984:407-10.
221. Frankel JP, Pirtosek Z, Kempster PA et al. Diurnal differences in response to oral levodopa. J Neurol Neurosurg Psychiatry 1990; 53:948-50.
222. Olesen OV, Thomsen K. Diurnal variations in serum lithium and renal lithium clearance in rats given lithium as a single small dose or as multiple high doses. Acta Pharmacol Toxicol Copenh 1985; 57:171-75.
223. Ohdo S, Nakano S, Ogawa N. Chronopharmacokinetics of valproic acid following constant-rate administration in mice. Chronobiol Int 1991; 8:35-43.

224. Halberg F. Protection by timing treatment according to bodily rhythms: an analogy to protection by scrubbing before surgery. Chronobiologia 1974; 1(Suppl.1):27-68.
225. Abernathy RS, Halberg F, Spink WW. Studies on the mechanism of chlorpromazine protection against brucella endotoxin in mice. J Lab clin Med 1957; 49:708-715.
226. Berendes HW, Marte E, Ertel RJ et al. Circadian physiologic rhythms and lowered blood 5-hydroxytryptamine in human subjects with defective mentality. Physiologist 1960; 3:20.
227. Halberg F, Anderson J, Ertel R et al. Circadian rhythm in serum 5-hydroxytryptamine of healthy men and male patients with mental retardation. Int J Neuropsychiatr 1967; 3:379-386.
228. Jones F, Haus E, Halberg F. Murine circadian susceptibility-resistance cycle to acetylcholine. Proc Minn Acad Sci 1963; 31:61-62.
229. Carlsson A, Serin F. Time of day as a factor influencing the toxicity of nikethamide. Acta Pharmacologica et Toxicologica 1950; 6:181-186.
230. Carlsson A, Serin F. The toxicity of nikethamide at different times of the day. Acta Pharmacologica et Toxicologica 1950; 6:187-193.
231. Woolley DE, Timiras PS. Estrous and circadian periodicity and electro-shock convulsions in rats. Am J Physiol 1962; 202:379-382.
232. Pizzarello DH, Witcofski RL, Lyons A. Variations in survival time after whole body radiation at 2 times of day. Science 1963; 139:349.
233. Rugh R, Castro V, Balter S et al. X-rays: are there cyclic variations in radiosensitivity? Science 1963; 142:53-56.
234. Pohle K, Matthies E, Meng K. Tagesperiodische Schwankungen der cancerostatischen Wirkungsstärke von N-oxyd-Lost beim Ehrlich-Ascites Carcinoma der Maus. Z Krebsforsch 1961; 64:215-218.
235. Dobrokhotov VN. On the importance of regularities governing the diurnal periodicity of cellular multiplication. Vestnik Akademii Meditsinskikh Nauk SSSR 1963; No. 7:50-62.
236. Halberg F. The 24-hour scale: A time dimension of adaptive function organization. Perspect Biol Med 1960; 3:491-527.
237. Levi F, Halberg F, Nesbit M et al. Chrono-oncology. In: Kaiser H, editor. Neoplasms-Comparative Pathology of Growth in Animals, Plants and Man. Baltimore: Williams and Wilkins, 1981:267-316.
238. Halberg F, Cornélissen G, Wall D et al. Engineering and governmental challenge: 7-day/24-hour chronobiologic blood pressure and heart rate screening: Part I. Biomedical Instrumentation & Technology 2002; 36:89-122.
239. Halberg F, Cornélissen G, Wall D et al. Engineering and governmental challenge: 7-day/24-hour chronobiologic blood pressure and heart rate screening: Part II. Biomedical Instrumentation & Technology 2002; 36:183-197.
240. Reinberg A. Circadian changes in psychological effects of ethanol. Neuropsychopharmacology 1992; 7:149-156.
241. Reinberg AE, Soudant E, Koulbanis C et al. Circadian dosing time dependency in the forearm skin penetration of methyl and hexyl nicotinate. Life Sciences 1995; 57:1507-1513.
242. Reinberg AE. Rythmes circadiens de la sensibilité des systèmes cibles aux médicaments: un phénomène sous-estimé. Bull Acad Natle Méd 1996; 180:533-547.
243. Reinberg A. Variations circadiennes et saisonnières des effets biologiques de l'ACTH 1-17 et de l'HCG. Applications cliniques. La revue Française d'Endocrinologie Clinique: nutrition et métabolisme 1999; 40:205-229.
244. Halberg F, Bingham C, Cornélissen G. Clinical trials: the larger the better? Chronobiologia 1993; 20:193-212.
245. Levine H, Halberg F, Taylor D. Circadian rhythms before and after removal of both eyes for bilateral retinoblastoma. Graefes Arch Ophthalmol 1973; 188:263-280.
246. Cornélissen G, Engebretson M, Johnson D et al. The week, inherited in neonatal human twins, found also in geomagnetic pulsations in isolated Antarctica. Biomedicine and Pharmacotherapy 2001; 55(Suppl 1):32-50.
247. Watanabe Y, Cornélissen G, Watanabe M et al. Effects of autogenic training and antihypertensive agents on circadian and circaseptan variation of blood pressure. Clin Exp Hypertens 2003; 25:405-412.
248. Watanabe Y, Cornélissen G, Katinas G et al. Case report: a drug for damping circannual and transannual amplitudes of blood pressure and heart rate. Abstract, Problems of Rhythms in Natural Sciences, Moscow: Russian People's Friendship University; 2004:18-20.
249. Shinagawa M, Kubo Y, Otsuka K et al. Impact of circadian amplitude and chronotherapy: relevance to prevention and treatment of stroke. Biomedicine and Pharmacotherapy 2001; 55 (Suppl 1):125-132.

Section II: Chronopharmacology

CHAPTER 8

Melatonin Interaction with BZ-Gaba$_A$ Receptors: Implications for Sleep Induction

Lennard P. Niles

Abstract

The pineal hormone, melatonin, can exert sedative/hypnotic, anxiolytic and other neuropharmacological effects in experimental animals. Usually, these effects are produced by large pharmacological doses of melatonin, which are known to interact with benzodiazepine (BZ) receptors in the central nervous system (CNS). There is evidence that flumazenil, a specific central-type BZ antagonist, can block some of these effects of melatonin. Therefore, it is thought that activation of central-type BZ receptors on the BZ-GABA$_A$ receptor complex, with consequent allosteric enhancement of GABAergic activity, is the primary mechanism underlying the neuropharmacological effects of melatonin. In addition, melatonin can interact with other BZ receptor subtypes to influence neurosteroidogenesis and cyclic AMP production, which can further modulate GABAergic activity in the CNS. However, in contrast to the high pharmacological doses of melatonin used in animal studies, the relatively low doses of this hormone, typically used in human sleep studies, are unlikely to reach the micromolar threshold required for binding to BZ receptors. In support of this view, flumazenil does not block the sedative/hypnotic effect observed in young adults following administration of a low pharmacological dose of 3 mg melatonin. Thus, while pharmacological sedation by high doses of melatonin is thought to involve enhancement of BZ-GABA$_A$ receptor signaling, it appears that physiological receptors and mechanisms mediate sleep induction by low doses of this psychotropic hormone in humans.

Melatonin

Considerable evidence indicates that melatonin influences the function of diverse neural, circadian, endocrine, immune, and other systems in mammals. Two G protein-coupled melatonin receptors, designated as MT_1 and MT_2 subtypes,[1] are present in the brain and peripheral organs of humans and other mammals.[2] Both of these receptors are coupled, via pertussis toxin-sensitive G-proteins, to inhibition of adenylyl cyclase activity, with a consequent decrease in cAMP production.[3] Melatonin also activates protein kinase C (PKC), via a pertussis toxin sensitive pathway,[4] or by a direct action on this intracellular enzyme.[5] In addition, melatonin has been linked to modulation of cGMP levels,[6] and activation of the mitogen-activated protein kinase (MAPK)-extracellular regulated kinase (ERK) pathway.[7] Thus, this hormone can interact with multiple cellular pathways to produce its diverse physiological effects, including neurotrophic factor upregulation in target cells.[8] In addition, when melatonin is administered in pharmacological doses, which greatly exceed its endogenous picomolar to low nanomolar levels, it can activate benzodiazepine receptors.[9,10] This chapter is focused on the mechanisms inderlying the interaction of melatonin with benzodiazepine receptors, with comments on the implications for the sedative/hynotic effects of this hormone.

Benzodiazepine Receptors

Benzodiazepine (BZ) drugs are extensively used clinically because of their anxiolytic, myorelaxant, anticonvulsant and sedative/hypnotic properties. The receptors for BZs have been generally classified as either central-type or peripheral-type but there is evidence that other distinct subtypes exist.[11-13]

Central-Type BZ Receptors

Initial binding studies using [^{3}H]diazepam and subsequently [^{3}H] flunitrazepam, identified high-affinity BZ receptors in rat and human brain.[14] The anxiolytic and anticonvulsant efficacy of BZs was matched by their binding affinity for these receptors, designated as central-type BZ receptors (CBRs). CBRs have been classified as either type I or II, based on their high or low affinity respectively, for triazolopyridazines such as CL218.872, β-carbolines and imidazopyridines like zolpidem.[15,16] The therapeutic actions of BZs are thought to be primarily mediated by CBRs which are coupled to type A γ–aminobutyric acid (GABA$_A$) receptors in the CNS.[17] The GABA$_A$ receptor is a pentameric protein consisting of diverse subunit classes ($\alpha_{1\text{-}6}$, $\beta_{1\text{-}3}$, $\gamma_{1\text{-}3}$, δ, ε, ϕ, π, $\rho_{1\text{-}3}$) which exhibit multiple isoforms.[18] Benzodiazepine-sensitive GABA$_A$ (BZ-GABA$_A$) receptors, which consist of α_1-, α_2-, α_3-, or α_5- subunits together with any of the β-subunits and the γ_2-subunit, are the most abundant in the brain, with the α_1, β_2, γ_2-containing subtype being predominant.[19] It is now evident that the type of subunit present in the BZ-GABA$_A$ receptor complex determines its pharmacological profile.[20,21] For example, α_1- BZ-GABA$_A$ receptors mediate the sedative and amnesic effects of BZs, such as diazepam, whereas their anxiolytic effects involve α_2-BZ-GABA$_A$ receptors.[19,22] Benzodiazepines and related ligands act as allosteric modulators of GABAergic activity by enhancing the binding of GABA to the BZ-GABA$_A$ receptor complex, thereby increasing the frequency at which the associated chloride channel is opened.[12,16]

Sleep and Sleep Disorders: A Neuropsychopharmacological Approach, edited by Malcolm Lader, Daniel P. Cardinali and S.R. Pandi-Perumal.

Peripheral-Type Receptors

In addition to central-type receptors, there are peripheral-type BZ receptors (PBRs) which exhibit a widespread distribution in the brain and peripheral tissues.[23] Cloning of a peripheral-type BZ receptor component responsible for the receptor binding properties showed no apparent sequence homology with any of the cloned $GABA_A$ receptor subunits, indicating the distinctness of the PBR.[14] Subcellular localization studies have shown that PBRs are primarily present on the outer mitochondrial membrane; therefore, these sites are also referred to as mitochondrial BZ receptors. Using the isoquinoline carboxamide, PK14105, for photoaffinity labelling, an 18 kDa isoquinoline binding protein (IBP) has been localized in diverse tissues that express PBRs. Expression studies utilizing the rat or bovine IBP cDNA, indicate that the binding domains for PBR ligands such as isoquinoline carboxamides (e.g., PK11195) and BZs are localized in the IBP. However, several lines of evidence indicate that functional receptor activity requires the multimeric PBR complex which consists of the IBP, a 30 kDa adenine nucleotide carrier and a 32 kDa voltage-dependent anion channel.[24,25] The human, bovine and rat IBPs lack a consensus mitochondrial insertion signal sequence, which suggests that these proteins are not exclusively localized in mitochondria.[14] This is consistent with increasing evidence that the PBR is also localized in nonmitochondrial membranes, such as the plasma membrane.[26,27] Indeed, it has been suggested that a PBR-like plasma membrane-localized G protein-coupled BZ receptor, mediates the inhibitory action of BZs on the adenylyl cyclase (AC)-cAMP pathway in the brain.[28,29]

Steroidogenesis

It is now known that PBRs play an essential role in steroidogenesis via regulation of cholesterol translocation from the outer to inner mitochondrial membranes. This rate-limiting step in steroid biosynthesis provides cholesterol to the cytochome P450 side-chain cleavage enzyme for conversion to pregnenolone, the precursor of various steroid hormones produced in diverse tissues including the brain.[25] Interestingly, the modulatory action on steroidogenesis mediated by the PBR provides a functional link between this receptor and the BZ-$GABA_A$ receptor. In the CNS, the PBR is primarily localized in glial cells which synthesize and release various neurosteroids, such as allopregnanolone and pregnenolone sulfate, which can exert positive or negative effects on the activity of the BZ-$GABA_A$ receptor.[30-32]

Melatonin Interaction with Benzodiazepine Receptors

Most studies of the interaction between melatonin and BZ-$GABA_A$ receptors have focused on the pharmacological effects of this hormone. However, there is also evidence for a physiological modulation of GABAergic activity by melatonin, acting on its high (picomolar)-affinity G protein coupled receptors in the mammalian CNS. The physiological effects of melatonin on neuroendocrine, circadian and other biological systems, are mediated by pertussis toxin-sensitive G protein-coupled receptors in the brain and peripheral target organs. As noted earlier, there are two distinct mammalian receptor subtypes, MT_1 and MT_2, which mediate inhibition of adenylate cyclase and are also linked to other signaling pathways.[2,3] Interestingly, electrophysiological studies indicate that these receptors mediate opposite effects on GABAergic function, with the MT_1 subtype producing enhancement whereas the MT_2 subtype mediates inhibition by melatonin.[33]

In contrast to the physiological (MT_1 and MT_2) targets for melatonin, there is substantial evidence that when administered in large pharmacological doses (e.g., ~10-20 mg/kg) melatonin can directly interact with BZ- $GABA_A$ receptors. For example, it competes for [^{3}H]diazepam binding sites in rat, human and bovine brain membranes with micromolar affinity.[34] Similarly, pharmacological doses of melatonin act on BZ- $GABA_A$ receptors to enhance both in vitro and in vivo binding of GABA, and to allosterically inhibit binding of the caged convulsant, t-butyl-bicyclophosphorothionate (TBPS), on GABA-gated chloride channels in rat brain.[35] In accordance with the foregoing, pharmacological studies have shown that the anxiolytic, anticonvulsant and other psychotropic actions of melatonin, which are similar to those exhibited by BZs, involve the enhancement of GABAergic activity.[9,10] The binding site for melatonin on the BZ-$GABA_A$ receptor complex is not known, but its ability to competitively inhibit [^{3}H]diazepam binding suggests a direct interaction within the BZ binding pocket, which is located at the α/γ subunit interface of the BZ- $GABA_A$ receptor complex.[17] Similarly, although the melatonin binding site on the PBR complex is not known, it is reasonable to assume an interaction with the IBP subunit, where both isoquinolines and BZs are thought to bind.[36]

Neuropharmacological Mechanisms of Melatonin Action

It is known that GABA binds to specific sites at the β/α subunit interface of the BZ/$GABA_A$ receptor[17] to induce opening of chloride channels, with consequent membrane hyperpolarization and neuronal inhibition.[12] Central-type benzodiazepines, such as clonazepam, bind to other sites on the BZ/$GABA_A$ receptor complex, as described above, resulting in an allosteric enhancement of GABA binding and thus GABAergic activity. Since micromolar concentrations of melatonin can interact with BZ central-type sites on the BZ/$GABA_A$ receptor, the neuropharmacological actions of this hormone are thought to primarily involve the allosteric enhancement of GABAergic activity in the CNS.[10,13] Some examples of the neuropharmacological effects of melatonin which have been linked to modulation of central GABAergic systems are listed in Table 1.

As discussed earlier, pharmacological concentrations of melatonin can also bind to PBRs which play an important role in neurosteroidogenesis. Neurosteroids can exert potent modulatory effects on ion-gated neurotransmitter receptors including central GABAergic systems.[37] Therefore, the activation of PBRs by melatonin, with potential changes in neurosteroid production, provides another pathway for GABAergic modulation by this hormone. Moreover, the ability of pharmacological concentrations of melatonin or BZs to inhibit the AC-cAMP pathway via putative G protein-coupled BZ receptors,[29] suggests yet another neuropharmacological mechanism for modulation of GABAergic activity. There is evidence that multiple kinases can phosphorylate various $GABA_A$ receptor subunits, to alter GABAergic activity in the brain.[38,39] Depending on the brain area and /or the receptor subunits involved, phosphorylation by various kinases can result in either activation or inhibition of GABAergic function. For example, tyrosine phosphorylation, induced by

Table 1. Neuropharmacological effects of melatonin which involve GABAergic activity

Effect	Melatonin Dose (ip)	Flumazenil Blockade	Reference
Anxiolytic activity in rats	1-20 mg/kg	Yes	9
	2.5 mg/kg	Yes	42
Anticonvulsant activity in rats and hamsters	200 mg/kg	No	43
	50 mg/kg	Yes	44
Sedative/Hypnotic activity in rats	10 mg/kg	Yes	45
	20 mg/kg	Not tested	46
Inhibitionof locomotor activity in hamsters	0.3 mg/kg	Yes	47
Inhibition of rotational behaviour in lesioned rats	10 mg/kg	Yes	10

intracellular application of protein tyrosine kinase (PTK), has been found to increase GABA$_A$-mediated currents in cultured CNS neurons.[40] In contrast, protein kinase A (PKA)-induced phosphorylation of the GABA$_A$ receptor is usually associated with GABA$_A$ receptor desensitization in various CNS areas, and results in decreased GABAergic activity.[38] Therefore, suppression of cAMP production by melatonin or BZs, which results in a decrease in PKA-induced phosphorylation, may produce an opposite potentiating effect on GABAergic activity. A schematic representation of the possible interaction of melatonin with BZ-GABA$_A$ receptors, PBRs and putative G protein-coupled BZ receptors in the CNS, is shown in Figure 1. The BZ-GABA$_A$ and G protein-coupled BZ receptors are depicted on the same neuron, which would allow the intracellular crosstalk proposed. The PBR is shown on a mitochondrion inside an astrocyte, where this receptor is predominantly expressed.[41]

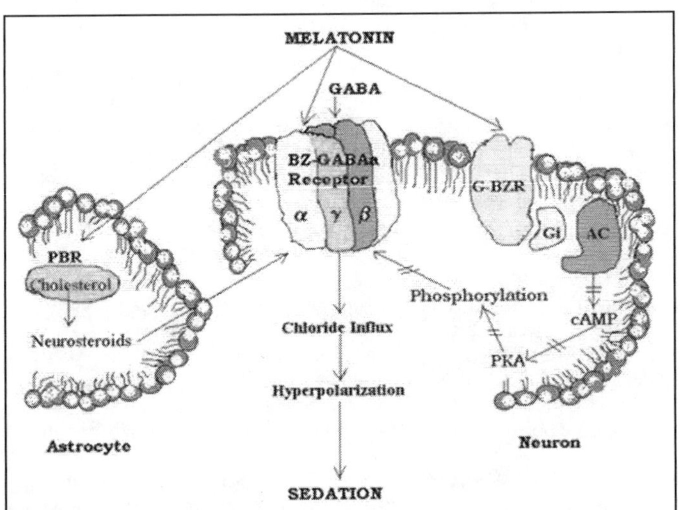

Figure 1. Model of the proposed interaction between melatonin and benzodiazepine receptors. Melatonin-induced activation of central-type BZ receptors on the BZ-GABA$_A$ receptor complex, peripheral-type BZ receptors on glial mitochondria and G_i protein-coupled BZ receptors (G-BZR) in the plasma membrane, results in enhancement of GABAergic activity, with associated neuropharmacological effects such as sedation, as described in the text. Double lines indicate inhibition of the cAMP pathway.

Sleep Modulation by Melatonin

Physiological

Although the focus of this chapter is on the pharmacological actions of melatonin, it would be worthwhile to briefly mention the physiological effects of this hormone on sleep, in order to provide a comparative perspective. The suggestion that the nocturnal increase in serum melatonin may provide a physiological signal for sleep onset,[48] is supported by the presence of decreased melatonin levels in long-term insomniacs.[49] Moreover, consistent with the age-related decline in circulating melatonin levels,[50,51] controlled-release melatonin replacement therapy has been found to significantly improve sleep quality in elderly insomniacs.[52] An important question concerns the mechanism(s) underlying the sedative action of endogenous melatonin. The effectiveness of relatively low melatonin doses, which produce normal nocturnal levels,[48] suggests that a physiological action mediated by high-affinity receptors in the CNS is involved. This view is supported by evidence that the sleep-like state induced by melatonin in zebrafish is blocked by luzindole, a nonselective melatonin receptor antagonist.[53] It is likely that the physiological effects of melatonin on sleep involve activation of its MT_1 and/or MT_2 receptors, with consequent modulation of GABAergic activity,[33] which plays an important role in sleep regulation.[54] Another interesting possibility involves the ability of physiological concentrations of melatonin to upregulate the expression of glial cell line-derived neurotrophic factor,[8] which has been reported to promote sleep in mammals.[55] In addition to its acute sedative effect, melatonin can influence sleep cycles via its circadian phase shifting effects, mediated by melatonin receptors in the circadian clock, which is localized in the suprachiasmatic nucleus of the hypothalamus.[4,56]

Pharmacological

Several investigators, utilizing pharmacological or lower doses of melatonin, have reported sedative/hypnotic effects in humans.[48,57-59] These findings, coupled with evidence that a GABAergic mechanism underlies the neuropharmacological effects of this hormone in experimental animals, suggest that a similar mechanism is involved in the sleep modulating effects of melatonin. Moreover, in a recent study, melatonin (10 mg/kg; ip) produced a significant sleep-promoting effect in rats, which was blocked by the CBR antagonist, flumazenil, the GABA binding site antagonist, bicuculline, or the BZ-GABA$_A$ chloride channel blocker, picrotoxin.[45] These observations support the view that pharmacological concentrations of melatonin may act directly on the BZ-$GABA_A$ receptor to enhance GABAergic activation of chloride influx with concomitant postsynaptic inhibition, which has been linked to sleep induction.[54,60] The involvement of GABA is well illustrated by studies of BZ hypnotics which act on BZ-$GABA_A$ receptors to increase nonrapid eye movement (NREM) sleep and to decrease REM sleep in humans. These drugs also alter the sleep electroencephalogram as shown by a reduction in delta activity and an increase in sigma activity in humans.[22] Thus, the enhancement of GABAergic activity by melatonin, via its interaction with BZ-$GABA_A$ receptors, may be a major mechanism underlying its sedative/hypnotic effects. In addition, the potential modulation of neurosteroidogenesis and cAMP production by melatonin could produce a three-way pharmacological convergence at the BZ-$GABA_A$ receptor complex resulting in enhanced central GABAergic activity, with concomitant neuropsychopharmacological effects including sedation. However, it is important to note that physiological modulation of GABAergic activity and/or other targets by melatonin, as discussed earlier, may also contribute to its sedative/hypnotic effects. Such physiological modulation would involve melatonin receptors but not BZ receptors, in keeping with evidence that the hypnotic effect of melatonin (3 mg) in young adults was not blocked by the central-type BZ receptor antagonist flumazenil.[61] Although this low pharmacological dose of melatonin produces supraphysiologic circulating levels of about 1370 pg/ml within one hour of ingestion in older adults,[62] this is equivalent to only a 5.9 nM concentration, which is well below the micromolar threshold for interaction with BZ receptors. Therefore, assuming that young adults receiving similar treatment would also exhibit a low nanomolar range of circulating melatonin, it is not surprising that flumazenil was ineffective in the above study, which presumably involved activation of melatonin receptors. In keeping with the foregoing, it is now known that low physiological doses of melatonin, which can interact with MT_1 and MT_2 receptors, are effective in treating insomnia.[48,62]

Summary and Conclusions

There is considerable behavioural, biochemical and pharmacological evidence that melatonin can interact with central-type and other BZ receptors in the CNS. Since this interaction mimics the effects of BZs, which are among the most potent and widely prescribed sedatives currently in use, it is thought that pharmacological sedation by melatonin involves the activation of BZ receptors. In particular, central-type BZ-$GABA_A$ receptors appear to play a major role in the sedative and other neuropsychopharmacological actions of melatonin, with the possible supplementary involvement of other BZ receptor subtypes. However, it is important to note that while high pharmacological doses of melatonin, as typically used in animal studies, can interact with BZ-$GABA_A$ and other BZ receptors, the relatively low doses used in human studies appear to induce sleep via other physiological pathways involving melatonin receptors. In view of increasing evidence that very low physiological doses of melatonin are effective in alleviating insomnia, it would seem reasonable to utilize this safer approach, especially for the long-term treatment of sleep disorders.

References

1. Dubocovich ML, Cardinali DP, Guardiola-Lemaitre B et al. Melatonin receptors. In: Girdlestone D, ed. The IUPHAR Compendium of Receptor Characterization and Classification. London: IUPHAR Media, 1998:187-193.
2. Vanecek J. Cellular mechanisms of melatonin action. Physiol Rev 1998; 78:687-721.
3. Niles LP. G protein-coupled melatonin receptors. In: Mishra RK, Baker GB, Boulton AA, eds. G Protein Methods and Protocols: Role of G Proteins in Psychiatric and Neurological Disorders. Neuromethods. New Jersey: Humana Press, 1997:31:223-281.
4. McArthur AJ, Hunt AE, Gillette MU. Melatonin action and signal transduction in the rat suprachiasmatic circadian clock: Activation of protein kinase C at dusk and dawn. Endocrinology 1997; 138:627-634.
5. Anton-Tay F, Ramirez G, Martinez I et al. In Vitro Stimulation of protein kinase C by melatonin. Neurochemical Res 1998; 23 601:606
6. Brydon L, Petit L, Delagrange P et al. Functional expression of MT2 (Mel1b) melatonin receptors in human PAZ6 adipocytes. Endocrinology 2001; 142:4264-4271.
7. Roy D, Belsham DD. Melatonin receptor activation regulates GnRH gene expression and secretion in GT1-7 GnRH neurons. J Biol Chem 2002; 277:251-258.
8. Armstrong KJ, Niles LP. Induction of GDNF mRNA expression by melatonin in rat C6 glioma cells. NeuroReport 2002; 13:473-475.
9. Golombek DA, Martinin M, Cardinali DP. Melatonin as an anxiolytic in rats: Time dependence and interaction with the central GABAergic system. Eur J Pharmacol 1993; 237:231-236.
10. Tenn CC, Niles LP. Central-type benzodiazepine receptors mediate the antidopaminergic effect of clonazepam and melatonin in 6-hydroxydopamine lesioned rats: Involvement of a GABAergic mechanism. J Pharmacol Exp Ther 1995; 274:84-89.
11. Sieghart W. Structure and pharmacology of γ aminobutyric $acid_A$ receptor subtypes. Pharmacol Rev 1995; 47:181-234.
12. Costa E. From $GABA_A$ receptor diversity emerges a unified vision of GABAergic inhibition. Annu Rev Pharmacol Toxicol 1998; 38:321-350.
13 Tenn CC, Niles LP. Modulation of dopaminergic activity in the striatum by benzodiazepines and melatonin. Pharmacol Rev Comm 2002; 12:171-178.
14. Parola AL, Yamamura HI, Laird HE. Peripheral-type benzodiazepine receptors. Life Sci 1993; 52:1329-1342.
15. Niddam R, Dubois A, Scatton B et al. Autoradiographic localization of [^{3}H] zolpidem binding sites in the rat CNS: Comparison with the distribution of [^{3}H]flunitrazepam binding sites. J Neurochem 1987; 49:890-899.
16. Bateson AN. Basic pharmacologic mechanisms involved in benzodiazepine tolerance and withdrawal. Curr Pharmaceut Design 2002; 8:5-21.
17. Sigel E, Buhr A. The benzodiazepine binding site of $GABA_A$ receptors. Trends Pharmacol Sci 1997; 18:425-429.
18. Sur C, Wafford KA, Reynolds DS et al. Loss of the major GABA(A) receptor subtype in the brain is not lethal in mice. J Neurosci 2001; 21:3409-3418.
19. Mohler H, Fritschy JM, Rudolph U. A new benzodiazepine pharmacology. J Pharmacol Exp Ther 2002; 300:2-8.
20. Pritchett DB, Seeburg PH. γ-Aminobutyric $acid_A$ receptor α_5 subunit creates novel type II benzodiazepine receptor pharmacology. J Neurochem 1990; 54:1802-1804.
21. Wisden W, Laurie DJ, Monyer H et al. The distribution of 13 $GABA_A$ receptor subunit mRNAs in the rat brain. I Telencephalon, diencephalon, mesencephalon. J Neurosci 1992; 12:1040-1062.
22. Tobler I, Kopp C, Deboer T et al. Diazepam-induced changes in sleep: Role of the $\alpha 1$ $GABA_A$ receptor subtype. Proc Natl Acad Sci 2001; 98:6464-6469.
23. Casellas P, Galiegue S, Basile AS. Peripheral benzodiazepine receptors and mitochondrial function. Neurochem Int 2002; 40:475-486.

24. McEnery MW, Snowman AM, Trifiletti RR et al. Isolation of the mitochondrial benzodiazepine receptor: Association with the voltage-dependent anion channel and the adenine nucleotide carrier. Proc Natl Acad Sci USA 1992; 89:3170-3174.
25. Papadopoulos V. Structure and function of the peripheral-type benzodiazepine receptor in steroidogenic cells. Proc Soc Exp Biol Med 1998; 217:130-142.
26. Oke BO, Suarez-Quian CA, Riond J et al. Cell surface localization of the peripheral-type benzodiazepine receptor in adrenal cortex. Mol Cell Endocrinol 1992; 87:R1-R6.
27. Woods MJ, Zisterer DM, Williams DC. Two cellular and subcellular locations for the peripheral-type benzodiazepine receptor in rat liver. Biochem Pharmacol 1996; 51:1283-1292.
28. Tenn CC, Neu JM, Niles LP. PK 11195 blockade of benzodiazepine-induced inhibition of forskolin-stimulated adenylate cyclase activity in the striatum. Brit J Pharmacol 1996; 119:223-228.
29. Tenn CC, Niles LP. Sensitization of G protein coupled benzodiazepine receptors in the striatum of 6-hydroxydopamine lesioned rats. J Neurochem 1997; 69:1920-1926.
30. Lambert JJ, Belelli D, Hill-Venning C et al. Neurosteroids and GABA$_A$ receptor function. Trends Pharmacol Sci 1995; 16:295-303.
31. Brot MD, Akwa Y, Purdy RH et al. The anxiolytic-like effects of the neurosteroid allopregnanolone: Intractions with GABA$_A$ receptors. Eur J Pharmacol 1997; 325:1-7.
32. Stoffel-Wagner B. Neurosteroid metabolism in the human brain. Eur J Endocrinol 2001; 145:669-679.
33. Wan Q, Man HY, Liu F et al. Differential modulation of GABAA receptor function by Mel1a and Mel1breceptors. Nat Neurosci 1999; 2:401-403.
34. Niles LP. Melatonin interaction with the benzodiazepine-GABA receptor complex in the CNS. In: Schwarcz R, Young SN, Brown RR, eds. Kynurenine and Serotonin Pathways: Progress in Tryptophan Research. Advances in Experimental Medicine and Biology. New York: Plenum Press, 1991:294:267-277.
35. Niles LP, Peace CH. Allosteric modulation of t-[^{35}S]butylcyclophosphorothionate binding in rat brain by melatonin. Brain Res Bull 1990; 24:635-638.
36. Joseph-Liauzun E, Farges R, Delmas P et al. The Mr 18,000 subunit of the peripheral-type benzodiazepine receptor exhibits both benzodiazepine and isoquinoline carboxamide binding sites in the absence of the voltage-dependent anion channel or of the adenine nucleotide carrier. J Biol Chem 1997; 272:28102-6.
37. Mellon SH, Griffin LD. Neurosteroids: Biochemistry and clinical significance. Trends Endocrinol Metab 2002; 13:35-43.
38. Browning MD, Endo S, Smith GB et al. Phosphorylation of the GABA$_A$ receptor by cAMP-dependent protein kinase and by protein kinase C: Analysis of the substrate domain. Neurochem Res 1993; 18:95-10.
39. Poisbeau P, Cheney MC, Browning M.D et al. Modulation of synaptic GABA$_A$ receptor function by PKA and PKC in adult hippocampal neurons. J Neurosci 1999; 19:674-683.
40. Wan Q, Man HY, Braunton J et al. Modulation of GABA$_A$ receptor function by tyrosine phosphorylation of beta subunits. J Neurosci 1997; 17:5062-5069.
41. Itzhak Y, Baker L, Norenberg MD. Characterization of the peripheral-type benzodiazepine receptors in cultured astrocytes: Evidence for multiplicity. Glia 1993; 9:211-8.
42. Pierrefiche G, Zerbib R, Laborit H. Anxiolytic activity of melatonin in mice: Involvement of benzodiazepine receptors. Res Comm Chem Pathol Pharmacol 1993; 82:131-142.
43. Green AR, Nutt DJ, Cowen PJ. Using Ro 15-1788 to investigate the benzodiazepine receptor in vivo: Studies on the anticonvulsant and sedative effect of melatonin and the convulsant effect of the benzodiazepine Ro 05-3663. Psychopharmacol 1982; 78:293-295.
44. Golombek DA, Duque DF, De Brito Sanchez MG et al. Time-dependent anticonvulsant activity of melatonin in hamsters. Eur J Pharmacol 1992; 210:253-258.
45. Wang F, Li J, Wu C et al. The GABA$_A$ receptor mediates the hypnotic activity of melatonin in rats. Pharmacol Biochem Behav 2003; 74:573-578.
46. Sugden D. Psychopharmacological effects of melatonin in mouse and rat. J Pharmacol Exp. Ther 1983; 227:587-591.
47. Golombek DA, Escolar E, Cardinali DP. Melatonin-induced depression of locomotor activity in hamsters: Time-dependency and inhibition by the central-type benzodiazepine antagonist Ro 15-1788. Physiol Behav 1991; 49:1091-1097.
48. Dollins AB, Zhdanova IV, Wurtman RJ et al. Effect of inducing nocturnal serum melatonin concentrations in daytime on sleep, mood, body temperature, and performance. Proc Natl Acad Sci USA 1994; 91:1824-1828.
49. Hajak G, Rodenbeck A, Staedt J et al. Nocturnal plasma melatonin levels in patients suffering from chronic primary insomnia. J Pineal Res 1995; 19:116-122.
50. Waldhauser F, Weiszenbacher G, Tatzer E et al. Alterations in nocturnal serum melatonin levels in humans with growth and aging. J Clin Endocrinol Metab 1988; 66:648-652.
51. Van Coevorden A, Mockel J, Laurent E et al. Neuroendocrine rhythms and sleep in aging men. Am J Physiol 1991; 260:E651-61.
52. Garfinkel D, Laudon M, Nof D et al. Improvement of sleep quality in elderly people by controlled-release melatonin. Lancet 1995; 346:541-544.
53. Zhdanova IV, Wang SY, Leclair OU et al. Melatonin promotes sleep-like state in zebrafish. Brain Res 2001; 903:263-268.
54. Gottesmann C. GABA mechanisms and sleep. Neurosci 2002; 111:231-239.
55. Kushikata T, Kubota T, Fang J et al. Glial cell line-derived neurotrophic factor promotes sleep in rats and rabbits. Am J Physiol Regul Integr Comp Physiol 2001; 280:R1001-1006.
56. Zhdanova IV, Lynch HJ, Wurtman RJ. Melatonin: A sleep-promoting hormone. Sleep 1997; 20:899-907.
57. MacFarlane JG, Cleghorn JM, Brown GM et al. The effects of exogenous melatonin on the total sleep time and daytime alertness of chronic insomniacs: A preliminary study. Biol Psychiatry 1991; 30:371-376.
58. Stone BM, Turner C, Mills SL et al. Hypnotic activity of melatonin. Sleep 2000; 23:663-669.
59. Satomura T, Sakamoto T, Shirakawa S et al. Hypnotic action of melatonin during daytime administration and its comparison with triazolam. Psychiat Clin Neurosci 2001; 55:303-304.
60. Mignot E, Taheri S, Nishino S. Sleeping with the hypothalamus: Emerging therapeutic targets for sleep disorders. Nat Neurosci 2002; 5:1071-1075.
61. Nave R, Herer P, Haimov I et al. Hypnotic and hypothermic effects of melatonin on daytime sleep in humans: Lack of antagonism by flumazenil. Neurosci Lett 1996; 214:123-126.
62. Zhdanova IV, Wurtman RJ, Regan MM et al. Melatonin treatment for age-related insomnia. J Clin Endocrinol Metab 2001; 86:4727-4730.

CHAPTER 9

Melatonin:
A Chronobiotic that Not Only Shifts Rhythms

Dieter Kunz and Richard Mahlberg

Abstract

Today, a chronobiotic is defined as a substance capable of shifting the phase of the CTS and reentraining circadian rhythms that have been dissociated in the short-term, or desynchronized in the long-term. The most widely recognized chronobiotic is melatonin. Its phase-shifting effects have been studied extensively. Surprisingly, the synchronizing effects of exogenous melatonin in patients suffering from long-term desynchronization (such as shift workers or the elderly) have not yet been tested. The chief goal of this manuscript is to bring this characteristic of melatonin back into the spotlight and emphasize that melatonin is a chronobiotic with multiple time-dependent effects. Another important goal of this paper is to stress that melatonin is a potent drug whose nonprescription use may have serious general health consequences.

Introduction

The circadian timing system (CTS) has been shown to be involved in the daily variation of almost every physiological and psychological variable evaluated thus far (hormones, neurotransmitters, receptor densities and affinities, gene expression, mood, motor activity, pharmacokinetics, pharmacodynamics, responses to pharmacological treatment, etc.).[1-6] Not surprisingly, this statement holds true for the sleep-wake cycle as well. As described in the two-process model of sleep-wake regulation, the circadian drive to sleep and wakefulness interacts with homeostatic sleep pressure.[7] This interaction allows humans to stay awake and perform well even in the early evening, when the homeostatic sleep pressure built up during prior wakefulness is high, and to maintain a good consolidated period of sleep in the early morning, when the homeostatic drive to sleep is low.[8,9]

Drugs that influence the CTS were first referred to as chronobiotics in 1973.[10] Earlier concepts of chronobiotics focused on compounds that had a direct therapeutic effect on the circadian pacemaker itself. However, subsequent research has shown that the CTS can be influenced indirectly and at a variety of levels, including the input level (retina), clock level, and output level (circadian hormones, peptides, neurotransmitters, etc.). Thus, newer concepts of chronobiotics also include substances that indirectly affect the CTS.[11]

Today, a chronobiotic is defined as a substance capable of shifting the phase of the CTS and reentraining circadian rhythms that have been dissociated in the short-term, or desynchronized in the long-term. The most widely recognized chronobiotic is melatonin. Its phase-shifting effects have been studied extensively.[12]

However welcome the phase-shifting effects of exogenous melatonin may be in times of transcontinental flights to treat jet lag symptoms or to align morning and evening types to the environmental light dark cycle, they are of minor importance when compared to the role played by endogenous melatonin. It has been proposed that the melatonin secreted during nighttime darkness serves as a kind of "circadian cement" that provides enough inertia to resist minor perturbations in the CTS.[13] If this is true, then endogenous melatonin may be able to prevent the internal dissociation or external desynchronization of circadian rhythms. Surprisingly, the synchronizing effects of exogenous melatonin in patients suffering from long-term desynchronization (such as shift workers or the elderly) have not yet been tested.

There are many excellent reviews on the phase-shifting effects of melatonin.[14-16] However, following the publication of Dawson and Armstrong's pivotal manuscript "Chronobiotics – drugs that shift rhythms" in 1996, melatonin's perhaps most fascinating mode of action seems to have been forgotten: namely, its *synchronizing* effect.[11]

The chief goal of this manuscript is to bring this characteristic of melatonin back into the spotlight and emphasize that melatonin is a chronobiotic with multiple time-dependent effects. Another important goal of this paper is to stress that melatonin is a potent drug whose nonprescription use may have serious general health consequences.

The Circadian Timing System

The rotation of the Earth is the source of the most reliably recurrent events in nature: the daily light-dark cycle. Internal clocks are the evolutionary result, and they drive the predictable part of daily physiological variations in a precise manner.[17,18] Many of the molecular mechanisms that govern clock function have recently been determined.[19,20] Clock genes are widely expressed throughout the body, but need cyclic input from a master clock in order to sustain rhythmicity.[21] In humans, the biological master clock is located in the suprachiasmatic nuclei (SCN) of the hypothalamus.[22]

Lesions of the SCN eliminate all circadian-driven rhythms.[23,24] Inversely, SCN transplants to animals whose own SCN have been ablated, can restore circadian activity rhythms by means of a diffusible signal.[25,26] Every single one of its 20,000 neurons exerts a

Sleep and Sleep Disorders: A Neuropsychopharmacological Approach, edited by Malcolm Lader, Daniel P. Cardinali and S.R. Pandi-Perumal.

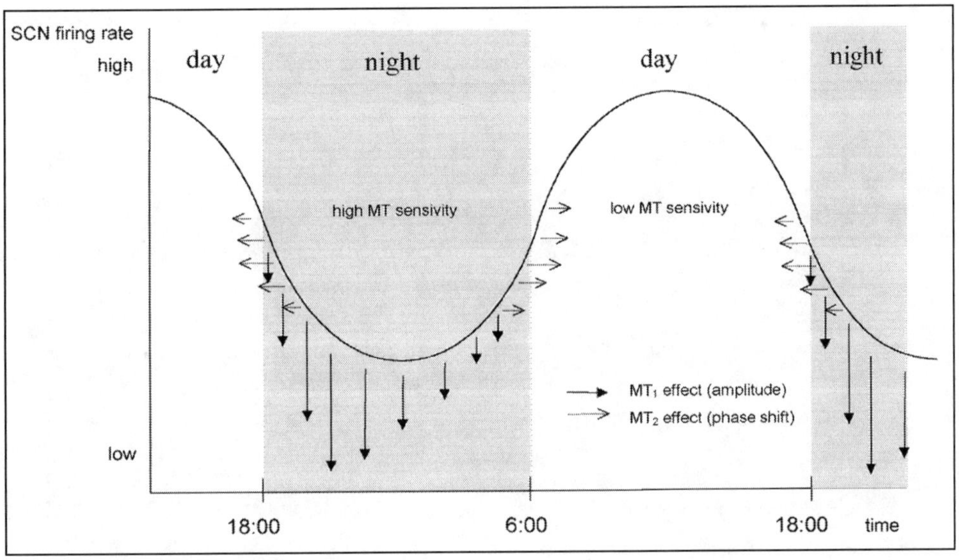

Figure 1. Schematic diagram for time dependent effects of exogenous melatonin on nucleus suprachiasmaticus neuronal activity (SCN). Neuronal firing rate is presented as the total of accumulated synchronized SCN neurons reflecting the output signal strength. Diagram is given for a 12 hour light / 12 hour dark day. Hours are given as clocktimes. Sensitivity of melatonin receptors to melatonin during the light day is low and high during nighttime darkness. Notice: Effect on amplitudes is most pronounced at early nighttime hours.

waxing and waning of the firing rate with a predictable circadian rhythm.[27] Synchronized by a yet unknown paracrine signal,[28] the SCN produces an output signal that "drives" endogenously generated daily oscillations in hormones, sleep state, alertness, performance, and other physiological functions.[22] Similar to a sinusoidal output signal, the signal produced by the SCN can be described by its period (cycle length), phase (position in the cycle), and amplitude (range between highest and lowest signal). The output amplitude reflects the "strength" or robustness of the CTS,[29] which can also be described as the drive to restore homeostasis in response to stimuli or the extent to which circadian behavior is separated into distinct periods of activity and rest within one cycle.

The SCN becomes synchronized, or *entrained*, to the environmental light-dark cycle by its primary *zeitgeber* ("time cue"): light.[30] This is true for humans as well, since the responsiveness to light and photoperiod are well conserved.[31,32] However, entrainment is highly time-dependant. In the fungus *Neurospora*, molecular mechanisms have been described by which the clock restricts its induction of behaviors in response to light at certain times during the day.[33]

Like the effects induced by the external zeitgeber light, effects by the internal zeitgeber melatonin are also time-dependent. The greatest density of high-affinity melatonin receptors in humans is located in the SCN.[34] Entraining free-running circadian rhythms by administering melatonin is only possible if the SCN is intact.[35] Within the SCN, melatonin reduces neuronal activity in a time-dependent manner.[36] In rodents, the effects of melatonin on SCN activity are mediated by at least two different receptors. They are insensitive during the day, but sensitive at dusk and dawn (MT2; causes phase shifts) and during early night period (MT1; decreases neuronal firing rate) (Fig. 1).[37,38] As a result, the phase and amplitude of SCN neuronal activity rhythm can be influenced differentially by the timing of stimuli (i.e., light and melatonin),[39] just like the pendulum of a clock.

Chronobiotic Effect on Phase

In humans, the phase-response curve (PRC) of melatonin predicts that melatonin can delay circadian rhythms when administered in the morning and advance them when administered in the afternoon or early evening. This is nearly opposite in phase to the PRCs for light exposure.[39] Accordingly, several studies have shown that melatonin can (1) phase-advance the sleep of patients suffering from a phase-delay syndrome,[40] (2) ameliorate symptoms and accelerate resynchronization of short-term dissociated circadian rhythms caused by jet lag,[12,41] and (3) reentrain the sleep-wake cycle to the environmental light-dark cycle in blind patients (long-term externally desynchronized individuals).[42]

In accordance with animal data,[37] the shifting effects caused by melatonin administration in the entrained individual (synchronized to a 12h light-12h dark cycle) are most pronounced between approximately 5-6pm for a phase advance and 5-6am for a phase delay (Fig. 2).[39] The periods during which no phase shift is observed are approximately 11-12am and 11-12pm. During the former time span (11-12am), the sensitivity of the SCN to changes in neuronal activity caused by melatonin is low.[28] During the later time span (11-12pm), the sensitivity of the SCN to melatonin is high.[38] Despite this high sensitivity, however, no phase shift occurs during the period between 11-12pm in humans.

Chronobiotic Effect on Amplitude

Although dusk and dawn are most suitable times for administering melatonin to shift the phase of the CTS, a decrease in the neuronal firing rate of SCN neurons can be achieved best during the early night period. During this time, MT1 receptors in the SCN are highly sensitive to melatonin.[37,38] Furthermore, during the early night period, the increase in endogenously secreted melatonin in entrained individuals is most rapid. Thus, this may be the most suitable time to administer exogenous melatonin in order to support the function of endogenous melatonin.

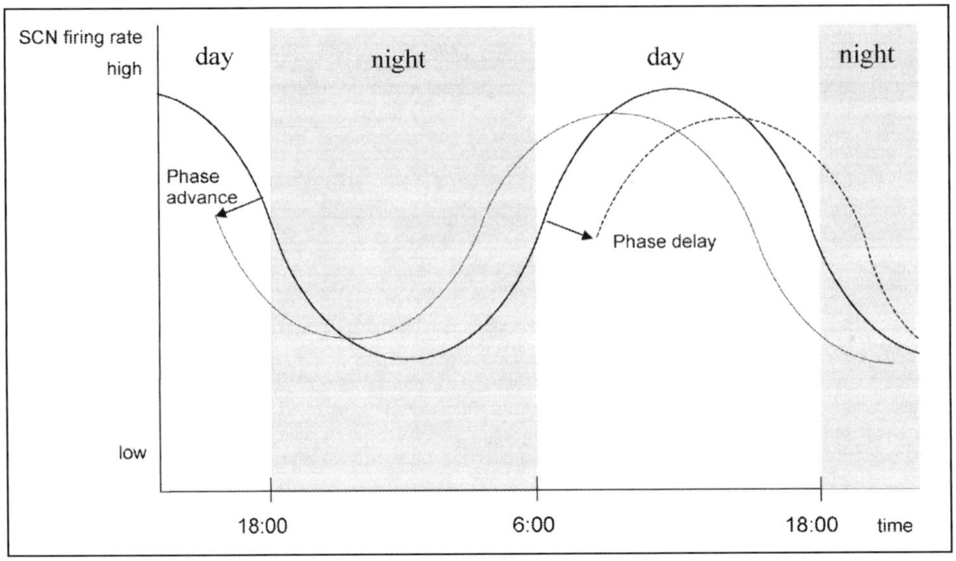

Figure 2. Schematic diagram for time dependent effects of exogenous melatonin on nucleus suprachiasmaticus neuronal activity (SCN). Neuronal firing rate is presented as the total of accumulated synchronized SCN neurons reflecting the output signal strength. Diagram is given for a 12 hour light / 12 hour dark day. Hours are given as clocktimes. At the beginning of the night, melatonin causes a phase advance (dashed line) and in the early morning hours causes a phase delay (dotted line)—MT2 receptors; note the acute reduction of output amplitude after a phase shift.

The term "internal desynchrony" was introduced to describe the dissociation of internally-generated circadian rhythms.[29] Under the natural conditions of a 24-hour day and with exposure to sufficiently strong zeitgebers, circadian rhythms are synchronized. However, internal rhythms can partly desynchronize as a result of temporal isolation, such as in forced desynchrony protocols or as the consequence of acute phase shifts caused, for example, by jet lag (Fig. 3). Earlier, it was assumed that circadian variations in physiological factors such as body temperature and activity in humans were driven by two different oscillators. Now, however, there is a general consensus that we are dealing with a peripheral multioscillator system governed by one master oscillator located in the SCN.[20,43]

When caused intentionally as part of a temporal isolation study, internal desynchrony is artificial and of short-term nature. The underlying circadian rhythmicity remains strong. The use of the term "internal desynchrony" was expanded to include its hypothesized, long-term consequences on the SCN level.[11] Within the SCN, every single neuron contributes to an increase and decrease in the firing rate in an approximately 24-hour rhythm.[27,44] When synchronized, these "clock cells" represent the human internal clock and have a strong output signal.[28,45] When partially desynchronized, SCN neurons may produce a weaker output signal. When completely uncoupled, however, they appear not to produce any output signal whatsoever. The strength of the biological clock determines the magnitude of the daily variation in circadian driven rhythms.[29] Thus, the consequence of a state of internal desynchrony is the cessation of the endogenously generated daily variation in physiological, biochemical, and behavioral parameters, which together can be described as the amplitude of circadian rhythms.[11,46]

Referring to the "use it or loose it" theory on the plasticity of SCN neurons,[47] long-term desynchronization can cause these neurons to become inactive, resulting in a decrease in the output amplitude of the circadian pacemaker. Because this state can persist even after shift work has been discontinued, it is hardly surprising that long-lasting sleep-related problems have been observed in this population.[48] However, because this inactivity is functional, it can also potentially be reversed.[47]

In order to counteract the dampening of CTS amplitude caused by internal desynchronization, it is necessary to reorganize or strengthen the circadian pacemaker. Increasing endogenous nighttime melatonin levels with well-timed photic stimulation or exogenous melatonin appears to be one possible means of doing so.[46] In animal studies, the amplitude of circadian driven parameters, such as body temperature and locomotive activity, was increased during melatonin application in elderly rats.[49] Similarly, the amplitude of locomotive activity in Syrian hamsters was increased following the application of bright light.[50] In another study, a melatonin agonist was shown to accelerate resynchronization in diurnal rodents.[51] In humans, direct and even indirect bright light improves circadian rest-activity rhythm disturbances in demented patients.[52,53] The application of bright light during the day to elderly insomniacs has been shown to improve sleep and increase nighttime melatonin secretion simultaneously.[54] Evening melatonin reduces nighttime agitation in Alzheimer's patients.[55] Nighttime melatonin administration in humans and rats increases endogenous melatonin secretion in the following nights only when administered at the day/night transition.[56,57] Most importantly, however, in a human jet lag experiment melatonin was found not only to accelerate resynchronization, but also to increase the amplitude of circadian rhythms.[41]

In conclusion, in an individual whose CTS output amplitude has been dampened, it may be possible to strengthen this signal by the well-timed administration of exogenous melatonin. Although the exact time for melatonin administration is still unknown, it is clear that this time needs to be fixed and be located after (!) the early evening phase-shifting period. Because melatonin acts on the SCN via a positive feedback loop, its synchronizing effects outlast the actual time of drug administration. Thus, like in the pendulum clock: Once the clock begins ticking properly, it will keep ticking.

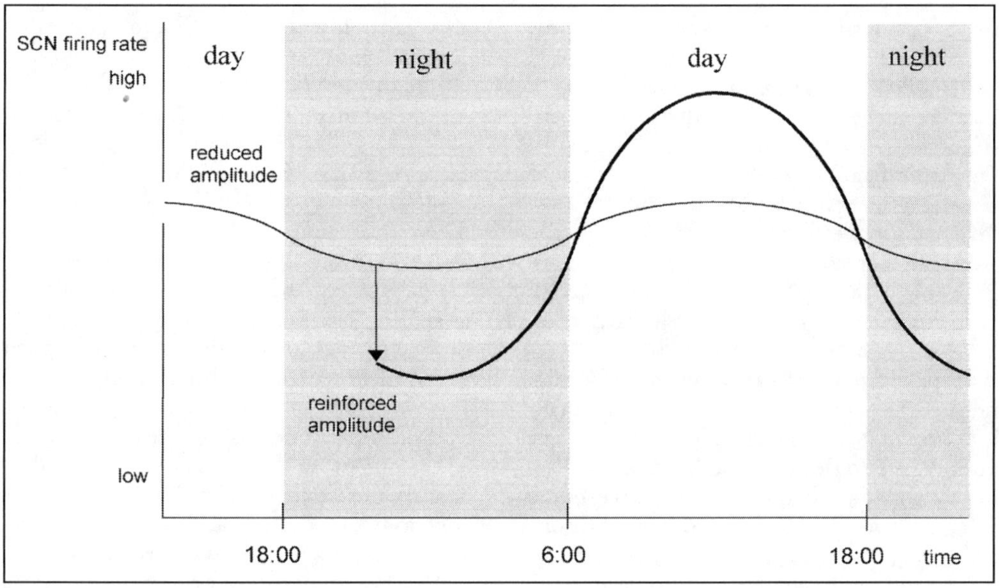

Figure 3. Schematic diagram for time dependent effects of exogenous melatonin on nucleus suprachiasmaticus neuronal activity (SCN). Neuronal firing rate is presented as the total of accumulated synchronized SCN neurons reflecting the output signal strength. Diagram is given for a 12 hour light / 12 hour dark day. Hours are given as clocktimes. In the case of reduced strength of the SCN output signal caused by internal desynchrony, melatonin administered during the early night period may resynchronize SCN neurons and thereby reinforce amplitude of SCN output signal without an effect on circadian phase – MT1 receptors.[37,38,90]

Causes of Reduced Amplitude

The invention of artificial light approximately 120 years ago enabled humans to place times of activity and inactivity at will – an achievement we have taken advantage of ever since. Due to the necessities of modern life, we change the length and the timing of our "day" by using bright artificial light. Furthermore, many of us consume beverages and administer medications that are known to shift the phase of the CTS, to influence the internal zeitgeber melatonin, or to change the sensitivity of the SCN to the external zeitgeber light.[58-63]

Some 30 years ago, Winfree reported that light stimulus of a critical strength and applied during a critical circadian phase could essentially stop the circadian clock, resulting in an amplitude of circadian oscillation equal to zero.[64] Similar observations of attenuated amplitude in response to critical stimuli were subsequently made in unicells, insects, plants, and, most recently, in humans.[65,66] Furthermore, prolonged exposure to periods of light or dark reduces amplitude: In one study, rats exposed to continuous light for a long period experienced complete suppression of circadian rhythms in body temperature and locomotor activity. This coincided with a 6-fold decrease in plasma melatonin and a marked dampening of the amplitude of its 24-hour variation.[67] In humans, experimental shifts in the sleep-wake cycle with respect to environmental time have led to a reduction in amplitude.[41,68] Thus, there seems to be solid evidence that the dampening of CTS output amplitude, which is considered to be a hallmark of aging,[46] is accelerated by a modern life style and can lead to a state of internal desynchrony.

Clinical Considerations

The two main populations with reduced CTS output amplitude are former shift workers and the elderly. Shift work involves repeated shifts of sleep-wake behavior within the light-dark cycle. The strength of the CTS—as indicated by the output amplitude (e.g., daily variations of body temperature, hormones, sleep propensity)—is dampened in elderly people and, to an even greater extent, in patients suffering from Alzheimer's dementia.[1,46,69,70]

Recently, the effects of exogenous melatonin on human sleep have been the subject of extensive investigations.[14,15] These have shown that melatonin exerts only minimal hypnotic effects on general sleep parameters, influencing instead specific circadian components of the sleep-wake cycle, such as sleep continuity and REM sleep.[71-73] In contrast, earlier sleep studies did not specifically address the influence of melatonin on circadian components in the sleep-wake cycle.[74] In our own study, we performed a randomized controlled trial, in which patients with reduced REM sleep duration showed improvements not only of clinical symptoms but also circadian-modulated REM sleep parameters and temperature decline during sleep.[72] Moreover, the effects of melatonin in our study outlasted the actual period of melatonin administration, and only diminished slowly over time. Because there was no evidence of phase-shift in our patient population, we attribute these results to an increase in CTS output amplitude.

The effects of melatonin in humans are time dependent,[75] so it follows that the time of melatonin administration needs to be fixed.[74] The greatest increase in endogenously secreted melatonin by the pineal gland in *entrained* humans (synchronized to the environmental light-dark cycle) occurs during the early night period. Only when exogenous melatonin is administered at that specific point in time can it support the functions of endogenous melatonin or act as a synchronizing chronobiotic. In contrast, improperly timing the administration of exogenous melatonin may lead to phase-shifts and, thus, reduced amplitude. In our melatonin and sleep studies, patients were instructed to take melatonin between 10-11pm.[71,72,76] Most interestingly, responders and nonresponders were distinguished best by sleep hygiene (stable vs. changing bedtimes and the time of melatonin administration). Thus, exogenous melatonin needs to be administered consistently within a very narrow time span – in contrast to the flexible administration (i.e., "at bedtime") often recommended on the bottles or package inserts of so-called "nutritional supplements" that contain melatonin.

What are the consequences of a reduced amplitude? High amplitude of the circadian driven rest-activity rhythm is positively associated with increased survival time in rats.[77] In hamsters with cardiomyopathic heart disease, chronic circadian desynchronization decreases survival.[78] Circadian amplitude of melatonin and body temperature as indicators of circadian strength are positively associated with increased sleep duration in humans (r = .87).[79] Most interestingly, in very healthy, elderly subjects, circadian amplitude, including the sleep-wake cycle, is not altered.[80]

The clinical phenomena of a reduced amplitude may mimic a chronic "jet lag" condition. Jet lag is often accompanied by malaise and fatigue. More importantly, jet lag and/or former shift work is associated with long-term sleep and health disturbances such as increased cortisol, impaired cognition, reduced temporal lobe volume, as well as gastrointestinal and cardiovascular disorders.[48,81-83] Due to repetitive phase shifts, shift workers internally desynchronize in the long run[84] and, consequently, tend to have shortened and impaired sleep.

Sleep has been shown to enhance cortical plasticity, memory consolidation, and learning processes.[85,86] Sleep deprivation in humans decreases nighttime interleukin-6 levels and may thus affect the integrity of immune system functioning.[87] Short-term sleep restriction in healthy young men leads to decreased carbohydrate tolerance, increased sympathetic tone, and increased cortisol blood concentrations.[88] These symptoms are all part of the "normal" aging process and are well-recognized risk factors for the development of insulin resistance, obesity, and hypertension. This implies that reduced amplitude may very well play a role in the metabolic syndrome. Thus, if melatonin, when administered flexibly "at bedtime" (rather than within a regular, fixed time span), does indeed reduce amplitude, the question arises: Does the uncontrolled use of melatonin contribute to disturbed sleep, obesity, diabetes, and hypertension?

Conclusions

Diurnal species that predominantly use their eyes in order to find their way in the world are at the mercy of their enemies during the hours of darkness. Evolution may have developed clock systems in order to promote inactivity and thus protect these species from harm. In humans, this state of inactivity is sleep. Most likely, the physiologic processes of which sleep is comprised developed secondarily. Nevertheless, now they are located within the period of time we designate as "sleep".

It has been proposed that sleep is involved in neuronal plasticity, learning processes, memory consolidation, the coordination of metabolic processes, and in the integrity of immune system functioning. The time for good quality sleep to occur is predetermined and anticipated by the circadian timing system, which is governed by an internal master clock located in the hypothalamic suprachiasmatic nuclei. One function of endogenous melatonin is to stabilize the CTS at night. Clearly, good quality sleep of sufficient length is a *conditio sine qua non* for the proper functioning of body and brain.

Though we seem to be independent of the environmental light dark cycle by means of the arbitrary application of artificial light and many other achievements of modern societies, man remains a diurnal species. Since the inception of artificial light 120 years ago, evolution has not overcome the necessity of timekeeping. Proper functioning of the CTS allows human physiology to anticipate behavior such as sleep and to fulfill its functions. The CTS is highly active in humans and a reduction of CTS output strength is considered a hallmark of biological aging.[46] The well-timed administration of the chronobiotic melatonin may stabilize the CTS and thereby normalize the temporal organization of human physiology.

There is one important limitation to the literature currently available on this topic: it represents a mixture of human and animal data. However, the molecular mechanisms behind biological time-keeping exhibit an extraordinary degree of evolutionary conservation in organisms as diverse as cyanobacteria, plants, fruit flies, and mammals. Fruit flies and mice even share some identical time-keeping proteins.[89] Nevertheless, there is still much work to be done before one can assume that melatonin receptor properties, among other things, are similar in rodents and humans.

Clearly, melatonin is an important hormone and a potent drug. As long as the precise mode of action with respect to physiology and pharmacology is not specified, however, taking melatonin without a prescription and a physician's guidance represents a significant risk. If administering melatonin at variable bedtimes, does indeed prove to destabilize the CTS, it appears that we may very well have a ticking time bomb on our hands.

References

1. Van Cauter E, Plat L, Leproult R et al. Alterations of circadian rhythmicity and sleep in aging: Endocrine consequences. Horm Res 1998; 49:147-152.
2. Wirz-Justice A. Circadian rhythms in mammalian neurotransmitter receptors. Prog Neurobiol 1987; 29:219-259.
3. Shearman LP, Sriram S, Weaver DR et al. Interacting molecular loops in the mammalian circadian clock. Science 2000; 288:1013-1019.
4. Boivin DB, Czeisler CA, Dijk DJ et al. Complex interaction of the sleep-wake cycle and circadian phase modulates mood in healthy subjects. Arch Gen Psychiatry 1997; 54:145-152.
5. Wyatt JK, Cecco AR, Czeisler CA et al. Circadian temperature and melatonin rhythms, sleep, and neurobehavioral function in humans living on a 20-h day. Am J Physiol 1999; 277:R1152-R1163.
6. Bruguerolle B. Chronopharmacology. In: Touitou Y, Haus E, eds. Biological Rhythms in Clinical and Laboratory Medicine. Paris: Springer-Verlag, 1992:114-137.
7. Borbély AA, Achermann P. Concepts and models of sleep regulation: An overview. J Sleep Res 1992; 1:63-79.
8. Lavie P. Ultrashort sleep-waking schedule. III. 'Gates' and 'forbidden zones' for sleep. Electroencephalogr Clin Neurophysiol 1986; 63:414-425.
9. Dijk DJ, Czeisler CA. Contribution of the circadian pacemaker and the sleep homeostat to sleep propensity, sleep structure, electroencephalographic slow waves, and sleep spindle activity in humans. J Neurosci 1995; 15:3526-3538.
10. Simpson HW, Bellamy N, Bohlen J et al. Double blind trial of a possible chronobiotic (Quiadon)r. Field studies in N.W. Greenland. Int J Chronobiol 1973; 1:287-311.
11. Dawson D, Armstrong SM. Chronobiotics-drugs that shift rhythms. Pharmacol Ther 1996; 69:15-36.
12. Herxheimer A, Petrie KJ. Melatonin for preventing and treating jet lag (Cochrane Review). Cochrane Database Syst Rev 1901; 1:CD001520.
13. Dawson D, Armstrong SM. Chronobiotics-drugs that shift rhythms. Pharmacol Ther 1996; 69:15-36.
14. Zhdanova IV, Lynch HJ, Wurtman RJ. Melatonin: A sleep-promoting hormone. Sleep 1997; 20:899-907.
15. Czeisler CA, Cajochen C, Turek FW. Melatonin in the regulation of sleep and circadian rhythms. In: Kryger MH, Roth T, Dement WC, eds. Principles and Practice of Sleep Medicine. Philadelphia: WB Saunders, 2000:400-406.

16. Cajochen C, Krauchi K, Wirz-Justice A. Role of melatonin in the regulation of human circadian rhythms and sleep. J Neuroendocrinol 2003; 15:432-437.
17. Pittendrigh CS. Temporal organization: Reflections of a Darwinian clock-watcher. Annu Rev Physiol 1993; 55:16-54.
18. Czeisler CA, Duffy JF, Shanahan TL et al. Stability, precision, and near-24-hour period of the human circadian pacemaker. Science 1999; 284:2177-2181.
19. Young MW. Life's 24-hour clock: Molecular control of circadian rhythms in animal cells. Trends Biochem Sci 2000; 25:601-606.
20. Cermakian N, Sassone-Corsi P. Multilevel regulation of the circadian clock. Nature Reviews 2000; 1:59-67.
21. Clayton JD, Kyriacou CP, Reppert SM. Keeping time with the human genome. Nature 2001; 409:829-831.
22. Klein DC, Moore RY, Reppert SM. The Suprachiasmatic Nucleus: The Mind,s Clock. New York: Oxford University Press, 1991.
23. Stephan FK, Zucker I. Circadian rhythms in drinking behavior and locomotor activity of rats are eliminated by hypothalamic lesions. Proc Natl Acad Sci USA 1972; 69:1583-1586.
24. Eastman CI, Mistlberger RE, Rechtschaffen A. Suprachiasmatic nuclei lesions eliminate circadian temperature and sleep rhythms in the rat. Physiol Behav 1984; 32:357-368.
25. Silver R, LeSauter J, Tresco PA et al. A diffusible coupling signal from the transplanted suprachiasmatic nucleus controlling circadian locomotor rhythms. Nature 1996; 382:810-813.
26. Earnest DJ, Liang FQ, Ratcliff M et al. Immortal time: Circadian clock properties of rat suprachiasmatic cell lines. Science 1999; 283:693-695.
27. Green DJ, Gillette R. Circadian rhythm of firing rate recorded from single cells in the rat suprachiasmatic brain slice. Brain Res 1982; 245:198-200.
28. Gillette MU, Tischkau SA. Suprachiasmatic nucleus: The brain's circadian clock. Recent Prog Horm Res 1999; 54:33-58.
29. Wever R. Order and disorder in human circadian rhythmicity: Possible relations to mental disorders. In: Kupfer DJ, Monk TH, Barchas JD, eds. Biological Rhythms and Mental Disorders. New York: Guilford, 1988:253-346.
30. Freedman MS, Lucas RJ, Soni B et al. Regulation of mammalian circadian behavior by nonrod, noncone, ocular photoreceptors. Science 1999; 284:502-504.
31. Lewy AJ, Wehr TA, Goodwin FK et al. Light suppresses melatonin secretion in humans. Science 1980; 210:1267-1269.
32. Wehr TA, Moul DE, Barbato G et al. Conservation of photoperiod-responsive mechanisms in humans. Am J Physiol 1993; 265:R846-R857.
33. Heintzen C, Loros JJ, Dunlap JC. The PAS protein VIVID defines a clock-associated feedback loop that represses light input, modulates gating, and regulates clock resetting. Cell 2001; 104:453-464.
34. Weaver DR, Stehle JH, Stopa EG et al. Melatonin receptors in human hypothalamus and pituitary: Implications for circadian and reproductive responses to melatonin. J Clin Endocrinol Metab 1993; 76:295-301.
35. Cassone VM, Chesworth MJ, Armstrong SM. Entrainment of rat circadian rhythms by daily injection of melatonin depends upon the hypothalamic suprachiasmatic nuclei. Physiol Behav 1986; 36:1111-1121.
36. Cassone VM, Roberts MH, Moore RY. Effects of melatonin on 2-deoxy-[1-14C]glucose uptake within rat suprachiasmatic nucleus. Am J Physiol 1988; 255:R332-R337.
37. Hunt AE, Al-Ghoul WM, Gillette MU et al. Activation of MT(2) melatonin receptors in rat suprachiasmatic nucleus phase advances the circadian clock. Am J Physiol Cell Physiol 2001; 280:C110-C118.
38. Liu C, Weaver DR, Jin X et al. Molecular dissection of two distinct actions of melatonin on the suprachiasmatic circadian clock. Neuron 1997; 19:91-102.
39. Lewy AJ, Bauer VK, Ahmed S et al. The human phase response curve (PRC) to melatonin is about 12 hours out of phase with the PRC to light. Chronobiol Int 1998; 15:71-83.
40. Dahlitz M, Alvarez B, Vignau J et al. Delayed sleep phase syndrome response to melatonin. Lancet 1991; 337:1121-1124.
41. Samel A, Wegmann HM, Vejvoda M et al. Influence of melatonin treatment on human circadian rhythmicity before and after a simulated 9-hr time shift. J Biol Rhythms 1991; 6:235-248.
42. Sack RL, Brandes RW, Kendall AR et al. Entrainment of free-running circadian rhythms by melatonin in blind people. N Engl J Med 2000; 343:1070-1077.
43. Lu J, Zhang YH, Chou TC et al. Contrasting effects of ibotenate lesions of the paraventricular nucleus and subparaventricular zone on sleep-wake cycle and temperature regulation. J Neurosci 2001; 21:4864-4874.
44. Welsh DK, Logothetis DE, Meister M et al. Individual neurons dissociated from rat suprachiasmatic nucleus express independently phased circadian firing rhythms. Neuron 1995; 14:697-706.
45. Liu C, Weaver DR, Strogatz SH et al. Cellular construction of a circadian clock: Period determination in the suprachiasmatic nuclei. Cell 1997; 91:855-860.
46. Myers BL, Badia P. Changes in circadian rhythms and sleep quality with aging: Mechanisms and interventions. Neurosci Biobehav Rev 1995; 19:553-571.
47. Swaab DF. Brain aging and Alzheimer's disease, "wear and tear" versus "use it or lose it". Neurobiol Aging 1991; 12:317-324.
48. Dumont M, Montplaisir J, Infante-Rivard C. Sleep Quality of Former Night-shift Workers. Int J Occup Environ Health 1997; 3:S10-S14.
49. Koster-van Hoffen GC, Mirmiran M, Bos NP et al. Effects of a novel melatonin analog on circadian rhythms of body temperature and activity in young, middle-aged, and old rats. Neurobiol Aging 1993; 14:565-569.
50. Labyak SE, Turek FW, Wallen EP et al. Effects of bright light on age-related changes in the locomotor activity of Syrian hamsters. Am J Physiol 1998; 274:R830-R839.
51. Van Reeth O, Olivares E, Turek FW et al. Resynchronization of a diurnal rodent circadian clock accelerated by a melatonin agonist. Neuroreport 1998; 9:1901-1905.
52. Satlin A, Teicher MH, Liebermann HR et al. Circadian locomotor activity rhythms in Alzheimer's disease. Neuropsychopharmacol 1991; 5:115-126.
53. Someren van EJW, Kessler A, Mirmiran M et al. Indirect bright light improves circadian rest-activity rhythm disturbances in demented patients. Biol Psychiatry 1997; 41:955-963.
54. Mishima K, Okawa M, Shimizu T et al. Diminished melatonin secretion in the elderly caused by insufficient environmental illumination. J Clin Endocrinol Metab 2001; 86:129-134.
55. Mahlberg R, Kunz D, Sutej I et al. Melatonin treatment of day-night rhythm disturbances and sundowning in Alzheimer,s disease: An open-label pilot study using actigraphy. J Clin Psychopharmacol 2003 -in revision
56. Zaidan R, Geoffriau M, Brun J et al. Melatonin is able to influence its secretion in humans: Description of a phase-response curve. Neuroendocrinology 1994; 60:105-112.
57. Bothorel B, Barassin S, Saboureau M et al. In the rat, exogenous melatonin increases the amplitude of pineal melatonin secretion by a direct action on the circadian clock. Eur J Neurosci 2002; 16:1090-1098.
58. Morin LP, Fitzgerald KM, Zucker I. Estradiol shortens the period of hamster circadian rhythms. Science 1977; 196:305-307.
59. Cowen PJ, Bevan JS, Gosden B et al. Treatment with beta-adrenoceptor blockers reduces plasma melatonin concentration. Br J Clin Pharmacol 1985; 19:258-260.
60. Ekman AC, Leppaluoto J, Huttunen P et al. Ethanol inhibits melatonin secretion in healthy volunteers in a dose- dependent randomized double blind cross-over study. J Clin Endocrinol Metab 1993; 77:780-783.
61. Wright KPJ, Badia P, Myers BL et al. Caffeine and light effects on nighttime melatonin and temperature levels in sleep-deprived humans. Brain Res 1997; 747:78-84.
62. Duncan WCJ, Johnson KA, Wehr TA. Decreased sensitivity to light of the photic entrainment pathway during chronic clorgyline and lithium treatments. J Biol Rhythms 1998; 13:330-346.
63. Arendt J. Melatonin and the Mammalian Pineal Gland. 1 ed. University Press: Cambridge, 1995.

64. Winfree AT. Integrated view of resetting a circadian clock. J Theor Biol 1970; 28:327-374.
65. Jewett ME, Kronauer RE, Czeisler CA. Light-induced suppression of endogenous circadian amplitude in humans. Nature 1991; 350:59-62.
66. Shanahan TL, Kronauer RE, Duffy JF et al. Melatonin rhythm observed throughout a three-cycle bright-light stimulus designed to reset the human circadian pacemaker. J Biol Rhythms 1999; 14:237-253.
67. Depres-Brummer P, Levi F, Metzger G et al. Light-induced suppression of the rat circadian system. Am J Physiol 1995; 268:R1111-R1116.
68. Monk TH, Buysse DJ, Carrier J et al. Inducing jet-lag in older people: Directional asymmetry. J Sleep Res 2000; 9:101-116.
69. Prinz PN, Christie C, Smallwood R et al. Circadian temperature variation in healthy aged and Alzheimer's disease. J Gerontol 1984; 39:30-35.
70. Swaab DF, Fliers E, Partiman TS. The suprachiasmatic nucleus of the human brain in relation to sex, age and senile dementia. Brain Res 1985; 342:37-44.
71. Kunz D, Bes F. Melatonin as a therapy in REM sleep behavior disorder patients: An open-labeled pilot study on the possible influence of melatonin on REM-sleep regulation. Mov Disord 1999; 14:507-511.
72. Kunz D, Mahlberg R, Müller LC et al. Melatonin in patients with reduced REM sleep duration: A randomized clinical trial. J Clin Endocrinol Metab 2004; 89:128-134.
73. Dijk DJ, Cajochen C. Melatonin and the circadian regulation of sleep initiation, consolidation, structure, and the sleep EEG. J Biol Rhythms 1997; 12:627-635.
74. Kunz D. Chronobiotic protocol and circadian sleep propensity index: New tools for clinical routine and research on melatonin and sleep. Pharmacopsychiatry 2003 - in revision
75. Lewy AJ, Ahmed S, Jackson JM et al. Melatonin shifts human circadian rhythms according to a phase-response curve. Chronobiol Int 1992; 9:380-392.
76. Kunz D, Bes F. Exogenous melatonin in periodic limb movement disorder: An open clinical trial and a hypothesis. Sleep 2001; 24:183-187.
77. Martin JR, Fuchs A, Bender R et al. Altered light-dark activity difference with aging in two rats strains. J Gerontol 1986; 41:2-7.
78. Penev PD, Kolker DE, Zee PC et al. Chronic circadian desynchronization decreases the survival of animals with cardiomyopathic heart disease. Am J Physiol 1998; 275:H2334-H2337.
79. Czeisler CA, Dumont M, Duffy JF et al. Association of sleep-wake habits in older people with changes in the output of circadian pacemaker. Lancet 1992; 340:933-936.
80. Monk TH, Buysse DJ. Circadian rhythms in the elderly: A comparison of field, laboratory and unmasked conditions. Sleep Res 1989; 18:433.
81. Knutsson A, Akerstedt T, Jonsson BG et al. Increased risk of ischaemic heart disease in shift workers. Lancet 1986; 2:89-92.
82. Cho K. Chronic ‚jet lag' produces temporal lobe atrophy and spatial cognitive deficits. Nat Neurosci 2001; 4:567-568.
83. Cho K, Ennaceur A, Cole JC et al. Chronic jet lag produces cognitive deficits. J Neurosci 2000; 20:RC66.
84. Weibel L, Brandenberger G. Disturbances in hormonal profiles of night workers during their usual sleep and work times. J Biol Rhythms 1998; 13:202-208.
85. Frank MG, Issa NP, Stryker MP. Sleep enhances plasticity in the developing visual cortex. Neuron 2001; 30:275-287.
86. Maquet P. Sleep on it! Nat Neurosci 2000; 3:1235-1236.
87. Redwine L, Hauger RL, Gillin JC et al. Effects of sleep and sleep deprivation on interleukin-6, growth hormone, cortisol, and melatonin levels in humans. J Clin Endocrinol Metab 2000; 85:3597-3603.
88. Spiegel K, Leproult R, Van Cauter E. Impact of sleep debt on metabolic and endocrine function. Lancet 1999; 354:1435-1439.
89. King DP, Takahashi JS. Molecular genetics of circadian rhythms in mammals. Annu Rev Neurosci 2000; 23:713-742.

CHAPTER 10

Melatonin and Human Sleep

Irina V. Zhdanova

Abstract

The major hormone of the circadian system, melatonin, plays an important role in human sleep regulation, acting via both circadian and homeostatic mechanisms. Melatonin treatment can be used to treat chronic insomnias of different origin and circadian sleep disorders. However, if administered in high doses or at inappropriate times, melatonin might induce the circadian rhythm disorders or sleep abnormalities, rather than cure them.

Introduction

Within the cells in the pineal gland, amino acid L-tryptophan is converted to melatonin (N-acetyl-5-methoxytryptamine) through series of enzymatic transformations. The hormone is released directly into the blood stream and the cerebrospinal fluid as it is synthesized and, being lipid soluble, has ready access to every cell of the body. The human fetus and newborn infant do not produce melatonin but rely on the hormone supplied via the placental blood, and, postnatally, via the mother's milk. After infants are 9-12 weeks old, rhythmic melatonin production increases rapidly; the highest nocturnal melatonin levels are attained in children under 5 years of age. Thereafter, melatonin levels decrease, with perhaps the most dramatic changes occurring during adolescence.[1] Though exceptionally healthy elderly sometimes maintain high melatonin production,[2] the majority of aged individuals have low circulating melatonin levels.[3-5] Typical nocturnal peak serum concentrations of melatonin in young humans are around 100 pg/ml, while daytime levels are as low as 1-3 pg/ml. Although the circulating melatonin pattern for a given individual tends to remain surprisingly constant from day to day, marked inter-individual variations in nocturnal melatonin levels are observed in all age groups. The hours of the increased melatonin secretion are concurrent with the habitual hours of sleep, and the onset of melatonin secretion correlates with the onset of evening sleepiness.[6-8] Studies of circadian phase shifts in humans also show that change in the timing of the onset of melatonin secretion correlates with change in the timing of evening sleepiness.

Melatonin Effects on Sleep via the Circadian Clock System

One of the most distinct features of melatonin production is its 24-hour pattern, characterized by a nocturnal raise. It is typically initiated about two hours before individual's habitual bedtime, unless blocked by environmental light, and declines in the morning. This rhythmic pattern of melatonin secretion persists in people maintained in constant darkness. Exposure to darkness during the daytime does not induce melatonin production, however the abrupt imposition of bright light at night can suppresses it.

A near-24-hour (circadian) rhythm of melatonin secretion from the human pineal gland depends on a periodic signal from suprachiasmatic nuclei (SCN) of the hypothalamus, the master biological "clock". The SCN neurons are capable of sustaining a circadian pattern of activity even in the absence of rhythmic environmental input. They are normally active during the day and slow down at night. The activation of SCN neurons has an inhibitory effect on the pineal gland, defining a nocturnal pattern of melatonin secretion. If SCN neurons are activated at night, e.g., by environmental light perceived by the retina, melatonin production declines. Melatonin, in turn, can attenuate the activity of SCN and produce a shift in the circadian phase of SCN activity, either advancing or delaying it. The direction of the phase-shift depends on the time of melatonin treatment, i.e., administration of melatonin in the late afternoon can advance the circadian clock, while early-morning treatment can cause a phase delay.[9] A temporal and functional interplay between melatonin and the SCN, and their response to environmental light, promote a temporal alignment of multiple circadian body rhythms with each other (internal synchronization) and with the periodic changes in the environment (external synchronization).

Sleep process is under the control of the biological clock. It is recognized now that many sleep disorders result from a misalignment of circadian rhythms. Some are related to abrupt advances or delays in the sleep period relative to the environmental light-dark cycle, e.g., due to transmeridian flight or shift work. Others reflect an intrinsic impairment of circadian and homeostatic regulation of sleep. The latter include a non24-h sleep-wake syndrome, rare in normally sighted persons but common in the blind; the delayed or advanced sleep phase syndromes (DSPS and ASPS), associated with much later (DSPS) or much earlier (ASPS) sleep onset and morning awakening than expected according to social cues, or individuals' circadian temperature patterns.[10] The most dramatic alteration of circadian regulation of sleep manifests as an irregular sleep-wake pattern, characterized by several daytime sleep periods and lack of a consolidated nighttime sleep episode. This circadian disorder is most commonly observed in patients with developmental or degenerative neurological conditions.

The ability of melatonin to entrain and phase shift the circadian clock makes it a good candidate for treating some circadian

Sleep and Sleep Disorders: A Neuropsychopharmacological Approach, edited by Malcolm Lader, Daniel P. Cardinali and S.R. Pandi-Perumal.

sleep disorders. There are contradictory reports regarding the ability of timely melatonin administration to phase shift and entrain the free-running rhythms in blind individuals. Some studies found improved sleep quality but not a circadian phase shift after melatonin treatment, while others documented both the entrainment of previously free-running circadian rhythms and an increase in sleep efficiency in blind subjects receiving melatonin.[11-14] Several studies described different degrees of success in treating DSPS patients with melatonin to achieve phase advance.[15-17] Interestingly, studies in adolescents, who frequently develop DSPS, suggest that a substantial decline in melatonin production at this age might be one of the predisposing factors for the onset of this sleep disorder.[18] Indeed, a rapidly developing deficit in both circadian and homeostatic effects of melatonin on sleep initiation could cause a delay of their nighttime sleep period.

Melatonin treatment might be also beneficial to individuals with neurological disorders, associated with the irregular sleep-wake pattern. The preliminary data suggest that severely altered sleep patterns in Alzheimer disease patients can be improved by timely melatonin treatment.[19] Similarly, nighttime melatonin treatment in children with Angelman syndrome (AS), a developmental neurological disorder, has been found to modify their irregular sleep-wake patterns by significantly increasing their nighttime sleep period and reducing sleep fragmentation.[20] In addition, we found a substantial delay in melatonin secretion onset in some of the AS children. Such phase delay may result from irregularities in sleep-wake and/or environmental light-dark cycles, or could be a sign of a weakened circadian clock function, as part of the general developmental disorder. Whatever the reason, daily melatonin treatment succeeded in readjusting patients' sleep-wake and melatonin rhythms and synchronized them with children's bedtimes.

Difficulty falling asleep at night and staying alert during the day are the major symptoms of a circadian desynchrony following a rapid transition between time zones (jet-lag syndrome). The ability of melatonin to produce a circadian phase shift suggests that it can help to facilitate a resetting of the circadian clock to match new environmental time after a transmeridian flight. Some studies have documented an improvement in subjective measures of sleep and performance when subjects received melatonin treatment after a real transmeridian flight or its simulation under laboratory conditions.[21-24] Others, however, did not find melatonin treatment to be efficacious in eliminating jet-lag symptoms compared to placebo.[25,26] Such discrepancies might result from low sensitivity of subjective measures of sleep and alertness employed in these studies, differences in the doses and preparations of melatonin used, and/or the masking effects of environmental light. Furthermore, the positive effects observed could be, to a large extent, explained by an acute sleep-promoting effect of the hormone, which would prevent an accumulation of "sleep debt" and thus improve daytime performance.

A chronic state of circadian desynchrony developed in response to shift work often leads to chronic sleep disruption during daylight off-shift hours and diminished alertness and performance during nighttime work hours. Most night-shift workers, even if they maintain a permanent night work schedule, fail to achieve complete physiological adaptation of their endogenous circadian rhythms.[27] Administration of exogenous melatonin has been reported to improve shift workers' tolerance of the night shift.[28] This may result from the entraining effects of melatonin and synchronization of sleep hours with a period of normal melatonin production. Alternatively, in addition to a phase-shifting effect, melatonin may help shift workers override the alerting signal of their intrinsic circadian system and improve their sleep during their subjective day, when their endogenous melatonin is low.

Acute Effects of Melatonin on Human Sleep

Aaron Lerner and his collaborators at Yale University, who isolated and identified N-acetyl-5-methoxytryptamine as an active compound of the pineal gland and named it melatonin in 1950s, were also the first to observe a sedative effect of melatonin administration in man. Melatonin treatment is typically found to produce subjective and/or objective sleep-promoting effects in healthy research subjects when administered during the day or to improve overnight sleep in insomniacs. Depending on the methods of sleep assessment used and the baseline sleep patterns characteristic of the various populations tested, the effects of melatonin treatment have been described as increased reaction time, diminished subjective alertness, increased fatigue or sleepiness, increased sleep propensity, reduction in latency to sleep onset, decrease in the number of nocturnal awakenings, or an increase in sleep efficiency. These effects are observed in response to doses that induce physiological (i.e., around 100 pg/ml) and pharmacological (over 200 pg/ml) circulating levels of the hormone. Remarkably, the effects of both physiological and pharmacological doses of melatonin are not accompanied by any dramatic changes in electrophysiological sleep architecture, a common complication encountered with many existing hypnotics and anxiolytics.

Melatonin treatment can improve sleep in aged insomniacs.[29,30] It should be noted, however, that the sleep promoting effects of melatonin demonstrate substantially higher variability in older individuals compared to young healthy adults, presumably due to age-related changes in sensitivity of melatonin receptors or alterations in melatonin metabolism. Studies in children with severe insomnia, associated with multiple neurological disorders, showed that the administration of melatonin could substantially improve their sleep patterns and increase sleep duration.[31] Similarly, our study of children with Angelman syndrome (AS), a rare genetic disorder characterized by severe mental retardation, hyperactivity, and disturbed sleep, found that timely administration of low melatonin doses both promoted nighttime sleep and advanced the circadian rhythm of melatonin production.[19] Furthermore, some AS children showed a reduction in hyperactivity and enhanced attention. Whether these are consequences of improved nighttime sleep, or represent additional results of melatonin treatment that could be beneficial to other populations suffering from attention deficits, needs further investigation.

Few studies have attempted to use melatonin to treat patients suffering from psychiatric disorders. Such treatment was found to improve sleep in patients with major depression, though it did not significantly affect their mood.[32,33] Both sleep and mood were improved by melatonin treatment in patients suffering from winter depression or from mania.[34,35] Sleep in schizophrenia patients, experiencing chronic insomnia, was also enhanced by melatonin administration.[36] Conversely, no changes after melatonin treatment in sleep or mood were documented in patients with rapid-cycling bipolar disorder.[37] All these studies involved relatively small numbers of patients, especially considering the heterogeneity of these disorders. Some studies in seasonal affective disorder (SAD) patients suggest that this condition is associated with, and perhaps results from, a circadian phase delay.[38,39] Melatonin treatment in the afternoon can improve both mood and sleep in patients with SAD.[40] Whether the effect on mood in this population results from an observed modest phase shift in patients' circadian body rhythms or from improved nighttime sleep, needs further clarification.

Melatonin Therapy: Doses and Timing

Whether melatonin is used to treat sleep or circadian rhythm disorder, it is important to administer it at the right circadian time, in doses that would induce physiological levels of the hormone, and under conditions of low environmental illumination. The benefit of treating a disorder using a normal physiologic agent in concentrations that are normal in humans is that we are less likely to disrupt the physiological harmony of the organism and cause side effects. Sensitive methods are now available for measuring melatonin levels in blood, saliva and urine, thus making it possible to detect melatonin deficiencies and to adjust the therapeutic dose of melatonin so as to not to raise its levels beyond the normal range during treatment. Available data show that physiological doses of melatonin (which raise plasma melatonin to levels within its normal nocturnal range, i.e., 60-200 pg/ml) can significantly promote daytime sleep onset in healthy individuals, improve overnight sleep in people suffering from age-related insomnia, shift the circadian phase of body rhythms or attenuate the subjective effects of drug withdrawal.[41] Such circulating melatonin levels are achieved by oral melatonin doses within a 0.1-0.3 mg range. Pharmacological doses of melatonin typically do not increase the effects of melatonin above those achieved by physiological doses, and might even be less effective.[42]

The circadian pattern of melatonin production, with high nighttime and low daytime levels, and the ability of the pineal hormone to affect the phase and amplitude of circadian body rhythms dictates that special attention should be paid to the timing of melatonin treatment. The effect of melatonin on sleep is typically manifest within 30-60 min after oral administration.[43] Although the acute effect of melatonin on sleep does not seem to depend on the time of administration, coordinating treatment with the time of melatonin's normal increase in secretion would minimize the possibility of an undesired circadian phase-shift. Thus, if melatonin is used to improve nighttime sleep, it should be taken about half an hour before bedtime. By contrast, the timing of the hormone's administration is crucial for the exploitation of its phase-shifting effect. Melatonin tends to produce a phase-advance when it is administered in the late afternoon, while early morning melatonin treatment tends to cause a phase-delay.[9] The mid-afternoon administration of melatonin does not pose a chronobiological effect. If melatonin is used therapeutically to induce phase advance, it should be administered 4-5 hours before the onset of the individual's nocturnal melatonin secretion. In a healthy person, this would correspond to about 5-6 hours before habitual bedtime. However, in a patient with DSPS, whose sleep propensity rhythm might be delayed relative to other circadian body rhythms, including that of melatonin, the timing of melatonin treatment should be determined using serum or salivary melatonin measurements.

If melatonin is administered to counteract the effect of an eastward transmeridian flight, resulting in a 3-6 hour time difference, it might be useful to administer a physiological dose (0.1-0.3 mg) at the local bedtime following the flight. Such treatment will restore the deficit in melatonin that the traveler will experience on his subjective afternoon, and might also promote a desired phase advance in his endogenous rhythmic pattern of melatonin secretion. Following a westward flight, it might be advisable not to ingest melatonin in the evening, when the endogenous level of the hormone during a subjective night is already increased and the person feels very sleepy. However, it might be helpful to take half-dose (e.g., 0.1 mg) immediately following mid-night or early morning awakening, typical of westward flights. This could help to acutely facilitate resumption of sleep and its maintenance for the next several hours, and to support a phase-delay of the circadian pacemaker by exploiting the phase-delaying effect of the hormone when administered during the subjective morning.

In undertaking the treatment of shift workers with melatonin, it is important to document the individual's initial endogenous melatonin pattern and its responsiveness to an actual work schedule. The variability between people in the capacity to adapt to shift work is pronounced, and this capacity is influenced by many factors, including the subject's age, sex, and the magnitude and direction of the shift work schedule, as well as the intensity and timing of the individual's light exposure. Thus, the specific treatment of a circadian phase disorder in a particular shift worker requires substantial knowledge of his physiologic responses and the environmental factors he is exposed to.

Summary

Several lines of evidence suggest that melatonin has a physiological role in human sleep initiation and maintenance, serving, perhaps, as a humoral link between circadian and homeostatic sleep mechanisms. Thus, combined sleep-promoting and chronobiological effects of melatonin treatment could be of substantial assistance to those suffering from insomnia of different origin, including age-related sleep disturbances or circadian rhythm sleep disorders characterized by advanced or delayed sleep phase. Importantly, melatonin might be a treatment of choice for blind individuals with "free-running" circadian rhythms and to patients with severe developmental or degenerative brain disorders associated with irregular sleep-wake patterns, such as children with Angelman syndrome or elderly with Alzheimer disease. The optimal use of melatonin treatment requires that the patient receives the correct dose of the hormone and at the proper time-of-day. Effective melatonin therapy should consider the characteristic individual melatonin-related parameters, including the endogenous levels, rate of metabolism and phase of its circadian rhythm. The circadian pattern of melatonin production, with high nighttime and low daytime levels, and the ability of the pineal hormone to affect the phase and amplitude of circadian body rhythms dictates that special attention should be paid to the timing of melatonin treatment. Finally, it is important to avoid bright light exposure during melatonin treatment. A combination of bright light and high circulating melatonin levels is likely to cancel or significantly attenuate the effect of melatonin treatment. More importantly, this could produce an adverse effect on the visual system, since melatonin has been reported to increase photoreceptor susceptibility to light-induced damage in animals.[44,45] and might have a similar effect in humans.[46]

References

1. Waldhauser F, Weiszenhacher G, Frisch H et al. Fall in nocturnal serum melatonin during prepuberty and pubescence. Lancet 1984; 1:362-365.
2. Zeitzer JM, Daniels JE, Duffy JF et al. Do plasma melatonin concentrations decline with age? Am J Med 1999; 107:432-436.
3. Iguchi H, Kato KI, Ibayashi H. Age-dependent reduction in serum melatonin concentrations in healthy human subjects. J Clin Endocrinol Metab 1982; 55:27-29.

4. Sack L, Lewy AJ, Erb DL et al. Human melatonin production decreases with age. J Pineal Res 1986; 3:379-388.
5. Sharma M, Palacios-Bois J, Schwartz G et al. Circadian rhythms of melatonin and cortisol in aging. Biol Psychiatry 1989; 25:305-319.
6. Tzischinsky O, Shlitner A, Lavie P. The association between the nocturnal sleep gate and nocturnal onset of urinary 6-sulfatoxymelatonin. J Biol Rhythms 1993; 8(3):199-209.
7. Akerstedt T, Froberg JA, Friberg Y et al. Melatonin excretion, body temperature and subjective arousal during 64 hours of sleep deprivation. Psychoneuroendocrinology 1979; 4:219-225.
8. Zhdanova IV, Wurtman RJ, Morabito C et al. Effects of low oral doses of melatonin, given 2-4 hours before habitual bedtime, on sleep in normal young humans. Sleep 1996; 19(5):423-431.
9. Lewy AJ, Ahmed S, Jackson JML et al. Melatonin shifts human circadian rhythms. J Nerv Ment Dis 1989; 177:300-303.
10. Coleman RM, Roffwarg HP, Kennedy SJ et al. Sleep-wake disorders based on a polysomnographic diagnosis. A national cooperative study. JAMA 1982; 247:997-1003.
11. Folkard S, Arendt J, Aldhous M et al. Melatonin stabilizes sleep onset time in a blind man without entrainment of cortisol or temperature rhythms. Neurosci Lett 1990; 113:193-198.
12. Arendt J, Aldhous M, Wright J. Synchronisation of a disturbed sleep-wake cycle in a blind man by melatonin treatment. Lancet 1988; 2(8588):772-773.
13. Palm L, Blennow G, Wetterberg L. Long-term melatonin treatment in blind children and young adults with circadian sleep-wake disturbances. Dev Med Child Neurol 1997; 39:319-325.
14. Sack RL, Brandes RW, Kendall AR et al. Entrainment of free-running circadian rhythms by melatonin in blind people. N Engl J Med 2000; 12;343(15):1070-1077.
15. Kayumov L, Zhdanova IV, Shapiro CM. Melatonin, sleep, and circadian rhythm disorders. Semin Clin Neuropsychiatry 2000; 5(1):44-55.
16. Nagtegaal JE, Kerkhof GA, Smits MG et al. Delayed Sleep Phase Syndrome: A placebo-controlled cross-over study on the effects of melatonin administered five hours before the individual dim light melatonin onset. J Sleep Res 1998; 7:135-143.
17. Kayumov L, Buttoo K, Shapiro CM. Delayed sleep phase syndrome: A randomized crossover placebo-controlled study of effect of exogenous melatonin administered at a partially fixed time. Sleep 1999; 22:160-161.
18. Carskadon MA, Vieira C, Acebo C. Association between puberty and delayed phase preference. Sleep 1993; 16(3):258-262.
19. Zhdanova IV, Wurtman RJ, Wagstaff J. Effects of low dose of melatonin on sleep in children with Angelman syndrome. J. Pediatric Endocrinology and Metabolism 1999; 12:57-67.
20. Brusco LI, Fainstein I, Marquez M et al. Effect of melatonin in selected populations of sleep-disturbed patients. Biol Signals Recept 1999; 8:126-131.
21. Arendt J, Aldhous M, Marks V. Alleviation of jet lag by melatonin: preliminary results of controlled double blind trial. Br Med J 1986; 3(6529):1170.
22. Nickelsen T, Demisch K. Influence of subchronic intake of melatonin at various times of the day on fatigue and hormonal levels: A placebo-controlled, double-blind trial. J Pineal Res 1989; 6:325-334.
23. Petrie K, Coraglen JV, Thompson L et al. Effect of melatonin on jet lag after long haul flights. Br Med J 1989; 298:705-707.
24. Petrie K, Dawson Ag, Thompson L et al. A double-blind trial of melatonin as treatment for jet lag in international cabin crew. Biol Psychiatry 1993; 33:526-530.
25. Spitzer RL, Terman M, Williams JB et al. Jet lag: Clinical features, validation of a new syndrome-specific scale, and lack of response to melatonin in a randomized, double-blind trial. Am J Psychiatry 1999; 156:1392-1396.
26. Edwards BJ, Atkinson G, Waterhouse J et al. Use of melatonin in recovery from jet-lag following an eastward flight across 10 time-zones. Ergonomics 2000; 43(10):1501-1513.
27. Sack RL, Blood ML, Lewy AJ. Melatonin rhythm in night shift workers. Sleep 1992; 15:434-441.
28. Folkard S, Arendt J, Clark M. Can melatonin improve shift workers' tolerance of the night shifts? Some preliminary findings. Chronobiol Int 1993; 10:315-320.
29. Haimov I, Lavie P, Laudon M et al. Melatonin replacement therapy of elderly insomniacs. Sleep 1995; 18:598-603.
30. Zhdanova IV, Wurtman RJ. Improvement of sleep quality by melatonin. Lancet 1995; 346(8988):1491.
31. Jan JE, Espezel H, Appleton RE. The treatment of sleep disorders with melatonin. Dev Med Child Neurol 1994; 36:97-107.
32. Dolberg OT, Hirschmann S, Grunhaus L. Melatonin for the treatment of sleep disturbances in major depressive disorder. Am J Psychiatry 1998; 155:1119-1121.
33. Dalton EJ, Rotondi D, Levitan RD et al. Use of slow-release melatonin in treatment-resistant depression. J Psychiatry Neurosci 2000; 25:48-52.
34. Robertson JM, Tanguay PE. Case study: The use of melatonin in a boy with refractory bipolar disorder. J Am Acad Child Adolesc Psychiatry 1997; 36:822-825.
35. Bersani G, Garavini A. Melatonin add-on in manic patients with treatment resistant insomnia.
36. Shamir E, Laudon M, Barak Y et al. Melatonin improves sleep quality of patients with chronic schizophrenia. J Clin Psychiatry 2000; 61:373.
37. Boivin DB, Czeisler CA, Dijk DJ et al. Complex interaction of the sleep-wake cycle and circadian phase modulates mood in healthy subjects. Arch Gen Psychiatry 1997; 54:145-152.
38. Lewy AJ, Bauer VK, Cutler NL et al. Morning vs evening light treatment of patients with winter depression. Arch Gen Psychiatry 1998; 55:890-896.
39. Thalen BE, Kjellman BF, Morkrid L et al. Melatonin in light treatment of patients with seasonal and nonseasonal depression. Acta Psychiatr Scand 1995; 92:274-284.
40. Lewy AJ, Bauer VK, Cutler NL et al. Melatonin treatment of winter depression: A pilot.
41. Zhdanova IV, Piotrovskaya VR. Melatonin reduces symptoms of acute nicotine withdrawal in humans. Pharm Biochem Behav 2000; 67:131-135.
42. Dollins AB, Zhdanova IV, Wurtman RJ et al. Effect of inducing nocturnal serum melatonin concentrations in daytime on sleep, mood, body temperature, and performance. Proc Natl Acad Sci USA 1994; 91:1824-1822.
43. Zhdanova IV, Wurtman RJ, Lynch HL et al. Sleep inducing effects of low melatonin doses ingested in the evening. Clinical Pharmacol Therapeut 1995; 57:552-558.
44. Wiechmann AF, Yang XL, Wu SM et al. Melatonin enhances horizontal cell sensitivity in salamander retina. Brain Res 1988; 453:377-380.
45. Wiechmann AF, O'Steen WK. Melatonin increases photoreceptor susceptibility to light-induced damage. Invest Ophthalmol Vis Sci 1992; 33:1894-1902.
46. Arushanian EB, Ovanesov KB. Melatonin lowers the threshold of light sensitivity of the human retina. Eksp Klin Farmakol 1999; 62:58-60.

CHAPTER 11

Melatonin Efficacy to Treat Circadian Alterations of Sleep in Alzheimer's Disease

Daniel P. Cardinali, Analía M. Furio, Luis I. Brusco and Cynthia Liberczuk

Summary

Alzheimer's disease (AD) patients show a greater disruption of the circadian sleep-wake cycle as compared to similarly aged non-demented controls. When this occurs demented patients spend their nights in a state of frequent restlessness and their days in a state of frequent sleepiness. These sleep-wake disturbances became increasingly more marked with the progression of the disease and may contribute to cognitive decay. Sleep architecture in AD is characterized by decreases of slow wave sleep and rapid eye movement sleep and increases of time and frequency of awakening compared to aged-matched control subjects. The sleep-wake disturbances in elderly people and AD patients appear to result from changes at different levels: reduction of environmental synchronizers, neurosensorial deficit or lack of mental and physical activity. However, increasing evidence exists about the loss of functionality of the hypothalamic suprachiasmatic nuclei (SCN), the principal oscillator in the circadian system. A chronobiological approach including melatonin, bright-light therapy, restricted time in bed and programmed diurnal activity is a promising therapeutic alternative in the management of sleep-wake disorders in AD patients. In elderly insomniacs melatonin treatment decreased sleep latency and increased sleep efficiency. The effect of melatonin on sleep is probably the consequence of increasing sleep propensity (by inducing a fall in body temperature) and of a synchronizing effect on the circadian clock (chronobiotic effect); typically the latter takes several weeks to occur. Generally, melatonin decreased sundowning in AD patients and reduced variability of sleep onset time, with improvement in mood and memory. The effect of melatonin was seen regardless of the concomitant medication employed to treat cognitive or behavioral signs of disease. The mechanisms accounting for the possible therapeutic effect of melatonin in AD patients remain unknown. Melatonin treatment could be beneficial in AD by augmenting the restorative phase of sleep. In addition, in vitro and in vivo data indicated that melatonin protects neurons against β amyloid toxicity, prevents β amyloid-induced lipid peroxidation and alters the metabolism of the β amyloid precursor protein. Melatonin prevented the chronobiological consequences of injecting β amyloid peptide 25-35 in the SCN of hamsters and cognitive decay in transgenic mice overexpressing β-amyloid.

The Circadian Clock Is One of the Most Indispensable Biological Functions

Circadian rhythms are daily cycles of physiology and behavior that are driven by an endogenous oscillator with a period of approximately (circa-) one day (diem). In mammals, the principal circadian oscillator is located in the suprachiasmatic nuclei (SCN) of the hypothalamus. The SCN circadian oscillator acts like a multifunctional timer to adjust the homeostatic system, including sleep and wakefulness, hormonal secretions and various other bodily functions, to the 24-h cycle.[18,48,82,105]

SCN neurons contain an internal pacemaker with an endogenous, rhythmic electrical activity. Thus, when the SCN is completely isolated,[54] or when individual SCN neurons are cultured in vitro, the cells from this hypothalamic region show high electrical activity during the subjective daytime.[43] A very important evidence showing the role of the SCN as a central clock was provided by transplantation experiments. If the SCN of a mutant hamster, with a rhythm faster than that of wild-type animals, was grafted into the hypothalamus of a wild-type, SCN-lesioned hamster, the recipient animal adopted the faster, mutant rhythm.[97]

During the past decade, enormous progress has been made in determining the molecular components of the biological clock. The molecular mechanisms that underlie the function of the clock are universally present in all the cells and consist of gene-protein-gene feedback loops in which proteins can down-regulate their own transcription and stimulate the transcription of other clock proteins.[30,135] The cellular oscillator consists of interlocked transcriptional and post-translational feedback loops: heterodimeric complexes encoded by the Clock and Bmal genes drive expression of Per and Cry genes during circadian day, leading to accumulation of Per/Cry protein complexes that enter the nucleus and suppress transcription of their cognate genes.

Although circadian rhythms are anchored genetically they are synchronized by (entrained) and maintain certain phase relationships to external (exogenous) factors, especially the sleep portion of the light-dark schedule.[48,82,105] These rhythms will persist with a period different from 24 h when external time cues are suppressed or removed, such as during complete social isolation or in constant light or darkness. Research in animals and humans has shown that only a few such environmental cues, like light dark cycles, are effective entraining agents for the circadian oscillator ("Zeitgebers"). An entraining agent can actually reset, or

Sleep and Sleep Disorders: A Neuropsychopharmacological Approach, edited by Malcolm Lader, Daniel P. Cardinali and S.R. Pandi-Perumal.

phase shift, the internal clock. Depending on when an organism is exposed to such an entraining agent, circadian rhythms may be advanced, delayed, or not shifted at all. Therefore, involved in adjusting the daily activity pattern to the appropriate time of day is a rhythmic variation in the influence of the Zeitgeber as a resetting factor.[18,48,82,105]

In humans, light exposure during the first part of the night delays the phase of the cycle; a comparable light change near the end of the night, advances it. At other times during the day light exposure have no phase-shifting influence.[63] Melatonin, a chemical code of the night, showed an opposite phase response curve to light, producing phase advances during the first half of the night and phase delays during the second.[63]

Pineal Melatonin Acts as an "Endocrine Arm" of the Circadian Clock

Melatonin synthesis occurs in the pineal gland as well as in many peripheral organs.[100] Yet, circulating levels of melatonin are mainly, if not exclusively, derived from the pineal gland.[57] Melatonin is primarily secreted from the pineal gland during the subjective night, its plasma levels showing in humans a characteristic profile with excretion rising in the early evening to a peak in the early morning and then declining to undetectable levels during daylight.[14]

While melatonin is produced in most organisms from algae to mammals, and its role varies considerably across the phylogenetic spectra,[101] in humans it seems to play a major function in the coordination of circadian rhythmicity, remarkable the sleep-wake cycle.[22,57] The circadian rhythm of melatonin is generated by the central pacemaker located in the SCN and it is synchronized to a 24-hour period largely by cues from the light-dark cycle received via the retino-hypothalamic pathway to the SCN.

The evening increase in melatonin secretion is associated with an increase in the propensity for sleep. Secretion of melatonin during the day, as occurs in diverse pathologic or occupational health situations, is strongly associated with daytime sleepiness or napping, Moreover, the administration of melatonin during the day induces sleepiness.[142]

Melatonin secretion is thus an "arm" of the biologic clock in the sense that it responds to signals from the SCN and in that the timing of the melatonin rhythm indicates the status of the clock, both in terms of phase (i.e., internal clock time relative to external clock time) and amplitude. From another point of view, melatonin is also a chemical code of night: The longer the night, the longer the duration of its secretion. In many species, this pattern of production serves as a time cue for seasonal rhythms.[22]

Melatonin (in a dose of 5 mg daily, timed to advance the phase of the internal clock) can maintain synchronization of the circadian rhythm to a 24-hour cycle in sighted persons who are living in conditions likely to induce a free-running rhythm, and it synchronizes the rhythm in some persons after a short period of free-running. In blind subjects with free-running rhythms, it has been possible to stabilize, or entrain, the sleep-wake cycle to a 24-hour period by giving melatonin, with resulting improvements in sleep and mood.[117] The phase shifting effects of melatonin were also sufficient to explain its utility as a treatment for circadian-related sleep disorders such as jet lag[19,49] or the delayed phase sleep syndrome.[57,116]

A certain degree of sleep disruption is needed for melatonin to show improvement of sleep. For example, Many times, old patients with minor or none sleep disturbance received, on a long-term basis, anxiolytic benzodiazepines in low doses for relief of a disturbance that has numerous, often concurrent etiologies, including medical conditions, medication or poor sleep hygiene.[88] Recently we carried out a study to assess whether melatonin could be useful to reduce low benzodiazepine dosage in this group of patients,[21] as it is in insomniac patients treated with hypnotic benzodiazepines.[35,115] A double blind-placebo controlled study on the efficacy of a 3 mg-melatonin dose p.o. was undertaken. The possible correlation of urinary excretion of 6-sulphatoxymelatonin before starting treatment and outcome of treatment was also examined. The results indicated that melatonin lacked to affect subjective assessment of wakefulness or sleep in this group of patients with minor sleep disturbance. The only effect that melatonin had in this group of patients was to advance sleep onset and to decrease significantly the variability of sleep onset time as compared to placebo.[21] Indeed, this melatonin efficacy to reduce variability of the sleep onset time was first described in demented patients exhibiting sundowning[31] and is the basis for the indication of melatonin as an effective therapy of sundowning in Alzheimer's disease (AD), discussed below.

Ascertaining further that a certain degree of sleep disruption is needed to see significant effects of exogenous melatonin, a recent publication by Baskett and co-workers can be cited. In a double blind randomized placebo controlled crossover trial in healthy older volunteers (20 normal and 20 problem sleepers), 5 mg of melatonin or matching placebo were given at bedtime for 4 weeks, separated by a 4-week washout period.[6] Sleep quality was measured using sleep diaries, the Leeds Sleep Evaluation Questionnaire, and actigraphy. Melatonin did not significantly improve any sleep parameter measured in either group.[6]

Sleep-Wake Cycle Is Severely Disturbed in Pathological Aging

Sleep architecture becomes modified with age both at the level of the macro and the microstructure of the sleep. The major macro structural characteristic of sleep in a 60 year-old subject is the reduction of the deep slow sleep (stages 2, 3 and 4).[102] This decrease is compensated by a relative increase of a lighter sleep (stage 1). As far as REM sleep, the average middle length of the episodes and the density of ocular movements remain steady during the night.[26] As regards to sleep microstructure, the delta waves ($<$ 4 Hz) decrease in amplitude and number, and during the stage 2, the time interval that separates the two spindles shortens.[85]

A number of studies have assessed sleep in AD. They report an increase in the sleep features commonly observed in normal aging: Loss of sleep continuity and a reduction in stages 3 and 4 and rapid eye movement (REM) sleep, shortening of REM sleep periods, trend towards a bimodal (afternoon and night) circadian partition of sleep. Modifications in sleep microstructure are also observed: Decrease in K complexes, vertex sharp waves and spindles.[81,96] These changes are part of a generalized disturbance in all the biological rhythms, including loss of normal temperature variations. Sleep anomalies increase with the degree of dementia.[81,132,133] Some anomalies may be related to the cholinergic dysfunction, which is particularly important in this disease and may underlie the slowing of the EEG rhythm during waking and REM sleep anomalies.

AD patients present numerous awakening episodes and an increased number of changes of sleep stages during the night.[95] Contrary to that observed in healthy aged subjects, AD patients show delayed latency to first REM as well as fragmentation of

REM and loss of the rhythmic ultradian alternation of slow sleep and REM sleep.[81] It is noteworthy that AD patients, in contrast to healthy controls, exhibit altered response to deprivation of sleep, i.e., selective deprivation of slow wave sleep and REM sleep are not followed by a sleep rebound in the subsequent nights.[13,122] This argues in favor of a loss of the capacity of CNS adaptation and recuperation in AD patients with major cognitive impairment.

Indeed, among demented aged subjects, the awakenings, agitation and the nocturnal wandering are frequently associated with major cognitive and functional deterioration that often leads to the institutionalization of the patient. Therefore, it is not surprising to see an increased use of sedative psychotropics in these patients.[38] However, about 80% of age-related insomnias are chronic and resistant to classic hypnotics.[52] Additionally, the sedative psychotropics like benzodiazepines can alter memory, more especially in AD, leading to a decrease of diurnal vigilance and to increase of risk of falls.[67]

It is well demonstrated that the deficit of the cholinergic system is one of the first and presumably most important biochemical changes in AD. AD is characterized especially by a reduction of the number of central cholinergic neurons. The disruptions of the continuity and architecture of the sleep, notably the anomalies of REM sleep could be explained in part by this cholinergic dysfunction, since induction and maintenance of REM sleep are under the control of cholinergic neurons located in the dorsolateral tegmentum, at the level of the pons.[87] Two other central structures that play an important role in the process of cortical activation during REM are the thalamus and the forebrain basal nucleus of Meynert. The latter sends numerous cholinergic projections to the cerebral cortex. In addition, AD patients show a loss of neurons in the locus coeruleus, the site of noradrenergic neurons involved in the genesis of slow sleep and inhibition of REM.[87]

Several behavioral and electrophysiological studies in man and experimental animals indicate that REM sleep is implicated in processing the information acquired during the waking state. In man, the total deprivation of sleep or the selective deprivation of REM sleep after training leads deficits of retention of complex logical tasks involving procedural or implicit memory.[81] Deletion of initial phases of REM leads to a similar mnesic disruptions as the whole deletion of REM or the total deletion of sleep.[119] Increases of the length of the REM phase and of the density of the fast ocular movements occur in the nights that follow the training.[50,120] Lately, studies with PET made it clear the cerebral hyperactivity during REM of the regions stimulated at the time of training sessions during the day.[50]

It is then clear that among aged subjects disruption of the internal organization of the sleep and possible interruptions in the harmonious succession of the cycles of the sleep could also disrupt the mnesic process.[70] In the case of AD patients, disruptions of the continuity of sleep in general and the anomalies of REM sleep in particular can alter the mnesic function very severely.

Since REM sleep could be one of the most sensitive targets for the deficit of cholinergic function that characterizes AD, acetycholinesterase inhibitors could improve memory processes in AD patients in part by stimulating REM sleep. Thus, the administration of the acetycholinesterase inhibitor donepezil 1 h before bedtime to healthy aged subjects led to an increase of the density of the fast ocular movements and the percentage of REM sleep and to a decrease in REM sleep latency.[112] Besides, a positive interrelationship between this increase in REM sleep and the mnesic performance was documented.[112] In another study, administration of the acetycholinesterase inhibitor rivastigmine also decreased significantly REM latency.[111]

The Circadian Apparatus Is Damaged in AD

The physiopathology of rhythm dysfunction in aging is not completely understood. Two major conditions appear to contribute; on one hand intrinsic alterations of the endogenous biologic clock and, on the other hand, the reduction of influence of the external synchronizers. During aging, morphological and neurochemical changes occurs in SCN.[98,124] With regard to the reduced influence of external synchronizers, one can theoretically distinguish those that are a matter for modifications of life style and that are reversible and readily accessible to prevention and those that depend on reduction of neurosensorial organ's sensitivity to the synchronizers. Indeed, aging is frequently accompanied by neurosensorial deficits (cataracts, retinopathies, hypoacousy, etc.) and by modifications of life style because of disappearance or reduction of socio-professional constraints.

A very remarkable chronobiological disturbance of the aged person is the advance of phase of the sleep-wake cycle that results in bedtime and awakening earlier than those of sociocultural habits.[88] Besides this advance of phase, the most important point is a decrease of amplitude of rhythm, i.e., poor sleep at night together with poor vigilance during the day. This results in the clinical expression of an ultradian rhythm of rest-activity, named Basic Rest-Activity Cycle (BRAC) many years ago).[60] With an advance age, BRAC turns to become a dominant rhythm that results in an increase of propensity to diurnal sleep with an increased number of naps during the day. Thus, the relation between the expression of circadian and ultradian rhythms of activity-rest radically unbalances with age.

As compared to age-matched controls, AD patients show a significant reduction of number and volume of neurons in SCN.[98,124] This is accompanied by a marked disruption of the rhythmic secretion of melatonin.[129] When the rhythm of plasma melatonin concentration was studied in 10 AD patients as compared to 10 age-matched controls, the AD group exhibited a reduction in amplitude of the secretory peaks and a global reduction in quantity of melatonin secreted.[79] There was a positive correlation between the severity of the disruption of activity-rest rhythm and the reduction of amplitude of melatonin peak and total quantity of the melatonin secreted. CSF melatonin was measured in postmortem specimens from 85 AD patients and 82 age-matched controls.[65] In the AD group the average melatonin concentration in CSF represented one-fifth of that of controls. Besides, in the AD group the lowest melatonin concentration was seen in subjects who had apolipoprotein E-epsilon 4/4 genotype.[65] The finding on up-regulated MT-1 receptors in hippocampus, retina and vascular tissue of AD patients suggests a widespread compensatory response to the markedly diminished melatonin levels.[110] In a recent study melatonin levels were determined in ventricular postmortem cerebrospinal fluid of 121 subjects.[143] Melatonin levels were significantly decreased in the aged individuals with early neuropathological changes in the temporal cortex, where the AD process starts. Indeed, the decrease in CSF melatonin levels may be an early event in the development of AD, possibly occurring even before the clinical symptoms.[143]

As AD evolves, a significant reduction of diurnal locomotive activity and an increase of the nocturnal activity with fragmentation of the sleep together with decrease of the amplitude of activity-rest rhythm are seen.[108] In addition, and contrary to that observed in normal aged individuals, AD patients present a delay of phase and internal desynchrony of rhythms, e.g., that of temperature and activity-rest rhythm.[108]

The stability of the sleep-wake cycle is bound to the existence of physical activity and a strong diurnal environmental light. Among AD patients disturbance of activity-rest rhythm are associated with a greater cognitive deficit and a more insane state.[77] Actigraphic recordings and melatonin determinations reveal the significant alteration of the sleep wake cycle with a positive interrelation between the evolutionary stage of AD and the disruptions of activity-rest rhythm.

As above mentioned, in AD patients the disrupted activity-rest rhythm and sleep are poorly or none affected by treatment with psychotropics. Moreover, the use of psychotropics can aggravate the existing unrest or can lead to more severe disruptions of sleep and behavior. As far as the benzodiazepines they can aggravate the respiratory disturbances bound to sleep, can lead the confusional states and that can impair mnesic performance.[71] Most of these effects are due to the modified pharmacokinetics of psychotropic drugs bound to age. In summary, and as far as possible, the psychotropics must be avoided or must be used parsimoniously.

Chronobiological Therapies Can Be Helpful in Circadian Alterations in AD

In controlled studies with AD patients it has been demonstrated that timely planned physical and social activities during the day improve the quality of the sleep,[58] increase slow wave sleep and improve the mnesic functions evaluated by neuropsychometric tests.[84] In a study on 21 aged healthy subjects the effects on sleep of two types of behavioral approaches were compared. These comprise either sleep hygienic advice (eviction of alcohol and of drinks containing caffeine, control of temperature of the bedroom, reduction of ambient noise, etc.) or 30 min of bed restriction in the evening plus one diurnal nap of 30 min. The authors noted a significant increase of the efficiency of sleep in the group with restriction of bed and practicing a diurnal nap.[51] In a retrospective survey on factors of risk in AD in Japan, 337 AD patients and 260 age-matched controls were asked for the habits concerning the diurnal naps.[3] The usual practice of nap (at least 3 per week) of a length lower than 60 min per day had a protective effect against the development of AD, more especially among the individuals who are carriers of the apolipoprotein E-epsilon 4/4 genotype. On the other hand, naps of a length superior to 60 min among the carriers of allele would tend to increase the risk of AD. When the behavioral chronobiological approach is complemented with the administration of the melatonin and phototherapy the effectiveness of the behavioral chronotherapy improves substantially.

As noted above, the circadian pacemaker in the SCN is synchronized with the 24 h day by "Zeitgebers", of which light is the most important. In elderly demented patients most Zeitgebers are reduced: social input is diminished, motor activity decreases, and in particular, there is less exposure to sufficient outdoor or bright light.[72,78] In addition to degeneration of SCN function with age, ocular light transmission may be impaired by different age-related diseases of the eye (i.e., macula degeneration, clouding of the ocular media, retinopathies, cataracts, etc.) or optic nerve degeneration. Thus, even if patients were exposed to adequate light intensities during the day, less light would actually reach the circadian clock.

Increasing Zeitgeber strength by augmenting light intensity has been suggested as a possible treatment for stabilizing disrupted sleep-wake rhythms in AD[17] and in recent years a number of studies have tested the effects of light in patients with dementia.[44,74,107,131] Light therapy was developed as a treatment for seasonal affective disorder (SAD).[93] Depressive symptoms are also found in early stages of dementia of the elderly. The sleep disturbance can be a direct consequence of dementia or related to the depressive symptoms. Although this hypothesis is plausible, there is no direct evidence that light therapy can improve depressive symptoms in dementia. The chronobiological treatment could improve the disrupted sleep-wake cycle, diminish the intake of hypnotics, improve cognitive function and/or reduce depressive symptoms.

Conventional light therapy is administered by sitting the patient for a minimum of 30 min per day in front of a light box. This procedure is troublesome in patients with dementia and requires supervision. As an alternative, to augment light intensity in the ambient surroundings of the main room where patients spend their days was proposed.[131] Another possibility is the use of dawn-dusk simulation mimicking outdoor twilight transitions. Here a gradual dusk and dawn is adapted to the patient's sleep time and does not require daily lightbox sessions or caregiver presence. Dawn-dusk simulation can be considered as a 'naturalistic' light therapy, using lower and gradual changes in light intensity.[127] It has been successfully used to treat SAD patients and for patients suffering circadian sleep-wake cycle disorders.[126] Low intensity dawn simulation has been shown to phase advance or prevent a delay drift in the circadian rhythm of melatonin secretion and in AD patients. Recently, a study of the feasibility of dawn-dusk simulation light therapy in a population of patients with different degrees of dementia, using actimetry to investigate their circadian rest-activity cycle and to quantify sleep disturbances.[36] This procedure induced an advance in the circadian rest-activity cycle by inducing an earlier onset of the most restful period of the night. An improved sleep was also observed.

Mishima et al[75] employed phototherapy in the morning (3000 to 5000 lux) for 2 h, during 4 weeks in 14 patients with vascular and AD dementia. During the whole period of phototherapy, there was a significant increase of the length of total sleep and of nocturnal sleep with a reduction of the length of diurnal sleep. In a subsequent study the same authors employed a crossed randomized protocol to compare the therapeutic effects of early light of strong intensity (5000 to 8000 lux, during 2 h) vs. plain light (300 lux, during 2 h) in 12 patients with vascular dementia and 10 patients with AD for periods of 2 weeks.[74] Activity-rest cycle was evaluated by wrist actigraphy. Compared to plain light, the strong light reduced nocturnal activity in the two groups of patients. The therapeutic effect of light was more marked among patients with vascular dementia.

In another study 10 AD patients with sundowning syndrome were exposed to a strong light in the evening for 2 h.[107] After one week of phototherapy the rest-sleep cycle increased in amplitude. Van Someren et al[131] used phototherapy during 4 weeks in a heterogeneous population of 22 patients with AD, vascular dementia and dementia attributed to alcohol or hydrocephaly. Bright light was administered during the day. Light intensity at the middle of the day was 1136 lux vs. 436 lux in regular light. Activity-rest rhythm evaluated by actigraphy became stabilized after 3 weeks of phototherapy. No effect on amplitude of rhythm was observed. Efficacy of treatment depended on the visual function of the patients.

Taken together these studies are in favor of the synchronizing effect of phototherapy to re-establish the disruptions of sleep and to reduce the frequency of unrest of behavior in demented patients. The modality of light exposure was extremely variable in terms of intensity and especially of periods of exposition (morning, evening, all day). Indeed, light in the morning should be given in the case of a phase delay and light in the evening should be given for phase advances.

The aim of these therapeutics is to improve sleep and diurnal activity and consequently to increase the quality of life in AD patients.[73,76] However, there is a very significant risk of retinal damage from repetitious exposure to the high intensities of visible light provided by bright light units in this population since a substantial number of studies have indicated that age related macular degeneration is the result of natural aging processes exacerbated by the cumulative effects of photo-oxidative damage.[5,7,9,15,46,103] Indeed, age-related maculopathy is associated with AD[59] and also the optic nerve shows degenerative changes in AD.[8] A recent development in phototherapy is the demonstration that the spectral sensitivity of the circadian system is very different than the spectral sensitivity of the retina used in visual activities such as reading and used to measure light in illuminating engineering.[10] Very short wavelength (blue) light is maximally effective at affecting the circadian system whereas middle wavelengths (yellow-green) are maximally effective for visual performance. By restricting light emission to wavelengths between 480 nm and 515 nm, one could reduce the intensity of light to levels normally used in artificially illuminated environments, thus providing effective stimulation in a comfortable environment and without the risk of retinal damage from high intensity light. In addition, the associate use of an antioxidant substance like melatonin may be useful to reduce potential photooxidative damage in AD patients.

Melatonin Is Effective to Treat Sleep Disorders and Sundowning in Demented Patients

The promoting effect of exogenous melatonin administration on sleep and sedation has been known since long. Initial studies addressing the effect of melatonin on sleep made use of the i.v. or the intranasal route[2,24,62,134] or administered very large doses of the methoxyindole by the oral route.[2,64] From these early studies it was concluded that melatonin reduces sleep latency and induces sleepiness and fatigue. Such effect of melatonin is probably the consequence of increasing sleep propensity (by inducing a fall in body temperature) and of a synchronizing effect on the circadian clock (chronobiotic effect). Typically, the thermoregulatory effect of melatonin arises immediately after melatonin administration while the second one needs some days (or weeks) to develop.[4,16,27,28,42,83,99,114,125,140,141]

Assessment of the hypnotic action of melatonin during daytime administration and its comparison with triazolam indicated that a 6 mg dose of melatonin demonstrated hypnotic effects that were nearly equal to those of triazolam at 0.125 mg. Rectal temperature was significantly decreased by melatonin.[109] In another placebo-controlled and double-blind with a cross-over design including temazepam (20 mg), the hypnotic activity of melatonin at early evening (presumably in the absence of endogenous melatonin) was similar to 20 mg temazepam.[37,123] These data in humans reproduced previous findings in rodents.[40]

In middle-aged and elderly insomniacs who made use of immediate-release (0.5 mg) and controlled release (0.5 mg) preparations of melatonin 30 min before bedtime, polysomnographic recordings and sleep actigraphy showed that melatonin shortened latencies to persistent sleep.[53] Administration of a 3-mg dose of melatonin during 14 nights to elderly patients with chronic primary insomnia brought about a significant reduction in wake time after sleep onset while total sleep time and sleep efficiency increased, with an increase of stage 2 sleep.[80] In studies monitoring sleep quality by wrist actigraphy in elderly insomniacs, controlled-release melatonin (2 mg) taken 2 h before the desired bedtime during 3 weeks reduced sleep latency and wake time after sleep onset and increased total sleep time and sleep efficiency.[34] Melatonin (3 mg) administered 30 min before the expected bedtime for 21 nights to patients with chronic insomnia significantly improved sleep quality and decreased the number of awakenings from day 2-3 of treatment.[31] A sustained-release preparation of melatonin (2 mg) improved sleep initiation, with further improvement of sleep initiation and sleep maintenance after 2 months.[45] Insomniac patients receiving 75 mg melatonin at 2200 h for 7 consecutive nights reported improved subjective sleep time and subjective daytime alertness.[66] Low amounts of melatonin (0.3 mg) given during 3 nights to middle-aged and elderly patients with chronic insomnia reduced sleep latency, the number of nocturnal awakenings and body movements per night, whereas core temperature remained unchanged.[142] In medically ill persons with initial insomnia receiving 5.4 mg melatonin or placebo, double-blind assessments of aspects of sleep indicated that melatonin significantly hastened sleep onset, improved quality and depth of sleep, and increased sleep duration.[1]

It is interesting to note that melatonin can facilitate discontinuation of benzodiazepine therapy while maintaining good sleep quality. This can be helpful in the elderly in view of the contraindications for a prolonged use of benzodiazepine at this age. A number of animal studies indicated that several behavioral effects of melatonin can be suppressed by inhibiting central benzodiazepine receptors.[39,41,136] Indeed melatonin and benzodiazepines share several properties, remarkable anxiolytic effects, and therefore melatonin could be useful to help patients to discontinue benzodiazepine treatment. In one study 14 out of 18 subjects under benzodiazepine therapy receiving melatonin (2 mg in a controlled-release formulation) discontinued benzodiazepine therapy and after a 6-month follow-up assessment of the 24 patients who discontinued benzodiazepine and received melatonin therapy, 19 maintained good sleep quality.[35] In another study, 13 of 20 patients taking benzodiazepines together with melatonin, benzodiazepine use could be stopped, and in another four patients, benzodiazepine dose could be decreased to 25-66% of the initial dose.[115]

Nighttime insomnia and nocturnal wandering in AD patients often poses unbearable problems for caregivers. As mentioned above, hypnotic or antipsychotic medications are only minimally effective[29,47,138] while sleep-wake cycle disturbances may even be aggravated by a classic neuroleptic like haloperidol.[137]

Several studies have shown that melatonin levels are lower in AD patients compared with aged matched controls.[32,65,79,86,118,128] Based on this evidence the supplementary administration of melatonin to treat sleep and behavior disorders in AD patients seemed to be a logical therapeutic approach. Indeed, initial studies of melatonin therapy in AD have had positive results. However, the majority of studies are open, and knowledge about side effects and long-term effects of melatonin is still limited.

In a first examination of the sleep-promoting action of melatonin (3 mg p.o. for 21 days) in a small non-homogenous group of elderly patients with primary insomnia and with insomnia associated with dementia or depression, 7 out of 10 dementia patients having sleep disorders and treated with melatonin (3 mg p.o. at bed time) showed a decreased sundowning and reduced variability of sleep onset time.[31] In another study, 10 individuals with mild cognitive impairment were given 6 mg of melatonin before bedtime. Improvement was found in sleep, mood and memory.[55]

Other studies include daily administration of 6-9 mg melatonin for longer periods of time to AD patients with sleep disorders and sundowning agitation. The retrospective account of 14 AD patients after a 2-3-year period of treatment with melatonin indicated that all improved sleep quality.[11] Sundowning, diagnosed clinically in all patients examined, was not longer detectable in 12 of them, and persisted attenuated in the other 2 patients. Another significant observation in this study was the halted evolution of the cognitive and mnesic alterations expected in comparable populations of patients not receiving. This should be contrasted with the significant deterioration of clinical conditions of the disease expected from patients after 1-3 years of evolution of AD.

Further support to the hypothesis that melatonin can be useful in AD patients was given by a case report study which included two 79 years old male monozygotic twins with AD diagnosed 8 years earlier.[12] The onset of the disease differed by about 6 months between twins. The patients exhibited similar cognitive and neuroimaging alterations at diagnosis. Patients were treated with vitamin E (800 I.U./day) and, starting about 3 years in advance to assessment, they received 50 mg/day thioridazine because of the behavioral and sleep disorder. Patient NN, who is still alive, has been treated with 6 mg /day of melatonin at bed time daily for more than 6 years. Three months after starting melatonin treatment, patient NN discontinued thioridazine. At the time of case report publication (36 months of treatment), neuropsychological evaluation by the Functional Assessment Tool For Alzheimer's Disease (FAST) indicated a 7b stage for patient ZZ and a 5 stage for patient NN, with Mini-Mental scores of 0/30 and 10/30, respectively. A generalized cortical atrophy in both patients, with a more important bitemporal atrophy and ventricular enlargement in patient ZZ, was seen in NMR.[12] Pacing in patient ZZ was intense and increased at the evening; he was an insomniac and exhibited sundowning episodes. Patient NN showed normal walking and only rudiments of primitive reflexes. Sleep-vigilance rhythm in patient NN was unimpaired. Six months after the case report publication, the twin not receiving melatonin died because of a complication of its AD while his brother was still alive, showing a normal sleep and severe impairment of cognitive function (Mini-Mental has changed from 10/30 to 0/30; FAST: 6b stage).

Other studies also support the efficacy of melatonin treatment in AD patients. Mishima and co-workers administered a 6 mg dose of melatonin for 4 weeks to 7 inpatients with AD who exhibited irregular sleep-waking cycle.[76] Melatonin significantly reduced percentage of nighttime activity compared to placebo. Cohen-Mansfield et al reported the efficacy of melatonin (3 mg/day at bed time) for improving sleep and alleviating sundowning in 11 elderly AD patients.[23] Analysis revealed a significant decrease in agitated behavior and a significant decrease in daytime sleepiness. In a case report observation on two AD patients, melatonin administration enhanced and stabilized the circadian rest-activity rhythm in one of them along with reduction of daytime sleepiness and improvement in mood.[56] Likewise, another observation on two AD patients given 6 mg melatonin daily for one year indicated appreciable improvement of the Mini-Mental score.[89]

We recently assessed the capacity of melatonin to improve sleep in 45 AD patients with sleep disturbances after 4 months of treatment with 6-9 mg melatonin/day.[20] A significant effect of treatment on global subjective evaluation of sleep was detected in this group. Moreover, sundowning, clinically diagnosed in all patients disappeared after 4 months of treatment with melatonin. The effect of melatonin was seen regardless of any concomitant medication employed to treat cognitive or behavioral signs of disease.

Observations in AD patients, however, are not always consistent. In a double blind randomized placebo controlled trial on the effect of 6 mg/day of melatonin for two weeks in demented patients with sleep disorders melatonin had no effect on median total time asleep, number of awakenings or sleep efficiency.[113] It must be noted that the doses employed and, particularly, the short term of observation might account for this discrepancy with earlier results. As noted above, melatonin effect in synchronizing the circadian clock needs sometimes weeks to develop.

Since there was no published study on the circadian consequences of injecting β amyloid peptide in experimental animals we recently assessed whether melatonin had the ability to protect against the circadian changes produced by β amyloid peptide 25-35 microinjection in SCN of golden hamsters.[33] Melatonin was given in the drinking water (25 μg/ml) starting 15 days in advance to the microinjection of β amyloid peptide into SCN. β Amyloid-treated hamsters exhibited a significant phase advance of onset of running activity as compared to saline-injected animals. They also showed a significantly greater variability in onset time of wheel running activity, mainly evident from 6 to 15 days of treatment. Melatonin administration prevented the phase advance of onset time and the increased variability of onset time brought about by β amyloid peptide. These results underlie the circadian consequences of injecting β amyloid peptide 25-35 in the SCN of hamsters.[33] The "fixing" effect of melatonin on onset time reported in hamsters administered with β amyloid peptide in their SCN somewhat resembles the reduction in variability of sleep onset time reported in demented[31] and non-demented patients.[21]

Melatonin May Act Both as a Chronobiotic and a Neuroprotective Agent in AD

The mechanisms accounting for the therapeutic effect of melatonin in AD patients remain unknown. Melatonin treatment promotes mainly non-REM sleep in the elderly,[80] and can be beneficial in AD by augmenting the restorative phases of sleep, including the augmented secretion of GH[130] and neurotrophins. In addition, in vitro experiments indicated that melatonin protects neurons against β-amyloid toxicity,[91,92] prevents β-amyloid-induced lipid peroxidation[25] and alters the metabolism of the β-amyloid precursor protein.[121] Melatonin given orally to rats was very effective to reduce β-amyloid-induced oxidative stress, and the neuroinflammatory response in CNS.[104] In addition, melatonin was able to reduce the free radical formation which follows the interaction between transition metal ions and β-amyloid.[139]

AD is considered a part of an emerging complex group of chronic and progressive neurodegenerative entities collectively known as disorders of protein folding.[68] In these diseases, normal molecules or their genetic variants self-assemble to form aggregates and/or fibrils that deposit in the cerebral vessels and/or brain parenchyma and that are associated with cognitive deficits, dementia and cerebellar and extrapyramidal signs. Among them, the most pathological conformer is rich in the β sheet conformation which can be stabilized by aggregation.

Thus melatonin may have two different ways of action to prevent pro-amyloidogenic environment that leads to AD amyloidosis: (1) anti-amyloidogenic and (2) antioxidant properties. The former causes the reduction of the β-sheet-rich conformer which becomes aggregated and neurotoxic, contributing to neuroprotection. Using AD transgenic mice the possibility that increases in microaggregated amyloid may be sufficient to produce oxidative stress was put forth.[90] Increased lipid peroxidation precedes amyloid plaque formation in transgenic mice, suggesting that brain oxidative stress contributes to amyloidosis before its deposition in affected brain.[94] The demonstration of the direct relationship between melatonin and the biochemical pathology of Alzheimer's disease was recently made in a transgenic mouse model of Alzheimer's amyloidosis by monitoring the effects of administering melatonin on brain levels of β-amyloid abnormal protein nitration and survival of the mice. The administration of melatonin partially inhibited the expected time-dependent elevation of β-amyloid, reduced abnormal nitration of proteins and increased survival in the treated transgenic mice.[69]

Since there is information indicating that supplementation with antioxidants delayed development of AD,[61,106] melatonin treatment may constitute a selection therapy to slow evolution of cognitive impairment in AD patients because it combines a potent neuroprotective activity with the amelioration of sundowning, the latter not observed with the use of regular antioxidants. Moreover, the combined treatment of melatonin together with bright light has the potential advantage of a preventing antioxidant effect of melatonin on harmful photooxidative processes at the retinal level.

Concluding Remarks

People with dementia are among the most vulnerable in our society. Symptoms often need to be treated expediently, and commonly employed drugs, although moderately effective, can be hazardous. On the other hand, sleep disorder and the associated behavioral and psychiatric problems exert a much greater toll than the cognitive impairment per se on caregivers of patients with AD.

Melatonin and light treatment seem to be safe and effective and may have an important role in managing sundowning in people with dementia. Indeed, initial studies of melatonin therapy in AD have had positive results. However, these studies are only a few and are open, and knowledge about side effects and long-term effects of melatonin is still limited. Indeed, melatonin may provide an innovative neuroprotective strategy against at least three known mechanism of neuronal death: Apoptosis, glutamate excitotoxicity, and oxyradical-mediated damage. In restoration of slow wave sleep after melatonin treatment may result in better regulation of neuronal metabolism.

Sleep can be considered as a neuroprotective strategy that can potentially improve the course and outcome of several brain disorders, and thus the quality of life of the affected individuals and their family members. To promote and protect an appropriate sleep can substantially reduce the costs of treatment and management, in particular, the enormous costs of lifetime treatment of some neuropsychiatric disorders. While a role for sleep in neuroprotection has not been conclusively established, several findings indicate at least that the idea merits serious consideration.

Acknowledgments

Studies in authors' laboratory were supported by the University of Buenos Aires, the Consejo Nacional de Investigaciones Científicas y Técnicas, Argentina, the Agencia Nacional de Promoción Científica y Tecnológica, Argentina (PICT 6153), Fundación Bunge y Born, Buenos Aires and Fundación Antorchas, Buenos Aires.

References

1. Andrade C, Srihari BS, Reddy KP et al. Melatonin in medically ill patients with insomnia: A double-blind, placebo-controlled study. J Clin Psychiatry 2001; 62:41-45.
2. Anton-Tay F. Melatonin: Effects on brain function. Adv Biochem Psychopharmacol 1974; 11:315-324.
3. Asada T, Motonaga T, Yamagata Z et al. Associations between retrospectively recalled napping behavior and later development of Alzheimer's disease: Association with APOE genotypes. Sleep 2000; 23:629-634.
4. Badia P, Hughes R, Murphy BD et al. Effects of exogenous melatonin on memory, sleepiness, and performance after a 4-hr nap. J Sleep Res 1996; 5(suppl 1):11.
5. Barron MJ, Johnson MA, Andrews RM et al. Mitochondrial abnormalities in ageing macular photoreceptors. Invest Ophthalmol Visual Sci 2001; 42:3016-3022.
6. Baskett JJ, Broad JB, Wood PC et al. Does melatonin improve sleep in older people? A randomised crossover trial. Age Ageing 2003; 32:164-170.
7. Beatty S, Koh H, Phil M et al. The role of oxidative stress in the pathogenesis of age-related macular degeneration. Surv Ophthalmol 2000; 45:115-134.
8. Blanks JC, Schmidt SY, Torigoe Y et al. Retinal pathology in Alzheimer's disease. II. Regional neuron loss and glial changes in GCL. Neurobiol Aging 1996; 17:385-395.
9. Boulton M, Dayhaw-Barker P. The role of the retinal pigment epithelium: Topographical variation and ageing changes. Eye 2001; 15:384-389.
10. Brainard GC, Hanifin JP, Greeson JM et al. Action spectrum for melatonin regulation in humans: Evidence for a novel circadian photoreceptor. J Neurosci 2001; 21:6405-6412.
11. Brusco LI, Marquez M, Cardinali DP. Melatonin treatment stabilizes chronobiologic and cognitive symptoms in Alzheimer's disease. Neuroendocrinol Lett 1998; 19:111-115.
12. Brusco LI, Marquez M, Cardinali DP. Monozygotic twins with Alzheimer's disease treated with melatonin: Case report. J Pineal Res 1998; 25:260-263.
13. Buysse DJ, Reynolds CF, Hoch C et al. Rapid eye movement sleep deprivation in elderly patients with concurre of depression and dementia. J Neuropsychiatry Clin Neurosci 1992; 4:249-256.
14. Cagnacci A. Melatonin in relation to physiology in adult humans. J Pineal Res 1996; 21:200-213.
15. Cai J, Nelson KC, Wub M et al. Oxidative damage and protection of the RPE. Progr Retinal Eye Rese 2000; 19:205-221.
16. Cajochen C, Krauchi K, von Arx MA et al. Daytime melatonin administration enhances sleepiness and theta/alpha activity in the waking EEG. Neurosci Lett 1996; 207:209-213.
17. Campbell SS, Dijk DJ, Boulos Z et al. Light treatment for sleep disorders: Consensus report. III. Alerting and activating effects. J Biol Rhythms 1995; 10:129-132.
18. Cardinali DP. The human body circadian: How the biologic clock influences sleep and emotion. Ciencia e Cultura 1998; 50:172-177.
19. Cardinali DP, Bortman GP, Liotta G et al. A multifactorial approach employing melatonin to accelerate resynchronization of sleep-wake cycle after a 12 time-zone westerly transmeridian flight in elite soccer athletes. J Pineal Res 2002; 32:41-46.
20. Cardinali DP, Brusco LI, Liberczuk C et al. The use of melatonin in Alzheimer's disease. Neuroendocrinol Lett 2002; 23(suppl 1):26-29.
21. Cardinali DP, Gvozdenovich E et al. A double blind-placebo controlled study on melatonin efficacy to reduce anxiolytic benzodiazepine use in the elderly. Neuroendocrinol Lett 2002; 23:55-60.

22. Cardinali DP, Pevet P. Basic aspects of melatonin action. Sleep Med Rev 1998; 2:175-190.
23. Cohen-Mansfield J, Garfinkel D, Lipson S. Melatonin for treatment of sundowning in elderly persons with demen. Arch Gerontol Geriatr 2000; 31:65-76.
24. Cramer H, Rudolph J, Consbruch U et al. On the effects of melatonin on sleep and behavior in man. Adv Biochem Psychopharmacol 1974; 11:187-191.
25. Daniels WM, van Rensburg SJ, van Zyl JM et al. Melatonin prevents beta-amyloid-induced lipid peroxidation. J Pineal Res 1998; 24:78-82.
26. Dijk DJ, Duffy JF, Riel E et al. Ageing and the circadian and homeostatic regulation of human sleep during forced desynchrony of rest, melatonin and temperature rhythms. J Physiol 1999; 516:611-617.
27. Dijk DJ, Roth C, Landolt HP et al. Melatonin effect on daytime sleep in men: Suppression of EEG low frequency activity and enhancement of spindle frequency activity. Neurosci Lett 1995; 201:13-16.
28. Dollins AB, Zhdanova IV, Wurtman RJ et al. Effect of inducing nocturnal serum melatonin concentrations in daytime on sleep, mood, body temperature, and performance. Proc Natl Acad Sci USA 1994; 91:1824-1828.
29. Dowling GA. Behavioral intervention strategies for sleep-activity disruption. Int Psychogeriatr 1996; 8(Suppl 1):77-86.
30. Dunlap JC. Molecular bases for circadian clocks. Cell 1999; 96:271-290.
31. Fainstein I, Bonetto A, Brusco LI et al. Effects of melatonin in elderly patients with sleep disturbance. A pilot study. Curr Ther Res 1997; 58:990-1000.
32. Ferrari E, Arcaini A, Gornati R et al. Pineal and pituitary-adrenocortical function in physiological aging and in senile dementia. Exp Gerontol 2000; 35:1239-1250.
33. Furio AM, Cutrera RA, Castillo Thea V et al. Effect of melatonin on changes in locomotor activity rhythm of Syrian hamsters injected with beta amyloid peptide 25-35 in the suprachiasmatic nuclei. Cell Mol Neurobiol 2002; 22:699-709.
34. Garfinkel D, Laudon M, Nof D et al. Improvement of sleep quality in elderly people by controlled-release melatonin. Lancet 1995; 346:541-544.
35. Garfinkel D, Zisapel N, Wainstein J et al. Facilitation of benzodiazepine discontinuation by melatonin: A new clinical approach. Arch Intern Med 1999; 159:2456-2460.
36. Gasio P, Krauchi K, Cajochen C et al. Dawn-dusk simulation light therapy of disturbed circadian rest-activity cycles in demented elderly. Exp Gerontol 2003; 38:207-216.
37. Gilbert SS, Van Den Heuvel CJ, Dawson D. Daytime melatonin and temazepam in young adult humans: Equivalent effects on sleep latency and body temperatures. J Physiol (Lond) 1999; 514(Pt 3):905-914.
38. Gilot B, Onen SH, Jalenques I. Hypnotic prescription in Alzheimer's disease patients. Int J Geriatric Psycopharmacology 1999; 2:10-12.
39. Golombek DA, Escolar E, Burin LJ et al. Chronopharmacology of melatonin: Inhibition by benzodiazepine antagonism. Chronobiol Int 1992; 9:124-131.
40. Golombek DA, Martini M, Cardinali DP. Melatonin as an anxiolytic in rats: Time-dependence and interaction with the central gabaergic system. Eur J Pharmacol 1993; 237:231-236.
41. Golombek DA, Pevet P, Cardinali DP. Melatonin effects on behavior: Possible mediation by the central GABAergic system. Neurosci Biobehav Rev 1996; 20:403-412.
42. Graw P, Werth E, Krauchi K et al. Early morning melatonin administration impairs psychomotor vigilance. Behav Brain Res 2001; 121:167-172.
43. Groos G, Mason R, Meijer J. Electrical and pharmacological properties of the suprachiasmatic nuclei. Fed Proc 1983; 42:2790-2795.
44. Haffmans P, Sival R, Lucius S et al. Bright light therapy and melatonin in motor restless behaviour in dementia: A placebo-controlled study. Int J Geriatr Psychiatry 2001; 16:106-110.
45. Haimov I, Lavie P. Potential of melatonin replacement therapy in older patients with sleep disorders. Drugs Aging 1995; 7:75-78.
46. Hall NF, Gale CR. Prevention of age related macular degeneration. BMJ 2002; 325:1-2.
47. Harper DG, Stopa EG, McKee AC et al. Differential circadian rhythm disturbances in men with Alzheimer disease and frontotemporal degeneration. Arch Gen Psychiatry 2001; 58:353-360.
48. Hastings MH, Best JD, Ebling FJ et al. Entrainment of the circadian clock. Prog Brain Res 1996; 111:147-174.
49. Herxheimer A, Petrie KJ. Melatonin for the prevention and treatment of jet lag. Cochrane Database Syst Rev 2002CD001520.
50. Hobson JA, Pace-Schott EF. The cognitive neuroscience of sleep: Neuronal systems, consciousness and learning. Nat Rev Neurosci 2002; 3:679-693.
51. Hoch CC, Reynolds CF, Buysse DJ et al. Protecting sleep quality in later life: A pilot study of bed restriction and sleep hygiene. J Gerontol B Psychol Sci Soc Sci 2001; 56:52-59.
52. Hohagen F, Kappler C, Schramm E et al. Prevalence of insomnia in elderly general practice attenders and the current treatment modalities. Acta Psychiatr Scand 1994; 90:102-108.
53. Hughes RJ, Sack RL, Lewy AJ. The role of melatonin and circadian phase in age-related sleep- maintenance insomnia: Assessment in a clinical trial of melatonin replacement. Sleep 1998; 21:52-68.
54. Inouye SIT, Kawamura H, Green DJ et al. Persistence of circadian rhythmicity in a mammalian hypothalamic 'island' containing the suprachiasmatic nucleus. Proc Natl Acad Sci USA 1979; 76:5962-5966.
55. Jean-Louis G, von Gizycki H, Zizi F. Melatonin effects on sleep, mood, and cognition in elderly with mild cognitive impairment. J Pineal Res 1998; 25:177-183.
56. Jean-Louis G, Zizi F, von Gizycki H et al. Effects of melatonin in two individuals with Alzheimer's disease. Percept Mot Skills 1998; 87:331-339.
57. Kennaway DJ, Wright H. Melatonin and circadian rhythms. Curr Top Med Chem 2002; 2:199-209.
58. King AC, Oman RF, Brassington GS et al. Moderate-intensity exercise and self-rated quality of sleep in older adults. A randomized controlled trial. JAMA 1997; 277:32-37.
59. Klaver CC, Ott A, Hofman A et al. Is age-related maculopathy associated with Alzheimer's Disease? The Rotterdam Study. Am J Epidemiol 1999; 150:963-968.
60. Kleitman N. Basic rest-activity cycle—22 years later. Sleep 1982; 5:311-317.
61. Kontush A, Mann U, Ant S et al. Influence of vitamin E and C supplementation on lipoprotein oxidation in patients with Alzheimer's disease. Free Rad Biol Med 2001; 31:345-354.
62. Lerner AB, Case MD. Melatonin. Fed Proc 1960; 19:590-592.
63. Lewy AJ, Ahmed S, Sack RL. Phase shifting the human circadian clock using melatonin. Behav Brain Res 1996; 73:131-134.
64. Lieberman HR, Waldhauser F, Garfield G et al. Effects of melatonin on human mood and performance. Brain Res 1984; 323:201-207.
65. Liu RY, Zhou JN, van Heerikhuize J et al. Decreased melatonin levels in postmortem cerebrospinal fluid in relation to aging, Alzheimer's disease, and apolipoprotein E-epsilon4/4 genotype. J Clin Endocrinol Metab 1999; 84:323-327.
66. MacFarlane JG, Cleghorn JM, Brown GM et al. The effects of exogenous melatonin on the total sleep time and daytime alertness of chronic insomniacs: A preliminary study. Biol Psychiatry 1991; 30:371-376.
67. Maggi S, Langlois JA, Minicuci N et al. Sleep complaints in community-dwelling older persons: Prevalence, associated factors, and reported causes. J Am Geriatr Soc 1998; 46:161-168.
68. Martin JB. Molecular basis of the neurodegenerative disorders. N Engl J Med 1999; 340:1970-1980.
69. Matsubara E, Bryant-Thomas T, Pacheco QJ et al. Melatonin increases survival and inhibits oxidative and amyloid pathology in a transgenic model of Alzheimer's disease. J Neurochem 2003; 85:1101-1108.
70. Mazzoni G, Gori S, Formicola G et al. Word recall correlates with sleep cycles in elderly subjects. J Sleep Res 1999; 8:185-188.
71. McCarten JR, Kovera C, Maddox MK et al. Triazolam in Alzheimer's disease: Pilot study on sleep and memory effect. Pharmacol Biochem Behav 1995; 52:447-452.

72. McCurry SM, Reynolds CF, Ancoli-Israel S et al. Treatment of sleep disturbance in Alzheimer's disease. Sleep Med Rev 2000; 4:603-628.
73. McGaffigan S, Bliwise DL. The treatment of sundowning. A selective review of pharmacological and nonpharmacological studies. Drugs Aging 1997; 10:10-17.
74. Mishima K, Hishikawa Y, Okawa M. Randomized, dim light controlled, crossover test of morning bright light therapy for rest-activity rhythm disorders in patients with vascular dementia and dementia of Alzheimer's type. Chronobiol Int 1998; 15:647-654.
75. Mishima K, Okawa M, Hishikawa Y et al. Morning bright light therapy for sleep and behavior disorders in elderly patients with dementia. Acta Psychiatr Scand 1994; 89:1-7.
76. Mishima K, Okawa M, Hozumi S et al. Supplementary administration of artificial bright light and melatonin as potent treatment for disorganized circadian rest-activity and dysfunctional autonomic and neuroendocrine systems in institutionalized demented elderly persons. Chronobiol Int 2000; 17:419-432.
77. Mishima K, Okawa M, Satoh K et al. Different manifestations of circadian rhythms in senile dementia of Alzheimer's type and multi-infarct dementia. Neurobiol Aging 1997; 18:105-109.
78. Mishima K, Okawa M, Shimizu T et al. Diminished melatonin secretion in the elderly caused by insufficient environmental illumination. J Clin Endocrinol Metab 2001; 86:129-134.
79. Mishima K, Tozawa T, Satoh K et al. Melatonin secretion rhythm disorders in patients with senile dementia of Alzheimer's type with disturbed sleep-waking. Biol Psychiatry 1999; 45:417-421.
80. Monti JM, Alvarino F, Cardinali D et al. Polysomnographic study of the effect of melatonin on sleep in elderly patients with chronic primary insomnia. Arch Gerontol Geriatr 1999; 28:85-98.
81. Montplaisir J, Petit D, Lorrain D et al. Sleep in Alzheimer's disease: Further considerations on the role of brainstem and forebrain cholinergic populations in sleepwake Mechanisms. Sleep 1995; 18:145-148.
82. Murphy PJ, Campbell SS. Physiology of the circadian system in animals and humans. J Clin Neurophysiol 1996; 13:2-16.
83. Nave R, Peled R, Lavie P. Melatonin improves evening napping. Eur J Pharmacol 1995; 275:213-216.
84. Naylor E, Penev PD, Orbeta L et al. Daily social and physical activity increases slow-wave sleep and daytime neuropsychological performance in elderly. Sleep 2000; 23:87-95.
85. Nicolas A, Petit D, Rompre S et al. Sleep spindle characteristics in healthy subjects of different age groups. Clin Neurophysiol 2001; 112:521-527.
86. Ohashi Y, Okamoto N, Uchida K et al. Daily rhythm of serum melatonin levels and effect of light exposure in patients with dementia of the Alzheimer's type. Biol Psychiatry 1999; 45:1646-1652.
87. Pace-Schott EF, Hobson JA. The neurobiology of sleep: Genetics, cellular physiology and subcortical networks. Nat Rev Neurosci 2002; 3:591-605.
88. Pandi-Perumal SR, Seils LK, Kayumov L et al. Senescence, sleep, and circadian rhythms. Ageing Res Rev 2002; 1:559-604.
89. Pappolla MA, Chyan Y, Poeggeler B et al. An assessment of the antioxidant and the antiamyloidogenic properties of melatonin: Implications for Alzheimer's disease. J Neural Transm 2000; 107:203-231.
90. Pappolla MA, Chyan YJ, Omar RA et al. Evidence of oxidative stress and in vivo neurotoxicity of beta-amyloid in a transgenic mouse model of Alzheimer's disease: A chronic oxidative paradigm for testing antioxidant therapies in vivo. Am J Pathol 1998; 152:871-877.
91. Pappolla MA, Simovich MJ, Bryant-Thomas T et al. The neuroprotective activities of melatonin against the Alzheimer beta-protein are not mediated by melatonin membrane receptors. J Pineal Res 2002; 32:135-142.
92. Pappolla MA, Sos M, Omar RA et al. Melatonin prevents death of neuroblastoma cells exposed to the Alzheimer amyloid peptide. J Neurosci 1997; 17:1683-1690.
93. Partonen T, Vakkuri O, Lamberg-Allardt C et al. Effects of bright light on sleepiness, melatonin, and 25-hydroxyvitamin D(3) in winter seasonal affective disorder. Biol Psychiatry 1996; 39:865-872.
94. Pratico D, Uryu K, Leight S et al. Increased lipid peroxidation precedes amyloid plaque formation in an animal model of Alzheimer amyloidosis. J Neurosci 2001; 21:4183-4187.
95. Prinz PN, Larsen LH, Moe KE et al. EEG markers of early Alzheimer's disease in computer selected tonic REM sleep. Electroencephalogr Clin Neurophysiol 1992; 83:36-43.
96. Prinz PN, Peskind ER, Vitaliano PP et al. Changes in the sleep and waking EEGs of non demented and demented elderly subjects. J Am Geriatr Soc 1982; 30:86-93.
97. Ralph MR, Foster RG, Davis FC et al. Transplanted suprachiasmatic nucleus determines circadian period. Science 1990; 247:975-978.
98. Ravid R, Swaab DF. The Netherlands brain bank-a clinico¡pathological link in aging and dementia research. J Neural Transm 1993; 39 suppl:143-153.
99. Reid K, Van den Heuvel C, Dawson D. Day-time melatonin administration: Effects on core temperature and sleep onset latency. J Sleep Res 1996; 5:150-154.
100. Reiter RJ. Functional pleiotropy of the neurohormone melatonin: Antioxidant protection and neuroendocrine regulation. Front Neuroendocrinol 1995; 16:383-415.
101. Reiter RJ, Tan DX, Burkhardt S et al. Melatonin in plants. Nutr Rev 2001; 59:286-290.
102. Reynolds CF, Spiker DG, Hanin I et al. Electroencephalographic sleep, aging, and psychopathology: New data and state of the art. Biol Psychiatry 1983; 18:139-155.
103. Roberts JE. Ocular phototoxicity. J Photochem Photobiol B 2001; 64:136-143.
104. Rosales-Corral S, Tan DX, Reiter RJ et al. Orally administered melatonin reduces oxidative stress and proinflammatory cytokines induced by amyloid-beta peptide in rat brain: A comparative, in vivo study versus vitamin C and E. J Pineal Res 2003; 35:80-84.
105. Rusak B, Zucker I. Neural regulation of circadian rhythms. Physiol Rev 1979; 59:449-526.
106. Sano M, Ernesto C, Thomas RG et al. A controlled trial of selegiline, alpha-tocopherol, or both as treatment for Alzheimer's disease. N Engl J Med 1997; 336:1216-1222.
107. Satlin A, Volicer L, Ross V et al. Bright light treatment of behavioral and sleep disturbances in patients with Alzheimer's disease. Am J Psychiatry 1992; 149:1028-1032.
108. Satlin A, Volicer L, Stopa EG et al. Circadian locomotor activity and core-body temperature rhythms in Alzheimer's disease. Neurobiol Aging 1995; 16:765-771.
109. Satomura T, Sakamoto T, Shirakawa S et al. Hypnotic action of melatonin during daytime administration and its comparison with triazolam. Psychiatry Clin Neurosci 2001; 55:303-304.
110. Savaskan E, Olivieri G, Meier F et al. Increased melatonin 1a-receptor immunoreactivity in the hippocampus of Alzheimer's disease patients. J Pineal Res 2002; 32:59-62.
111. Schredl M, Weber B, Braus D et al. The effect of rivastigmine on sleep in elderly healthy subjects. Exp Gerontol 2000; 35:243-249.
112. Schredl M, Weber B, Leins M et al. Donepezil-induced REM sleep augmentation enhances memory performance in elderly, healthy persons. Exp Gerontol 2001; 36:353-361.
113. Serfaty M, Kennell-Webb S, Warner J et al. Double blind randomised placebo controlled trial of low dose melatonin for sleep disorders in dementia. Int J Geriatr Psychiatry 2002; 17:1120-1127.
114. Shirakawa SI, Sakamoto T, Uchimura N et al. Effect of melatonin on sleep and rectal temperature of young healthy evening types. Psychiatry Clin Neurosci 2001; 55:301-302.
115. Siegrist C, Benedetti C, Orlando A et al. Lack of changes in serum prolactin, FSH, TSH, and estradiol after melatonin treatment in doses that improve sleep and reduce benzodiazepine consumption in sleep-disturbed, middle-aged, and elderly patients. J Pineal Res 2001; 30:34-42.
116. Skene DJ. Optimization of light and melatonin to phase-shift human circadian rhythms. J Neuroendocrinol 2003; 15:438-441.
117. Skene DJ, Lockley SW, Arendt J. Melatonin in circadian sleep disorders in the blind. Biol Signals Recept 1999; 8:90-95.
118. Skene DJ, Vivien-Roels B, Sparks DL et al. Daily variation in the concentration of melatonin and 5- methoxytryptophol in the human pineal gland: Effect of age and Alzheimer's disease. Brain Res 1990; 528:170-174.
119. Smith C. Sleep states and memory processes. Behav Brain Res 1995; 69:137-145.

120. Smith C, Lapp L. Increases in number of REMS and REM density in humans following an intensive learning period. Sleep 1991; 14:325-330.
121. Song W, Lahiri DK. Melatonin alters the metabolism of the beta-amyloid precursor protein in the neuroendocrine cell line PC12. J Mol Neurosci 1997; 9:75-92.
122. Spiegel R, Koberle S, Allen SR. Significance of slow wave sleep: Considerations from a clinical viewpoint. Sleep 1986; 9:66-79.
123. Stone BM, Turner C, Mills SL et al. Hypnotic activity of melatonin. Sleep 2000; 23:663-669.
124. Stopa EG, Volicer L, Kuo-Leblanc V et al. Pathologic evaluation of the human suprachiasmatic nucleus in severe dementia. J Neuropathol Exp Neurol 1999; 58:29-39.
125. Terlo L, Laudon M, Tarasch R et al. Effects of low doses of melatonin on late afternoon napping and mood. Biol Rhythm Res 1997; 28:2-15.
126. Terman JS, Terman M, Lo ES et al. Circadian time of morning light administration and therapeutic response in winter depression. Arch Gen Psychiatry 2001; 58:69-75.
127. Terman M, Schlager D, Fairhurst S et al. Dawn and dusk simulation as a therapeutic intervention. Biol Psychiatry 1989; 25:966-970.
128. Uchida K, Okamoto N, Ohara K et al. Daily rhythm of serum melatonin in patients with dementia of the degenerate type. Brain Res 1996; 717:154-159.
129. Uchida K, Okamoto N, Ohara K et al. Daily rhythm of serum melatonin in patients with dementia of the degenerate type. Brain Res 1996; 717:154-159.
130. van Coevorden A, Mockel J, Laurent E et al. Neuroendocrine rhythms and sleep in aging men. Am J Physiol 1991; 260:E651-E661.
131. van Someren EJ. Circadian and sleep disturbances in the elderly. Exp Gerontol 2000; 35:1229-1237.
132. Vitiello MV, Prinz PN. Alzheimer's disease, sleep and sleep/wake patterns. Clin Geriatr Med 1989; 5:289-299.
133. Vitiello MV, Prinz PN, Williams DE et al. Sleep disturbances in patients with mild-stage Alzheimer's disease. J Gerontol Med Sci 1990; 45:M131-M138.
134. Vollrath L, Semm P, Gammel G. Sleep induction by intranasal administration of melatonin. Adv Biosci 1981; 29:327-329.
135. Wager-Smith K, Kay SA. Circadian rhythm genetics: From flies to mice to humans. Nature Genet 2000; 26:23-27.
136. Wang F, Li J, Wu C et al. The GABA(A) receptor mediates the hypnotic activity of melatonin in rats. Pharmacol Biochem Behav 2003; 74:573-578.
137. Wirz-Justice A, Werth E, Savaskan E et al. Haloperidol disrupts, clozapine reinstates the circadian rest-activity cycle in a patient with early-onset Alzheimer disease. Alzheimer Dis Assoc Disord 2000; 14:212-215.
138. Witting W, Kwa IH, Eikelenboom P et al. Alterations in the circadian rest-activity rhythm in aging and Alzheimer's disease. Biol Psychiatry 1990; 27:563-572.
139. Zatta P, Tognon G, Carampin P. Melatonin prevents free radical formation due to the interaction between beta-amyloid peptides and metal ions [Al(III), Zn(II), Cu(II), Mn(II), Fe(II)]. J Pineal Res 2003; 35:98-103.
140. Zhdanova IV, Wurtman RJ, Lynch HJ et al. Sleep-inducing effects of low doses of melatonin ingested in the evening. Clin Pharmacol Ther 1995; 57:552-558.
141. Zhdanova IV, Wurtman RJ, Morabito C et al. Effects of low oral doses of melatonin, given 2-4 hours before habitual bedtime, on sleep in normal young humans. Sleep 1996; 19:423-431.
142. Zhdanova IV, Wurtman RJ, Regan MM et al. Melatonin treatment for age-related insomnia. J Clin Endocrinol Metab 2001; 86:4727-4730.
143. Zhou JN, Liu RY, Kamphorst W et al. Early neuropathological Alzheimer's changes in aged individuals are accompanied by decreased cerebrospinal fluid melatonin levels. J Pineal Res 2003; 35:125-130.

CHAPTER 12

Pharmacotherapy for Seasonal Affective Disorder

Timo Partonen

Summary

The current first-line treatment for patients with winter seasonal affective disorder (SAD) is bright-light therapy, that is exposure to strong artificial light visible to the eye. Patients at risk of light-induced eye damage, such as those using photosensitizing medication, need to consult an ophthalmologist before the bright-light therapy is started, and eventually at regular intervals thereafter. Antidepressant drugs have not gained much attention in the treatment of winter SAD. They are still a relatively unexplored mode of and best regarded as the second-line treatment. Preliminary data from randomized controlled trials suggest that several antidepressants may be effective. Continuing treatment throughout the winter season, or in some cases all the year round, is advisable to prevent relapses.

Introduction

This chapter is meant to give a guideline on biological modes of treatment for patients with the winter pattern of recurrent depressive episodes, how to treat them and what to prescribe on the basis of current evidence. To guide the reader further, there are several overviews on the treatment of seasonal affective disorder (SAD) that provide additional information.[1-5]

Diagnosis

Seasonal affective disorder is a recurrent affective disorder whose episodes are bound to the changes of seasons. The basic concepts, diagnostic criteria and clinical picture are presented more in detail in the following.

Concepts

The syndrome coined in 1984 as SAD was originally seen as a condition in which depressive episodes develop during autumn or winter and remit the following spring or summer for at least two successive years.[6] In addition, the SAD patient had to show a history of major depressive or bipolar disorder. Since then, two subtypes of SAD have been described in the literature: winter SAD and summer SAD, of which the former is far more frequent. Hence, the concept of SAD is of higher degree, being not a synonym for the winter pattern but including two subtypes that have been verified thus far. So, concerning people living in regions of temperate zone spring SAD and fall SAD are in principle potential subtypes but infrequent in clinical practice.

Subsyndromal SAD is a condition with similar but milder symptoms that do not impair functioning to a major degree.[7] The tendency to experience seasonal changes in mood and behavior, also known as seasonality, is manifested to a different degree in individuals, ranging from the extreme and pathological end of the spectrum, namely patients with SAD, through the mildly pathological, as in subsyndromal SAD, to the normal.

Classification

The original, operational conceptualizations of SAD were eventually transformed into diagnostic criteria based on the Diagnostic and Statistical Manual of Mental Disorders, in the latest version (DSM-IV) of which SAD is regarded as a specifier of either bipolar or recurrent major depressive disorder, with a seasonal pattern of major depressive episodes (codes 296.3x to 296.8x).

The ICD-10 Classification of Mental and Behavioural Disorders (ICD-10) gives only provisional diagnostic criteria for SAD on the grounds that its status is best regarded as uncertain. Subject to these reservations, SAD is recognized as a form of bipolar affective or recurrent depressive disorder, with episodes varying in degrees of severity (codes F30.*xx* to F33.*xx*).

Symptoms and Signs

The onset of winter SAD typically occurs between 20 and 30 years of age, but the affected subject does not usually seek psychiatric attention for some years. Depressive episodes are usually mild to moderate in severity. Clinical features associated with winter SAD are rather consistent across series of patients from industrialized cultures (Table 1). So-called atypical depressive symptoms (increased duration of sleep, increased appetite, weight gain, and carbohydrate craving) often precede impaired functioning. Somatic symptoms are frequently the presenting complaint at visits to general practice. On recovery, improvement in tests of cognitive function, except in those of visual or spatial memory, have been observed that can be attributed to abnormal cognitive sensitivity to light.

Interestingly, whereas healthy subjects report sedation after ingestion of carbohydrates, depressed winter SAD patients experience activation and are less sensitive to the sweet taste. Resting metabolic rates may also be increased in depressed winter SAD patients secondary to changes in appetite and caloric intake. The high intake of sweets late in the day is known to predict both a rapid and persistent response to the first-line treatment, bright-light therapy, among patients with winter SAD.[8] By similarity, medications that alleviate the high intake of sweets, via the suppression of insulin secretion or some other mechanism of action, may be of use in their treatment.

Sleep and Sleep Disorders: A Neuropsychopharmacological Approach, edited by Malcolm Lader, Daniel P. Cardinali and S.R. Pandi-Perumal.

Table 1. Symptom profile of patients[a]

Frequent	Percent
Sadness	96
Decreased activity	96
Social misfortune[b]	92
Anxiety	86
Irritability[b]	86
Occupational misfortune	84
Daytime tiredness[b]	81
Fairly Frequent	**Percent**
Increased sleep	76
Poor quality of sleep[b]	75
Increased weight	74
Carbohydrate craving	70
Decreased libido	68
Increased appetite	65
Infrequent	**Percent**
Mixed or no change in appetite[b]	17
Mixed or no change in weight[b]	17
Decreased appetite	15
Decreased weight	7

[a] Data are derived from the NIMH Seasonal Studies Program (n=662, except for [b] n=366).

Mixed Disorders

Mixed disorders often compromise the search for winter SAD, and each may require specific intervention. First, the lifetime prevalence of anxiety disorders in winter SAD is relatively high, though not different from the rate seen in nonseasonal recurrent major depression. Generalized anxiety disorder, simple phobias and social phobia are the most common comorbid disorders. Second, avoidant personality disorder is also a usual concurrent disorder in patients with winter SAD. Third, a considerable number of bulimic patients suffers from seasonal depressive symptoms. Some data indicate that patients with winter SAD and those with bulimia nervosa have similar attitudes towards eating that are reflected as distorted perceptions of body size and shape, but opposite styles of eating with low versus high scores of restraint eating behaviors respectively.

Pathogenesis

The decreasing daylight period as winter approaches is thought to trigger a depressive episode in subjects predisposed to winter SAD, in particular. However, no causal relationship can be drawn between the incidence of winter SAD and the northerly climate with extensive changes in levels of illumination and ambient temperatures over the year. Although bright-light exposure has been used in treatment, the cause of winter SAD is not inevitably the shortage of light. Winter SAD may well be sensitive to factors that are common to various forms of recurrent affective disorder. It could best be seen as a disorder that is driven by endogenous fluctuations with a period of about a year, or so-called circannual rhythms. As for example, winter SAD may be characterized by disturbed metabolism of melatonin and serotonin in winter and spring respectively.

Melatonin

Melatonin is regarded as important for the timing of the circadian rhythm phase position and the sleep-wake cycle. Winter SAD was earlier believed to be related to abnormal melatonin metabolism, but recent findings support the view that there are no abnormalities in melatonin production. It appears that depressed patients with winter SAD generate an abnormal signal of change of season in the form of circulating concentrations of melatonin, the problem being caused rather by slow elimination than abnormal production of melatonin.[9] This may be a key to and give an explanation for the disappointing results from trials with melatonin in the treatment of winter SAD.[10-12] Melatonin administration may not form the appropriate time-giving signal to the circadian pacemaker in conditions where the elimination of the hormone is compromized and its half-life lengthened.

Circadian Pacemaker

Information about the direction and velocity of change of the length of daylight (the photoperiod) is being transformed into the production of melatonin. Fixed phase-delays in the timing of circadian rhythms, such as those in melatonin production, were previously thought to be related to winter SAD. Recent data show that the circadian rest-activity cycle appears to be more elastic in winter SAD patients compared to healthy subjects across the days, deviating more from 24 hours and peaking at less regular times.[13]

This may be due to compromized passage of light via the retinohypothalamic tract, leading to changes in the stability and precision in function of the circadian clock. Few studies do suggest that depressed winter SAD patients tend to be insensitive to illumination during the day, as measured with the electro-oculographic ratios, but to show a high degree of sensitivity to light exposure at night, as assessed with the light-induced suppression of melatonin production.

It is likely that there are separate time-givers for circadian oscillation in the morning and evening hours, modulating the circadian clockwork in a complex way. The disordered input to the circadian clock together with the inaccurate feedback on it through the actions of melatonin may reset these two time-givers discordantly, predisposing to irregularities in the daily rest-activity cycles.

Sleep

Since mood is influenced by a complex interaction of circadian phase and the duration of prior wakefulness, even moderate changes in the timing of the sleep-wake cycle may have profound effects on mood.[14] The extent of prolonged sleep seen in depressed winter SAD patients does not differ markedly from that reported by the general population, but their sleep-related complaints are accompanied by abnormal findings in the structure of sleep, including decreased slow-wave sleep, increased rapid-eye-movement (REM) density, and impaired sleep efficiency.

In winter SAD, there is no evidence of abnormal homeostatic regulation of sleep, as assessed with constant routine protocols. However, sleep-related events, such as regulation of core body or brain temperatures at night, may in part influence the emergence of disturbances that are common in winter SAD. In these patients, the frequencies of electroencephalograph during nonREM (NREM) sleep tend to resemble the profiles of individuals who have been deprived of sleep. In addition, brain temperatures during sleep are in part determined by the level of brain cooling activity, which oscillates across the NREM-REM sleep cycle and may be reflected by facial skin temperatures. Facial skin temperatures during sleep are abnormally low and do not match with rectal temperatures in depressed patients with winter SAD, showing that brain cooling activity is reduced.[15]

Treatment

Treatment of winter SAD is similar to any recurrent affective disorder, albeit the exception that bright-light therapy has been established as the first-line treatment. Particular care needs to be taken as far as continuation of treatment and prevention of recurrence are concerned.

Everyday Tips

Regular walks or outdoor exercises, such as skiing and skating, can be of great benefit for patients with winter SAD, thanks to both physical exercise and exposure to light that appear to have an additive effect on mood regulation. However, it is also important to try to receive sunlight while spending time indoors. One could examine, if it is possible to work in brightly lit surroundings and, for instance, to move one's desk close to a window. At home, those suffering from winter SAD may also try to spend as much time as possible in the brightest areas of apartment, for instance by a window facing the sun. It is often a relief for many to travel to locations with a lot of sunshine during the dark period of the year. Some individuals may be in the position to negotiate with their employee to take vacation during winter rather than to have a summer holiday. They may then stay in sunny locations for prolonged times in winter. However, these adjustments may be difficult to achieve in modern societies. Furthermore, in regions at northerly latitudes and those with heavy overcast, the adjustments listed above may not provide enough light from the habitat to relieve winter SAD, but the administration of bright-light therapy together with other options for treatment, such as medication, needs to be considered.

Bright-Light Therapy

Bright-light therapy is the treatment of choice for the management of patients with winter SAD. The response to bright-light therapy is often good or excellent, and no additional treatment is usually required. Bright-light therapy may induce adverse effects, but they are usually mild and few. However, manic behavior and suicidal tendencies have although seldom been reported during the initiation of light therapy. It is therefore preferable to contact a psychiatrist or a physician before light treatment is started. This is especially important for those who have a history of severe mental illness such as mania or severe depression, or suicidal thoughts.

It is important to consult a doctor before starting the bright-light therapy, if the individual is taking any medication. Patients taking medications that have photosensitizing effects need ophthalmologic examination. The following medications are examples of drugs that may have these effects: chloroquine (antimalarial), hematoporphyrins (used in the photodynamic therapy for cancer), hypericum (St. John's wort), lithium (mood stabilizer), melatonin (pineal hormone), methoxypsoralens (used in the ultraviolet phototherapy for psoriasis), and phenothiazines (antiemetics and antipsychotics). Animal studies show retinal changes with drugs such as beta blockers, tricyclic antidepressants and L-tryptophan, but ophthalmologic monitoring for patients on these drugs is not required, according to the current consensus, unless they have other ocular risk factors.

Some experts advice that an ophthalmologic check-up will be routinely performed on all individuals taking light treatment, whereas others do not. Ophthalmologic consultation is recommended for patients with the following known risk factors for retinal toxicity to light exposure: a preexisting retinal or eye disease (for example, retinal detachments, retinitis pigmentosa and glaucoma), a systemic illness that affects the retina (for example, diabetes mellitus), previous cataract surgery and lens removal, and older age because of a greater risk of age-related degeneration. However, it is advised to consult a doctor before starting the bright-light therapy, if the individual has any eye-related problems.

It is essential to monitor the initial effects of bright-light therapy, for example by administering it at a well-equipped outpatient facility. Rating scales measuring the severity of depression are encouraged to be used before and after the treatment and periodically thereafter. It is better for the clinician to assess the benefits before the patient is advised to purchase a device for personal use. Best would be, if the response to light exposure can be assessed before the patient starts the self-administration at home. Patients who are using a light-therapy device at home without close supervision must be informed about health hazards and encouraged to keep in contact. It might bear a problem that individuals can nowadays buy these devices from an e-shop at the Internet or local stores, without having adequate instructions for use, and carry out the treatment with no prior consultation or supervision.

Medication

When there appears to be no response to bright-light therapy, or the patient prefers another mode of treatment, a prescription of an antidepressant drug needs to be considered. Data from randomized, controlled trials suggest that antidepressants are effective in the treatment of winter SAD. Based on the current evidence, the best choice would then be one of the selective serotonin reuptake inhibitors or reversible inhibitors of monoamine oxidase A. However, other new antidepressants may show the efficacy as well, similar to their general efficacy in the treatment of depressive disorders, but this remains to be discovered.

Three double-blind, placebo-controlled trials using a parallel design on a total of 289 winter SAD patients have been conducted with antidepressant drugs. First, sertraline, a selective serotonin reuptake inhibitor (SSRI), was more effective than placebo in a multicenter trial of 8 weeks on 187 patients, although the abbreviated report of this study has yet to be published as a formal journal article.[16] Second, fluoxetine, another SSRI, produced a slightly higher response rate than placebo in a multicenter trial of 5 weeks on 68 patients.[17] Third, moclobemide, a reversible inhibitor of monoamine oxidase A (RIMA), appeared no more effective than placebo in a one-center trial of 3 weeks on 34 patients.[18] In addition, a double-blind, active-control trial of 6 weeks with fluoxetine and moclobemide indicated good efficacy with both drugs.[19] However, the last three trials mentioned here need to be evaluated with caution because of a short duration, small number of study subjects, or both.

No evidence-based data can be derived from and are available on the basis of most of the drug trials (see the list of further reading). Only the trial with sertraline has included a fairly large number of subjects thus far. Many of the trials have not been designed as formal pharmacological randomized controlled trials, but instead of analyzing the clinical efficacy they have been applied for

Table 2. Recommended length of treatment for SAD

Randomized controlled trials, with smaller numbers or insufficient power, show that:

1. A therapeutic trial of antidepressants should be at least 6 weeks in length.
2. Because of risk of relapse, patients should continue with treatment for the entire winter season, until the time of their natural spring or summer remission. Treatment is not generally recommended during the summer.

Evidence that expresses the opinion of the committee members who have reviewed the literature and guidelines, following discussion with peers, shows that:

1. When possible, antidepressants should be tapered instead of abruptly discontinued.
2. Treatment should be restarted in the autumn, either with onset of mild symptoms, or in advance of the usual onset of symptoms.
3. Preventative year-round antidepressant treatment inclusive the summertime need to be considered in situations where:
   - patients are poorly compliant or motivated.
   - they take a long time to taper off and on medications.
   - they are unable to recognize early signs and symptoms of a depressive episode.
   - they have very early onset or very late offset of symptoms.
   - they experience symptoms during the summer.

testing the hypothesized mechanisms of action in the pathogenesis of the disorder.

The dosages of antidepressants should be similar to those used in the treatment of major depressive disorder, but the duration of treatment in patients with winter SAD can often be shorter than that required for other conditions. If the patient suffers from insomnia, an antidepressant (for example, mirtazapine or nefazodone) that rapidly improves sleep quality may be preferred for prescription.

Recently, the association of winter SAD with a polymorphism of 471 Leu/Ser in the neuronal PAS domain protein 2 (NPAS2) gene was reported, showing for the first time a recessive effect of the leucine allele on disease-specific susceptibility for winter SAD.[20] Whereas the transcription factor CLOCK controls circadian oscillation in the suprachiasmatic nucleus, NPAS2 has a similar function in other brain regions plus a humoral mechanism of action to reset peripheral clocks of the body. NPAS2 plays a substantive role in maintaining circadian behaviors, rest-activity cycles as well as adaptation to a shortage of food. Parallel to novel antidepressants already under development, molecules that have an effect on the function of the NPAS2 gene, or its associates, will be of high priority in terms of drug development and medication for mood and sleep disorders in general terms, using winter SAD as a leading example.

Poor Response

Patients showing a limited response to the treatment need to be evaluated to guarantee that they have an adequate dosing of treatment, are fully compliant with the treatment and do not have unrecognized conditions. The fact that the patient has two or more concurrent, or comorbid, disorders is likely to influence the clinical picture by modulating the course of illness, and each of the disorders may require specific intervention. For patients with refractory illness, it is important to take a detailed medical history and to examine previous treatment responses.

Factors to consider when deciding on a first-line treatment for patients with winter SAD include the severity of depressive symptoms, patient preference, safety, patient compliance, adverse effects, and costs. As a rule of thumb, the patient needs to start with a single treatment only, and this would help with the evaluation of treatment response and eventual adverse events. If needed, however, versatile strategies for care may need to be tailored to the patient's needs using appropriate options.

There are several options for improving responses to medications when the clinical outcome is not good enough due to insufficient efficacy. They include switching to another antidepressant, combining with other antidepressants, augmenting with another psychotropic drug, and combining with bright-light or psychological therapy.

Each therapeutic trial with antidepressants need to be at least 6 to 8 weeks in length (see Table 2). Combinations of two antidepressants do necessitate careful monitoring for potential adverse events. Although there is a rationale for combining psychotropic drugs of different classes, the efficacy of augmentation strategies with another psychotropic drugs appear to remain modest and in practice they may be of limited use only. Harmful drug-light interactions have been reported for two individual cases in the context of bright-light therapy. Currently, there is only a preliminary evidence of the efficacy of cognitive therapy for winter SAD.

Conclusion

The first-line treatment for patients with winter SAD is bright-light therapy. Patients at risk of light-induced eye damage, such as those having photosensitizing medication, must consult an ophthalmologist before the bright-light therapy is started, and eventually at regular intervals thereafter.

Antidepressant drugs are a relatively unexplored mode of treatment for winter SAD, but preliminary data from randomized controlled trials suggest that several antidepressants may be effective. Combined treatment needs to be considered whenever the initial response is not good enough.

References

1. Partonen T, Lönnqvist J. Seasonal affective disorder. Lancet 1998; 352:1369-1374.
2. Partonen T, Lönnqvist J. Seasonal affective disorder: A guide to diagnosis and management. CNS Drugs 1998; 9:203-212.
3. Lam RW. Seasonal affective disorder: Diagnosis and management. Primary Care Psychiatry 1998; 4:63-74.
4. In: Lam RW, Levitt AJ, eds. Canadian consensus guidelines for the treatment of seasonal affective disorder: A summary of the report of the Canadian Consensus Group on SAD. The Canadian Journal of Diagnosis Oct 1998; (Suppl):1-16.

5. Zulman DM, Oren DA. Seasonal affective disorder. Curr Opinion Psychiatry 1999; 12:81-86.
6. Rosenthal NE, Sack DA, Gillin JC et al. Seasonal affective disorder: A description of the syndrome and preliminary findings with light therapy. Arch Gen Psychiatry 1984; 41:72-80.
7. Kasper S, Rogers SLB, Yancey A et al. Phototherapy in individuals with and without subsyndromal seasonal affective disorder. Arch Gen Psychiatry 1989; 46:837-844.
8. Kräuchi K, Wirz-Justice A, Graw P. High intake of sweets late in the day predicts a rapid and persistent response to light therapy in winter depression. Psychiatry Res 1993; 46:107-117.
9. Wehr TA, Duncan Jr WC, Sher L et al. A circadian signal of change of season in patients with seasonal affective disorder. Arch Gen Psychiatry 2001; 58:1108-1114.
10. Wirz-Justice A, Graw P, Krauchi K et al. Morning or night-time melatonin is ineffective in seasonal affective disorder. J Psychiatr Res 1990; 24:129-137.
11. Lewy AJ, Bauer VK, Cutler NL et al. Melatonin treatment of winter depression: A pilot study. Psychiatry Res 1998; 77:57-61.
12. Leppämäki S, Partonen T, Vakkuri O et al. Effect of controlled-release melatonin on sleep quality, mood, and quality of life in subjects with seasonal or weather-associated changes in mood and behaviour. Eur Neuropsychopharmacol 2003; 13:137-145.
13. Teicher MH, Glod CA, Magnus E et al. Circadian rest-activity disturbances in seasonal affective disorder. Arch Gen Psychiatry 1997; 54:124-130.
14. Boivin DB, Czeisler CA, Dijk DJ et al. Complex interaction of the sleep-wake cycle and circadian phase modulates mood in healthy subjects. Arch Gen Psychiatry 1997; 54:145-152.
15. Schwartz PJ, Rosenthal NE, Kajimura N et al. Ultradian oscillations in cranial thermoregulation and electroencephalographic slow-wave activity during sleep are abnormal in humans with annual winter depression. Brain Res 2000; 866:152-167.
16. Moscovitch A, Blashko C, Wiseman R et al. A double-blind, placebo-controlled study of sertraline in the treatment of outpatients with seasonal affective disorder [abstract]. 148th Annual Meeting of the American Psychiatric Association. Miami May 1995; 20-25.
17. Lam RW, Gorman CP, Michalon M et al. Multicenter, placebo-controlled study of fluoxetine in seasonal affective disorder. Am J Psychiatry 1995; 152:1765-1770.
18. Lingærde O, Reichborn-Kjennerud T, Haggag A et al. Treatment of winter depression in Norway: II. A comparison of the selective monoamine oxidase A inhibitor moclobemide and placebo. Acta Psychiatr Scand 1993; 88:372-380.
19. Partonen T, Lönnqvist J. Moclobemide and fluoxetine in treatment of seasonal affective disorder. J Affect Disord 1996; 41:93-99.
20. Johansson C, Willeit M, Smedh C et al. Circadian clock-related polymorphisms in seasonal affective disorder and their relevance to diurnal preference. Neuropsychopharmacology 2003; 28:34-739.

Further Reading

1. Childs PA, Rodin I, Martin NJ et al. Effect of fluoxetine on melatonin in patients with seasonal affective disorder and matched controls. Br J Psychiatry 1995; 166:196-198.
2. Dilsaver SC, Jaeckle RS. Winter depression responds to an open trial of tranylcypromine. J Clin Psychiatry 1990; 51:326-329.
3. Dilsaver SC, Del Medico VJ, Quadri A et al. Pharmacological responsiveness of winter depression. Psychopharmacol Bull 1990; 26:303-309.
4. Dilsaver SC, Qamar AB, Del Medico VJ. The efficacy of bupropion in winter depression: Results of an open trial. J Clin Psychiatry 1992; 53:252-255.
5. Garcia-Borreguero D, Jacobsen FM, Murphy DL et al. Hormonal responses to the administration of m-chlorophenylpiperazine in patients with seasonal affective disorder and controls. Biol Psychiatry 1995; 37:740-749.
6. Ghadirian A-M, Murphy BEP, Gendron M-J. Efficacy of light versus tryptophan therapy in seasonal affective disorder. J Affect Disord 1998; 50:23-27.
7. Gloth 3rd FM, Alam W, Hollis B. Vitamin D vs broad spectrum phototherapy in the treatment of seasonal affective disorder. J Nutr Health Aging 1999; 3:5-7.
8. Hesselmann B, Habeler A, Praschak-Rieder N et al. Mirtazapine in seasonal affective disorder (SAD): A preliminary report. Hum Psychopharmacol Clin Exp 1999; 14:59-62.
9. Hilger E, Willeit M, Praschak-Rieder N et al. Reboxetine in seasonal affective disorder: An open trial. Eur Neuropsychopharmacol 2001; 11:1-5.
10. Jacobsen FM, Mueller EA, Rosenthal NE et al. Behavioral responses to intravenous meta-chlorophenylpiperazine in patients with seasonal affective disorder and control subjects before and after phototherapy. Psychiatry Res 1994; 52:181-197.
11. Joseph-Vanderpool JR, Jacobsen FM, Murphy DL et al. Seasonal variation in behavioral responses to m-CPP in patients with seasonal affective disorder and controls. Biol Psychiatry 1993; 33:496-504.
12. Lam RW, Levitan RD, Tam EM et al. L-tryptophan augmentation of light therapy in patients with seasonal affective disorder. Can J Psychiatry 1997; 42:303-306.
13. Levitan RD, Kaplan AS, Brown GM et al. Hormonal and subjective responses to intravenous m-chlorophenylpiperazine in women with seasonal affective disorder. Arch Gen Psychiatry 1998; 55:244-249.
14. Levitt AJ, Brown GM, Kennedy SH et al. Tryptophan treatment and melatonin response in a patient with seasonal affective disorder. J Clin Psychopharmacol 1991; 11:74-75.
15. Lingærde O, Haggag A. Moclobemide in winter depression: Some preliminary results from an open trial. Nord J Psychiatry 1992; 46:201-203.
16. Lingærde O, Føreland AR, Magnusson A. Can winter depression be prevented by Ginkgo biloba extract? A placebo-controlled trial. Acta Psychiatr Scand 1999; 100:62-66.
17. Martinez B, Kasper S, Ruhrmann S et al. Hypericum in the treatment of seasonal affective disorders. J Geriatr Psychiatry Neurol 1994; 7(Suppl 1):S29-S33.
18. McGrath RE, Buckwald B, Resnick EV. The effect of L-tryptophan on seasonal affective disorder. J Clin Psychiatry 1990; 51:162-163.
19. Oren DA, Moul DE, Schwartz PJ et al. A controlled trial of levodopa plus carbidopa in the treatment of winter seasonal affective disorder: A test of the dopamine hypothesis. J Clin Psychopharmacol 1994; 14:196-200.
20. Oren DA, Teicher MH, Schwartz PJ et al. A controlled trial of cyanocobalamin (Vitamin B12) in the treatment of winter seasonal affective disorder. J Affect Disord 1994; 32:197-200.
21. O'Rourke DA, Wurtman JJ, Brzezinski A et al. Serotonin implicated in etiology of seasonal affective disorder. Psychopharmacol Bull 1987; 23:358-359.
22. O'Rourke D, Wurtman JJ, Wurtman RJ et al. Treatment of seasonal depression with d-fenfluramine. J Clin Psychiatry 1989; 50:343-347.
23. Pande AC. Pharmacological treatments of SAD. Can J Psychiatry 1990; 35:721-722.
24. Rosenthal NE, Jacobsen FM, Sack DA et al. Atenolol in seasonal affective disorder: A test of the melatonin hypothesis. Am J Psychiatry 1988; 145:52-56.
25. Ruhrmann S, Kasper S, Hawellek B et al. Effects of fluoxetine versus bright light in the treatment of seasonal affective disorder. Psychol Med 1998; 28:923-933.
26. Schlager DS. Early-morning administration of short-acting β blockers for treatment of winter depression. Am J Psychiatry 1994; 151:1383-1385.
27. Schwartz PJ, Turner EH, Garcia-Borreguero D et al. Serotonin hypothesis of winter depression: Behavioral and neuroendocrine effects of the 5-HT(1A) receptor partial agonist ipsapirone in patients with seasonal affective disorder and healthy control subjects. Psychiatry Res 1999; 86:9-28.
28. Sherer MA, Weingartner H, James SP et al. Effects of melatonin on performance testing in patients with seasonal affective disorder. Neurosci Lett 1985; 58:277-282.
29. Teicher MH, Glod CA. Seasonal affective disorder: Rapid resolution by low-dose alprazolam. Psychopharmacol Bull 1990; 26:197-202.

30. Thorell LH, Kjellman B, Arned M et al. Light treatment of seasonal affective disorder in combination with citalopram or placebo with 1-year follow-up. Int Clin Psychopharmacol 1999; 14(Suppl 2):S7-S11.
31. Turner EH, Schwartz PJ, Lowe CH et al. Double-blind, placebo-controlled study of single-dose metergoline in depressed patients with seasonal affective disorder. J Clin Psychopharmacol 2002; 22:216-220.
32. Wheatley D. Hypericum in seasonal affective disorder (SAD). Curr Med Res Opin 1999; 15:33-37.
33. Wirz-Justice A, van der Velde P, Bucher A et al. Comparison of light treatment with citalopram in winter depression: A longitudinal single case study. Int Clin Psychopharmacol 1992; 7:109-116.
34. Yamadera H, Okawa M, Takahashi K. Open study of effects of alprazolam on seasonal affective disorder. Psychiatry Clin Neurosci 2001; 55:27-30.

CHAPTER 13

Use and Discontinuation of Hypnosedative Medications

Mirko Petrovic

Abstract

Benzodiazepines (BZDs) constitute the most widely used symptomatic treatment of insomnia and anxiety. Many of these drugs are associated with adverse effects, such as daytime sedation and dependence with continued use. There is a concern about the rationale for and extent of benzodiazepine (BZD) use. There is a general awareness that BZD use is inappropriate in many patients, and therefore discontinuation should be recommended whenever possible. Moreover, long-term use of these drugs should be actively discouraged. Although no unanimous recommendations concerning the optimal duration of the withdrawal process exist, BZDs may easily be withdrawn during a short period in most patients who are habituated to a low dose, if an initial phase with dose reduction and psychological support are provided. Alternative approaches involve sleep hygiene guidelines, behavioural treatment and psychotherapy tailored to the needs of the individual patient.

Introduction

For almost five decades, benzodiazepines (BZDs) have been used in clinical practice. Following their introduction into pharmacotherapy these drugs were widely welcomed because they were a valuable substitute for tranquillizers or hypnosedatives with a small therapeutical safety margin, such as bromides and barbiturates. Due to their clinical efficacy BZDs became by far the most frequently prescribed psychotropic drugs in the world.[1]

Pharmacology

The term benzodiazepine (BZD) has been related to the portion of the structure composed of a benzene ring fused to a seven-membered diazepine ring. Since almost all the BZDs contain a 5-aryl substituent and a 1,4-diazepine ring, the term actually denotes the 5-aryl-1, 4-benzodiazepines. Although variations in the structure of the ring systems yield different chemical compounds, pharmacological activity may remain similar.[2] Three nonbenzodiazepine sedative-hypnotic agents are now available: zaleplon (a pyrazolpyrimidine), zolpidem (an imidazopyridine) and zopiclone (a cyclopyrrolone). Although termed nonBZD and widely referred to as such in the literature, zolpidem, zopiclone and zaleplon interact with BZD receptors.[2,3]

Nearly all the effects of the BZDs result from actions on the central nervous system (CNS). Only coronary vasodilatation and neuromuscular blockade appear to result from actions on peripheral tissues.

Molecular targets for BZD actions in the CNS are inhibitory neurotransmitter receptors directly activated by the amino acid gamma-amino butyric acid (GABA).[2] The main type of GABA receptor in the brain, named $GABA_A$ receptor, is an integral membrane chloride channel that mediates the rapid, inhibitory neurotransmission in the CNS. According to the $GABA_A$ receptor hypothesis, BZDs bind to the receptor/ion channel complex, allosterically modulate its activity and thus facilitate GABA-ergic neurotransmission.[2,4] The local or systemic administration of BZDs appears to reduce spontaneous or evoked electrical activity of large neurons both of the brain and the spinal cord.[2]

Two central BZD receptor subtypes (BZ_1 and BZ_2) and one peripheral BZD receptor have been identified. BZ_1 (also called omega 1) receptors are preferentially located in the cerebellum. Anxiolytic and hypnosedative actions seem to be mediated mostly through the BZ_1 receptor subtype. BZ_2 (omega 2) receptors, on the other hand, are located predominantly in the spinal cord and striatum. These receptors may be involved in mediating the muscle relaxant actions of BZDs.[4] Most BZDs interact nonselectively with both receptor subtypes, which results in a variety of inhibitory central nervous system effects.[5] The peripheral BZD receptor is abundant in the kidney. Its role in anxiolytic and hypnosedative actions remains unclear.[4]

Different BZD-like effects have been classified as full agonistic (faithfully mimicking agents such as diazepam and demanding relatively low fractional occupancy of binding sites), partial agonistic effects (producing less intense maximal effects and demanding high fractional occupancy in comparison to agents such as diazepam) or inverse agonistic effects (opposite to those of diazepam in the absence of BZD-like agonists).[6,7]

There are marked differences in potency between different BZDs. The equivalent dose may vary as much as 20-fold.[8,9] We should keep this in mind when substituting one BZD by another. These differences in potency relate to differences in affinity for various receptor subtypes.

Zolpidem and zaleplon, in contrast to most BZDs and zopiclone, are believed to bind selectively to the BZ_1 receptor. They both exhibit sedative effects similar to those of the BZDs, but with a lower probability of undesirable side effects as memory loss and abuse potential.[10]

Almost all of the BZDs are completely absorbed; some of them reach the systemic circulation only in the form of active

Sleep and Sleep Disorders: A Neuropsychopharmacological Approach, edited by Malcolm Lader, Daniel P. Cardinali and S.R. Pandi-Perumal.

metabolites. Accordingly, three categories are identified based on the elimination half-lives ($t_{1/2}$): short-acting agents ($t_{1/2}$ 2-5 hours), intermediate-acting agents ($t_{1/2}$ 6-24 hours) and long-acting agents ($t_{1/2}$ more than 24 hours).[2]

The BZDs and their metabolites have a high affinity for binding to plasma proteins. The extent of binding correlates with lipid solubility and varies from 70 to 99%. Most BZDs have a large volume of distribution due to their high lipid solubility. The concentration in cerebrospinal fluid is nearly equal to the concentration of free drug in plasma.[11]

BZDs are metabolised in the liver primarily by oxidation, nitro-reduction and glucuronidation.[2,7] Most of these compounds can be classified as low clearance drugs.

Clinical Effects

All BZDs are characterized by, in slightly varying degrees, 5 major effects: hypnosedative, anxiolytic, anticonvulsant, muscular relaxant and amnesic. In the short term, BZDs may be used safely in certain clinical conditions. With long-term use, tolerance, dependence and withdrawal effects may prove to be major drawbacks.[12,13]

Hypnotic effects. BZDs accelerate sleep onset, decrease nocturnal arousals, and increase total sleep time.[14] Nevertheless, they change the normal sleep pattern: light sleep is prolonged, while the duration of slow wave sleep and rapid eye movement sleep is reduced. The onset of the first rapid eye movement sleep episode may be delayed.[9] Although some long-term users of BZDs are confident that sleep quality and duration is improved as a consequence of the treatment, the results of polysomnographic studies show that the sleep pattern remains clearly different from normal sleep.[15] The aberrant sleep profile possibly results from unselective depression of both arousal and sleep mechanisms in the brainstem.[16,17]

Anxiolytic effects are present in doses that cause minimal sedation, although the hypnotic, muscular relaxant and amnesic actions may all provide relief of associated tension and insomnia.[18] The effect on anxiety is probably related to suppressive activity in limbic and other brain areas involved in anxiogenesis. The main clinical attribute of BZDs prescribed for anxiety is the rapid onset of action, usually visible after a single dose. BZDs provide only symptomatic treatment for anxiety. Nevertheless they may be indicated in the initial management of distressing anxiety, while awaiting enduring clinical effects from more specific nondrug measures.[4]

Anticonvulsant effects. BZDs are effective in the treatment of status epilepticus and convulsions due to drug poisoning. These drugs can only be used in emergency situations and are not appropriate for the extended treatment of epilepsy, because of the development of tolerance in the majority of patients.[19,20]

Muscular relaxant effects of BZDs can sometimes be used in a variety of motor disorders (i.e., dystonias and involuntary movements, myoclonus, restless limbs syndrome) and muscle spasm associated with pain.[21]

Amnesic effects. BZDs also cause dose related anterograde amnesia. These amnesic properties may be clinically significant, particularly in the elderly and in those with coexisting medical problems.[14,22]

Side Effects

Although BZDs initially induce and prolong sleep, tolerance develops quickly. Sleep latency and duration regress to pretreatment levels after a few weeks of continued treatment. Sleep quality, however, does not improve, since deep NREM sleep and REM sleep stages are partially replaced by stage 2 light NREM sleep.[14,23] Tolerance to the anxiolytic properties of BZDs develops more gradually than to the hypnotic effects. Nevertheless, the extended use over years helps little to control and may even worsen anxiety. Therefore, BZD use in most anxiety states should be restricted to short term (not more than 4 weeks) or intermittent courses.[24,25] BZDs may reduce the efficiency of psychological therapies because of impaired learning.[26]

Rebound insomnia refers to an increase in the original symptom beyond the baseline level after withdrawal from BZDs. There have been inconsistent reports of rebound insomnia with short-acting BZDs. Population surveys and results from large treatment effectiveness studies show rebound insomnia in 14-20 % of patients treated with BZDs, a rate indistinguishable from that seen with over-the-counter drugs or placebo.[27,28] Rebound insomnia is claimed to be less frequent with slowly eliminated BZDs, as well as zolpidem and zaleplon.[29,30]

BZDs frequently give rise to subjective hangover. Even those BZDs that are quickly eliminated may cause the impairment of psychomotor performance and memory the next day. The effects depend on type of BZD, dosage and duration of use.[31,32]

Residual effects occur mostly with slowly eliminated BZDs, particularly if used in the long term and when administered to the elderly.[31] Increased volume of distribution due to a gain in body fat, in combination with reduced clearance leads to accumulation of BZDs in the elderly and a marked prolongation of $t_{1/2}$.[33]

Dependence on BZD hypnotics is a psychological or physical need to continue taking these drugs. It may be psychological, physical or both. Psychological dependence, also referred to as habituation, is characterized by an intermittent or continuous craving for BZDs.[34,35] Physical dependence on BZDs is characterized by a need to take these drugs to prevent the occurrence of a withdrawal or abstinence syndrome. Dependence on BZD hypnotics can develop if the drugs are taken systematically for some time. Some individuals are susceptible to dependence, even after a few weeks of treatment.[36] It is known that all drugs, including placebo, can produce psychological dependence, as some patients will attribute beneficial effects to the medication despite the lack of evidence for their pharmacological value. The need to continue taking a drug, as reported by the patient, is therefore not synonymous with pharmacological dependence. It is part of a complex clinical situation, composed by the patient's perception, negative conditioning, the personality structure, past and current psychopathology and also the pattern of chronic BZD use.[37] However, if the continuation of a drug is required because it prevents discomfort, this can be a sign of psychological dependence on a pharmacological basis.[34,38] Dependence to BZDs is often limited to patients with preexisting or other substance abuse. Although the risk for developing dependence to hypnosedative medication is reported to be lower in zolpidem, cases of abuse have been reported.[39]

Sporadically, the BZDs may provoke paradoxical stimulation, including excitement, irritability and even furious reactions.[40] In patients with mixed anxiety and depression the extended use of BZDs may exacerbate depression with suicidal inclination.[41]

The elderly in particular are vulnerable to adverse effects of hypnotic drugs. These patients are more sensitive to CNS depression, states of confusion and ataxia that can result in falls and fractures.[42,43] They are as well susceptible to respiratory depression, diminished ventilatory response, hypercapnia and increased hypopnoeic episodes during sleep.[43]

Epidemiology of Insomnia and Anxiety

The prevalence of insomnia increases steadily with age. A literature review mentioned a gradually rise from approximately 5% in the age group between 18 and 30 years to a prevalence of 40-60% in the subjects aged over 65 years.[44] Prevalence rates of anxiety differ between studies and age ranges under observation.[30,45] It has been estimated that the prevalence is about 3% in the general population. Among the patients seen by general practitioners a prevalence of 15-20% has been reported.[30,45] Most studies in regard to both disorders only report data for individuals over 65 years as one group. Moreover, different sampling strategies and different ways of case definition are used, which makes comparisons between studies difficult.[46]

Often, benzodiazepines are prescribed on a long-term basis; 1 to 3% of the population in the Western world have received continuous benzodiazepine therapy for more than 1 year. Frequently, inappropriately large doses of benzodiazepines are prescribed with minimal physician follow-up, especially among elderly patients.

Individuals over 65 receive 30% of all prescriptions for BZDs and nonBZD minor sedatives. One-year exposure to BZD use for the elderly averages 32% (range 9-54%).[47] Current prevalence of BZD use in the institutional setting varies between 11 and 42%.[47-49] In comparison, the prevalence of BZD use in the community setting among the elderly varies between 10 and 37%.[47,50] Long-term users of BZDs constitute a heterogeneous group.[51] Most studies consistently find that rates of use of benzodiazepines are substantially higher among older than among younger subjects and are higher among female than among male subjects.

A prototype of a long-term user is an aged widowed female with various health problems and a moderate psychiatric disorder.[34,52] Widespread BZD use is particularly common in nursing homes. This is a matter of concern especially when the reason for therapy is not known or no obvious attempts have been made to taper BZD treatment.[53]

Guidelines for Rational Use

Since the beginning of the eighties more attention has been paid to the problems arising from the widespread use of benzodiazepines. Apart from the risk of abuse and primary dependency, there is also the risk of "low-dose dependency", which is of special importance because of the high rate of long-term benzodiazepine treatment.[54]

Treatment with benzodiazepines typically is initiated during periods of acute stress, medical illness, or hospitalisation, or simply when a person can no longer cope with the daytime sequelae of chronic sleep disturbance. Despite the initial intent to limit their use to a short period, some people continue using hypnotics for prolonged periods of time. The onset of this pattern of prolonged usage is often insidious, with both psychological and physiological factors contributing to its maintenance. Some individuals continue using medications because of chronic insomnia or anxiety, but other may do so even after their sleep or mood disturbance have subsided. Sometimes, prescriptions of hypnotic drugs are renewed without adequate evaluations of continued need or sustained efficacy of the medication.[55]

Nonpharmacological Alternatives

Instead of ordinarily prescribing BZDs for these symptoms, the causes and types of insomnia and anxiety, the pharmacological effects of the drugs and the needs of the individual patient should be taken into consideration.[56,57] A nonpharmacological therapeutic approach may often be equally or even more effective, as the poor sleep may result from adjustable external factors, such as uncomfortable sleeping conditions, irregular arousal or bed-times, daytime naps, lack of physical activity, underlying medical or psychiatric conditions, use of alcohol, caffeine, tobacco or other concurrent drugs, and last but not least unrealistic expectations of sleep quality.[58] Attempts should be made to ensure good sleep behaviour before prescribing BZDs in subjects with insomnia. Patients should be interviewed about habitual food intake, and about particular habits that may impair falling or staying asleep at night. Naps taken during the day should be restricted to half an hour in the early afternoon.[57]

Behavioural therapy was recently shown to be as or even more effective as pharmacological management of insomnia, particularly in the elderly. Cognitive-behavioural therapy consists of behavioural, cognitive and educational components.[59] The behavioural component includes sleep restriction-therapy and stimulus control procedures.[60] Sleep restrictions consist of limiting time in bed to the actual sleep time. The stimulus control procedures regulate the sleep-wake schedules and help subjects to reassociate the bed/bedroom and bedtime stimuli with sleep rather than with the irritation and anxiety associated with lying in bed trying to sleep.[60] Pharmacotherapy alone, although effective in short-term, may not be satisfactory for long-term management of chronic insomnia. Sleep improvements are more constant over time with behavioural treatment.[58]

Rationale for Prescription

If cognitive-behavioural therapy alone is not sufficient in reducing insomnia, initiation of drug therapy should be considered. The decision should be founded on the severity of symptoms and on the patient's quality of life. Another important issue is whether untreated acute insomnia may develop into long-term insomnia due to increased anxiety at bedtime.[57] Hypnotics in low doses for 1 or 2 weeks only may be indicated for short-term insomnia due to temporary stress. Treatment of chronic insomnia secondary to physical, psychological or psychiatric conditions with hypnotics is not always, or perhaps rarely, indicated.[14]

Before choosing a hypnotic agent the types of sleep abnormalities experienced by the patient should be taken into consideration and matched to the properties of different available drugs. An ideal hypnotic agent should have a rapid onset of action, a duration of action that lasts through the night with no residual and no adverse effects.[61]

The primary treatment options include BZD receptor agonists. Unlike the BZDs, zolpidem at therapeutic doses at 5 to 10 mg has little effect on the stages of sleep.[62] The drug is as effective as BZDs in shortening sleep latency and prolonging total sleep time in patients with insomnia. Following discontinuation of zolpidem, the beneficial effects on sleep have been reported to persist up to one week. Zaleplon is newly available hypnotic agent. Its primary clinical advantage over other medications for insomnia is its short elimination half-life of approximately one hour.[63] Zaleplon, particularly at the 5 and 10 mg doses, demonstrates short-term efficacy in individuals with insomnia.[10] It is effective in helping subjects fall asleep, and usually does not cause residual effects the next day. In comparison with long-acting BZDs it causes less psychomotor and cognitive impairment.[10,64] Zolpidem is currently approved only for the short-term treatment of insomnia.[62] Zaleplon should also be considered only in the short-term management of transient insomnia.[10] So far, there are no published data of clinical trials examining its efficacy in the long-term treatment of insomnia. Zopiclone, at therapeutic doses at 3.75 to 7.5 mg has a slightly later peak drug concentration than zaleplon but a more rapid peak than zolpidem. However, its half-life (3.5-6 hours) is much longer than that of either zaleplon of zolpidem. Residual effects the next day after use of zopiclone have been reported.[65]

BZDs might be considered for a time-limited, stressful emotional state (i.e., acute grief reactions) or the short-term (2-4 weeks) relief of severe and disabling anxiety. As in insomnia they provide only symptomatic treatment and do not cure the underlying disorder. Other indications[47] for short-term BZD prescription include:

1. Adjunctive treatment for psychotic disorders with hyperexcitability and aggressiveness
2. Treatment of panic and agoraphobia together with antidepressants and psychological therapies
3. Alcohol withdrawal
4. Agitation associated with delirium
5. Movement disorders (i.e., periodic limb movement disorder or REM behaviour disorder)
6. Anaesthetic practice, not only because of the sedative, but also because of the amnesic effects
7. Procedures such as colonoscopy

Alternative Treatment

There are various medication classes that are suitable for substitution of BZDs, especially antipsychotic, antidepressant, phytotherapeutic, analgesic and anti-inflammatory, spasmolytic and muscle relaxant and other sedative drugs.

Anxiety states and psychosomatic instability are mainly treated with antidepressants as BZD replacement. Buspirone could also be used as a possible alternative anxiolytic. It is well tolerated and has no significant pharmacokinetic drug interactions.[66] Trazodone, a nontricyclic sedative antidepressant, is frequently used in low doses as a hypnotic. It has also been proposed as a convenient agent for the substitution of BZDs in withdrawal programmes.[67-69] Trazodone is particularly effective for patients with depression-associated insomnia. But, its therapeutic efficacy as a hypnotic in the nondepressed insomniacs remains unknown.[57]

Antipsychotic drugs are more effective than BZDs in aggressive states and agitation.[70]

Analgesics and anti-inflammatory drugs are a reasonable therapeutic alternative for BZDs in myocardial infarction, pain and muscle tension syndromes.[70]

The utilization of BZDs is often habitual and frequently not based on a prescription for any therapeutic reason. Therefore to request for "something for sleep or nerves" could be answered by prescribing phytotherapeutics (over-the-counter and/or prescription drugs like herbal or natural medications).[70] Valerian (Valeriana officinalis) is an herb that has long been advocated as a sleep promoter.[71,72] Reports of adverse events with valerian administration are scarce and cases that were reported were mild and similar to those experienced with placebo. Addiction to valerian preparations has never been reported.[73] The evidence available from randomised, placebo-controlled, double blind trials of the efficacy of valerian for improving sleep is promising but not fully conclusive.[71]

Guidelines for Withdrawal

Very few people abuse benzodiazepines in the same manner as narcotics or stimulants. Instead, most people tend to misuse their sleep medications not in terms of exceeding the recommended therapeutic dosage, but more in using them for long periods. Patients may have gradually increased the amount and frequency of medications, but it is seldom that dosage is higher than the upper recommended limit. As such, dependency on hypnotic medications is often more psychological than physical in nature.[55]

Therefore, it is essential that the practitioner develops a treatment plan when prescribing benzodiazepines. There is a growing agreement with the view that most patients currently taking benzodiazepines should discontinue them whenever possible.[74] If physical dependence exists without other adverse effects, the clinician and the patient can decide whether to detoxify from benzodiazepines and attempt alternative treatments. If benzodiazepine may be a factor in exacerbating the patient's problem, withdrawal is preferred, and observation of benzodiazepine therapy should be continued for at least 6 weeks to allow the possible effects of withdrawal or rebound to be evaluated. Before initiating discontinuation of hypnotics, it is essential to evaluate the patient's readiness and motivation concerning the undertaking of this program. Discontinuation of a hypnotic drug is more likely to be successful if motivation is intrinsic, rather than extrinsic.[55,74]

Self-monitoring and goal setting should be integral components of withdrawal program. A diary including sleep parameters and type, frequency and dosage of sleep aids completed before and during withdrawal program is a helpful instrument for monitoring progress and the compliance. Setting a weekly goal for reduction in dosage is another useful procedure.[55]

Many patients who have taken BZDs for years can have these drugs withdrawn successfully. The factors presumed to affect withdrawal are personality profile, dose and half-life of BZDs, duration of treatment and mode of withdrawal.[34,75] Men under the age of 50 without personality disturbance show particularly good response after withdrawal from BZDs. On the contrary, older women with anxious symptoms and personality disturbance respond less successfully to withdrawing.[34] In contrast, there are reports sharing the observation that with the exception of age, withdrawal outcome is not related to any particular variable.[76]

Withdrawal symptoms, if any, occur most often after 3 months of habitual use. They may include anxiety, restlessness, sleep disturbance, headache, muscle cramps, nausea, delirium or convulsions.

The treatment of BZD withdrawal includes suitable psychological support together with a gradual dosage tapering. The degree of psychological support may range from simple encouragement to anxiety management or behavioural therapy. Chronic insomnia can be effectively treated with structured and sleep-focused interventions designed to change poor sleep habits.[56] In most of the cases, sleep symptoms progressively improve after withdrawal. It has been demonstrated that the elderly tolerate withdrawal as well as, if not better than young individuals.[77]

There are no unanimous recommendations in the literature regarding the optimal duration of the withdrawal process. Hence, a variable withdrawal period ranging from 4 to 16 weeks may be allowed in outpatient settings. The size of the stepwise lowering in dosage that should be utilized is arguable as well. Certain investigators use a fixed tapering schedule (i.e., 25% at intervals of 1-2 weeks), while others claim that the reduction rate should be titrated against the patient's withdrawal symptoms.[67]

An inpatients program will provide a safer environment to reduce the medication and respond promptly to severe withdrawal symptoms. It has been shown that a short-term BZD withdrawal program is possible in the hospital setting. Most patients who are habituated to a low-dose of BZD may wean during a short period, if an initial phase with dose reduction and psychological support is included in the withdrawal program. A faster taper should therefore be encouraged, as it may fit in a short-term admission to the ward.[67,68,78]

The tapering principles are similar for younger and older adults. The only exception may be for subjects using very high dosages of hypnosedative medications. Because of greater risk of toxicity and withdrawal effects, it is important to go on slowly and under close supervision.

References

1. Shorr RI, Robin DW. Rational use of benzodiazepines in the elderly. Drugs Aging 1994; 4:9-20.
2. Hobbs WR, Rall TW, Verdoorn TA. Hypnotics and sedatives; ethanol. In: Hardman JG, Gilman AG, Limbird LE, eds. The pharmacological basis of therapeutics. New York: Goodman & Gilman's Mc Graw-Hill Health Profession Division, 1996:361-397.
3. Heydorn WE. Zaleplon a review of a novel sedative hypnotic used in the treatment of insomnia. Exp Opin Invest Drugs 2000; 9:841-858.
4. Stahl SM. Anxiolytic and sedative hypnotics. In: Stahl SM, ed. Essential Psychopharmacology. Cambridge: Cambridge University Press. 2000:297-333.
5. Kirkwood CK. Management of insomnia. J Am Pharm Assoc 1999; 39:688-696.
6. Greenblatt DJ. Pharmacology of benzodiazepine hypnotics. J Clin Psychiatry 1992; 53(Suppl 6):7-13.
7. Breiner DD. Clinical pharmacokinetics of hypnotics. Clin Pharmacokinet 1997; 2:93-109.
8. Roth T, Roehrs TA. Issues in the use of benzodiazepine therapy. J Clin Psychiatry 1992; 53 (Suppl 6):14-18.
9. Ashton H. Guidelines for rational use of benzodiazepines - when and what to use. Drugs 1994; 48:25-40.
10. Weitzel KW, Wickman JM, Augustin SG et al. Zaleplon: A pyrazolopyrimidine sedative-hypnotic agent for the treatment of insomnia. Clin Ther 2000; 22:1254-1267.
11. Hämmerlein A, Derendorf H, Lowenthal DT. Pharmacokinetic and Pharmacodynamic changes in the elderly. Clin Pharmacokinet 1998; 35:49-64.
12. Swift CG, Swift MR, Hamley J et al. Side-effect 'tolerance' in elderly long-term recipients of benzodiazepine hypnotics. Age and Ageing 1987; 13:335-343.
13. Petursson H, Lader MH. Benzodiazepine dependence. Br J Addiction 1984; 76:133-145.
14. Buysse DJ, Reynolds CF. Pharmacologic treatment. In: Lichstein KL, Morin CM, eds. Treatment of Late-Life Insomnia. Thousand Oaks: Sage Publications. 2000:231-267.
15. Schneider-Helmert D. Why low-dose benzodiazepine-dependent insomniacs can't escape their sleeping pills. Acta Psychiatr Scand 1988; 78:706-711.
16. Grad RM. Benzodiazepines for insomnia in community dwelling elderly: A review of benefit and risk. J Fam Practice 1995; 41:473-481.
17. Wheatley D. Effects of drugs on sleep. In: Wheatley D, ed. Psychopharmacology of Sleep. New York: Raven Press, 1981:153-176.
18. Tyrer P. Choices of Treatment in Anxiety. In: Tyrer P, ed. Psychopharmacology of anxiety. Oxford: Oxford Medical Publications, 1989:255-282.
19. Olivier H, Fitz-Gerald MJ, Babiak B. Benzodiazepines revisited. J La State Med Soc 1988; 150:483-485.
20. Brodie MJ. Status epilepticus in adults. BMJ 1990; 336:551-552.
21. Herrington RN. The use of benzodiazepines in neuropsychiatry. In: Hindmarch I, Bauamont G, Brandon S, Leonard BE, eds. Benzodiazepines: Current concepts. Chichester: John Wiley & Sons Ltd., 1990:95-110.
22. Buysse DJ. Drugs affecting sleep, sleepiness and performance. In: Monk TH ed. Sleep, sleepiness and performance. London: John Wiley & Sons, 1991:249-306.
23. Lucki I, Rickels K, Geller AM. Chronic use of benzodiazepines and psychomotor and cognitive test performance. Psychopharmacol 1986; 88:426-433.
24. Wheatley D. The new alternatives. In: Wheatley D ed. The anxiolytic jungle: Where next? Chichester: John Wiley & Sons, 1990:163-184.
25. Consensus Conference. Guidelines for the management of patients with generalized anxiety. Psychiatric Bull 1992; 16:560-565.
26. Gray JA. The neuropsychology of motion and personality. In: Stahl SM, Iversen SD, Goodman ED, eds. Cognitive neurochemistry. Oxford: Oxford University Press, 1987:171-190.
27. Balter MB, Uhlenhuth EH. The beneficial and adverse effects of hypnotics. J Clin Psychiatry 1991; 52:16-23.
28. Hajak G, Clarenbach P, Fischer W et al. Rebound insomnia after hypnotic withdrawal in insomniac outpatients. Eur Arch Psychiatry Clin Neurosci 1998; 248:148-156.
29. Tyrer P. Withdrawal from hypnotic drugs. Brit Med J 1993; 306:706-708.
30. Ancoli-Israel S, Walsh JK, Mangano RM et al. Zaleplon, a novel nonbenzodiazepine hypnotic, effectively treats insomnia in elderly patients without causing rebound effects. Primary care Companion to the J of Clin Psychiatry 1999; 1:114-120.
31. Hindmarch I. Human psychopharmacological differences between benzodiazepines. In: Hindmarch I, Beaumont G, Brandon S, Leonard BE, eds. Benzodiazepines: Current Concepts. Chichester: John Wiley & Sons, 1990:73-92.
32. Dement WC, Mitler MM. It's time to wake up to the importance of sleep disorders. JAMA 1993; 269:1548-1550.
33. Greenblatt DJ, Harmatz JS, Shader RI. Clinical pharmacokinetics of anxiolytics and hypnotics in the elderly: Therapeutic considerations. Clin Pharmacokinet 1991; 21:165-177.
34. Holton A, Riley P, Tyrer P. Factors predicting long-term outcome after chronic benzodiazepine therapy. J Affect Disord 1992; 24:245-252.
35. Ashton H. Benzodiazepine withdrawal: An unfinished story. BMJ 1984; 288:1135-40.
36. Livingston MG. Benzodiazepine dependence. Br J Hosp Med 1994; 51:281-286.
37. Marriott S, Tyrer P. Benzodiazepine dependence: Avoidance and withdrawal. Drug Saf 1993; 9:93-103.
38. Owen RT, Tyrer P. Benzodiazepine dependence: a review of the evidence. Drugs 1983; 25:385-398.
39. Madrak LN, Rosenberg M. Zolpidem Abuse Am J Psychiatry 2001; 158:1330-1331.
40. Dietch JT, Jennings RK. Aggressive dyscontrol in patients treated with benzodiazepines. J Clin Psychiatry 1988; 49:184-188.
41. Lader MH, Petursson H. Benzodiazepine derivates - side effects and dangers. Biol Psychiatry 1981; 16:1195-1212.
42. Herings RMC, Stricker BHC, de Boer A et al. Benzodiazepines and the risk of falling leading to femur fractures. Arch Int Med 1995; 155:1801-1807.
43. Morgan K. Hypnotics in the elderly: What cause for concern? Drugs 1990; 40: 688-696.
44. Gillin JC, Byerley WF. The diagnosis and management of insomnia. N Engl J Med 1990; 322: 239-48.
45. Halstrom C. Coping with anxiety: The patient's predicament. In: Wheatley D, ed. The anxiolytic jungle: Where next? Chichester: John Wiley & Sons, 1990: 99-111.
46. Schaub RT, Linden M. Anxiety and anxiety disorders in the old and very old - results from the Berlin aging study (BASE). Comprehensive Psychiatry 2000; 41: 48-54.
47. Llorente MD, David D, Golden AG et al. Defining patterns of benzodiazepine use in older adults. J Geriatr Psychiatry Neurol 2000; 13: 150-160.
48. Svarstad BL, Mount JK. Chronic benzodiazepine use in nursing homes: Effects of federal guidelines, resident mix, and nurse staffing. J Am Geriatr Soc 2001; 49:1673-1678.
49. Ramesh M, Roberts G. Use of night-time benzodiazepines in an elderly inpatient population. J Clin Pharm Ther 2002; 27:93-97.

50. Fourrier A, Letenneur L, Dartigues JF, Moore N, Bégaud B. Benzodiazepine use in an elderly community-dwelling population. Eur J Clin Pharmacol 2001; 57:419-425.
51. Barter G, Cormack M. The long-term use of benzodiazepines: Patients' views, accounts and experiences. Fam Practice 1996; 13:491-497.
52. Petrovic M, Van Dierendonck A, Mariman A. et al. Personality traits and socio-epidemiological status of hospitalised elderly benzodiazepine users. Int J Geriat Psychiatry 2002; 17:733-738.
53. Monane M. Insomnia in the elderly. J Clin Psychiatry 1992; 53(Suppl 6):23-28.
54. Juergens SM. Problems with benzodiazepines in elderly patients. Mayo Clin Proc 1993; 68:818-820.
55. Morin CM, Baillargeon L, Bastien C. Discontinuation of sleep medications. In: Lichstein KL, Morin CM, eds. Treatment of late-life insomnia. Thousand Oaks-London-New Delhi: Sage Publications, 2000:271-296.
56. Ancoli-Israel S. Sleep problems in older adults: Putting myths to bed. Geriatrics 1997; 52:20-30.
57. Ancoli-Israel S. Insomnia in the elderly: A Review For The Primary Care Practitioner Sleep 2000; 23(Suppl 1):S23-S29.
58. Morin CM, Colecchi C, Stone J et al. Behavioural and pharmacological therapies for late-life insomnia: A randomised controlled trial. JAMA 1999; 281:991-999.
59. Engle-Friedman M, Bootzin RR, Hazlewood L et al. An evaluation of behavioural treatments for insomnia in the older adult. J Clin Psychol 1992; 48:77-90.
60. Morin CM, Kowatch RA, Barry T et al. Cognitive behaviour therapy for late-life insomnia. J Consult Clin Psychol 1993; 61:137-146.
61. Mendels J. Criteria for selection of appropriate benzodiazepine hypnotic therapy. J Clin Psychiatry 1991; 52:42-46.
62. Hanly P, Powles P. Hypnotics should never be used in patients with sleep apnoea. J Psychosom Res 1993; 37:59-65.
63. Beer B, Clody DE, Mangano R et al. A review of the preclinical development of zaleplon, a novel nonbenzodiazepine hypnotic for the treatment of insomnia. CNS Drug Res 1997; 3:207-224.
64. Walsh JK, Fry J, Richardson GS et al. Short-term efficacy of zaleplon in older patients with chronic insomnia. Clin Drug Invest 2000; 20:143-149.
65. Noble S, Langtry HD, Lamb HM. Zopiclone: An update on its pharmacology, clinical efficacy and tolerability in the treatment of insomnia. Drugs 1998; 55:277-392.
66. Sramek JJ, Zarotsky V, Cutler NR. Generalised anxiety disorder: Treatment options Drugs 2002; 62:1635-1648.
67. Petrovic M, Pevernagie D, Van Den Noortgate N et al. A programme for short-term withdrawal from benzodiazepines in geriatric hospital inpatients: Success rate and effect on sleep quality Int J Ger Psychiatry 1999; 14:754-760.
68. Petrovic M, Pevernagie D, Mariman A et al. Fast withdrawal from benzodiazepines in geriatric inpatients: A randomised double-blind, placebo-controlled study Eur J Clin Pharmacol 2002; 57:759-764.
69. Ansseau M, De Roeck J. Trazodone in benzodiazepine dependence. J Clin Psychiatry 1993; 54:189-191.
70. Linden M, Gothe H. Benzodiazepine substitution in medical practice. Analysis of pharmacoepidemiologic data based on expert interviews. Pharmacopsychiat 1993; 26:107-113.
71. Stevinson C, Ernst E. Valerian for insomnia: A systematic review of randomised clinical trials. Sleep Medicine 2000; 1:91-99.
72. Wagner J, Wagner ML, Hening WA. Beyond benzodiazepines: Alternative pharmacologic agents for the treatment of insomnia. Ann Pharmacother 1998; 32:680-691.
73. Bos R, Woerdenbag HJ, De Smet PAGM et al.Valeriana species. In: De Smet PAGM, Keller K, Hänsel R, Chandler RF, eds. Adverse Effects of Herbal Drugs. Berlin: Springer-Verlag, 1997:3:165-180.
74. Salzman C, Fisher J, Nobel K et al. Cognitive improvement following benzodiazepine discontinuation in elderly nursing home residents. Int J Ger Psychiatry 1992; 7:89-93.
75. Roy-Byrne PP, Hommer D. Benzodiazepine withdrawal: Oerview and implications for the treatment of anxiety Am J Med 1988; 84:1041-1052.
76. Ashton H. Benzodiazepine withdrawal: Outcome in 50 patients. Br J Addiction 1987; 82:665-671.
77. Schweizer E, Case WG, Rickels K. Benzodiazepine dependence and withdrawal in elderly patients. Am J Psychiatry 1989; 146:529-531.
78. Petrovic M, Mariman A, Warie H et al. Is there a rationale for prescription of benzodiazepines in the elderly? Review of the literature. Acta Clin Belg 2003; 58:27-36.

Section III: Clinical Pharmacology

CHAPTER 14

Long-Term Use of Sleeping Pills in Chronic Insomnia

Milton Kramer

Elderly patients, among others, will often describe the benefits they derive from taking sleeping pills.[1] Their physicians feel obliged to reduce or take them off such medication. Most medical experts will support such efforts, especially in the elderly fearing the increased risk of falls and the development of tolerance, dependence and cognitive impairment.[2] Some consultants feel "withholding effective medication because of generalizations about their potential hazards may deprive a patient of rapid and safe symptomatic relief. Under-treatment may be as harmful as over-treatment".[3] Physicians who withdraw or decrease sleeping pills may feel they have protected a patients physiological status while decreasing their psychological well being.[4]

Expert Advice Is Challenged by Clinical Practice

Tsoi[5] in a review and update on hypnotics for the general physician concluded that benzodiazepines suffer from "unwanted effects which included tolerance, dependence, withdrawal symptoms, rebound insomnia, hangover effects, alterations of memory processes, and synergy with ethanol". Kupfer and Reynolds[6] recommend that if hypnotic medication is used, especially in treating chronic insomnia, that its use be limited to 3-4 weeks, as there are no efficacy or safety data for any longer time periods. Kripke et al[7] warn that the use of hypnotics is associated with an excess death rate. And Ashton[8] echoes these voices of concern in proposing that perhaps pharmacological treatment for insomnia is rarely indicated.

There is however, a significant number of patients with insomnia who have had the problem for a year or longer,[9,10] who use medications frequently, if not every night, for protracted periods.[11-16] They report they have been helped, at least to some degree, by the medication.[9,10,12-18] This population of chronic insomniacs is primarily older, over 65, and made up mainly of women.

Insomniacs Are a Source of Concern

Insomnia warrants the interest and concern of the medical community as moderate to severe insomnia is quite common and occurs in 13-19% of people.[19] The most frequent complaint of insomniacs is awakening in the morning unrefreshed.[9,20] The insomniac then has increased difficulty the next day with concentration, drowsiness, memory, irritability, and accidents.[21] They have poorer social relations and lower moods and their quality of life is compromised.[21,22] Insomnia is more frequent in women, older people, the unmarried, the less occupationally successful, the less well educated, the unemployed, the lower socioeconomic class and is both a covariant and a risk factor in various medical and psychiatric illnesses e.g., cardiovascular disease and depression.[19] The effect of insomnia treatment on these problems remains to be demonstrated.

The Medication Treatment of Insomnia

There are a number of cases of insomnia which are not due to another sleep disorder, a mental or physical illness or the effect of a medication.[6] These primary insomnias respond equally well, short-term, to behavioral[23] or medicinal[24] treatment, but there is the clear recognition, even by advocates of behavioral therapy, that behavioral approaches are unlikely to become the main treatment modality for insomnia.[25] An understanding of the limitations and usefulness of hypnotic medications in the long term treatment of chronic insomnia is imperative, particularly as we have no laboratory based, randomized, double blind, parallel groups, placebo controlled trials of the efficacy of these medicines beyond 35 days.[24] Two additional studies of sleeping pills have been reported in abstract form, one of an open label extension for 6-12 months in the elderly of a double blind trial,[26] and the other a 6 month randomized double blind, placebo controlled study of a new hypnotic.[27] Problems with selection criteria, bias, drop out rates (40%), and lack of follow up data, understandable in an abstract, make interpretation difficult.

Potential Problems in Using Medication to Treat Insomnia

The concerns about the long term treatment of chronic insomnia with hypnotics, almost all of which currently are benzodiazepines or benzodiazepine agonists, include: (1) residual daytime effects e.g., sedation, (2) memory impairment, (3) falling, (4) respiratory depression, (5) rebound insomnia, (6) medication abuse, (7) tolerance development, (8) dose escalation, (9)dependency and withdrawal difficulties, and (10) an increased risk of death. Awareness of these concerns is necessary to appreciate the risk patients are potentially exposed to if these medications are used long term.

Residual Daytime Effects

Hangover and other residual effects e.g., psychomotor impairments, memory problems, and difficulties maintaining balance, have all been demonstrated following the use of sleeping pills.[28] Interestingly, with time evidence of improvement, adaptation, occurs.[29]

Sleep and Sleep Disorders: A Neuropsychopharmacological Approach, edited by Malcolm Lader, Daniel P. Cardinali and S.R. Pandi-Perumal.

A confounding issue in assessing hangover is the high frequency of hangover type complaints in untreated insomniacs,[18] e.g., 74% report feeling tired in the morning and 53% report problems with their memory.[21] Improving the patient's sleep with an hypnotic leads to their reporting feeling clear headed and less hung over.[30] Langer et al[29] concluded "... currently hypnotics have relatively rare adverse effects, which are generally mild. There is little daytime residual sedation with the shorter-acting benzodiazepines and the newer nonbenzodiazepine agents".

Memory Impairment

All benzodiazepines can affect memory, particularly causing anterograde amnesia and interfering with memory consolidation.[31] The problem is considered particularly an issue for the elderly who may already have compromised memory function.[32,33] In a large group of elderly chronic insomniacs who have used benzodiazepines for years, no cognitive impairment was demonstrated.[12] Question has been raised about how significant the medication induced memory problems are when memory problems are so common in untreated insomniacs.[17,21]

Falling

A relationship in the elderly between falls, hip fractures and benzodiazepine use has been described and is explained by the negative effect these drugs have on balance.[33] However, not all reports confirm that chronic benzodiazepine use in the elderly impairs balance.[12] A population study of older individuals found a relationship between hip fractures and depression but not with insomnia.[34] Falls are relatively rare events in the hospital.[29,35] And, only 1/3 of all falls could be related to benzodiazepine use.[35] Reducing benzodiazepine use with the goal of demonstrating an effect on the rate of hip fractures would be a horrendous undertaking.[33]

Respiratory Depression

Using benzodiazepine hypnotics might for some patients create respiratory difficulties.[36] If the patient does not have primary pulmonary disease i.e., chronic obstructive pulmonary disease (COPD) or obstructive sleep apnea (OSA), the use of hypnotics is unlikely to cause difficulty.[29] Hypnotics can be used in mild to moderate COPD. The problem is in making the diagnosis of OSA. In a very careful screening procedure to select older subjects for an insomnia study, 20% of the subjects had OSA and were not identified until they had a sleep study.[37] Continuing consideration must be given to this problem in the long -term hypnotic therapy of insomnia in the elderly because of the high incidence of OSA in the older person.

Rebound Insomnia

Rebound insomnia has been described as the primary discontinuation symptom from hypnotic use[38] and is more likely to be associated with short and intermediate acting benzodiazepine compounds,[6] but not with the newer benzodiazepine agonists.[36] It is most marked when the drugs have been taken regularly for long periods.[6] There is individual variation in the susceptibility to experiencing rebound based on the severity of the preexisting insomnia[29]and the personality of the insomniac.[38] Gradual tapering of the medication is an effective technique to minimize any rebound.[31]

It is noteworthy that 2/3 of the patients on benzodiazepine hypnotics can stop using them without experiencing rebound insomnia.[38] And prevalence rates for rebound insomnia were found to be low and not related to the half-life of the medication.[18]

Medication Abuse

The long-term users of benzodiazepines are older adults and not the stereotypical "drug abusers".[33] The long term users are older adults "with multiple physical health problems and in moderate psychiatric distress".[33] Addiction to benzodiazepines need not be a major source of concern, as abuse with hypnotic agents is generally not seen, as recreational use is limited to poly-drug abusers.[18,28,36]

Tolerance

Sleep laboratory studies of the nightly use of benzodiazepines described the rapid development, within a few weeks, of tolerance to their hypnotic effects.[8] This conclusion, that tolerance developed to the benzodiazepine hypnotics, has been repeated in review articles for clinicians.[5,28] In contrast, surveys of long term users of hypnotics[12-16] reported no evidence of tolerance development. The dissociation of changes in the sleep electroencephalogram of insomniacs who used benzodiazepines for a short time from the clinical improvement in chronic insomniacs who used the medication long term raises the question that perhaps the sleep disturbance in insomnia is more complex than what we are able to capture with the clinical polysomnogram.

Dose Escalation

The frequent adaptation to the sedative side effects,[29] the low abuse potential,[18] the relative absence of tolerance development and the satisfaction expressed by hypnotic drug users,[18] all suggest that dose escalation is unlikely to be much of a problem in those who take hypnotic drugs and that is indeed the case. Patients dose escalation is no more frequent with hypnotics than with other psychotropic medications.[17] Poor sleepers report continued efficacy without dose escalation.[8] Benzodiazepine users in Great Britain[13] and Wales[12] showed no tendency to increase their medication.

Dependency and Withdrawal Difficulties

There is a widespread concern about the development of dependence, both physiological and psychological,[39] with the use of psychotropic drugs, particularly the benzodiazepines.[33] Of those who use psychotropic medication, 27-45% are dependent on them.[37] A population survey in Great Britain[13] estimated that 1.1% of the population may be dependent on benzodiazepines. It is thought that 1/3 of long term benzodiazepine users have difficulty in stopping the medication[38]and, 71% of an elderly group of long term regular users of benzodiazepine hypnotics said they would be unable to sleep without their medication.[12] Among general practice patients, 80% of them who tried to stop using their benzodiazepines got worse.[14] Some 13.5% of older subjects who volunteered for a study for the treatment of their insomnia were unwilling or unable to stop their hypnotic medication.[37] Only 1/3 to 1/2 of patients who had stopped their benzodiazepine were able to stay drug free.[33] Nevertheless, many patients who have taken benzodiazepines nightly for years can be successfully withdrawn; if it is done slowly enough,[8] as abrupt withdrawal runs the risk, among other things, of seizures.[28,36] The unmotivated elderly, it is suggested, should not be pushed to withdraw from their benzodiazepine hypnotic.[8]

Responses to the cessation of the use of benzodiazepine hypnotics can be categorized as relapse, rebound, or withdrawal.[38] Relapse would be the reappearance of the insomnia and the associated symptom complex ie. daytime fatigue, irritability, memory and performance problems.[21] Chronic insomnia is a fluctuating condition and relapse may occur immediately on medication cessation or after some stressful life event.[38] Distinguishing relapse from rebound or withdrawal is a problem. Long term benzodiazepine treatment, it is recognized, has often been used appropriately to control long term symptoms.[13]

Rebound is a worsening of the patient's symptoms to levels beyond what was originally experienced.[38] It has been discussed earlier in this report. The comparison depends on the adequacy of the characterization of the patient's baseline and post-medication condition as sleep parameters are highly variable night to night.[40] Some reports have failed to find rebound[30,41-43] and others have reported it without an appropriate analysis.[44] An alteration, which has little or no clinical consequence, is of limited concern.[18]

Withdrawal syndromes are potentially of the greatest concern.[38] The benzodiazepine withdrawal syndrome is "the emergence of a well defined syndrome with predictable onset, direction and offset of action, containing psychological and bodily symptoms not previously complained of by the patient".[38] It may be difficult to distinguish the withdrawal symptoms from underlying insomnia and anxiety.[33] The specificity of the so-called "Triazolam Syndrome " has been called into question.[45] These withdrawal syndromes are most likely to occur in dependent personalities[38] and the severity of the withdrawal may be related to personality variables.[33] Serious problems with withdrawal are relatively rare.[28]

Increased Risk of Death

Kripke et al[7] have shown a correlative relationship between prescription sleeping pill use and an increased risk, 22-35%, of death within 6 years. They are careful to point out that the association they found is not proof of causality. They acknowledge the lack of data about concomitant psychiatric illness and sleep disorders that could have lead to overestimation. There was uncertainty, they note, about what sleeping pills were being used as the data was collected during a period, 1982-1988, when many types of nonbenzodiazepine sleeping pills were still being used. They suggest as possible psychopathologic risks, which might help explain the increase in mortality, exacerbation of sleep apnea, sedation, suppression of self-care functions, confusion and amnesia and disinhibition. As the hazard ratio for prescription sleeping pills was significantly higher than for Valium and Librium, it suggested the possibility of a specific risk factor for some subclass of hypnotic compounds. Kripke et al[7] call for long term controlled trials to determine if any of the currently available hypnotic drugs have favorable risk/benefit ratios. It seems unlikely that the increase in hazard ratio is related to benzodizepine hypnotics, as the hazard ratio for Valium and Librium is lower than for all other hypnotics. Kripke[67] in 2000 again calls attention to his many concerns about the use of sleeping pills and the distortion that the funding of hypnotic drug studies from what he calls "the hypnotic industrial-sales complex" creates.

The Lack of Long Term Controlled Trials of Sleeping Pills

There is the recognized need for long term randomized, double blind, parallel group, controlled trials of hypnotics.[6,28,29,45] The necessity for a placebo control is of particular importance as 10% to 50% of the effect of hypnotic drugs may be a placebo response.[28,46] In a controversial report, 75% of the effect of anti-depressant medication was attributed to a placebo response.[47,48] Questions have been raised, not about the value of long term controlled trials, but of the practicality of such trials.[48] Most of our knowledge about the long-term risks of hypnotic medications comes from voluntary reporting. Unfortunately, voluntary reporting of emergent risks is unreliable and it has been suggested that the development of large clinical data bases may offer the best opportunity to identify emerging risks with long term use during post-marketing survailance.[48] As noted above, we have two attempts reported in abstract form on longer term, 6 months, treatment of insomnia. One in the elderly which is open label[26] and the other[27] with a 40% drop out rate, both of which report clinical improvement and limited side effect problems, but both without post treatment follow up.

We are left with examining clinical data-bases i.e., clinical case series, to provide some clue as to the risks as well as the possible benefits of the long term use of hypnotic medication. It even has been suggested that, " Optimal trial conditions (efficacy) misrepresents the real world (effectiveness) where variations in clinical skills, the intensity and duration of interventions, patient adherence, and local resources influence outcomes".[42,49] Sadly, case series data is open and uncontrolled, but it may tell us "what to do until the double blind arrives" and what to watch for after it does, if it does.

What Do We Know from Clinical Practice?

The population survey data points to there being a significant number of patients with chronic insomnia who have taken medication regularly for years, are well adapted, unwilling or unable to stop taking the medication, and who are of the opinion that the medication helps.[9,10,12-16,18] In a "natural history" study of a small group of carefully characterized primary insomniacs, who had only been given sleep hygiene advice when seen in consultation, 79% at 5 year follow-up[50] were using pharmacologically active agents to help them sleep. In a group of older volunteer insomniacs who had been treated with either behavior therapy, medication, or both; it was found that, at the 2 year follow up, 30%, 58%, and 50% respectively of each group were using sleeping aids.[37] In a clinical report seven patients were described who suffered from an intractable and disabling insomnia whom after many trials were found to respond to only one medication which, in most cases, was of an addicting nature.[50]

There are a limited number of open studies treating chronic insomniacs with benzodiazepines or benzodiazepine agonists for a year or more and they all report positive results. Pakes, et al[51] report positively on a study of 21 chronic psychiatric in-patients who took 0.5 mgm to 1.0 mgm of Triazolam for 1 year. Clark et al[52] reviewed a work in which 37 adult chronic psychiatric patients were given Loprazolam nightly for 12 months and whose sleep was improved. Maarek et al[53] described the effective treatment for 1 year of 49 out-patients treated by their general practitioners with 10-20mgm of Zolpidem. Schenck et al[54] treated 25 chronic insomniacs with Clonazepam or Alprazolam for over a year and obtained substantial improvement. None of these trials reported significant, disturbing side effects.

All of these studies[51-54] are of open uncontrolled treatment and most contain explicit cautions about generalizing from these reports to the long-term treatment of chronic insomnia with hypnotic medication. Nevertheless, coupled with the observations in the survey studies[9,10,12-18] and more recent long-term studies,[26,27] they encourage a further look at the clinical practice of treating chronic insomnia with hypnotic medication.

Clinical Case Series: Treating Insomnia Long Term

To provide a clinical case series of medication treated chronic insomniacs, Kramer et al[55] searched the records of the 5,460 patients they had treated between 1984 and 1992. They selected those patients (1) who had a final diagnosis of chronic insomnia, (2) who had had a clinical polysomnogram that confirmed the diagnosis and ruled out any concomitant sleep disorder, and (3) who were only treated with medication . One hundred fifty cases were found that met the criteria.

The group of patients was 56% women, 38% over age 50, 89% Caucasian, and 47% of a lower socio-economic class. The chief complaint in 52% was of insomnia e.g., can't sleep or trouble sleeping, while in 37.5% the complaint was of sleepiness or snoring. Seventy five percent of the patients had a history of insomnia for over a year. The descriptive polysomnographic diagnosis was of an interruption insomnia in 65.7%, onset insomnia in 0.7% and both in 32.6%.

At the time of last contact, 61.3% of the patients were rated as improved, similar to what others have reported.[56] The length of treatment varied from 1 to 90 months with 79% being treated for less 1 year. At the time of the review 16.7% of the patients were still in treatment.

The last medication used by class was a sedative anti-depressant, e.g., Doxepin, Trazadone, in 47.2%, 18.3% were on a benzodiazepine, e.g., Clonazepam, Triazolam, 21.2% were on other medications, e.g., Tryptophan, Thioridazine, and 13.3% were on a stimulating medications, e.g., Protriptyline, Methylphenidate. The percent improved at last contact was 52.2% for sedative anti-depressants, 73.1% for those on benzodiazepines, 43.4% for these on other medications, and 68.4% for those on stimulating medications.

Only one medication had been tried in 56% of the patients, two were tried in 20.7%, three in 8%, four in 7.3%, and 7.9% had been tried on five or more. It had been suggested that multiple medication trials might be necessary to find a medication effective for a particular patient.[51]

Along with many others,[57] Kramer et al[55] had reduced their use of benzodiazepines and increased their use of sedative anti-depressants. In effect they substituted one set of unknown risks for another,[29] while decreasing their improvement percentage from 73.1% to 52.2%. They had potentially decreased the numerator while increasing the denominator and thereby negatively effecting the risk/ benefit ratio for the patients they were treating.

With an interest in the long-term medicinal treatment of chronic insomnia, Kramer et al[58] attempted to contact the 125 patients who were no longer in treatment with them. Two attempts resulted in 30 responses (24%). Thirteen patients indicated they were improved and still on hypnotic medication; 15 were improved but no longer taking any medication; and 2 were worse than when last seen but still on medication. One patient had died. The over all improvement rate was 36% for the 149 patients. The very low response rate may be attributed to the high percentage of patients from the lowest socio-economic class, who have a high incidence of insomnia[59] but a poor rate of treatment compliance.

In a clinical behavioral treatment program that treated 105 chronic insomniacs, it was found, at a 5-year mean follow-up time, that 39% of the patients had improved. But 78% of these improved patients had needed concomitant hypnotic medication.[60] In a clinical behavioral program that treated 72 psychiatric out-patients with chronic insomnia, 72% were rated as improved 6 months after the last behavioral treatment, but 74% of them had been on concomitant hypnotic medication.[61]

There Is a Reluctance to Examine the Use of Drugs in the Long Term Treatment of Insomnia

There is the strong impression from the present review that there is a role for hypnotic medications in the long-term treatment of chronic insomnia. An attitude of reluctance exists to exploring the use of medications, currently the benzodiazepines primarily, in the treatment of sleep problems and anxiety disorders. In a survey of internists and psychiatrists[62] "a persistent negative bias against benzodiazepines" was found and this"... misinformation results in under treatment". Trainees were found to be more conservative in prescribing benzodiazepines than experienced practitioners,[63] and primary care physicians who said they never or only sometimes prescribe benzodiazepines for insomnia actually prescribed them in over 90% of such cases.[64] "...We do not always do what we mean to do, and... we do not always do what we think we do".[64]

The negative perception about the use of benzodiazepine hypnotics affects clinical practice.[2] Regenstein and Reich[56] provide a list of cautionary expert opinion on the use of sedative hypnotics that illustrates the negative view. The experts advise (1) that hypnotic drugs grant no long term relief, (2) that the risks out weigh the benefits, (3) that they work largely through placebo effects, (4) that they should never be prescribed for more than a few days nor for more than "occasional" use, and (5) that there is no indication for chronic use and such is always contraindicated.

Assumptions about insomnia and the problems associated with the drug treatment of insomnia influence our attitudes about and efforts at assessment. Dunner[65] asks "...whether sleep disorders are true medical conditions". In a national survey,[20] respondents were asked whether they agreed or disagreed with the statements," Sleeping pills should not be taken unless you have had difficulty sleeping for more than a month" and " Sleeping pills are addictive". Sixty-six percent agreed with the first statement and 83% with the second. Are these statements worded to seek information or to find confirmation with the preexisting beliefs of experts? In commenting on the subjects screened for a study who elected not to participate, the commentators described them as unable to stop the use of hypnotics,[68] however in the methods section of the article they were described as uninterested or unable to avoid taking their sleeping pills.[37]

The warnings against the long-term use of benzodiazepines pervade aspects of our data interpretation. The authors conclude, in summarizing the results of the 1991 National Sleep Foundation survey, that chronic insomniacs "...were likely to try different treatments, but rarely found any of them effective".[9] However, 60% of the chronic insomniacs in the survey report that prescription medication, other than sleeping pills, were effective. It was concluded in another study,[14] that as only 22% of severe insomniacs reported significant improvement that "...the long-term administration of benzodiazepine hypnotics seems to be an inadequate treatment strategy in chronic insomnia". Yet 34% of the patients reported being somewhat better, so that 56% were improved compared to 44% who were unchanged or worse.

Schenck et al[54] and Swift et al[12] both found positive effects of long term benzodiazepine hypnotics in chronic insomnia but warn "...data from this study of the treatment of chronic, severe insomnia, a subset of all insomnia, are not generalizeable to the typical insomnia patient"[54] and " inspite of our observations the desirability of protracted benzodiazepine use in the elderly must be seriously questioned...".[12]

The description of insomniacs as overestimating their baseline degree of sleep disturbance and the efficacy of hypnotic drugs[8] raises a concern about taking the observations of insomniacs seriously. This fits the "negative perspective toward the long term use of sleeping medications...(as) these medications are often subject to abuse by patients...and they might be used for recreational rather than medical use".[65]

Morality and the Use of Drugs for Insomnia

The use of benzodiazepines for anxiety and insomnia in the United States has declined across the past 20 years.[33,55] An increasingly stoic attitude towards the treatment of these disorders has developed.[33] Hohagen et al[14] have pointed out that the greater use of hypnotics in Germany compared to the United States might well reflect a more conservative attitude in American society towards the medicinal treatment of insomnia. This conservative stance in the United States has been described by Klerman as "Pharmacological Calvinism".[41] It is an attitude towards drugs which perceives taking them as morally wrong or something for which one has to pay for with dependence. "If a drug makes you feel good, it must be bad".

Conclusion

Chronic primary insomnia is a recurrent condition that negatively effects the daily functioning of patients diminishing the quality of their lives. It is associated with and in some situations is a risk factor in both psychiatric (depression) and physical illness (cardio-vascular). Treatment effectiveness in insomnia has been shown short term for both drug (benzodiazepine and benzodiazepine agonists) and behavioral treatment. Expert opinion has strongly advised against long term drug treatment because of concerns about residual sedative effects, memory impairment, falls, respiratory depression, rebound insomnia, medication abuse, dose escalation, dependency and withdrawal difficulties, and an increased risk of death possibly associated with the current hypnotic medications. Many of these concerns could be made against using these agents at all. Worries about these potential problems are challenged by the widespread clinical practice of using hypnotic drugs long term without any of these difficulties developing and with patients who feel their sleep and daily function is improved with the nightly use of their sleeping pill. The ability to mount a randomized, placebo controlled, parallel group, double blind trial of hypnotic medication in primary insomnia may not be possible. We may have to develop large systematic clinical databases, a number of case series in effect, to monitor both emergent symptoms and possible clinical effectiveness. There is the additional concern that there is a reluctance to examine the long - term drug treatment of insomnia. This reluctance may reflect a negative moral judgement about treating primary insomnia with drugs, a sort of "PharmacologicalCalvinism", rather than just a data based judiciousness.

Acknowledgement

The present report is a revision of a previous report: Kramer M. Hypnotic medication in the treatment of chronic insomnia: Non nocere! Doesn't anyone care? Sleep Medicine Reviews 2000; 4:529-541.

References

1. Salzman C. An 87-year-old woman taking a benzodiazepine. JAMA 1999; 281:1121-1125.
2. Portnoi VA. Should benzodiazepines in an 87-year-old woman be tapered and discontinued? JAMA 1999; 282:1128.
3. Salzman C. In reply: Should benzodiazepines in an 87-year-old woman taking a benzodiazepine be tapered? JAMA 1999; 282:1128.
4. Parker RA, Hartman EE. An 87-year-old woman taking a benzodiazepine. 1 year later. JAMA 1999; 282:1960
5. Tsoi WF. Insomnia: Drug treatment. Ann Acad Med Singapore 1991; 20:269-272.
6. Kupfer DJ, Reynolds CF. Current concepts: Management of insomnia. NEJM 1997; 336:341-346.
7. Kripke DF, Klauber MR, Wingard DL et al. Mortality hazard associated with prescription g a benzodiazepine be tapered and discontinued? JAMA 1999; 282:1128.
8. Ashton H. Guidelines for the rational use of benzodiazepines. When and what to use. Drugs 1994; 48:25-40.
9. Ancoli-Israel S, Roth T. Characteristics of insomnia in the united states: Results of the 1991 national sleep foundation survey.1. Sleep 1999; 22(Suppl):S347-353.
10. The gallop organization. Sleep in America. Washington DC: National sleep foundation, 1995:1-78.
11. Mellinger GD, Balter MB, Uhlenhuth EH. Insomnia and its treatments, prevalence and correlates. Arch Gen Psychiatry 1985; 42:225-32.
12. Swift CG, Swift MR, Hamley J et al. Side effect 'tolerance' in elderly recepients of benzodiazepine hypnotics. Age Ageing 1984; 13:335-43.
13. Dunbar GC, Perera MH, Jenner FA. Patterns of benzodiazepine use in Great Britain as measured by a population survey. Br J Psychiatry 1989; 155:836-41.
14. Hohagen F, Rink K, Kappler C et al. Prevalence and treatment of insomnia in general practice. A longitudinal study. Eur Arch Psychiatry Clin Neurosci 1993; 242:329-36.
15. Rayon P, Serrao-Castro M, del-Barrio H et al. Hypnotic drug use in Spain: A national study based on a network of community pharmacies. Spanish group for the study of hypnotic drug utilization. Ann Pharmacother 1996; 30:1092-100.
16. Ohayon M. Epidemiological study of insomnia in the general population. Sleep 1996; 19(Suppl 3):S7-15.
17. Balter MB, Uhlenhuth EH. The beneficial and adverse effects of hypnotics. J Clin Psychiatry 1991; 52(Suppl):16-23.
18. Balter MB, Uhlenhuth EH. New epidemiologic findings about insomnia and its treatment. J Clin Psychiatry 1992; 53(Suppl):34-9.
19. Walsh J, Ustun TB. Prevalence and heath consequences of insomnia. Sleep 1999; 22(Suppl 3):S427-36.
20. Johnson EO. Sleep in America. Washington DC: National Sleep Foundation 1999:1-122.
21. Roth T, Ancoli-Israel S. Daytime consequences and correlates of insomnia in the united states: Results of the 1991 national sleep foundation survey. II Sleep 1999; 22(Suppl 2):S354-58.
22. Zammit GK, Weiner J, Damato N et al. Quality of life in people with insomnia. Sleep 1999; 22(Suppl 2):S379-85.
23. Morin CM, Culbert JP, Schwartz SM. Nonpharmacological interventions for insomnia: A metaanalysis of treatment efficacy. Am J Psychiatry 1994; 151:1172-80.
24. Nowell PD, Mazumdar S, Buysse DJ et al. Benzodiazepines and zolpidem for chronic insomnia: A metaanalysis of treatment efficacy. JAMA 1997; 278:2170-77.
25. Morin CM. Behavioral and pharmacological treatment for insomnia. In reply: Letter to the editor. JAMA 1999; 282:1130-31.

26. Ancoli-Israel S, Richardson GS, Mangano RM et al. Long term exposure to zaleplon is safe and effective in younger- elderly and older elderly patient with primary insomnia. Sleep (abst.) 2003; 26:A-77.
27. Krystal A, Walsh J, Roth T et al. The sustained efficacy and safety of eszopiclone over six months of nightly treatment: A placebo-controlled study in patients with chronic insomnia. Sleep (abst.) 2003; 26:A310.
28. Gillen JC, Byerley WF. The diagnosis and management of insomnia. NEJM 1990; 322:339-48.
29. Langer S, Mendelson W, Richardson G. Symptomatic treatment of insomnia. Sleep 1999; 22(Suppl 3):S437-45.
30. Kesson CM, Lawson DH, Ankier SI. Long-term efficacy and tolerability of anew hypnotic- loprazolam. Br J Clin Pract 1984; 38:306-12.
31. Vogel G. Clinical uses and advantages of low doses of benzodiazepine hypnotics. J Clin Psychiatry 1992; 53(Suppl):19-22.
32. Reynolds CF, Regestein Q, Nowell PD et al. Treatment of insomnia in the elderly. In: Salzman C, ed. Clinical Geriatric Psychopharmacology. Baltimore: Williams and Wilkens, 1998:395-416.
33. Shorr RI, Robin DW. Rational use of benzodiazepines in the elderly. Drugs Aging 1994; 55:192-9.
34. Foley DJ, Monjan AA, Izmirlian G et al. Incidence and remission of insomnia among elderly adults: An epidemiologic study of 6,800 persons over three years. Sleep 1999; 22(Suppl 2):S366-72.
35. Kramer M, Svinte M. Falls, hypnotics and altering physician prescribing practice. Sleep Res 1992; 21:58.
36. Erman M. Efficacy and safety of hypnotic medications. In: Goldberg JR, ed. The Pharmacological Management of Insomnia. Washington DC: National Sleep Foundation, 1996:23-34.
37. Morin CM, Colecchi G, Stone J et al. Behavioral and pharmacological therapies for late-life insomnia. JAMA 1999; 281:991-99.
38. Lader M. Anxiety or depression during withdrawal of hypnotic treatments. J Psychosom Res 1994; 38(Suppl 1):113-23.
39. Ettorre E, Klaukka T, Riska E. Psychtropic drugs: Long term use, dependency and the gender factor. Soc Sci Med 1994; 39:1667-73.
40. Roth T, Kramer M, Roehrs T. The consistency of sleep measures. In: Koella WP, Levin P, eds. Sleep 1976: Proceedings of the Third European Sleep Congress. New York: S. Karger, 1977:286-8.
41. von Delbruck O, Goetzke E, Nagel C. Tolerance studies with brotizolam in hospitalized patients. Br J Clin Pharmacol 1983; 16(Suppl 2):S385-89.
42. Mittler MM, Carscadon MA, Phillips RL et al. Hypnotic efficacy of temazepam: A long-term sleep laboratory evaluation. Br J Clin Pharmacol 1979; 8:S63-8.
43. Leibovitz M, Sunshine A. Long-term hypnotic efficacy and safety of triazolam and flurazepam. J Clin Pharmacol 1978; 18:302-9.
44. Oswald I, French C, Adam K et al. Benzodiazepine hypnotics remain effective for 24 weeks. Br Med J 1982; 284:860-3.
45. Bunney Jr WE, Azarnoff DL, Brown Jr BW et al. Report of the institute of medicine committee on the efficacy and safety of halcion. Arch Gen Psychiatry 1999; 56:349-52.
46. Hartellius H, Larsson AK, Lepp M et al. Acontrolled study of flunitrazepam,nitrazepam and placebo, with special regard to withdrawal effects. Acta Psychiatr Scand 1978; 58:1-15.
47. Kirsch I, Sapirstein G. Listening to prozac but hearing placebo: A metaanalysis of antidepressant medication. Prevention & Treatment 1998; 1:Article 0002a. Available on the World Wide Web: http:// journals.apa.org/prevention/ volume1/ pre0010002a.html.
48. Klein DF. Listening to meta-analysis but hearing bias. Prevention & Treatment Amer Psychol Assoc 1998; 1:Article 0006c.
49. Woolf S. The need for perspective in evidence based medicine. JAMA 1999; 282:2358-65.
50. Regestein Q. Specific effects of sedative/hypnotic drugs in the treatment of incapacitating chronic insomnia. Am J Med 1987; 83:909-16.
51. Pakes GE, Brogden RN, Heel RC et al. Triazolam: A review of its pharmacological properties and therapeutic efficacy in patients with insomnia. Drugs 1981; 22:81-110.
52. Clark BG, Jue SG, Dawson GW et al. Lopazolam. A preliminary review of its pharmacodynamic properties and therapeutic efficacy in insomnia. Drugs 1986; 31:500-16.
53. Maarek L, Cramer P, Attali P et al. The safety and efficacy of zolpidem in insomniac patients: A long –term open study in general practice. J Int Med Res 1992; 20:162-70.
54. Schenck CH, Mahwold MW. Long-term nightly benzodiazepine treatment of injurious parasomnias and other disorders of disrupted nocturnal sleep in 170 adults. Am J Med 1996; 100:333-7.
55. Kramer M, Bailey S, Sepate M et al. The effectiveness of sleeping medication in chronic insomnia: A clinical case series. Sleep Res 1993; 22:219.
56. Regestein QR, Reich P. Asleep clinic within a general hospital psychiatry services. Gen Hosp Psychiatry 1980; 2:112-17.
57. Walsh J, Engelhardt C. Trends in the pharmacological treatment of insomnia. J Clin Psychiatry 1995; 53(12 Suppl):10-7.
58. Kramer M, Bailey S, Sepate M. Longterm medicinal treatment of insomnia: A follow up study. Sleep Res 1995; 24:266.
59. Ford DE, Kamerow DB. Epidemiologic study of sleep disturbances and psychiatric disorders: An opportunity for prevention. JAMA 1989; 262:1479-84.
60. Kramer M, Dashevsky B, Bailey S et al. Behavioral treatment of insomnia :A follow up study. Sleep Res 1996; 25:271.
61. Dashevsky B, Kramer M. Behavioral treatment of chronic insomnia in psychiatrically ill patients. J Clin Psychiatry 1998; 59:693-9.
62. Silberman EK. Psychiatrists and internists beliefs about benzodiazepines. Primary Psychology 1998; 5:1-8.
63. Zwar NA, Gordon JJ, Sanson-Fisher RW. Antibiotic and benzodiazepine prescribing by general practice trainees. Med J Aust 1994; 161:491-3.
64. Yeo GT, deBurgh SP, Letton T et al. Educational visiting and hypnosedative prescribing in general practice. Fam Pract 1994; 11:57-61.
65. Dunner DL. Commentary: Long-term use of sedative and hypnotic medication. Arch Gen Psyychiatry 1999; 56:355.
66. Mendelson WB.Long term follow-up of chronic insomnia. Sleep 1995; 18:698-701.
67. Kripke DF. Chronic hypnotic use: Deadly risks, doubtful benefit. Sleep Medicine Reviews 2000; 4:5-20.
68. Reynolds CF, Buysee DJ, Kupfer DJ. Treating insomnia in older adults: Taking a long term view. JAMA 1999; 281:1034-1035.

CHAPTER 15

Risks of Chronic Hypnotic Use

Daniel F. Kripke

Abstract

Hypnotic drugs are consumed mainly by chronic users, who often take hypnotics for years continuously. Two studies of the American Cancer Society, each involving over 1 million subjects followed prospectively for 6 years, showed that hypnotic use predicted increased mortality after control for comorbidities. At least 8 other studies have also noted that hypnotic use predicts increased mortality. There is an epidemiologic consensus that hypnotic use is associated with increased risk of death. Chronic hypnotic use is also strongly associated with insomnia, poor function, and poor quality of life. There is considerable evidence that hypnotics may in part cause this increased morbidity, as well as automobile accidents and falls. In contrast, there is no persuasive evidence that long-term use of hypnotics produces any benefit. Available evidence is that the risks of chronic hypnotic use outweigh the benefits.

Risks of Chronic Hypnotic Use

Most prescriptions of hypnotic drugs are consumed by chronic consumers, who often take hypnotics nightly for years at a time. The greatest sales, the greatest consumption, the greatest health impact and the greatest economic impact are with the chronic users. From U.S. data, it has been estimated that perhaps 2/3 or 3/4 of prescriptions go to the chronic user.[1,2] In the European setting, the percentage of chronic use might be rather higher. Ohayon found that 80% of those in France and 76% of those in Quebec who were taking a sleep-promoting drug had consumed it for at least 6 months.[3] In Italy, 73% taking a hypnotic had done so for a year or more.[4] In Spain, 72% had used the hypnotic they received for over 3 months.[5] Until recently, the dispensing of hypnotic drugs had more economic impact in American medicine than all other aspects of sleep medicine combined. That may still be so. It is a shame that we have no random clinical trials of hypnotics tested objectively over several months or years. Thus, we must make our best judgement of the risks and benefits of chronic hypnotic use from other evidence.

This discussion will focus on the risks of chronic hypnotic use, because substantial evidence is available about such risks. Little can be said about the benefits of chronic hypnotic use, for there is no persuasive scientific evidence that such benefits occur. Accordingly, this discussion will review the mortality risks and the morbidity risks of chronic hypnotic use, followed by a brief critique of the information on long-term benefits.

Mortality Associated with Prescription Hypnotics

In questionnaires collected from more than 1 million people in 1959, the American Cancer Society asked participants if they took "sleeping pills" "never," "seldom" or "often." Men who indicated that they took sleeping pills "often" had 57% increased mortality and women had 54% increased mortality after prospective 6-year follow-up, controlling for age and for reported sleep duration.[6] The ratios for "seldom" using sleeping pills were 15% and 13% increased, respectively, also statistically significant. The risk for sleeping pill users was more than 100% greater among those who took sleeping pills "often", if reported insomnia and sleep duration were not controlled. There were four important limitations to this study. First, the study did not distinguish between prescription versus over-the-counter "sleeping pills." Second, when the questionnaires were distributed in 1959, barbiturates were the predominant hypnotics, but in the next decade they were rapidly replaced by benzodiazepines with less acute overdose toxicity. Third, there was uncertainty of what frequency of use participants signified when they indicated that they took sleeping pills "often." Fourth, because of limitations on computer facilities at the time the data were analyzed, it was not possible to control extensively for comorbidities.

Most of the limitations of this study were overcome in the American Cancer Society's Cancer Prevention Study II (CPSII), which received questionnaires from 1.1 million participants in 1982, at a time when benzodiazepines dominated the hypnotics market.[7] Participants were asked "how many times in the last month" they used "prescription sleeping pills." Before controlling for comorbidities, as in the previous study, use of sleeping pills 30 or more times per month (essentially every night) was associated with more than doubled mortality hazard.[2] Controlling simultaneously for 32 risk factors and comorbidities, the hazard associated with sleeping pill use 30+ times per month was increased 24% for women and 25% for men.[7] For use 1-29 times per month (median 3), the hazard was increased 10% for women and 15% for men. In all these analyses, the adjusted hazard of sleeping pill use was statistically significant. The risk was particularly elevated for suicide, but was also elevated for cancer.[8] A significantly elevated risk for cancer was found, even excluding those sleeping pill users who said that they had cancer at the time they completed the questionnaire (unpublished).

Sleep and Sleep Disorders: A Neuropsychopharmacological Approach, edited by Malcolm Lader, Daniel P. Cardinali and S.R. Pandi-Perumal.

Strengths of this study were the extensive control for concomitant risk factors and the questionnaire's distinction of "prescription sleeping pills" from "Valium" and "Librium," which were the most popular minor tranquilizers in 1982. Limitations were the inability to specify exactly which "prescription sleeping pills" each participant was taking, and the fact that some popular contemporary hypnotics were not marketed at the time. The most popular hypnotics in 2002 were benzodiazepine agonists such as zolpidem and zaleplon which are not actually benzodiazepines. These contemporary agents bind somewhat selectively to the various benzodiazepine receptors. Since neither Valium or Librium was associated with excess mortality hazard,[7] one might be skeptical that the long-acting hypnotics such as flurazepam and quazepam, which are very similar pharmacologically, would be at fault, but shorter-acting benzodiazepines such as temazepam and triazolam might pose a higher risk. The kinetics of zolpidem and zaleplon at the receptor are rather similar to those of triazolam.

Several much-smaller studies have also examined the mortality hazards of hypnotics with a variety of control methods for confounding risk factors. Allgulander and Nowak found that men and women who took hypnotics and tranquilizers had about 7-fold increases in suicide.[9] Allgulander also noticed a high rate of suicides among those admitted for sedative-hypnotic dependence.[10] Thorogood and colleagues found that benzodiazepine and barbiturate use significantly predicted 4-fold and 6-fold risks of fatal myocardial infarction.[11] Rumble and Morgan found that in Nottingham, use of sleep medication significantly predicted mortality.[12] The effect was stronger for medications not usually classified as hypnotics, but the numbers were small. Brabbins et al found no significant relationship of hypnotic use to mortality in Liverpool, but the small sample included only 86 hypnotics users.[13] Sundquist et al observed that municipalities with higher sedative/hypnotic use had higher mortality.[14] Merlo et al found an increased risk of cancer mortality in men using anxiolytics-hypnotics.[15] Hays, Blazer, and Foley found no significant relationship of sleep medications to survival.[16] Kojima and colleagues found that use of drugs for sleep was associated with total mortality with adjusted risk ratios of 1.28 for men (not significant) and 1.81 for women (marginally significant).[17] Mallon et al found habitual sleeping pill use significantly related to total mortality, with risk ratios (RR) of 3.0 in men and 3.8 in women.[18] After multivariate adjustment, the relationships were significant for cancer deaths in men (RR=5.3) and noncancer noncoronary deaths in women (RR=3.3).[18] In conclusion, a variety of studies from different countries have found a preponderance of evidence for increased mortality risk associated with the use of hypnotics. Lack of significant evidence of hypnotics risk in some of these studies could be attributed to insufficient power in small samples.

It appears that hypnotics may cause deaths through several mechanisms. The lethal effects of barbiturates and benzodiazepine agonists in overdose are extremely well understood on molecular, neuronal, and organismic levels. Hypnotics in high doses are reliably used to put animals to sleep and to execute the death sentence. In lower doses, hypnotics may be synergistic with alcohol and other drugs. Whether hypnotics occasionally cause respiratory arrest in recommended doses is not known, but there is evidence that hypnotics exacerbate sleep apnea in at least a percentage of patients. This might lead to chronic hypertension, cardiac ischemia, and consequent demise. Evidently, hypnotic use is associated with increased suicide, but it is unclear whether this is mediated through depressive effects of hypnotics, through heightened disinhibition or confusion, or through confounding with depression. A mechanism by which hypnotics would increase cancer deaths is not currently known, but the suggestion for some specificity in cancer deaths deserves more study.

Figure 1. In epidemiologic evaluations of hypnotic risks for mortality, the evidence of 10 studies tilts heavily on the side associating hypnotics with increased mortality. There are two negative studies, but there is no suggestion that hypnotics lessen mortality.

Available studies indicate that not all mortality related to sleeping pill use can be explained by comorbidities. Although control for risk factors substantially reduced the mortality associated with use of prescription use in several studies, there may have been over control. For example, if sleeping pills cause deaths by increasing sleep apnea, thus increasing hypertension and heart disease, control for hypertension and heart disease may have biased the analysis to underestimate mortality caused hypnotics. Lapane et al found that benzodiazepine users had a 2-fold relative risk of ischemic heart disease.[19] Pratt et al found that barbiturate use was associated with a 2.11 odds ratio for myocardial infarction; the odds ratio for benzodiazepines, 1.33, was not significant.[20] On the other hand, control may have been insufficient for comorbid factors such as depression. Depression was directly controlled in only a minority of the studies. In my opinion, the likelihood of over control and the likelihood of under control were about equal in the epidemiologic literature concerning hypnotic risks.

I know of no epidemiologic evidence that use of any hypnotic prolongs survival or improves general health in any way. If one looks at the weight of epidemiologic evidence (Fig. 1), the weight is certainly on the side of hypnotics increasing mortality.

It is true that controlled trials of long-term hypnotic administration would be necessary for absolute proof of causal effects of hypnotics (either negative or positive), and as new hypnotics are introduced, epidemiologic data on the latest drugs will always be sparse. In the absence of controlled trials, the epidemiologic evidence is the best evidence available on the mortality risks-to-benefits ratio of hypnotics. The burden is on advocates of chronic hypnotic prescribing to prove that any hypnotic might be safe for long term use.

Morbidity Associated with Chronic Hypnotic Use

As might be expected, the use of hypnotics "often" is associated with having insomnia "often".[6] In the 2001 Sleep In America Poll (national data kindly loaned by the National Sleep Foundation), use of prescription medicines for sleep was significantly correlated with trouble falling asleep, awakening during the night, awakening early, and awakening unrefreshed. Use of medications was not significantly correlated with self-reported total sleep time in the Sleep In America Poll or in French data concerning insomniacs.[21,22] We expect people to take hypnotics because they have insomnia symptoms, so it is even surprising that there is a substantial portion of those taking hypnotics who do not report insomnia.[6,23,24] Converse causality, that taking hypnotics causes insomnia symptoms, has rarely been considered. There are no data to exclude that hypnotics causing insomnia is one pathway contributing to the association of hypnotic use and insomnia. Indeed, there are preliminary reports that when chronic hypnotics users undergo withdrawal from these drugs, they eventually sleep better,[25,26] which would suggest that hypnotics have caused insomnia.

Certainly, hypnotics with short half lives cause early awakening when administered at bedtime, and they cause withdrawal insomnia and anxiety on nights when they are withheld.[27,28] It would appear that these withdrawal mechanisms are active with contemporary hypnotics. Patients taking zolpidem sleep significantly worse subjectively than patients taking placebo on nights when the drug is skipped.[29] Zaleplon does not consistently increase total nightly sleep time,[30] so its benefit of reducing sleep latency is seemingly counterbalanced by more awakening later in the night.[31] These withdrawal effects of hypnotics demonstrate a complex interplay between their capacity to slightly increase sleep and their capacity to harm sleep.

There are some indications that the withdrawal effects and other sleep-damaging effects of hypnotics are more persistent than their benefits for sleep, to which rapid tolerance may develop. For example, the withdrawal effects of triazolam were worse than those of placebo at a time when, because of tolerance, the benefits did not exceed placebo.[32] Early awakening effects of triazolam seemingly increased over several weeks.[27] In chronic hypnotic users, tolerance to flurazepam and midazolam developed within 2 weeks, as compared to placebo.[33] Zolpidem was ineffective in its recommended dose at 5 weeks.[34] After 8 weeks, intermittent use of zolpidem did not increase weekly subjectively-reported total sleep.[29] Although the subjective quality of sleep was not significantly improved by two benzodiazepines as compared to placebo after 24 weeks, significant deterioration of quality was demonstrated when the benzodiazepines were withdrawn.[35] This strange evidence that the sleep-harming effects of hypnotics might be more persistent than their benefits suggests a very negative assessment of the chronic effects of these drugs. The implication of these studies is that with chronic use, hypnotics may be of no benefit at all for sleep, but patients may continue dosing themselves because of the distress which is caused by drug withdrawal. This drug withdrawal is certainly a factor in the widely-reported habituation and dependence on hypnotics which contributes to the chronicity of hypnotic use.

Although hypnotics with short half-life cause much less impairment of daytime performance than those which are persistent during the day, even with short-half-life drugs, there is more evidence that they impair performance than that they improve performance, at least when the indication is insomnia rather than jet lag or shift work.[36,37] In a recent study, 8 weeks of intermittent use of zolpidem caused no improvement in health or performance, though the study had 80% power to detect a small-to-medium effect of 0.4 SD.[38] On the other hand, the more persistent hypnotics, such as flurazepam and quazepam, cause considerable impairment of daytime performance which augments with continuing daily use as the drug accumulates.[36] It also appears that the long-half-life hypnotics cause more problems with automobile accidents and falls, (but see data on zoplicone).[39-45] Both short and long-half-life hypnotics can impair memory.[46]

People who take hypnotics report an excess of a variety of somatic conditions and psychological symptoms.[47-50] Moreover, after adjusting for initial levels and confounding factors, Byles et al found that use of sleep medication predicted poor quality of life (SF-36), increased falls, increased doctor consultations, and increased days in the hospital upon 3-year follow-up.[50] The latter evidence would suggest that hypnotics caused these multiple aspects of morbidity and poor health.

Benefits of Chronic Hypnotic Use

In clinical trials of hypnotics, a frequent design has been to compare an initial drug-free or placebo baseline with results during several weeks of hypnotic administration. These sequential designs without cross-over are completely flawed and useless for judging drug effects, because parallel-placebo studies show that improvement occurs sequentially among patients given placebo. Therefore, the only scientifically meaningful designs would be counter-balanced cross-overs or better, parallel placebo-hypnotic contrasts.

There have been few parallel-placebo contrasts of hypnotic administration for more than 4 weeks, and no counter-balanced cross-overs of which I am aware. Of the parallel designs, as has been mentioned, several suggested that after 2-8 weeks of administration, hypnotics produce no more benefit than placebo at recommended doses, that is, there was no long-term benefit.[32-34] In 8 weeks of intermittent use, the benefits of zolpidem on drug nights was balanced by the harm on drug-withdrawal nights, resulting in no significant increase in weekly sleep.[38] The heroic study of Oswald and colleagues gave somewhat uncertain results for 24-week administration of two benzodiazepines, because only subjective measurements were made, because the benefit for sleep quality did not remain significant for 24 weeks, and because the apparent small benefit for sleep latency was not fully consistent.[35] In treatment of late-life insomnia for 8 weeks, temazepam was not significantly better than placebo on polysomnographic variables, though temazepam was superior on sleep diary scores.[51] However, upon follow-up, patients receiving temazepam deteriorated and were slightly worse than the placebo group at 24 months, when some had resumed taking medication. To summarize the meager evidence of parallel studies of longer-term use of hypnotics, use of hypnotics for more than 4 weeks produces no objective improvement in sleep and little (if any) subjective benefit. I am not aware of any placebo-controlled parallel studies which show benefits of long-term use of hypnotics for performance, mood, health, or general quality of well-being.

Need for Research

To my knowledge, there has been no parallel-placebo-controlled trial of giving a hypnotic for more than 8 weeks which used objective sleep recording. Objective recording is important, because subjective and objective reports of sleep are so frequently discordant, particularly when addictive and memory-impairing properties of hypnotics may distort patient reports. Elsewhere, I have discussed the need for hypnotics trials in more detail.[52,53] In the absence of conclusive trials demonstrating the causal effects of chronic use, our best evidence comes from hypnotics epidemiology.

Conclusions

We must recognize that chronic use of hypnotics is associated with poor sleep, poor health, and reduced survival. We are faced with considerable evidence that hypnotics may create sleep and performance problems, memory disturbance, driving accidents,

Figure 2. For chronic hypnotic use, the balance of evidence tilts strongly toward risks, there being only a tiny cloud of evidence for subjective benefits for sleep.

and falls. Lacking controlled evidence that chronic hypnotic use improves health or function in any way, and with evidence that there is no objective benefit even for sleep, we see that the weight of evidence favors risks over benefits in chronic hypnotic use (Fig. 2). Several consensus conferences in the U.S. have reached the same conclusion,[54-57] along with the U.S. Food and Drug Administration and the regulatory agencies of several other countries.

Acknowledgment

Supported by National Institutes of Health AG12364, HL61280, and AG15763 and by the Sam and Rose Stein Institute for Research on Aging.

Note Added in Proof

Krystal and colleagues (Krystal AD, Walsh JK, Laska E et al. Sustained efficacy of eszopiclone over 6 months of nightly treatment: results of a randomized, double-blind, placebo-controlled study in adults with chronic insomnia. Sleep 2003; 26:793-799) found subjective evidence for improvements in average health in a placebo-controlled 6-month trial of eszopiclone. However, the eszopliclone group had almost 3 times the severe adverse effects, leaving the benefits/risks ratio uncertain.

References

1. Kripke DF. Chronic hypnotic use: He neglected problem. In: Koella WP, Ruther E, Schulz H, eds. Sleep '84. Stuttgart: Gustav Fischer Verlag, 1985:338-340.
2. Kripke DF, Klauber MR, Wingard DL et al. Mortality hazard associated with prescription hypnotics. Biol Psychiatry 1998; 43:687-693.
3. Ohayon MM, Caulet M. Psychotropic medication and insomnia complaints in two epidemiological studies. Can J Psychiatry 1996; 41:457-464.
4. Balestrieri M, Bortolomasi M, Galletta M et al. Patterns of hypnotic drug prescription in Italy. Brit J Psychiat 1997; 170:176-180.
5. Rayon P, Càstro-Serrano M, del Barrio H et al. Hypnotic drug use in Spain: A cross-sectional study based on a network of community pharmacies. Pharmacoepidemiology 1996; 30:1092-1100.
6. Kripke DF, Simons RN, Garfinkel L et al. Short and long sleep and sleeping pills: Is increased mortality associated? Arch Gen Psychiatry 1979; 36:103-116.
7. Kripke DF, Garfinkel L, Wingard DL et al. Mortality associated with sleep duration and insomnia. Arch Gen Psychiatry 2002; 59:131-136.
8. Kripke DF, Marler MR. Specific causes of mortality associated with prescription sleeping pill usage [abstract]. Sleep Research Online 1999; 2:144.
9. Allgulander C, Nowak J, Rice JP. Psychopathology and treatment of 30,344 twins in Sweden. I. The appropriateness of psychoactive drug treatment. Acta Psychiatrica Scandinavica 1990; 82:420-426.
10. Allgulander C, Ljungberg L, Fisher LD. Long-term prognosis in addiction on sedative and hypnotic drugs analyzed with the Cox regression model. Acta Psychiatrica Scandinavica 1987; 75:521-531.
11. Thorogood M, Cowen P, Mann J et al. Fatal myocardial infarction and use of psychotropic drugs in young women. Lancet 1992; 340:1067-1068.
12. Rumble R, Morgan K. Hypnotics, sleep, and mortality in elderly people. J Am Geriatr Soc 1992; 40:787-791.
13. Brabbins CJ, Dewey ME, Copeland RM et al. Insomnia in the elderly: Prevalence, gender differences and relationships with morbidity and mortality. Int J Ger Psych 1993; 8:473-480.
14. Sundquist J, Ekedahl A, Johansson S-E. Sales of tranquillizers, hypnotics/sedatives and antidepressants and their relationship with underprivileged area score and mortality and suicide rates. Eur J Clin Pharmacol 1996; 51:105-109.
15. Merlo J, Hedblad B, Ogren M et al. Increased risk of ischaemic heart disease mortality in elderly men using anxiolytics-hypnotics and analgesics. Eur J Clin Pharmacol 1996; 49:261-265.
16. Hays JC, Blazer DG, Foley DJ. Risk of napping: Excessive daytime sleepiness and mortality in an older community population. J Am Geriatr Soc 1996; 44:693-698.
17. Kojima M, Wakai K, Kawamura T et al. Sleep patterns and total mortality: A 12-year follow-up study in Japan. J Epidemiol 2000; 10:87-93.
18. Mallon L, Broman J-E, Hetta J. Sleep complaints predict coronary artery disease mortality in males: A 12-year follow-up study of a middle-aged Swedish population. J Int Med 2002; 251:207-216.
19. Lapane KL, Zierler S, Lasater TM et al. Is the use of psychotropic drugs associated with increased risk of ischemic heart disease? Epidemiology 1995; 6:376-381.
20. Pratt LA, Ford DE, Crum RM et al. Depression, psychotropic medication, and risk of myocardial infarction. Prospective data from the Baltimore ECA follow-up. Circulation 1996; 94:3123-3129.
21. Ohayon MM, Caulet M, Arbus L et al. Are prescribed medications effective in the treatment of insomnia complaints? J Psychosom Res 1999; 47:359-368.
22. Leger D, Janus C, Pellois A et al. Sleep, morning alertness and quality of life in subjects treated with zopiclone and in good sleepers. Study comparing 167 patients and 381 good sleepers. Eur Psychiatry 1995; 10:99s-102s.
23. Ohayon M, Caulet M, Lemoine P. Sujets agés, habitudes de sommeil et consommation de psychotropes dan la population française. L'Encephale 1996; 22:337-350.
24. Ohayon MM, Lader MH. Use of psychotropic medication in the general population of France, Germany, Italy, and the United Kingdom. J Clin Psychiatry 2002; 63:817-825.
25. Morin CM, Bastien CH, RadoucoThomas M. Late-life insomnia and chronic use of benzodiazepines: Medication tapering with and without behavioral intervention [abstract]. Sleep 1998; 21S:99
26. Parrino L, Smerieri A, Boselli M et al. Discontinuation of elevated doses of benzodiazepines used as hypnotics: Polysomnographic assessment of sleep parameters [abstract]. Sleep 1998; 21S:101
27. Kales A, Soldatos CR, Bixler EO et al. Early morning insomnia with rapidly eliminated benzodiazepines. Science 1983; 220:95-7.
28. Kales A, Soldatos CR, Bixler EO et al. Rebound insomnia and rebound anxiety: A review. Pharmacology 1983; 26:121-37.
29. Walsh JK. Reply to Kripke. Sleep Medicine Reviews 2003; 7:195-196.
30. Elan Pharmaceuticals: Sonata (zaleplon) Capsules. 11/26/2001 [prescribing information]. Philadelphia, PA: Wyeth Laboratories, 2002.
31. George CFP. Pyrazolopyrimidines. Lancet 2001; 357:1623-1626.
32. Mitler MM, Seidel WF, van den Hoed J et al. Comparative hypnotic effects of flurazepam, triazolam, and placebo: A long-term simultaneous nighttime and daytime study. J Clin Psychopharmacol 1984; 4:2-15.

33. Kripke DF, Hauri P, Ancoli-Israel S et al. Sleep evaluation in chronic insomniacs during 14-day use of flurazepam and midazolam. J Clin Psychopharmacol 1990; 10(Supplement 4):32S-43S.
34. Scharf MB, Roth T, Vogel GW et al. A multicenter, placebo-controlled study evaluating zolpidem in the treatment of chronic insomnia. J Clin Psychiatry 1994; 55:192-199.
35. Oswald I, French C, Adam K et al. Benzodiazepine hypnotics remain effective for 24 weeks. Bristish Medical Journal 1982; 284:860-863.
36. Judd LL, Ellinwood E, McAdams LA. Cognitive performance and mood in patients with chronic insomnia during 14-day use of flurazepam and midazolam. J Clin Psychopharmacol 1990; 10:56S-67S.
37. Johnson LC, Chernik DA. Sedative-hypnotics and human performance. Psychopharmacology (Berlin) 1982; 76:101-113.
38. Walsh JK, Roth T, Randazzo A et al. Eight weeks of nonnightly use of zolpidem for primary insomnia. Sleep 2000; 23:1087-1096.
39. Barbone F, McMahon AD, Davey PG et al. Association of road-traffic accidents with benzodiazepine use. Lancet 1998; 352:1331-1336.
40. Greenblatt DJ, Allen MD. Toxicity of nitrazepam in the elderly: A report from the Boston collaborative drug surveillance program. Br J Clin Pharmacol 1978; 5:407-13.
41. Hemmelgarn B, Suissa S, Huang A et al. Benzodiazepine use and the risk of motor vehicle crash in the elderly. JAMA 1997; 278:27-31.
42. Oster G, Huse DM, Russell MW et al. Benzodiazepine tranquilizers and the risk of accidental injury. Am J Public Health 1990; 80:1467-1470.
43. Ray WA, Griffin MR, Schaffner W et al. Psychotropic drug use and the risk of hip fracture. N Engl J Med 1987; 316:363-9.
44. Tinetti ME, Speechley M, Ginter SF. Risk factors for falls among elderly persons living in the community. N Engl J Med 1988; 319(26):1701-1707.
45. Ray WA, Griffin MR, Downey W. Benzodiazepines of long and short elimination half-life and the risk of hip fracture. JAMA 1989; 262(23):3303-3307.
46. Greenblatt DJ, Harmatz JS, Engelhardt N et al. Pharmacokinetic determinants of dynamic differences among three benzodiazepine hypnotics: Flurazepam, temazepam, and triazolam. Arch Gen Psychiatry 1989; 46:326-332.
47. Allgulander C, Nasman B. Regular hypnotic drug treatment in a sample of 32,679 Swedes: Associations with somatic and mental health, inpatient psychiatric diagnoses and suicide, derived with automated record-linkage. Psychosom Med 1991; 53(1):101-108.
48. Quera-Salva MA, Orluc A, Goldenberg F et al. Insomnia and use of hypnotics: Study of a French population. Sleep 1991; 14(5):386-391.
49. Roehrs T, Hollebeek E, Drake C et al. Substance use for insomnia in Metropolitan Detroit. J Psychosom Res 2002; 53:571-576.
50. Byles JE, Mishra GD, Harris MA et al. The problems of sleep for older women: Changes in health outcomes. Age Ageing 2003; 32:154-163.
51. Morin CM, Colecchi C, Stone J et al. Behavioral and pharmacological therapies for late-life insomnia. A randomized controlled trial. JAMA 1999; 281:991-999.
52. Kripke DF. Chronic hypnotic use: Deadly risks, doubtful benefit. Sleep Medicine Reviews 2000; 4:5-20.
53. Kripke DF. http://www.DarkSideOfSleepingPills.com 2002.
54. NIOM: Sleeping Pills, Insomnia, and Medical Practice. Washington DC: National Academy of Sciences, 1979:1-198.
55. Consensus Conference: Drugs and insomnia. The use of medications to promote sleep. JAMA 1984; 251(18):2410-2414.
56. National Institutes of Health: Consensus development conference statement: The treatment of sleep disorders of older people. Sleep 1991; 14(2):169-177.
57. Bunney Jr WE, Azarnoff DL, Brown Jr BW et al. Report of the institute of medicine committee on the efficacy and safety of Halcion. Arch Gen Psychiatry 1999; 56:349-352.

CHAPTER 16

Effects of Psychotropics on Driving Performance

Henry J. Moller, Colin M. Shapiro and Leonid Kayumov

Introduction

Driving a motorized vehicle is a complex task of information processing and motor skills, requiring a variety of cognitive and psychomotor performance abilities to be intact: alertness, attention, multitasking, memory, co-ordination and visuospatial perception, to name a few. Numerous prescription drugs acting within the central nervous system (CNS) have the potential to affect both sleep and daytime functioning.[1] While this phenomenon is well known to clinicians and researchers alike, it is sometimes discussed as if it was an abstract concept lacking any consequence to the functional abilities and consequences on the day-to-day lives of our patients. Furthermore, some performance situations, such as operation of a vehicle or other complex mechanical device represent real risks to both patients and the public, creating potential medicolegal liabilities for the prescribing physician.[2] As current medical research increasingly occupies itself with topics of ergonomics, human error and human-machine interface, effects of psychotropic medications on specific daytime performance tasks such as driving have become a more relevant topic.

Medical researchers, and specifically psychiatrists, have historically shown only passing interest in driving ability as an example of human interaction with machines. An article entitled 'The accident prone driver" (a term borrowed from the psychoanalytical literature's exploration of parapraxes in "the accident-prone individual") appeared in the American Journal of Psychiatry in 1949 which attempted to correlate individual psychopathology with poor driving records in two groups of taxi drivers.[3] Since then, the term 'accident' has long fallen out of favor in the medical and human factors literature, as this term would see a crash/error as a random/unpredictable event, as opposed to a possible predictable and/or preventable entity, based on particular risk factors.[4] These risk factors can be driver/operator-specific, machine/vehicle-specific, or relate to the human-machine interface.

While vehicle safety has increased dramatically since the introduction of the automobile, and psychopharmacological interventions have generally become more sophisticated in recent decades, current trends in telematic devices such as mobile phones and Global Positioning System (GPS) units have introduced an additional multi-tasking element to the driving task.[5] Thus an individual with reduced concentration due to mildly sedating effects of a medication may have had little trouble driving while attending exclusively to this task, but may struggle with competing attentional demands of conversing on a cellular telephone and navigating through urban rush-hour traffic.

Studies of drug consumption by patients involved in crashes in the United Kingdom have shown that between 11-20% are taking psychotropic drugs, however, as has been true of other epidemiological studies, the extent to which the drug(s) may have contributed remains difficult to ascertain.[6,7]

Perception, Cognition and Driving

This raises the issue of perceptual context in driving. The capacity for human attention and cognitive capacity in relation to bandwidth has been described as one of the most important economic commodities of the 21st century.[8] Research into the field of perceptual immersion or 'presence' has been progressing in parallel with research in the field of simulated environments.[9,10] Presence can be defined as the feeling a conscious organism experiences when immersed in a concrete external world, i.e., that one is "really there" in the environment. Converse to this exists a state of 'absence', in which an individual is preoccupied or engaged with an internally constructed world.[9] In fact, this spectrum coincides well with the neuroscience of sleep and wakefulness, which would attempt to correlate these fluctuations between 'absent' and 'present' states through electrophysiological patterns. When an individual is fully alert, i.e., 'on task' he would perceive a state of full presence/immersion, whereas increasing states of absence would likely be a precursor towards actual lapses into an unconscious state. Often this will involve a feedback loop between individual and environment. A lack of external stimulation, which is cognitively experienced as boredom, leads to metabolic lowering of brain activity. On the other end of the spectrum, neurocognitive overstimulation/overarousal results when mental tasks are excessively complex and sustained.[11]

Applying this to the study of automotive crashes, this can be exemplified by the contrasting (but both hazardous scenarios) of monotonous nighttime highway driving (Type 1) vs. rush hour "mid-town madness" (Type 2) (see Fig. 1). This implies that medications that impart a sedating effect on the patient's CNS may be hazardous in some driving environments requiring prolonged vigilance, while medicines with the potential to cause anxiety or agitation may be specifically dangerous in other situations. Returning to the previous topic of the 'distracting' effects of communication devices, long-haul truck drivers have long known that they can improve their wakefulness on long soporific drives through the use of CB radio, in an attempt to optimize their level of perceptual stimulation. While overall crash rates tend to be dependent on vehicle density and peak at rush hour, crashes where drowsiness due to medication or sleep disturbance plays a role have been shown to have a circadian variation correlating with

Sleep and Sleep Disorders: A Neuropsychopharmacological Approach, edited by Malcolm Lader, Daniel P. Cardinali and S.R. Pandi-Perumal.

Two major MVA categorizations

Type I	Type II
Single vehicle	Multi-vehicle
Soporific situation	Multi-taking/ overarousal
Rural/ Long mononotonous stretch of highway	Urban/ intersection
Circadian component	More often related to traffic volume.

Figure 1. Stratification of motor vehicle "accidents" by interaction of environment and cognitive state.

periods of maximum sleepiness, between 2 and 4 a.m. and, to a lesser degree, the 'siesta period' during mid-afternoon.[12,13]

Classification and Measurement Issues in Quantifying Medication Effects

Classifying centrally acting medications is a complex endeavor. While some classifications use a system based on desired therapeutic action (e.g., sedative, antidepressant, anxiolytic, etc), others tend to base categorizations on pharmacodynamic properties (e.g., GABA- or serotonin- agonists, antihistamines, etc). In fact both classification systems are imperfect. Frequently, both therapeutic and neurochemical properties of CNS agents are mixed. As an example, the sedating antidepressant trazodone has both a serotonin 5HT-2 agonist as well as histamine antagonist properties. Furthermore, are the benzodiazepines best classified as anticonvulsants, hypnotics, sedatives or anxiolytics? Rather than adhering to simple-minded categorizations, it behooves the clinician to understand the unique biochemical and therapeutic properties of medications used to treat their patients.

Particularly with psychotropics, the prescribing physician needs to be aware of the drug's pharmacodynamic and pharmacokinetic characteristics. Factors such as hepatic interactions at the cytochrome p450 level and medication half-life must be monitored for; with respect to the latter, it is important to consider the individual neurochemical target receptor properties, as the wash-out period from the CNS is often significantly longer than from serum.

Another key clinical and research consideration relates to the terminology used to define phenomenology of these altered states of consciousness imparted upon the patient by psychotropics, and the methodology used to assess these. For example, drowsiness, lethargy, fatigue and somnolence are sometimes used interchangeably. While for one patient, these descriptors may seem interchangeable, others will be more specific in their descriptions of drug-related sedative effects. Researchers have used measurements of subjective states of consciousness, such as the Epworth Sleepiness Scale[14] (a more global subjective rating instrument) and the Stanford Sleepiness Scale (intended as a more instantaneous approximation of a subject's mental state. Alternately, physiological measures have been employed to study this problem, such as the Multiple Sleep Latency Test (MSLT) and the Maintenance of Wakefulness Test (MWT),[15] and more recently, pupillometrics.[16]

While such tests may accurately measure a subject's propensity to fall asleep or experience impaired alertness under controlled laboratory conditions, a difficulty remains regarding the ecological validity, i.e., relevance to 'real-world' situations. Alternately, in vivo testing centers exist that can perform on-road testing. This method, while more ecologically valid, is often more costly, and more difficult to standardize. Driving simulator studies have been used,[17,18] occasionally in conjunction with polygraphic monitoring[19,20] to approximate subject's driving impairment in a laboratory environment with increased ecological validity. The fact that such tests may yield discrepant results from traditional tests of daytime alertness[18] raises the question of which test is the most accurate approximation of an individual's mental state; ultimately, the medicated patient's behavior and subjective experience in their activities of daily living outside the hospital remains the gold standard, providing the clinician is able to obtain a reliable history. For this reason, the conservative clinician may need to gather collateral sources of information such as from the spouse or employer.

Alcohol and Illicit Substances

Alcohol and drugs of abuse have been studied in simulated driving environments. While the literature is near-unanimous on the impairing effects of alcohol in a dose-dependent fashion, legal standards in different countries vary in terms of acceptable breathalyzer sample readings, based on departure of test drivers' reaction times from a standardized normative reaction time and lane obedience in simulated driving tasks. The fact that states of sleepiness can result in severe driving impairment, analogous to states of intoxication with alcohol has been well described[21-23] However, it is also important to note that fatigue and sleepiness can exaggerate driving impairment exponentially, even in situations with conventionally acceptable blood alcohol concentrations. While studies comparing marijuana to alcohol have generally found both to impair driving abilities[24-26] patterns of driving behavior appear to differ for drivers intoxicated with the respective substance and the literature appears divided regarding the risk of driving while under the influence of cannabis. This may be due to the relative effect on insight into impairment that is generally lost with alcohol intoxication compared with cannabis intake. Combined intake of marijuana and alcohol has been shown to be more impairing than with either substance alone.[24,25] It has been reported that misuse of substances such as amphetamines, cocaine and other stimulants is a common coping strategy used by truck drivers to self-medicate their perceived risk of driving impairment due to sleep deprivation and periods of shift work.[27] Military pilots have also reportedly used medications such a methylphenidate or dextroamphetamine as provigilants.[28] While many of these stimulant substances do increase alertness in the short term, unsupervised and illicit use invariable increases crash risk. Firstly, there is a risk of rebound sleepiness during times of withdrawal; secondly, increased proneness to erratic and/or risk-taking behavior may accompany intake of excessively stimulating substances, especially if overtitrated. Thirdly, some studies have reported the induction of 'tunnel vision' with use of stimulant substances.[29] Use of stimulants is generally prohibited in commercial pilots. MDMA, a currently popular drug of abuse, has been shown to produce a disinhibited driving pattern elevating risks of errors of commission and crash risk.[30]

Therapeutic Medications

Not all centrally acting drugs are physician-prescribed. It is recognized even among the general public that commonly used over-the counter (OTC) drugs such as opiate-containing antitussives or anticholinergic antinauseants can impair alertness relevant to driving.[1] Classic first-generation antihistamines, which far more readily enter the CNS than the liphophobic second-generation H1-receptor antagonists, are also well known to produce performance decrements and drowsiness.[1,31] Caffeine is the most commonly used centrally acting performance enhancing agent. A typical cup of American-style coffee has been estimated to contain about 100 mg of caffeine, and its provigilant effect has been estimated to last about 1 to 3 hours.[32] Tea and chocolate are among other caffeine-containing foodstuffs. However, if used as a countermeasure to physiologically occurring sleepiness, there appears to be a tendency towards diminishing alertness with repeated use. In fact, Rayner and Horne[33] demonstrated that a brief therapeutic 'powernap', followed by caffeine consumption, was far more effective than caffeine alone in improving driving performance.

A large literature already exists on the adverse effects of hypnotics and sedatives,[12] but there is also significant interest in the effects of widely prescribed psychotropic medications like anti-anxiety and antidepressant medications.[4,12,22] While it is known that conditions such as depression and anxiety disorders significantly affect cognitive functioning, the pharmacological mode of action of drugs used to combat these disorders may vary vastly in their therapeutic and adverse effects on the spectrum from highly sedating to activating, depending on the neurotransmitters affected by the medication in question.

The key issue with respect to sedatives and hypnotics regards medication half-life.[34,35] This issue is discussed more extensively elsewhere in this book. Virtually all benzodiazepine-like substances are known to cause drowsiness and decreased fine-motor coordination, and patients should generally expect impairments comparable to alcohol on driving ability. Less well known to the public is the danger of residual 'hang-over' sedation of medications of various classes taken during the night, or even at bed-time the night before a daytime driving task, a clinically common scenario in patients using CNS-active agents for sleep neuropsychiatric disturbance. In particular, the use of long-or medium-acting sedatives for middle insomnia can create significant driving impairment the morning following use, and for this reason, if patients need to use middle-of-the night hypnotics on an occasional p.r.n. basis, an ultra-short acting sedative such as zaleplon is recommended.[35]

Depressed patients show impaired information processing, learning, memory and tracking skills.[36-38] While all these are potential target symptoms of medication and psychological treatment, a judicious clinician should be aware of both risks and benefits of treatment options. The effects of antidepressants on driving ability have probably best been described by Hindmarch, using both 'in vivo' road tests and computer-based performance tests such as the critical flicker fusion frequency and choice reaction time tests.[39] A recent excellent review of antidepressant effects on driving ability by Ramaekers[37] has attempted to broadly divide antidepressant medicines into sedating and non-sedating groups, and found that on an actual road test using a car equipped with a telematic performance monitoring tracker, the measure standard deviation of lateral position (SDLP), an index of a driver's propensity to 'weave', was most sensitive in distinguishing differential performance due to various antidepressants. While an exhaustive discussion of specific drugs is beyond the scope of this article, tricyclic medications, and those novel antidepressants with antagonism of central histaminergic, muscarinic or adrenergic receptors do demonstrate a dose-dependent increase in a medicated test subject's SDLP. Furthermore, an important co-factor was the combined use of benzodiazepines (a clinically common scenario), particularly when there was a likely cytochrome p450 inhibitory interaction of the two drugs.

This is not to suggest that sedating antidepressants should be avoided in patients depending on their driving privileges. In fact, sleep disturbance is an extremely common clinical manifestation of depression, typically taking form in some combination of initial and middle ('early morning' insomnia) with anxious or depressive ruminations. Often, the sleeplessness associated with the psychiatric condition will pose more of a hazard than the possible residual effect of the antidepressant treatment. Sedating antidepressant agents such as mirtazapine or trazodone can be effective treatments for a combination of sleep and mood disturbance. Mirtazapine has been shown in at least one study of administration over a 5 day period to potentially impair driving performance at bed-time doses of 15 to 30 mg.[41] For more refractory mood and agitation states, it is reasonable to combine antidepressants with benzodiazepine-like sedatives; furthermore, sedating atypical antipsychotics such as olanzapine or quetiapine may be effective add-ons, and possibly less impairing than older neuroleptics.[42] What is important in the use of such medications is a clear discussion of risks and benefits with the patient. It is generally advised to avoid driving altogether within the initial titration period of treatment, as CNS-effects that impair co-ordination and mental alertness are usually most prominent then. Similarly, in the days following dosing adjustments, additional caution is warranted. For example, while Robbe and Ohanlon[43] found paroxetine showed no evidence of elevated psychomotor impairment on a divided attention task compared to placebo at 20mg, a mild increase in impairment began to appear at higher doses, although performance on a driving task remained preserved, illustrating the subtlety of performance testing. Similarly, acute administration of the now dicontinued antidepressive nefazodone showed impairment on lateral position control in an on-road driving test, at 200mg twice daily, but not 100 mg twice daily, in conjunction with dose-related mild cognitive and memory decrements.[44]

When used therapeutically, stimulant medicines have been shown to improve psychomotor performance on an acute basis in adolescent drivers with attention deficit disorder.[45] Alerting or provigilant medications may also come from the antidepressant class, where they activate combinations of serotonergic (e.g., fluoxetine), noradrenergic (venlafaxine) or dopaminergic (buproprion) most prominently.[1] As with stimulants, caution should be exercised to avoid inducing states of hypervigilance or agitation through overly aggressive dosing. Furthermore, it is not uncommon for individual psychomotor performance to vary significantly for different antidepressant agents.[37,39,40]

Not uncommonly, individuals with more serious mental disorders such as bipolar disorder or psychotic illnesses possess valid driver's licenses, and may even depend on driving as part of their activities of daily living. The effects of psychotropics used to treat these conditions is less well understood. There are reports of potentially impairing effects of lithium[46,47] as well anticonvulsants,[48] whether used as mood stabilizers or to control seizure disorders. The use of antipsychotics has been less extensively studied. Similarly to the treatment of depression, severe cognitive and psychomotor impairments are often a component of the illness state, and may improve with treatment. While acute administration of most antipsychotics in drug-naïve subjects results in psychomotor impairment, a review by Judd[49] points out that generally, patients suffering from psychotic illness tend to demonstrate improved psychomotor performance during chronic treatment with these medications. Similarly to tricyclic antidepressant agents, sedating effects due to the anti-histaminergic, anti-α 1 adrenergic or anticholinergic systems, often quite prominent in acute antipsychotic administration, may abate with prolonged treatment.

Schweitzer's review of antihypertensives points out that beta-blockers, particularly those with α1-blocking activity have been associated with fatigue and somnolence.[1] However, there is a paucity of literature in the domain of alertness testing, and specifically, driving performance. In general, for cardiovascular medications, it is appropriate to be aware of potential idiosyncratic cognitive and/or performance decrements.

A recent controversy regarding psychotropic agents and driving has followed a report by Frucht,[50] alerting clinicians of the possibility of "sleep attacks" secondary to the novel anti-parkinsonian agents ropinirole and pramipexole. It has been pointed out that rather than dealing with an impairing risk relating to these specific drugs, a broader issue may reside in the frequently compromised driving ability of often elderly individuals with advanced Parkinson's disease. Furthermore, the potential class-effect of other dopaminergic agents including levo-dopa to at least have the potential to cause idiosyncratic and/or dose-related impairments of alertness has been noted.[51]

In general, elderly individuals are more likely to carry multiple medical diagnoses, to be treated with multiple medications, including psychotropics, and are more likely to have decrements in psychomotor skills. Clinicians must be cautious of the altered pharmacokinetics and pharmacodynamics of the elderly. While a recent large-scale study of nearly 1000 drivers revealed an increase in odds ratio for benzodiazepine use in at-fault crashes for seniors,[7] other medicines such as nonsteroidal anti-inflammatories, ACE inhibitors and anticoagulants also showed elevated odds ratios, possibly due to medical morbidity (muskuloskeltal and cardiovascular) associated with these drugs. This illustrates the challenge of examining causative drug effects on driving in epidemiological studies.

Conclusions

It is intuitive that CNS-active drugs have the potential to affect perceptual and psychomotor skills relevant to driving. Yet, variability in testing paradigms and difficulty in controlling performance-testing conditions continue to create challenges in achieving a more exact understanding of pharmacological influences on driving behavior. Advances in technology have created opportunities for better measurement, both in simulator and on-road testing. Controlled prospective, rather than retrospective epidemiological studies will be more likely to shed light on cause and effect of motor vehicle crashes in relation to drug use. Alternative comparison data for driving performance of healthy unmedicated individuals is urgently needed to further knowledge in this field.

Psychotropic medications have the potential to improve aspects of driving performance, however, this is dependent on factors such as perceptual driving context and trip duration. Cognitive and psychomotor aspects of psychiatric disorders often respond to pharmacological treatment, but care must be taken to avoid acute or persistent adverse effects such as sedation or agitation, which could equally elevate crash risk through separate mechanisms. The performance of driving will continue to evolve in coming years, along with sociological and technological developments, making this a topic of continued interest to researchers interested in psychopharmacology and human performance.

References

1. Schweitzer PK. Drugs that disturb sleep and wakefulness. In: Kryger MH, Roth T, Dement WC, eds. Principles and Practice of Sleep Medicine. 3rd Ed. Saunders 2000:441-61
2. Determining Medical Fitness to Drive: A Guide for Physicians (6th ed). 2000; Canadian Medical Association.
3. Tillman WA, Hobbs GE. The Accident-prone Automobile Driver, American J Psychiatry 1949; 321-331.
4. O'Neill D. Benzodiazepines and driver safety (editorial) Lancet. 1998; 352:1324-5
5. Haigney D, Westerman SJ. Mobile (cellular) phone use and driving: a critical review of research methodology. Ergonomics 2001; 44:132-43.
6. Beeley L. Drugs and Medicines. In: Medical Aspects of fitness to drive: Medical commission on accident prevention. 4th ed. London: HMSO 1985:65-9.
7. McGwinG, Sims RV, Pulley L et al. Relations among chronic medical conditions, medications and automobile crashes in the elderly: a population-based case-control study. Am J Epidemiology 2000; 152:424-31.
8. Dodge M, Kitchin R. Introducing Cyberspace. In: Dodge, Kitchen, eds. Mapping Cyberspace. Routledge, 2001.
9. Waterford JA, Waterford El. Presence and Absence in Educational VR: The Role of Perceptual Seduction in Conceptual Learning, Themes in Education. 2000; 1:7-38.
10. Chultheis M, Mourant R. Virtual reality and driving: The road to better assessment for cognitively impaired populations. Presence 2001; 10:431-9.
11. Moller H, Kayumov L, Shapiro CM. Microsleep Episodes, Attention Lapses and Circadian Variation in Psychomotor Performance in a Driving Simulation Paradigm. Proceedings of 2nd International Conference on Driving Assessment. Park City, Utah, 2003.
12. Horne JA, Reyner LA Sleep-related vehicle accidents. BMJ 1995; 310:565-567.
13. Kecklund, Akerstedt. Time of day and Swedish road accidents. Shiftwork Int Newsletter 1995; 12:31.
14. Johns MW. A new method for measuring daytime sleepiness: the Epworth Sleepiness Scale. Sleep 1991; 14:540-545.
15. Mitler M, Carskaddon M, Hirschkowitz M. Daytime Tests. In: Kryger MH, Roth T, Dement WC, eds. Principles and Practice of Sleep Medicine. 3rd Ed. Saunders, 2000:1251-8.
16. Schmidt HS. Pupillometric assessment of disorders of arousal, Sleep. 1982; 5(Suppl 2):S157-64.
17. Arnedt JT, Acebo C, Seifer R et al. Assessment of a Simulated Driving Task for Sleep Research. Sleep 24S 2001; A413.
18. Arnedt JT, Wilde GJ, Munt PW et al. Simulated driving performance following prolonged wakefulness and alcohol consumption: separate and contributions to impairment. J Sleep Res 2000; 9:233-241.
19. George CF, Smiley A. Sleep apnea & automobile crashes. Sleep 1999; 22:790-5.
20. Moller H, Lowe A, Kayumov L et al. Can impaired alertness be detected more sensitively using a computerized driving simulator? Sleep 2002; A252-253
21. Powell NB, Schectman KB, Riley RW et al. The road to danger: comparative risks of driving while sleepy. Laryngoscope 2001; 111:887-93.
22. Lyznicki JM. Sleepiness, driving and motorvehicle crashes. JAMA 1998; 279:1908-13.
23. Dawson D, Reid K. Fatigue, alcohol and performance impairment. Nature 1997; 388:235.
24. Ramaekers JG, Robbe HW, O'Hanlon JF. Marijuana, alcohol and actual driving performance. Hum Psychopharmacol 2000; 7:551-558.
25. Lamers CT, Ramaekers JG. Visual search and urban driving under the influence of marijuana and alcohol. Hum Psychopharmacol 2001; 5:393-401
26. Liguori A, Gatto CP, Jarrett DB. Separate and combined effects of marijuana and alcohol on mood, equilibrium and simulated driving. Psychopharmacology (Berl). 2002; 163:399-405

27. Couper FJ, Pemberton M, Jarvis A et al. Prevalence of drug use in commercial tractor-trailer drivers. J Forensic Sci 2002; 47:562-7
28. Caldwell JA, Caldwell JL. An in-flight investigation of the efficacy of dextroamphetamine for sustaining helicopter pilot performance. Aviat Space Environ Med 1997; 68:1073-80.
29. Mills KC, Spruill SE, Kaune RW et al. The influence of stimulants, sedatives and fatigue on tunnel vision: risk factors for driving and piloting. Hum Factors 2001; 43:310-27.
30. Logan BK, Couper FJ. 3,4-Methylenedioxymethamphetamine (MDMA, ecstasy) and driving J Forensic Sci 2001; 46:1426-33.
31. Meltzer EO:Performance effects of antihistamines. J Allergy Clin Immunol 1990; 86:613-619.
32. Lisper HO, Tornos J, Van Loon J. Effects of caffeine or diazepam on subsidiary reaction time in a long term driving task. In: L Goldberg, ed. Alcohol Drugs and Traffic Safety. Stockholm: Almqvist & Wiksell: 1986:1024-1049.
33. Reyner LA, Horne JA. Suppression of sleepiness in drivers: combination of caffeine with a short nap. Psychophysiology 1997; 34:721-72
34. Hemmelgarn B, Suissa S, Huang A. Benzodiazepine use and the risk of motor-vehicle crash in the elderly. JAMA 1997; 278:27-31.
35. Vermeeren A, Riedel W, van Boxtel MP et al. Differential residual effects of zaleplon and zopiclone on actual driving: a comparison with low dose of alcohol. Sleep 2002; 25:224-31.
36. Linnoila M, Seppala T, Antidepressants and driving. Accid Anal & Prev 17 1985; 297-301.
37. Ramaekers JG. Antidepressants and driver impairment: Empirical Evidence from a standard on-the-road-test. J Clin Psychiatry 2003; 64:20-29
38. Linnoila M, Seppala T, Mattila M et al. Clomipramine and doxepin in depressive neurosis. Arch Gen Psychiatry 1980; 37:1295-1299.
39. Hindmarch I. The behavioural toxicity of the selective serotonin reuptake inhibitors. Int Clin Psychopharmacol. 9 S 1995; 9(Suppl 4):13-7.
40. O'Hanlon JF, Freeman H. Categorizing the behavioural toxicities of antidepressants. B J Psychiaty 1995; 166:421-423.
41. Ridout F, Meadows R, Johnsen S et al. A placebo controlled investigation into the effects of paroxetine and mirtazapine on measures related to car driving performance. Hum Psychopharmacol. 2003; 18:261-9.
42. Kagerer S, Winter C, Moller HJ et al. Effects of haloperidol and atypical neuroleptics on psychomotor performance and driving ability in schizophrenic patients. Results from an experimental study. Neuropsychobiology 2003; 47:212-8.
43. Robbe HW, O'Hanlon JF. Acute and subchronic effects of paroxetine 20 and 40 mg on actual driving, psychomotor performance and subjective assessments in healthy volunteers. Eur Neuropsychopharmacology 1995; 5:35-42.
44. van Laar M,Willigenburg AP and Volkerts ER. Acute and subchronic effects of nefazodone and imipramine on highway driving, cognitive functions, and daytime sleepiness in healthy adult and elderly subjects. J Clin Psychopharmacology 1994; 15:30-40
45. Cox DJ, Merkel L, Kovatchev B et al. Effect of stimulant medication on driving performance of young adults with attention-deficit hyperactivity disorder. J Nerv Mental Dis 2000; 188:230-234.
46. Hatcher S, Sims R, Thompson D The effects of chronic lithium treatment on psychomotor performance related to driving B J Psychiaty 1990; 157: 275-8
47. Linnoila M, Saario I, Maki M. Effect of treatment with lithium and alcohol on psychomotor skills related to driving Eur J Clin Psychoparmacol 1974; 7:337-42.
48. Aldenkamp AP, De Krom M, Reijs R. Newer antiepileptic drugs and cognitive issues. Epilepsia 2003; 44(Suppl 4):21-9.
49. Judd LL. The effect of antipsychotic drugs on driving and driving related psychomotor functions. Accid Anal Prev 1985; 17:319-22
50. Frucht S, Rogers JD, Greene PE et al. Falling asleep at the wheel: Motor vehicle mishaps in people taking pramipexole and ropinorole. Neurology 1999; 52:1908-10
51. Moller HJ. Antiparkinsonian Drugs and Sleep Attacks. CMAJ 2001; 164:1038-39.

CHAPTER 17

Diagnosis, Pathophysiology and Treatment of Hypersomnias

Sebastiaan Overeem and Michel Billiard

Abstract

Besides the obstructive sleep apnea/hypopnea syndrome, the most prevalent cause of hypersomnia, other disorders may be the cause of more severe hypersomnia. These include primary disorders of the central nervous system and hypersomnia associated with various medical disorders. Among the first ones are narcolepsy-cataplexy, idiopathic hypersomnia and recurrent hypersomnia. Narcolepsy-cataplexy is the one disorder that has benefited from the most extensive pharmacological, biochemical, pathological and genetic findings, mostly due to the availability of an animal model, the narcoleptic dog. In comparison, idiopathic hypersomnia and recurrent hypersomnia are much less known. Treatment of these disorders used to be limited to amphetamines and their derivatives with non negligible adversed effects. Today a new compound with awakening properties, modafinil, has become available. Its mode of action is not yet fully understood, but it has the great advantage of very limited adverse effects. However it is still a purely symptomatic treatment. For narcolepsy-cataplexy, there are high hopes that the recently discovered central role played by the hypocretin (orexin) system in both canine and human narcolepsy will result in a causal treatment of the condition. Hypersomnias associated with various medical disorders have received less attention. However hypersomnias associated with neurologic and infectious disorders may open the door to a better understanding of the anatomic basis of hypersomnia.

Introduction

Falling asleep at the wheel, while reading a newspaper or a book, or in conversation with friends during the daytime are all abnormal situations testifying to hypersomnia. However, in contrast to insomnia which is a widespread phenomenon, familiar to patients and physicians alike, hypersomnia is often overlooked and considered as rare or even exceptional. Nevertheless, numerous studies have shown prevalence of 4 to 6% of the population being afflicted by severe hypersomnia and 15 to 20% by moderate hypersomnia.[1,2] Hypersomnia is not free of consequences. Subjects with sleep apnea or narcolepsy have a higher rate of accidents than control subjects.[3] Cognitive functions are affected, especially memory.[4] Mortality rate has been shown to be higher in subjects reporting daytime sleep episodes than in controls in an older community population.[5] The main cause of hypersomnia is the obstructive sleep apnea/hypopnea syndrome. However there are other causes of hypersomnia including primary disorders of the central nervous system, namely narcolepsy-cataplexy, idiopathic hypersomnia, recurrent hypersomnias, and hypersomnias secondary to various medical conditions, neurological, psychiatric, infectious and endocrine. Among the primary disorders of the central nervous system, narcolepsy-cataplexy is the one which has benefited from the most extensive pharmacological, biochemical, pathological and genetic findings. Treatment of hypersomnias used to be limited to amphetamines and their derivatives with non negligible undesirable effects. Fortunately, the last twenty years have seen the development of a new compound, modafinil, with awakening properties and very limited adverse effects. Moreover other compounds are either in development or being conceived given the progress in understanding the pathophysiology of narcolepsy-cataplexy.

Primary Disorders of the Central Nervous System

Narcolepsy-Cataplexy

Introduction

The term "narcolepsy" was coined by Gélineau in 1880, to describe a syndrome characterized by the combination of unwanted episodes of sleep and episodes of muscle weakness later called cataplexy.[6] Afterwards, it was recognized that hypnagogic hallucinations and sleep paralysis occurred in many narcoleptic patients as well. This notion resulted in the description of the "narcoleptic tetrad", encompassing these four symptoms.[7] Shortly after the initial discovery of REM sleep in 1953, it was observed that narcoleptic patients tend to enter directly into REM sleep.[8] This and other observations led to the hypothesis that cataplexy, sleep paralysis and hypnagogic hallucinations are dissociated features of REM sleep. In this view, narcolepsy is a combination of excessive daytime sleepiness and disordered REM sleep regulation. Observing patients with different lesions of the hypothalamus, von Economo predicted as early as the 1930s, that narcolepsy is a disorder of that brain region.[9] Remarkably, von Economo was proved right by a sequence of discoveries commencing in 1999, which pinpointed a damaged hypothalamic hypocretin system as the principal cause of narcolepsy.

Sleep and Sleep Disorders: A Neuropsychopharmacological Approach, edited by Malcolm Lader, Daniel P. Cardinali and S.R. Pandi-Perumal.

Epidemiology

The most recent epidemiological studies have estimated the prevalence of narcolepsy to be around 2.5 per 10,000 in Western countries.[10] Differences in prevalences have been reported in various countries, the highest in Japan and the lowest in Israel. The exact meaning of these differences are unclear; they could be due to methodological differences or differences in criteria to diagnose narcolepsy. However, they may also point to genetic differences influencing susceptibility to acquire narcolepsy. Men are affected somewhat more often than women.

In the majority of patients, the age of onset lies between 15 and 30 years.[11] Typically, the presenting symptom is excessive daytime sleepiness, with or without hypnagogic hallucinations or sleep paralysis. Cataplexy usually occurs later on, on average six years after the onset of sleepiness. However, in 5-8% of patients, cataplexy is the presenting symptom. In the end, only 10-15% of patients develop the full narcoleptic tetrad. About 30% of patients have hypnagogic hallucinations and 25% sleep paralysis.

Clinical Features

Excessive daytime sleepiness (EDS) is the principal symptom of narcolepsy. Clinically, it can be expressed in different ways. Patients may report a continuous feeling of sleepiness throughout the day. Alternatively, they may present with sudden episodes of irresistible sleepiness, with relatively few complaints between these episodes. When questioned in detail, most patients experience both forms, although one may dominate. These differences may have implications for the treatment of choice. Sleepiness must be distinguished from fatigue, which is a very nonspecific complaint. It is therefore important to specifically ask a patient if he or she "ever unintentionally has fallen asleep". While the severity of EDS may vary from day to day, narcoleptic patients will never be completely free of sleepiness. EDS in narcolepsy is best described as an "inability to stay awake" rather than an increased amount of sleep. Sleep episodes are much more fragmented, but over 24 hours, the total amount of sleep is not increased.[12]

Naps typically occur in situations of lowered alertness, for example during reading, watching television or as a passenger in a car. In contrast to the "archetypical narcoleptic", naps seldom occur under highly unusual circumstances like while riding a bicycle. Physical activity may sometimes even postpone the moment of falling asleep, a reason why many patients tend to force themselves to be active. Naps last relatively short in most instances. Most of the times, naps are described to be refreshing. It is important to realize that no amount of sleep will prevent the occurring of a new sleep episode for more than a few hours. Furthermore, every narcoleptic patient will have difficulties staying awake in a dark quiet room, regardless of the previous amount of sleep, previous amount of physical activity or the use of stimulant or other medication.

Cataplexy is the most specific symptom of narcolepsy. It is defined as a sudden loss of voluntary muscle tone, with preserved consciousness, triggered by strong emotions.[13] Narcoleptic patients may in fact literally become "weak with laughter". All striated muscles may be affected, except the extraocular and respiratory musculature. Cataplectic attacks may be complete, leading to a fall to the ground. More often however, they are only partial. A partial attack usually affects the neck (resulting in dropping of the head) or the lower limbs (leading to buckling of the knees). Sometimes, cataplexy is limited to a mere feeling of weakness, without an apparent loss of muscle tone. Most patients have an individual pattern of attacks, although complete and partial attacks may occur in the same patient. Attacks are short-lived, lasting between several seconds and one or two minutes. Sometimes an attack may gradually shift into (REM) sleep, making it seem that the attack lasted longer. Attack frequency may vary widely, from less than one per month, to tens of attacks a day.

Several emotions may evoke a cataplectic attack. Laughing or telling a joke is most often reported as a trigger.[13] Other triggers include the sudden meeting of an aquaintance, scoring a goal during sports or merely thinking about the punchline of a joke and less often anger or stress. Interestingly, it seems necessary that a patient is in a somewhat relaxed, familiar surroundings in order to get an attack. Consequently it is seldom possible to evoke an attack during a consultation in the hospital. If an attack is witnessed, neurological examination will show atonia and a loss of tendon reflexes.

Consiousness is always preserved during cataplexy, so patients are fully aware of their surroundings during an attack. It is important to verify this, especially to differentiate cataplexy from attacks mimicking the condition.

Hypnagogic hallucinations are mental imagery at the moment of falling asleep, that may be so real that they are mistaken for something that actually happened. Visual imagery is predominant, but auditory and tactile components are frequently present. The content of hypnagogic hallucinations is rarely pleasant, most patients report them to be bizarre and frightening. This may be one of the reasons that hypnagogic hallucinations are frequently not reported spontaneously by a patient, so their presence should be specifically asked for.

Sleep paralysis is the inability to move during the transition from sleep to wakefulness or vice versa, while the patients is subjectively awake and conscious. In this respect, sleep paralysis resembles cataplexy, but is different with respect to its timing, the absence of an emotional trigger and its duration (up to 10 minutes). People are usually frightened during an attack, especially the first time. Sometimes, sleep paralysis is accompanied by hypnagogic hallucinations, making the experience even worse.

Nowadays, disturbed nocturnal sleep is considered to be an integral part of narcolepsy. Although patients fall asleep very quickly at night, most have frequent awakenings during the night. For some patients, this fragmented nighttime sleep is even the most troublesome symptom. Interestingly, there is no relation between the (lack of) quality of sleep at night, and the severity of daytime sleepiness.[14]

Other symptoms that are frequently reported by narcoleptics are: automatic behaviour (semi-purposeful activity during a state of drowsiness) and memory disturbances. Furthermore, the majority of patients is overweight. Obesity in narcolepsy is not caused by inactivity during the day or an increase in caloric intake, and is probably a direct consequence of the pathophysiological mechanisms involved (see below).[15]

Positive Diagnosis

Narcolepsy is primarily diagnosed based on the patient's history. However, the exact diagnostic criteria have been subject of much debate. It is generally accepted that the combined presence of EDS and cataplexy is diagnostic for narcolepsy. When cataplexy is not present, or when its presence is doubted, difficulties arise. The criteria of the International Classification of Sleep Disorders (ICSD) then requires the use of ancillary polysomnographic studies to make the diagnosis (Table 1).[16] Currently, the most important diagnostic test is the Multiple Sleep Latency Test (MSLT). The MSLT measures the mean sleep latency during four

Table 1. ICSD criteria for narcolepsy[16]

A The patient has a complaint of excessive sleepiness or sudden muscle weakness
B Recurrent daytime naps or lapses into sleep occur almost daily for at least 3 months
C Sudden bilateral loss of postural muscle tone occurs in association with intense emotion (cataplexy)
D Associated features include:
   1. Sleep paralysis
   2. Hypnagogic hallucinations
   3. Automatic behaviors
   4. Disrupted major sleep episode
E Polysomnography demonstrates one or more of the following:
   1. Sleep latency less than 10 minutes
   2. REM sleep latency less than 20 minutes and
   3. An MSLT that demonstrates a mean sleep latency of less than 5 minutes and
   4. Two or more sleep-onset REM periods
F HLA typing demonstrates DQB1*0602 or DR2 positivity
G No medical or mental disorder accounts for the symptoms
H Other sleep disorders (e.g., periodic limb movement disorder or central sleep apnea syndrome) may be present but are not the primary cause of the symptoms

Minimal Criteria: B plus C, or A plus D plus E plus G.

or five 20-minute periods divided over the day, in which a subject is asked to try to fall asleep in a quiet and dark room. In addition, the occurence of REM sleep is noted. The ICSD criteria require a mean sleep latency of less than 5 minutes together with two or more sleep-onset REM periods (SOREMPs) to make the diagnosis of narcolepsy. This may even be the case without the presence of cataplexy, and some authors prefer to call this condition specifically 'narcolepsy without cataplexy'.

Recent studies question both the sensitivity and specificity of the current MSLT criteria for narcolepsy. For example, up to 15% of patients with clear-cut cataplexy do not fulfill these criteria.[17] Conversely, patients without any sleep complaints may test postive for narcolepsy based on the MSLT. HLA-typing has been advocated by some as an aid in diagnosis, but has limited value. Although more than 90% of narcoleptics are carrier of the HLA subtype DQB1*0602, this holds true for around 25% of the general population. So, HLA typing may only be useful to exclude a diagnosis of narcolepsy.

The finding that the vast majority of patients with narcolepsy have no detectable levels of the neurotransmitter hypocretin-1 in their cerebrospinal fluid, may give rise to a new diagnostic test.[18,19] Currently, the ICSD criteria are being revised according to these new insights, for example to include hypocretin-1 measurements in the diagnostic trajectory. Interestingly, the absence of hypocretin-1 is almost completely associated with the presence of cataplexy (Fig. 1), giving additional support to the opinion that cataplexy is an obligatory feature of narcolepsy.

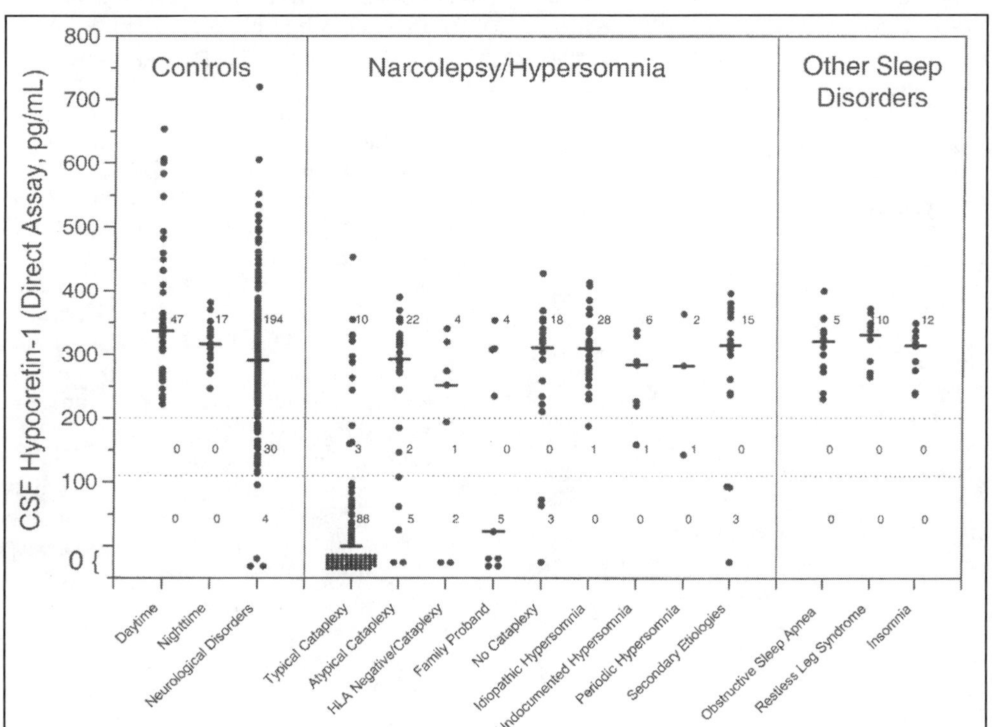

Figure 1. CSF Hypocretin-1 levels across various disease categories. Each dot represents a single patient. Hypocretin-1 values below 110 pg/ml were determined to be diagnostic for narcolepsy. Concentrations above 200 pg/ml best determine healthy control values. The number of subjects with hypocretin values below or equal to 110 pg/ml, above 200 pg/ml and between these two values is indicated for each category. Reprinted with permission from ref. 19.

Differential Diagnosis

As mentioned, the combination of EDS and cataplexy is pathognomonic for narcolepsy. Several differential diagnostic possibilities exist for the separate symptoms.[20] Excessive daytime sleepiness is a feature of other intrinsic sleep disorders such as idiopathic hypersomnia and recurrent hypersomnia, which are discussed later in this chapter. In addition, disturbed nocturnal sleep due to sleep apnea or periodic limb movement disorder is associated with EDS. Excessive daytime sleepiness may be confused with fatigue, and syndromes associated with fatigue.

Cataplexy may occur without sleepiness in rare diseases such as Niemann Pick disease type C, the Prader-Willy syndrome and Norrie disease. Furthermore, a familial form of isolated cataplexy has been described. Cataplexy must be distinguished from attacks that merely resemble the condition, such as syncope, drop attacks, atonic seizures and psychogenic disorders. The preservation of consciousness during cataplexy and the emotional trigger are important diagnostic clues.

Both sleep paralysis and hypnagogic hallucinations occur in isolation in a part of the normal population. The latter may also be confused with nightmares, seizures or psychiatric conditions, especially schizophrenia.

Pathophysiology/Neuropharmacology

On a conceptual level, the symptoms of narcolepsy have been explained as a 'loss of state boundary control'.[21] This results in two key disturbances of sleep/wake regulation. First, narcoleptics are unable to sustain any given sleep or wakefulness state: patients can be awake but do not stay awake, and can be asleep but do not stay asleep. Second, various characteristics of sleep stages that normally occur together (e.g., eye movements and muscle atonia during REM sleep) may be expressed out of their context and at inopportune moments. Cataplexy is one of these so-called dissociation phenomena and is portrayed as REM-sleep atonia occurring at the wrong time. While this theoretical framework makes it easier to describe the symptoms of narcolepsy, it has never been proven and in fact, recent data may disagree with the state-boundary-control hypothesis.[13]

The neurobiology of human sleep is discussed elsewhere in this book and in recent reviews.[22,23] Neuropharmacological data on human narcolepsy has been derived from patients with a long history of drug treatment, so data may certainly have been influenced by this. Neuroanatomical information has been gained from case-reports of symptomatic narcolepsy secondary to brain lesions, and points to an important role of the diencephalon and upper brain stem.[24] Fortunately, much has been learned from the naturally occurring animal model: canine narcolepsy. The narcoleptic dog model has a remarkable resemblance to the human disease, including cataplexy (Fig. 2).[25] A huge amount of neuropharmacological data has been derived using the canine model, as described in the landmark review of Nishino and Mignot.[25]

EDS in narcolepsy is primarily caused by a hypoactivity of the dopaminergic system. Indeed, most stimulant medications act by increasing dopaminergic tone. Cataplexy is caused by an imbalance between pontine aminergic and cholinergic systems. Cataplexy is aggravated by cholinergic activation and/or deactivation of monoaminergic, particularly adrenergic, systems. The various medications with alleviating effect on cataplexy increase norepinephrine availability. Conversely, the alpha-1 adrenergic blocker prazosine, used as an antihypertensive agent, severely worsens cataplexy.

Figure 2. Attack of cataplexy, here elicited by a piece of palatable food, in a dog with the hereditary form of narcolepsy.

The genetic basis of human narcolepsy is complex. Although genetic factors do influence the susceptibility of narcolepsy, most human cases do not occur in families. Furthermore, only 20-30% of monozygotic twins is concordant for the disease. However, for first-degree relatives of a patient, the risk of acquiring narcolepsy is about 40 times higher than the general population.[26] The best studied genetic locus in narcolepsy is the HLA system. Early studies found an association with the serologically determined HLA subtype DR2 which varied with ethnicity. High-resolution DNA-based mapping of this HLA region identified the strongest association with the subtype DQB1*0602. Over 90% of patients with narcolepsy are positive for DQB1*0602. There are several other genetic loci that may be associated with narcolepsy, of which the TNF-alpha gene is the studied in greatest detail. Polymorphisms in this gene influence the susceptibility to narcolepsy, although plasma TNF-alpha levels are unaltered in narcolepsy.[27]

As most autoimmune disorders are linked to specific HLA haplotypes, narcolepsy is hypothesized to be an autoimmune disease as well. Extensive studies have been performed looking for general markers of (auto)immune activation, such as oligoclonal bands in the cerebrospinal fluid, serum immunoglobulin levels and lymphocyte subset rations. To date, no direct evidence for an autoimmune process has been identified.[28]

In contrast to human narcolepsy, narcolepsy in dogs is transmitted as an autosomal recessive trait. In 1999, a decade long search for the responsible gene was concluded when mutations in the hypocretin-receptor 2 gene were identified in two strains of narcoleptic dogs.[29] This discovery initiated a host of new research which provided key information about the pathophysiology of the disorder. Only two weeks after the gene for canine narcolepys was found, the group of Yanagisawa reported on the phenotype of hypocretin ligand knockout mice.[30] These mice have several sleep abnormalities including a decreased REM latency and SOREMPs, as well as brief episodes of behavioural arrests which seem to be the murine equivalent of cataplexy. Very soon the first study appeared which confirmed that the hypocretin system plays a critical role in human narcolepsy as well. Nishino and colleagues measured CSF hypocretin-1 levels in 9 patients and 8 controls, and found that 7 patients had no detectable hypocretin-1 at all, while the other 2 patients and all the controls had stable, detectable concentrations.[18] These findings have now been replicated in over 150 narcolepsy patients and several hundreds of subjects with other sleep disorders and various neurological diseases (Fig. 1). It is clear that hypocretin-1 deficiencies are highly specific for narcolepsy with cataplexy; most narcoleptic patients without cataplexy have normal levels as do all subjects with other sleep disorders.[19]

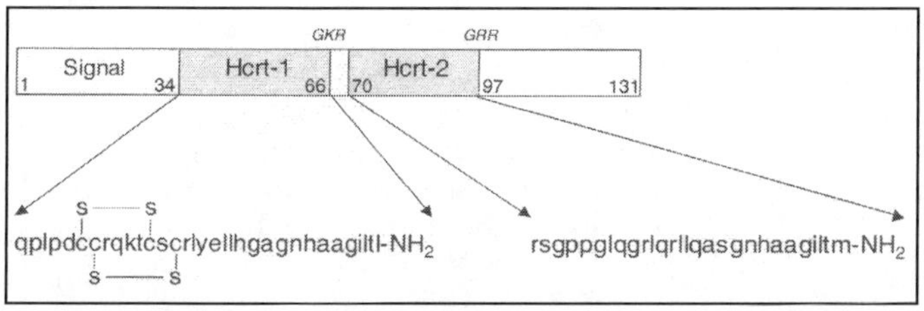

Figure 3. Diagram of the preprohypocretin from which the hypocretin peptides are cleaved. *GKR* and *GRR* depict dibasic residues, that are potential cleavage sites for prohormone convertases. The derived aminoacid sequences for hypocretin-1 and hypocretin-2 are shown as well.

Two neuropathological studies using post-mortem brain tissue from several narcoleptic patients showed that both hypocretin mRNA and peptide are not detectable in the hypothalamus and in hypocretin projection areas.[31,32] However, no consistent evidence was found for gliosis, or other signs of neuronal damage. So the cause of the hypocretin deficiency remains unknown. However, the findings fit well with the autoimmune hypothesis, and many current studies are searching for evidence of an autoimmune attack on the hypocretin neurons.

The hypocretin system consists of two neuropeptides, hypocretin-1 and -2 (also known as orexin-A and -B) and two receptors (Fig. 3).[33] The hypocretin producing neurons are located in the perifornical part of the hypothalamus. Despite the confined area of the hypocretin cell bodies, the neurons project widely throughout the brain. Projection areas include the brain regions known to be involved in narcolepsy, with densest connections to brain stem aminergic nuclei. In addition, the hypocretin system plays a role in the regulation of appetite, energy metabolism and various neuroendocrine systems. It is therefore not surprising that narcoleptic patients show alterations in these functions as well. Alterations in various hormonal ensembles, including the leptin system, the pituitary-adrenal axis and the growth hormone axis have been documented.[34]

Treatment

Narcolepsy is a chronic disease. Once established, the symptoms of narcolepsy remain more or less stable, although cataplexy may sometimes improve in time. At this time, there is no causal treatment available. Therefore, narcolepsy requires long-term, sometimes life-long, treatment. The various symptomatic treatment options will never completely abolish symptoms. Treatment goal is therefore to achieve an optimal balance between beneficial effects and side effects. Most treatments aim at specific symptoms of narcolepsy, so combination therapy will often be necessary. In practice, the most disabling symptom is treated first, often this will be the EDS. When this symptom is sufficiently controlled, it can be decided if the other symptoms need treatment. Treatment effects are monitored clinically, i.e., guided by the patient's report and not by means of polysomnography or other measurements. Nonpharmacological treatment such as life-style advices may be of some help in a subset of patients. However, most patients will require some form of pharmacological intervention.

It is important that patients try to live a regular life, go to bed around the same time each night, and get up around the same time every morning. In most patients, several scheduled naps during the day will alleviate sleepiness to some extent.[35] The optimal frequency, duration and timing of these naps may differ between individuals. Narcoleptics may be more sensitive to the sleep-inducing properties of carbohydrates. Therefore, they should not eat large carbohydrate-rich meals.

Table 2. Stimulants commonly used in the treatment of EDS

Compound	Total Daily Dose (mg)	Typical Number of Doses
amphetamine-like		
methamphetamine	10-50	1
dextroamphetamine	5-60	1-2
methylphenidate	10-60	2-3
non amphetamine-like		
modafinil	200-400	1-2

Stimulants are the most important pharmacological group in the treatment of EDS.[36] Several drugs are available in this category (Table 2), which differ in effectiveness, number of side-effects and duration of action. Most stimulants are amphetamine-like drugs, including methylphenidate, dextroamphetamine and methamphetamine. The main mode of action is to increase aminergic signalling, most important dopaminergic transmission. Side effects are common and include nervousness, palpitations, irritability, agitation and headache. These side effects and the occurence of tolerance are major drawbacks in the use of stimulants. Modafinil is a nonamphetamine like wake-promoting drug that has recently been approved in the USA, after having been used for a longe time in Europe specially in France. Its exact mechanism of action is unclear as of yet. The efficacy of modafinil seems to be comparable to the other stimulants. The main advantage is the lower frequency and severity of side effects. Long acting stimulants (including modafinil) are often better tolerated than short acting drugs (such as methylphenidate). However, the latter have the advantage that they can be used 'on demand', targeted at specific times during the day when a high level of attention is needed most.

Generally, the treatment of cataplexy is associated with a reduction of hypnagogic hallucinations and sleep paralysis.[25] The various drugs used increase the availability of norepinephrine and serotonin and are potent REM-sleep inhibitors (Table 3). The most effective treatment of the REM-dissociation symptoms are the tricyclic antidepressants. Several authors consider clomipramine the treatment of choice, but imipramine and protryptiline have been used as well. The tricyclics are often effective in very low doses, clomipramine may be sufficient in a dose of only 10 mg a day. Low dosages may partly prevent the occurrence of side effects, an important disadvantage of the tricyclics. The most common side effects are anticholinergic in nature and include dry mouth, increased sweating, weight gain, delayed orgasm or erectile dysfunction. These side effects triggered the study of alternative treatments, especially the selective serotonin reuptake inhibitors. Several drugs in this class have been studied, including fluoxetine and fluvoxamine. Importantly, SSRIs must be used in relative higher dosages than the tryciclics. Venlafaxine is being advocated by some as treatment of first choice. However, there are no published studies on its efficacy and it does not seem to be better tolerated than for example low doses

Table 3. Drugs used in the treatment of cataplexy

Compound	Total Daily Dose (mg)	Typical Number of Doses
Tricyclic antidepressants		
clomipramine	10-150	1-3
imipramine	10-150	1-3
protryptiline	2.5-50	1-2
SSRIs		
fluoxetine	20-60	1
fluvoxamine	100-300	1-2
venlavaxine	37.5-150	1-2
Gammahydroxybutyrate	3-9 gram	2 doses, given at bedtime and 3-4 hours later

of clomipramine. Antidepressants should never be abruptly withdrawn, as that may severely aggravate cataplexy.

Gammahydroxybutyrate (GHB, sodium oxybate) was first proposed as a treatment of REM-based symptoms by Broughton and Mamelak.[37] It has recently been approved in the USA for the treatment of cataplexy. GHB is a sedating drug with a very short half life. It should be taken at night in two doses, one at bedtime and one 3-4 hours later. In a recent large study, 2 times 4.5 grams of GHB substantially reduced the frequency of cataplexy.[38] Side effects include dizziness and nausea, and urinary incontinence. GHB gained attention as a drug of abuse, and as a 'date-rape' drug. Therefore, its use must be closely monitored.

In some patients, disturbed nocturnal sleep may be a major complaint. Treatment options are limited and often unsatisfactory however. GHB may improve the (subjective) quality of nocturnal sleep. Furthermore, benzodiazepine hypnotics may be effective in consolidating nocturnal sleep, although tolerance may develop and side effects occur. Interestingly, treatment of fragmented nighttime sleep is not associated with a decrease of EDS during the day.[14]

Idiopathic Hypersomnia

Introduction

In comparison with narcolepsy or the Kleine-Levin syndrome, idiopathic hypersomnia stands as a recently identified entity. Dement et al proposed that subjects with excessive sleepiness but neither cataplexy, sleep paralysis nor SOREMPs should not be considered as narcoleptics and should be relegated to another diagnostic category.[39] Three years later Rechtschaffen and Roth gave a sound description of a disease that they simply called "hypersomnia", characterized by daytime sleep not as irresistible as that of narcolepsy but usually lasting much longer; normal sleep except for its profound extension; and major difficulty in coming to full wakefulness following nocturnal sleep.[40] In 1972 Roth et al published a paper entitled "hypersomnia with sleep drunkenness"[41] in which they made a masterpiece description of the major difficulty waking up of these subjects and in 1976 Roth coined the word "idiopathic hypersomnia".[42] This disorder should be divided into two forms, monosymptomatic characterized by excessive daytime sleepiness only and polysymptomatic characterized by excessive daytime sleepiness, prolonged nocturnal sleep and major difficulty in awakening. In the revised version of the ICSD now in preparation, the polysymptomatic and the monosymptomatic forms will be considered as separate entities. They will be referred to as idiopathic hypersomnia with long sleep time and idiopathic hypersomnia without long sleep time

Epidemiology

Due to the nosological uncertainty and the relative rarity of the condition, prevalence studies have not been performed so far. However some predictions can be made. Various sleep disorders center populations including both narcolepsy and idiopathic hypersomnia patients have been reported, allowing a calculation of the ratio of idiopathic hypersomnia to narcolepsy. Interestingly the ratio decreases in the successive reports from an initial 76.9%[42] to a present 13.9%,[43] due to the identification of new sleep disorders such as the upper airway resistance syndrome or hypersomnia associated with mental disorders and more stringent criteria of the disease. The age of onset varies from childhood to young adulthood with very few cases occurring after the age of 25. There is apparently no gender predominance.

Clinical Features

Idiopathic hypersomnia with long sleep time is remarkable for three symptoms: a complaint of constant or recurrent daily excessive sleepiness and unwanted naps, usually longer and less irresistible than in narcolepsy, and non refreshing irrespective of their duration; night sleep is sound, uninterrupted and prolonged; morning awakening is laborious. Subjects do not awaken to the ringing of a clock, a telephone and often rely on their family members who must use vigorous and repeated procedures to wake them up. Even then, patients remain confused, unable to react adequately to external stimuli, a state referred to as "sleep drunkenness". Associated symptoms suggesting a dysfunction of the autonomic system are not uncommon. They include headaches which may be migrainous in quality; fainting episodes (syncopes); peripheral vascular complaints.

Idiopathic hypersomnia without long sleep time stands as isolated excessive daytime sleepiness.

Course

In contrast with narcolepsy, onset of idiopathic hypersomnia is much more progressive, so that a precise date of onset cannot be given. The disorder is generally stable and long lasting. Complications are mostly social and professional.

Positive Diagnosis

Diagnosis is mostly clinical. However laboratory tests are necessary to rule out other hypersomniac conditions.

Polysomnographic monitoring of nocturnal sleep demonstrates normal sleep, except for its prolonged duration in the case of idiopathic hypersomnia with long sleep time. NREM and REM sleep are in normal proportions. There are no sleep onset REM periods. Sleep apneas and periodic limb movements should theoretically be absent, but may be acceptable in the case of an early onset of idiopathic hypersomnia and of their late occurrence. Monitoring of oesophageal pressure is mandatory to rule out upper airway resistance syndrome.

The MSLT demonstrates a mean sleep latency less than 10 minutes, which could be longer than in narcolepsy in the form with long sleep time and in the range of narcolepsy in the form without long sleep time.

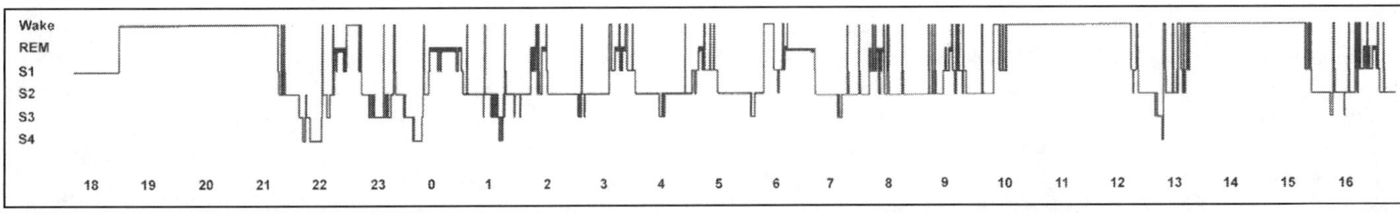

Figure 4. Continuous 24-hour recording in a 21 year old woman with idiopathic hypersomnia. The patient slept from 21h 29 to 10h16, from 12h29 to 13h31 and from 15h33 to 17h21. Total sleep time over the 24 hour period was 14hours and 56 minutes.

In the case of idiopathic hypersomnia with long sleep time, the diagnostic value of the MSLT is somewhat questionable as awakening the subject in the morning in view of the later MSLT sessions precludes documenting the abnormally prolonged night sleep, and the MSLT sessions preclude recording of prolonged unrefreshing daytime sleep episode(s) of major diagnostic value. Thus another diagnostic procedure, 24 or 48 hour continuous polysomnography, on an ad-lib sleep/wake protocol, still awaiting standardization and validation, seems of potential interest in documenting a major sleep episode (> 10 hours) and at least one nap of long duration (> 1 hour) (Fig. 4).

CT scan or MRI of the brain is always indicated to rule out a brain lesion. Psychological interview and tests are advisable to exclude hypersomnia associated with mental disorder.

Differential Diagnosis

Idiopathic hypersomnia is frequently overdiagnosed due to an unfortunate tendency to label as such all hypersomnias that do not fit the criteria of either narcolepsy or obstructive sleep apnea/hypopnea syndrome.

The first diagnosis to consider is the upper airway resistance syndrome. Excessive daytime sleepiness, snoring loudly without obvious apneas, and fatigue on awakening are the main clinical features. In any event the presence of multiple brief arousals occurring during polysomnography must draw attention and call for monitoring oesophageal pressure (Pes). The typical pattern is a sequence of breaths characterized by increasing respiratory effort leading to an arousal from sleep, which does not meet criteria for an apnea or hypopnea. This pattern is referred to as respiratory effort-related arousal (RERA) event.[44]

Narcolepsy without cataplexy is a clinical variant of narcolepsy without cataplexy but associated REM abnormalities including significant symptoms of dissociated REM sleep (sleep paralysis, hypnagogic hallucinations) and/or several SOREMPs during the MSLT.

Hypersomnia associated with psychiatric disorder should be considered in a subject with abnormal personality features. The complaint of excessive sleepiness and prolonged sleep is rather similar to that of patients with idiopathic hypersomnia, except that it may vary from day to day and is often associated with poor night sleep. The MSLT does not demonstrate an abnormally short mean sleep latency.[45]

Post-traumatic hypersomnia may mimic idiopathic hypersomnia. Past medical history including an initial coma after head trauma is revealing. Hypersomnia usually develops within 6 to 18 months after the head trauma.

Hypersomnia following a viral infection such as pneumonia, mononucleosis or Guillain-Barré syndrome usually develops within months after the infection, when the subject realizes that he is not only fatigued but abnormally sleepy.

Chronic fatigue syndrome is characterized by persistent or relapsing fatigue that does not resolve with sleep or rest. Polysomnography shows reduced sleep efficiency and may include alpha intrusion into sleep EEG.[46]

Pain or other medical symptoms responsible for fragmented night sleep may result in excessive sleepiness as may occur in subjects with ankylosing spondilytis or rheumatoid arthritis.

Insufficient sleep syndrome is associated with excessive daytime sleepiness, impaired concentration and lowered energy level. A detailed history of the subject's current sleep schedule may point to this condition.

Pathophysiology

A genetic basis for idiopathic hypersomnia has been suggested by two reports, one by Nevsimalova and Roth[47] and one by Roth[48] with data in favour of an autosomal dominant mode of inheritance. Despite a few reports, no consistent association with HLA has been documented.

In contrast with narcolepsy no natural model of idiopathic hypersomnia is available. However destruction of norepinephrine neurons of the rostral third of the locus coeruleus complex or of the norepinephrine bundle at the level of the isthmus in the cat leads to hypersomnia with a proportional increase of NREM sleep and REM sleep suggestive of idiopathic hypersomnia. In this situation telencephalic norepinephrine has been shown to be decreased and 5-HIAA and tryptophan to be increased.[49]

Neurochemical studies have been performed. According to one study assessing mean CSF concentrations of monoamine metabolites and using principal component analysis, all four monoamine metabolites (DOPAC, MHPG, HVA and 5-HIAA) were highly intercorrelated in normal volunteers. In contrast HVA and DOPAC, the dopamine (DA) metabolites, did not correlate with the other two metabolites in narcoleptic subjects, and MHPG, the norepinephrine (NE) metabolite, did not correlate with the other three metabolites in idiopathic hypersomnia subjects. Accordingly an alteration of the DA system in narcolepsy and an alteration of the NE system in idiopathic hypersomnia were suggested.[50]

It was interesting in the context of recently discovered CSF hypocretin-1 deficiency in narcolepsy-cataplexy to measure CSF hypocretin-1 levels in subjects with idiopathic hypersomnia. In a first study none of the 12 subjects with idiopathic hypersomnia had undetectable low levels[51] whereas in a second one low levels were reported.[52]

Treatment

Despite different types of excessive sleepiness in patients with idiopathic hypersomnia and narcolepsy, treatment of idiopathic hypersomnia has relied on the same drugs as for sleepiness and unwanted sleep episodes of narcolepsy. Stimulant drugs including dextroamphetamine, methamphetamine, methylphenidate, pemoline, have been used with some success on excessive sleepiness but with less success on the difficulty in morning awakening. Moreover, unwanted effects such as headache, tachycardia or irritability have been commonly reported, may be more often than in narcoleptics. Modafinil has yielded good results in patients with idiopathic hypersomnia.[53] However awakening difficulties are not improved and a double-blind controlled study or a comparative study with conventional stimulants has not yet been performed.

Behavioral treatment possibilities are limited. Naps are of no help as they are both lenghty and nonrefreshing. "Saturating" the patient with sleep on week-ends (recommending him or her to sleep for as long as possible) does not seem to have a sustained effect.

Recurrent Hypersomnia

Introduction

This refers to a group of rare disorders characterized by episodes of more or less continuous sleep, with an average duration of one week, recurring at highly variable intervals of between one and several months. The most classic picture is that of the Kleine-Levin syndrome in which sleep episodes are associated with other symptoms. The first descriptions of the syndrome are attributed to Kleine[54] and Levin[55] and the eponymous term Kleine-Levin syndrome was coined by Critchley and Hoffman.[56] Later on Critchley[57] emphasised four complementary clinical features: the syndrome essentially or even exclusively affects the male sex; onset occurs in adolescence; overeating is of a compulsory nature and is not due to pathological hunger; recovery occurs spontaneously.

Epidemiology

The Kleine-Levin syndrome is an uncommon disorder with roughly 200 cases published in the world literature. The male to female ratio is about 4/1. Early adolescence is the most usual age of onset.

Clinical Features

The syndrome is typically characterized by recurrent episodes of hypersomnia associated with behavioral disorders including binge eating (rapid consumption of a large amount of food on a compulsory manner), oversexuality in the form of sexual advances, shamlessly expressed sexual fantasies or masturbation in public, irritability, aggressivity, odd behaviours (like standing on the head, singing loudly, talking in a childish manner) and cognitive disorders, feeling of unreality, confusion, visual or auditory hallucinations. Simultaneous occurrence of all these symptoms is more the exception than the rule however. The hypersomniac episode occurs in a matter of hours or gradually over several days, sometimes marked by headache. During the episode the patient may sleep as long as 14 to 18 hours per day, waking or getting up only to eat and void. Urinary incontinence does not occur. Body weight gain of a few kilograms can be observed during the episode. Physical examination is sometimes remarkable for reddish face, severe perspiration, stinking hair or urines. Amnesia of the episode, transient dysphoria or elation with insomnia for one or two days may follow the episode itself. During asymptomatic intervals patients sleep normally and do not experience behavioral or cognitive disorders.

Course

Long-term follow-up studies of patients with the Kleine-Levin syndrome are lacking. Case report review suggests a frequent benign course, with episodes lessening in frequency, duration and severity. However, in a number of cases, follow-up is lacking after the last episode and cases are described of episodes recurring over a period of 10 to 20 years. Complications are mainly social and occupational.

Positive Diagnosis

Diagnosis of the Kleine-Levin syndrome is essentially clinical and laboratory tests merely serve to exclude the possibility of recurrent hypersomnia of secondary origin, organic or psychiatric.

Routine EEG obtained during episodes show general slowing of the background activity and often paroxysmal bursts of bi-synchronous, generalized, moderate to high voltage 5 to 7 cycles/sec waves of 0.5 to 2 second duration. Polysomnography may indicate a low sleep efficiency and increased wake time after sleep onset. Results of multiple sleep latency test are higly dependent on the subject's willingness to comply with the specific procedures of the test. 24-hour polysomnography is of more interest in demonstrating prolonged total sleep time up to 16 hours or more (Fig. 5).

Neuroimaging, either CT scan or MRI, is normal, during or after the episode. SPECT, performed both during a symptomatic and an unsymptomatic period in a limited number of cases has demonstrated hypoperfusion of the left or right hippocampus. Cerebrospinal fluid is normal. Hormonal levels are normal as well as 24 hour secretory patterns, except in a few case reports. Psychological testing is difficult to achieve during the episode and generally normal between episodes.

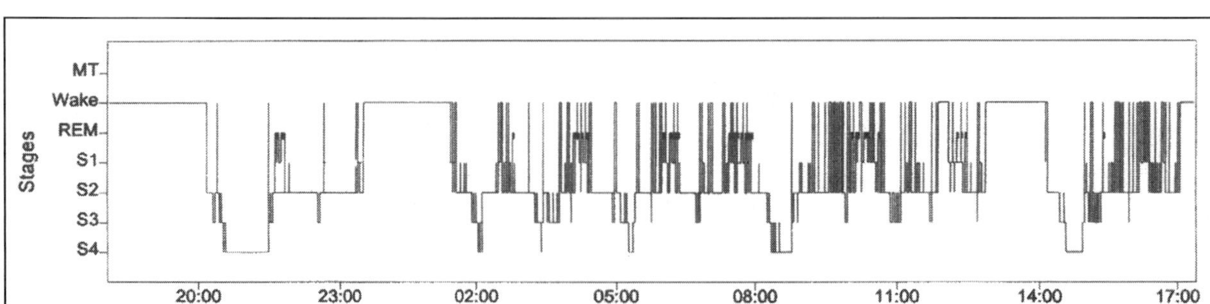

Figure 5. Continuous 24-hour recording in a 16 year old boy with the Kleine-Levin syndrome. Total sleep time over the 24 hour period was 16 hours and 11 minutes. Sleep was interrupted by multiple brief awakenings.

Clinical Subtypes

Monosymptomatic recurrent hypersomnia refers to cases in which symptomatology is limited to recurrent episodes of hypersomnia.

Periodic episodes of hypersomnia, plus or minus behavioral and cognitive symptoms, occurring in association with menstruation are indicative of the menstruation associated periodic hypersomnia. The disease occurs within the first months after menarche. Hormonal unbalance is likely since oral contraceptive therapy will usually lead to prolonged remission.[58]

Recurrent episodes of hypersomnia may depend on an organic insult to the CNS. Tumours proliferating inside the 3rd ventricle, such as colloid cyst, pediculated astrocytoma and sometimes craniopharyngioma, may be responsible for intermittent obstruction of the cavity of the ventricle, leading to headache and a series of associated symptoms such as vomiting, vague sensorial disturbance and paroxysmal impairment of alertness. Tumours of other location may also, although less frequently, account for recurring hypersomnia. Other instances of organic recurrent hypersomnia may rarely develop after an encephalitis, a head traumatism or a stroke.

Recurrent episodes of hypersomnia are described in psychiatric disorders, such as major depression of the recurrent type, and somatoform disorder.

Differential Diagnosis

The major risk is misdiagnosing the Kleine-Levin syndrome for psychosis, given disordered behavioural features, especially hypersexuality. Along this line it is not rare that patients are first hospitalized on psychiatric wards and given antipsychotic drugs.

Pathophysiology

HLA-DQB1*0201 allele frequency was recently found to be increased in patients with the Kleine-Levin syndrome.[59]

Precipitating factors are found in more than half of the cases, most frequently a flue-like episode or an infection of the upper airway occurring immediately before the onset of the first episode. Less frequent circumstances include sea sickness, acute drunkenness, blow to the face, sunstroke, general anesthesia.

Four postmortem neuropathological case reports are available. One subject showed a considerable degree of lymphocytic cuffing of the small vessels in the hypothalamus, in the amygdala and in the grey matter of the temporal lobes resembling those of a mild localized encephalitis.[60] A second and a third one showed the same type of lesion in two different locations, thalamus in one case,[61] diencephalon and midbrain in the other.[62] As for the 4th case the only finding in the brain was a small locus coeruleus with decreased pigmentation in the substantia nigra. Thus generalization appears hazardous.

The special combination of the clinical features of the Kleine-Levin syndrome has raised the possibility of recurrent hypothalamic dysfunction. However no consistent endocrinological data allow confirmation of a pituitary-hypothalamic disorder.[63] Magnetic resonance imaging does not reveal morphological changes, and out of 4 neuropathological studies, only one showed inflammatory lesions in the hypothalamus.

On the other hand the recurrence of episodes, the young age of onset, the frequent infectious diseases preceding the first episode and the recently evidenced association with HLA-DQB1*0201 may suggest an autoimmune disorder.[59]

Treatment

No controlled trial of pharmacological treatments of the Kleine-Levin syndrome has ever been and will ever be performed given the rarity of the condition. Thus we are limited to case reports and personal experience.

It is logical to use a stimulant drug such as methylphenidate or modafinil at the time of hypersomniac episodes. However results must be treated with caution to the extent that the methods of evaluation are purely subjective and that the episodes vanish spontaneously within a few days. Neuroleptics are often prescribed before diagnosis, when symptoms are interpreted as manifestations of a psychotic state. There is no proof that these drugs relieve the symptoms or affect the course of the episode.

Of greater interest is the prophylactic use of mood stabilizers. Positive results, that is absence of recurrence of the symptoms throughout the period of administration and recurrence with discontinuation, has been reported in some cases with the use of carbamazepine, valpromide, lithium carbonate and more recently valproic acid. On the other hand no result has been obtained in other cases. In practice these drugs should not be resorted to systematically, but in accordance with the frequency and severity of the abnormal episodes and their socio-professional impact.

In the case of menstruation associated periodic hypersomnia oestroprogestatives which inhibit ovulation seem to be effective.[58]

Hypersomnias Associated with Various Medical Diseases

Associated with Neurological Diseases

Hypersomnia may occur in any intracranial pressure syndrome, but it may also result, although more rarely, from tumours affecting the diencephalon, specially the ventro-lateral posterior part of the hypothalamus or the peduncular region.[64] Among the former are suprasellar craniopharyngiomas, extending posteriorly and superiorly, compressing the floor of the third ventricle; tumours of the pineal region; teratomas affecting the posterior part of the third ventricle; colloid cysts exerting a valve action in the third ventricle. Tumours affecting the peduncular region lead more often to more serious disorders of alertness, like obnubilation or stupor. Cases of narcolepsy secondary to brain tumours affecting the hypothalamus or the midbrain region have been reported.

Uni- or bilateral paramedian thalamic infarcts and paramedian pedunculo-thalamic infarcts are the most typical causes of hypersomnia of vascular origin. In addition hypersomnia is often a transient state marking the initial period of a stroke (Fig. 6).

A nonnegligible fraction of patients with Parkinson's disease present excessive sleepiness.[65,66] In a recent questionnaire survey, a significant correlation between heavy snoring and daytime sleepiness was found in Parkinson's disease patients.[67] This is even more the case of patients with multiple system atrophy which is often responsible for sleep-related breathing abnormalities. In addition "sleep attacks" have been reported as a specific unwanted effect of nonergoline dopamine agonists.[68] However it has been later recognised that this effect was observed with a steadily increasing list of dopaminergic agents including levodopa.

Normal pressure hydocephalus, Arnold-Chiari malformation, neuromuscular diseases, myotonic dystrophy may also lead to excessive sleepiness.

Treatment is uncertain. Surgery in the case of tumours, management of sleep breathing abnormalities, shunting CSF from

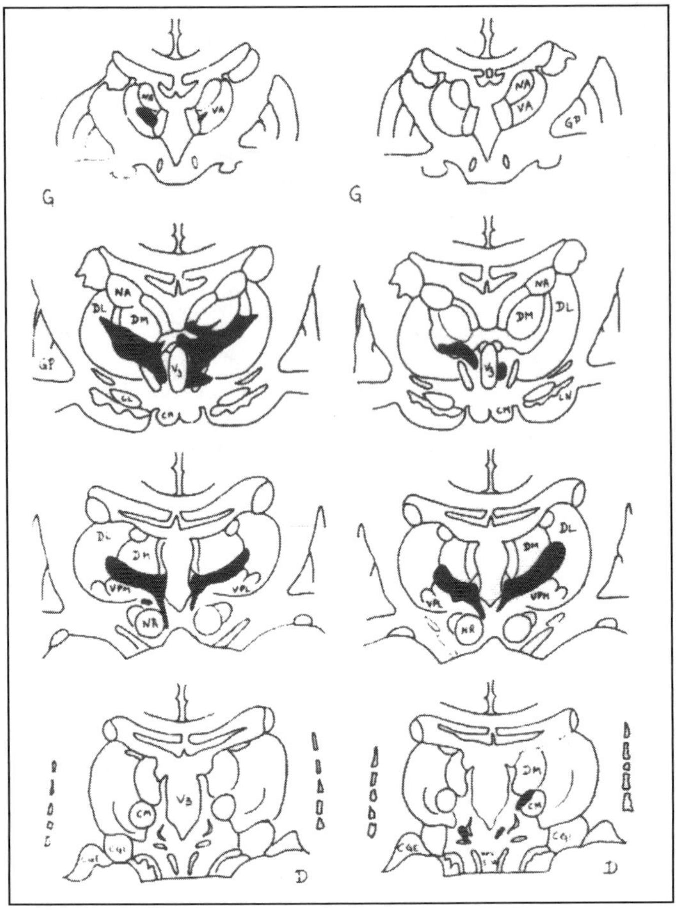

Figure 6. Topography of thalamic lesions in two subjects with excessive sleepiness consecutive to stroke. Left: thalamic infarct involving the retro-mammillary pedicle territory, extending foreward and on the left to the ventral anterior nucleus and on the right to the medial formations. Moderate lesion of the posterior hypothalamus extending to the superior internal pole of the red nucleus specially on the left side. Right: Cystic infarct of the thalamic territory of the perforated thalamic artery. Lesions are more extended foreward and on the left, and more extended laterally in the dorso-lateral nuclei, mainly backward in the centromedian nucleus on the left (from P. Castaigne and R. Escourolle, Rev Neurol (Paris), 1967; 116: 547-584). Reprinted with permission from Masson.

the dilated ventricles in the case of normal pressure hydrocephalus, may alleviate the symptomatology in some cases.

Associated with Psychiatric Conditions

Hypersomnia is often referred to in the literature on depressive patients. However objective studies show that most subjects spend considerable time in bed, not only at night but also during the day (clinophilia) with no significant increase in their degree of sleepiness or total sleep time[45,69] (Fig. 7). Thus the complaint of sleepiness is likely related to lack of interest, withdrawal, decreased energy, or psychomotor retardation inherent in the anergic depressed condition rather than an increase in true sleep propensity. Antidepressants are often unsuccessful while modafinil may result in clear improvement even at low dosage.

Seasonal affective disorder is characterized by episodes of depression beginning in the autumn or winter and ending toward the spring equinox. Anergy, hypersomnia, hyperphagia (carbohydrates) are the main clinical features. Phototherapy is the elective treatment.

Associated with Infectious Diseases

Intense fatigue and severe excessive sleepiness may develop in the months following Epstein-Barr disease.[70] The same holds true of atypical viral pneumonia, hepatitis B viral infection and the Guillain-Barré syndrome. Hypersomnia tends to go into gradual remission after several months or years.

Disorders of alertness and/or consciousness are found in virtually all patients affected by viral encephalitis.

Human African Trypanosomiasis (sleeping sickness) is a subacute or chronic parasitic disease caused by the inoculation of a protozoan, *Trypanosoma brucei* transmitted by the tsé-tsé fly. It is endemic to certain regions of tropical Africa. The form found in West and Central Africa is due to *Trypanosoma gambiense* hosted by the fly *Glossina palpalis*. The invasion of the nervous system is characterized by meningoencephalitis with headache, trembling, dyskenesia, choreoathetosis, mood changes, sleep disorder and epileptic seizures. Polysomnography has shown episodes of sleep occurring randomly day and night and sleep onset REM periods.[71]

Diagnosis is evoked by epidemiological setting, clinical symptoms and the presence of severe inflammatory syndrome in blood and CSF. It is confirmed by the demonstration of the trypanosome in smears of peripheral blood or in fluid aspirated from an

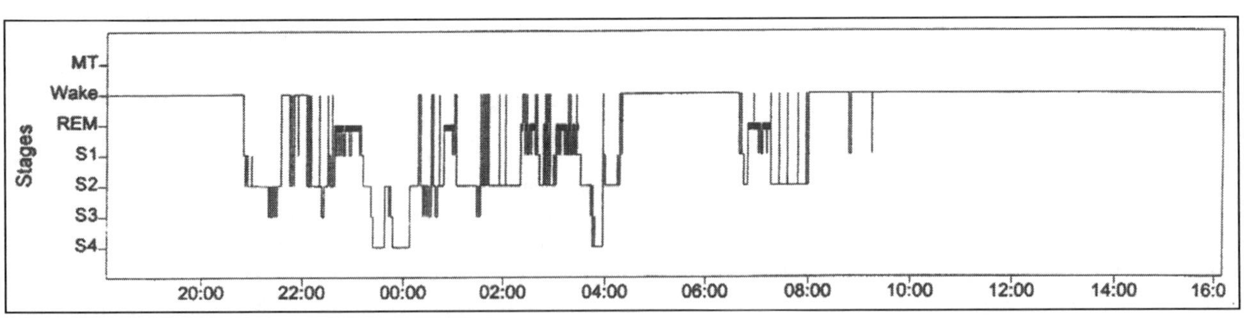

Figure 7. Continuous 24-hour recording in a 41 year-old woman complaining of severe sleepiness. Score on the Epworth sleepiness scale was 19. Mean sleep latency on the multiple sleep latency test was 20 minutes. Total sleep time at night was 8 hours and 5 minutes. Sleep was interrupted by frequent awakenings. No sleep period was documented during the daytime. Psychiatric interview concluded in favour of a somatoform disorder.

enlarged lymph node. In advanced stages they may be found only in the CSF.

So far the treatment for the early stages of Human African trypanosomiasis involves the drugs pentamidine and suramin which have been very successful. In the second stage of the disease, during which the trypanosomes reside in the CSF, treatment is dependent exclusively on the arsenical compound melarsoprol in spite of its toxicity. This is largely due to the inability to find compounds that can cross the blood brain barrier and kill the CSF-residing trypanosomes.

Associated with Endocrine Disorders

Hypothyroidism and acromegaly are the two main sources of hypersomnia, usually due to obstructive sleep apnea/hypopnea syndrome. Continuous positive airway pressure is indicated until etiological treatment is successful.

Conclusion and Perspectives

The hypersomnias are an invalidating group of sleep disorders, which are more prevalent than many people, physicians included, realize. Narcolepsy is the 'prototype' disease, and therefore very well studied. In the recent years crucial advances have been made in the understanding of narcolepsy. These new insights in the pathophysiology have also increased our knowledge in the regulation of normal sleep. Hopefully, similar breakthroughs will follow for the other hypersomnias. In order to achieve this, it will be important to carefully diagnose and classify the various disorders. This will enable to employ the powerful genetic, immunological and neurobiological techniques that are available today. At this time, the treatment options available are purely symptomatic. For narcolepsy, there are high hopes that replacing the lost hypocretin signalling will result in the first real causal treatment. Perhaps that modulating hypocretin neurotransmission will be of benefit in treating idiopathic hypersomnia and recurrent hypersomnias as well. As for hypersomnias associated with various medical diseases they are certainly of interest to clear up the anatomical basis of hypersomnia. On the other hand their treatment is still problematical.

Acknowledgment

We like to thank Prof. J.G. van Dijk for drawing Figure 1.

References

1. Gislason T, Almqvist M. Somatic diseases and sleep complaints: An epidemiological study of 3201 Swedish men. Acta Med Scand 1987; 221:474-481.
2. Ohayon MM, Caulet M, Philip P et al. How sleep and mental disorders are related to complaints of daytime sleepiness. Arch Intern Med 1997; 157(22):2645-2652.
3. Findley L, Unverzagt M, Guchu R et al. Vigilance and automobile accidents in patients with sleep apnea or narcolepsy. Chest 1995; 108(3):619-624.
4. Schulz H, Wilde-Frenz J, Grabietz-Kurfurst U. Cognitive deficits in patients with daytime sleepiness. Acta Neurol Belg 1997; 97(2):108-112.
5. Hays JC, Blazer DG, Foley DJ. Risk of napping: Excessive daytime sleepiness and mortality in an older community population. J Am Geriatr Soc 1996; 44:693-698.
6. Gélineau JB. De la narcolepsie. Gaz Hôp (Paris) 1880; 53:626-628.
7. Yoss RE, Daly DD. Criteria for the diagnosis of the narcoleptic syndrome. Proc Staff Meet Mayo Clin 1957; 32:320-328.
8. Vogel G. Studies in Psychophysiology of dreams. III. The dream of narcolepsy. Arch Gen Psychiatry 1960; 3:421-428.
9. von Economo C. Sleep as a problem of localization. J Nerv Ment Dis 1930; 71(3):249-259.
10. Hublin C, Partinen M, Kaprio J et al. Epidemiology of narcolepsy. Sleep 1994; 17(8 Suppl):S7-12.
11. Okun ML, Lin L, Pelin Z et al. Clinical aspects of narcolepsy-cataplexy across ethnic groups. Sleep 2002; 25(1):27-35.
12. Broughton R, Krupa S, Boucher B et al. Impaired circadian waking arousal in narcolepsy-cataplexy. Sleep Res Online 1998; 1(4):159-165.
13. Overeem S, Lammers GJ, van Dijk JG. Cataplexy: 'Tonic immobility' or 'REM-sleep atonia'? Sleep Medicine 2002; 3(6):471-477.
14. Broughton R, Dunham W, Weisskopf M et al. Night sleep does not predict day sleep in narcolepsy. Electroencephalogr Clin Neurophysiol 1994; 91(1):67-70.
15. Kok SW, Overeem S, Visscher TL et al. Hypocretin deficiency in narcoleptic humans is associated with abdominal obesity. Obes Res 2003; 11:1147-1154.
16. ICSD—International Classification of Sleep Disorders, revised: Diagnostic and Coding Manual. American Sleep Disorders Association, 1997.
17. Aldrich MS, Chervin RD, Malow BA. Value of the multiple sleep latency test (MSLT) for the diagnosis of narcolepsy. Sleep 1997; 20(8):620-629.
18. Nishino S, Ripley B, Overeem S et al. Hypocretin (orexin) deficiency in human narcolepsy. Lancet 2000; 355(9197):39-40.
19. Mignot E, Lammers GJ, Ripley B et al. The role of cerebrospinal fluid hypocretin measurement in the diagnosis of narcolepsy and other hypersomnias. Arch Neurol 2002; 59(10):1553-1562.
20. Overeem S, Mignot E, van Dijk JG et al. Narcolepsy: Clinical features, new pathophysiologic insights, and future perspectives. J Clin Neurophysiol 2001; 18(2):78-105.
21. Broughton R, Valley V, Aguirre M et al. Excessive daytime sleepiness and the pathophysiology of narcolepsy- cataplexy: A laboratory perspective. Sleep 1986; 9(1 Pt 2):205-215.
22. Saper CB, Chou TC, Scammell TE. The sleep switch: Hypothalamic control of sleep and wakefulness. Trends Neurosci 2001; 24(12):726-731.
23. Pace-Schott EF, Hobson JA. The neurobiology of sleep: Genetics, cellular physiology and subcortical networks. Nat Rev Neurosci 2002; 3(8):591-605.
24. Scammell TE. The neurobiology, diagnosis, and treatment of narcolepsy. Ann Neurol 2003; 53(2):154-166.
25. Nishino S, Mignot E. Pharmacological aspects of human and canine narcolepsy. Prog Neurobiol 1997; 52(1):27-78.
26. Mignot E. Genetic and familial aspects of narcolepsy. Neurology 1998; 50(2 Suppl 1):S16-S22.
27. Hohjoh H, Terada N, Kawashima M et al. Significant association of the tumor necrosis factor receptor 2 (TNFR2) gene with human narcolepsy. Tissue Antigens 2000; 56:446-448.
28. Lin L, Hungs M, Mignot E. Narcolepsy and the HLA region. J Neuroimmunol 2001; 117(1-2):9-20.
29. Lin L, Faraco J, Li R et al. The sleep disorder canine narcolepsy is caused by a mutation in the hypocretin (orexin) receptor 2 gene. Cell 1999; 98(3):365-376.
30. Chemelli RM, Willie JT, Sinton CM et al. Narcolepsy in orexin knockout mice: Molecular genetics of sleep regulation. Cell 1999; 98(4):437-451.
31. Peyron C, Faraco J, Rogers W et al. A mutation in a case of early onset narcolepsy and a generalized absence of hypocretin peptides in human narcoleptic brains. Nat Med 2000; 6(9):991-997.
32. Thannickal TC, Moore RY, Nienhuis R et al. Reduced number of hypocretin neurons in human narcolepsy. Neuron 2000; 27(3):469-474.
33. Willie JT, Chemelli RM, Sinton CM et al. To eat or to sleep? Orexin in the regulation of feeding and wakefulness. Annu Rev Neurosci 2001; 24:429-458.
34. Overeem S, Scammell TE, Lammers GJ. Hypocretin/orexin and sleep: Implications for the pathophysiology and diagnosis of narcolepsy. Curr Opin Neurol 2002; 15(6):739-745.
35. Mullington J, Broughton R. Scheduled naps in the management of daytime sleepiness in narcolepsy- cataplexy. Sleep 1993; 16(5):444-456.

36. Mitler MM, Aldrich MS, Koob GF et al. Narcolepsy and its treatment with stimulants. ASDA standards of practice. Sleep 1994; 17(4):352-371.
37. Broughton 5, Mamelak M. The treatment of narcolepsy-cataplexy with nocturnal gamma-hydroxybutyrate. Can J Neurol Sci 1979; 6:1-6.
38. Xyrem US. Multicenter Study Group. A randomized, double blind, placebo-controlled multicenter trial comparing the effects of three doses of orally administered sodium oxybate with placebo for the treatment of narcolepsy. Sleep 2002; 25(1):42-49.
39. Dement W, Rechtschaffen A, Gulevich G. The nature of the narcoleptic sleep attack. Neurology 1966; 16(1):18-33.
40. Rechtschaffen A, Roth B. Nocturnal sleep of hypersomniacs. Activ Nerv 1969; (Suppl 11):229-233.
41. Roth B, Nevsimalova S, Rechtschaffen A. Hypersomnia with "sleep drunkenness". Arch Gen Psychiatry 1972; 26(5):456-462.
42. Roth B. Narcolepsy and hypersomnia: Review and classification of 642 personally observed cases. Schweiz Arch Neurol Neurochir Psychiatr 1976; 119(1):31-41.
43. Billiard M, Besset A. Idiopathic hypersomnia. In: Billiard M, ed. Sleep, Physiology, Investigations and Medicine. New York: Kluwer Academic, Plenum Publishers, 2003:429-35.
44. The report of an American Academy of Sleep Medicine Task Force. Sleep-related breathing disorders in adults: Recommendations for syndrome definition and measurement techniques in clinical research. Sleep 1999; 22:667-689
45. Nofzinger EA, Thase ME, Reynolds CF 3rd et al. Hypersomnia in bipolar depression: A comparison with narcolepsy using the multiple sleep latency test. Am J Psychiatry 1991; 148:1177-1181.
46. Fukuda K, Strauss SE, Hickie I et al. The chronic fatigue syndrome: A comprehensive approach to its definition and study. Ann Int Med 1994; 121:953-959.
47. Nevsimalova S, Roth B. Genealogical study of hypersomnia and narcolepsy. Cesk Neurol 1972; 35(1):1-8.
48. Roth B. Narcolepsy and Hypersomnia. In: Karger S, ed. Basel: 1980.
49. Petitjean F, Jouvet M. Hypersomnie et augmentation de l'acide 5-hydroxy-indolacétique cérébral par lésion isthmique chez le chat. C R Acad Sci (Paris) 1970; 164:2288-2293.
50. Faull KF, Thiemann S, King RJ et al. Monoamine interactions in narcolepsy and hypersomnia: A preliminary report. Sleep 1986; 9(1 Pt 2):246-249.
51. Kanbayashi T, Inoue Y, Chiba S et al. CSF hypocretin-1 (orexin-A) concentrations in narcolepsy with and without cataplexy and idiopathic hypersomnia. J Sleep Res 2002; 11(1):91-93.
52. Ebrahim IO, Sharief MK, De Lacy S et al. Hypocretin (orexin) deficiency in narcolepsy and primary hypersomnia. J Neurol Neurosurgery Psychiatry 2003; 74:127-130.
53. Bastuji H, Jouvet M. Successful treatment of idiopathic hypersomnia and narcolepsy with modafinil. Prog Neuropsychopharmacol Biol Psychiatry 1988; 12(5):695-700.
54. Kleine W. Periodische Schlafsucht. Msch Psychiat Neurol 1925; 57:285-320.
55. Levin M. Narcolepsy (Gelineau's syndrome) and other varieties of morbid somnolence. Arch Neurol Psychiatr 1929; 22:1172-1200.
56. Critchley M, Hoffman HL. The syndrome of periodic somnolence and morbid hunger (Kleine-Levin syndrome). BMJ 1942; 1:137-139.
57. Critchley M. Periodic hypersomnia and megaphagia in adolescent males. Brain 1962; 85:627-656.
58. Billiard M, Guilleminault C, Dement WC. A menstruation-linked periodic hypersomnia. Neurology 1975; 25:436-443.
59. Dauvilliers Y, Mayer G, Lecendreux M et al. Kleine-Levin syndrome: An autoimmune hypothesis based on clinical and genetic analyses. Neurology 2002; 59(11):1739-1745.
60. Takrani LB, Cronin D. Kleine-Levin syndrome in a female patient. Can Psychiatr Assoc J 1976; 21:315-318.
61. Carpenter S, Yassa R, Ochs R. A pathological basis for Kleine-Levin syndrome. Arch Neurol 1982; 39:25-28.
62. Fenzi F, Simonati A, Crosato F et al. Clinical features of Kleine-Levin syndrome with localized encephalitis. Neuropediatrics 1993; 24:292-295.
63. Mayer G, Leonhard E, Krieg J et al. Endocrinological and polysomnographic findings in Kleine-Levin syndrome. Sleep 1998; 21(3):278-284.
64. Castaigne P, Escourolle R. Etude topographique des lésions anatomiques dans les hypersomnies. Rev Neurol (Paris) 1967; 116:547-584.
65. Rye DB, Bliwise DL, Dihenia B et al. Daytime sleepiness in Parkinson's disease. J Sleep Res 2000; 9(1):63-69.
66. Arnulf I, Konofal E, Merino-Andreu M et al. Parkinson's disease and sleepiness: An integral part of PD. Neurology 2002; 58(7):1019-1024.
67. Högl B, Seppi K, Brandauer E et al. Increased daytime sleepiness in Parkinson's disease: A questionnaire survey. Mov Disord 2003; 18(3):319-323.
68. Frucht S, Rogers JD, Greene PE et al. Falling asleep at the wheel: Motor vehicle mishaps in persons taking pramipexole and ropinirole. Neurology 1999; 52(9):1908-1910.
69. Billiard M, Dolenc L, Aldaz C et al. Hypersomnia associated with mood disorders: A new perspective. J Psychosom Res 1994; 38(Suppl 1):41-47.
70. Guilleminault C, Mondini S. Infectious mononucleosis and excessive daytime sleepiness. A long-term follow-up study; Arch Intern Med 1986; 146:1333-1335.
71. Buguet A, Bert J, Tapie P et al. Sleep-wake cycle in human African Trypanosomiasis. J Clin Neurophysiol 1993; 10:190-196.

CHAPTER 18

Chronic Disease and Sleep Architecture: Pharmacotherapeutic Considerations

James J. Herdegen

Abstract

This chapter details the pattern of sleep disturbances associated with chronic medical conditions. It illustrates the disturbances in sleep architecture manifested by a number of medical conditions as detailed by EEG or polysomnography. Sleep disturbances are common with a number of medical conditions and abnormal health states may also lead to primary sleep disorders such as insomnia, sleep apnea, or restless legs syndrome. Disturbances in sleep architecture caused by pharmaceuticals used in the management of chronic medical conditions are also described. Understanding the complex interaction between sleep, medical disease and medications may have important implications in 24/7 approach to clinical disease management.

Introduction

It has been increasingly recognized that a number of chronic medical conditions are associated with primary or secondary sleep disorders. In addition, a number of medications used to treat these medical disorders can also contribute to sleep disorders or abnormal sleep architecture when polysomnography is performed. This chapter will detail the medical conditions associated with secondary sleep disorders along with the subsequent sleep/EEG abnormalities observed as a result of pharmaceutical intervention. It is now clear that the prevalence of sleep disorders is high among patients with chronic medical conditions and the pharmaceutical disruption of sleep architecture is often unavoidable. Understanding the interaction between disease states and pharmacology as it relates to sleep medicine will help clinicians choose treatment options that consider a 24 hour day.

Medical Conditions and Sleep

General

A number of studies have found an association between sleep disturbances and comorbid conditions. Additional factors influencing sleep complaints include age, female gender, poor education and socioeconomic status, recent stress, and alcohol or drug abuse. In a random sample of Swedish men aged 30 to 69 years,[1] patients with hypertension, chronic obstructive pulmonary disease (COPD), diabetes mellitus, rheumatic diseases including low back pain and other musculoskeletal disorders, and obesity had more sleep complaints than the rest of the population. The primary sleep complaints included difficulty initiating or maintaining sleep, too little sleep followed by excessive daytime somnolence (EDS), and too much sleep. Several other studies support these findings. Weiss et al[2] in 1962 reported that 22% of patients with somatic complaints had sleep disorders. McGhie and Russell[3] found that patients with chronic headache and cardiovascular problems had the greatest number of sleep complaints.

Another questionnaire study explored sleep complaints in patients with type II diabetes mellitus, recent myocardial infarction, chronic paraplegia due to spinal cord injury, affective disorders, and musculoskeletal complaints.[4] Each group included sex- and age-matched control subjects. Patients with diabetes did not differ from controls in terms of sleep complaints. Patients with myocardial infarction had more difficulties initiating and maintaining sleep than controls. Patients with paraplegia and affective disorders had the greatest number of sleep complaints. Patients with paraplegia had sleep initiation and maintenance problems, whereas those with affective disorders and rheumatic complaints had sleep initiation problems and EDS. Patients with musculoskeletal symptoms without associated depression had fewer sleep complaints than those with affective disorders. The authors concluded that having an organic disease does not necessarily explain the high prevalence of sleep disorders in chronically ill patients. They felt that psychological and social factors contribute significantly to sleep complaints. The study by Ford and Kamerow[5] supports this observation. This study of 7,594 adult patients found a high cooccurrence of insomnia and mental disorder (40%) and a high risk for development of depression in patients with insomnia within 1 year. In a study of 1,722 subjects aged 15 to 100 years, Ohayon et al[6] reported a high prevalence of psychiatric illness among insomnia sufferers in the Montreal metropolitan area. A French and United Kingdom survey by the same authors found that insomniacs also demonstrated a high prevalence of psychiatric disorders.[7,8]

Insomnia features or excessive sleepiness are not unique to psychiatric illnesses although depression may be a unifying theme. In a survey of 9,282 subjects $\geq$ 65 years, Foley et al[9] found that complaints of insomnia were associated with respiratory disease, physical disability, use of over-the-counter medications, depression, and perception of poor health. Asplund[10] investigated EDS in 6,143 community-dwelling elderly in Sweden and found that EDS and daytime napping were associated with poor health status and somatic diseases. The next section will focus on specific medical conditions, their associated sleep disturbances along with supporting evidence from polysomnographic studies (see also Table 1).

Sleep and Sleep Disorders: A Neuropsychopharmacological Approach, edited by Malcolm Lader, Daniel P. Cardinali and S.R. Pandi-Perumal.

Cardiovascular Disease

Sleep disturbances in patients with cardiovascular disease can result from nocturnal angina, myocardial infarction, cardiac arrhythmias, and congestive heart failure. It is known that nonfatal myocardial infarction and myocardial ischemic episodes are more likely to occur in the morning. In an analysis of 2,203 patients with sudden cardiac death, there was a low incidence during the night and a high incidence from 7-11 AM.[11] This may be related to increased sympathetic activity in the morning, possibly causing increased myocardial electrical instability. Myocardial ischemia appears to be more common in REM sleep. In a study of 12 patients with nocturnal angina using polygraphic testing, ischemic ECG changes occurred more frequently in REM.[12] In a study of sleep apnea patients with or without coronary heart disease, sleep architecture showed a reduction of slow wave and REM sleep as well as increased and more severe arousals during periods with myocardial ischemia than during control episodes.[13] They also noted that approximately 78% of ischemic episodes occurred during REM sleep.

Patients with congestive heart failure demonstrate a high prevalence of sleep disturbances including obstructive or central sleep apnea along with periodic limb movements (PLM). In a small study comparing patients with stable congestive heart failure (CHF) to healthy controls, 52% of CHF patients demonstrated PLM greater than 25/hour compared to 11% of control subjects.[14] In two case series of patients with CHF undergoing polysomnography, obstructive sleep apnea (OSA) was detected in 37% of 450 subjects and 11% of 81 subjects.[15,16] These same two studies described a prevalence of central sleep apnea as being 33% and 40% respectively. The principal risk factors for central sleep apnea are male sex, hypocapnia, atrial fibrillation, and increasing age, but not obesity. For obstructive sleep apnea, the main risk factor for men is obesity and in women is older age. The pattern of arousal is different among patients with primarily obstructive or central sleep apnea. In OSA, the arousal is an important defense mechanism that leads to apnea termination by triggering opening of the upper airway. In central apneas, arousal often occur after airflow has resumed. If fact, theses arousals usually trigger and propagate central apneas. In another distinguishing feature, central sleep apnea is more pronounced during nonREM sleep, where chemical-metabolic factors are the predominant influence on ventilatory control, than during REM sleep where behavioral nonmetabolic factors predominate.[17,18] Thus CHF can lead to sleep fragmentation and reduction of slow wave and REM sleep both as a direct result of CHF and through secondary sleep disorders such as sleep-disordered breathing.

Table 1. Medical conditions associated with sleep architecture changes

Medical Condition	Sleep Architecture Change
Coronary artery disease	Increased arousals Reduced slow wave and REM sleep
Congestive heart failure	Increased arousals
Chronic obstructive lung disease	Increased arousals Increased wake after sleep onset
Asthma	Increased awakenings Early morning awakenings Reduced slow wave sleep
Restrictive lung disease	Increased arousals with sleep stage shifts Increased stage I and reduced REM sleep
Hypothyroid/Myxedema	Reduced slow wave sleep
Acromegaly	Increased stage 1 and 2 sleep, reduced stage 4
Seizure disorder	Increased stage shifts Decreased total sleep time Reduced REM sleep
Renal failure	Decreased sleep efficiency, increased sleep fragmentation Decreased slow wave sleep
Chronic pain	Increased arousals Increased alpha in NREM sleep
Peptic ulcer disease	Awakening 2-3 hours after sleep onset
Gastroesophageal reflux	Increased arousals
Depression	Reduced REM latency, increased REM density Reduced total sleep time Decreased slow wave sleep Shift of slow wave sleep from the first nonREM cycle to the second
Anxiety disorder	Prolonged sleep-onset latency Increased stage 1 and 2 sleep Reduced slow wave sleep
Schizophrenia	Prolonged sleep onset latency Increased wake time
Cancer	Increased awakenings
Parkinson's disease	Reduced sleep duration and sleep efficiency Reduced slow wave and REM sleep

Respiratory Diseases

Patients with chronic obstructive pulmonary disease (COPD) are susceptible to sleep disturbances from several causes. Physiologic changes in respiration, respiratory muscles, and control of breathing during sleep adversely affect breathing. In COPD patients, two basic mechanisms worsen hypoxemia during sleep: alveolar-hypoventilation, which is worse during REM sleep, and ventilation-perfusion mismatching. This results in a fall of oxygen saturation and PaO2 during sleep and a rise of PaCO2, all of which worsen to a greater degree in REM sleep. A number of sleep architecture abnormalities have been reported in COPD patients.[19-24] The predominant findings include a reduction of sleep efficiency, delayed sleep onset, increased wake after sleep onset (WASO), frequent stage shifts, and frequent arousals. COPD patients are a heterogeneous group of patients. However, they are sometimes divided into "pink puffers" and "blue bloaters". Pink puffers generally have normal blood gases, hyperinflated lungs, no hypoxemia or hypercapnia, and no cardiomegaly or cor pulmonale. On the other hand, blue bloaters are usually hypoxemic and hypercapnic and have cor pulmonale, polycythemia, enlarged heart, and reduced ventilatory response to hypoxemia and hypercapnia. In general, blue bloaters have more severe hypoxemia of longer duration than pink puffers.[25,26]

Asthma, characterized by dyspnea, wheezing, and cough can be a chronic, but reversible lung disease. Asthma attacks may occur at any hour of the day, and while nocturnal attacks are distributed at random without any relationship to a particular sleep stage, symptoms will often demonstrate circadian variation. A number of studies have found a high incidence of sleep disturbances in asthma patients, including early morning awakening, difficulty in maintaining sleep, and excessive daytime sleepiness.[27-29] In one study of 7,729 asthmatics, 74% admitted to waking up at least one night a week, 64% at least three nights a week, and 39% every night.[27] In one study of 93 asthma attacks, no relationship between asthma attacks and sleep stage or time of night was found.[30] The sleep pattern showed less total sleep time, frequent periods of wake time after sleep onset, early final awakenings, and reduced stage IV sleep. The circadian variability of asthma is supported by several observations. The peak expiratory flow (PEFR) is highest at 4 PM and lowest at 4AM. This variation is usually around 5-8% but can reach as high as 50% in some asthmatics which increases the risk of respiratory arrest. This circadian variation in PEFR is related to sleep and not to body position.[31] Airway resistance in asthmatics increases at night and is related to the duration of sleep and not to sleep stages.[32] Finally, similar to sleep apnea in COPD patients, nocturnal asthma is associated with OSA (hypopneas more than apneas, worse during REM sleep), leading to sleep fragmentation.[33,34]

Restrictive lung disease is characterized functionally by a reduction of total lung capacity, functional residual capacity, vital capacity, expiratory reserve volume, and diffusion capacity. This pattern can be seen in patients with interstitial lung disease, neuromuscular disease, and kyphoscoliosis. Sleep abnormalities in patients with interstitial lung disease include repeated arousals with sleep fragmentation and multiple sleep stage shifts, increased stage I and reduced REM sleep accompanied by oxygen desaturation during REM and nonREM sleep.[35,36]

Endocrine Diseases

Advanced hypothyroidism (myxedema) has been associated with both obstructive[37] and central[38] sleep apneas which disappeared after thyroxine treatment. Mechanisms include deposition of mucopolysaccharides in the upper airways as well as central respiratory dysfunction supported by impaired hypercapnic and hypoxic ventilatory responses seen in myxedema patients.[39] In a study of myxedema patients, slow wave sleep was reduced and normalized after treatment.[40] Sleep architecture changes in patients with thyrotoxicosis is controversial. One study found an increased amount of slow wave sleep which returned to normal after treatment.[41] Another study did not find changes in slow wave sleep but found an increase in sleep-onset latency[42] while yet another study found no relationship between stages of sleep and alteration of thyroid function.[43]

Excessive growth hormone release (acromegaly) has been associated with sleep apnea,[44] thought to be a result of enlargement of the tongue and pharyngeal wall leading to upper airway narrowing. In a small series of 8 adults with isolated growth hormone deficiency, stage IV sleep was reduced with increases in stage I and II NREM sleep leading to an increase in total sleep time.[45] In a follow-up study after 6 months of growth hormone treatment, there was a decrease in total sleep time due mainly to a reduction in stage II sleep, unchanged slow wave, and an increase in REM sleep time.[46]

Neurologic Diseases

Neurologic diseases can cause a variety of sleep disturbances including both insomnia and hypersomnolence. Neurologic causes for insomnia can include cerebral hemispheric and brainstem strokes, neurodegenerative diseases, brain tumors, post-traumatic insomnia, neuromuscular disorders, headache syndromes, fatal familial insomnia, and epilepsy. Significant overlap exists in disorders that cause hypersomnolence and include brain tumors, supratentorial and infratentorial strokes, trauma, multiple sclerosis, encephalitides, encephalopathies, neurodegenerative disorders, neuromuscular disorders, cerebral trypanosomiasis, and epilepsy.

Sleep architecture in epileptic patients can be affected by frequent awakenings, increase in the number of stage shifts, decrease in total sleep time, and reduction of up to 50% of REM sleep time.[47-49] Less common observations include increase in NREM, S1, and S2,[49] decrease in NREM, S3, and S4, reduction in the density of sleep spindles,[50] and increase in sleep onset and REM latency.[51]

Patients with Parkinson's disease demonstrate a number of sleep disturbances including a reduced sleep duration and sleep efficiency. The number of arousals is increased and may be related to an increased risk of sleep-disordered breathing, periodic limb movements, or bradykinesia and rigor that prevents patients from changing position leading to discomfort.[52] Sleep disorders commonly associated with Parkinson's disease include sleep apnea, periodic limb movements, and REM sleep behavior disorder.

Renal Disorders

Patients with chronic renal failure with or without dialysis commonly have sleep disturbances. The sleep complaints range from insomnia, excessive daytime sleepiness, and day-night reversal with disturbed nocturnal sleep. Several polysomnographic studies have demonstrated reduced sleep efficiency, increased sleep fragmentation, frequent awakenings with difficulty in maintenance of sleep, decreased slow wave sleep, and disorganization of sleep cycles.[53-55] Several studies have described no change in sleep disturbances in patients on dialysis or those who underwent renal transplantation. Another study, however, found that the uremic form of restless legs syndrome causing sleep disturbance can be cured after successful kidney transplantation.[56] Two other primary sleep disorders commonly associated with chronic renal failure and leading to significant sleep fragmentation include sleep apnea and periodic limb movements.

Chronic Pain Disorders

A number of studies have estimated that between 50% and 88% of patients with chronic nonmalignant pain disorders have significant sleep complaints.[57-60] This sleep disturbance my increase pain sensitivity and create a self-perpetuating cycle of sleep disruption, increased pain and depression.[61] Fibromyalgia syndrome, a syndrome characterized by diffuse muscle aches and pains not related to diseases of the joints, bones, or connective tissues, has been estimated to occur in 3-6 million Americans.[62] Sleep disturbance is common with the characteristic polysomnographic finding being intermittent alpha activity during NREM sleep giving rise to the characteristic alpha-delta or alpha-NREM sleep pattern. However, one study found that less than 40% of patients demonstrating this alpha-EEG pattern complained of pain.[63] Alpha-NREM sleep has been reported in other rheumatic disorders,[64] febrile illness, postviral fatigue syndrome,[65] psychiatric patients, and even normal individuals.[66] In general, associated nonspecific polysomnographic abnormalities of nonrestorative sleep (such as fibromyalgia) include increased awakenings, decreased sleep efficiency, and alpha-NREM sleep.

Gastrointestinal Disorders

Peptic ulcer disease affects about 10% of the adult population with a male predominance. The most common presentation is episodic pain localized to the epigastrium that is relieved by food, antacids, or other acid suppressants. This pain often awakens patients 2-3 hours after going to bed leading to disturbed sleep. Several studies have found that patients with duodenal ulcer fail to inhibit gastric acid secretion during the first 2 hours after sleep onset.[67,68] The nocturnal epigastric pain causes arousals and repeated awakenings leading to significant sleep fragmentation. Gastroesophageal reflux disease (GERD) frequently occurs in middle-aged and elderly women, and sometimes in younger women during pregnancy. The characteristic symptom is heartburn, described as retrosternal burning pain exacerbated by lifting or straining or when the patient lies down at night.[69,70] The burning pain causes difficulty in initiating sleep, frequent awakenings, and fragmentation of sleep. The nocturnal pain is relieved by sitting up, ingesting food, or acid-suppressant medications. Another potential complication of GERD is exacerbation of nocturnal asthma.

Psychiatric Disorders

A number of EEG findings have been described in patients with major depression. The primary changes in sleep architecture include shortened REM sleep-onset latency, increased REM density, reduced total sleep time, reduced sleep efficiency, increased awakenings, decreased slow-wave sleep, and a shift of slow wave sleep from the first nonREM cycle to the second. High-amplitude fast-frequency EEG activity has also been suggested as a marker for depression.[71] Several authors have suggested that sleep architecture changes found in depression may serve as a marker for the development of depression in those genetically predisposed to depression.[72]

Patients with generalized anxiety disorder typically have prolonged sleep-onset latency, increased stage I and II sleep, less slow wave sleep, a smaller REM sleep percentage, and usually an increased or normal REM sleep latency.[73,74] Patients with social phobia show increased sleep-onset latency, awakening after sleep onset, and reduced total sleep time.[75]

The EEG sleep profile in schizophrenia is often reported to be characterized by an impaired sleep continuity (prolonged time to fall asleep and frequent nocturnal awakenings), a decrease in slow wave sleep, an early onset of the first rapid eye movement, and an increased density of rapid eye movements during REM sleep.[76-78] However, most of the EEG sleep studies performed on patients with a schizophrenic disorder are limited by the fact that the patients were studied either during chronic neuroleptic treatment or after a short drug washout period. In a study of 22 drug-naïve patients with paranoid schizophrenia, slow wave and REM sleep were similar to controls while a prolonged sleep onset latency, increased wake time and decreased stage 2 sleep was seen.[79]

Malignancy

Sleep disturbances in cancer patients appear common and may be related to the primary malignancy, treatment for the malignancy, associated pain, discomfort, or psychosocial disturbance. Sleep fragmentation and insomnia features appear to be among the more common symptoms. Cancer patients reported more difficulty falling asleep and staying asleep compared to other medical and surgical patients.[80,81] A study of 14 lung cancer patients found that their perception of whether they slept well or poorly was directly related to the amount of deep sleep they obtained in the laboratory.[82]

Table 2. Medications associated with sleep architecture changes

Medical Condition	Sleep Architecture Change
Beta blocker	Increased wake after sleep onset Decreased REM sleep
Clonidine	Increased wake after sleep onset Increased stage shift changes
Alpha Methyldopa	Increased REM during first half of sleep night Decreased slow wave sleep
Diuretics	Increased wake after sleep onset
Theophylline	Increased arousals
Prednisone	Increased sleep onset latency Increased wake after sleep onset
Diphenhydramine	Decreased REM sleep
Tricyclic antidepressant	Decreased REM sleep Increased stage 2 sleep
SSRI (selective serotonin reuptake inhibitors)	Decreased REM sleep Eye movements in NREM sleep
Benzodiazepines	Decreased REM sleep Decreased sleep latency onset, Increased total sleep time Increased spindle activity
Lithium	Decreased REM sleep Increased slow wave sleep
Phenobarbital	Decreased wake after sleep onset Decreased REM sleep Increased stage 2 sleep Increased spindle activity
Gabapentin	Increased REM sleep percentage Reduced awakenings Reduced stage 1 sleep
Phenytoin	Decreased sleep latency Decreased sleep efficiency
Narcotics	Increase wake after sleep onset Decreased REM sleep
Dopamine agonists	Sleep attacks Increased REM sleep latency Decreased REM sleep

Pharmacological Effects on Sleep

Chronic medical conditions contribute greatly to sleep disturbances with polysomnographic studies demonstrating increased arousals, altered sleep architecture such as reduced slow wave sleep and reduced or delayed REM sleep. Primary sleep disorders may also be associated with a number of medical conditions and can include sleep-disordered breathing (central or obstructive), insomnia, parasomnias (such as REM behavior disorder), and restless legs/periodic limb movement disorder. In addition, medications used in the treatment of chronic medical disorders can also have a significant impact on normal sleep architecture. To follow is an illustration of just some of these affects (see also Table 2).

Cardiovascular Medications

Some of the beta blockers (such as propranolol or metoprolol) have been associated with depression or insomnia features,[83] increasing wakefulness after sleep onset and decreasing REM sleep.[84] Clonidine increases wakefulness after sleep onset and stage shift changes, maintains total sleep time and decreases REM sleep.

Clonidine may cause insomnia and daytime sedation. Alpha-methyldopa increases REM during the first half of the night, cause nightmares in some individuals, and decrease delta sleep. Diuretics increase wakefulness after sleep onset, especially if taken too late in the day.

Respiratory Medications

Methylxanthines (theophylline) have a sleep-reducing effect, primarily by causing insomnia features.[85,86] However, it can also increase the frequency of gastroesophageal reflux, leading to frequent arousals.[87,88] Systemic steroids contribute to frequent insomnia features, especially when taken for more than -5 days with the EEG demonstrating increased wake time after sleep onset (WASO). The older antihistamine agents cause significant sedation and decrease REM sleep.[89,90]

Hypnotic/Psychiatric Medications

The psychiatric medications are some of the best studied medications for their affects on sleep architecture. The tricyclic and monoamine oxidase inhibitor antidepressants decrease REM sleep and increase stage 2 and delta sleep. Upon withdrawal of medication, wake after sleep onset and REM sleep is increased. The selective serotonin reuptake inhibitors (SSRI's) will decrease REM sleep while not affecting total sleep time or delta sleep. SSRI's can cause insomnia or hypersomnia and may cause eye movements in nonREM sleep that make sleep study interpretation more difficult. The effects of benzodiazepines on sleep architecture vary based on their duration of action. In general they decrease sleep latency, wakefulness after sleep onset, and REM sleep while increasing total sleep time and stage 2 sleep. Some agents may increase spindle activity, especially in stage 2 sleep. The benzodiazepine receptor agents (such as zolpidem) will decrease sleep latency and increase total sleep time but not have a significant affect on REM or delta sleep. Lithium has been shown to decrease REM and increase delta sleep. Phenothiazine increases total sleep time and delta sleep with inconsistent effects on REM sleep. The neuroleptics, tricyclic antidepressants, selective serotonin reuptake inhibitors, and lithium have all been associated with secondary restless legs syndrome.

Neurologic Medications

Barbiturates increase total sleep time, decrease wakefulness after sleep onset and REM sleep while increasing stage 2 sleep and spindle activity. Daytime sedation, rapid tolerance development and insomnia symptoms after drug withdrawal is common. The acute administration of carbamazepine increases the number of stage shifts, reduces REM sleep and increased REM sleep fragmentation.[91] These effects appeared to be reversed after chronic treatment. Lamotrigine and gabapentin were also evaluated in the same study. Lamotrigine increased REM sleep, reduced the number of REM sleep cycles, decreased the number of phase shifts, and decreased the percentage of slow-wave sleep. Gabapentin increased the REM sleep percentage, increased mean duration of REM periods, reduced the number of awakenings, and reduced stage 1 sleep percentage. Phenytoin has been reported to decrease sleep efficiency, decrease sleep latency, decrease stage 1 and 2 sleep, cause a small decrease in REM sleep, increase or have no effect on stage 3 and 4 sleep, and increase wakefulness and arousals.[92] Phenobarbital causes an increase in total sleep time with short-term therapy, but no change with chronic therapy. There is a decrease in sleep onset latency, increase in stage 1, 2, and sleep spindles, decrease in REM sleep percentage, decrease in number and duration of REM cycles, decrease in movement-related arousals, less wakefulness after sleep onset, and no change in stage 3 or 4 sleep.[92]

Table 3. Medical conditions associated with primary sleep disorders

Medical Condition	Sleep Disorder
Congestive heart failure	Obstructive sleep apnea Central sleep apnea Periodic limb movements
Hypertension	Obstructive sleep apnea
Myxedema	Obstructive sleep apnea
Chronic renal failure	Obstructive sleep apnea Restless legs/periodic limb movements
Diabetes	Obstructive sleep apnea Restless legs syndrome
Depression	Insomnia
Parkinsons disease	Restless legs syndrome REM behavior disorder Sleep-disordered breathing
Alzheimers disease	REM behavior disorder
Hyperthyroidism	Insomnia
Lumbosacral radiculopathies	Restless legs syndrome
Anemia	Restless legs syndrome

Dopamine agonists used to treat Parkinson's disease have been associated with sleep attacks (sudden uncontrollable sleep). Reduced sleep onset latency, increased REM sleep latency, and reduced REM sleep have been described with use of these medications.[93]

Pain Medications

Acutely, narcotics increase wakefulness after sleep onset and decrease REM sleep however, on a chronic basis do not appear to affect wake after sleep onset or delta sleep. Upon withdrawal, hypersomnolence may occur, decreasing wakefulness after sleep onset. Acute aspirin intake may decrease delta sleep, possibly acting via prostaglandin inhibition and temperature effects. However it has also been shown to decrease slow wave sleep and increase stage 2 sleep.[94] Withdrawal of misused pain medications may result in improved sleep as shown in a study of 26 women with chronic daily headaches due to medication misuse.[95] Polysomnography performed 3 months after medication withdrawal showed significant improvement in total sleep time, sleep efficiency, and number of arousals.

Summary

This chapter has reviewed the broad range of sleep disturbances associated with chronic medical conditions. For a number of medical conditions, sleep onset is delayed or fragmented leading to insufficient sleep and possibly hypersomnolence. In addition, chronic pain or poor sleep can significantly contribute to insomnia features. Primary sleep disorders (see Table 3) such as sleep apnea are also frequently seen associated with medical conditions such as congestive heart failure, hypertension, hypothyroidism, and diabetes. Restless legs syndrome/PLMD are seen in a variety of neurological disorders and can be induced or worsened by a number of medications used to treat psychiatric conditions such as depression. A greater knowledge of how medications interfere with normal sleep architecture can help guide management of the underlying medical condition. For example, use of medications that limit sleep disruption or do not worsen coexisting sleep conditions can be preferentially chosen. Improved health outcomes such as quality of life and blood pressure control have been demonstrated with treatment of primary sleep conditions such as sleep apnea. The integration of pharmacologic, medical and sleep therapeutics that improve sleep quality may not only improve a patient's health state perception but also other measurable health outcomes

References

1. Gislason T, Almqvist M. Somatic diseases and sleep complaints: An epidemiological study of 3201 Swedish men. Acta Med Scand 1987; 221:475-481.
2. Weiss H, Kasinoff B, Bailey M. An exploration of reported sleep disturbance. J Nerv Ment Dis 1962; 134:528-534.
3. McGhie A, Russell SM. The subjective assessment of normal sleep pattern. J Ment Sci 1962; 108:642-654.
4. Hyyppa MT, Kronholm E. Quality of sleep and chronic illnesses. J Clin Epidemiol 1989; 42:633-638.
5. Ford DE, Kamerow DB. Epidemiologic study of sleep disturbances and psychiatric disorders. An opportunity for prevention? JAMA 1989; 262:1479-1484.
6. Ohayon MM, Caulet M, Guilleminault C. How a general population perceives its sleep, and how this relates to the complaint of insomnia. Sleep 1997; 20:715-723.
7. Ohayon MM, Caulet M, Priest RG et al. DSM-IV and ICSD-90 insomnia symptoms and sleep dissatisfaction. Br J Psychiatry 1997; 171:382-388.
8. Ohayon MM. Epidemiological study on insomnia in the general population. Sleep 1996; 19:S7-S15.
9. Foley DJ, Monjan AA, Brown SL et al. Sleep complaints among elderly persons: An epidemiologic study of three communities. Sleep 1995; 18:425-432.
10. Asplund R. Daytime sleepiness and napping amongst the elderly in relation to somatic health and medical treatment. J Intern Med 1996; 239:261-267.
11. Muller JE, Stone PH, Turi ZG et al. Circadian variation in the frequency of onset of acute myocardial infarction. NEJM 1985; 313:1315-1322.
12. Murao S, Harumi K, Katayama S et al. All-night polygraphic studies of nocturnal angina pectoris. Jpn Heart J 1972; 13:295-306.
13. Schafer H, Koehler U, Ploch T et al. Sleep-related myocardial ischemia and sleep structure in patients with obstructive sleep apnea and coronary artery disease. Chest 1997; 111:387-393.
14. Hanly PJ, Zuberi-Khokhar N. Periodic limb movements during sleep in patients with congestive heart failure. Chest 1996; 109(6):1497-1502.
15. Javaheri S, Parker TJ, Liming JD et al. Sleep apnea in 81 ambulatory male patients with stable heart failure: Types and their prevalences, consequences, and presentations. Circulation 1998; 97:2154-2159.
16. Sin DD, Fitzgerald F, Parker JD et al. Risk factors for central and obstructive sleep apnea in 450 men and women with congestive heart failure. AJRCCM 1999; 160:1101-1106.
17. Bradley TD, Phillipson EA. Central sleep apnea. Clin Chest Med 1992; 13:493-505.
18. Naughton M, Benard D, Tam A et al. Role of hyperventilation in the pathogenesis of central sleep apneas in patients with congestive heart failure. Am Rev Respir Dis 1993; 148:330-338.
19. Calverley PMA, Brezinova V, DN J et al. The effect of oxygenation on sleep quality in chronic bronchitis and emphysema. Am Rev Respir Dis 1982; 126:206-210.
20. Arand DL, McGinty DJ, Littner MR. Respiratory patterns associated with hemoglobin desaturation during sleep in chronic obstructive pulmonary disease. Chest 1981; 80:183-190.
21. Fleetham J, West P, Mezon B et al. Sleep, arousals and oxygen desaturation in chronic obstructive pulmonary disease: The effect of oxygen therapy. Am Rev Respir Dis 1982; 126:429-433.
22. Fletcher EC, Martin RJ, Monlux RD. Disturbed EEG sleep patterns in chronic obstructive pulmonary disease. Sleep Res 1982; 11:186.
23. Brezinova A, Catterall JR, Douglas NJ et al. Night sleep of patients with chronic ventilatory failure and age matched controls: Number and duration of the EEG episodes of intervening wakefulness and drowsiness. Sleep 1982; 5:123-130.
24. Leitch AG, Clancy LJ, Leggett RJ et al. Arterial blood gas tensions, hydrogen ion, and electroencephalogram during sleep in patients with chronic ventilatory failure. Thorax 1976; 31:730-735.
25. DeMarco FJJ, Wynne JW, Block AJ et al. Ocygen desaturation during sleep as a determinant of the "blue and bloated" syndrome. Chest 1981; 79:621-625.
26. Flenley DC, Claverly PM, Douglas NJ et al. Nocturnal hypoxemia and long-term domiciliary oxygen therapy in "blue and bloated" bronchitics. Physiopathological correlations. Chest 1980; 77:305-307.
27. Turner-Warwick M. Epidemiology of nocturnal asthma. Am J Med 1988; 85:6-8.
28. Janson C, Gislason T, Bowman G et al. Sleep disturbances in patients with asthma. Respir Med 1990; 84:37-42.
29. Deegan PC, McNicholas WT. Continuous noninvasive monitoring of evolving acute severe asthma during sleep. Thorax 1994; 49:613-614.
30. Kales A, Beall GN, Bajor GF et al. Sleep studies in asthmatic adults: Relationship of attacks to sleep stage and time of night. J Allergy 1968; 41:164-173.
31. Clark TJH, Hetzel MR. Diurnal variation of asthma. Br J Dis Chest 1977; 71:87-92.
32. Ballard RD, Saathoff MC, Patel DK et al. The effect of sleep on nocturnal bronchoconstriction and ventilatory patterns in asthmatics. J Appl Physiol 1989; 67:243-249.
33. Catterall JR, Douglas NJ, Calverley PM et al. Irregular breathing and hypoxemia during sleep in chronic stable asthma. Lancet 1982; 1:301-304.
34. Chan CS, Woolcock AJ, Sullivan CE. Nocturnal asthma: Role of snoring and obstructive sleep apnea. Am Rev Respir Dis 1988; 137:1502-1504.
35. Perez-Padilla RR, West P, Lertzman M et al. Breathing during sleep in patients with interstitial lung disease. Am Rev Respir Dis 1985; 132:224-229.
36. Bye PT, Issa F, Berthan-Jones M et al. Studies of oxygenation during sleep in patients with interstitial lung disease. Am Rev Respir Dis 1984; 129:27-32.
37. Skatrud J, Iber C, Ewart R et al. Disordered breathing during sleep in hypothyroidism. Am Rev Respir Dis 1981; 124:325-329.
38. Millman RP, Bevilacqua J, Peterson DD et al. Central sleep apnea in hypothyroidism. Am Rev Respir Dis 1983; 127:504-507.
39. Zwillich CW, Pierson DJ, Hofeldt FD et al. Ventilatory control in myxedema and hypothyroidism. NEJM 1975; 292:662-665.
40. Kales A, Heuser G, Jacobson A et al. All-night sleep studies in hypothyroid patients before and after treatment. J Clin Endocrinol & Metabolism 1967; 27:1593-1599.
41. Dunleavy DLF, Oswald I, Brown P et al. Hyperthyroidism, sleep and growth hormone. Electroencephalogr Clin Neurophysiol 1974; 36:259-263.
42. Passouant P, Passouant-Fountaine T, Cadilhac J. The influency of hyperthyroidism on sleep. Clinical and experimental study [French]. Rev Neurol (Paris) 1966; 115:353-366.
43. Johns MW, Rinsler MG. Sleep and throid function: Further studies in healthy young men. J Psychosom Res 1977; 21:161-166.
44. Grunstein RR, Ho KY, Sullivan CE. Sleep apnea in acromegaly. Ann Intern Med 1991; 115:527-532.
45. Astrom C, Lindholm J. Growth hormone-deficient young adults have decreased deep sleep. Neuroendocrinology 1990; 51:82-84.
46. Astrom C, Pedersen SA, Lindholm J. The influence of growth hormone on sleep in adults with growth hormone deficiency. Clin Endocrinol 1990; 33:495-500.
47. Touchon J, Baildy-Moulinier M, Billiard M et al. Sleep organization and epilepsy. Epilepsy Res 1991; suppl 2:73-81.
48. Baldy-Moulinier M. Temporal lobe epilepsy and sleep organization. In: Sterman MB, Shouse MN, Passouant P, eds. Sleep and epilepsy. New York: Academic Press, 1982:347-360.

49. Besset A. Influence of generalized seizures on sleep organization. In: Sterman MB, Shouse MN, Passouant P, eds. Sleep and epilepsy. New York: Academic Press, 1982:339-346.
50. Montplaisir J, Saint-Hilaire JM, Walsh J et al. Localization of primary epileptic foci by all night polygraphic recording: A study of 13 patients chronically implanted with multiple depth electrodes (abstract). Sleep Res 1979; 8:239.
51. Montplaisir J, Laverdiere M, Saint-Hilaire JM. Sleep and epilepsy. In: Gotman J, Ives J, Gloor P, eds. Long-term monitoring in epilepsy. EEG supplement 37. Amsterdam: Elsevier, 1985:215-239.
52. Clarenbach P. Parkinson's disease and sleep. J Neurol 2000; 247(suppl 4):20-23.
53. Williams RL. Sleep disturbances in various medical and surgical conditions. In: Williams RL, Karacan I, Moore CA, eds. Sleep disorders. New York: Wiley, 1988:265.
54. Fraser CL, Arieff AI. Nervous system complications in uremia. Ann Intern Med 1988; 109:143-153.
55. Kimmel PL, Gavin C, Miller G et al. Disordered sleep and noncompliance in a patient with end-stage renal disease. Adv Ren Replace Ther 1997; 4:55-67.
56. Yasuda T, Nishimura A, Katsuki Y et al. Restless legs sndrome treated successfully by kidney transplantation: A case report. Clin Transplants 1986; 138:138.
57. Atkinson JH, Anconi-Israel S, Slater MA et al. Subjective sleep disturbance in chronic back pain. Clin J Pain 1988; 4:225-232.
58. Morin CM, Gibson D, Wade J. Self-reported sleep and mood disturbance in chronic pain patients. Clinical Journal of Pain 1998; 14(4):311-314.
59. Pilowsky I, Crettenden I, Townley M. Sleep disturbance in pain clinic patients. Pain 1985; 23:27-33.
60. Smith MT, Perlis ML, Smith MS et al. Sleep quality and presleep arousal in chronic pain. Journal of Behavioral Medicine 2000; 23(1):1-13.
61. Moldofsky H. Psychogenic rheumatism or the fibrositis syndrome. In: Hill O, ed. Modern trends in psychosomatic medicine. London: Butterworths, 1976:3:187-195.
62. Goldenberg DL. Fibromyalgia syndrome: An emerging but controversial condition. JAMA 1987; 257:2782-2787.
63. Rains JC, Penzien DB. Sleep and chronic pain: Challenges to the alpha-EEG sleep pattern as a pain specific sleep anomaly. Journal of Psychosomatic Research 2003; 54(1):77-83.
64. Moldofsky H, Lue FA, Smythe H. Alpha EEG sleep and morning symptoms of rheumatoid arthritis. J Rheumatol 1983; 10:373-379.
65. Moldofsky H, Saskin P, Lue FA. Sleep and symptoms in fibrositis syndrome after a febrile illness. J Rheumatol 1988; 15:1701-1704.
66. Hauri P, Hawkins DR. Alpha-delta sleep. Electroencephalogr Clin Neurophysiol 1973; 34:233-237.
67. Watanabe M, Nakazawa S, Yoshino J et al. A study of the relationship between nocturnal intragastric pH and sleep stages of peptic ulcer. Nippon Shokakibyo Gakkai Zasshi 1995; 92:1241-1249.
68. Orr WC, Hall WH, Stahl ML et al. Sleep patterns and gastric acid secretion in duodenal ulcer disease. Arch Intern Med 1976; 136:655-660.
69. Pope CEI. Acid-reflux disorders. NEJM 1994; 331:656-660.
70. Richter JE, Castell DO. Gastroesophageal reflux. Pathogenesis, diagnosis, and therapy. Ann Intern Med 1982; 97:93-103.
71. Armitage R. Microarchitectural findings in sleep EEG in depression: Diagnostic implications. Biological Psychiatry 1995; 37:72-84.
72. Giles DE, Roffwarg HP, Kupfer DF et al. Secular trend in unipolar depression: A hypothesis. J Affect Disorders 1989; 14:71-75.
73. Reynolds CF, Shaw DH, Newton TF et al. EEG sleep in outpatients with generalized anxiety: A preliminary comparison with depressed outpatients. Psychiatry Research 1983; 8:81-89.
74. Papadimitriou GN, Kerkhofs M, Kempenaers C et al. EEG sleep studies in patients with generalized anxiety disorder. Psychiatry Research 1988; 26:183-190.
75. Stein MB, Kroft CD, Walker JR. Sleep impairment in patients with social phobia. Psychiatry Research 1993; 49:251-256.
76. Zarcone VP. Sleep and schizophrenia. In: Williams RL, Karacan J, Moore CA, eds. Sleep Disorders: Diagnosis and Treatment. 2nd ed. New York: Wiley & Sons, 1988:165-188.
77. Thaker GK, Wagman AM, Kirkpatrick B et al. Alterations in sleep polygraphy after neuroleptic withdrawal: A putative supersensitive dpaminergic mechanism. Biological Psychiatry 1989; 25:75-86.
78. Keshavan MS, Reynolds CF, Kupfer DF. Electroencephalographic sleep in schizophrenia: A critical review. Compr Psychiatry 1990; 30:34-47.
79. Lauer CJ, Schreiber W, Pollmacher T et al. Sleep in schizophrenia: A polysomnographic study on drug-naive patients. Neuropsychopharmacology 1997; 16(1):51-60.
80. Kaye J, Kaye K, Madow L. Sleep patterns in patients with cancer and patients with cardiac disease. J Psychol 1983; 114:107-113.
81. Beszterczey A, Lipowski ZJ. Insomnia in cancer patients. Can Med Assoc J 1997; 116:135.
82. Silberfarb PM, Hauri PJ, Oxman TE et al. Factors affecting couples' adjustment to recurrent breast cancer. Soc Sci Med 1985; 20:849-850.
83. Uchiumi M, Murasaki M, Matsumoto J et al. The effects of beta-blockers on the psychomotor function with sepcial reference to multiple sleep latency test. Psychopharmacology 1988; 96(suppl):320.
84. Kostis JB, Rosen RC. Central nervous system effects of beta-blockers. A study with objective measures. Clin Pharmacol Ther 1986; 39:203.
85. Janson C, Gislason T, Almqvist M et al. Theophylline disturbs sleep mainly in caffeine-sensitive persons. Pulm Pharmacol 1989; 2:125-129.
86. Rhind GB, Connaughton JJ, McFie J et al. Sustained release choline theophyllinate in nocturnal asthma. British Med J Clin Res 1985; 291:1605-1607.
87. Stein MR, Towner TG, Weber RW et al. The effect of theophylline on the lower esophageal sphincter pressure. Ann Allergy 1980; 45:238-241.
88. Berquist WE, Rachelefsky GS, Kadden M et al. Effect of theophylline on gastroesophageal reflux in normal adults. J Allergy Clin Immunol 1981; 67:407-411.
89. Soldatos CR, Dikeos DG. Neuroleptics, antihistamines and antiparkinsonian drugs: Effects on sleep. In: Kales A, ed. The Pharmacology of Sleep. Berlin-Heidelberg: Springer, 1995:443-464.
90. Nicholson AN. Antihistamines and sedation. Lancet 1983; 2:211-212.
91. Placidi F, Diomedi M, Scalise A et al. Effect of anticonvulsants on nocturnal sleep in epilepsy. Neurology 2000; 54(Supp 1):S25-S32.
92. Sammaritano M, Sherwin A. Effect of anticonvulsants on sleep. Neurology 2000; 54(Supp 1):S16-S24.
93. Schafer D, Greulich W. Effects of parkinsonian medication on sleep. J Neurol 2000; 247(suppl 4):24-27.
94. Horne JA, Percival JE, Traynor JR. Aspirin and human sleep. Electroencephalogr Clin Neurophysiol 1980; 49:409-413.
95. Hering-Hanit R, Yavetz A, Dagan Y. Effect of withdrawal of misused medication on sleep disturbances in migraine sufferers with chronic daily headache.[comment]. Headache 2000; 40(10):809-812.

CHAPTER 19

Benzodiazepines for Sedation in Infants and Children

Eugene Ng and Vibhuti Shah

Abstract

Benzodiazepines are commonly used to provide sedation for infants and children undergoing intensive care or diagnostic and therapeutic procedures in a variety of clinical settings. This chapter focuses on Midazolam as representative of this class of drug. Midazolam provides sedation by altering the neuroinhibitory pathway mediated by gamma-aminobutyric acid. It is primarily metabolized by the hepatic cytochrome P450 enzyme subfamily and eliminated via the renal route. Plasma clearance of midazolam is affected by the degree of hepatic and renal immaturity in the newborn period. In addition, there is a large inter-individual variability in midazolam metabolism in neonates and children, leading to heterogeneity in drug handling.

Pediatric patients, especially neonates, are therefore susceptible to adverse effects associated with the use of midazolam. Transient neurologic, respiratory and cardiovascular reactions have been reported. Although most appeared to be transient, some studies suggest that neonates exposed to midazolam may have longer-term adverse neurodevelopmental effects. Furthermore, in review of literature to date, even though midazolam infusion is efficacious in sedating critically ill infants and children undergoing intensive care, the use of midazolam for procedural sedation in the pediatric population may be less efficacious than alternatives such as ketamine and may not be appropriate in all clinical circumstance. Therefore, until further research is done on the safety and efficacy of midazolam administration in infants and children, cautious use of this medication in this population is recommended.

Introduction

The provision of medical care for infants and children often requires diagnostic and therapeutic procedures, which are increasingly performed outside the operating room setting.[1] These procedures are frequently associated with pain and stress. Stress is defined as "a physical, chemical, or emotional factor that causes bodily or mental tension and may be a factor in disease causation", and the response to stress can be directly related to the event causing stress, or can be non-specific.[2] Despite the advances in pain assessment and management, the use of sedation to alleviate non-painful stress in infant and children continues to be relatively limited. This is partly due to the subjectivity of measuring stress, the relative inability of young children to verbalize stress, and the fear of untoward complications associated with sedative use.[1]

However, the completion of most diagnostic and therapeutic procedures in infants and children often depends heavily on the success of sedation. In neonates undergoing intensive care, evidence also suggests that adequate sedation during invasive procedures such as mechanical ventilation may decrease stress[3] and prevent complications such as pneumothoraces and intracranial hemorrhages.[4,5] The goals of providing sedation to infants and children are to (a) minimize discomfort directly caused by the procedure; (b) decrease anxiety and negative psychological effects associated with the procedure, and (c) facilitate the performance of the procedure by controlling behavior. These goals must be achieved without compromising the safety of the patients during the procedure, and must return them to the state of health prior to the administration of sedation.[6-8] The American Academy of Pediatrics has published guidelines for proper management and monitoring of children during sedation.[9,10] These guidelines suggest the universal principles of pre-sedation medical evaluation with special attention to airway examination and preparation prior to the procedure, skilled personnel to administer appropriate drugs and monitoring during sedation, and observation post-sedation prior to discharge to ensure the safety and effectiveness of the sedation.

In older infants and children, a variety of pharmacologic agents have been used to provide sedation. These include benzodiazepines, barbiturates, chloral hydrate and phenothiazines.[2] Of the benzodiazepines, midazolam and, to a lesser degree, lorazepam are most commonly used for sedation of pediatric patients. In this article, the pharmacological properties of benzodiazepines, and, more specifically, the developmental responses to benzodiazepines will be considered. The occurrence of various adverse effects associated with benzodiazepine use will also be discussed. Due to the vast amount of literature on the use of midazolam in infants and children, this chapter will consider the clinical applications of midazolam as representative of this class of drug in a variety of clinical settings, such as mechanical ventilation in intensive care units, preoperative sedation, sedation for ambulatory and emergency procedures, and treatment of opioid withdrawal.

Pharmacology of Benzodiazepines

The benzodiazepines modulate cerebral function in a dose-dependent fashion, producing varying degrees of neuronal inhibition ranging from anxiolysis, sedation, to anesthesia. This class of drug is also known for its effects of anterograde amnesia, muscle relaxation, and anti-convulsion; however, it does not provide analgesia.[1,11] Considerable work has been done in order to understand the pharmacological properties and the mechanisms of cerebral effects of the benzodiazepines. Evidence suggests that benzodiazepines act on specific receptors and alter the pathway

Sleep and Sleep Disorders: A Neuropsychopharmacological Approach, edited by Malcolm Lader, Daniel P. Cardinali and S.R. Pandi-Perumal.

mediated by gamma-aminobutyric acid (GABA), the major inhibitory neurotransmitter in the central nervous system (CNS).[11] Once activated, the benzodiazepine receptors enhance the effectiveness of GABA-ergic neurotransmission by facilitating ion movement across the subsynaptic chloride channels, leading to a dampening of neuronal excitation.[12]

The central benzodiazepine receptors are high affinity binding sites, which are associated with GABA receptors in the neuronal membranes. The areas of the CNS with the highest density of benzodiazepine receptors include the cerebral cortex, the cerebellum, the limbic system, and the basal ganglia.[11,12] The functionality of each receptor relies on the presence of three classes of subunits: alpha, beta, and gamma. Variants exist within each class of subunit, leading to a certain degree of heterogeneity of benzodiazepine receptors within the CNS. This may partly explain the differences in clinical effects of the various benzodiazepines.[12]

The Effect of Age

As with many other classes of drugs, the pharmacodynamic and pharmacokinetic properties of benzodiazepines vary with age. Studies comparing the responses to midazolam between elderly and younger adult patients showed that, even though there is no significant age-related difference in pharmacokinetic properties, the more pronounced clinical effects of midazolam in elderly patients could be explained by differences in pharmacodynamic response.[13,14] Similarly, in clinical studies, it has been shown that neonates, particularly those born preterm, also have a higher neurologic sensitivity toward the benzodiazepines compared with older infants and children.[15-25] However, from limited data, it has been shown that such difference in response to benzodiazepines may be related to the developmental ontogeny of benzodiazepine metabolism, leading to observed variations in pharmacokinetic parameters.[11]

Benzodiazepine Pharmacokinetics in Infants and Children

In neonates and children, particularly those receiving intensive care, midazolam, amongst all benzodiazepines, has been most often prescribed and studied.[26] Midazolam has long been considered the ideal sedative for neonates and children, particularly those receiving intensive care. Its water solubility presents a unique advantage over other benzodiazepines in this population for its fast onset and short duration of action, and rapid rate of receptor association-dissociation and elimination.[6,26-28] The dosage of benzodiazepines administered to neonates and children should be calculated based on milligram per kilogram of body weight.

Pharmacokinetics

After administration, midazolam is extensively metabolized through hepatic microsomal oxidation and glucuronidation; its major active metabolite, 1-hydroxymidazolam, is then eliminated via the renal route.[27,29] In the liver, midazolam is biotransformed by the cytochrome P450 3A (CYP 3A) enzyme subfamily.[30,31] In adults, it has been shown that plasma clearance of midazolam correlates with level of CYP 3A 4 and CYP 3A 5 activities. Within the first two weeks of birth, as a result of the ontogeny of the hepatic CYP 3A 4 and 5 activity, a considerable reduction in plasma clearance of midazolam, in the order of 1.5 to 5 times, has been demonstrated.[30] In addition, a significant heterogeneity within individual neonates in the maturational rate of CYP 3A activity exists. Coinciding with a reduced hepatic metabolism is the variability in renal maturation, which also shows inter-individual variations, and could potentially lead to a decrease in renal elimination of the active metabolite of midazolam.[27,31] Together, these two age-dependent factors contribute significantly to the delay in plasma clearance of midazolam in this population compared with older infants, children, and adults. These findings were confirmed by population pharmacokinetic studies of midazolam in the neonates.[27,32]

Furthermore, inter-individual variability in midazolam metabolism has been demonstrated. The heterogeneity in illness severity, and the multiple comorbid conditions, such as sepsis and hypotension, may have partly contributed to such variability. As midazolam clearance is dependent on adequate perfusion of the major organs, changes in hemodynamic status, as seen in conditions such as sepsis or patent ductus arteriosus, and the correction of these conditions, may significantly affect the clearance of midazolam. The prenatal exposure of betamethasone, which may itself induce activity of the CYP 3A activities, may influence the clearance of midazolam as well.[27,30]

In children, there also appeared to be a large inter-individual variability in midazolam pharmacokinetic parameters.[33-35] Furthermore, the uncertain contribution of the active metabolite, and the nature of the critical illnesses may also contribute to the heterogeneity in drug handling in the population studied. Contrary to neonates, however, studies have consistently shown that the plasma clearance of midazolam and lorazepam in children is higher than that in adults, which was thought to be related to a higher metabolic turnover and/or a smaller volume of distribution in the former age group.[36-38]

Routes of Administration

Midazolam is currently available in intravenous (IV) and intramuscular (IM) forms, and oral syrup has recently become available for use in ambulatory settings. In addition, there have been reports of the IV preparation being administered through other transmucosal routes, such as rectal (PR), sublingual (SL) and intranasal (IN) routes in children.

The IM route of midazolam administration is quite comparable to the IV administration, owing to the over 90% bioavailability of midazolam by this route.[36] Whereas lorazepam has similar characteristics, other more lipophilic benzodiazepines such as diazepam are much more variable in systemic absorption after IM injections compared to midazolam.[11]

In a dose-finding study, the dose of oral midazolam required for sedation was at least three times the usual intramuscular dose of 0.15 mg/kg.[36] Overall, the time to reach peak midazolam serum concentration is significantly longer after an oral dose, compared with that given by IM or rectal routes. However, serum concentration stayed at above the therapeutic level for up to 2 hours with the oral dose of 0.45 mg/kg. In addition, adolescents, similar to young adults, were found to have a much slower absorption rate compared with young children. Because of first pass effect, the bioavailability of oral midazolam is considerably lower than that of IM midazolam, although results from studies were highly variable. The clearance rate and elimination half-life in children seemed to be shorter than that of adults, although these disposition characteristics were also found to be highly variable and not necessarily related to age or dose of midazolam administered.[26,36]

Despite the effects of first pass metabolism and the variability of rectal pH affecting absorption of midazolam in children,[39,40] PR administration of midazolam has been shown to have a relatively fast onset of action. At a dose of 0.35mg/kg, plasma concentration reaches an adult sedative level within minutes, and the effect was further accentuated with higher doses.[36,41] Furthermore, the duration of action is prolonged, with serum concentration remaining at therapeutic level after 2 hours,[41] which may be related to ongoing reabsorption of midazolam from the rectum.

Comparative studies showed that IN and SL routes of midazolam administration at equal dosing could achieve a much higher plasma concentration than the PR route at a comparable rate of action onset.[37,38,40,42] The bioavailability of IN midazolam as measured in two studies was considerably higher than rectal and oral midazolam, presumably because absorption via the IN route does not require first pass metabolism. Half-lives of IN midazolam is comparable to that of IV midazolam.[37,38]

Drug Interactions

The absorption of midazolam after oral administration may be affected by the gastrointestinal pH environment.[36,40] In a randomized controlled study, Lammers et al[43] demonstrated that raising the gastric pH by using an antacid increased the absorption of midazolam, as suggested by a significantly faster onset of sedation. The mechanism for the enhanced absorption may be related to the increase in lipid solubility of midazolam by closure of its imidazole ring in an alkaline environment, thus facilitating its absorption by the lipid bilayer gastric mucosa.[43]

Another type of drug interaction relates to the effects on hepatic first pass metabolism. Grapefruit juice[44] is known to inhibit intestinal and hepatic CYP4 3A4, and has been reported to delay hepatic metabolism of midazolam, with the potential of increasing plasma midazolam concentration to an excessive level. Similar effects have been shown with the co-administration of midazolam and erythromycin.[45,46]

As benzodiazepine is frequently used in conjunction with an analgesic in different clinical settings, Hase et al[47] studied the effects of fentanyl on midazolam pharmacokinetics in adults undergoing orthopedic surgery. In a randomized controlled trial, the plasma clearance of midazolam was found to be delayed by 30%, and serum half-life increased by 49% when fentanyl was co-administered. This may again be related to a reduction in hepatic CYP4 3A metabolism due to competitive inhibition by fentanyl.[47]

Summary

- There is a developmentally related heterogeneity in metabolism of benzodiazepines amongst individuals in the neonatal and pediatric population.
- In children, despite the significant first pass effect of oral and rectal routes of administration, midazolam appears to be quickly absorbed after administration, reaching therapeutic concentrations in a relative short time, and the duration of action is adequate (albeit shorter than that in adults).
- In neonates, however, hepatic and renal immaturity may lead to significant delay in clearance of benzodiazepines, which appears to be birth weight and/or gestational age dependent. Further influences of other co-morbid conditions predispose this population to a much higher risk of adverse effect from benzodiazepine administration.
- Therefore, from a pharmacokinetic viewpoint, whereas it is a relatively safe and effective sedative for children, the safety of midazolam administration in neonates is uncertain.
- The absorption and metabolism of midazolam is also affected by the pH environment and co-administration of drugs that affects the hepatic cytochrome p450 enzymes, with potential to cause toxicity.

Adverse Reactions Associated with Benzodiazepine Use

A variety of adverse events (AE) related to benzodiazepine use in infants and children have been reported. The majority of these events were transient, including neurologic and neuropsychological disturbances, cardiovascular abnormalities and respiratory depression. The majority of reports on these AE over the past decade were largely anecdotal in nature, making it difficult to estimate the frequency of occurrence of these AE. A recent study[48] was performed to determine the incidence of AE associated with benzodiazepine use in infants in a neonatal intensive care unit, and to characterize these AE. Over a period of 16 months, 10 (16%) of the 63 infants studied had 14 documented AE associated with the use of benzodiazepines, including seizures, hypotension, and respiratory depression. The majority of the AE occurred immediately after the first dose or after the initiation of an infusion of benzodiazepine, and the occurrence of AE was not correlated with the number of doses and the dosage of benzodiazepine administered, or the concomitant use of analgesia. These results demonstrate that, at least in the neonatal population, midazolam and lorazepam were frequently associated with adverse reactions.

Adverse Reaction Related to the Route of Administration

There have been a few reports on the administration of intranasal midazolam causing significant burning and irritation of the nasal mucosa in children, which, in some cases, were more intense than the pain caused by the procedures themselves, leading to poor acceptance of the medication.[49-51] A case of osmotic diarrhea was reported in an infant given oral midazolam, which was thought to be caused by the propylene glycol and polyethylene glycol in the intravenous formulation used orally.[52]

Hiccups

Hiccups have been reported in children receiving midazolam either intravenously[53] or rectally.[54] The occurrence of this transient, benign adverse reaction appears to be age dependent, with a higher incidence in younger children, but is not dose related. It is of quick onset, short duration, and self-resolving, although successful treatment using intranasal administration of ethyl chloride has been reported.[54]

Transient CNS Reactions

In neonates, isolated cases and case series on a variety of adverse neurologic movements such as seizures,[15,20,22-24] myoclonus,[17,19-21,25] hypertonia[16] or hypotonia,[18,22] and extrapyramidal movements[16] have been reported (Table 1).

In children, midazolam premedication significantly delays emergence from general anesthesia, although no delay in discharge post-operatively has been reported.[55,56] Furthermore, even though low doses (0.1-0.2mg/kg) of midazolam has been shown to reduce the incidence and severity of emergence delirium after general anesthesia,[57,58] it has also been reported as a potential cause of emergence delirium in the pediatric population.[59]

Table 1. Reported adverse neurologic effects with benzodiazepine administration in neonates

Study	N	FT/Prem	Drug	Adverse Drug Reactions
Adams (1997)[15]	11	Prem	Midazolam	Twitching, jerking, rhythmic movements
Bergman (1991)[16]	3	FT	Midazolam	Choreic movement, stiff posture, dyskinesis
Chess (1998)[17]	2	FT	Lorazepam	Clonic jerks
Collins (1991)[18]	1	FT	Midazolam	Unresponsive, hypotonic
Cronin (1992)[19]	3	Prem	Lorazepam	Myoclonic jerk
Lee (1994)[20]	3	Prem	Lorazepam	Rhythmic jerks, myoclonus
Magny (1994)[21]	6	Prem	Midazolam	Myoclonus
Reiter (1993)[22]	1	Prem	Lorazepam	Seizure, unresponsive, hypotonia
Sexson (1995)[23]	4	Prem	Lorazepam	Overactive, rhythmic jerks
van den Anker (1992)[24]	NR	Prem	Midazolam	Epileptiform movement
Waisman (1999)[25]	3	Prem	Midazolam	Myoclonic jerks

FT= full term infants; NR= not reported; Prem= preterm infants

Other adverse psychological effect termed "paradoxical reaction" has been reported in children.[60-63] This reaction may manifests as agitation, combativeness, restlessness, tachycardia, inconsolable crying, hallucinations, disorientation, and the inability to recognize parents or family members. Massanari et al[62] published a large series of such reactions, and found that incidence to be 1.4%. The reaction occurred at a mean of 17 minutes after administration of midazolam, and could last up to 4 hours if not treated. The onset, duration, or time to recovery was not correlated with the patient's age or the dose of midazolam used. Such reactions has been successfully treated with physostigmine,[62] and flumazenil,[62-64] which may not totally reverse the sedative effect of midazolam, allowing for continuation of the procedure under adequate sedation.[63] Although such adverse events were transient and short-lived, similar behavioural reactions have been reported in children up to one week after conscious sedation with midazolam.[65] The mechanism of paradoxical reaction is not known, but may be related to effects of benzodiazepines on neurotransmitters such as acetylcholine, 5 hydroxytryptamine and catecholamines.[64]

Long-Term Neurologic Sequalae

Although the majority of reported neurologic AE appears to be transient with full recovery, it remains unclear whether there could be more long-term sequelae. The potential long-term deleterious effect of benzodiazepines has been examined in a series of studies on the teratogenic and neurodevelopmental effects on infants after in-utero exposure to benzodiazepine from regular maternal use during pregnancies.[66-70] The findings were summarized in Table 2. These concerns were echoed by Ng et al.[71] In a systematic review of midazolam infusion use in the neonates, the largest included study by Anand et al[72] reported a higher incidence of poor neurologic outcome [death, severe intraventricular hemorrhage (IVH), periventricular leukomalacia (PVL)] in the midazolam group compared to the placebo group and the morphine group (32% vs. 24% vs. 4%, respectively, p = 0.03), even though meta-analyses of these results from all included studies did not demonstrate any significance. However, midazolam treated infants were found to have a statistically significantly longer length of stay in the NICU than the placebo group.

The reason for the predilection of adverse neurologic effects in the developing brain is unclear, and may be related to immaturity of their neuronal inhibitory pathways.[15] Specifically, term and preterm neonates express far less GABA receptors than adults,[17] so that they are predisposed to neuronal excitation.[17,23] Moreover, the specific benzodiazepine-binding $GABA_A$ receptors have different subunit composition at various developmental stages,[73] so that the neuronal inhibitory property of benzodiazepines mediated by GABA may have different effects on the neonatal brain compared with adults.

Another possible mechanism of neurologic effects may be a benzodiazepine-induced decrease in cerebral perfusion. Two studies of cerebral blood flow velocity in preterm infants receiving an intravenous bolus of midazolam[74,75] showed a significant but transient decrease in cerebral blood flow velocity from baseline within 5 to 15 minutes of drug administration. Such significant albeit transient effects on cerebral perfusion may have explained a trend toward an increased incidence of PVL in preterm infants exposed to midazolam infusion in our systematic review[71] and may have long-term neurodevelopmental implications in this patient population. Finally, the mechanism of benzodiazepine-induced hypotension has been shown in animal studies to be secondary to vasodilatation mediated by prostanoids (PGE_2) and calcium.[76] A decrease in circulating catecholamines and a direct effect on the myocardium may also contribute to benzodiazepine-induced hypotension.[77]

Respiratory Depression

Just as other sedatives and opioid analgesics may suppress respiration, there have been many reports of benzodiazepines, when used alone or in combination with other analgesics or anesthetics, to cause significant respiratory depression. Yaster et al[78] first reported on a case of respiratory arrest in a young child who received a combination of midazolam and fentanyl for sedation during a bone marrow aspiration that required brief manual positive pressure ventilation. Various studies demonstrated that midazolam, used alone or in combination with other sedatives was associated with significant hypoxia manifesting as arterial oxygen desaturation.[79-82] The degree of hypoxia has been shown to be more profound than when general anesthesia was administered.[83]

Table 2. Reported long-term adverse events from exposure to benzodiazepines in utero

Study	N	Drug Exposed In Utero	Duration of Exposure	Adverse Long-Term Events
Whitelaw (1989)[66]	53	Lorazepam	Peripartum	Delayed establishment of oral feeds over first 5 days
Sanchis (1991)[67]	2	Lorazepam	Throughout pregnancy	Feeding problems and abnormal muscle tone starting 1 wk of life
Laegreid (1989)[68]	8	Benzodiazepine	Throughout pregnancy	Teratogenic effects resembling fetal alcohol syndrome. One death showed abnormal neuronal migration and heterotopias
Laegreid (1992a)[69]	17	Benzodiazepine	Throughout pregnancy	Neurologic investigation on day 2, compared with controls: Lower birth weight for birth length; higher frequency of perinatal complications; higher frequency of abnormal neurologic examinations (symptoms of intoxication or withdrawal)
Laegreid (1992b)[70] – same series of infants as in 1992a	17	Benzodiazepine	Throughout pregnancy	Compared with controls at follow up: Decreased head circumference and poor catch up head growth; craniofacial abnormalities in 5 infants; delay in gross motor development (catch up by 18 months); delay in fine motor development (persistent to 18 months)

This dose-dependent respiratory depression may be related to the level of sedation causing a reduction in tidal volume and peak inspiratory flow,[84] in addition to the suppression of ventilatory drive in the face of hypoxia.[78] Furthermore, the co-administration of opioids may have additive effects on respiratory depression,[85] and may cause a decrease of upper airway smooth muscle tone causing obstruction.[86]

The types of benzodiazepine used may have variable effects on respiratory depression. Chiulli et al[87] compared lorazepam and diazepam in the management of children with status epilepticus. Whether in combination with phenytoin or not, patients who were treated with diazepam is significantly more likely to require endotracheal intubation than those treated with lorazepam.

Withdrawal from Benzodiazepines

Withdrawal syndrome from benzodiazepine use has been described in adults and is characterized by a variety of neurologic, neuropsychological, and somatic symptoms.[88,89] Similar withdrawal reactions have also been described in children. In two earlier case reports, five children who received midazolam infusion during intensive care for duration of 7 to 29 days experienced withdrawal reaction within 12 to 24 hours after discontinuation of the infusion.[90,91] Symptoms including agitation, aggression, confusion, visual hallucination, failure to recognize parents, generalized or subtle seizures, tachycardia, fever, aerophagia, and vomiting were noted. All were treated with reintroduction of benzodiazepines, leading to resolution of symptoms within 12 hours to 7 days. Data from larger cohort studies determined that the incidence of withdrawal from benzodiazepines in children is between 11 to 29%.[92,93]

Summary

- The use of benzodiazepines in infants and children may be associated with frequent occurrence of neurologic, respiratory, and cardiovascular AE
- The combination of benzodiazepines and opioids may be associated with serious adverse effects; caution should be exercised if both drugs are co-administered
- Although most of the reported AE were transient, studies of longer term, in-utero exposure suggest potential neurologic abnormalities that may be persistent
- The use of benzodiazepines in this population must therefore be cautious, with close monitoring for AE, especially in infants with medical conditions and/or those receiving medications that can predispose to benzodiazepine-induced AE
- Prospective studies with longitudinal follow up are necessary to confirm these findings

Clinical Application of Benzodiazepines in Neonates and Children

Use of Benzodiazepines in Neonatal Intensive Care Settings

Since the introduction of midazolam into the neonatal intensive care units (NICUs) in the 1980s, little information has been published on its effectiveness and safety when administered to critically ill neonates. The majority of reports to date are case series and studies of midazolam use in diverse age groups (including pediatric and neonatal patients).[94-100]

In a systematic review[71] three randomized controlled trials[72,101,102] reported on the effectiveness of midazolam infusion in a total of 146 infants (Table 3). In all three studies, midazolam infusion appears to be an effective sedative for neonates undergoing intensive care. Compared with placebo groups, infants in the midazolam group were more adequately sedated according to various measurements used. However, pooling the results by meta-analysis to obtain an overall estimate of effect size was not possible because different sedation scores were used, and the methods of reporting were different amongst studies. Moreover, the various sedation scores used were adapted from scores for older

Table 3. Randomized studies of midazolam infusion for neonates undergoing intensive care

Study	**Jacqz-Aigrain 1994**[101]	**Anand 1999**[72]	**Arya 2001**[102]
Comparison	Midazolam vs. Placebo	Midazolam vs. Morphine vs. Placebo	Midazolam vs. Placebo
Number	46	67	33
GA or BW criteria	None	24-32 weeks	< 2000 gm
Age at enrollment	First 48 hours	First 72 hours	First wk
Dose of Midazolam	60 μg/kg/hr (> 33 weeks GA); 60 μg/kg/hr on day 1 then 30 μg/kg/hr (< 33 weeks GA)	200 μg/kg loading dose then 20-60 μg/kg/hr depending on GA	200 μg/kg loading dose then 60 μg/kg/hr
Co-administration of morphine	No	Yes	Yes
Duration of treatment	up to 5 days	up to 14 days	48 hours
Outcome measures	Sedation score (Barrier score[170]) Physiologic changes (HR, BP)	Adverse neurologic events* sedation score (COMFORT score[171]); Neurodevelopmental outcome at 36 weeks (NAPI score[172])	Sedation score (Barrier score)
Midazolam effective?	Yes	Yes (Better than placebo)	Yes

* Adverse neurologic events include neonatal death, Papile[173] grade III/VI intraventricular hemorrhage, and periventricular leukomalacia. BP= Blood pressure; BW= Birth weight; GA= Gestational age; HR= Heart rate

infants and children and none had been validated in neonates. Therefore, a definite conclusion on the effectiveness of midazolam as sedative for neonates could not be drawn.

Use of Benzodiazepines for Narcotic Withdrawal

Neonatal abstinence syndrome (NAS) from exposure to narcotics during fetal life manifests in the neonatal period with symptoms of tremors, irritability, hyperactivity, inconsolable cry, poor feeding and vomiting, hypertonia, yawning, sneezing, nasal congestion, sweating, loose stools, and seizures.[103-108] The administration of pharmacologic agents to manage NAS must be individualized based on the severity of symptoms and weighing the risks and benefits of initiating treatment. The American Academy of Pediatrics recommends that pharmacologic therapy for NAS be given to those with seizures, feeding difficulties, diarrhea and vomiting causing dehydration or significant weight loss, sleep disturbance, and fever.[109]

Approximately 30% of infants with NAS can be managed by conservative, non-pharmacologic measures.[110] There is limited comparison data from the literature to conclude on the best pharmacologic agent for NAS.[111] Although the American Academy of Pediatrics recommended tincture of opium for NAS from opiate exposure,[109] sedatives such as benzodiazepines have been studied and used. It may have the potential benefit of avoiding further exposure to opiates and may therefore be more efficacious than tincture of opium. In a systematic review, Osborn et al[112] reviewed the effectiveness of various sedatives in managing NAS. Meta-analyses of results from 3 studies[113-115] comparing the use of phenobarbital versus diazepam for neonates withdrawing from in-utero exposure to opiates showed that treatment failure is significantly more likely when infants are treated with diazepam compared to phenobarbital. The results hold true even when data from infants of mothers using opiates alone were analyzed.[112] Clearly, the methodological qualities of the individual studies prevented a firm conclusion to be made; however, with the available evidence to date, benzodiazepines are not the drug of choice for management of NAS.

Use of Benzodiazepines in Pediatric Intensive Care Settings

Similar to neonates, critically ill children in pediatric intensive care units (PICUs) are subjected to invasive and uncomfortable procedures. In addition, the intense level of light and noise can induce a significant amount of stress, with the potential to affect clinical outcomes. Proper sedation may also facilitate mechanical ventilation, preventing adverse events such as dislodgement of endotracheal tube.[116] Studies on the use of continuous midazolam infusion for children during mechanical ventilation and post-cardiac surgery intensive care reported success in achieving sedation. Occasionally, prolonged sedation with midazolam was found, which is most likely due to accumulation of midazolam or its active metabolite.[117,118] Adequate sedation with midazolam can usually be achieved at an infusion rate between 100-400 μg/kg/hr and serum concentration of 100-500 μg/l. A much higher infusion rate (up to 480 μg/kg/hr) may be required if not used in combination with an opioid such as fentanyl.[94,119,120]

A more recent study[121] examined the quality of sedation in children undergoing mechanical ventilation. The adequacy of sedation was objectively evaluated in 28 children using a sedation score based on responses to tracheal suctioning. A combination of morphine 10-20 μg/kg/hr and midazolam 120 μg/kg/hr infusions were initiated, and then titrated as needed to a maximum of 50 μg/kg/hr of morphine and 480 μg/kg/hr of midazolam, respectively. Additional sedatives used include chloral hydrate and/or antihistamines given orally. Over 81 assessments in all patients, 79% were considered well sedated, 11% were acceptably sedated, and only 10% were considered inadequately sedated. The majority of the inadequately sedated assessments were on one individual, showing some inter-patient variability in drug handling.

Use of Benzodiazepines in Burn Patients

For acute and long-term care of intubated or nonintubated pediatric burn patients, and for those burn patients who require reconstructive surgery, the management of background pain and anxiety is of utmost importance. Although many other factors, such as the type of wound care and support of family, may modulate the degree of pain and anxiety in these patients, the use of analgesia and sedative is usually warranted.[122] Midazolam infusion has been used in conjunction with an opioid infusion to maintain comfort in burned children undergoing mechanical ventilation.[122,123] However, in cases where prolonged period of ventilation is required, the dose of midazolam required may need to be increased. The increase in requirement may be related to development of tolerance,[122] although pharmacokinetic studies in adult burn patients showed that benzodiazepine clearance and elimination may actually be augmented.[124] Intranasal midazolam use has also been reported in sedating a burn child successfully during venous catheter cannulation.[125]

Peri-Operative Applications of Benzodiazepines

The ideal premedication to relieve anxiety before induction of anesthesia in pediatric patients must be easy to use, with a relatively fast onset, short duration of action, and devoid of adverse effects.[126] Benzodiazepines, particularly midazolam, possess these qualities, and therefore have been used extensively for pediatric patients in this setting. In addition, the administration of benzodiazepines after induction of anesthesia has also been shown to have antiemetic properties.[127-130]

Earlier studies suggest that an oral dose of midazolam was better able to achieve anxiolysis for children in the preoperative period compared with a longer-acting benzodiazepine such as diazepam or placebo.[131,132] Subsequently, a number of randomized studies compared oral midazolam to placebo in children and adolescents in this setting showed a marginal and inconsistent effect of midazolam on anxiolysis.[133-138] Jones et al[139] further demonstrated that midazolam 0.5 mg/kg given orally 2 hours prior to anesthesia induction was able to provide sedation, as shown by a decline in psychomotor function with standardized tests. Physiologically, the preoperative administration of midazolam may modulate the hormonal stress response, as evident by a lower catecholamine and cortisol levels compared with controls.[140]

As to the timing of premedication administration, various studies demonstrated the onset of sedation within 10-45 minutes of midazolam administration, as evident by the ability to be separated from parents, cooperation with anesthesia induction, and anterograde amnesia.[141,142]

Although midazolam premedication appears to be effective in providing anxiolysis in children, two studies reported delay in recovery due to its use.[136,138] In addition, many studies also showed that children may be calmed preoperatively without midazolam, thus raising the question of its necessity.[133-137,143,144]

Benzodiazepines Sedation for Invasive Procedures

In ambulatory and hospital inpatient settings, painful procedures are frequently performed in children. The need for analgesia as well as sedation for these procedures, such as tooth extraction, laceration repair in the emergency department, and bone marrow aspiration for oncology patients, has been widely reported. Conscious sedation can be defined as a "medically controlled state of depressed consciousness that (1) allows protective reflexes to be maintained; (2) retains the patient's ability to maintain a patent airway independently and continuously; and (3) permits appropriate response by the patient to physical stimulation or verbal command."[9] The objectives for conscious sedation of pediatric patients during these procedures not only include facilitating the successful performance of these procedures by preventing movement, but the relief of distress caused by such procedures as well.[145]

Sedative use in these settings has been considered adjunctive to analgesics. However, a national survey of pediatric dentists

Table 4. Studies of midazolam sedation for invasive procedures in children

Study	Sedative	Route & Dose	Patient Settings	Effective Sedation?	Remarks
Friedman 1991[149]	Midazolam	0.2mg/kg IV vs. Placebo	Bone marrow aspiration/ LP in oncology patients	Yes	Good amnestic property
Elder 1991[150]	Midazolam	0.6mg/kg IV vs. Placebo	VCUG	Yes	
Shane 1994[151]	Midazolam	0.45mg/kg PR vs. Placebo	Laceration repair in ER	Yes	
Theroux 1993[152]	Midazolam	0.4mg/kg IN vs. Placebo	Laceration repair in ER	Yes	Cried and struggled less during the procedure
Connors 1994[153]	Midazolam	0.25mg/kg IN vs. 0.5mg/kg PO	Laceration repair in ER	Yes	Both regimens equally effective
Everitt 2002[154]	Midazolam	0.4mg/kg IN vs. 0.5mg/kg PO vs. 1mg/kg PO	Laceration repair in ER	Yes	All effective, but IN route sedates faster although less well tolerated

ER= Emergency room; IN= Intranasal; LP= Lumbar puncture; PO= Oral; PR= Rectal; VCUG= Voiding cystourethrogram

Table 5. Summary of studies on benzodiazepines alone for sedation of infants and children in a variety of clinical settings

Studies	Clinical Setting	Method of Administration	Dosage	Effective Sedation?	Remarks
Neonates and Infants					
Jacqz-Aigrain 1994[101] Anand 1999[72] Arya 2001[102]	Neonatal intensive care	Continuous infusion	See Table 3	Yes	Sedation scores used not validated for neonates
Finnegan 1984[113] Madden 1977[114] Kaltenbach 1986[115]	Narcotic withdrawal	Oral	Not reported	No	Diazepam more likely to cause treatment failure than Phenobarbital
Children					
Hartwig 1991[94] Moore 1993[119] Marx 1993[120] Playfor 2000[121]	Pediatric intensive care	Continuous infusion	100-480 μg/kg/hr	Yes	Seems more effective in combination with opioid
Sheridan 1994[123] Sheridan 2001[122]	Burn	Continuous infusion	40-140 μg/kg/hr	Yes	
Rice 1990[125]	Burn	Intranasal	NA	Yes	
Fell 1985[131] Parnis 1992[132] McMillan 1992[133] Cray 1996[134] Riva 1997[135] Bevan 1997[136] Mitchell 1997[137] Brosius 2002[138] Jones 1994[139]	Pre-anesthetic anxiolysis	Oral	0.25-0.5 mg single dose	Marginal and inconsistent effectiveness	Question raised as to whether pharmacological anxiolysis in this setting is necessary
See Table 4	Sedation for invasive procedures	Intravenous, intranasal, oral, rectal	Variable doses: IV 0.2-0.6mg/kg; IN 0.25-0.4 mg/kg; PO 0.5-1.0 mg/kg; PR 0.45mg/kg	Yes	Uncertain whether it is the ideal drug in this setting (see Table 4)

IV= Intravenous; IN= Intranasal; PO= Oral; PR= Rectal

showed inconsistency of its use.[146] Benzodiazepines, particularly midazolam, are one of the most commonly used agents for conscious sedation in children.[147] The reported efficacy of midazolam is approximately 60-70%.[148] The effectiveness of midazolam for sedation during invasive procedures were studied in a number of randomized trials comparing midazolam to placebo or comparing different routes of administration of midazolam in various settings.[149-154] The findings were summarized in Table 4. Overall, midazolam administered intravenously, intranasally, orally or rectally, are equally effective in providing sedation in these settings, with no reported adverse effect noted.

In terms of the optimal dose of midazolam for conscious sedation, one retrospective review of children who received midazolam in additional to meperidine for gastrointestinal endoscopy found equal efficacy and safety using low (mean 0.18 mg/kg) or high (mean 0.49 mg/kg) doses of intravenous midazolam.[155] However, another study[156] comparing midazolam 0.2 mg/kg and 0.5 mg/kg given orally as the single agent for minor procedures in the emergency department found significant differences between the two regimens in level of sedation during the procedure. The higher dose of 0.5 mg/kg was found to be effective and safe for sedation in most children for minor procedures.

Question remains, however, as to whether benzodiazepines are the ideal agent in these settings. Many studies aimed to answer this question by comparing benzodiazepines to other commonly used agents, most notably ketamine,[157,158] chloral hydrate,[159-161] and propofol.[162,163]

Three studies comparing ketamine given IM or orally to midazolam given IN, IM, or orally found that sedation was better and faster achieved using ketamine compared to midazolam in children undergoing laceration repair.[157,158,164] Studies comparing midazolam to chloral hydrate given orally for sedation during diagnostic procedures showed that chloral hydrate is superior to midazolam, as evident by a significantly lower rate of sedation failure[159] and an overall deeper level of sedation.[161] However, recovery time after the procedure may also be prolonged with chloral hydrate.[161] In a retrospective study, chloral hydrate, in combination with meperidine and hydroxyzine, was able to provide better sedation than midazolam alone, as evident by the higher percentage of sleeping and quiet behavior in children on the former regimen.[160] Two recent studies compared the efficacy of midazolam versus propofol in achieving sedation in children for invasive procedures showed that propofol is superior to midazolam in achieving sedation. In addition, propofol has less

Table 6. Comparative studies of midazolam with other sedative/anesthetic or combination of sedatives/anesthetics for sedation in children

Study	Drug Compared		Which One Is More Effective?
Ketamine			
McGlone 1998[158]	Ketamine 2.5 mg/kg IM	Midazolam 0.5 mg/kg IN	Ketamine
Mcglone 2001[164]	Ketamine 2.5 mg/kg IM	Midazolam 0.4 mg/kg IM	Ketamine
Younge 2001[157]	Ketamine 10 mg/kg PO	Midazolam 0.7 mg/kg PO	Ketamine
Chloral hydrate			
D'Agostino 2000[159] and Wheeler 2001[161]	Chloral hydrate 75 mg/kg PO	Midazolam 0.5 mg/kg PO	Chloral hydrate
Wilson 2000[160]	Chloral hydrate + hydroxyzine +/- meperidine (variable doses)	Midazolam (variable doses)	Chloral hydrate combinations
Propofol			
Arya 2002[163] and Krugliak 2000[162]	Propofol IV	Midazolam IV	Propofol
Midazolam combinations			
Klein 2002[148]	Midazolam 0.5 mg/kg PO + Transmucosal fentanyl 5-10 mg/kg	Midazolam 0.5 mg/kg PO	No difference in sedation level but more adverse reaction with combination
Acworth 2001[165]	Midazolam 0.1 mg/kg IV + Ketamine 1 mg/kg IV	Midazolam 0.4 mg/kg IN	Midazolam-ketamine combination (but time to discharge significantly longer)
Mason 2001[166]	Midazolam 0.1 mg/kg IV + Pentobarbital 2-6 mg/kg IV	Pentobarbital 2-6 mg/kg IV	Pentobarbital alone

IM= Intramuscular; IN= Intranasal; IV= Intravenous; PO= Oral

hemodynamic effects, and recovery time is significantly shorter than midazolam.[162,163]

A number of studies examined the benefit of combining various agents such as fentanyl[148] or remifentanyl[82] to improve the quality of sedation in children undergoing invasive procedures. No additional benefit has been shown with the combination. However, a higher incidence of adverse effects including vomiting, hypoxia, pruritis, and respiratory depression were found. When combined with ketamine, midazolam was shown to achieve better sedation than when given alone, although the length of time from recovery to discharge was also significantly higher.[165] Another study comparing IV pentobarbital with a combination of IV pentobarbital and midazolam during radiological procedures found patients in the combination regimen group more difficult to sedate and require longer to recover after the procedure.[166] Other studies examined the role of adding midazolam to improve the quality of recovery after ketamine sedation and found no benefit in reducing post-procedure agitation.[95,167,168] Again, the combination of ketamine and midazolam was associated with some adverse effects such as transient oxygen desaturation.[167]

From available studies, it remains uncertain whether benzodiazepines, alone or in combination, is an effective agent to provide conscious sedation in children. Indeed, it is not clear whether conscious sedation is adequate for certain procedures, particularly when they are more invasive, such as bone marrow aspiration. A recent study by Crock[145] examined that question from the perspective of family members of pediatric oncology patients. Ninety-six children with various types of neoplasms were randomized to receive general anesthesia (GA) or sedation with midazolam (0.4-0.6 mg/kg PO or 0.2-0.4 mg/kg IN) prior to lumbar puncture or bone marrow aspiration. Compared to the GA group, children receiving midazolam more frequently required physical restraint during the procedure, reported more distress before, during, and after the procedure, and experienced more pain than those receiving GA. Parents also perceived more distress in their children in the midazolam group, and the majority wanted a GA for similar procedures in the future.

Summary

The clinical applications of benzodiazepines, alone, or in combination with other sedative/anesthetics in infants and children are summarized in Tables 5 and 6.

Conclusion

The increasing popularity of benzodiazepine use in infants and children over the past two decades stems from the awareness that this group of patients do experience pain and stress during medical procedures. These procedures, whether invasive or noninvasive, are such commonplace in the delivery of even the most routine pediatric and neonatal care. As an extreme example, in one neonatal intensive care unit, each infant undergoes a documented average of invasive 61 invasive procedures throughout the hospital stay, with the highest of 488 procedures in one infant.[169] While the use of analgesics and sedative should not be overlooked, a thorough understanding of the properties of these potent medications is necessary to ensure the safety and efficacy of their use. This review aimed at an understanding of the pharmacokinetics and pharmacodynamics of the sedative benzodiazepines, consideration of its application in various clinical settings, and review of the potential short and long-term adverse effects from its use in this diverse population. After appraisal of the literature to date, some concerns were raised as to the safety of benzodiazepine use in the neonatal population, particularly since there are potential adverse long-term neurodevelopmental implications after its exposure. Moreover, its use in the pediatric population, although deemed safe, may not be appropriate in all clinical circumstance. It is clear, however, that further research will be necessary to define the limit of appropriate use of the benzodiazepines in infants and children. For now, it appears that judicious use of this medication is warranted.

References

1. Ferguson S, Ball AJ. Sedation and sedative drugs in paediatrics. Br J Hosp Med 1996; 55:611-5.
2. Canadian Paediatric Society. Prevention and management of pain and stress in the neonate. Paediatr Child Health 2000; 5:31-38.
3. Quinn M, Wild J, Dean H et al. Randomised double-blind controlled trial of effect of morphine on catecholamine concentrations in ventilated preterm babies. Lancet 1993; 342:324-7.
4. Greenough A, Morley C, Davis J. Interaction of spontaneous respiration with artificial ventilation in preterm babies. J Pediatr 1983; 103:769-73.
5. Perlman J, Goodman S, Kreusser K et al. Reduction in intraventricular hemorrhage by elimination of fluctuating cerebral blood-flow velocity in preterm infants with respiratory distress syndrome. N Engl J Med 1985; 312:1353-7.
6. Noerr B. Midazolam (Versed). Neonatal Netw 1995; 14:65-7.
7. Kaplan RF, Yang CI. Sedation and analgesia in pediatric patients for procedures outside the operating room. Anesthesiol Clin North America 2002; 20:181-94.
8. Schwengel D. Pain management. In: Rudolph CD, Rudolph AM, eds. Rudolph's Pediatrics. 21st ed. New York: McGraw-Hill, 2002:341-348.
9. American Academy of Pediatrics, committee on drugs. Guidelines for monitoring and management of pediatric patients during and after sedation for diagnostic and therapeutic procedures. Pediatrics 1992; 89:1110-5.
10. American Academy of Pediatrics, committee on drugs. Guidelines for monitoring and management of pediatric patients during and after sedation for diagnostic and therapeutic procedures: Addendum. Pediatrics 2002; 110:836-8.
11. Blumer JL. Clinical pharmacology of midazolam in infants and children. Clin Pharmacokinet 1998; 35:37-47.
12. Gaillard JM. Benzodiazepines and GABA-ergic transmission. In: Kryger MH, Roth T, Dement WC, eds. Principles and Practice of Sleep Medicine. 2nd ed. Toronto: WB Saunders, 1994:349-54.
13. Platten HP, Schweizer E, Dilger K et al. Pharmacokinetics and the pharmacodynamic action of midazolam in young and elderly patients undergoing tooth extraction. Clin Pharmacol Ther 1998; 63:552-60.
14. Klotz U. Effect of age on pharmacokinetics and pharmacodynamics in man. Int J Clin Pharmacol Ther 1998; 36:581-5.
15. Adams MM, Hahn JS, Benitz WE. A series of neonatal patients with paradoxical seizure-like reactions to bolus intravenous injections of midazolam. Pediatr Res 1997; 41:134A (Abstract 790).
16. Bergman I, Steeves M, Burckart G et al. Reversible neurologic abnormalities associated with prolonged intravenous midazolam and fentanyl administration. J Pediatr 1991; 119:644-9.
17. Chess PR, D'Angio CT. Clonic movements following lorazepam administration in full-term infants. Arch Pediatr Adolesc Med 1998; 152:98-9.
18. Collins S, Carter JA. Resedation after bolus administration of midazolam to an infant and its reversal by flumazenil. Anaesthesia 1991; 46:471-2.
19. Cronin CM. Neurotoxicity of lorazepam in a premature infant. Pediatrics 1992; 89:1129-30.
20. Lee DS, Wong HA, Knoppert DC. Myoclonus associated with lorazepam therapy in very-low-birth-weight infants. Biol Neonate 1994; 66:311-5.
21. Magny JF, d'Allest AM, Nedelcoux H et al. Midazolam and myoclonus in neonate. Eur J Pediatr 1994; 153:389-90.
22. Reiter PD, Stiles AD. Lorazepam toxicity in a premature infant. Ann Pharmacother 1993; 27:727-9.
23. Sexson WR, Thigpen J, Stajich GV. Stereotypic movements after lorazepam administration in premature neonates: A series and review of the literature. J Perinatol 1995; 15:146-9.
24. van den Anker JN, Sauer PJ. The use of midazolam in the preterm neonate. Eur J Pediatr 1992; 151:152.
25. Waisman D, Weintraub Z, Rotschild A et al. Myoclonic movements in very-low-birth-weight premature infants associated with midazolam intravenous bolus administration. Pediatrics 1999; 104:579.
26. Reed MD, Rodarte A, Blumer JL et al. The single-dose pharmacokinetics of midazolam and its primary metabolite in pediatric patients after oral and intravenous administration. J Clin Pharmacol 2001; 41:1359-69.
27. Burtin P, Jacqz-Aigrain E, Girard P et al. Population pharmacokinetics of midazolam in neonates. Clin Pharmacol Ther 1994; 56:615-25.
28. Marshall J, Rodarte A, Blumer J et al. Pediatric Pharmacodynamics of midazolam oral syrup. J Clin Pharmacol 2000; 40:578-89.
29. Jacqz-Aigrain E, Wood C, Robieux I. Pharmacokinetics of midazolam in critically ill neonates. Eur J Clin Pharmacol 1990; 39:191-2.
30. de Wildt SN, Kearns GL, Hop WC et al. Pharmacokinetics and metabolism of intravenous midazolam in preterm infants. Clin Pharmacol Ther 2001; 70:525-31.
31. Jacqz-Aigrain E, Daoud P, Burtin P et al. Pharmacokinetics of midazolam during continuous infusion in critically ill neonates. Eur J Clin Pharmacol 1992; 42:329-32.
32. Lee TC, Charles BG, Harte GJ et al. Population pharmacokinetic modeling in very premature infants receiving midazolam during mechanical ventilation. Anesthesiology 1999; 90:451-7.
33. Tolia V, Brennan S, Aravind MK et al. Pharmacokinetic and pharmacodynamic study of midazolam in children during esophagogastroduodenoscopy. J Pediatr 1991; 119:467-71.
34. Mathews HM, Carson IW, Lyons SM et al. A pharmacokinetic study of midazolam in paediatric patients undergoing cardiac surgery. Br J Anaesth 1988; 61:302-7.
35. Nahara MC, McMorrow J, Jones PR et al. Pharmacokinetics of midazolam in critically ill pediatric patients. Eur J Drug Metab Pharmacokinet 2000; 25:219-21.
36. Payne K, Mattheyse FJ, Liebenberg D et al. Pharmacokinetics of midazolam in paediatric patients. Eur J Clin Pharmacol 1989; 37:267-72.
37. Rey E, Delaunay L, Pons G et al. Pharmacokinetics of midazolam in children: Comparative study of intranasal and intravenous administration. Eur J Clin Pharmacol 1991; 41:355-7.
38. Walbergh EJ, Wills RJ, Eckhert J. Plasma concentrations of midazolam in children following intranasal administration. Anesthesiology 1991; 74:233-5.
39. Reves JG, Fragen RJ, Vinik HR et al. Midazolam: Pharmacology and uses. Anesthesiology 1985; 62:310-24.
40. Malinovsky JM, Lejus C, Servin F et al. Plasma concentrations of midazolam after i.v., nasal or rectal administration in children. Br J Anaesth 1993; 70:617-20.
41. Kraus GB, Gruber RG, Knoll R et al. Pharmacokinetic studies following intravenous and rectal administration of midazolam in children. Anaesthesist 1989; 38:658-63.
42. Geldner G, Hubmann M, Knoll R et al. Comparison between three transmucosal routes of administration of midazolam in children. Paediatr Anaesth 1997; 7:103-9.
43. Lammers CR, Rosner JL, Crockett DE et al. Oral midazolam with an antacid may increase the speed of onset of sedation in children prior to general anesthesia. Paediatr Anaesth 2002; 12:26-8.
44. Goho C. Oral midazolam-grapefruit juice drug interaction. Pediatr Dent 2001; 23:365-6.
45. Hiller A, Olkkola KT, Isohanni P et al. Unconsciousness associated with midzaolam and erythromycin. Br J Anaesth 1990; 65:826-8.
46. Olkkola KT, Aranko K, Luurila H et al. A potentially hazardous interaction between erythromycin and midazolam. Clin Pharmacol Ther 1993; 53:298-305.

47. Hase I, Oda Y, Tanaka K et al. I.v. fentanyl decreases the clearance of midazolam. Br J Anaesth 1997; 79:740-3.
48. Ng E, Klinger G, Shah V et al. Safety of benzodiazepines in newborns. Ann Pharmacother 2002; 36:1150-5.
49. Lugo RA, Fishbein M, Nahata MC et al. Complication of intranasal midazolam. Pediatrics 1993; 92:638.
50. Zedie N, Amory DW, Wagner BK et al. Comparison of intranasal midazolam and sufentanil premedication in pediatric outpatients. Clin Pharmacol Ther 1996; 59:341-8.
51. Griffith N, Howell S, Mason D. Intranasal midazolam for premedication of children undergoing day-case anaesthesia: Comparison of two delivery systems with assessment of intra-observer variability. Br J Anaesth 1998; 81:865-9.
52. Marshall JD, Farrar HC, Kearns GL. Diarrhea associated with enteral benzodiazepine solutions. J Pediatr 1995; 126:657-9.
53. Rodríguez-Núñez A, Redondo L, Martinón JM. Hiccups due to midazolam in children. Eur J Pediatr 1993; 153:271.
54. Marhofer P, Glaser C, Krenn CG et al. Incidence and therapy of midazolam induced hiccups in paediatrc anaesthesia. Paediatr Anaesth 1999; 9:295-8.
55. Viitanen H, Annila P, Viitanen M et al. Midazolam premedication delays recovery from propofol-induced sevoflurane anesthesia in children 1-3 yr. Can J Anaesth 1999; 46:766-71.
56. Horgesheimer JJ, Pribble CG, Lugo RA. The effect of midazolam premedication on discharge time in pediatric patients undergoing general anesthesia and dental restorations. Pediatr Dent 2001; 23:491-4.
57. Ko YP, Huang CJ, Hung YC et al. Premedication with low-dose oral midazolam reduces the incidence and severity of emergence agitation in pediatric patients following sevoflurane anesthesia. Acta Anaesthesiol Sin 2001; 39:169-77.
58. Kulka PJ, Bressem M, Wiebalck A et al. Prevention of "post-sevoflurane delirium" with midazolam. Anaesthesist 2001; 50:401-5.
59. Doyle WL, Perrin L. Emergence delirium in a child given oral midazolam for conscious sedation. Ann Emerg Med 1994; 24:1173-5.
60. Marcus A, Ammermann C, Bahro M et al. Benzodiazepine administration induces exogenic psychosis: A case of child abuse. Child Abuse Negl 1995; 19:833-6.
61. van den Berg AA. Hallucinations after oral lorazepam in children. Anaesthesia 1986; 41:330-1.
62. Massanari M, Novitsky J, Reinstein LJ. Paradoxical reactions in children associated with midazolam use during endoscopy. Clin Pediatr 1997; 36:681-4.
63. Nur Saltik I, Özen H. Role of flumazenil for paradoxical reaction to midazolam during endoscopic procedures in children. Am J Roentgenol 2000; 95:3011-2.
64. Thakker P, Gallagher TM. Flumazenil reverses paradoxical reaction to midazolam in a child. Anaesth Intensive Care 1996; 24:505-7.
65. McGraw T. Oral midazolam and postoperative behaviour in children. Can J Anaesth 1993; 40:682-3.
66. Whitelaw AG, Cummings AJ, McFadyen IR. Effect of maternal lorazepam on the neonate. Br Med J 1981; 282:1106-8.
67. Sanchis A, Rosique D, Catala J. Adverse effects of maternal lorazepam on neonates. Ann Pharmacother 1991; 25:1137-8.
68. Laegreid L, Olegard R, Walstrom J et al. Teratogenic effects of benzodiazepine use during pregnancy. J Pediatr 1989; 114:126-31.
69. Laegreid L, Hagberg G, Lundberg A. The effect of benzodiazepines on the fetus and the newborn. Neuropediatrics 1992; 23:18-23.
70. Laegreid L, Hagberg G, Lundberg A. Neurodevelopment in late infancy after prenatal exposure to benzodiazepines - A prospective study. Neuropediatrics 1992; 23:60-7.
71. Ng E, Taddio A, Ohlsson A. Intravenous midazolam infusion for sedation of infants in the neonatal intensive care unit (Cochrane review). In: The Cochrane Library, issue 1, 2003. Oxford: update software 2003.
72. Anand K, McIntosh N, Lagercrantz H et al. Analgesia and sedation in preterm neonates who require ventilatory support - results from the NOPAIN trial. Arch Pediatr Adolesc Med 1999; 153:331-8.
73. Brooks-Kayal AR, Pritchett DB. Developmental changes in human γ-aminobutyric $acid_A$ receptor subunit composition. Ann Neurol 1993; 34:687-93.
74. Harte GJ, Gray PH, Lee TC et al. Haemodynamic responses and population pharmacokinetics of midazolam following administration to ventilated, pterterm neonates. J Paediatr Child Health 1997; 33:335-8.
75. van Straaten HL, Rademaker CM, de Vries LS. Comparison of the effect of midazolam or vecuronium on blood pressure and cerebral blood flow velocity in the premature newborn. Dev Pharmacol Ther 1992; 19:191-5.
76. Modanlou HD, Beharry, K. Mechanism of midazolam-induced hypotension: Possible role of prostanoids and Ca2+. Pediatric Research 1997; 41:57A (Abstract #329).
77. Burtin P, Daoud P, Jacqz-Aigrain E et al. Hypotension with midazolam and fentanyl in the newborn. Lancet 1991; 337:1545-6.
78. Yaster M, Nichols DG, Deshpande JK et al. Midazolam-fentanyl intravenous sedation in children: Case report of respiratory arrest. Pediatrics 1990; 86:463-7.
79. Wathen JE, Roback MG, Mackenzie T et al. Does midazolam alter the clinical effects of intravenous ketamine sedation in children? A double-blind, randomized, controlled, emergency department trial. Ann Emerg Med 2000; 36:579-88.
80. Roelofse JA, de V Joubert JJ. Arterial oxygen saturation in children receiving rectal midazolam as premedication for oral surgical procedures. Anesth Prog 1990; 37:286-9.
81. Roelofse JA, Roelofse PG. Oxygen desaturation in a child receiving a combination of ketamine and midazolam for dental extraction. Anesth Prog 1997; 44:68-70.
82. Litman RS. Conscious sedation with remifentanil and midazolam during brief painful procedures in children. Arch Pediatr Adolesc Med 1999; 153:1085-8.
83. Lamireau T, Dubreuil M, Daconceicao M. Oxygen saturation during esophagogastroduodenoscopy in children: General anesthesia versus intravenous sedation. J Pediatr Gastroenterol Nutr 1999; 28:455.
84. Fosel T, Cartarius R, Gruness V et al. The effect of midazolam after intranasal administration on spontaneous respiratory control in young children. Anasthesiol Intensivmed Notfallmed Schmerzther 1996; 31:22-5.
85. Bailey PL, Andriano KP, Goldman M et al. Variability of the respiratory response to diazepam. Anesthesiology 1986; 64:460-5.
86. Negus BH, Street NE. Midazolam-opioid combination and postoperative upper airway obstruction in children. Anaesth Intensive Care 1994; 22:232-3.
87. Chiulli DA, Terndrup TE, Kanter RK. The influence of diazepam or lorazepam on the frequency of endotracheal intubation in childhood status epilepticus. J Emerg Med 1991; 9:13-7.
88. MacKinnon GL, Parker WA. Benzodiazepine withdrawal syndrome: A literature review and evaluation. Am J Drug Alcohol Abuse 1982; 9:19-33.
89. Busto U, Sellers EM, Naranjo CA et al. Withdrawal reaction after long-term therapeutic use of benzodiazepines. N Engl J Med 1986; 315:854-9.
90. Sury MR, Billingham I, Russell GN et al. Acute benzodiazepine withdrawal syndrome after midazolam infusions in children. Crit Care Med 1989; 17:301-2.
91. van Engelen BG, Gimbrere JS, Booy LH. Benzodiazepine withdrawal reaction in two children following discontinuation of sedation with midazolam. Ann Pharmacother 1993; 27:579-81.
92. Hughes J, Gill A, Leach HJ et al. A prospective study of the adverse effects of midazolam on withdrawal in critically ill children. Acta Paediatr 1994; 83:1194-9.
93. Fonsmark L, Rasmussen YH, Carl P. Occurrence of withdrawal in critically ill sedated children. Crit Care Med 1999; 27:196-9.
94. Hartwig S, Roth B, Theisohn M. Clinical experience with continuous intravenous sedation using midazolam and fentanyl in the paediatric intensive care unit. Eur J Pediatr 1991; 150:784-8.
95. Pellier I, Monrigal J, Le Moine P et al. Use of intravenous ketamine-midazolam association for pain procedure in children with cancer. A prospective study. Paediatr Anaesth 1999; 9:61-8.
96. Rosen D, Rosen K. Midazolam for sedation in the paediatric intensive care unit. Intensive Care Med 1991; 17:S15-9.
97. Stenhammer L, Högberg L, Lewander P et al. Intravenous midazolam in small bowel biopsy. Arch Dis Child 1994; 71:558.

98. McCarver-May D, Kang J, Aouthmany M et al. Comparison of chloral hydrate and midazolam for sedation of neonates for neuroimaging studies. J Pediatr 1996; 128:573-6.
99. Kawakami K, Ohata J, Kadosaki M et al. Midazolam for anesthetic induction in neonates. Masui 1998; 47:570-5.
100. Parkinson L, Hughes J, Gill A et al. A randomized controlled trial of sedation in the critically ill. Paediatr Anaesth 1997; 7:405-10.
101. Jacqz-Aigrain E, Daoud P, Burtin P et al. Placebo-controlled trial of midazolam sedation in mechanically ventilated newborn babies. Lancet 1994; 344:646-50.
102. Arya V, Ramji S. Midazolam sedation in mechanically ventilated newborns: A double blind randomized placebo controlled trial. Indian Pediatr 2001; 38:967-72.
103. Neumann LL, Cohen SN. The neonatal narcotic withdrawal syndrome: A therapeutic challenge. Clin Perinatol 1975; 2:99-109.
104. Herzlinger RA, Kandall SR, Vaughan HG. Neonatal seizures associated with narcotic withdrawal. J Pediatr 1977; 91:638-41.
105. Kandall SR, Albin S, Gartner LM et al. The narcotic-dependent mother: Fetal and neonatal consequences. Early Hum Dev 1977; 1:159-69.
106. Ostrea EM, Chavez CJ, Strauss ME. A study of factors that influence the severity of neonatal narcotic withdrawal. J Pediatr 1976; 88:642-5.
107. Kron RE, Litt M, Finnegan LP. Narcotic addiction in the newborn: Differences in behavior generated by methadone and heroin. Int J Clin Pharmacol Ther Toxicol 1975; 12:63-9.
108. American Academy of Pediatrics, committee on drugs. Neonatal drug withdrawal. Pediatrics 1998; 101:1079-88.
109. Neonatal drug withdrawal. Pediatrics. American Academy of Pediatrics, Committee on Drugs, 1998:101:1079-88.
110. Zelson C. Acute management of neonatal addiction. Addict Dis 1975; 2:159-68.
111. Theis JG, Selby P, Ikizler Y et al. Current management of the neonatal abstinence syndrome: A critical analysis of the evidence. Biol Neonate 1997; 71:345-56.
112. Osborn DA, Jeffery HE, Cole MJ. Sedatives for opiate withdrawal in newborn infants (Cochrane Review). In: The Cochrane Library, issue 1, 2003. Oxford: update software 2003.
113. Finnegan LP, Michael H, Leifer B et al. An evaluation of neonatal abstinence treatment modalities. NIDA Res Monogr 1984; 49:282-8.
114. Madden J, Chappel J, Gumpel J et al. Observation and treatment of neonatal narcotic withdrawal. Am J Obstet Gynecol 1977; 127:199-201.
115. Kaltenbach K, Finnegan LP. Neonatal abstinence syndrome, pharmacotherapy and developmental outcome. Neurobehav Toxicol Teratol 1986; 8:353-5.
116. Notterman DA. Sedation with intravenous midazolam in the pediatric intensive care unit. Clin Pediatr 1997; 36:449-54.
117. Lloyd-Thomas AR, Booker PD. Infusion of midazolam in paediartric patients after cardiac surgery. Br J Anaesth 1986; 58:1109-15.
118. Booker PD, Beechey A, Lloyd-Thomas AR. Sedation of children requiring artificial ventilation using an infusion of midazolam. Br J Anaesth 1986; 58:1104-8.
119. Moore A, Hughes J, Turner S et al. Midazolam infusion in pediatric intensive care unit. Ann Pharmacother 1993; 27:791.
120. Marx CM, Reed MD. Sedative efficacy of midazolam in pediatric ICU. Ann Pharmacother 1993; 27:1543-4.
121. Playfor SD, Thomas DA, Choonara I. Quality of sedation during mechanical ventilation. Paediatr Anaesth 2000; 10:195-9.
122. Sheridan R, Stoddard F, Querzoli E. Management of background pain and anxiety in critically burned children requiring protracted mechanical ventilation. J Burn Care Rehabil 2001; 22:150-3.
123. Sheridan RL, McEttrick M, Bacha G et al. Midazolam infusion in pediatric patients with burns who are undergoing mechanical ventilation. J Burn Care Rehabil 1994; 15:515-8.
124. Martyn J Greenblatt DJ. Lorazepam conjugation is unimpaired in burn trauma. Clin Pharmacol Ther 1987; 43:250-5.
125. Rice TL, Kyff JV. Intranasal administration of midazolam to a severely burned child. Burns 1990; 16:307-8.
126. Funk W, Jakob W, Riedl T et al. Oral preanaesthetic medication for children: Double-blind randomized study of a combination of midazolam and ketamine vs midazolam or ketamine alone. Br J Anaesth 2000; 84:335-40.
127. van Hoff J, Olszewski D. Lorazepam for the control of chemotherapy-related nausea and vomiting in children. J Pediatr 1988; 113:146-9.
128. Khalil SN, Berry JM, Howard G et al. The antiemetic effect of lorazepam after outpatient strabismus surgery in children. Anesthesiology 1992; 77:915-9.
129. Splinter W, Noël L, Roberts D et al. Antiemetic prophylaxis for strabismus surgery. Can J Ophthalmol 1994; 29:224-6.
130. Splinter W, MacNeil H, Menard E et al. Midazolam reduces vomiting after tonsillectomy in children. Can J Anaesth 1995; 42:201-203.
131. Fell D, Gough MB, Northan A et al. Diazepam premedication in children. Anaesthesia 1985; 40:12-7.
132. Parnis SJ, Foate JA, van der Walt JH et al. Oral midazolam is an effective premedication for children having day-stay anaesthesia. Anaesth Intensive Care 1992; 20:9-14.
133. McMillan CO, Spahr-Schopfer IA, Sikich N et al. Premedication of children with oral midazolam. Can J Anaesth 1992; 39:545-550.
134. Cray SH, Dixon JL, Heard CM et al. Oral midazolam premedication for paediatric day case patients. Paediatr Anaesth 1996; 6:265-70.
135. Riva J, Lejbusiewicz G, Papa M et al. Oral premedication with midazolam in paediatric anaesthesia. Effects on sedation and gastric contents. Paediatr Anaesth 1997; 7:191-6.
136. Bevan JC, Veall GR, Macnab AJ et al. Midazolam premedication delays recovery after propofol without modifying involuntary movements. Anesth Analg 1997; 85:50-4.
137. Mitchell V, Grange C, Black A et al. A comparison of midazolam with trimeprazine as an oral premedicant for children. Anaesthesia 1997; 52:416-21.
138. Brosius KK, Bannister CF. Oral midazolam premedication in preadolescents and adolescents. Anesth Analg 2002; 94:31-6.
139. Jones RD, Visram AR, Kornberg JP et al. Premedication with oral midazolam in children—an assessment of psychomotor function, anxiolysis, sedation and pharmacokinetics. Anaesth Intensive Care 1994; 22:539-44.
140. Burkhardt U, Wild L, Vetter B et al. Modulation of the stress response in children in the preoperative preparation. Anaesthesist 1997; 46:850-5.
141. Weldon BC, Watcha MF, White PF. Oral midazolam in children: Effect of time and adjunctive therapy. Anesth Analg 1992; 75:51-5.
142. Kain Z, Hofstadter M, Mayes L et al. Midazolam: Effects on amnesia and anxiety in children. Anesthesiology 2000; 93:676-84.
143. Vetter TR. A comparison of midazolam, diazepam, and placebo as oral anesthetic premedicants in younger children. J Clin Anesth 1993; 5:58-61.
144. Ong BC, Ng AS, Chew SL. Oral premedications in paediatric day surgery. Singapore Med J 1996; 37:139-42.
145. Crock C, Olsson C, Phillips R et al. General anaesthesia or conscious sedation for painful procedures in childhood cancer: The family's perspective. Arch Dis Child 2003; 88:253-7.
146. Houpt M. Project USAP 2000—use of sedative agents by pediatric dentists: A 15-year follow-up survey. Pediatr Dent 2002; 24:289-94.
147. Krauss B, Green SM. Sedation and analgesia for procedures in children. N Engl J Med 2000; 342:938-45.
148. Klein EJ, Diekema DS, Paris CA et al. A randomized, clinical trial of oral midazolam plus placebo versus oral midazolam plus oral transmucosal fentanyl for sedation during laceration repair. Pediatrics 2002; 109:894-7.
149. Friedman AG, Mulhern RK, Fairclough D et al. Midazolam premedication for pediatric bone marrow aspiration and lumbar puncture. Med Pediatr Oncol 1991; 19:499-504.
150. Elder JS, Longenecker R. Premedication with oral midazolam for voiding cystourethrography in children: Safety and efficacy. Am J Roentgenol 1995; 164:1229-32.
151. Shane SA, Fuchs SM, Khine H. Efficacy of rectal midazolam for the sedation of preschool children undergoing laceration repair. Ann Emerg Med 1994; 24:1065-73.
152. Theroux MC, West DW, Corddry DH et al. Efficacy of intranasal midazolam in facilitating suturing of lacerations in preschool children in the emergency department. Pediatrics 1993; 91:624-7.
153. Connors K, Terndrup TE. Nasal versus oral midazolam for sedation of anxious children undergoing laceration repair. Ann Emerg Med 1994; 24:1074-9.

154. Everitt IJ, Barnett P. Comparison of two benzodiazepines used for sedation of children undergoing suturing of a laceration in an emergency department. Pediatr Emerg Care 2002; 18:72-4.
155. Gremse DA, Kumar S, Sacks AI. Conscious sedation with high-dose midazolam for pediatric gastrointestinal endoscopy. South Med J 1997; 90:821-5.
156. Davies FC, Waters M. Oral midazolam for conscious sedation of children during minor procedures. J Accid Emerg Med 1998; 15:244-8.
157. Younge PA, Kendall JM. Sedation for children requiring wound repair: A randomised controlled double blind comparison of oral midazolam and oral ketamine. Emerg Med J 2001; 18:30-3.
158. McGlone RG, Ranasinghe S, Durham S. An alternative to "brutacaine": A comparison of low dose intramuscular ketamine with intranasal midazolam in children before suturing. J Accid Emerg Med 1998; 15:231-6.
159. D'Agostino J, Terndrup TE. Chloral hydrate versus midazolam for sedation of children for neuroimaging: A randomized clinical trial. Pediatr Emerg Care 2000; 16:1-4.
160. Wilson S, Easton J, Lamb K et al. A retrospective study of chloral hydrate, meperidine, hydroxyzine, and midazolam regimens used to sedate children for dental care. Pediatr Dent 2000; 22:107-12.
161. Wheeler DS, Jensen RA, Poss WB. A randomized, blinded comparison of chloral hydrate and midazolam sedation in children undergoing echocardiography. Clin Pediatr 2001; 40:381-7.
162. Krugliak P, Ziff B, Rusabrov Y et al. Propofol versus midazolam for conscious sedation guided by processed EEG during endoscopic retrograde cholangiopancreatography: A prospective, randomized double-blind study. Endoscopy 2000; 32:677-82.
163. Arya VS, Damle SG. Comparative evaluation of midazolam and propofol as intravenous sedative agents in the management of uncooperative children. J Indian Soc Pedod Prev Dent 2002; 20:6-8.
164. McGlone R, Fleet T, Durham S et al. A comparison of intramuscular ketamine with high dose intramuscular midazolam with and without intranasal flumazenil in children before suturing. Emerg Med J 2001; 18:34-8.
165. Acworth JP, Purdie D, Clarke RC. Intravenous ketamine plus midazolam is superior to intranasal midazolam for emergency paediatric procedural sedation. Emerg Med J 2001; 18:39-45.
166. Mason KP, Zurakowski D, Karian VE et al. Sedative used in pediatric imaging: Comparison of IV pentobarbital with IV pentobarbital with midazolam added. Am J Roentgenol 2001; 177:427-30.
167. Parker RI, Mahan RA, Giugliano D et al. Efficacy and safety of intravenous midazolam and ketamine as sedation for therapeutic and diagnostic procedures in children. Pediatrics 1997; 99:427-31.
168. Sherwin TS, Green SM, Khan A et al. Does adjunctive midazolam reduce recovery agitation after ketamine sedation for pediatric procedures? A randomized, double-blind placebo-controlled trial. Ann Emerg Med 2000; 35:229-38.
169. Barker D, Rutter N. Exposure to invasive procedures in neonatal intensive care unit admissions. Arch Dis Child Fetal Neonatal Ed 1995; 72:47-8.
170. Barrier G, Attia J, Mayer M et al. Measurement of post-operative pain and narcotic administration in infants using a new clinical scoring system. Intensive Care Med 1989; 15(Suppl 1):S37-9.
171. Ambuel B, Hamlett K, Marx C et al. Assessing distress in pediatric intensive care environments: The COMFORT scale. J Pediatr Psychol 1992; 17:95-109.
172. Korner A, Constantinou J, Dimiceli S et al. Establishing the reliability and developmental validity of a neurobehavioural assessment for preterm infants: A methodological process. Child Development 1991; 62:1200-8.
173. Papile L, Burstein J, Burstein R et al. Incidence and evolution of subependymal and intraventricular hemorrhage: A study of infants with weights less than 1500 grams. J Pediatr 1978; 92:529-34.

CHAPTER 20

The Pharmacological Management of Fatigue and Sleepiness in Affective Disorders

Karl Doghramji

Abstract

Fatigue, daytime sleepiness, and hypersomnia are commonly associated with affective illnesses. In major depressive disorder, they not only accompany the acute episode, but also frequently precede its onset. These symptoms also often complicate the recovery phase itself. In this chapter, the longitudinal course, manifestations, and treatment of sleepiness and fatigue in depression are reviewed. Although many approaches to management exist, this chapter will focus on pharmacological measures, including monotherapy and polytherapy.

Definitions

The literature in affective disorders that serves as the database of this chapter utilizes a diverse terminology. While not universally accepted, I will adhere to the following terms and definitions:

1. Excessive daytime sleepiness (somnolence, EDS): Propensity to fall asleep during the day.[1]
2. Hypersomnia: Increased time spent asleep or in bed.
3. Fatigue: Weariness, tiredness, exhaustion, loss of energy.[2]

Other terms used historically to describe this area include asthenia and sedation.

It is possible that these terms actually describe distinct clinical states. This is evident in the inventories and tests used to measure each. In affective disorders, a third area of importance is the degree of psychomotor retardation, commonly measured by the Hamilton Depression Retardation Factor.[2a] Each also seems to be associated with its characteristic set of disorders. For example, EDS is a hallmark symptom of sleep deprivation, shiftwork sleep disorder, narcolepsy, and obstructive sleep apnea syndrome, among others; fatigue is a hallmark symptom of multiple sclerosis, depression, chronic fatigue syndrome, and others. Hypersomnia characterizes the daytime complaint of bipolar depression and seasonal affective disorder. Nevertheless, the relationships between these states, and their differential diagnoses have been largely unexplored.

Prevalence and Impact

Alterations in sleep and wakefulness are some of the most common complaints in patients suffering from affective disorders; an estimated 80% of patients with major depression, for example, voice these complaints.[3] Over the past few decades, the focus of attention in this area, both in research and clinical settings, has been on nature of sleep and its disturbances, and little attention has been paid to the daytime and waking complaints. This is quite surprising, when one considers the large body of data suggesting that these complaints are at least as commonly voiced in depressives. Sampled populations indicate that 94% of depressed patients complain of fatigue and loss of energy, 84% of impaired concentration, 73% of tiredness, and 97% of reduced energy.[4] Additionally, symptoms of anhedonia, in the presence of psychomotor retardation, hypersomnia, fatigue/loss of energy, and diminished ability to think/concentrate/decide are sufficient to meet core criteria for a major depressive episode.[5] Although they may occur in the context of any depressive episode, these symptoms seem to be more likely among patients with childhood unipolar depression, seasonal affective disorder, and bipolar disorder, which often tend to involve "anergic" states (characterized by fatigue, hypersomnolence, psychomotor slowing, increased appetite, and weight gain).

The negative impact of these symptoms on quality of life and daytime performance has been well established, if not in affective disorders, then certainly in other disease states such as narcolepsy, obstructive sleep apnea syndrome and sleep deprivation/restriction. Data from controlled settings have documented impairments such as slower response time, instability of attention, cognitive slowing on subject-paced tasks, decline in both short-term recall and working memory performance, reduced learning (acquisition) of cognitive tasks, and neglect of nonessential activities. Demographic studies have clearly documented the real-life impairments associated with sleepiness, including traffic accidents, occupational performance decrements, and breaches in relationships, among others. In contrast, cause-and-effect relationships between insomnia and similar impairments have been much more difficult to clarify.

Longitudinal Relationships

Premorbid

The association between current hypersomnia and depression was documented in a recent survey of 1200 randomly selected young adult members of a health maintenance organization; patients with complaints of either hypersomnia or both insomnia and hypersomnia were more likely to exhibit current major depression (25.3% and 54.3%, respectively) than patients without

Sleep and Sleep Disorders: A Neuropsychopharmacological Approach, edited by Malcolm Lader, Daniel P. Cardinali and S.R. Pandi-Perumal.

sleep complaints (2.7%).[6] Studies further indicate an association between excessive sleepiness and depression. For example, in the Finnish Twin Cohort study, approximately one fourth of the patients who reported daytime sleepiness also displayed symptoms of moderate to severe depression and, conversely, half of the patients with major depression reported daily sleepiness.[7] Longitudinal studies also suggest that hypersomnia can be a premonitory symptom in depressives. As part of the National Institute of Mental Health Epidemiologic Catchment Area Study (1981-1985), respondents were interviewed at baseline and 1 year later to determine the extent of insomnia and hypersomnia and the relative probability of having or developing a new psychiatric disorder.[8] The investigators observed a community prevalence rate of 3.2% for hypersomnia. At baseline, only 0.9% of those with no sleep complaints had major depression, whereas 9.9% of those with hypersomnia had this disorder. Remarkably, 28.6% of community residents with complaints of hypersomnia that persisted for one year developed new major depressive disorder (MDD), compared to only 1% of their counterparts who did not have evidence of hypersomnia during that same year. The rate of development of new psychiatric disorders in general was even higher in persistent hypersomnia residents, 50%, compared to 8.8% their noncomplaining counterparts. There is compelling evidence, therefore, that hypersomnia strongly predicts the emergence of future psychiatric conditions, including depression. This temporal relationship has led to speculation of a bi-directional relationship between hypersomnia and MDD, raising the interesting notion that hypersomnia, secondary to other causes, might foment affective instability. This is certainly evident in the case of bipolar depression, where externally imposed sleep curtailment has been shown to lead to a switch to mania in anecdotal reports. Such speculation raises, in turn, intriguing possibilities regarding the provision of prophylactic benefit by the efficacious management of chronic hypersomnia in the context of affective disorders. There is the equal likelihood that hypersomnia represents an early and sensitive symptom of an emerging depressive disorder. Clinically, this suggests that patients who have a vulnerability for MDD, but who are euthymic, should be carefully followed for emerging hypersomnia, and those exhibiting this should be managed for MDD itself early in the course of their recurrence of relapse.

Residual

In general, antidepressants, especially serotonin specific reuptake inhibitors (SSRIs), significantly diminish depression-related fatigue by an order of 35-50% following 8 weeks of treatment. Despite such beneficial effects, however, a considerable degree of residual fatigue seems to remain in the acute treatment phase of MDD and appears to be the least responsive to treatment. Of 21 Beck Depression Inventory items examined in one study of MDD patients undergoing psychotherapy, least improvement was noted on the fatigue item following 6 to 16 sessions.[9] Fatigue also commonly complicates the clinical picture in the context of antidepressant management, despite achievement of remission. In a study of patients with MDD, 215 patients received a fixed dose of fluoxetine 20 mg for 8 weeks. In the subgroup of 108 patients who achieved remission, 35% had threshold, or subthreshold, residual fatigue, and fatigue represented the second most common residual symptom, second only to sleeplessness.[10] Additionally, in this study, 93% of patients with post-treatment fatigue had the same pretreatment symptom, suggesting that persistent fatigue is more commonly secondary to an incomplete resolution of the depressive diathesis, rather than a direct adverse effect of the antidepressant itself.

Regardless of cause, the persistence of fatigue in the context of antidepressant management is clearly a difficult problem, since it affects quality of life in patients and their families, and predicts greater healthcare visits and longer functional impairment. Residual symptoms are also strong predictors of subsequent early relapse; in one study, relapse occurred in 76% (13/17) of those with residual symptoms and 25% (10/40) of those without.[11]

Treatment-Related

The persistence of fatigue and sleepiness during antidepressant management raises the possibility that these symptoms can be secondary to the antidepressants themselves. There are no studies into the comparative sedative potential of antidepressants. However, Physician Desk Reference data, which are noncomparative since they have been gathered utilizing diverse methodologies and under different experimental conditions, suggest that sedation rates may be higher for certain antidepressants than for others. As determined by subtracting placebo rates of sedation from active drug rates, the order of most sedating to least is as follows: Mirtazepine, venlafaxine, paroxetine, nefazodone, citalopram, sertraline, fluoxetine, escitalopram, and bupropion. Therefore, if fatigue is prominent, consideration may be given to antidepressants that are less sedating to avoid further intensification of this impactful symptom.

Pharmacologic Management of Excessive Sleepiness, Fatigue, and Hypersomnia in Depressed Patients

The comments presume that the depressed patient has received appropriate treatment for the underlying depressive disorder, such as psychotherapy or antidepressants, yet fatigue and related symptoms continue to pose a clinical problem. Additionally, in keeping with the topic of this chapter, only pharmacological techniques will be considered. However, it should be noted that a variety of behavioral measures, such as sleep hygiene techniques, circadian realignment, phototherapy, and sleep restriction, among others, might also be appropriate. Only unipolar depressions will be considered. Additionally, it should be borne in mind that fatigue might be the product of comorbid psychiatric conditions, miscellaneous medical conditions, sleep disorders such as sleep apnea syndrome, sedating medications, and personality traits or disorders. Therefore, a thorough clinical evaluation with attention to such issues should precede any direct intervention.

The point at which fatigue and related symptoms are directly managed following antidepressant management is a matter of clinical judgment. Clearly, if sleepiness is so profound that it interferes with daytime functioning or poses a threat to the safety of the patient or others, independent management of fatigue may be relevant at the outset, at the point when antidepressant or psychotherapeutic treatment is instituted. However, in most cases, treatment will be most strongly considered a few weeks following the institution of management for MDD, when the determination is made that the treatment modality of the depressive

Table 1. Adjunctive therapy with CNS stimulants

Study Author, Design, Subjects (N)	Medications and Dose (mg/day)	Efficacy Results	Most Common Adverse Events
Fawcett J, et al[13] Retrospective chart review of patients with treatment-resistant MDD (16)	MAOI + MP 10-15 or DEX 5-20	9 of 16 (55%) patients had very much improved or much improved Clinical Global Impression (CGI) scores	Orthostatic hypotension (5), edginess/irritability/agitation (3), insomnia, dizziness, nausea, impaired short-term memory (2 each)
Feighner JP, et al[14] Case series of patients with treatment-refractory MDD (32)	MAOI + PEM 18.75–112.5 (n = 28) or DEX 5–40 (n = 17) (13 pts had both courses)	25 of 32 (78%) patients had a very much improved or much improved CGI scores	PEM = mood cycling (5), orthostatic hypotension (2) DEX = memory problems (2), mood cycling (1)
Masand PS, et al[15] Case series of patients with treatment-refractory MDD (7)	Various SSRIs + stimulant	Marked improvement in clinical symptoms of depression in all 7 cases; particular improvement in feelings of apathy/fatigue	None reported
Lavretsky H, et al[16] 8-week open-label study of elderly patients (mean = 79.8 years old) (10)	Citalopram + MP	8 of 10 (80%) patients with statistically significant improvement in clinical symptoms of depression by Week 8	Well tolerated, no discontinuations; insomnia (3), anxiety (3) (2 pts reported both AEs)

AE= Adverse event; DEX= Dexedrine; MAOI= Monoamine oxydase inhibitor; MP= Methylphenidate; PEM= Pemoline; SSRI= Serotonin specific reuptake inhibitor

disorder itself is not likely to provide sufficient symptomatic benefit for fatigue, sleepiness, and hypersomnia.

Single-Medication Strategies

While all of these strategies are appropriate in the context of an incomplete response, their specific effects on fatigue and related symptoms have not been adequately explored.

1. Wait. Extending treatment duration at the same dose of antidepressant, beyond 8 weeks, has been shown to add to the rate of remission. However, this would me most appropriate for patients with mild symptoms.

2. Raise antidepressant dose. This is the most commonly employed technique. The most common cause of treatment failure is sub-optimal dosing.

3. Decrease the antidepressant dose. This should be considered if fatigue intensifies following introduction of the antidepressant, suggesting that it is related to the antidepressant itself. Of course, this strategy is risky since it introduces the risk of suboptimal response.

4. Switch to another antidepressant. This would be an appropriate choice for patients who have experienced a minimal response to optimal doses of the original agent in terms of fatigue and related symptoms. Although trials have not been conducted to support this practice, it seems reasonable to proceed to an antidepressant with fewer sedating effects, as discussed above.

Multiple-Medication Strategies

Although a wide variety of agents have been utilized, we will consider those which have been considered to be more energizing, as opposed to those which tend to be sedating, such as lithium carbonate, hypnotic agents, antipsychotic agents, etc. Multiple medication strategies have the advantage, over the single-medication strategies, of being able to treat specific, undesired, symptoms with in a focused manner. There is greater flexibility in making dosage adjustments without compromising the treatment of the depressive disorder itself. However, they also raise the potential for additional cost and may increase the likelihood of adverse effects, due not only to the added medication itself, but also to the possible pharmacodynamic and pharmacokinetic interations between medications. Although adverse effects of various agents are not fully discussed in this chapter, readers are urged to carefully consider such side effect possibilities prior to engaging in this strategy.[11a]

1. Antidepressants. The most popular strategy in this category is adjunctive therapy with bupropion, widely considered to be the most activating antidepressant. In a retrospective case series,[12] the addition of bupropion to 21 partial responders on SSRI resulted in improvement in energy levels in 57% of patients and no change in 43%. However, this practice, as well as the practice of bupropion augmentation in general, has not been subjected to controlled trials with validated instruments for the assessment of fatigue and related symptoms. Bupropion's main adverse effects are agitation, nausea and vomiting; seizures, headache, sleep disturbance, and rash.[11a]

Table 2. Adjunctive therapy with Modafinil

Study Design, Subjects (N)	Medications and Dose (mg/day)	Efficacy Results	Adverse Events
DeBattista et al[4] 6-week, double blind, placebo-controlled study (136)	Various antidepressants + modafinil 100–400	Significant improvement at Weeks 1 and 2 in ESS and FSS scores; changes from baseline to final visit not statistically different from placebo	Headache, nervousness, insomnia
Markovitz PJ, et al[17] Prospective, open-label study (27)	Various antidepressants (primarily SSRIs) + modafinil 200–400	Significant improvements in Global Assessment of Functioning (GAF) scores from baseline	Headache, nausea caused 1 discontinuation each

HDRS= Hamilton Depression Rating Scale; BDI= Beck Depression Inventory; CGIS= Clinical Global Impression (Severity); VASF= Visual Analog Scale for Fatigue; FSI= Fatigue Severity Inventory.

2. Psychostimulants, such as methylphenidate, pemoline, and d-amphetamine, all exert their behavioral activating effects primarily via potentiation of central dopaminergic activity. Altnough utilized mainly for attention deficit hyperactivity disorder, as a class, psychostimulants have been shown to be highly effective for relief of excessive sleepiness.[12a] Additionally, data from uncontrolled case studies suggest that they may help relieve depression when used as adjuncts to antidepressant medication, including TCAs, MAOIs, and SSRIs (Table 4). However, these studies suffer from methodological limitations including small sample sizes, the lack of placebo control groups and inadequate assessment measures. For example, validated measures of fatigue, EDS, or hypersomnia were not utilized. In MDD, the stimulants promise to provide rapid relief from fatigue and related symptoms. Pemoline, which has not been investigated in this area, has been linked to hepatic failure, which has limited its utility in clinical settings. Modafinil is a wakefulness promoting agent which has been shown to be effective in various clinical models, including narcolepsy, shift work sleep disorder, and obstructive sleep apnea syndrome. The mechanism of action of modafinil has not yet been fully characterized. Nevertheless animal studies suggest that it does not exert its effects through enhancements of dopaminergic activity. Drawbacks to the stimulants are the development of insomnia, irritability, decreased appetite, gastrointestinal upset, elevated heart rate and blood pressure and agitation. Modafinil's main drawbacks are headache, nausea, nervousness, rhinitis, diarrhea, back pain, anxiety, insomnia, dizziness, and dyspepsia. Amphetamine and methylphenidate are schedule II agents, whereas pemoline and modafinil are schedue IV agents. Therefore, these agents should be utilized with caution in individuals with a history of addiction or drug abuse, and patients should be regularly monitored for dose escalation and compulsive use (Physicians' Desk Reference, 2005).

3. Wake promoting agents. Modafinil is the only available compound in this class. It has been shown to be effective for improving wakefulness in various clinical models, including narcolepsy, shift work sleep disorder, and obstructive sleep apnea syndrome. The mechanism of action of modafinil has not yet been fully characterized. Nevertheless it does not appear to exert its effects through enhancements of dopaminergic activity. In animal studies modafinil's initial activity seems to be localized to the hypothalamic regions. It is categorized as a Schedule IV drug.

Modafinil has been examined as an adjunctive medication in depressives on current antidepressants. A retrospective case series revealed a diminution of residual feelings of excessive sleepiness, fatigue, and depression ratings on the Hamilton Depression Rating Scale (HAM-D) and Beck Depression Inventory.[18] More recent studies, one controlled and the other not, are summarized in Table 2. The former represents the largest controlled series in this area. Collectively, these studies suggest that modafinil provides a reasonably safe alternative in the management of residual fatigue and EDS in the depressed patient.

Dose-response studies have not been performed for these adjunctive strategies in depressives; therefore, the most prudent clinical recommendation seems to be to start with the lowest dose feasible and to proceed upwards gradually. This guideline is especially relevant in depressives, who may have a high level of background anxiety and tension, which can be fomented by any activating compounds. The interval between dose changes should be made in consideration of the patient's response, emerging adverse effects, and the medication's elimination half-life, with longer intervals being necessary for the longer half-life medications. Since long-term data in depressives are not available for these agents when utilized adjunctively, the most prudent guideline for long-term use in depressives seems to be to monitor carefully for side effects and to taper their use episodically to determine if continued treatment is necessary.

Conclusions

Excessive sleepiness, hypersomnia, and fatigue are common and often clinically consequential in depressives. These symptoms may be associated with decreased attention, concentration, and other cognitive abilities. They may also be associated with increased morbidity, decreased quality of life, and an increased risk of accidents. Longitudinal studies suggest that these symptoms often precede the onset of depressive episodes, and that they are common as residual symptoms in the context of antidepressant and psychotherapeutic management. Regarding management, a variety of pharmacological options are available, including single antidepressant pharmacotherapy, and adjunctive use of antidepressants, stimulants, and wake promoting agents.

Statement of Off-Label Usages

To the author's best knowledge, the following medications are NOT indicated for the corresponding disorders by the US Food and Drug Administration:

1. Waking promoting agents (modafinil) and psychostimulants (methylphenidate, dextroamphetamine, pemoline, among others) for depression
2. Bupropion for adjunctive management of depression

Acknowledgements

This work was supported through a grant from Cephalon.

References

1. Johns MW. A new method for measuring daytime sleepiness: The epworth sleepiness scale. Sleep 1991; 14:540-545.
2. Krupp LB, LaRocca NG, Muir-Nash J et al. The fatigue severity scale. Application to patients with multiple sclerosis and systemic lupus erythematosus. Arch Neurol 1989; 46:1121-1123.

2a. Hamilton M. Development of a rating scale for primary depressive illness. Br J Soc Clin Psychol 1967; 6:278-296.

3. Reynolds CF, Kupfer DJ. Sleep research in affective illness: State of the art circa 1987. Sleep 1987; 10:199-215.
4. DeBattista C, Doghramji K, Menza MA et al. Adjunct modafinil for the short-term treatment of fatigue and sleepiness in patients with major depressive disorder: A preliminary, double-blind placebo-controlled study. J Clin Psychiatry 2003; 23:27-30.
5. DSM IV: Diagnostic and Statistical Manual of Mental Disorders. 4th ed. American Psychiatric Association, 1994.
6. Breslau N, Roth T, Rosenthal L et al. Sleep disturbance and psychiatric disorders: A longitudinal epidemiological study of young adults. Biol Psychiatry 1996; 39:411-418.
7. Hublin C, Kaprio J, Partinen M et al. Daytime sleepiness in an adult, Finnish population. J Intern Med 1996; 239:417-423.
8. Ford DE, Kamerow DB. Epidemiologic study of sleep disturbances and psychiatric disorders: An opportunity for prevention? JAMA 1989; 262:1479-1484.
9. Barkham M, Rees A, Shapiro DA et al. Outcomes of time-limited psychotherapy in applied settings: Replicating the Second Sheffield Psychotherapy Project.[comment]. J Consult Clin Psychol 1996; 64(5):1079-85.

10 Nierenberg AA, Keefe BR, Leslie VC et al. Residual symptoms in depressed patients who respond acutely to fluoxetine. J Clin Psychiatry 1999; 60:221-225.

11 Paykel ES, Ramana R, Hayhurst H et al. Residual symptoms after partial remission: An important outcome in depression. Psychol Med 1995; 25(6):1171-1180.

11a. Physicians' Desk Reference. 59 ed. Montvale, New Jersey: Thomson Healthcare, 2005.

12 Bodkin JA, Lasser RA, Wines JD et al. Combining serotonin reuptake inhibitors and bupropion in partial responders to antidepressant monotherapy. J Clin Psychiatry 1997; 58:137-145.

12a. Mitler MM, Aldrich MS, Koob GF et al. Narcolepsy and its treatment with stimulants. ASDA standards of practice. Sleep 17(4):352-71,199.

13 Fawcett J, Kravitz HM, Zajecka JM et al. CNS stimulant potentiation of monoamine oxidase inhibitors in treatment-refractory depression. J Clin Psychopharmacol 1991; 11:127-132.

14 Feighner JP, Herbstein J, Damlouji N. Combined MAOI, TCA, and direct stimulant therapy of treatment-resistant depression. J Clin Psychiatry 1985; 46(6):206-9.

15 Masand PS, Anand VS, Tanquary JF. Psychostimulant augmentation of second generation antidepressants: A case series. Depress Anxiety 1998; 7:89-91.

16 Lavretsky H, Kumar A. Methylphenidate augmentation of citalopram in elderly depressed patients. Am J Geriatr Psychiatry 2001; 9(3):298-303.

17 Markovitz PJ, Wagner S. An open-label trial of modafinil augmentation in patients with partial response to antidepressant therapy. J Clin Psychopharmacol 2003; 23:207-208.

18 Menza MA, Kaufman KR, Castellanos AM. Modafinil augmentation of antidepressant treatment in depression. J Clin Psychiatry 2000; 61:378-381.

CHAPTER 21

Next-Day Residual Effects of Sleeping Medications on Driving Ability:

A Review of the Literature

Joris C. Verster, Marinus N. Verbaten and Edmund R. Volkerts

Abstract

Poor sleep quality often results in decreased alertness, drowsiness and sleepiness the following day. Pharmacological treatment of sleep complaints can aggravate these effects, resulting in impaired performance at work and during daily activities such as driving a car. The first hypnotics, the barbiturates, have a limited safety profile and produce significant daytime impairment. They were replaced by the benzodiazepines during the 1970s, which showed to be efficient in the treatment of insomnia but are much safer than the barbiturates. However, bedtime use of benzodiazepines also produces sleepiness and significantly impairs driving performance the following day. The presence and severity of driving impairment varies between benzodiazepines, depending on their half-life, dosage and time after administration. Nonbenzodiazepines such as zopiclone, zolpidem and zaleplon were developed to overcome the residual hypnotic effects interfering with daytime performance. Zolpidem showed little to no daytime driving impairment when administered at bedtime, but middle-of-the-night administration is not recommended for this drug. In contrast, zaleplon (10 mg and 20 mg), administered either at bedtime or in the middle of the night, does not affect driving ability. Therefore, zaleplon is a safe alternative for patients suffering from insomnia that want treatment as needed (during the night when symptoms occur), and are willing to drive a car the following morning.

Introduction

Main complaints of patients suffering from insomnia include sleep initiation problems, and nocturnal or early-morning awakenings, all resulting in poor sleep quality. Sleep disorders not only affect nighttime sleep, but they also interfere with daytime functioning. Approximately 35% of the population experiences sleep problems, and 17% of these patients reported that insomnia affects their daily living negatively.[1] Thus, sleep disturbances are commonly reported and must be viewed as a health problem with a great impact on society. Sleep disturbances can be categorized according to their duration as transient insomnia (< 1 week), short-term insomnia (1-3 weeks) or chronic insomnia (months). Further, insomnia can be a primary disorder, or sleep disturbances can occur secondary to another medical condition or psychiatric disease, such as depression, anxiety or panic disorder.

Ideally, treatment of insomnia will produce adequate relief of sleep disturbances, without residual hangover effects the day following treatment. That is, patients should feel refreshed after a hypnotic-induced sleep. In reality, hangover effects often accompany the use of hypnotics. Hangover effects include daytime fatigue, concentration problems, impaired memory functioning and psychomotor effects that can compromise the performance of daily activities such as driving a car.

Since barbiturates were abandoned from practice due to their narrow therapeutic span and unfavorable adverse effect profile, benzodiazepines became the first-choice treatment for those suffering from insomnia. Drugs from the benzodiazepine family showed to be therapeutically effective, but next day hangover effects often accompany their use. In addition, abuse potential, withdrawal and rebound effects limit the clinical use of benzodiazepines. Nonbenzodiazepine drugs such as zopiclone, zolpidem and zaleplon were developed to overcome these effects. This review will focus on the residual effects of hypnotics on driving, one of the most common daily activities.

Although residual effects of hypnotics are sometimes preferred in hospitalized patients (e.g., after surgery), most patients suffering from insomnia are ambulatory patients and presumably continue their daily activities including driving a car. Driving a car is part of almost everyone's life, promoting personal freedom and independence, and thus an important determinant of our quality of life. On the other hand, participating in traffic also includes the risk of becoming involved in traffic accidents. Traffic accidents occur on a daily basis and are accepted as a possible risk that's taken while driving. However, some drivers such as the elderly and young novice drivers have increased traffic accident risks. Other high-risk groups include those driving under the influence of alcohol and psychoactive drugs.

The implication of pharmacological treatment of sleep disorders for driving performance becomes evident from epidemiological data. A classic epidemiological study performed by Borkenstein and colleagues[2] established a relationship between increased traffic accident risk and the use of alcohol. That is, with increasing blood alcohol concentration (BAC) the risk of being involved in traffic accidents increases exponentially. Various studies replicated their finding and more recent research concentrated on the relationship between medicinal drugs and traffic accident risks. A relationship has also been reported between traffic

Sleep and Sleep Disorders: A Neuropsychopharmacological Approach, edited by Malcolm Lader, Daniel P. Cardinali and S.R. Pandi-Perumal.

BOX 1. The Dutch Road Traffic Act, 1994, Article 8.1.

"It is forbidden for everyone to drive a vehicle, or as the driver, allow a vehicle to operate, while he is impaired under the immediate influence of any substance, which he knows or should reasonably know, that the use of – alone or in combination with another substance – can diminish his driving ability so that it can be assumed he is unfit for driving". (Italics added)

accident risks and the use of benzodiazepines. However, results have to be interpreted with caution since they generally do not take into account presumable differences in drug dosages and time after administration. In addition, these studies generally do not differentiate between the specific benzodiazepine drugs. Nevertheless, these studies showed that increased number of traffic accidents were significant only for long-acting benzodiazepines (> 24 h), but not for short-acting benzodiazepines (< 24 h).[3-7]

These epidemiological findings illustrate that the effect of hypnotics on driving ability is a matter of ongoing concern. To enhance overall traffic safety, preclusion of these patients from driving would be favorable, however, serious socioeconomic consequences make this an unrealistic solution. Moreover, patients often perceive that driving is their right, and preclusion from driving would presumably result in poor therapeutic compliance or illegal driving.

During the previous decades, epidemiological evidence and experimental findings have resulted in legislative changes regarding the use of alcohol and psychoactive drugs while participating in traffic. For example, in The Netherlands driving is prohibited at or above a blood alcohol concentration (BAC) of 0.05%. Concerning the use of hypnotics (and other psychoactive drugs), the Dutch law is less clear (see Box 1).

That is, the decision whether to drive or not is the responsibility of the patient. Driving is prohibited, if the patients experience that the drug negatively affects their driving ability. For psychoactive drugs that probably impair driving ability, the package labeling at the Dutch pharmacies contains a general warning: "This drug may decrease your reaction speed. (Driving a car—operating machinery—playing in the street.) Be cautioned with alcohol!" Although it is sometimes suggested that a patient is expert on his own bodily feelings, approximately 80% of all drivers report their driving ability as above average.[8,9] Also, Dutch surveys showed that the majority of patients using psychoactive drugs do not adjust their driving habits with the current warning labeling. Hence, it would be more helpful if the package labeling would clearly state whether and/or when it is safe to drive while using the respective drug. During the last two decades, much research effort has focused on the effects of hypnotics on driving ability. Results from these studies enable us to be more accurate in formulation of guidelines for the usage of hypnotics.

Measurement of Driving Ability

Up to now, the most accurate and direct way to measure driving ability is the on-the-road driving test during normal traffic. The on-the-road driving test was developed in the nineteen eighties[10,11] and is applied in over 50 studies with both healthy volunteers and patients.

In the driving test, subjects operate an instrumented vehicle over a highway circuit during normal traffic. Subjects are instructed to operate the instrumented vehicle with a constant speed (usually 90 or 95 km/h) and steady lateral position within the right (slower) traffic lane over a long distance (usually 100 km). A camera, mounted on the roof of the car, continuously records the actual position of the car within the traffic lane, by tracking the relative distance of the car from delineated stripe in the middle of the road. This is illustrated in Figure 1.

Figure 1. Illustration of the test vehicle. Note that the camera for lateral position measurements is equipped with two infrared lights, to enable recording during the night and dark weather circumstances. All data are continuously recorded on a board computer with a sampling rate of 2 Hz and edited off-line to remove data that are disturbed by extraneous events (e.g., overtaking, traffic jams, road maintenance). The experimenter, seated in the back seat of the car, monitors the board computer.

Primary parameter measuring vehicle control in the driving test is the Standard Deviation of Lateral Position (SDLP), i.e., the amount of weaving of the car. In placebo conditions, SDLP of experienced drivers (driving at least 5000 km/year) generally ranges between 18 and 22 cm. However, drug effects can produce SDLP increment to 35 cm or more. This excessive weaving results in unsafe driving characterized by repeated excursions out-of-lane into both the road shoulder and the adjacent traffic lane. As illustrated by Figure 2, SDLP can thus be regarded as an index of driving safety.

During the driving test, which lasts approximately 1 hour, performance worsens gradually. Performance decrement in tests requiring sustained attention has been reported for many benzodiazepines[12] and is caused by the monotonous vigilance character of these tests.[13] In the driving test, performance decrement (SDLP increment) over distance traveled is observed after both placebo

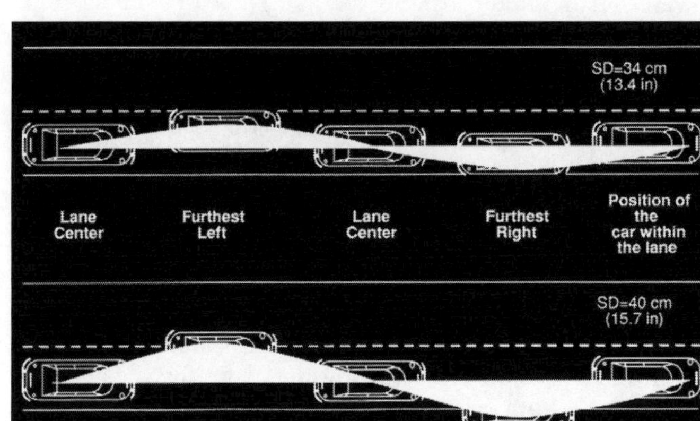

Figure 2. Standard deviation of the lateral position, SDLP.

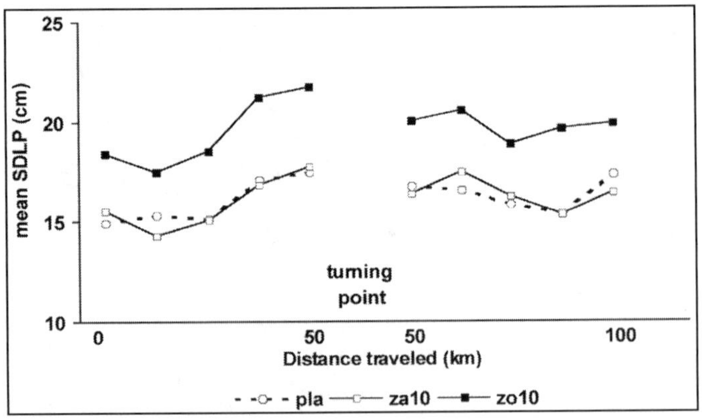

Figure 3. Mean SDLP over distance traveled, computed for subsequent 10-km segments, for placebo (pla), zaleplon 10 mg (za10) and zolpidem 10 mg (zo10), administered in the middle-of-the-night, 4h before the driving test. The driving test is a vigilance test over a 50 km circuit between the two Dutch cities of Utrecht (Start- and endpoint) and Arnhem (turning point). Note that the gradual performance decrement (SDLP elevation) over distance traveled, caused by the monotonous characteristic of the driving test, is interrupted at the turning point.

and drug treatments. Typical examples of SDLP over distance traveled are shown in Figure 3.

A secondary parameter of the driving test giving insight in the amount of vehicle control is the Standard Deviation of Speed (SD Speed). Speed is measured from a pulse generator triggered by magnetic induction at a rate proportional to the revolutions of the drive wheels. Although speed variability clearly expressed the amount of vehicle control, the relationship with driving safety is less clear when compared to the weaving index (SDLP). Mean lateral position and mean speed are control parameters to infer whether the subjects performed the driving test according to the given instructions.

In the right front seat, a licensed driving instructor accompanies the subject. His main responsibility is to guard safety during the driving test, and is equipped with a brake and clutch system. If the subject or the driving instructor judges that it is unsafe to continue driving, the test is terminated before completion and the driving instructor transports the subject back to the Institute.

In addition to the on-the-road driving test, several alternative driving-related tests have been developed such as driving simulators and closed road tests. The advantage of the driving simulator comprises the controlled testing circumstances in the laboratory. However, an important characteristic of driving is the interaction with other drivers in sometimes unexpected or even risky situations. Since both closed-road tests and driving simulators lack these elements, the validity and reliability of test results may be doubtful. Further, the tasks (e.g., maneuvering along a pilot circuit) often do not represent real driving. It is therefore not surprising that results from closed-road studies and driving simulators do not correlate well with results from on-the-road studies, and are sometimes even contradictory. A comparative study[14] showed that the on-the-road driving test was more sensitive to drug-induced impairment relative to a driving simulator. The predictive validity of results from the driving simulator to SDLP was only 23%. It must be noted, however, that driving simulator technology has improved significantly over the years. That is, the simulator studies performed during the 1980s must be considered as complex tracking tests in which attention has to be relocated simultaneously to visual stimuli rather than the more sophisticated driving simulators used today. Nevertheless, the presence of other traffic and the occurrence of unexpected events, which are sometimes regarded as problematic, are in fact necessary components of driving, and a perquisite of the ecological validity of a driving test. In this context, the most advanced driving simulator cannot resemble these real-life circumstances.

In numerous studies, the effects of psychoactive drugs on driving ability are assessed by using laboratory tests. The advantage of laboratory tests lies in the fact that they are easy to apply, relatively cheap, and can be conducted under well-controlled circumstances. Unfortunately, the relationship between actual driving and many laboratory tests is unclear. That is, popular simple tests such as finger tapping tests, the Digit Symbol Substitution Test (DSST) and the Critical Flicker Fusion Test (CFFT) showed to be sensitive to drug effects, but they do not resemble driving (or most other real life activities) and their parameters represent a mixture of several psychological constructs. The lack of standardization, i.e., the use of different tests and methodologies, make comparisons between studies difficult and it seems that the use of many laboratory tests is based upon their frequent use rather than their acknowledged validity or reliability.[15-17] In addition, time-on-task of the on-the-road driving test is approximately 1 hour whereas that of most laboratory tests varies between 5 and 10 minutes.

Since sustained attention is generally accepted to be involved in driving, the short time-on-task of these laboratory tests does not represent this important psychological function.

Driving a car is an example of skilled but complex behavior. For example, driving requires proper functioning of working memory (e.g., remembering road signs and make use of them), motor control (e.g., road tracking and vehicle control) and division of attention (e.g., interacting with other traffic and performing tasks simultaneously). Laboratory tests designed to examine these so-called "driving-related skills" should at least have a theoretical rationale, be sensitive and reliable to drug-induced impairment, and their outcome should be clearly and unequivocally represent a psychological construct or skill.

A recent comparative analysis[18] using data from three placebo-controlled, double blind randomized studies showed that even laboratory tests that meet these criteria poorly predict on-the-road driving performance. Placebo data from ninety-eight healthy volunteers was analyzed who performed the on-the-road driving test and a test battery including a Sternberg memory scanning test (representing time-limited judgment and decision making), a tracking test (representing lane keeping), and a divided attention test (performing the tracking and memory test simultaneously). Together, the predictive validity of these laboratory tests to SDLP was only 33%.

In conclusion, the on-the-road driving test during normal traffic is the golden standard to determine driving ability. Other measurements to determine driving ability should rather be viewed as supportive evidence.

Effects of Hypnotics on Driving Ability

Insomnia is an under diagnosed health problem. This is probably caused by the fact that complaints are often transient and most people do not regard sleeping problems as a medical problem. Results from the 1991 National Sleep Foundation Survey[19] showed that only 5% of all patients with insomnia have consulted their physician to discuss these problems. Treatments used by patients with insomnia include over-the-counter drugs (23%), alcohol (28%), and prescription drugs (21%), often combined with lifestyle adaptations and relaxation techniques. Pharmacological treatment of insomnia is generally of short duration, since prolonged use can produce unwanted effects such as dependency and recurrence of sleep disturbances.

The first drugs used during the 1950s for the treatment of insomnia include acetylcarbromal, ethchlorvynol, glutethimide, chloral hydrate and paraldehyde. Nowadays, these drugs are out of use, because of their relatively low clinical efficacy, and high risks of toxicity, abuse potential, and next-day hangover effects. They were replaced by the barbiturates, which showed similar mechanism of action. That is, a nonselective binding to $GABA_A$ (Gamma Amino Butyric Acid, type A) receptors were they enhance the inhibitory effects of GABA by increasing the duration of chloride ion channel opening. In addition, they inhibit the excitatory action at glutamate receptors. Together, these actions induce CNS (Central Nervous System) depression and promote sleep. Today, barbiturates are rarely prescribed since they have the same disadvantages as their precursors. Tolerance to their hypnotic effects develops within 2 weeks, enhancing dose increments. Abuse potential and withdrawal effects further limited their clinical use, and next day hangover effects were commonly reported by patients using barbiturates. Of the most frequently prescribed barbiturates for the treatment of insomnia (i.e., phenobarbital, butabarbital, secobarbital and pentobarbital), only the effects of secobarbital on driving ability have been investigated.[11,20] Although the 200 mg dose used in this study was rather high (secobarbital was used as a verum), driving ability was significantly impaired up to 17h after bedtime administration.

Since barbiturates were replaced by benzodiazepines in the 1970s, benzodiazepines are the number one therapeutic choice in the pharmacological treatment of insomnia. Relative to barbiturates, benzodiazepines have a significantly improved safety profile, especially in case of overdose. Benzodiazepines are more efficient in the treatment of insomnia, and unfavorable adverse effects such as dependency, withdrawal effects, rebound and next day hangover effects were less pronounced when compared to barbiturates.

Benzodiazepines

Benzodiazepines are the most prescribed drugs in the pharmacological treatment of insomnia. They act as full agonists at the $GABA_A$ receptor complex and bind nonselective to benzodiazepine receptor subtypes 1 and 2. Major differences between benzodiazepines are their pharmacokinetics. More specifically, benzodiazepines differ in half-life, the presence or absence of active metabolites and their onset of action. These pharmacokinetic properties determine the amount of next-day sleepiness and other residual adverse effects. An overview of benzodiazepine drugs is shown in Table 1.

It is evident from Table 1 that benzodiazepines can be classified according to their half-life, being short (< 8 h), moderate (8h-24 h), or long (> 24 h). In addition to half-life, residual effects may cumulate after repeated (chronic) use and depend on the administered dose. Further, some benzodiazepines (such as flurazepam) have active metabolites that extend the duration of hypnotic effects into daytime. Therefore, half-life values presented in Table 1 include those of the active metabolites.

Six on-the-road driving studies investigating the residual effects of benzodiazepines were performed during the 1980s. In these studies (referred to as Study 1-6), treatments were administrated at bedtime in a double blind, placebo-controlled crossover design. Next-day driving ability was assessed in a morning driving test (10-11 h after administration) and afternoon driving test (16-17 h after administration) on the road during normal traffic. In Study 1-5, subjects were female patients with a history of insomnia and experienced with benzodiazepine treatment. Study 6 was performed in 18 healthy male volunteers. Results from these studies are summarized in Table 2. An overview of SDLP values (changes from placebo after two days of nocturnal treatment) from the six benzodiazepine studies is shown in Figure 4.

The clinical relevance of the observed SDLP increment, relative to placebo, becomes evident from comparisons with alcohol. In a calibration study of the on-the-road driving test in 24 social drinkers, significant and dose-related SDLP increments were recorded after acute alcohol intoxication.[29] The correlation between BAC and SDLP obtained in this study was so strong (r=0.98) that the results have been used often as a historical control. Hence, SDLP increments obtained after administration of different dosages of alcohol (0.05%, 0.08%, and 0.10%), corresponding to the most common legal limits for driving, are depicted in Figure 4 as well.

It is evident from the results summarized in Table 2 and Figure 4 that driving a car is not recommended the day following nocturnal administration of benzodiazepines with an intermediate or long half-life. Driving after using benzodiazepines with a short half-life is relatively safe if they are administered at bedtime. However, half-life is not the only determinant of driving impairment. Higher doses of benzodiazepines with a short half-life show similar SDLP increment of that observed with long-acting benzodiazepines. Therefore, patients using any kind of benzodiazepine

Table 1. Benzodiazepine drugs used in the treatment of insomnia

Hypnotics	Dose (mg)	$T_{1/2}$ (h)	T_{MAX} (h)	Active Metabolite(s)
Triazolam	0.25	1.5-5.5	1	+
Midazolam	15	2.1-3.5*	0.5-1.5	+
Brotizolam	0.25	3-6	1-2	+
Temazepam	20	7-11	0.8	-
Loprazolam	1	8*	2-5	-
Estazolam	1-2	10-24	2	-
Lormetazepam	1	10	1-2.5	-
Flunitrazepam	2	16-35	1.2	+
Nitrazepam	5	18-34	2	+
Flurazepam	30	47-100*	0.5-2	+
Anxiolytics	**Dose (mg)**	**$T_{1/2}$ (h)**	**T_{MAX} (h)**	**Active Metabolite(s)**
Oxazepam	50	4-15	2-3	-
Alprazolam	1	12-15	1-2	-
Lorazepam	2.5	12-16	2	-
Bromazepam	10	15-22	0.5	+
Quazepam	15	15-40	0.5	+
Medazepam	20	42-100	1	+
Prazepam	30	42-100	2.5	+
Ketazolam	60	42-100*	3	+
Diazepam	10	20-100*	1-2	+

$T_{1/2}$ (h) includes those of the active metabolites; + = active metabolites, - = no active metabolites

Table 2. Summary of the six on-the-road driving studies with benzodiazepines

Study	Subjects	Nights	Treatment	Outcome
1a (11,20)	24 females with a history of insomnia and hypnotic treatment	2	Flurazepam 15 en 30 mg Secobarbital 200 mg placebo	SDLP was significantly increased after flurazepam (15 and 30 mg) and secobarbital 10-11 h and 16-17 h after administration.
1b (11,20)	4 females from Study 1a	8	Flurazepam 30 mg placebo	SDLP was significantly increased 10-11 h and 16-17 h after administration.
2 (11,21)	16 females with a history of insomnia and hypnotic treatment	2	Loprazolam 1 and 2 mg Flunitrazepam 2 mg placebo	SDLP was significantly increased 10-11 h and 16-17 h after Flunitrazepam and loprazolam 2 mg administration. Loprazolam 1 mg significantly increased SDLP 10-11 h after administration, but not in the afternoon.
3 (22,23)	16 females with a history of insomnia and hypnotic treatment	2	Zopiclone 7.5 mg Nitrazepam 5 mg Flunitrazepam 2 mg placebo	SDLP was significantly increased 10-11 h and 16-17 h after Flunitrazepam 2 mg. SDLP was significantly increased 10-11 h after zopiclone 7.5 mg administration, but not in the afternoon. Nitrazepam 5 mg did not impair driving ability.
4 (24,25)	12 females with a history of insomnia and hypnotic treatment	8	Temazepam 20 mg (caps.) Nitrazepam 10 mg placebo	Temazepam did not significantly impair driving ability. SDLP was significantly increased after 2 nights of Nitrazepam 10 mg, but not on day 5 and 8. The effects of Nitrazepam were most pronounced in the afternoon test.
5 (26,27)	16 females with a history of insomnia and hypnotic treatment	8	Lormetazepam 1 mg and 2 mg (caps.) Flurazepam 30 mg placebo	Driving ability was not significantly impaired after lormetazepam 1 mg and 2 mg. Flurazepam 30 mg significantly impaired driving ability after 2, 4 and 7 nights of treatment.
6 (14,28)	18 healthy male volunteers	2	Lormetazepam 1 mg Oxazepam 50 mg placebo	The morning following the first treatment night, lormetazepam 1 mg and oxazepam 50 mg significantly impaired performance on the driving test, but not in a driving simulator. After 2 nights, in the morning test, SDLP was significantly increased after both lormetazepam 1 mg and oxazepam 50 mg. Driving was not significantly impaired in the afternoon.

hypnotic must always be cautioned while driving a car the following day.

In this context it must be stressed that the results depicted in Figure 4 are group averages. These averages do not take into account individual differences in driving performance. That is, subjects participating in the six benzodiazepine studies sometimes showed great variance in SDLP after drug treatment. For example, in Study 3 SDLP ranges in the morning test showed substantial variation after administration of zopiclone (15-28 cm), flunitrazepam (18-36 cm) and nitrazepam (16-31 cm), in contrast to placebo (17-22 cm). At the same time, group mean effects of these drugs on driving ability seem only moderate, corresponding to SDLP values of BAC between 0.05% and 0.08%. Also, in Study 5 lormetazepam 2 mg elevated mean SDLP approximately 2-3 cm in the morning driving test. However, this effect did not reach significance. This was probably caused by great inter-individual variability in SDLP, showing some individual SDLP increments up to 10 cm whereas other subjects had SDLPs comparable to those obtained after placebo.

In other words, after receiving the same drug treatment SDLP increment of some individual subjects was much more pronounced than that of others. This is illustrated by data from a subject participating in Study 2. This female subject performed a driving test, 10-11 h after administration of loprazolam 2 mg. In contrast to the other participants in study 2, her driving test was terminated after approximately 40 km. SDLP values after loprazolam over the successive 10-km segments increased progressively (27.1, 37.3, 46.5 and 53.8 cm, respectively), resulting in repeated excursions into both the adjacent traffic lane and the road shoulder, accompanied by significant variations in mean speed. In comparison, her SDLP after placebo varied between 16 and 19 cm. The mean SDLP after loprazolam 2 mg of other participants was approximately 26 cm. The data of her driving tests after placebo and loprazolam (2 mg) is shown in Figure 5.

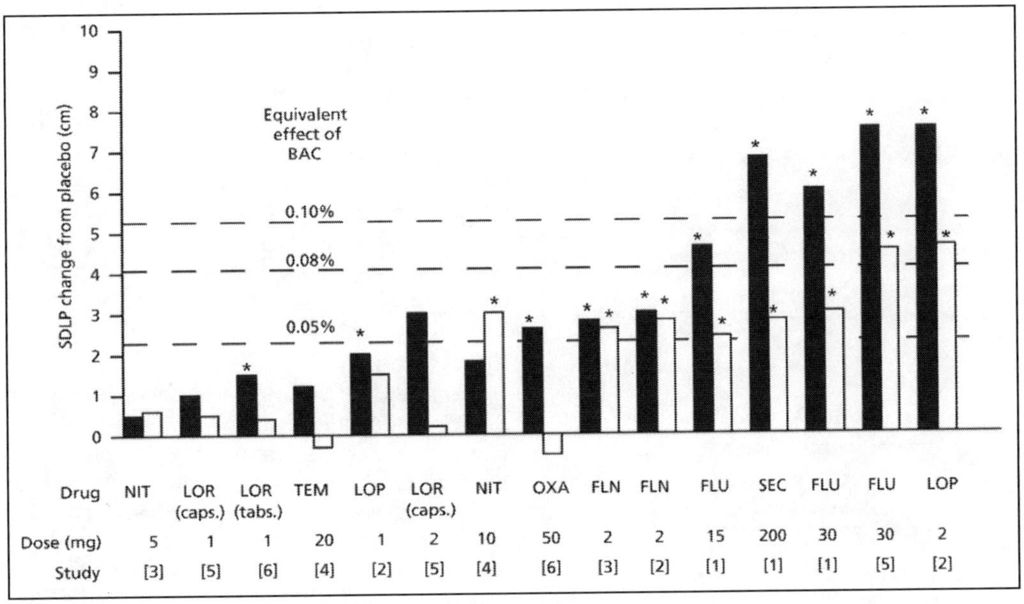

Figure 4. Effects of benzodiazepines on actual driving. SDLP changes from placebo (cm) were determined 10-11h (black bars) and 16-17h (white bars) after two consecutive nights of nocturnal treatment. SDLP changes from placebo after administration of ethanol (BAC 0.05%, 0.08% and 0.10%) are depicted for comparison. * = significant ($p < 0.05$) from placebo. Drugs: NIT= nitrazepam, LOR= lorazepam, TEM= temazepam, LOP= loprazolam, OXA= oxazepam, FLN= flunitrazepam, FLU= flurazepam. Studies: [1] = reference 20, [2] = reference 21, [3] = reference 22, [4] = reference 24, [5] = reference 26, [6] = reference 28.

Flurazepam

Flurazepam significantly impairs driving ability up to 17 hours after bedtime administration. On-the-road driving studies showed that flurazepam 15 mg produced driving impairment comparable to blood alcohol concentrations between 0.05% and 0.08% (Study 1A), whereas after flurazepam 30 mg SDLP increments were equal to those observed with blood alcohol concentrations above 0.10% (Study 1A and Study 5). Sub-chronic treatment with flurazepam 30 mg showed that further deteriorated until 4 nights of administration (Study 1B). After 7 nights SDLP increment was less pronounced but driving was still significantly impaired.

A closed road study[30] examining driving skills 12 hours after bedtime administration of flurazepam (20 mg) to 12 female subjects also reported impairment. Maneuvering through a number of passable gaps produced significant more collisions with the sides than after placebo. However, driving speed and correct judgment of nonpassable gaps were not affected by flurazepam. Performance was also impaired the morning following flurazepam (30 mg) administration in a driving simulator.[31] In conclusion, driving the day following flurazepam treatment is not recommended.

Nitrazepam

Most females that participated in the benzodiazepine studies (Study 1-5) were housewives, formerly treated with nitrazepam. In real life approximately 75% of them reported driving a car, whether they experienced residual drug effects during daytime or not. This is of concern, since it is presumably illustrative for the fact that the majority of ambulatory patients receiving hypnotic treatment do not adjust their lifestyle and driving habits accordingly.

In Study 4, on-the-road driving ability was significantly impaired up to 17 hours after bedtime administration of a 10 mg dose, comparable to driving impairment observed with a blood alcohol concentration of 0.05%. Driving impairment did not reach significance after 4 and 7 nights of treatment with the 10 mg dose. Nitrazepam 5 mg did not significantly impair driving ability (Study 3). In line, absence of significant performance impairment the morning after 3 consecutive nights of nitrazepam 5 mg administration was reported in studies using a driving simulator and a closed road test.[32,33] After a single night of nitrazepam 5 mg administration significant impairment was reported on the simulator test, but maneuvering a closed-road circuit (obstacle avoidance test) was not significantly affected after nitrazepam 5 mg. In conclusion, driving a car is relatively safe on the day after bedtime administration of nitrazepam 5 mg, but should be avoided when using a 10 mg dose.

Flunitrazepam

On-the-road driving ability was significantly impaired up to 17 hours after bedtime administration of flunitrazepam (Study 2 and Study 3). Flunitrazepam caused SDLP elevations comparable to those observed with a BAC between 0.05% and 0.08%. Another on the road study[34] examined speed, lateral acceleration and steering velocity recorded over a 1-km stretch (combined in a composite measure). The authors reported impaired performance the morning after a single night and 7 nights of consecutive flunitrazepam (2 mg) administration in 16 outpatients with insomnia. The on-the-road observations are also in line with results from laboratory studies[35] and observations after a 1 mg dose in a flight simulator[36,37] and a driving simulator.[38] Further, Scandinavian studies examining blood samples of drugged drivers showed that flunitrazepam intoxication was the cause of death in 130 out of 641 fatal accidents.[39] Also, flunitrazepam concentrations were significantly related to their level of driving impairment.[40] In conclusion, flunitrazepam produces significant residual effects the day following bedtime administration. Hence, driving a car during flunitrazepam treatment is unsafe.

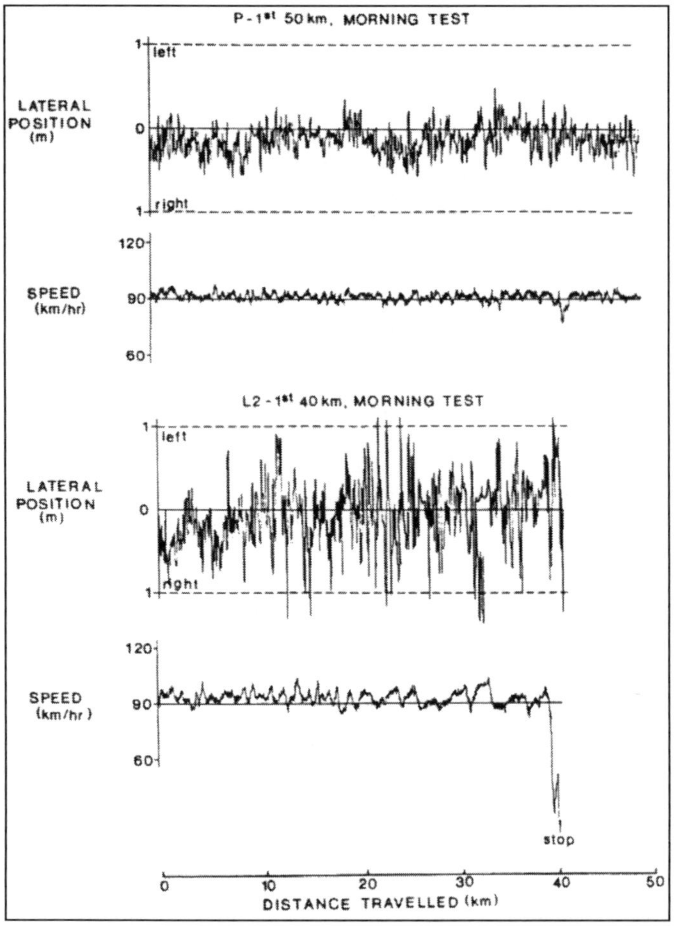

Figure 5. Examples of lateral position and speed changes over distance traveled by a single subject during morning driving tests 10h after placebo (P; top) and loprazolam 2 mg (L2; bottom). Lateral position is measured as the distance between vehicle-center and lane center. One meter left indicates left wheel excursion into the adjacent traffic lane and one meter right indicates right wheel excursion into the road shoulder. In the loprazolam condition, the driving test was stopped after 40 km. Adapted from Volkerts et al, 1984 (ref. 21).

Loprazolam

On-the-road driving ability was significantly impaired after loprazolam 1 mg 10-11h after administration, but less than that observed with a BAC of 0.05%. The 1 mg does did not significantly impair driving ability in the afternoon (Study 2). Driving performance after loprazolam 2 mg (twice the recommended dose) was significantly impaired up to 17 h after administration. SDLP increment was worse than with blood alcohol concentrations of 0.10% in the morning session, and greater than SDLP increments observed after a BAC of 0.08% in the afternoon (Study 2). Hence, driving a car is not recommended following nocturnal treatment with loprazolam.

Lormetazepam

Lormetazepam 1 mg significantly impaired driving ability in one study (Study 6), but not in another study (Study 5). This difference is possibly caused by the fact that lormetazepam was administered in soft gel capsules in Study 5, whereas in Study 6 lormetazepam was administered in a tablet formulation. Nevertheless, in both studies SDLP increment was moderate, and less than the SDLP increment observed after a BAC of 0.05%. Driving performance 10-11 h after administration of lormetazepam 2 mg was worse than after BAC 0.05%, but this effect also did not reach significance (Study 5). In the afternoon, driving performance after lormetazepam (1 and 2 mg) was comparable to placebo. In a driving simulator, lormetazepam (2 mg) did not significantly impair performance in 12 healthy volunteers, the morning following seven days of bedtime administration.[31] In conclusion, driving a car seems relatively safe the day following bedtime administration of lormetazepam.

Temazepam

Temazepam (20 mg) did not significantly impair on-the-road driving ability (Study 4). Accordingly, SDLP increment after temazepam was less than that observed after a BAC of 0.05%. An on-the-road driving study[34] examining speed, lateral acceleration and steering velocity reported improved driving performance (relative to placebo) the morning after a single nocturnal dose of temazepam (20 mg) in 16 outpatients with insomnia. After 7 consecutive nights this improvement did not reach significance. Similar development of tolerance to the effects temazepam (30 mg) was reported after subchronic (7 day) administration in a variety of laboratory tests and the multiple sleep latency test.[41] On a closed road circuit, driving performance of 12 healthy female subjects performing a weaving task (maneuvering between bollards) was not significantly impaired 12 hours after bedtime administration of temazepam (20 mg).[30] However, maneuvering a car over a circuit comprising passable and nonpassable gaps produced a significant increased number of side-hits while crossing passable gaps. In conclusion, the impairing effects of temazepam are mild, suggesting that driving is relatively safe the morning following bedtime administration of temazepam.

Oxazepam

Oxazepam (50 mg) significantly impaired on-the-road driving performance 10-11 h after administration, but not in the afternoon (Study 6). SDLP increment after oxazepam was moderate and comparable to that observed after a BAC of 0.05%.

Anxiolytic Benzodiazepines

Diazepam, alprazolam and lorazepam are benzodiazepines prescribed in the treatment of anxiety disorders. Since these compounds also possess sedative properties, they can be helpful in promoting sleep in anxious patients, but they are not regularly prescribed if complaints are limited to insomnia.

On-the-road driving tests performed 1-2 h after single-dose daytime administration have shown that both diazepam (10 mg)[10] and alprazolam (1 mg)[42] severely impair vehicle control, expressed in both significant SDLP increment (comparable to that observed with BAC greater than 0.15%) and significantly increased speed variability. Chronic administration of diazepam (5 mg, 3 times daily) to anxious outpatients showed that tolerance developed slowly, and SDLP elevations differed significantly from placebo up to 3-4 weeks of daily treatment.[43]

An on-the-road driving study[44-45] performed in anxious outpatients receiving lorazepam (2 mg, twice a day) for 8 successive days showed that driving performance was significantly impaired after administration of the morning dose on day 1 (SDLP increment of 18 cm). Although driving impairment was less on day 8, SDLP was still significantly increased (SDLP increment of 10 cm). At day one, patients recognized their driving impairment, but of serious concern was the fact that patients judged their driving performance on day 8 as unimpaired. An on-the-road driving study in 18 healthy male volunteers also reported significant driving impairment after 7 days of lorazepam treatment (1.5 mg, twice a day).[46]

Other Benzodiazepines

Research concerning the hangover effects of some other benzodiazepines on driving ability is either limited to closed road tests and driving simulators or does not exist.

In a driving simulator test no significant effects were found in morning test sessions after three subsequent nights of brotizolam (0.25 mg) treatment in healthy volunteers.[47]

Triazolam concentrations have been found in blood plasma samples of drivers arrested for impaired driving or involved in traffic accidents.[48] On a closed road circuit, 18 healthy volunteers performed a avoidance maneuvering test and a simulated driving test in the laboratory the morning following 1 and 3 nights of triazolam (0.25 mg) treatment. On both tests, performance was not significantly different from placebo.[32]

The effects of midazolam (15 mg) were investigated on a closed road circuit in 8 healthy female volunteers.[49] The morning after bedtime administration no significant performance impairment was observed in the maneuvering task.

Ten outpatients suffering from anxiety participated in a driving study with bromazepam (1.5 mg, three times a day for two weeks).[50] A licensed driving instructor rated the subjects' driving skills. After two weeks of treatment the number of items scored as insufficient was significantly increased relative to the pretreatment scores. Another study showed that performance on braking tests and simulated driving of 14 anxious inpatients was not significantly impaired after subchronic administration of medazepam (16.5 mg daily).[51]

Although these benzodiazepine drugs have not been examined with the on-the-road driving tests or laboratory tests measuring driving related skills, it is reasonable to assume that they produce similar impairing effects on driving ability as observed with other benzodiazepines.

Antihistamines

Approximately 23% of people with sleep complaints self-medicate with over-the-counter (nonprescription) drugs.[19] Most used nonprescription drugs for the promotion of sleep are diphenhydramine and doxylamine. These antihistamine drugs were developed for the treatment of allergic symptoms. However, their adverse effects (sedative properties producing drowsiness and sleepiness) make them very popular for the treatment of mild sleep problems. Histamine plays an important role in the sleep-wake cycle. In the brain, blockade of H_1-receptors by antihistamine drugs induces sedation and promotes sleepiness. Unfortunately, antihistamines have little effect on sleep architecture and may even worsen sleep complaints. Thus, these drugs may increase sleep duration, but do not improve sleep quality. Therefore, the usefulness of antihistamines in the treatment of serious sleep disorders is limited. Moreover, first-generation antihistamines are known to impair daytime performance, even when taken the night before.[52]

Antidepressants

Depression is sometime accompanied by sleep complaints. Some antidepressant drugs have shown to improve sleep quality in patients suffering from depression. In particular, trazodone, doxepin, amitriptyline, mitrazapine, and mianserin are useful if insomnia is a symptom of depression. Doxepine (25 mg, three times a day) significantly impaired on-the-road driving on day 1, but not after 8 days of treatment in depressive patients.[53] Sleep duration after doxepine was prolonged, but sleep quality was not improved. Acute and sub-chronic bedtime administration of mitrazapine (15 mg or 30 mg) and mianserin (30 mg and 60 mg) administrated at bedtime did not impair on-the-road driving ability the following morning after 1, 8 and 15 days of treatment.[54] Amitriptyline, administrated for 8 consecutive days (50 mg at bedtime and 25 mg in the morning), significantly impaired on-the-road driving ability 1.5 hours after the morning dose following the first treatment night.[55] However, driving impairment was absent after 8 days of treatment. In conclusion, these results suggest that antidepressants do not affect next morning driving ability after subchronic use. However, after acute administration and during the first days of treatment patients should be cautioned when driving a car.

Nonbenzodiazepine Hypnotics

The profile of unwanted residual adverse effects of benzodiazepines stimulated the development of nonbenzodiazepine hypnotics. Aim of this effort was to design short acting hypnotics with at least equal therapeutic efficacy as benzodiazepines, but devoid of residual sleepiness the following day. In addition, these drugs should not produce tolerance, dependency, withdrawal or rebound effects after discontinuation. The search for such an improved drug profile resulted in the subsequent introduction of zopiclone (a cyclopyrrolone; half life of 3.5-6.5 h), zolpidem (an imidazopyridine; half life of 1.2-4.0 h), and zaleplon (a pyrazolopyrimidine; half life of 0.8-1.4 h).

Zopiclone has no active metabolites and its recommended dose is 7.5 mg. The recommended dose of both zaleplon and zolpidem is 10 mg. A zolpidem dose should not be higher than 10 mg. Zaleplon doses may vary from 5 mg up to 20 mg. These nonbenzodiazepine hypnotics act as agonists at the benzodiazepine $GABA_A$ receptor complex, but unlike the benzodiazepines which bind nonselectively to both type 1 and type 2 benzodiazepine $GABA_A$ receptors, zopiclone, zolpidem and zaleplon bind more selectively to the type 1 benzodiazepine receptor. Most $GABA_A$ receptors are composed of three classes of subunits with several variants (α_{1-6}, β_{1-3}, γ_{1-3}). Benzodiazepine receptors can be differentiated upon their subunit structure: type 1 receptors are composed of $\alpha_1\beta_{1-3}\gamma_2$ subunits whereas type 2 receptors are $\alpha_{2,3,5}\beta_{1-3}\gamma_2$ subunits. The functional significance of these subunits is still under investigation, but it has been established that sedation and anterograde amnesia are mediated by α-$_1$ containing receptors, whereas α_2 receptors mediate anxiolytic effects.[56]

On-the-road driving studies with zopiclone, zolpidem and zaleplon are summarized in Table 3.

Zopiclone

Results from on-the-road driving studies with zopiclone are summarized in Figure 6.

On-the-road driving studies reported driving impairment the following morning comparable to a BAC between 0.05% and 0.08%. In the afternoon driving tests, no significant effects on driving ability were found. Further, epidemiological evidence[6] showed that bedtime administration of zopiclone is associated with a 4-fold increased traffic accident risk the following day.

Table 3. On-the-road driving studies performed after a single night administration of zopiclone, zolpidem and zaleplon

Study	Subjects	Treatment	Outcome
7 (57)	17 female subjects with a history of insomnia and hypnotic treatment	zolpidem 10 mg flunitrazepam 2 mg partial sleep deprivation placebo	10-11 hours after bedtime administration, SDLP did not differ significantly from placebo after zolpidem and flunitrazepam. Also, partial sleep deprivation did not affect driving performance.
8 (58)	28 healthy volunteers	zaleplon 10 mg and 20 mg zopiclone 7.5 mg placebo	Zopiclone significantly increased SDLP and SD speed 6h after administration; SDLP was also significantly elevated 11h after administration. Driving performance after both doses of zaleplon matched placebo.
9 (59)	30 healthy volunteers	zopiclone 7.5 mg zaleplon 10 mg placebo alcohol (BAC< 0.05%)	Zopiclone significantly impaired driving ability, 10 h after bedtime administration. Alcohol also significantly impaired driving performance. However, mean SDLP elevation after zopiclone, relative to placebo, was twice that observed after alcohol. Zaleplon did not affect driving performance.
10 (60)	30 healthy volunteers	zaleplon 10 mg and 20 mg zolpidem 10 mg and 20 mg placebo alcohol (BAC< 0.05%)	Zolpidem 10 and 20 mg significantly impaired driving ability in a dose-dependent manner, 4h after middle-of-the-night administration. SD speed was significantly increased after zolpidem 20 mg. SDLP increment after both doses of zolpidem were much higher than after alcohol. Zaleplon 10 mg and 20 mg did not affect driving ability.

On the other hand, results from laboratory tests (e.g., CFFT, choice reaction time tests, DSST) assessing residual effects of zopiclone on cognitive and psychomotor performance are inconclusive, some studies reported significant impairment the day following nocturnal treatment,[61-63] whereas other studies did not.[64-70]

Taking together, these results imply that patients should be cautioned when driving a car the day following nocturnal treatment with zopiclone, especially during the morning hours.

Zolpidem

Results from on-the-road driving studies with zolpidem are summarized in Figure 7.

On-the-road driving studies showed that zolpidem 10 mg administered at bedtime produces minimal daytime residual effects. Next-morning performance after nocturnal zolpidem administration was also unimpaired on a simulated driving test,[38] a flight simulator 7h following administration,[71] and a variety of laboratory tests.[35,72] These finding underline that after bedtime administration of the recommended dose of zolpidem (10 mg), next morning driving is safe and daytime residual effects are minimal. Thus, relative to the benzodiazepines, zolpidem is a safer alternative when used at bedtime.

However, when administered in the middle of the night, zolpidem significantly impaired driving ability, producing SDLP increments that were 4-fold (zolpidem 10 mg) and 11-fold (zolpidem 20 mg) of those observed after a BAC of 0.05%.[60] Driving after zolpidem 20 mg was severely impaired. This was illustrated by the fact that three female subjects had to stop their driving test before completion. Excessive weaving into both the road shoulder and the adjacent traffic lane resulted in serious unsafe driving behavior in these individuals. After zolpidem 10 mg, weaving was less pronounced and all subjects completed their driving tests. In line, several laboratory studies consistently reported significant performance impairment on psychomotor and memory tests within the first 5 hours after zolpidem administra-

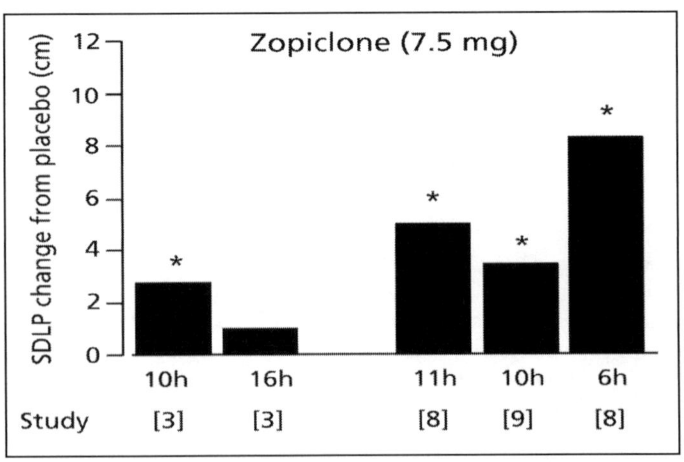

Figure 6. Effects of zopiclone on actual driving at different times after administration. Note that the subjects participating in Study 3 were patients with a history of insomnia and benzodiazepine treatment, whereas subjects in Study 8 and Study 9 were healthy volunteers. * = Significantly ($p< 0.05$) different from placebo.

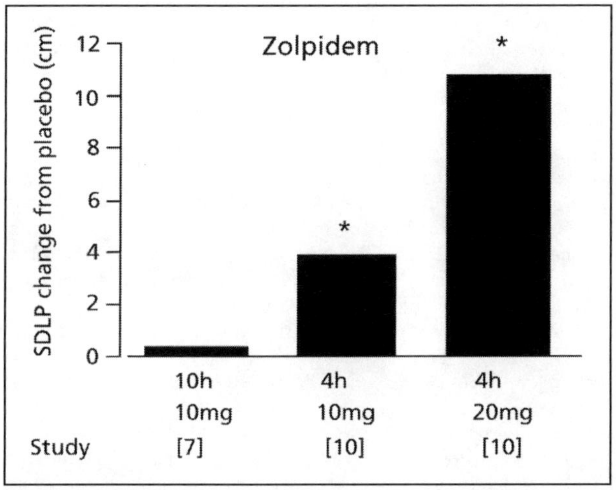

Figure 7. Effects of zolpidem on driving ability at different times after administration. * = Significantly ($p < 0.05$) different from placebo.

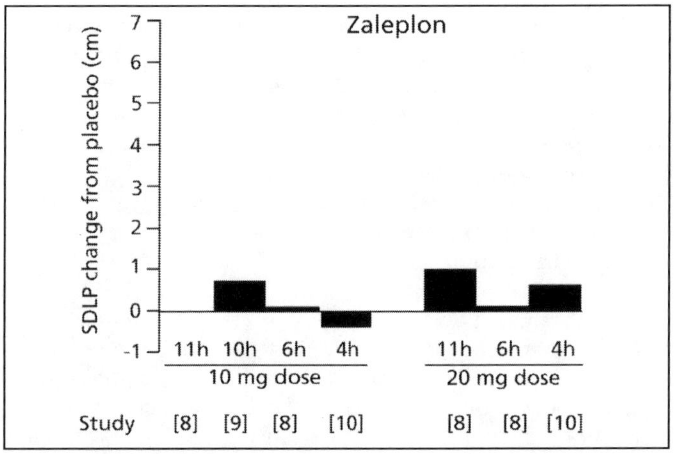

Figure 8. Effects of zaleplon on actual driving. None of the differences from placebo were significant.

tion.[73-82] In conclusion, middle-of-the-night administration of zolpidem is not recommended.

Zaleplon

Results from on-the-road driving studies with zaleplon are shown in Figure 8.

On-the-road driving studies showed that zaleplon does not impair driving ability when administered 6-12h before driving.[58,59] Also, laboratory test performance was unimpaired after bedtime administration.[59,83] These findings show that bedtime administration of zaleplon 10 mg and 20 mg does not affect performance the following day. Concerning middle-of-the-night administration, it was found that on-the-road driving was not impaired as shortly as 4 h after intake of zaleplon 10 mg and 20 mg.[60] Laboratory tests showed that zaleplon 10 mg and 20 mg did not affect performance 2-4 h after administration.[80,84] However, unlike zaleplon 10 mg, zaleplon 20 mg (twice the recommended dose) did impair performance 1-1.5 h after administration.[83,84]

In conclusion, zaleplon is the first hypnotic drug that can be used at bedtime and at flexible times during the night, without producing impaired driving the following morning.

Discussion

The prescription rates of psychoactive drugs including hypnotics increase every year. On the other hand, approximately 96% of the Dutch adult population drives a car. The latter is not unique; it probably equals that observed in most modern Western countries. In this context, safety and favorable adverse effect profiles are increasingly important in the development of new psychoactive drugs. The demands of our 24-hour economy requires that sleep disturbances are treated adequately, but also that treatment does not interfere with performance during daytime. It is therefore not surprising that lawmakers and other (non) governmental organizations are increasingly interested in the effects of psychoactive drugs on driving ability and there is a lively debate whether to adjust the present laws and guidelines concerning driving a car during treatment.

As became evident from the introduction, reliable conclusions regarding driving safety depend highly on the merits of the tests that were used to determine driving ability. Laboratory tests are insufficient to provide this information, since in general they do not measure the vigilance characteristic (sustained attention), which is an important aspect of real driving. In addition, the relationship of the parameters of these tests to driving is often unclear. Finally, the absence of normal traffic in laboratory settings and simulator studies lower their ecological validity. Hence, to change laws and guidelines regarding driving ability while using psychoactive drugs, one should rely primarily on the results from on-the-road driving studies during normal traffic.

Benzodiazepines are still the most prescribed drugs in the pharmacological treatment of insomnia. On-the-road driving studies showed that benzodiazepines, when administered at bedtime, produce significant impairment on the driving test the following morning. For long-acting benzodiazepines driving impairment was still evident in the afternoon, up to 17 h after administration. Residual effects are reported for both short-acting and long-acting benzodiazepines, since—in addition to half-life—the impact on driving ability depends also on the administered dose. Higher doses generally produce more pronounced driving impairment. In addition, individual differences have been reported on the impact of benzodiazepines on driving ability. Although benzodiazepine hypnotics have proven their clinical efficacy, residual daytime effects greatly limit their usefulness in the treatment of ambulatory patients.

As an alternative to benzodiazepine hypnotics, zopiclone and zolpidem were developed. Both are efficient hypnotics, but they have a more specific action at the $GABA_A$-receptor complex and a relative short half-life. When administered at bedtime, these drugs produce minimal residual effects the following day. However, effects on driving ability are not completely absent and patients using these drugs should be especially cautioned during the morning hours.

In the Introduction it was concluded that the ideal hypnotic should be safe and effective at the moment sleep disturbances occur (for example in the middle-of-the-night or after early morning awakenings), without producing unwanted daytime sedation.[85,86] These aims are in sharp contrast to the drug profiles of benzodiazepines, zopiclone and zolpidem, of which the use can be characterized as a preventive strategy at bedtime. Since many patients experience nightly or early morning awakenings or next-day hangover effects it is not surprising that they show poor compliance to the instructions that these hypnotics should only be used at bedtime. A recent study[60] revealed significantly impaired driving performance 4 hours after middle-of-the-night administration of zolpidem (10 mg and 20 mg). In addition, although prescription labeling clearly state that zolpidem should be used at bedtime only, a study examining blood samples of stopped drivers[87] showed that patients had used higher doses than the recommended dose of zolpidem (> 10 mg) and at inappropriate times during the night. The latter illustrates the need to develop hypnotics that can be used as needed during the night without producing performance impairment the following morning.

Zaleplon has been suggested to be one of those new hypnotics, devoid of next-day residual effects. Zaleplon has a high selectivity and high affinity to the α_1 receptor subtype, but a shorter half-life when compared to zopiclone and zolpidem. The finding of unimpaired driving ability 4 hours after middle-of-the night administration of zaleplon (10 mg and 20 mg) show that this is the first hypnotic drug that can be used during the night without producing driving impairment the following morning.[60] Relative to the other benzodiazepines this is a great advantage for those patients who want to participate fully in society and engage in activities such as driving a car.

Abbreviations

SDLP = Standard deviation of lateral position
BAC = Blood alcohol concentration
DSST = Digit symbol substitution test
CFFT = Critical flicker fusion test
GABA = Gamma amino butyric acid
CNS = Central nervous system

References

1. Freeman HL. Is there a need for a pure hypnotic? Approaches to the codiagnosis of insomnia and anxiety. J Drug Dev Clin Pract 1996; 7:289-302.
2. Borkenstein RF, Crowther RF, Shumate RP et al. The role of the drinking driver in traffic accidents. Bloomington: Indiana University, 1964.
3. Skegg DCG, Richards SM, Doll R. Minor tranquillisers and road accidents. Br Med J 1979; 1:917-919.
4. Neutel CI. Risk of traffic accident injury after a prescription for a benzodiazepine. AEP 1995; 5:239-244.
5. Hemmelgarn B, Suissa S, Huang A et al. Benzodiazepine use and the risk of motor vehicle crash in the elderly. JAMA 1997; 278:27-31.
6. Barbone F, McMahon AD, Morris AD et al. Association of road-traffic accidents with benzodiazepine use. Lancet 1998; 352:1331-1336.
7. Thomas RE. Benzodiazepine use and motor vehicle accidents. Systematic review of reported association. Can Fam Physician 1998; 44:799-808.
8. Svenson O. Are we all less risky and more skillful than our fellow drivers? Acta Psychologica 1981; 47:143-148.
9. McCormick IA, Walkey FH, Green DE. Comparative perception of driving ability – a confirmation and expansion. Accid Anal & Prev 1986; 18:205-208.
10. O'Hanlon JF, Haak TW, Blaauw GJ et al. Diazepam impairs lateral position control in highway driving. Science 1982; 217:79-81.
11. O'Hanlon JF. Driving under the influence of drugs: Rationale for, and application of, a new test. Br J Clin Pharmacol 1984; 18:121S-129S.
12. Koelega HS. Benzodiazepines and vigilance performance: A review. Psychopharmacol 1989; 98:145-156.
13. MackWorth NH. The breakdown of vigilance during prolonged visual search. Q J Exp Psychol 1948; 1:6-21.
14. Volkerts ER, van Laar MW, van Willigenburg APP et al. A comparative study of on-the-road and simulated driving performance after nocturnal treatment with lormetazepam 1 mg and oxazepam 50 mg. Hum Psychopharmacol 1992; 7:297-309.
15. Parrott AC. Performance tests in human psychopharmacology (1): Test reliability and standardization. Hum Psychopharmacol 1991; 6:1-9.
16. Parrott AC. Performance tests in human psychopharmacology (2): Content validity, criterion validity, and face validity. Hum Psychopharmacol 1991; 6:91-98.
17. Parrott AC. Performance tests in human psychopharmacology (3): Construct validity and test interpretation. Hum Psychopharmacol 1991; 6:197-207.
18. Verster JC. Measurement of the effects of psychoactive drugs on driving ability and related psychological processes. Thesis. Utrecht, The Netherlands: ISBN 90:393-3132-4, 2002.
19. Ancoli-Israel S, Roth T. Characteristics of insomnia in the United States: Results of the 1991 National Sleep foundation Survey. Sleep 1999; 22:22(Suppl 2):S347-S353.
20. O'Hanlon JF, Volkerts ER, de Vries G et al. Flurazepam HCl's residual ('hangover') effects upon actual driving performance. Groningen, The Netherlands: VSC, Report VK 83-02, Traffic Research Centre, 1983.
21. Volkerts ER, de Vries G, Meijer T et al. Driving performance the day after use of loprazolam, flunitrazepam and placebo. Groningen, The Netherlands: VSC, Report VK 83-04, Traffic Research Centre, 1984.
22. Volkerts ER, Louwerens JW, Gloerich ABM et al. Zopiclone's residual effect upon actual driving performance versus those of nitrazepam and flunitrazepam. Groningen, The Netherlands: VSC, Report 84-10, Traffic Research Centre, 1984.
23. O'Hanlon JF, Volkerts ER, Louwerens JW et al. Zopiclone's residual effect upon actual driving performance versus those of nitrazepam and flunitrazepam. In: The cycloperrolones non benzodiazepine hypnotics and tranquillizers. Florence: CINP Congress, 1984.
24. O'Hanlon JF, Volkerts ER, Brookhuis KA et al. Repeated dose effects of nitrazepam and temazepam upon actual driving performance. Groningen, The Netherlands: VSC, Report VK 85-02, Traffic Research Centre, 1985.
25. O'Hanlon JF, Volkerts ER. Hypnotics and actual driving ability. Acta Psychiatr Scand 1986; 74(Suppl 332):95-104.
26. Brookhuis KA, Volkerts ER, O'Hanlon JF. Repeated dose effects of lormetazepam 1 and 2 mg (in soft gelatine capsules) and flurazepam 30 mg upon driving performance. Groningen, The Netherlands: VSC, Report VK 86-18, Traffic Research Centre, 1986.
27. Brookhuis KA, Volkerts ER, O'Hanlon JF. Repeated dose effects of lormetazepam and flurazepam upon driving performance. Eur J Clin Pharmacol 1990; 38:1-5.
28. Volkerts ER, Abbink F, van Laar MW et al. A double-blind study to compare the acute residual effects of lormetazepam 1 mg, oxazepam 50 mg and placebo on driving performance in an over-the-road driving test. NIDDR: University of Utrecht, 1989.
29. Louwerens JW, Gloerich ABM, de Vries G et al. The relationship between drivers' blood alcohol concentration (BAC) and actual driving performance during high speed travel. In: Noordzij PC, Roszbach, eds. Alcohol, Drugs and traffic safety. Proceedings of the 10th International Conference on Alcohol, Drugs and Traffic Safety. Amsterdam: Exerpta Medica, 1987:183-192.
30. Betts TA, Birtle J. Effects of two hypnotics on actual driving performance next morning. Br Med J 1982; 285:852.
31. Willumeit HP, Neubert W, Ott H et al. Driving ability following subchronic application of lormetazepam, flurazepam and placebo. Ergonomics 1983; 26:1055-1061.
32. Laurell H, Tornros J. The carry-over effects of triazolam compared with nitrazepam and placebo in acute emergency driving situations and in monotonous simulated driving. Acta Pharmacol Toxicol 1986; 58:182-186.
33. Törnros J, Laurell H. Acute and carry-over effects of brotizolam compared to nitrazepam and placebo in monotonous simulated driving. Pharmacol Toxicol 1990; 67:77-80.
34. Schmidt U, Brendemuhl D, Ruther E. Aspects of driving after hypnotic therapy with particular reference to temazepam. Acta Psychiatr Scand 1986; 332(Suppl):112-118.
35. Bensimon G, Foret J, Warot D et al. Daytime wakefulness following a bedtime oral dose of zolpidem 20 mg, flunitrazepam 2 mg and placebo. Br J Clin Pharmacol 1990; 30:463-469.
36. Seppala T, Nuotto E, Dreyfus JF. Drug-alcohol interactions on psychomotor skills: Zopiclone and flunitrazepam. Pharmacology 1983; 27(Suppl 2):127-135.

37. Sicard BA, Trocherie S, Moreau J et al. Evaluation of zolpidem on alertness and psychomotor abilities among aviation ground personel and pilots. Aviat Space Environ Med 1993; 64:371-375.
38. Bocca ML, le Doze F, Etard O et al. Residual effect of zolpidem 10 mg and zopiclone 7.5 mg versus flunitrazepam 1 mg and placebo on driving performance and ocular saccades. Psychopharmacol 1999; 143:373-379.
39. Druid H, Holmgren P, Ahlner J. Flunitrazepam: An evaluation of use, abuse and toxicology. Forensic Sci Int 2001; 122:136-141.
40. Bramness J, Skurtveit S, Morland J. Clinical impairment of benzodiazepines – relation between benzodiazepine concentration and impairment in apprehended drivers. Drug Alcohol Depend 2002; 68:131.
41. Roehrs T, Kribbs N, Zorick F et al. Hypnotic residual effects of benzodiazepines with repeated administration. Sleep 1986; 9:309-316.
42. Verster JC, Volkerts ER, Verbaten MN. Effects of alprazolam on driving ability, memory functioning and psychomotor performance: A randomized, placebo-controlled study. Neuropsychopharmacol 2002; 27:260-269.
43. Van Laar MW, Volkerts ER, van Willigenburg APP. Therapeutic effects and effects on actual driving performance of chronically administered buspirone and diazepam in anxious outpatients. J Clin Psychopharmacol 1992; 12:86-95.
44. O'Hanlon JF, Vermeeren A, Uiterwijk MM et al. Anxiolytics' effects on the actual driving performance of patients and healthy volunteers in a standardized test. An integration of three studies. Neuropsychobiol 1995; 31:81-88.
45. Vermeeren A, Swijgman HF, O'Hanlon JF. Effects of alpidem, lorazepam and placebo on actual driving performance of anxious patients. Eur Psychiatry 1996; 11(Suppl 4):241S.
46. Van Laar M, Volkerts E, Verbaten M. Subchronic effects of the GABA-agonist lorazepam and the 5-$HT_{2A/2C}$ antagonist ritanserin on driving performance, slow wave sleep and daytime sleepiness in healthy volunteers. Psychopharmacol 2001; 154:189-197.
47. Törnros J, Laurell H. Acute and carry-over effects of brotizolam compared to nitrazepam and placebo in monotonous simulated driving. Pharmacol Toxicol 1990; 67:77-80.
48. Joynt BP. Triazolam blood concentrations in forensic cases in Canada. J Anal Toxicol 1993; 17:171-177.
49. Hindmarch I, Subhan Z. The effects of midazolam in conjunction with alcohol on sleep, psychomotor performance and car driving ability. Int J Clin Pharmacol Res 1983; 3:323-329.
50. De Gier JJ, 't Hart BJ, Nelemans FA. The effects of lorazepam and bromazepam on actual driving and psychomotor performance of patients. In: O'Hanlon JF, de Gier JJ, eds. Drugs and driving. London and Philadelphia: Taylor & Francis, 1986.
51. Moore NC. Medazepam and the driving ability of anxious patients. Psychopharmacol 1977; 52:103-106.
52. Kay GG. The effects of antihistamines on cognition and performance. J Allergy Clin Immunol 2000; 105(6 Pt. 2):S622-S627.
53. Louwerens JW, Brookhuis JKA, O'Hanlon JF. The effects of antidepresants oxaprotiline, mianserin, amitriptyline and doxepin upon actual driving performance. Groningen, Netherlands Traffic Research Center, 1984.
54. Ramaekers JG, Muntjewerff ND, O'Hanlon JF. Acute and subchronic effects of mitrazapine (15/30 mg nocte) and mianserin (30/60 mg nocte) on psychomotor and actual driving performance and sleep in healthy, young volunteers. Eur Neuropsychopharmacol 1995; 5:294-295.
55. Robbe HWJ, O'Hanlon JF. Acute and subchronic effects of paroxetine 20 and 40 mg on actual driving, psychomotor performance and subjective assessments in healthy volunteers. Eur Neuropsychopharmacol 1995; 5:35-42.
56. Dämgen K, Lüddens H. Zaleplon displays a selectivity to recombinant $GABA_A$ receptors different from zolpidem, zopiclone and benzodiazepines. Neurosci Res Commun 1999; 25:139-148.
57. Vermeeren A, O'Hanlon JF, DeClerck AC et al. Acute effects of zolpidem and flunitrazepam on sleep, memory and driving performance, compared to those of partial sleep deprivation and placebo. Acta Therapeutica 1995; 21:47-64.
58. Vermeeren A, Danjou PE, O'Hanlon JF. Residual effects of evening and middle-of-the-night administration of zaleplon 10 and 20 mg on memory and actual driving performance. Hum Psychopharmacol Clin Exp 1998; 13:S98-S107.
59. Vermeeren A, Riedel WJ, van Boxtel MPJ et al. Differential residual effects of zaleplon and zopiclone on actual driving: A comparison with a low dose of ethanol. Sleep 2002; 25:224-231.
60. Verster JC, Volkerts ER, Schreuder AHCML et al. Residual effects of middle-of-the-night administration of zaleplon and zolpidem on driving ability, memory functions, and psychomotor performance. J Clin Psychopharmacol 2002; 22:576-583.
61. Lader M, Denney SC. A double-blind study to establish the residual effects of zopiclone on performance in healthy volunteers. Int Pharmacopsychiatr 1982; 17(Suppl 1):98-108.
62. Nicholson AN, Stone BM. Zopiclone: Sleep and performance studies in healthy man. Int Psychopharmacol 1982; 17(Suppl 2):92-97.
63. Billiard M, Besset A, de Lustrac C et al. Dose-response effects of zopiclone on night sleep on nighttime and daytime functioning. Sleep 1987; 10(Suppl 1):27-34.
64. Nicholson AN, Stone BM. Efficacy of zopiclone in middle age. Sleep 1987; 10(Suppl 1):35-39.
65. Harrison C, Subhan Z, Hindmarch I. Residual effects of zopiclone and benzodiazepine hypnotics on psychomotor performance related to car driving. Drugs Exp Clin Res 1985; 11:823-829.
66. Broadhurst A, Cushnaghan RC. Residual effects of zopiclone (Imovane). Sleep 1987; 10(Suppl 1):48-53.
67. Wickstrøm E, Gotlibsen OB, Bredesen JE et al. Performance and mood following partial sleep deprivation: A randomized, double-blind, cross-over study of zopiclone, triazolam, flunitrazepam, ethanol and placebo. Hum Psychopharmacol 1988; 3:3-11.
68. Hindmarch I. Immediate and overnight effects of zopiclone 7.5 mg and nitrazepam 5 mg with ethanol on psychomotor performance and memory in healthy volunteers. Int Clin Psychopharmacol 1990; 5(Suppl 2):105-114.
69. Moon CAL, Hindmarch I, Holland RL. The effects of zopiclone 7.5 mg on the sleep, mood and performance of shift workers. Int Clin Psychopharmacol 1990; 5(Suppl 2):79-83.
70. Tafti M, Besset A, Billiard M. Effects of zopiclone on subjective evaluation of sleep and daytime alertness and on psychomotor and physical performance tests in athletes. Prog Neuro-Psychopharmacol Biol Psychiat 1992; 16:55-63.
71. Sicard B, Trocherie S, Moreau J et al. Assessment of zolpidem's residual effects on alertness and psychomotor abilities in navy pilots and air force personnel. J Sleep Res 1992; 1(Suppl 1):212.
72. Quera-Salva MA, McCann C, Boudet J et al. Effects of zolpidem on sleep architecture, night time ventilation, daytime vigilance and performance in heavy snorers. Br J Clin Pharmacol 1994; 37:539-543.
73. Berlin I, Warot D, Hergueta T et al. Comparison of the effects of zolpidem and triazolam on memory functions, psychomotor performances, and postural sway in healthy subjects. J Clin Psychopharmacol 1993; 13:100-106.
74. Richens A, Mercer AJ, Jones DM et al. Effects of zolpidem on saccadic eye movements and psychomotor performance: A double-blind, placebo-controlled study in healthy volunteers. Br J Clin Pharmacol 1993; 36:61-65.
75. Roehrs T, Merlotti L, Zorick F et al. Sedative, memory, and performance effects of hypnotics. Psychopharmacol 1994; 116:130-134.
76. Wesensten NJ, Balkin TJ, Belenky GL. Effects of daytime administration of zolpidem versus triazolam on memory. Eur J Clin Pharmacol 1995; 48:115-122.
77. Wesensten NJ, Balkin TJ, Belenky GL. Effects of daytime administration of zolpidem and triazolam on performance. Aviat Space Environ Med 1996; 67:115-120.
78. Mattila MJ, Vanakoski J, Kalska H et al. Effects of alcohol, zolpidem, and some other sedatives and hypnotics on human performance and memory. Pharmacol Biochem Behav 1998; 59:917-923.
79. Rush CR, Armstrong DL, Ali JA et al. Benzodiazepine-receptor ligands in humans: Acute performance-impairing, subject-rated and observer-rated effects. J Clin Psychopharmacol 1998; 18:154-166.

80. Danjou P, Paty I, Fruncillo R et al. A comparison of the residual effects of zaleplon and zolpidem following administration 5 to 2 h before awakening. Br J Clin Pharmacol 1999; 48:367-74.
81. Mintzer MZ, Griffiths RR. Selective effects of zolpidem on human memory functions. J Psychopharmacol 1999; 13:18-31.
82. Mintzer MZ, Griffiths. Triazolam and zolpidem: Effects on human memory and attentional processes. Psychopharmacol 1999; 144:8-19.
83. Troy SM, Lucki I, Unruh MA et al. Comparison of the effects of zaleplon, zolpidem, and triazolam on memory, learning, and psychomotor performance. J Clin Psychopharmacol 2000; 20:328-337.
84. Hindmarch I, Patat A, Stanley N et al. Residual effects of zaleplon and zolpidem following middle of the night administration five hours to one hour before awakening. Hum Psychopharmacol Clin Exp 2001; 16:159-167
85. Doghramji K. The need for flexibility in dosing of hypnotic agents. Sleep 2000; 23(Suppl 1):S16-S20.
86. Scharf MB. Individualizing therapy for early, middle-of-the-night and late night insomnia. Int J Clin Pract 2001; 116(Suppl):20-24.
87. Logan BK, Couper FJ. Zolpidem and driving impairment. J Forensic Sci 2001; 46:105-110.

CHAPTER 22

Sleep and Pain

Wilfred R. Pigeon, Hyung Park and Michael J. Sateia

Overview of the Chapter

This chapter focuses primarily on the co-occurrence of sleep disturbance and chronic, non-malignant pain. As might be expected, in chronic painful conditions such as rheumatoid arthritis poor sleep becomes a significant, pronounced, and enduring patient complaint. While there may be some increased risk for the development of other sleep disorders in chronic pain conditions (e.g., obstructive sleep apnea due to weight gain and loss of muscle tone), this chapter attends more closely to the most common reasons for sleep complaints in chronic pain patients: namely, insomnia and nonrestorative sleep. We characterize the objective and subjective nature of sleep disturbance in several representative populations of chronic pain patients. In addition, the evidence for the mediating roles of depression, anxiety, and trauma exposure in the sleep-pain relationship is explored. The most likely, potential underlying mechanisms in this relationship are briefly outlined. Finally, we address the surprisingly limited research informing the pharmacological or behavioral treatment of secondary insomnia and/or nonrestorative sleep in chronic pain patients. Overall, while some seminal work has been completed, the sleep-pain relationship awaits greater attention from the entire spectrum of science, from basic pathophysiology to targeted treatment interventions.

Background

Chronic pain is a common complaint in primary care settings and impacts 40-70 million people.[1] Among patients with chronic pain, a significant number complain of poor sleep with rates of sleep disturbance ranging from 42 to 88%.[2-8] A recent national survey projected that 56 million Americans have pain that interferes with sleep.[9] The direct and indirect economic costs of these conditions are enormous.[1,10] The patient burden is also significant for each disorder and is highlighted by demonstrated decrements in quality of life,[11-13] and increased medical[14] and psychiatric[15,16] morbidity. Furthermore, sleep disturbance is reported to be a significant quality of life issue by chronic pain patients[17,18] and reduces pain coping.[19] The relationship between sleep and pain, however, is complex. A clear consensus on contributing factors is lacking, although depression and anxiety are implicated as mediators.[2,7] Post-traumatic stress disorder (PTSD) is prevalent in both pain[20] and disturbed sleep[21] populations, but has not been examined as a mediator in the pain-sleep relationship.

The Bi-Directional Influences of Pain and Sleep

Patients and health care providers alike often implicate pain in creating sleep problems. However, like many phenomena, the relationship between pain and sleep is bi-directional. While pain severity and sleep disturbance are correlated in a variety of chronic pain populations,[2,4,5,7,22,23] the direction of this association is not always clear. In a longitudinal study of fibromyalgia patients spanning 30 days, Affleck et al analyzed daily sleep and pain diaries to find that pain and sleep had a bi-directional influence on each other.[24]

While pain severity has been found to be a predictor of sleep disturbance in some clinical studies,[6,24] several others found that pain severity was not the best predictor of sleep disturbance in patients with chronic pain.[5,7,17-19,25] Instead, factors such as depression, decreased activity and cognitive arousal can be more robust predictors of sleep disturbance.[22,25] In our own sample, chronic pain patients who reported decreased pain, increased activity and improved mood still continued to have sleep complaints.[26] This relates to a model of insomnia, which proposes that insomnia is perpetuated by behavioral and cognitive factors even after precipitating factors, such as pain or mood disturbance, are removed.[27] For example, Morin, Gibson and Wade divided a sample of 105 heterogeneous chronic pain patients into good and poor sleepers and found that poor sleepers did not have any higher prevalence of depression or anxiety than good sleepers.[4]

Experimentally, the nature of the pain-sleep relationship has been tested by inducing pain in subjects and monitoring effects on sleep. Animal studies have demonstrated that induced pain decreased REM sleep and SWS in cats.[28] Landis et al compared temporarily induced arthritis in rats and found decreased length of REM sleep and SWS periods as well as increased sleep fragmentation compared to controls.[29,30] In human experimental studies in which a variety of painful stimuli were administered to participants during sleep, there is consistent evidence that such stimuli lead to some level of sleep disruption. For instance, Drewes et al found that stimuli administered during the SWS of 10 healthy subjects produced decrements in delta activity and corresponding elevations in alpha and beta EEG activity when PSG EEG data were spectrally analyzed.[31] Lavigne et al introduced painful stimuli across sleep stages in nine subjects free of pain or sleep problems.[32] Visually scored arousals were higher in stage 2 sleep than in either SWS or REM sleep. Frank awakenings and stage shifts did not, however, differ by stage of sleep. These findings suggest that the macrostructure of sleep is less affected by nociception, than the microstructure of sleep.

Conversely, there is also experimental evidence that supports a causative role of sleep disturbance on the experience of pain, although it is not as robust as the evidence for the effect of pain on sleep. In animal models, REM sleep disruption has clearly been shown to increase various types of pain sensitivity.[33-35] In

Sleep and Sleep Disorders: A Neuropsychopharmacological Approach, edited by Malcolm Lader, Daniel P. Cardinali and S.R. Pandi-Perumal.

humans, a classic study by Moldofsky et al demonstrated that disruption of stage 4 sleep in healthy subjects resulted in muscle pain and tenderness that resembled the diffuse pain in fibromyalgia.[36,37] Two recent studies have shown increased morning pain sensitivity following SWS disruption.[38,39] Others,[40] however, failed to find an effect on masseter muscle pressure pain sensitivity following disruption of SWS for 3 consecutive nights. The method of disruption, auditory tone, may not have been adequate as 5 of the 10 subjects did not have a reduction of SWS. The remaining 5 "responders" who did have reduced SWS showed a trend ($p= .06$) towards increased pain sensitivity. Onen and colleagues[33] compared the effects of total sleep deprivation, REM sleep deprivation, SWS deprivation and recovery sleep on pain tolerance in an eloquent, two-way crossover design counterbalancing SWS and REM sleep interruption. Compared to a baseline night following an adaptation night, total sleep deprivation significantly decreased pain tolerance. Neither the selective interruption of REM sleep or SWS produced significantly lower pain tolerance scores, though they trended in that direction. An interesting outcome occurred on the recovery night in that recovery from SWS interruption, but not from REM interruption, a heightened analgesic effect. This suggests that increased SWS has a protective effect on sensitivity to pain.

Taken together these experimental findings especially implicate SWS in contributing to pain tolerance/pain sensitivity both at night and during the day. It is important to note that this is based on a limited number of studies only one of which directly compared the disruption of SWS to REM disruption. Furthermore, comparison of these experimental findings to clinical paradigms are premature since (1) sleep deprivation is not at all similar to microarousals or even to extended awakenings seen in insomnia and (2) experimentally and acutely induced pain in healthy subjects does not necessarily compare to the experience of chronic pain. Nonetheless, both the clinical and experimental data support the oft-stated notion of the bi-directional influence in the pain-sleep relationship. From a theoretical perspective, one conceptualization of this relationship is that the experience of chronic pain can disturb sleep acutely. It is then plausible that the insomnia that occurs in association with chronic pain has at its root the same etiologic factors that contribute to the occurrence and severity of insomnia in patients with primary insomnia (poor sleep hygiene, excessive time in bed, and increased worry and rumination). As the sleep disturbance persists, it then decreases the ability to tolerate and cope with chronic pain.

Potential Mediators of the Pain-Sleep Relationship

The Role of Depression

Depression has been investigated as a mediating variable in the relationship between pain and sleep, but studies have yielded inconsistent findings about its role. Although depression is implicated as a major contributor to sleep disturbance in pain patients in some studies,[2,18] others have not found depression to be predictive of poor sleep.[4,8,17,41] Moreover the relationship between depression and pain, in studies that did not consider sleep, is also somewhat equivocal. One review[42] concludes that chronic pain is clearly associated with depression, and that there is more evidence to support depression as a consequence of chronic pain, than as an antecedent but that this evidence is not overwhelming. Similarly, depression is highly co-morbid with sleep disturbance and sleep in depression is marked by increased arousals, among other sleep changes.[43] Meanwhile, insomnia can precede and be an early marker of depression and other psychiatric conditions.[14] Thus, it seems that depression itself may have a bi-directional relationships with both pain and insomnia. It is difficult to discern from the literature whether depression is a true mediator in the pain-sleep relationship. These findings are further complicated by a diagnostic dilemna faced in assessing chronic pain patients. That is, as a result of the pain experience itself and pharmacotherapy, these patients can meet many of the criteria for depression (poor sleep, irritability, fatigue, change in weight, decreased appetite, etc.) in the absence of depressed mood.

The Role of Anxiety

Anxiety is associated with both insomnia and chronic pain[44] and pain-related anxiety is especially prevalent in patients with chronic pain.[45] Some, who looked directly at the association of anxiety and the sleep of pain patients, found anxiety did contribute to sleep disturbance,[41,46] while others did not.[4-6] Anxiety as manifested in the form of presleep cognitive arousal has predicted sleep disturbance independent of depression and pain severity in a group of 51 patients with benign, chronic pain.[7] Smith et al also reported that thoughts related to pain had a stronger negative association to sleep continuity than did other arousing thoughts in a group of 31 patients with heterogeneous chronic pain.[25] This may explain why some measures of anxiety have not supported an anxiety-sleep relationship in chronic pain, as most instruments do not probe for pain-related ruminations.

The Role of PTSD

Exposure to traumatic events is associated with physical health complaints.[47] Furthermore, both trauma[48] and subsequent post-traumatic stress disorder (PTSD)[49] are highly co-morbid with chronic pain. Zayfert and Deviva reported that sleep complaints continue to manifest after successful treatment of PTSD.[50] PTSD is also associated with poorer adjustments to chronic pain.[51] In regards to sleep, PTSD is also consistent with overall sleep disturbance[52] and specific disturbances including excessive motor activity,[53] sleep disordered breathing,[54] REM sleep disruptions,[55] insomnia,[56,57] and negative dream content.[58] Despite these relationships only one account of the impact of PTSD on sleep of pain patients was identified. In unpublished work, Zayfert found that PTSD scores associated with significantly higher sleep disturbance in chronic pain patients, independent of depression. Therefore, while the role of PTSD in the pain-sleep relationship remains to be established; it could explain some of the inconsistent findings reviewed.

To summarize, studies examining the role of the most likely mediators in the pain-sleep relationship lend more complexity than clarity to the picture. This is in part due to the still nascent literature that faces us on this relationship. Though studies need to be well-powered to do so, any research on sleep in pain populations would do well to include measures of depression, anxiety, and PTSD. There is also enough evidence to consider these potential mediators in the clinical setting.

Sleep in Representative Chronic, Non-Malignant Pain Conditions

Fibromyalgia and Sleep

Sleep disturbance has long been associated with fibromyalgia (FM). In a survey of 188 FM patients, 80% reported sleep disturbance.[59] In a path analysis involving 105 FM patients, Nicassio et al reported that poor sleep quality accounted for the majority of the relationship between pain and fatigue.[60] The nature of the sleep disturbance in FM has undergone some scrutiny in the past three decades.

In 1973 Hauri and Hawkins first described the alpha-delta sleep pattern, wherein alpha EEG activity intruded in the slow wave sleep (SWS) in a 9 patient case series.[61] In 1975 Moldofsky identified this alpha-delta pattern in FM patients.[62] In 1976 Moldofsky et al experimentally induced this pattern of alpha EEG intrusion in the sleep of healthy subjects by disturbing SWS with a noxious auditory stimuli.[36] Alpha-delta intrusion became a signature feature of FM.[63,64] Subsequently, Moldofsky and others showed that alpha intrusion was not confined to SWS, but that alpha sleep was prominent throughout the NREM sleep of patients with FM.[65] Moreover, alpha sleep is not specific to FM, but is found in other disorders as well. For instance, Rains and Penzien in an archival study of 1,000 consecutive sleep patients undergoing polysomnography found that approximately 5% of their sample evidenced the alpha intrusion in their visually interpreted sleep records.[66] Within this sub-sample of 54 patients, the primary presenting diagnoses were chronic pain (n=21; with 18 of those with FM), psychiatric disorder (n=18), and other sleep disorder (n=18). This suggests that the alpha sleep anomaly is not specific to FM, but this study did not assess the prevalence of alpha sleep in FM patients.

Drewes et al used power spectral analysis of EEG activity in FM patients and healthy controls. They found increased power in the alpha band (8-12 Hz) across the night[67] as well as increased power in the higher frequency sigma domain (12-14.5 Hz) and decreased delta power (0.5-3.5 Hz).[68] Such findings support SWS decrements as a putative cause of daytime fatigue complaints in FM.

In one of the few longitudinal studies of sleep in FM, Affleck et al had 50 FM patients record sleep quality, pain intensity, and attention to pain for 30 days.[24] Two significant findings were that (1) the group of poorest sleepers reported more pain and (2) for all patients there was a bi-directional influence between sleep and pain, such that poor sleep led to a day of increased pain and days with high pain levels led to poorer sleep.

There is conflicting data on the association between FM and sleep apnea. Some have reported no association.[69,70] Jennum et al compared the PSG scored sleep of 20 FM patients and 10 healthy controls and found no differences in sleep architecture between groups. Of primary interest is that the only difference between groups was that FM patients had more arousals and that these arousals were related to respiratory events.[71] In a small case series FM was reported in 3 of 11 patients with established sleep apnea.[72] In May et al sample of 117 FM patients, 44% of males, but only 2% of females were diagnosed with sleep apnea following PSG.[73] All the patients had their sleep apnea successfully treated and 50% of them also experienced pain reduction, suggesting that in males with FM, sleep disruption related to obstructive events may be a causative agent in the diffuse pain associated with FM.

In sum, there is evidence that FM is associated with nonrestorative sleep, though the number of controlled trials in this area is limited. The nature of the sleep disturbance has been related to increased arousal and particularly to increased levels of alpha activity. The original findings of the visually identified alpha sleep pattern has been corroborated by quantitative EEG analysis. It is appealing to attach a causative role to disturbed and/or reduced SWS in FM as contributing to pain and tenderness since pain and tenderness have been created experimentally with SWS disruption. However, the clinical evidence does not yet consistently support this contention.

Arthritic Pain and Sleep

Arthritic disease is associated with heightened levels of fatigue and sleep disturbance. Fifty to seventy-five percent of patients with rheumatoid diseases complain of poor sleep.[74-78] Polysomnographically recorded sleep in 14 osteoarthritis patients was characterized by increased stage 1 and decreased stage 2 sleep compared to healthy controls.[79] Moldofsky, Lue and Saskin reported that 8 osteoarthritic patients with morning hand pain and stiffness had higher indices of periodic limb movements (PLM's) and arousals than 7 similar patients without morning hand pain and stiffness.[80] Sleep architecture between the groups was not significantly different and although there was no healthy control group, both groups appeared to have fairly normal sleep architecture compared to normative data. This study suggests that a portion of pain complaints in this pain population may be related to nonrestorative sleep caused by a primary sleep disorder.

Nicassio and Wallston found that initial self-reports of poor sleep were significantly associated with current depression in rheumatoid arthritis (RA) patients, while initial pain levels did not have a statistical effect on depression in 242 patients with RA.[19] When re-evaluated by self-report two years later, multiple regression analyses revealed that prior pain exacerbated poor sleep, but that prior poor sleep did not exacerbate subsequent pain levels. It was also determined that both pain and sleep problems contributed to subsequent depression. In an actigraphy study, Lavie et al demonstrated that RA patients have more PLM's than either low back pain patients or healthy controls.[81]

There have also been a number of uncontrolled studies in RA, in which PSG recordings demonstrated elevated alpha activity or sleep fragmentation.[76,82-84] Drewes at al performed two consecutive nights of home sleep recordings on 41 RA patients and included 19 matched controls. They also found PLM's to be significantly elevated in the RA group. In contrast to previously reviewed findings on SWS in chronic pain patients, their RA group had more stage 3 sleep (45 vs. 36 minutes; p= .05). The authors contend that this may be related to the more normalized sleep captured by home recordings compared to the laboratory-based recordings that found reduced SWS in prior work. Under quantitative EEG analysis, no differences in delta power were noted between groups. However, alpha-EEG sleep was found to be elevated throughout the night under spectral analysis in the RA group. Finally, graphical chain models of multiple variables supported direct and indirect interactions between pain and sleep.

Taken together, the findings in rheumatic disease support the general observation that chronic pain is associated with sleep fragmentation as evidenced by either increased arousal, visually scored alpha sleep, or quantitatively characterized elevations of alpha activity. Thus, this pattern is not limited to fibromyalgia, but may be endemic across at least several chronic pain conditions. The interaction between pain and sleep continues to be highlighted by the studies reviewed in this section. The contribution of both pain and sleep to depression is also noted. The higher prevalence of PLM's in arthritic disease is noteworthy. The contradictory findings with regard to pain and SWS remain to be empirically settled, although potential mechanisms for these disparities are explored in a later section.

Headache and Sleep

Several survey studies of varying methodological rigor have found that fatigue and lack of sleep are causally implicated in headache onset by among 30-70% of headache sufferers.[86-89] Paiva et al[90] subjected a heterogeneous group of patients with headache (n= 25) to polysomnography and, as a result, 13 of the patients had their diagnosis changed to a primary sleep disorder. In a later study, Paiva and colleagues[91] found that 17% of consecutive patients at a headache clinic (n= 288) complained of nocturnal or morning headaches and that approximately half of these headache sufferers were diagnosed with a primary sleep disorder. Epidemiological studies highlight snoring and sleep apnea as correlates of morning headaches[92,93] and unrefreshing sleep being associated with migraine headaches.[94] Treatment of sleep apnea has mitigated headaches in two controlled trials.[95,96]

Recently, Neau et al completed a prospective study of the association of headaches to snoring and sleep apnea in 312 patients undergoing clinical PSG for a referral of snoring.[97] All patients completed a headache questionnaire prior to PSG and assigned to a post-hoc group based on their apnea hypopnea index ((AHI); AHI < 15: snorer group, AHI >= 15: apnea group). Thirty-six percent of the snorer group and 32% of the apnea group endorsed headaches. However, when migraine type headaches were excluded, the apnea group had significantly higher rates of headache. There were few associations between headache and sleep parameters such as sleep architecture, AHI, or oxygen desaturations. Headaches were positively associated with the presence and severity of depression. Finally, 83 of the patients with headache were available at follow-up 1 to 4 years after PSG. 70% of apnea patients had improved headaches with apnea treatment, while 48% of the snorers improved regardless of treatment. Complete relief of headaches was significantly higher in the apnea group and correlated with treatment and treatment compliance within that group.

A recent survey of 100 confirmed narcoleptics with cataplectic attacks revealed that 37% (approximately half of the women and one fourth of the men) met IHS (International Headache Society) criteria for migraine.[98] Medication type, which can contribute to headache, did not alter these findings significantly. This study replicated earlier findings by the same group[99] and corresponds to a two- to four-fold increase in the risk for migraines in patients with narcolepsy.

Little additional work has been done that can link headache type to variations or disturbances in sleep architecture. Dexter[100] found an increase in SWS and REM sleep in patients with migraine. Drake et al[101] utilized a 4 channel EEG cassette (with one EEG lead) to study 10 patients each with tension, migraine and mixed type headaches and compared their sleep to normative data. There is no statistical analysis of the data, the authors simply reported means. All headache groups showed increased stage 1 and decreased stage 2 sleep. The migraine group had elevated SWS compared to norms (18.7% vs. 15%) while the other headache types had reduced SWS.

In summary, evidence to date in headache patients supports the observations that primary sleep disorders, especially obstructive sleep apnea and narcolepsy, are associated with headaches. In both migraine and tension type headache, depression may be mediating the relationship with sleep disturbances. The rather limited number of polysomnographic studies consistently fail to detect any striking differences in sleep architecture associated with headaches, except that there is a small body of literature pointing to increased SWS in patients with migraine type headache.

Back Pain and Sleep

In an actigraphic study, patients with low back pain did not differ from those with RA in sleep-wake activity.[81] In a controlled PSG study, Harman and Pivik[102] found that the alpha EEG anomaly and less restful sleep characterized low back pain patients compared to controls.

More recently Harman and colleagues[103] completed an important polysomnographic study that addresses the pain-sleep relationship in depressed (n= 40) and nondepressed (n= 6) chronic low back pain patients compared to age and gender-matched controls (n= 11). Rigid exclusion criteria for psychiatric, medical or occult sleep disorders, for additional pain conditions, and for use of medications that alter sleep were maintained. There were several interesting findings. First, the alpha sleep anomaly was present in members of each of the three groups and was found across NREM sleep stages and across occipital, central and frontal sites. Second, there were no sizable differences between groups on standard measures of sleep architecture and no evidence to corroborate the sleep fragmentation that was subjectively reported by the pain patients. Third, findings from quantitative EEG analysis again revealed fewer between group differences than expected. Increased beta activity, as a measure of arousal, did not differentiate between combined pain patients and controls. However, the depressed pain patients alone did have elevated beta activity compared to the nondepressed pain patients and the controls. Perhaps the most alluring spectral finding was that sigma activity (12-14 Hz), which is associated with more effective muting of sensory processing during sleep, was the only power spectral finding that distinguished both pain groups from the control group. The pain subgroups individually, and combined, demonstrated lower sigma power than the controls. As the authors suggest, this can explain the self-reports of poorer quality sleep in these pain patients.

Potential Mechanisms Underlying the Sleep-Pain Relationship

At the cortical level, there is consistent evidence that there is disruption of SWS and increased high frequency EEG activity in the sleep of patients with chronic pain. Several mechanisms for these disturbances await further investigation. For instance, in migraine headaches, serotonergic effects have been implicated in both the development and resolution of migraines, which are now widely treated with serotonergic agents.[104,105] Moldofsky has pointed to circadian alterations of various interdependent biological systems in fibromyalgia and chronic fatigue syndrome.[106] Vgontzas and colleagues have found alterations of cytokines mediating fatigue and sleepiness in several primary sleep disorders,[107] while Opp has found cytokines to be involved in dysregulation of sleep in response to infection.[108] Disruption of the neuroendocrine axes can also impact sleep. For instance Gillin showed that administration of glucocorticoids suppresses REM sleep.[109] Patients with insomnia have elevated HPA axis activity.[110] To the extent that pain patients have been exposed to trauma, which is marked by HPA dysregulation, this may be a path to sleep disturbance. Finally, for some patients the experience of pain itself becomes a chronic stressor.

Turning to the behavioral perspective, as originally set forth by Spielman and colleagues,[27] primary insomnia occurs acutely in relation to both trait (such as perfectionism or trait anxiety) and precipitating factors (such as a stressful event). In insomnia that is secondary to medical conditions what begins as acute insomnia becomes chronic when it is reinforced by maladaptive coping strategies that condition somatic and cognitive arousal.[111]

These mechanisms, including behavioral conditioning, are not mutually exclusive in modeling the pain-sleep relationship. Much work at both the basic and clinical research levels obviously remains in order to further elucidate these relationships.

Treatment

Antidepressants for Sleep in Pain Patients

Sedating antidepressants are often used to treat sleep complaints in chronic pain patients although there has been little direct research on the benefits of antidepressants for sleep in patients with pain. Antidepressants may improve sleep disturbance in chronic pain by three major mechanisms: (1) direct sedative effect; (2) antidepressant activity; (3) analgesic activity. Most research focuses on sleep as part of a group of target symptoms and thus makes direct correlation difficult. Therefore, it is difficult to tease out whether the benefits of antidepressants are due to direct effects on sleep or by diminishing pain, which then helps patients sleep better. The obvious benefits to depressed mood by antidepressants can further complicate the assessment of these medications directly on sleep. Many pain patients treated with antidepressants may have improved sleep due to effective treatment of underlying mood symptoms regardless of direct effect on sleep or pain. Finally, the evidence is mostly related to older tricyclic antidepressants (TCA) which carry significant side effects and risks.[113,114]

Nonetheless, there is some evidence to show that antidepressants might help certain pain patients sleep better. For instance, there exists modest evidence for the benefit of antidepressants for sleep in patients with fibromyalgia. O'Malley et al, in their review of 13 studies which included amitriptyline, clomipramine, maprotiline, S-adenosylmethionine, fluoxetine and citalopram, reported specific improvements in sleep for patients with fibromyalgia when antidepressants were used.[115] There were no differences between drug classes but only one study tracked depression. Other types of sedating medications, such as chlorpromazine[114] and cyclobenzaprine,[116] have also been shown to benefit sleep in these patients. The evidence for effectiveness of antidepressants for sleep disturbance in chronic fatigue syndrome is more mixed, with one study showing no benefit with fluoxetine[117] in any target symptoms, including sleep, and one study showing some benefit in sleep with phenelzine.[118] Ventafridda et al reported improvement in deafferentation pain, including cancer pain, with the use of trazodone or amitriptyline with concurrent improvement in hours of sleep.[119] Patients with other pain syndromes (acute pain, chronic low back pain, headache, chronic pelvic pain, facial pain, rheumatoid arthritis, pain associated with irritable bowel syndrome)[120] have benefited from TCAs or SSRIs with respect to pain level. By extension, it seems likely that such medications would also improve sleep in these populations. Antidepressants such as trazodone and amytriptyline have been used specifically for sleep initiation due to their sedative properties.[121] Sedating antidepressants like trazodone have improved subjective sleep quality in patients with primary insomnia,[122-124] although in one of these studies[122] objective sleep as measured by electrocephalogram was not improved and in another study of depressed patients objective sleep was worsened.[125]

Thus, the reader is cautioned to use careful assessments of the patient to weigh the risks and benefits and to determine if and how antidepressants should be used for sleep. Antidepressants should be considered when they benefit not only sleep but concomitantly occurring pain and mood symptoms. When these other symptoms need not be considered, sedative-hypnotics might be a better option. Furthermore, nonpharmacological treatments such as good sleep hygiene and cognitive therapy can help by themselves or when used with medication.[126]

Sedative and Analgesics/Narcotics for Sleep in Pain Patients

Effective control of pain through narcotics/analgesics has been demonstrated to improve sleep.[127] Improvements were found for sleep initiation, sleep maintenance and subjective sleep quality. Sedatives such as benzodiazepines, anti-histamines and non-benzodiazepine sleep aids are often used for sleep disturbances in pain patients and there is support for their use.[128-130] There is also a suggestion that the combination of analgesics and sedatives might act synergistically to improve sleep greater than can be accounted by the additive effects of each alone. A multipharmacy approach which includes various combination of oral and epidural opioids, peripheral acting analgesics, antidepressants, benzodiazepines, anticonvulsants, and nerve blocks produced an almost 80 percent in the number of cancer patients whose sleep was interrupted by pain.[131] As is the case for any pain patient, it is important to use appropriate pain medications to ensure that pain is adequately controlled. Chronic, long-term use of sedative medications lead to tolerance as well as carry risk for dependence and should be used cautiously. Sedatives and narcotics can worsen obstructive sleep apnea and patients should be monitored for worsening symptoms.

There has been evidence to show that the use of controlled release opiates results in better sleep in patients with chronic pain. Hanks et al described improved sleep in cancer patients with control released morphine compared to morphine sulfate administered every four hours.[132] In addition to the above cautions, opiates can cause acute changes in sleep, such as REM suppression and decrease total sleep time, as well as changes related to chronic use, such as decreased sleep onset latency and decreased sleep efficiency. However, these considerations must be weighed against benefits gained, including better sleep, by effective pain management.

Nonpharmacological Treatments

Cognitive-behavioral therapy (CBT) for pain involves training patients in relaxation, increasing physical activity, pacing activity to avoid pain flare-ups and/or deconditioning, and restructuring negative or ruminative thoughts. CBT for pain decreases both pain severity and depression,[133] yet there are few reports assessing the generalization of these improvements to sleep. Basler et al studied 45 patients with chronic pain associated with ankylosing spondylitis in a waiting list controlled trial of CBT for pain (48) and found improvements in self-reported measures of pain intensity, sleep, physical functioning, and psychological distress at both post-treatment and a 3-month follow-up.[134] Becker et al (2000) conducted a controlled clinical trial in a heterogeneous sample of 189 chronic pain patients randomized to 6 months of (1) a multidisciplinary pain treatment, (2) treatment in a General Medicine Practice with a pain specialist consultation, or (3) a waitlist control group.[135] Self-reported pain, quality of life, depression, and physical functioning were improved compared to the two control groups at 3 months, but sleep quality was not different. At 6-months, however, a modest improvement in sleep quality was found.

Others, meanwhile, have found no improvements in sleep following CBT for pain. For instance, Wigers et al, found that no impact on sleep was evidenced in 60 patients with fibromyalgia receiving treatment as usual plus random assignment to (1) CBT/Stress management, (2) aerobic exercise, or (3) symptom monitoring.[136] However, pain intensity, depression, and anergia did improve in the two treatment groups, suggesting that benefits from CBT for pain or from exercise did not generalize to improving sleep. Finally, some of our data bear directly on the issue of whether CBT for pain has significant effects on insomnia secondary to chronic pain. In a preliminary study,[26] 183 heterogeneous chronic pain patients referred to a behavioral medicine service completed a standardized 10-session treatment regimen. The sample exhibited significant pre-post change across a broad range of pain and mood measures, but did not exhibit significant change with respect to sleep complaints on two Likert scale items. A subsample of 12 patients demonstrated a similar lack of improvement in sleep as evidenced by a validated insomnia severity index. This suggests that (1) CBT for pain while effective for pain management does not reduce insomnia symptoms and that (2) chronic insomnia in the context of pain may be maintained by factors other than pain severity. These results are consistent with the above literature review in terms of the acute effects of CBT for pain and preliminarily suggest that targeted treatment for insomnia in pain patients is warranted.

Most of the studies summarized above used sleep measures of unknown validity, where most utilized single item measures instead of either validated sleep measures or sleep diaries. One study employed a validated measure on a subsample of only 12 participants and it was not a controlled trial.[26] Therefore, there is little basis to ascertain whether CBT for pain is effective for sleep disturbances.

The few published accounts of the direct treatment of insomnia in chronic pain patients do report success in treating sleep complaints with improvement in depression found in one study. Two, however, are case studies with one and three cases respectively.[137,138] Bruni reported that sleep hygiene alone was effective in treating both sleep disturbance and the primary complaint of migraine.[139] The only clinical trial for the treatment of insomnia secondary to chronic pain with CBT for insomnia (CBT-I) was conducted by Currie et al with 60 chronic pain patients.[140] It was found that sleep latency and WASO were reduced, while sleep efficiency and total sleep time increased. All the pre-post changes were significant.

None of these studies delivered the insomnia treatment in conjunction with or following pain management treatment. There is preliminary evidence for the use of sequential treatment of insomnia secondary to Generalized Anxiety Disorder (GAD).[141] In a cross-over design delivering CBT-I and CBT for GAD, it was determined that the most robust gains on insomnia measures were derived when patients received treatment for GAD first and insomnia treatment second. Accordingly, until further work tells us otherwise, it is expected that treatment of insomnia in chronic pain may best be delivered in conjunction with or after CBT for pain, though this remains to be tested.

Summary

The high prevalence of sleep disturbance in chronic pain conditions is due in part to bi-directional influences in which pain disturbs sleep and poor sleep reduces pain thresholds and aggravates pain. Other variables, especially depression and anxiety, may mediate this relationship, although supporting data has been equivocal. Trauma exposure and resulting PTSD symptoms have not been adequately evaluated in the pain-sleep relationship. The hyperarousal symptoms of PTSD are conceived to impact sleep directly in the manner that hyperarousal maintains insomnia. Even when pain severity, depression, and anxiety are decreased, sleep disturbance may persist due to the perpetuating factors of insomnia not being fully addressed in most patients. The importance of ruling out primary sleep disorders other than insomnia (sleep apnea and PLMD) must be underscored as they can contribute to sleep disruption that leads to increased pain sensitivity. Several lines of inquiry in regard to underlying sleep-pain mechanisms promise to unravel this complex relationship. Treatment of sleep complaints in chronic pain patients with antidepressant is best approached with caution. Nonpharmacological approaches to insomnia warrant further study and a tempered call for increased utilization based on a growing literature of successful treatment of insomnia secondary to medical conditions.

References

1. Latham J, Davis BD. The socio-economic impact of chronic pain. Disabil Rehabil 1994; 16:39-44.
2. Atkinson JH, Ancoli-Israel S, Slater MA et al. Subjective sleep disturbance in chronic back pain. Clin J Pain 1988; 4:225-232.
3. Becker N, Thomeson AB, Olsen AK et al. Pain epidemiology and health relted quality of life in chronic non-malignant pain patients referred to a Danish multidisciplinary pain center. Pain 1997; 73:393-400.
4. Morin CM, Gibson D, Wade J. Self-reported sleep and mood disturbance in chronic pain patients. Clin J Pain 1998; 14:311-314.
5. Pilowsky I, Crettenden I, Townley M. Sleep disturbance in pain clinic patients. Pain 1985; 23:27-33.
6. Riley JL, Benson MB, Gremillion HA et al. Sleep disturbance in orofacial pain patients: pain related or emotional distress? Cranio 2001; 19:106-113.
7. Smith MT, Perlis, ML, Smith MS et al. Sleep quality and presleep arousal in chronic pain. J Beh Med 2000; 23:1-13.
8. Wittig RM, Zorick FJ, Blumer D et al. Disturbed sleep in patients complaining of chronic pain. J Nerv Mental Dis 1982; 170:429-431.
9. National Sleep Foundation Gallup Poll on Adult Public's Experience with Nighttime Pain. Washington DC, 1995.
10. Stoller, M. Economic effects of insomnia. Clin Ther 1994; 16:873-897.
11. Zammit GK, Weiner J, Damato N et al. Quality of life in people with insomnia. Sleep 1999; 22:S379-S385.
12. Skevington SM, Carse MS, Williams AC. Validation of the WHOQOL-100: pain management improves quality of life for chronic pain patients. Clin J Pain 2001; 17:264-275.
13. Smith MT, Carmody TP, Smith MS. Quality of well-being scale and chronic low back pain. J Clin Psych in Med Settings 2000; 7:175-184.
14. Ford DE, Kamerow DB. Epidemiologic study of sleep disturbances and psychiatric disorders. An opportunity for prevention? JAMA 1989; 262:1479-1484.

15. Wilson KG, Mikail SF, D'Eon JL et al. Alternative diagnostic criteria for major depressive disorder in patients with chronic pain. Pain 2001; 91:227-234.
16. Katz DA, McHorney CA. Clinical correlates of insomnia in patients with chronic illness. Arch Intern Med 1998; 158:1099-1107.
17. Menefee LA, Frank ED, Doghramji K et al. Self reported sleep quality and quality of life for individuals with chronic pain conditions. Clin J Pain 2000; 16:290-297.
18. Follick M, Smith T, Ahern D. The sickness impact profile: a global measure of disability in chronic low back pain. Pain 1985; 21:6-76.
19. Nicassio PM, Wallston KA. Longitudinal relationships among pain, sleep problems, and depression in rheumatoid arthritis. J Abnorm Psychol 1992; 101:514-520.
20. Asmundson GJG, Coons MJ et al. PTSD and the experience of pain: research and clinical implications of shared vulnerability and mutual maintenance models. Can J Psychiatry 2002; 47:930-937.
21. Boiset-Fioro MH, Esdaile JM, Fitzcharles MA. Sexual and physical abuse in women with fibromyalgia. Arthr Rheum 1995; 38:235-241.
22. Wilson KG, Watson ST, Currie SR. Daily diary and ambulatory activity monitoring of sleep in patients with insomnia associated with chronic musculoskeletal pain. Pain 1998; 75:75-84.
23. Wilson KG, Eriksson MY, D'Eon JT et al. Major depression and insomnia in chronic pain. Clin J Pain 2002; 77-83.
24. Affleck G, Urrows S, Tennen, H, Sequential daily relations of sleep, pain intensity, and attention to pain among women with fibromyalgia. Pain 1996; 68:363-368.
25. Smith MT, Perlis, ML, Carmody TP et al. Presleep Cognitions in patients with insomnia secondary to chronic pain. J Beh Med 2001; 24:93-114.
26. Pigeon, WR, Seville, JL, Flood L. Impact of a cognitive-behavioral pain management group on sleep complaints in a chronic pain population. Sleep 2003; 26S:381.
27. Spielman A, Caruso L, Glovinsky P. A behavioral perspective on insomnia treatment. Psychiatr Clin North Am 1987; 10:541-553.
28. Carli G. Neuroplasticity and clinical pain. Progress in Brain Res 2000; 129:325-30.
29. Landis CA, Levine JD, Robinson CR. Decreased slow-wave and paradoxical sleep in a rat chronic pain model. Sleep 1989; 12:167-77.
30. Landis CA, Robinson CR, Levine JD. Sleep fragmentation in the arthritic rat. Pain 1988; 34:93-99.
31. Drewes AM, Nielsen KD, Nielsen-Arendt L et al. The effect of cutaneous and deep pain on the electroencephalogram during sleep-an experimental study. Sleep 1997; 20:623-640.
32. Lavigne G, M Zucconi, C Castronovo et al. Sleep arousal response to experimental thermal stimulation during sleep in human subjects free of pain and sleep problems. Pain 2000; 84:283-290.
33. Onen SH, Alloui A, Gross A et al. The effects of total sleep deprivation, selective sleep interruption, and sleep recovery on pain tolerance thresholds in healthy subjects. J Sleep Res 2001;10:35-42.
34. Onen SH, Alloui A, Jourdan D et al. Effects of rapid eye movement (REM) sleep deprivation on pain sensitivity in the rat. Brain Res 2001; 900:261-7.
35. Hicks RA, Coleman DD, Ferrante F et al. Pain thresholds in rats during recovery from REM sleep deprivation. Percept Mot Skills 1979; 48:687-690.
36. Moldofsky H, Scarisbrick P. Induction of neurasthenic musculoskeletal pain syndrome by selective sleep stage deprivation. Psychosom Med 1976; 38:35-44.
37. Moldofsky H, Scarisbrick P, England R et al. Musculoskeletal symptoms and Non-REM sleep disturbance in patients with "fibrositis syndrome" and healthy subjects. Psychosom Med 1975; 37:341-351.
38. Lentz MJ, Landis CA, Rothermel J et al. Effects of selective slow wave sleep disruption on musculoskeletal pain and fatigue in middle aged women. J Rheumatol 1999; 26:1586-1592.
39. Older SA, Battafarano DF, Danning CL et al. The effects of delta wave sleep interruption on pain thresholds and fibromyalgia-like symptoms in healthy subjects; correlations with insulin-like growth factor I. J Rheumatol 1998; 25:1180-1186.
40. Arima TP, Svensson C, Rasmussen KD et al. The relationship between selective sleep deprivation, nocturnal jaw-muscle activity and pain in healthy men. J Oral Rehabil 2001; 28:140-148.
41. Haythornthwaite JA, Hegel MT, Kerns RD. Development of a sleep diary for chronic pain patients. J Pain Sympt Manage 1991; 6:65-72.
42. Fishbain DA, Cutler R, Rosomoff HL et al. Chronic pain-associated depression: antecedent or consequence of chronic pain? A review. Clinical J Pain 1997; 13:116-137.
43. Benca RM, Obermyer WH, Thisted RA et al. Sleep and psychiatric disorders: a meta-anlysis. Arch Gen Psychiatry 1992; 49:651-668.
44. Katz DA, McHorney CA. Clinical correlates of insomnia in patients with chronic illness. Arch Intern Med 1998; 158:1099-1107.
45. McCracken LM, Zayfert C, Gross RT. The pain anxiety symptoms scale: development and validation of a scale to measure fear of pain. Pain 1992; 50:67-73.
46. Moffit PF, Kalucy EC, Kalucy RS et al. Sleep difficulties, pain and other correlates. J Int Med 1991; 230:245-249.
47. Phifer JF. Psychological distress and somatic symptoms after natural disaster: differential vulnerability among older adults. Psych Aging 1990; 5:412-420.
48. Toomey TC, Hernandez JT, Gittleman DF et al. Relationship of physical and sexual abuse to pain and psychological assessment variables in chronic pelvic pain patients. Pain 1993; 53:105-109.
49. Rappaport MH. Chronic pain and post traumatic stress disorder. Am J Psychiatry 1987; 144:120.
50. Zayfert C, DeViva JC. Residual sleep disturbance following cognitive-behavioral therapy for PTSD. J Trauma Stress 2004; 17(1):69-73.
51. Zayfert C, Seville J, O'Donnell K et al. PTSD is related to poorer adjustment to chronic pain. Presentation: International Society of Traumatic Stress Studies 1997; Montreal, Quebec.
52. Breslau N, Davis GC. Posttraumatic stress disorder in an urban population of young adults: risk factors for chronicity. Am J Psychiatry 1992; 149:671-676.
53. Ross RJ, Ball WA, Dinges DF et al. Motor dysfunction during sleep in posttraumatic stress disorder. Sleep 1994; 17:723-732.
54. Krakow BJ, Lowy C, Germain A et al. A retrospective study on improvements in nightmares and posttraumatic stress disorder following treatment for co-morbid sleep-disordered breathing. J Psychosom Res 2000; 49:291-298.
55. Mellman TA, Bustamante V, Fins AI et al. Rapid eye movement sleep and the early development of posttraumatic stress disorder. Amer J Psychiatry 2002; 159:1696-1701.
56. Deviva J, Zayfert C, Pigeon W. Treating residual sleep difficulties: The effects of insomnia treatment following exposure therapy for posttraumatic stress disorder. Proc Assoc of Beh Ther, 2002.
57. Krakow BJ, Melendrez DC, Johnston LG et al. Sleep dynamic therapy for Cerro grande fire evacuees with posttraumatic stress symptoms: a preliminary report. J Clin Psychiatry 2002; 63:673-684.
58. Pigeon WR, Mellman T. Dream content in recently hospitalized trauma patients. Proceedings Intnl Soc Trauma Stress Studies 2002.
59. Goldenberg DL. Fibromyalgia syndrome: an emerging but controversial condition. JAMA 1987; 257:2782-2787.
60. Nicassio PM, Moxham EG, Schuman CE et al. The contribution of pain, reported sleep quality, and depressive symptoms to fatigue in fibromyalgia. Pain 2002;100:271-279.
61. Hauri P, Hawkins DR. Alpha-delta sleep. Electroencephalogr Clin Neurophysiol 1973; 34:233-237.
62. Moldofsky H, Scarisbruck P, England R et al. Muskuloskeletal symptoms and nonREM sleep disturbance in patients with fibrositis syndrome and healthy subjects. Psychosom Med 1975; 37:341-351.
63. Moldofsky H, Lue FA. The relationship of alpha and delta EEG frequencies to pain and mood in 'fibrositis' patients treated with chlorpromazine and L-tryptophan. Electroencephalogr Clin Neurophysiol 1980; 50:71-80.
64. Branco J, Atalalaia A, Paiva T. Sleep cycles and alpha-delta sleep in fibromyalgia syndrome. J Rheumatol 1994; 21:1113-1117.
65. Moldofsky H. The contribution of sleep-wake physiology to fibromyalgia. Adv Pain Res Ther 1990; 17:227-240.
66. Rains JC, Penzien DB. Sleep and chronic pain: challenges to the alpha-EEG sleep pattern as a pain specific anomaly. J Psychosom Res 2003; 54:77-83.
67. Drewes AM, Gade K, Nielsen KD et al. Clustering of sleep electroencephalographic patterns in patients with the fibromyalgia syndrome. Br J Rheumatol 1995; 34:1151-6.

68. Drewes AM, Nielsen KD, Taagholt SJ et al. Sleep intensity in fibromyalgia: focus on the microstructure of the sleep process. Br J Rheumatol 1995; 34:629-635.
69. Fiona D, Esdaile JM, Kimoff JR et al. Musculoskeletal complaints and fibromyalgia in patients attending a respiratory sleep disorders clinic. J Rheum 1996; 23:1612-1616.
70. Lario BA, Teran J, Alonso JL et al. Lack of association between fibromyalgia and sleep apnea syndrome. Ann Rheum Dis 1992; 51:108-111.
71. Jennum P, Drewes AM, Andreasen A et al. Sleep and other symptoms in primary fibromyalgia and in healthy controls. J Rheumatol 1993; 20:1756-1759.
72. Malony RR, Macpeek DM, Schiffman PL et al. Sleep, sleep apnea and the fibromyalgia syndrome. J Rheumatol 1986; 13:797-800.
73. May KP, West SG, Baker MR et al. Sleep apnea in male patients with the fibromyalgia syndrome. Am J Med 1993; 94:505-508.
74. Belza BL. Comparison of self-reported fatigue in rheumatoid arthritis and controls. J Rheum 1995; 22:639-43.
75. Drewes AM, Nielsen KD, Hansen B et al. A longitudinal study of clinical symptoms and sleep parameters in rheumatoid arthritis. Rheumatol 2000; 39:1287-1289.
76. Moldofsky H, Lue F, Smythe HA. Alpha EEG sleep and morning symptoms in rheumatoid arthritis. J Rheumatol 1983; 10:373-379.
77. Pincus T, Swearingen C, Wolfe F. Towards a multidimensional health questionnaire(MDHAQ): assessment of advances activities of daily living and psychological status in the patient-friendly health assessment questionnaire format. Arth Rheum 1999; 42:2220-2230.
78. Wolfe F, Hawley DJ, Wilson K. The prevalence and meaning of fatigue in rheumatic disease. J Rheumatol 1997; 23:1407-1417.
79. Leigh TJ, Bird HA, Hindmarch I. A comparison of sleep in rheumatic and non-rheumatic patients. Ann Rheum Dis 1988; 47:40-42.
80. Moldofsky H, Lue FA, Saskin P. Sleep and morning pain in primary osteoarthritis. J Rheumatol 1987; 14:124-128.
81. Lavie P, Epstein R, Tzschinsky O et al. Actigraphic measurements of sleep in rheumatoid arthritis: comparison with low back pain and healthy controls. J Rheumatol 1992; 19:362-365.
82. Crosby LJ. Factors which contribute to fatigue associated with rheumatoid arthritis. J Adv Nurs 1991; 16(8):974-81.
83. Lavie P, Nahir M, Lorber M et al. Nonsteroidal antiinflamatory drug therapy in rheumatoid arthritis. Arth Rheum 1991; 34:665-659.
84. Mahowald MW, Mahowald ML, Bundlie Sr et al. Sleep fragmentation in rheumatoid arthritis. Arth Rhem 1989; 32:974-983.
85. Drewes AM, Svendsen L, Taagholt SJ et al. Sleep in rheumatoid arthritis: a comparison with healthy subjects and studies of sleep/wake interactions. Br J Rheumatol 1998; 37:71-81.
86. Blau JN. Sleep deprivation headache. Cephalgia 1990; 10:157-160.
87. Paiva T, Esperanca P, Martins I et al. Sleep disorders in headache patients. Headache Quarterly 1992; 3:438-442.
88. Paiva T, Martins I, Tellas J. Sleep disturbances in patients with morning headaches. J Sleep Res 1994; 3:190.
89. Spierings EL, van Hoof MJ. Fatigue and sleep in chronic headache sufferers: an age- and sex-controlled questionnaire study. Headache 1997;37(9):549-52.
90. Paiva T, Batista A, Martins P et al. The relationship between headaches and sleep disturbances. Headache 1995; 35(10):590-6.
91. Paiva T, Farinha A, Martins A et al. Chronic headaches and sleep disorders. Arch Intern Med 1997; 157:1701-1705.
92. Jennum P, Hein HO, Suadicani P et al. Headache and cognitive dysfunction in snorers. A cross-sectional study of 3323 men aged 54 to 74 years: the Copenhagen male study. Arch Neurol 1994; 51:937-942.
93. Ulfberg J, Carter N, Talbach M et al. Headache, snoring and sleep apnea. J Neurol 1996; 243:621-625.
94. Rasmussen BK. Migraine and tension-type headache in a general population: precipitating factors, female hormones, sleep pattern and relation to lifestyle. Pain 1993; 53:65-72.
95. Kiely JL, Murphy M, McNicholas WT. Subjective efficacy of nasal CPAP therapy in obstructive apnoea syndrome: a prospective controlled study. Eur Respir J 1999; 13:1086-1090.
96. Loh NK, Dinner DS, Foldvary N et al. Do patients with obstructive sleep apnea wake up with headaches? Arch Int Med 1999; 159:1765-1768.
97. Neau JP, Paquereau J, Bailbe M et al. Relationhsip between sleep apnoea syndrome, snoring and headaches. Cephalgia 2002; 22:333-339.
98. Dahmen N, Kasten M, Wieczorek S et al. Increased frequency of migraine in narcoleptic patients: a confirmatory study. Cephalgia 2003; 23:14-19.
99. Dahmen N, Querings K, Grun B, Bierbrauer J. Increased frequency of migraine in narcoleptic patients. Neurology 1999; 52:1291-1293.
100. Dexter JD. The relationship between stage III+IV+REM sleep and arousals with migraine. Headache 1979; 19:364-369.
101. Drake ME, Pakalnis A, Andrews JM et al. Nocturnal sleep recording with cassette EEG in chronic headaches. Headache 1990; 30:600-603.
102. Harman K, Matsunga L, Pivik RT. Non-restorative sleep complaints and sleep depth: an auditory arousal threshold study in chronic low back pain patients. Sleep 1998; 21:63.
103. Harman K, Pivik RT, D'eon JL et al. Sleep in depressed and nondepressed participants with chronic low back pain: electroencephalographic and behaviour findings. Sleep 2002; 25:775-783.
104. Saper JR, Silberstein SD, Lake AE et al. Double-blind trials of fluoxetine: chronic daily headache and migraine. Headache 1994; 34:497-502.
105. Dodick DW, Eross EJ, Parish JM. Clinical, anatomical and physiologic relationship between sleep and headache. Headache 2003; 43:282-292.
106. Moldofsky H. Sleep, neuroimmune and neuroendocrine functions in fibromyalgia and chronic fatigue syndrome. Adv Neuroimmunol 1995; 5:39-56.
107. Vgontzas AN, Papanicolaou DA, Bixler EO et al. Elevation of plasma cytokines in disorders of excessive daytime sleepiness: role of sleep disturbance and obesity. J Clin Endocrin Metab 1997; 82:1313-1316.
108. Opp MR. Fever, body temperature, and levels of arousal. In: Lydic R and Baghdoyan HA, eds. Handbook of behavioral control: cellular and molecular mechanisms. CRC Press, Boca Raton 1999:623-640.
109. Gillin JC, Jacobs LS, Fram DH et al. Acute effect of glucocorticoid on normal human sleep, Nature 1972; 237:398-399.
110. Vgontzas AN, Bixler EO, Lin E-M et al. Chronic insomnia is associated with nyctohemeral activation of the hypothalamus-pituitary-adrenalk axis: clinical implications. J Clin Endocrinol Metab 2001; 86:3787-3794.
111. McCrae CS, Lichstein KL. Secondary insomnia: diagnostic challenges and intervention opportunities. Sleep Med Reviews 2001; 5:47-61.
112. Morin CM. Measuring outcomes in randomized clinical trials of insomnia treatments. Sleep Med Reviews 2003; 7:263-279.
113. Moldofsky H, Lue FA. The relationship of alpha and delta EEG frequencies to pain and mood in 'fibrositis' patients treated with chlorpromazine and L-tryptophan. Electroencephalogr Clin Neurophysiol 1980; 50:71-80.
114. Carette S, Oakson G, Guimont C et al. Sleep electroencephalography and the clinical response to amitriptyline in patients with fibromyalgia. Arth Rheum 1995; 38(9):1211-7.
115. O'Malley PG, Balden E, Tomkins G et al. Treatment of fibromyalgia with antidepressants. A meta-analysis. J Gen Intern Med 2000;15:659-666.
116. Reynolds WJ, Moldofsky H, Saskin P et al. The effects of cyclobenzaprine on sleep physiology and symptoms in patients with fibromyalgia. J Rheumatol 1991; 18:452-454.
117. Vercoulen J, Swanink C, Zitman F et al. Randomised, double-lbind, placebo-controlled study of fluoxetine in chronic fatigue syndrome. Lancet 1996; 347:858-861.
118. Natelson B, Cheu J, Pareja J et al. Randomized, double-blind, controlled placebo-phase in trial of low dose phenelzine in the chronic fatigue syndrome. Psychopharmacology 1996; 124:226-230.
119. Ventafridda V, Caraceni A, Saita L et al. Trazodone for deafferentation pain. Comparison with amitriptyline. Psychopharmacology 1988; 95:S44-S49.
120. Fishbain D. Evidence-based data on pain relief with antidepressants. Ann Med 2000; 32:305-316.

121. Walsh JK, Schweitzer PK. Ten-year trends in the pharmacological treatment of insomnia. Sleep 1999; 22: 371-375.
122. Montgomery I, Oswald I, Morgan K et al. Trazodone enhances sleep in subjective quality but not in objective duration. Br J Clin Pharm 1983; 16:139-144.
123. Reimann D, Voderholzer U, Cohrs S et al. Trimipramine in primary insomnia: results of a polysomnographic double-blind controlled study. Pharmacopsychiatry 2002; 35:165-174.
124. Walsh JK, Erman M, Jamieson A et al. Subjective hypnotic efficacy of trazodone and zolpidem in DSMIII-R primary insomnia. Hum Psychopharm 1998; 13:191-198.
125. Nowell PD, Buysee DJ. Treatment of insomnia in patients with mood disorders. Depress Anx 2001; 14:7-18.
126. Bennett RM. Multidisciplinary group programs to treat fibromyalgia patients. Rheum Dis Clin North Am 1996; 22:351-367.
127. Ventafridda V, Tamburini M, Caraceni A et al. A validation study of the WHO method for cancer pain relief. Cancer 1987; 59:850-856.
128. Smith GM, Smith PH. Effects of doxylamine and acetaminophen on postoperative sleep. Clin Pharm Theraputics 1985; 37:549-557.
129. Drewes AM, Bjerregard K et al. Zopiclone as night medication in rheumatoid arthritis. Scand J Rheumatol 1998; 27:180-7.
130. Gronblad M, Nykanen J et al. Effect of zopiclone on sleep quality, morning stiffness, widespread tenderness and pain and general discomfort in primary fibromyalgia patients: a double-blind randomized trial. Clin Rheum 1993; 12:186-191.
131. Banning A, Sjogren P, Henriksen H. Treatment outcomes in a multidisciplinary cancer pain clinic. Pain 1991; 47:129-134.
132. Hanks GW et al. Controlled release morphine tablets: a double-blind trial in patients with advanced cancer. Anaesthesia 1987; 42:840-844.
133. McCracken LM, Turk DC. Behavioral and cognitive-behavioral treatment for chronic pain: outcome, predictors of outcome and treatment process. Spine 2002; 27:2564-2573.
134. Basler HD, Rehfisch HP. Cognitive-behavioral therapy in patients with ankylosing spondylitis in a German self-help organization. Psychosom Res 1991; 35:345-354.
135. Becker NP, Sjogren P, Bech AK et al. Treatment outcome of chronic non-malignant pain patients managed in a danish multidisciplinary pain centre compared to general practice: a randomised controlled trial. Pain 2000; 84:203-211.
136. Wigers SH, Stiles TC, Vogel PA. Effects of aerobic exercise versus stress management treatment in fibromyalgia. A 4.5 year prospective study. Scand J Rheumatol 1996; 25:77-86.
137. Morin CM, Kowatch RA, Wade JB. Behavioral management of sleep disturbances secondary to chronic pain. J Beh Ther Exper Psychiatry 1989; 20:295-302.
138. Morin CM, Kowatch RA, O'Shanick G. Sleep restriction for the inpatient treatment of insomnia. Sleep 1990: 13:183-186.
139. Bruni O, Galli F, Guidetti V. Sleep hygiene and migraine in children and adolescents. Cephalgia 1999; 19:57-59.
140. Currie SR, Wilson KG, Pontefract AJ et al. Cognitive-behavioral treatment of insomnia secondary to chronic pain. J Clin Consult Psych 2000; 68:407-416.
141. Blais FC, Mimeault V, Morin CM. Treatment of comorbid insomnia and generalized anxiety disorder. Proc Assoc Advancement Beh Ther 2003.

CHAPTER 23

Head Injuries and Sleep

Chanth Seyone and Babita Kara

Introduction

We do not remember days, we remember moments.

-Unknown

The above quote crystallizes who we, as individuals, are. We are a species with the ability to remember, especially those moments in life that stand out and that act as guides in our path through the maze of life. In patients with head injuries, these guides become absent and the maze unforgiving due to difficulties with being able to remember. This becomes even worse in head injured patients who subsequently develop sleep disorders. The sleep disorder then disrupts their lives even further, impedes rehabilitation and wrecks havoc on a persons sense of his own worth, and his or her sense of ever coming to terms with the damage sustained. It also impacts adversely on family dynamics by also interfering with the sleep patterns of caregivers thus leading to further problems.

It is not the intent of this chapter to discuss extensively either head injury (or the more recent usage, acquired brain injury (ABI), or sleep disorders but a brief overview will aid in the understanding of the pathophysiology and presentation of sleep disturbances in the head injured population and make easier the comprehension of management strategies.

Acquired Brain Injury

Head Injury and Acquired Brain Injury are terms becoming almost synonymous with traumatic brain injuries. However, in this chapter the terms are used more broadly to denote injuries sustained through trauma as well as other means such as lesions due to cerebrovascular accidents (CVA), abscesses, infections etc.

Head trauma leading to brain injury and neuropsychiatric sequelae including sleep disturbances has been described as a "disease of modernity", with the majority being due to motor vehicle accidents, gunshot wounds or occupational mishaps.This is attributed to the fact that with improved medical care, a larger proportion of patients survive injuries that would have been fatal in the past.

Head injuries leads to a large number of casualties each year in all countries. The incidence is estimated at greater than 400-600 cases per 100,000 populations if minimal injury is included. The peak age incidence is between 15-24 years in whites and in the 25-40 year range in blacks, with males being injured two to four times more frequently than females.[1]

Accidental head injury is the developed world's leading cause of death in people under the age of 45 years. Only the incidence of death due to Stroke and Cancer are greater than that from Head Injuries. Each year about 13,600 people in Ontario sustain a brain injury severe enough to be admitted to hospital. Outcome surveys conducted in the United States, United Kingdom and in the Netherlands indicate overall that after head injury, 5% remain in a vegetative state, 15% are still severely disabled six months after injury, 20% have psychiatric problems, and only 60% make a full recovery.[4] Thus, a full 40% have some form of neurological or psychiatric sequelae. This is especially tragic considering the age group of patients involved, usually in the prime of life between 15-24 years and thus the complications are all the more disabling not only physically but also emotionally, and financially. This is not only to the person involved, but also to his or her family and ultimately to the society. It is more of a problem now that dramatic reductions in mortality after head injuries over the last 10 years have led to young, otherwise healthy individuals having to live with chronic neuropsychiatric disabilities.

Head trauma causes brain injury through both direct (coup, contracoup and compression injuries) and indirect effects (cerebral edema, hypoxia, shock etc.). Trauma to the head can be broadly classified as penetrating or non-penetrating depending on whether the dura has been invaded. The vast majority are non-penetrating and result from MVA (> 50%), falls (21%), violence (12%), and recreational activities (10%).[3] Non-penetrating injury can lead a variety of injuries from simple, non-life threatening injuries from which patients may recover with little or no sequelae to those that are almost always accompanied by some form of neuropsychiatric difficulties assuming that the patient survives the initial injury.

The end result of head trauma and brain injury may run the gamut from full recovery to delirium, dementia, amnestic disorders or secondary psychiatric syndromes. These includes cognitive disorders (decreased efficiency or speed of processing, decreased attention and vigilance, problems with memory and verbal processing), post-traumatic amnesia, postconcussional disorder, personality changes or adjustment disorders. There may also be a depressive illness, difficulties with anxiety and even psychosis. Psychosocially, patients may have behavioral difficulties including substance use problems, inappropriate sexual expression and aggression, poor impulse control and dyscontrol, an inability to return to pre-morbid functioning, and concerns with family, shelter and finances. And finally, brain injuries may lead to difficulties with sleep patterns or even the development of significant secondary sleep disorders.

The time spent unconscious and post-traumatic amnesia (PTA) is a good predictor of prognosis. In general, the longer the period of unconsciousness and PTA and the deeper the level of coma, the greater the likelihood that the patient has suffered brain

Sleep and Sleep Disorders: A Neuropsychopharmacological Approach, edited by Malcolm Lader, Daniel P. Cardinali and S.R. Pandi-Perumal.

Table 1. Estimates of risk disability post head injury

Degree of Concussion	Length of PTA	Estimated Time Before Returning to Work
Slight	< 1 hour	4-6 weeks
Moderate	1-24 hours	6-8 weeks
Severe	1-7 days	2-4 months
Very severe	> 7 days	4-8 months

Adapted from Roberts GW, Leigh PN, Weinberger DR. Neuropsychiatric Disorders. Figure 5.2, p 5.3: Wolfe: Toronto, 1993.

damage and will suffer neuropsychological and psychiatric sequelae (Table 1). However, this has not been shown to be true for all sequelae. Sleep disorders do not seem to have as good a correlation to the length of coma. This could possibly be explained in part by the impact of multiple forces on the sleep patterns of head injured patients that bias the pathway between length of coma and the development of sleep disorders.

Radiographic examination of the head is used to identify and delineate fractures of the skull. CT and MRI allow for the detection of contusions, hematomas, and other intracranial pathologies in a noninvasive fashion. Despite its diagnostic usefulness in patients with post-traumatic epilepsy, EEG has not proved to be a better predictor of the development of post-traumatic seizures than such clinical features as early epilepsy, acute hematoma and depressed skull fracture. The degree of abnormality in evoked potentials is predictive of outcome and has been shown to be related to other neurological indexes of the severity of injury. Neuropsychological testing is a good way of delineating a persons cognitive deficits and of assessing a persons personality structure.

Sleep Assessment

Before discussing the sleep disorders that may accompany head injuries, it is worthwhile briefly summarizing the assessment of sleep disorders especially in those with brain injuries. Sleep physiology is discussed elsewhere in this book. Sleep assessment has undergone considerable advances in the past decade with the advent of new technologies and materials for incorporation into specialized sleep laboratories for the evaluation of a multitude of sleep disturbances.

A standard clinical interview combined with a detailed functional analysis remains the most effective method for gathering preliminary information on the sleep complaint and it's potential contributing factors. This is particularly so in patients with an acquired brain injury who without inquiry may not divulge information about brain injury, especially if it was "minor". Functional inquiry is also important to quantify the extent of limitations due to the development of secondary problems in ABI patients. Therefore, these patients should be assessed for:

- major psychiatric syndromes, including behavioral abnormalities
- substance use/abuse
- pain syndromes
- mobility issues
- psychosocial factors
- medications

It is also important to determine the onset and duration of symptoms to differentiate between transient, short-term and long-term insomnia as management strategies would differ. Patients often exaggerate the difficulty they have falling asleep or underestimate the amount of time they are able to sleep. It is therefore preferable to assess the severity of sleep disruption by evaluating daytime functioning especially daytime somnolence.

Notwithstanding, a history from the patient should include key questions as outlined by Kales et al[11,12] which question the patient about the characteristics of the sleep difficulty, duration, sleep hygiene, dietary habits, medical/psychiatric condition and treatments. The history should also include a thorough family history to rule out hereditary sleep disorders (e.g., narcolepsy) and should question the bed-partner to determine the presence of certain sleep disorders such as sleep bruxism, somnambulism, abnormal movements, snoring, and sleep apnea.[11] Concomitant medical and psychiatric conditions should be identified and a review conducted of alcohol, drug and medication use including the use of tobacco. Additional evaluation, if required, to either elicit or confirm diagnostic impressions can be achieved by using a daily sleep/wake diary for two to three weeks, doing a nocturnal polysomnographic recording and in certain patients ambulatory monitoring. Finally, other specialized measures have been developed and can be useful assessment tools in certain situations. These include the Pre-Sleep Arousal Scale,[13] the Insomnia Symptom Questionnaire,[14] the Sleep Hygiene and Practice Scale,[15] the Self Efficacy Scale,[16] Daytime Sleepiness Scale and the Multiple Sleep Latency Test.[17] A number of standardized measures of mood and psychological functioning are also available and include the MMPI, Beck Depression Inventory, the State-Trait Anxiety Inventory and the Profile of Mood States.[17] Specific questionnaires for some of the sequelae of ABI, in addition to the above, may include the Structured Clinical Interview for DSM-IV Diagnoses (SCID).[18] Neurobehavioral Rating Scale (NBRS),[19] the Positive and Negative Symptom Scale (PANSS),[20] the Overt Aggression Scale (OAS), and the Overt Agitation Severity Scale (OASS).[21]

Coleman et al[22] estimated the etiologies of insomnia diagnoses as mostly due to psychiatric disorders (15%) followed by psycho-physiologic (15%), drugs and alcohol dependency (12%), periodic limb movement disorder (12%), and medical, toxic, environmental factors (12%). Most treatment efforts are thus focused on interventions for conditioned insomnia, drug and/or alcohol dependent insomnia, and insomnia associated with medical/psychiatric conditions. The underlying medical/psychiatric problem should of course be treated and treatments as much as possible should be tailored to the specific diagnostic group.[17]

Pathophysiology and Presentation of Sleep Disorders in ABI

Sleep disturbances are a prevalent symptom in patients with ABI.[23,24] The rate of insomnia in traumatic brain injury patients

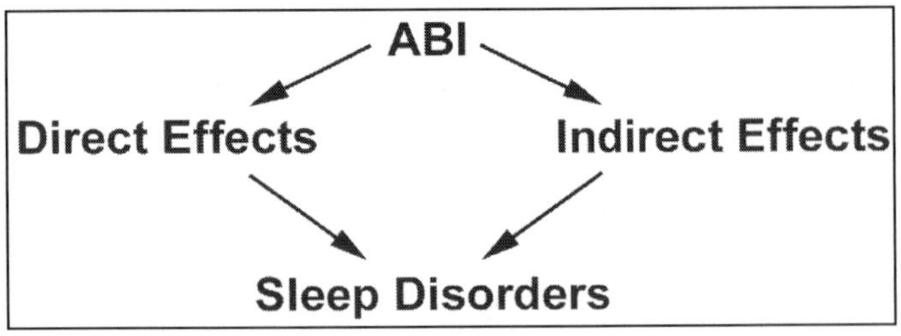

Figure 1. Pathophysiology of sleep disorders post head injury – main.

is almost twice the frequency as that found in the general population.[25] 70 to 90% of brain injuries that occur are classified as mild.[26] However, patients with mild brain injury seem to have more frequent complaints of sleep difficulty than patients with more severe injuries. Loss of consciousness does not seem to be a necessary prerequisite for the development of hypersomnolence.[27] Furthermore, whiplash, a relatively mild form of trauma has been strongly linked to the development of sleep apnea.[28] The patient's age has not been found to be a predictor of the development of sleep complaints in patients with head injury. While more males tend to be injured, however, it appears that females may present with a relatively higher incidence of sleep complaints.[29]

Sleep disorders remain a difficult entity to identify, quantify and manage, especially in the acquired brain injured population. This is in large measure due to the large number of other complaints that may overshadow sleep difficulties including by and large, mood, anxiety and psychotic symptomatology, significant cognitive deficiencies, behavioral abnormalities and neurological deficits including epilepsy. Medication use for any of the above may also confound the picture. In these cases, sleep patterns, as can be easily imagined, may be overlooked.

If not, insomnia is usually presented as an afterthought while in the physician's office for an unrelated complaint.[30] In addition, patients often do not receive the treatment they need due to provider concerns about the safety of treatment, or because providers lack awareness of effective treatment strategies. They may also regard insomnia as a minor or peripheral complaint and not review the patients history of sleep impairment or acquire the necessary information to determine the etiology of the insomnia.[31]

The pathophysiology of sleep disturbances in the acquired brain injured population are many and can be visualized graphically as shown in Figures 1-3.

The high correlation between head injury and sleep disorders most likely reflect a physiological sleep alteration as a consequence of central nervous system lesions.[29] With the variety of midbrain, basal forebrain and brainstem structures involved in generation of sleep rhythm and pattern, focal or diffuse injuries resulting in neurotransmitter or neuroendocrine imbalance may directly alter sleep.[32,33] Disorganization of the sleep-wake pattern, by itself, can have a detrimental effect on mental performance.[34]

The sleep-wake cycle is closely linked to other circadian rhythms such as temperature which follow an approximate 24 hr rhythm. The suprachiasmatic nucleus is the "clock" that controls these rhythms, in addition to the reticular activating system and interaction between several neurotransmitters, neuroendocrine axis and alterations in these functions. Serotonin, norepinephrine, dopamine, γ-aminobutyric acid (GABA) and acetylcholine have all been postulated to be keys in the possible production of disordered sleep rhythms in ABI patients.[33] Serotonin is a major neurotransmitter involved in sleep onset and maintenance. Infarction or damage to the brainstem involving the raphe nucleus that controls the production of serotonin may thus decrease the total sleep time without affecting REM sleep.

Norepinephrine on the other hand controls the shift to REM sleep. It is produced in the locus ceruleus but acts via modulating activity in the raphe nucleus. Markedly elevated levels of norepinephrine have been noted after acute ABI.[35,36] It seems to be correlated to more severe injury and poorer outcome. However, specific lesions may deplete norepinephrine by interrupting the nerve tracts that course anteriorly from the brain stem to curve around the hypothalamus, the basal ganglia, and the frontal cortex.[37] This, therefore, can lead to impairments in REM sleep as well. By interrupting these tracts, serotonin is depleted as well, and this may explain why administering l-tryptophan, a natural amino acid precursor of serotonin may stabilize some patients sleep rhythms. REM sleep is also impacted upon by acetylcholine which causes the characteristic physiologic changes of REM sleep; activating the cerebral cortex leading to EEG waves similar to that in the awake state, stimulating cranial nerves III and VI leading to the characteristic REM of REM sleep,

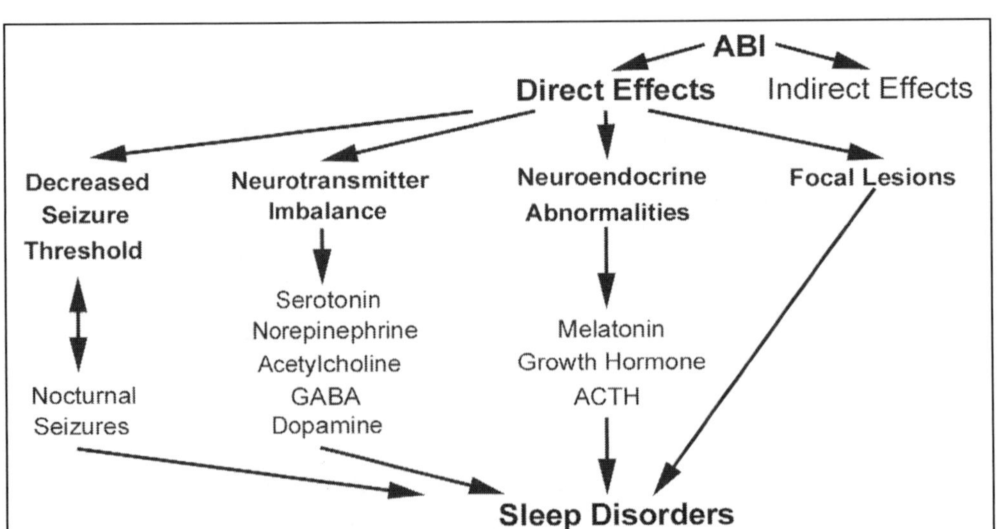

Figure 2. Pathophysiology of sleep disorders post head injury – direct.

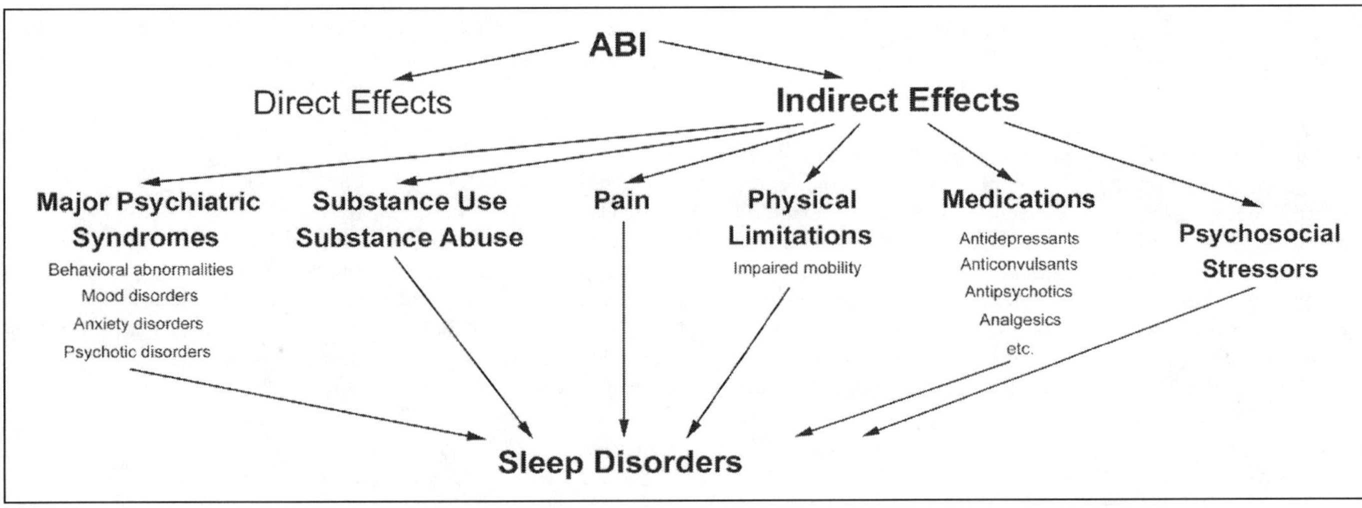

Figure 3. Pathophysiology of sleep disorders post head injury – indirect.

and decreasing muscle tone via the bulbar inhibitor reticular formation.[33]

The exact mechanism of action of GABA and dopamine on sleep phases is unclear even though they are known to lead to sleep pattern disturbances.

Melatonin, a neurohormone, seems to have a rhythm setting function closely related to darkness and light. It is secreted in large quantities during sleep from the pineal gland. Other hormones secreted from the anterior pituitarty also seem to have a circadian pattern. Growth hormone is linked to actual sleep time, whereas ACTH is linked to clock time and requires up to two weeks to readjust if sleep times are altered.[31] Thus, damage to the anterior pituitary or close to it could lead to sleep disturbances.

Direct damage of the brain can also lead to specific sleep disorders (Fig. 2), especially post-traumatic narcolepsy,[38-40] delayed sleep phase disorder,[41] and Kleine-Levin syndrome.[27] Rarer conditions such as paroxysmal nocturnal hemidystonia during NREM sleep have also been reported.[42] Facial trauma has been shown to produce obstructive sleep apnea.[43] Additionally, the occurrence of post-traumatic central sleep apnea has also been reported, presumably directly resulting from brain trauma.[44] However, in head injury patients without facial or cervical fractures, 40% of those complaining of excessive daytime sleepiness were found to have sleep apnea of mixed[5] origin.[45,46]

All of the above pre-supposes actual damage to the brain in ABI, either focal or diffuse, that leads via the mechanisms mentioned to altered sleep patterns. However, in ABI patients, sleep disturbances can also be produced by indirect effects as shown in Figure 3. Brain injury can lead to major psychiatric and neurologic syndromes including chronic pain that have as a consequence sleep disturbances. Impaired mobility and the ability to turn may disturb sleep as patients have to consciously wake and make the effort to turn in the middle of the night which can fragment their sleep leading to non-refreshing sleep and excessive daytime sleepiness. Medication prescribed for a variety of conditions may also disturb sleep; antipsychotics through dopamine blockade, antidepressants through REM suppression, anticonvulsants through increasing slow wave sleep and pain medication such as morpine through REM suppression. Finally psychosocial stressors may cause significant insomnia.

Anxiety and depression are all common sequelae of brain trauma[47,48] and are major causes of sleep disturbances.[29] Depression is the most common sequelae of traumatic brain injury,[47,49] often persisting for years after the injury. The frequency of post-head injury depression is reported to vary in the range from 10 to 60%,[50] but a more recent study using structured psychiatric interviews and defined diagnostic criteria has found an incidence of 27% of patients meeting the criteria for major depression. This is still well above the incidence of 4% reported in the general population.[51] Anxiety symptoms, concentration difficulty and thought disorders are also likely to accompany this depression.[52] Depression, by itself, has profound effects on sleep quality and duration. Symptoms such as difficulty falling asleep, early awakening, reduced amounts of slow wave sleep and reduced REM latency are all typically observed.[53]

Sleep Abnormalities Noted in ABI

Sleep disturbances in the ABI population frequently presents as excessive daytime sleepiness.[23,24] Difficulty falling asleep, a decrease in the amount of restorative (slow wave) sleep and rapid eye movement (REM) sleep, an increase in the number of awakenings during the NREM and REM sleep period, and decreased sleep duration can all occur.[54-57] Patients may also present with symptomatology characteristic of sleep apnea,[44-46] narcolepsy,[38-40] delayed sleep phase disorder,[41] Kleine-Levin syndrome,[42] or rarer conditions such as paroxysmal nocturnal hemidystonia during NREM.[43] Sleep problems can and more often than not present itself as daytime sleepiness; 20% vs. 5% in the general population.[55,58] In adolescents and teenagers Segalowitz et al in 1995 noted a higher degree of sleep difficulty than normal controls. One-third of them have been estimated to have suffered a mild head injury. In children diagnosed with learning disabilities, this incidence increases to about 40%.[26]

The question also arises as to when ABI patients manifest their sleep disturbances. Cohen et al,[29] noted that 72.7% of a cohort of 22 inpatients with ABI manifested sleep disorders 3-5 months post injury while another 51.9% of 77 patients had sleep complaints even after 29.5 months since the ABI. A distinction was present in that early post injury patients had difficulty initiating and maintaining sleep, while late post-injury patients had a preponderance of excessive somnolence during the day. This presence of sleep problems years after the ABI has been corroborated by Prigatano et al.[56]

Management

Management of these unfortunate individuals is two fold. First, make a diagnosis by keeping in mind that sleep disorders more often than not occur in ABI patients. Therefore, inquire and investigate for the possibility of such conditions. Second, an attempt should be made to manage the major sequela of the ABI especially any concomitant depression, anxiety, psychosis, behavioral difficulties, pain, mobility issues, medication issues or psychosocial factors prior to initiating any treatment of the sleep disorder itself as all of the above can adversely affect sleep. Thus, attempting to manage the sleep problems prior to optimizing the other manifestations will generally be futile. Notwithstanding, if a patient is diagnosed as having a major sleep disorder such as sleep apnea, consideration should be given to treating this condition as a first instance as without improvement in this condition, the other manifestation may never completely resolve and rehabilitation becomes delayed and frustrating not only to the patient but also to caregivers and health care professionals. It has been recognized that the cognitive and personality disturbances resulting from severe head injuries in young adults detrimentally impair their psychosocial functioning and their ability to become gainfully employed.[59]

Treating ABI patient with sleep disturbances or disorders is therefore a multidisciplinary and multimodal approach. The sequelae of ABI should be managed with rehabilitation program, therapies and medications, while sleep disorders need to be treated as necessary with pharmacological and/or non-pharmacological approaches depending on the presentation and complexity. It is beyond the scope of this chapter to discuss in detail the numerous approaches available to managing either of these conditions.

All programs that attempt to rehabilitate ABI patients need to be individualized and should have a combination of medical, cognitive, behavioral and supportive therapies. Basic steps should include initially a thorough case review (medical, educational, social), and assessment of current and past level of functioning (competency-treatment and financial, daily living skills, cognitive ability, social skills, behavioral issues, family and social relationships, recreational activities, employment interests and abilities). Follow up management will include development of strategies, both medical and therapeutic, to deal effectively with aggression and agitation (teach alternative acceptable behavior, develop contingencies, medications), disinhibition and sexual inappropriateness (develop acceptable outlets, recreational workshops), passivity (address fatigue, if any, reinforce social groups and activities, medications), attentional disorders (instructions, reinforcements and prompts, medications), memory deficits (address methods to both improve memory such as the spaced-retrieval technique and cope with deficits as with daily planners), psychiatric syndromes of depression, psychosis, anxiety (medications), poor patient-family interaction (encourage social skills training, family therapy), finances (obtain financial assistance from appropriate sources) and housing (help patient get and keep accommodation). Last but not least, sleep disorders should be identified and managed appropriately.

Cognitive recovery can accompany a normalization of the sleep pattern,[60] and the restoration of normal REM sleep has been shown to generally improve higher mental function in this patient population.[57] In turn, this improved sleep can facilitate physical and psychological recovery.[61] Screening for such disorders, however, is crucial as the patient may not always be aware of their sleep disorder.[62] Unfortunately, treatment of more the overt sleep disorder symptoms does not always ameliorate the daytime sleepiness.[40]

Conclusion

In conclusion, in rehabilitation of the ABI patient, not only must the patient overcome a number of cognitive, psychiatric and physical difficulties, but also sleep difficulties which are often difficult to treat. Notwithstanding, an attempt is certainly required as without it it is difficult to provide a better psychosocial outcome for these patients following an ABI.

References

1. Seyone C, Shapiro C. What happens after a blow to the head. Can J Diagnosis 1997; May:87-97.
2. Capruso DX, Levin HS. Neuropsychiatric aspects of head trauma. In: Kaplan HI, Sadock BJ, eds. Comprehensive textbook of psychiatry, 6th ed. Baltimore: Williams & Wilkins, 1995:207-220.
3. McAllister TW. Neuropsychiatric sequelae of head injuries. Psychiatr Clin North Am 1992; 15(2):395-413.
4. Roberts GW, Leigh PN, Weinberger DR. Neuropsychiatric disorders. Toronto:Wolfe, 1993:5.1-5.16.
5. Frazee JG. Head trauma. emergency medicine clinics of north america 1986; 4(4):859-874.
6. Ontario Brain Injury Association. 1996.
7. Lishman WA. Organic psychiatry: the psychological consequences of Cerebral Disorders. 2nd ed. London: Blackwell, 1987.
8. Prigatano GP, Fordyce D, Zeiner HK et al. Neuropsychological rehabilitation after closed head injury in young adults. J Neurol Neurosurg Psychr 1984; 47:505-513.
9. Bootzin RR, Engle-Friedman M. The assessment of insomnia. Behavioral Assessment 1981; 3:107-126.
10. Lacks P, Morin CM. Recent advances in the assessment and treatment of insomnia. J Consult Clin Psych 1992; 60(4):586-594.
11. Czeisler CA, Richardson GS. Detection and Assessment of Insomnia. Clinical Therapeutics 1991; 13(6):663-679.
12. Kales A, Soldatos CR, Kales JD. Taking a sleep history. Am Fam Phys 1980. 22:101-108
13. Nicassio PM, Mendlowitz DR, Fussell JJ et al. The phenomenology of the pre-sleep state: the development of the pre-sleep arousal scale. J Beh Ther Res 1985; 23:263-271.
14. Spielman AJ, Saskin P, Thorpy MJ. Treatment of chronic insomnia by restriction of time in bed. Sleep 1987; 10:45-56.
15. Lacks P, Bertelson AD, Gan L et al. The effectiveness of three behavioral treatments for different degrees of sleep onset insomnia. Behaviour Ther 1983 14:593-605.
16. Lacks P. Behavioral treatment of persistent insomnia. 1987. New York, NY: Pergamon
17. Lillie JK, Rosenberg RP. Behavioral treatment of insomnia. Progress in Behavioral Modification 1990; 25:152-177.
18. Spitzer RL, Williams JBW, Gibbon M et al. Structured Clinical Interview for DSM-IV (SCID) User's guide. Washington, DC, American Psychiatric Press 1997.
19. Levin HS, High WM, Goethe KE. The Neurobehavioral Rating Scale: assessment of the behavioral sequelae of head injury by the clinician. J Neurol Neurosurg Psychiatry 1987; 50:183-193.
20. Kay SR, Fiszbein A, Opler LA. The positive and negative syndrome scale (PANSS) for schizophrenia. Schizophr Bull 1987; 13:261-276.
21. Yudofsky SC, Silver JM, Jackson W et al. The Overt Aggression Scale for the objective rating of verbal and physical aggression. Am J Psychiatry 1986; 143:35-39.
22. Coleman RM, Roffwarg HP, Kennedy SJ et al. Sleep-wake disorders based on a polysomnographic diagnosis: a national cooperative study. JAMA 1982; 247:997-1003.
23. Binder L. Persisting symptoms after mild head injury: a review of the post-concussive syndrome. J Clin Exp Neuropsych 1986; 8:323-346.

24. Evans RW. The post-concussive syndrome and the sequelae of mild head injury. Neurological Clinics 1992; 10:815-847.
25. Beetar JT, Guilmette TJ, Sparadeo FR. Sleep and pain complaints in symptomatic traumatic brain injury and neurologic populations. Arch Phys Med Rehabil 1996; 77:1298-1302.
26. Segalowitz SJ, Lawson S. Subtle symptoms associated with self-reported mild head injury. J Learn Disabil 1995; 28(5):309-319.
27. Will RG, Young JPR, Thomas DJ. Klein-Levin syndrome: report of two cases with onset of symptoms precipitated by head trauma. Br J Psychiatry 1988; 152:410-412.
28. Guilleminault C, Faull KF, Miles L, van den Hoed J. Posttraumatic excessive daytime sleepiness: a review of 20 patients. Neurology 1983; 33:1584-9.
29. Cohen M, Oksenberg A, Snir D et al. Temporally related changes of sleep complaints in traumatic brain injured patients. J Neurol Neurosurg Psychiatry 1992; 55:313-5.
30. Dement WC, Seidel W, Carskadon M. Issues in the diagnosis and treatment of insomnia. In: Hindmarch I, Ott H, Roth T, eds. Psychopharmacology Supplementum I. Berlin: Springer-Verlag 1984; 11-43.
31. Everell DE, Avorn J, Baker MW. Clinical decision-making in the evaluation and treatment of insomnia. Am J Med 1990; 89:357-362.
32. Giubilei F, Formisano R, Fiorini M et al. Sleep abnormalities in traumatic apallic syndrome. J Neurol Neurosurg Psychiatry 1995; 58:484-486.
33. Reimer M. Sleep pattern disturbances related to neurological dysfunction. Axon 1989; 10(3):65-68.
34. Beutler LE, Ware JC, Karacan I et al. Differentiating psychological characteristics of patients with sleep apnea and narcolepsy. Sleep 1981; 4:39-47.
35. Clifton GL, Ziegler MG, Grossman RG. Circulating catecholamine and sympathetic activity after head injury. Neurosurgery 1981; 8:10-14.
36. Hamill RW, Woolf PD, McDonald JV et al. Catecholamines predict outcome in traumatic brain injury. Ann Neurol 1987; 21:438-443.
37. Morrison JH, Molliver ME, Grzanna R. Noradrenergic innervation of cerebral cortex: widespread effects of local cortical lesions. Science 1979; 205:313-316.
38. Good JL, Barry E, Fishman PS. Posttraumatic narcolepsy: the complete syndrome with tissue typing. J Neurosurg 1989; 71:765-767.
39. Maccario M, Ruggles KH, Meriwether MW. Post-traumatic narcolepsy. Military Medicine 1987; 152(7):370-371.
40. Wing Y-K, Lee S, Chiu HFK et al. A patient with coexisting narcolepsy and morbid jealousy showing a favorable response to fluoxetine. Postgrad Med J 1994; 70:34-36.
41. Patten SB, Lauderdale WM. Delayed sleep phase disorder after traumatic brain injury. J Am Acad Child Adolesc Psychiatry 1992; 31(1):100-102.
42. Biary N, Singh B, Bahou Y et al. Posttraumatic paroxysmal nocturnal hemidystonia. Movement Disorders 1994; 9(1):98-99.
43. Mossaz CF, Richter M, Oehlrich S. Case report: surgical-orthodontic management of posttraumatic obstructive sleep apnea syndrome. The Angle Orthodontist 1997; 67(2):155-161.
44. Quera-Salva MA, Guilleminault C. Post-traumatic central sleep apnea in a child. J Pediatrics 1987; 110(6):906-909.
45. Guilleminault C. Clinical features and evaluation of obstructive sleep apnea. In: Kryger MH, Roth T, Dement WC, eds. Principles and Practice of Sleep Medicine, 2nd ed. Philadelphia: WB Saunders Company, 1994:667-677.
46. Guilleminault C, Faull KF, Miles L et al. Posttraumatic excessive daytime sleepiness: a review of 20 patients. Neurology 1983. 33:1584-9.
47. Fedoroff JP, Starkstein SE, Forrester AW et al. Depression in patients with acute traumatic brain injury. Am J Psychiatry 1992; 149:918-923.
48. Jorge RE, Robinson RG, Arndt S. Are there symptoms that are specific for depressed mood in patients with traumatic brain injury? J Nervous Mental Disorders 1993; 181(2):91-99.
49. Kinsella G, Moran C, Ford B et al. Emotional disorder and its assessment within the severe head injured population. Psychol Med 1988; 18:57-63.
50. O'Shanick GJ. Neuropsychiatric complications in head injury. Adv Psychosom Med 1986; 16:173-193.
51. Judd LL, Paulus MP, Wells KB et al. Socioeconomic burden of sybsyndromal depressive symptoms and major depression in a sample of the general population. Am J Psychiatry 1996; 153(11):1411-1417.
52. Jorge RE, Robinson RG, Arndt S. Are there symptoms that are specific for depressed mood in patients with traumatic brain injury? J Nerv Mental Disorders 1993; 181(2):91-99.
53. Benca RM. Mood Disorders. In: Kryger MH, Roth T, Dement WC, eds. Principles and Practice of Sleep Medicine. 2nd ed. WB Saunders Company, Philadelphia, 1994:899-913.
54. Hicks JA, Shapiro CM. Sleep architecture of head injured patients. In: Cognitive Rehabilitation: Advances in the Rehabilitation of Acute and Age-Related Brain Disorders, 5th Rotman Research Institute Conference. 1995.
55. Perlis ML, Artiola L, Giles DE. Sleep complaints in chronic postconcussion syndrome. Perceptual and motor skills 1997; 84:595-599.
56. Prigatano GP, Stahl ML, Orr WC et al. Sleep and dreaming disturbances in closed head injury patients. J Neurol Neurosurg Psychiatry 1982 45:78-80.
57. Ron S, Algom D, Hary D et al. Time-related changes in the distribution of sleep stages in brain injured patients. EEG Clin Neurophysiol 1980; 48: 432-41.
58. Roth T, Roehrs TA, Carskadon MA et al. Daytime sleepiness and alertness. In: Kryger MH, Roth T, Dement WC, eds. Principles and Practice of Sleep Medicine. 2nd ed. Philadelphia: WB Saunders Company, 1994:40-49.
59. Bond MR, Brooks DN. Understanding the process of recovery as a basis for the investigation of rehabilitation for the brain injured. Scand J Rehab Med 1976; 8:127-33.
60. Alexandre L, Nertempi P, Farinello C. Sleep alterations during post-traumatic coma as a possible predictor of cognitive deficits. Acta Neursurgica 1997 Suppl 28; 188-192.
61. Shapiro CM, Devins GM, Hussain MRG. Sleep problems in patients with medical illness. Br Med J 1993; 306:1532-1535.
62. Askenasy JJM, Rahamani L. Neuropsycho-social rehabilitation of head injury. Am J Physical Med 1988; 66(6):315-327.

CHAPTER 24

Psychopharmacological Management of Restless Legs Syndrome and Periodic Limb Movements in Sleep

Raed Hawa, Leonid Kayumov, Alan Lowe and Colin M. Shapiro

Restless Legs Syndrome

Restless leg syndrome (RLS) was described as a distinctive sensory-motor phenomenon by Willis in 1672. Ekbom is credited for giving it its current name and identifying its broad clinical presentation.[7,8]

Clinical Features

Restless legs syndrome (RLS) together with comorbid periodic limb movements in sleep (PLMs) is the third leading cause of insomnia after psychiatric disorders and drug abuse. It affects 5-15% of the general population.[12,21] It may start at any age but there is a demonstrated increase in the elderly – up to 25-30%.[14] The International RLS study group[31] developed the following criteria for diagnosis RLS:

1. an urge to move the limbs with uncomfortable sensation (paresthesia or dysesthesia);
2. motor restlessness;
3. worsening of symptoms at rest and relieved by activity;
4. worsening of symptoms in the evening and early at night.

Associated features include: difficulty in sleep initiation, involuntary periodic limb movements during sleep and/or wakefulness, a chronic progressive clinical course with occasional remissions and positive family history.

Diagnosis

The diagnosis of RLS is based on history taking of the patient and the patient's partner. There is no laboratory test to confirm the diagnosis. Many patients describe a sense of formication-feeling as if ants are crawling on their legs. They may say that trying to push their feet into the floor provides some relief. Others use terms such as creeping, tingling, burning or aching. Polysomnographic studies can be useful to exclude sleep apnea or other sleep related disorders. More specific tests such as Multiple Sleep Latency Test (MSLT) are helpful if the clinical presentation includes excessive daytime sleepiness or if there are concerns regarding driving or operating machinery.[19]

Differential Diagnosis

It is important to perform a full medical history and physical examination to exclude secondary causes of the syndrome, such as pregnancy (which affects 12-20% of pregnant women), anemia (ferritin, iron, folate and B12 deficiency states have been associated with RLS), uremia (50% of hemodialysis patients suffer from RLS), and diabetes.[10,15,17] Suspicion of neuropathy or nerve root damage should be suggestive of performing appropriate EMG or nerve conduction studies.[22] Medications such as Dilantin, Amitriptyline, Paroxetine, Lithium and neuroleptics have been implicated to cause or worsen RLS.[15,17] Alcohol, caffeine, and tobacco may aggravate symptoms of RLS.[20]

Periodic Limb Movement Disorder (PLMD)

PLMs are stereotypic, repetitive movements of the leg that occur primarily in the nonREM sleep. These movements were first described by Symonds in 1953. They were first recorded polygraphically by Lugaresi et al in 1965. PLMs refer to brief muscle contractions that occur every 5-90 seconds, each lasting 0.5-5 seconds. Dorsiflexion of the big toe, ankle, knee and some times the hip are involved. The arms could be implicated at times.

Asking the patient and bed partner often yields incidence of kicking, restlessness, cold feet, disrupted beds' sheets, insomnia or excessive daytime sleepiness. The patient usually does not complain of these movements. However, the bed partner frequently complains of the kicking and restlessness. Sleep studies of PLMD and RLS indicate an increase in stage 1 and 2, decrease in SWS and REM sleep, with frequent arousals and awakenings.[16] Examples of the typical patterns of leg jerks during sleep are shown in Figure 1.

PLMD increase in prevalence with age – about 30% of the population older than 50 years is affected.[18] It is frequently associated with RLS. In fact, it is reported that 80% of RLS patients suffer from PLMD and 30% of PLMD patients have RLS.[9]

These conditions, RLS and PLMD, are extremely distressing for patients. The symptoms may fluctuate dramatically. One high powered executive we treated purchased an actigraph and plotted his PLMs for several months. He observed that his worst night in terms of leg twitches was Sunday night. We concluded that the positive stress of anticipating money making day in the office on Monday was a trigger for his PLMs and poor sleep. He rearranged his schedule to have no important meetings on Monday mornings and found a dramatic resolution of his PLMs rate on Sunday nights. A few years later he exited from a stressful marriage and found an overall decline in his PLMs rate (after an initial increase at the time of the separation and divorce).

Severity of PLMD is assessed in our clinic according to the criteria shown in Table 1.

Sleep and Sleep Disorders: A Neuropsychopharmacological Approach, edited by Malcolm Lader, Daniel P. Cardinali and S.R. Pandi-Perumal.

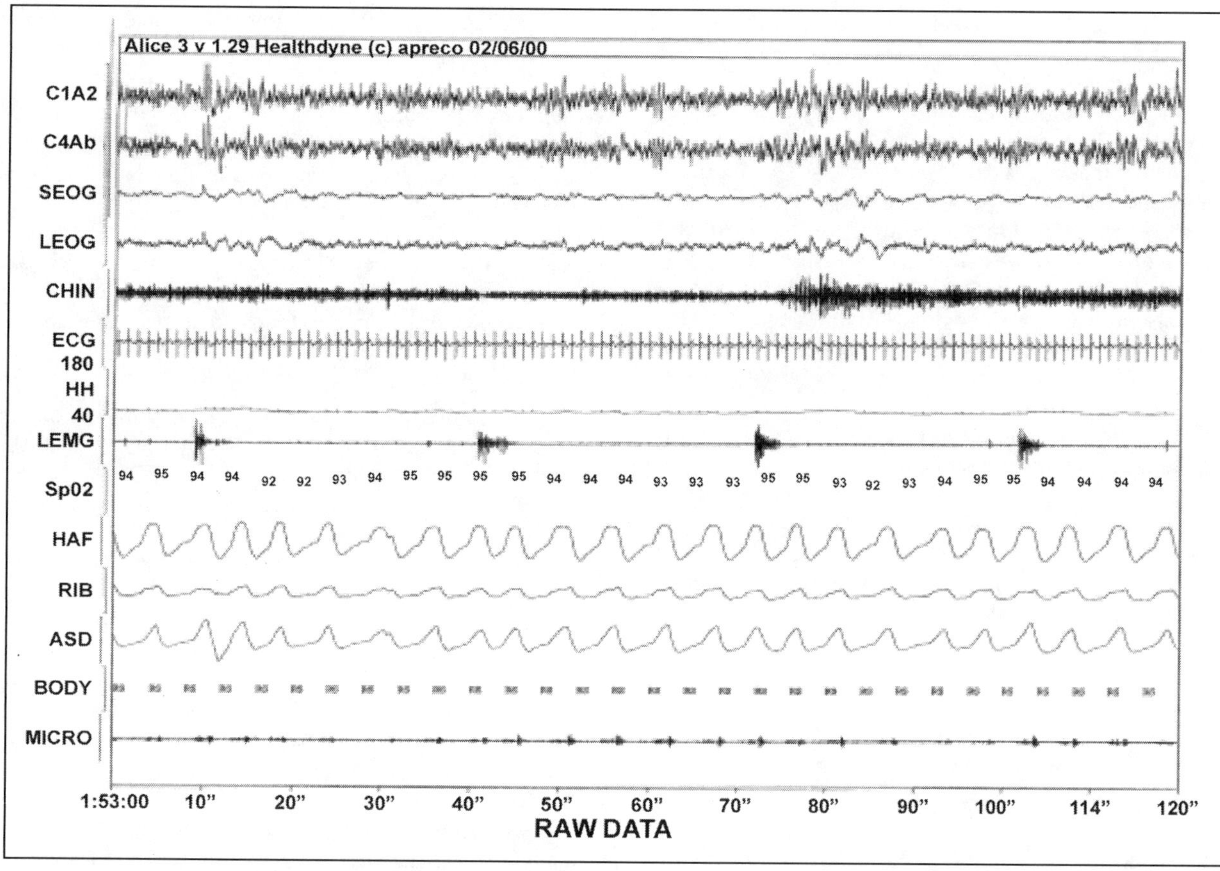

Figure 1. Periodic leg movements in stage 2 sleep (LEMG lead) with the periodicity of approximately 30 sec.

Treatment

1. Treat underlying conditions. It goes without saying that treatment of any medical condition contributing to RLS/PLMD is essential. The choice of what treatment modality should follow the rules of any proper medical management including weighing the benefits and outlining the risks.
2. Patients who present with RLS/PLMD have learnt what usually relieve their symptoms to some extent prior to presenting to a medical office. For some, it is walking or exercising, for others, it is having a very hot or very cold bath. Avoiding certain stimulants such as tobacco, alcohol and caffeinated beverages is helpful in certain cases. There has been suggestion in the literature of diet supplementation, vitamins and minerals to which there is no solid scientific support.
3. Certain medications should be avoided, if possible because of their propensity to worsen symptoms of RLS. These medications include some calcium channel blockers, phenytoin, anti-nausea medications and allergy/cold medications. The use of certain antipsychotics (Haldol and Phenothiazines) and antidepressants (SSRI and TCA) could increase RLS symptoms.
4. Maintaining good sleep hygiene practices such as regular sleep and waking times, comfortable environment and appropriate number of sleeping hours is helpful to some degree.
5. Pharmacological management.

Active involvement of the patient in the decision making regarding available pharmacological treatment is highly desirable. Medications that are available have variable side effects and treatment response rate. The following drugs have been suggested as possible therapies:

Dopaminergic Agents

Dopamine Precursors

Sinemet (Carbidopa/Levo-dopa in 1 to 4 ratio) and Sinemet CR (Sustained release carbidopa/levo-dopa) are used in RLS patients to improve sleep quality, decrease PLMs and sleep latencies. The typical dose is 25/100 to 100/400 of carbidopa/levo-dopa taken half an hour to one hour prior to bedtime. Another dose might be needed in the middle of the night. The carbidopa component helps preventing the peripheral metabolism of levo-dopa ensuring more brain delivery. Sinemet has been reported to benefit up to 80% of RLS patients over 2 to 3 year period.[9]

Table 1. Criteria for Evaluating the Severity of PLMD

	PLM Index	Sleep Latency* (in Minutes)	Insomnia or EDS
Mild	5-25	10-13	Mild
Moderate	25-50	5-9	Moderate
Severe	> 50	< 5	Severe

Sinemet can be started at a half pill, 30 to 60 minutes before bedtime. It can be increased up to 3 tablets before bed time and will last about 5-6 hours. If early morning awakenings occur, another half a pill can be added. This medication generally works better for night time RLS than for daytime RLS symptoms. The slow release tablet comes to peak action in two hours so it is often combined with short acting Sinemet to get relief within 30 minutes. Sinemet CR can be used in the morning for sustained daytime relief. Side effects include nausea, lightheadedness and headaches. The side effects are temporary and usually subside within few days of treatment. Ingesting the medication with food might help alleviate the nausea. Patients should be cautioned about avoiding high protein diets that might interfere with absorption of Sinemet.

Patient should be also made aware about the possibility of developing rebound and augmentation. Augmentation refers to the development of increasingly worsened restlessness earlier in the day. This usually occurs in patients on higher doses of Sinemet and those who have severe pretreatment symptoms. Treatment dictates reduction or discontinuation of Sinemet and substitution of another medication. Although this treatment is highly effective, we usually reserve it as a "last resort".

Dopamine Agonists

Pergolide (Permax) which is a mixed D1 and D2 agonist. Pergolide is usually started at 0.05 to 0.1 mg taken 1-2 hours before bedtime. Usually the dose is increased by 0.05 mg per day to avoid hypotension. Typical doses of Pergolide are 0.1-0.6 mg. Side effects of Pergolide include nasal stuffiness, nausea, hallucinations and constipation. There has been a recent article cautioning of valvular heart disease associated with Pergolide.[23]

In a head to head comparison, Pergolide versus Sinemet in RLS patients, the former was found to be as efficacious with lesser side effects.[27] This medication may work better than Sinemet for daytime RLS symptoms and for more severe RLS. Pergolide has been shown to help patients up to 73% who failed Sinemet. It has a longer duration of action of up to 10-12 hours. This drug does not usually cause significant rebound or augmentation. However some patients find the side effects intolerable.

Bromocriptine is a more typical D2 receptor agonist.[3] The typical dose is 5 to 7.5 mg. Bromocriptine is usually started at 1.75 to 21.5 mg before bed time and increased with caution over 1-2 weeks period to decrease possibility of side effects. The side effects include arterial hypotension, nausea and vertigo. Bromocriptine has been used for treatment of idiopathic RLS as well as RLS with narcolepsy and PLMD.[5] Both Pergolide and Bromocriptine are less likely to cause augmentation.

Newer Dopamine Agonists

Mirapex (Pramixole)

This medication has a more complete and specific binding to the D2 receptors. It is a full dopamine agonist. Its half-life is 8-12 hours. Mirapex is excreted renally. The side effect profile is similar to other dopamine agonists including nasal stuffiness, leg edema, insomnia, sleepiness, fatigue, malaise and nausea. The side effects generally occur mostly with the higher doses of the drug. Cimetidine can increase the blood levels of Mirapex if prescribed concomittantly. Mirapex is started at 0.125 mg 2 hours before sleep and increased slowly by no more than 0.125 mg every 3-5 days up to 2 mg a day. Most patients are usually controlled by doses between 0.25 to 2 mg.[15,17,26]

Requip (Ropinirole)

This is a more specific dopamine agonist similar to Mirapex.[2,24] Its half-life is 4-6 hours and is hepatically metabolized. The starting dose should be 0.25 mg which can be given twice daily with an earlier dose in late afternoon or early evening. The dose can be increased by 0.25 mg every 3-5 days up to 2 mg. If necessary, the dose can be further increased up to 4 mg. Experience is limited with the use of this medication for the treatment of RLS/PLMD.

Facilitating Agents

Seligiline Hidrochloride often used in Parkinson's disease has been shown to be helpful.[8A] It is a MAOI-B with very little propensity to show the food reactions of MAOI-A's. We have found that a regimen of 2.5 mg morning and mid-day for 2 weeks, increased to 5 mg bid for 2 weeks and then further increased to 10 mg bid for 2 weeks will establish an effective dosage level and induce very few side effects. Many patients describe a benefit of increased alertness. The main drawback is the cost of the drug. The dose which the patients find the most improvement is usually selected and maintained. No dietary restriction is required.

There are reports of using clonidine (0.1-1 mg) and baclofen (20-40 mg) for treatment of RLS symptoms.[6,32]

Benzodiazepines

Clonazepam at doses of 0.5 mg titrated up to 4 mg a day is the most widely used benzodiazepine for treatment of RLS and PLMS. Clonazepam has been shown in some studies to significantly decrease sleep latency, leg movements associated with arousals and improved sleep efficiency.[25] Other studies have stated that benzodiazepines lose their efficacy to treat PLMD after long term use.[4] Caution should be used in prescribing benzodiazepines for the elderly due to the possible confusion and excessive daytime drowsiness. Benzodiazepines should not be the first choice of treatment for RLS/PLMS in a drug dependent individual.

Opioids

Narcotics have been used to successfully treat RLS and PLMD. Oxycodone at doses ranging from 5 mg up to a maximum of 25 mg given in divided doses two hours prior to bed time, at bed time, and in the middle of the night, was efficacious in reducing PLMs per hour of sleep, number of arousals per hour of sleep, and improving subjective ratings.[30]

Propoxyphene in doses ranging from 130 to 520 mg a day was found to improve subjective ratings on motor activity but not PLMs.[11] Other narcotics have been used including codeine (50 mg to 240 mg) and methadone (5 mg to 30 mg). Drawbacks to the use of these medications include constipation, sedation, and development of tolerance.

Anti-Seizure Medications

Neurontin or Gabapentin

This is the best studied of the anti-convulsants in treating RLS. The dose of neurontin varies from 300 to 2000 mg per day with the average dose being 800 to 900 mg.[1] Neurontin appears to be helpful in decreasing painful sensations associated with the urge to move.

Tegretol or Carbamazepine

Doses between 100 to 300 mg per day were reported to be effective but the side effects are notorious. Younger patients seem to have a better response.[29,34]

Conclusions

Psychological (e.g., stress), physiological (e.g., pregnancy), pharmacological (e.g., caffeine and SSRIs) and pathological (e.g., renal failure and Parkinson's disease) factors all play a role in RLS and PLMD. Sleep specialists debate whether continuous long-term treatment for an undulating (but at times very disruptive condition) is merited. In "quiet" times, there may be the unnecessary ingestion of regularly prescribed medication. However, the disruption of sleep is unpredictable for most patients and an overall improvement of symptoms and quality of life may be achieved with continuous treatment. We generally favor the latter approach but have observed (especially with patients on Seligiline) that there may be several months of symptom free respite after inadvertently stopping the medication. This implies that some patients may be well on published regimens.

References

1. Adler CH. Treatment of restless legs syndrome with gabapentin. Clin Neuropharmacol 1997; 20:148-51.
2. Ahmed I. Ropinirole in restless leg syndrome. Mo Med 2002; 99:500-1.
3. Becker PM, Jamieson AO, Brown WD. Dopaminergic agents in restless legs syndrome and periodic limb movements of sleep: Response and complications of extended treatment in 49 cases. Sleep 1993; 16:713-6.
4. Boghen D, Lamothe L, Elie R et al. The treatment of the restless leg syndrome withclonazepam: A prospective controlled study. Can J Neurol Sci 1986; 13:245-247.
5. Boivin DB, Lorrain D, Montplaisir J. Effects of bromocriptine on periodic limb movements in human narcolepsy. Neurology 1993; 43:2134-6.
6. Brown LK, Heffner JE, Obbens EA. Transverse myelitis associated with restless legs syndrome and periodic movements of sleep responsive to an oral dopaminergic agent but not to intrathecal baclofen. Sleep 2000; 23:591-4.
7. Ekbom KA. Restless legs: A clinical study. Acta Med Scand 1945; 158(suppl):1-122.
8. Ekbom KA. Restless legs syndrome. Neurology 1960; 10:868-73.

8A. Grewal M, Hawa R, Shapiro CM. Treatment of periodic limb movements in sleep with selegiline HCl. Mov Disord 2002; 17(2):398-401.

9. Hening W, Allen R, Earley C et al. The treatment of restless legs syndrome and periodic limb movement disorder. Sleep 1999; 22:970-99.
10. Hui DS, Wong TY, Li TS et al. Prevalence of sleep disturbances in chinese patients with end stage renal failure on maintenance hemodialysis. Med Sci Monit 2002; 8:CR331-6.
11. Kaplan PW, Allen RP, Buchholz DW et al. A double-blind, placebo-controlled study of the treatment of periodic limb movements in sleep using carbidopa/levodopa and propoxyphene. Sleep1993; 16:717-23.
12. Lavigne GJ, Montplaisir JY. Restless legs syndrome and sleep bruxism: Prevalence and association among Canadians. Sleep 1994; 17:739-43.
13. Lugaresi E, Tassinari CA, Coccagna G et al. Particularities cliniques et polygraphiques du syndrome d'impatience des membres inferieurs. Rev neurol 1965; 113:545-555.
14. Milligan SA, Chesson AL. Restless legs syndrome in the older adult: Diagnosis and management. Drugs Aging 2002; 19:741-51.
15. Montplaisir J, Nicolas A, Godbout R et al. Restless leg syndrome and periodic limb movement disorder. In: Kryger M, Roth T, Dement IIIrd WC, eds. Principles and practice of sleep medicine. WB Sounders Co., 2000:742-752.
16. Montplaisir J, Boucher S, Gosselin A et al. Persistence of repetitive EEG arousals (K-alpha complexes) in RLS patients treated with L-DOPA. Sleep 1996; 19:196-9.
17. Montplaisir J, Denesle R, Petit D. Pramipexole in the treatment of restless legs syndrome: A follow-up study. Eur J Neurol 2000; 1(Suppl):27-31.
18. Mosko SS, Dickel MJ, Paul T et al. Sleep apnea and sleep-related periodic leg movements in community resident seniors. J Am Geriatr Soc 1988; 36:502-8.
19. Nicolas A, Lesperance P, Montplaisir J. Is excessive daytime sleepiness with periodic leg movements during sleep a specific diagnostic category? Eur Neurol 1998; 40:22-6.
20. Paulson GW. Restless legs syndrome. How to provide symptom relief with drug and non drug therapies. Geriatrics 2000; 55:35-48.
21. Phillips B, Young T, Finn L et al. Epidemiology of restless legs symptoms in adults. Arch Intern Med 2000; 160:2137-2141.
22. Polydefkis M, Allen RP, Hauer P et al. Subclinical sensory neuropathy in late-onset restless legs syndrome. Neurology 2000; 55:1115-21.
23. Pritchett AM, Morrison JF, Edwards WD et al. Valvular heart disease in patients taking pergolide. Mayo Clin Proc 2002; 77:1280-6.
24. Saletu B, Gruber G, Saletu M et al. Sleep laboratory studies in restless legs syndrome patients as compared with normals and acute effects of ropinirole. 1. Findings on objective and subjective sleep and awakening quality. Neuropsychobiology 2000; 41:181-9.
25. Saletu M, Anderer P, Saletu-Zyhlarz G et al. Restless legs syndrome (RLS) and periodic limb movement disorder (PLMD): Acute placebo-controlled sleep laboratory studies with clonazepam. Eur Neuro psycho pharmacol 2001; 11:153-61.
26. Saletu M, Anderer P, Saletu-Zyhlarz G et al. Acute placebo-controlled sleep laboratory studies and clinical follow-up with pramipexole in restless legs syndrome. Eur Arch Psychiatry Clin Neurosci 2002; 252:185-94.
27. Staedt J, Wassmuth F, Ziemann U et al. Pergolide: Treatment of choice in restless legs syndrome (RLS) and nocturnal myoclonus syndrome (NMS). A double-blind randomized crossover trial of pergolide versus L-Dopa. J Neural Transm 1997; 104:461-8.
28. Symonds CP. Nocturnal myoclonus. J Neurol Neurosurg psychiatr 1953; 16:166-171.
29. Telstad W, Sorensen O, Larsen S et al. Treatment of the restless legs syndrome with carbamazepine: A double blind study. Br Med J (Clin Res Ed) 1984; 288:444-6.
30. Walters AS, Wagner ML, Hening WA et al. Successful treatment of the idiopathic restless legs syndrome in a randomized double-blind trial of oxycodone versus placebo. Sleep 1993; 16:327-32.
31. Walters AS. Toward a better definition of the restless legs syndrome. The International Restless Legs Syndrome Study Group. Mov Disord 1995; 10:634-42.
32. Wetter TC, Pollmacher T. Restless legs and periodic leg movements in sleep syndromes. J neurol 1997; 244(4Suppl1):S37-45.
33. Willis T. De animae brutorum. London: Wells and Scott;1672:339.
34. Zucconi M, Coccagna G, Petronelli R et al. Nocturnal myoclonus in restless legs syndrome effect of carbamazepine treatment. Funct Neurol 1989; 4:263-71.

CHAPTER 25

A Comparison of Visual Analog Scale and Categorical Ratings in Assessing the Patient's Estimate of Sleep Quality

Nava Zisapel, Ricardo Tarrasch and Moshe Laudon

Abstract

Study Objectives

Visual analog scales (VAS) and diary cards are used to measure changes in patients' sleep quality in clinical trials and practice. In this study we compared the 2 methods for assessing changes in sleep quality in patients with insomnia.

Methods

A VAS method of assessing sleep quality (QOS-means of questions 4 and 5 in the Leeds Sleep Evaluation Questionnaire) was compared with a standard categorical five-grade scale (5GS) in a single blind clinical trial involving 394 treated patients with insomnia aged 55 years and older who received placebo for 2 weeks, prolonged release melatonin 2 mg for 3 weeks and then placebo for 2 weeks for improvement of quality of sleep.

Results

Changes in VAS and 5GS scores were concordant and highly correlated. In patients with severely impaired sleep at baseline the 5GS showed a slightly greater sensitivity than the VAS when reporting a change whereas the contrary was found for patients with minimally impaired sleep.

Conclusion

The 5GS is comparable to VAS in capturing changes in sleep quality. The sensitivity of the two methods appeared similar with some advantage for the 5GS in patients with more severe baseline pathology.

Introduction

The growing prevalence of complaints concerning poor quality of sleep with age has prompted the development of methods to measure a patient's sleep quality state and its changes resulting from disease or therapy. The all-night polysomnographic (PSG) sleep studies assess the quantitative aspects of a patient's sleep and provide reliable measurements of latency to consolidated sleep, total sleep time, sleep efficiency, sleep fragmentation and architecture.[1] However, PSG variables do not predict subjective patient estimates of sleep satisfaction.[2] This lack of useful correlation between objective measurements and subjective complaints is a particularly intractable problem when measuring sleep quality in insomnia patients.[2] Based on the currently available data and critical reviews of the scientific literature, several expert groups have recommended that PSG should not be used routinely in the diagnosis of primary insomnia.[3] As a consequence, instruments (sleep diaries and sleep questionnaires) have been designed to assess the patient's sleep quality state in terms of perception, emotional symptoms or function, and activities of living. To make the data comparable across clinical trials, patients are asked to complete scales at different times, to yield serial data similar to that used in physical measurement, and the difference between 2 points in time is considered to represent the change.

The Leeds Sleep Evaluation Questionnaire (LSEQ) is a widely used standardized instrument for the measurement of sleep difficulties in clinical settings and like other standardized questionnaires for the same purpose consists of several well defined components.[4] It comprises ten individual visual analogue scales (100 mm) which have been shown by factor analysis to assess four discrete, independent domains of sleep and daytime behavior: Getting To Sleep (Questions 1-3), Quality Of Sleep (Questions 4,5), Awakening From Sleep (Questions7,8) and Behavior Following Wakening (Questions 8-10).[5]

As is the case with most widely used sleep quality questionnaires, such as the St. Mary's Hospital Sleep Questionnaire, the Pittsburgh Sleep quality index and the Karolinska Sleep impairment index, the Leeds Sleep Evaluation Questionnaire (LSEQ) is a retrospective instrument by which the patient is asked to recall events over a period of time (e.g., 4 weeks). In general, it has been assumed that change inferred from serial measurement is more accurate than the patient's retrospective perceptions of the change.[6] Thus, the LSEQ was used in a repetitive manner, yielding serial measurements for use in the field of drug evaluation and it is the only instrument specifically developed and validated to measure medication effects.[7]

Another way of assessing sleep aspects is by use of categorical scale rating in sleep diaries. These instruments are simple, user friendly and easy to use in general medical practice, allowing repeated accurate sampling of quality of sleep with enhanced reliability of the measure.[7] Four and five-grade severity rating scales are commonly used in medical practice and clinical trials[8,9] and it

Sleep and Sleep Disorders: A Neuropsychopharmacological Approach, edited by Malcolm Lader, Daniel P. Cardinali and S.R. Pandi-Perumal.

is generally accepted that any change noted by the patient on such scales is of definite clinical relevance.[10,11]

Intuitively, it seems that the VAS would allow for greater sensitivity in determining treatment effects than a categorical scale such as the 5GS, since there is a greater range of possible scores (0-100 for a 100 mm VAS). Furthermore, for analysis purposes, the 5GS is usually analyzed by nonparametric statistics, which might be expected to further reduce its sensitivity compared with a continuous scale. On the other hand, instruments such as the LSEQ contrasting current sleep and wake aspects with those at the time before the trial are retrospective. They have the advantage of being able to quickly summarize events occurring over a long period of time (3-4 weeks), but are at the same time prone to distortions that are inevitably introduced when collapsing and distilling descriptions. Memories are often incomplete or selective. Insomniacs have a natural tendency to focus on the worst experiences and perhaps to amplify their importance.[7] It could also be argued that the larger range of possible scores on the VAS might lead to greater variability in response, and could thereby result in reduced sensitivity for determining treatment effects.

There has been no comprehensive evaluation of the relative utility of the VAS and 5GS scales for assessing differences between treatments in an insomnia trial. In this paper we report the results of analyses that were undertaken to compare the 5GS and VAS in assessment of sleep quality. Both instruments were used concurrently in a large single-blind clinical trial which investigated the efficacy of prolonged release melatonin 2 mg. Each patient served as its own control.

Methods

Patients

A total of 488 men and women aged 55 years and over who met the DSM-IV criteria for primary insomnia were enrolled. Of these, 427 took treatment and 394 provided efficacy data on both 5GS and VAS measures.

All patients had at least a 6-month history of insomnia. Patients with anxiety, depression or dementia symptoms or suffering from other sleep disturbances or severe mental disorders were excluded.

Approval was obtained from the necessary regulatory body and Ethical Review Committee, and each patient who participated in the trials signed an informed consent. The study was conducted in accordance with the Declaration of Helsinki and conformed to Good Clinical Practice.

Study Design and Procedure

The study was single-blind design and was conducted in 1997 at 130 general practitioners clinics in France. Following 2 weeks placebo treatment (baseline) they received prolonged release melatonin (Circadin™) 2 mg tablets for 3 weeks (treatment) and finally placebo for 2 weeks (washout).

Patients were instructed to take the blinded study medication once daily in the evening between 9-11 PM.

Outcome Measures

Patients rated their sleep quality the previous night daily using a standard 5GS, and on the last 3 days of each period also using the LSEQ. In the latter, subjects were asked to contrast aspects of their current sleep and daytime functioning with those at the time before they joined the study. The French versions of the scales were used.

The mean of VAS questions 4 and 5 of the LSEQ related to the question "How would you compare the quality of sleep using the medication with nonmedicated (your usual) sleep?)" was taken as an index of the contrast for sleep quality QOS.[12] The scales for questions 4 and 5 were not bounded and the extreme descriptors for question 4 were "More restless than usual" at the left end and "More restful than usual" at the right end and for question 5 "More periods of wakefulness than usual" at the left end and "Fewer periods of wakefulness than usual" at the right end. QOS before the study is thus assumed to be 50 mm ("no change"). The results of the 3 nights at each treatment period were averaged to yield 2 contrasts for each patient (i.e., when changing from baseline to drug and when changing from drug to placebo washout).

Patients were instructed to regard the VAS as a continuous dimension (as opposed to an 'either/or' categorical scale), and to make a single vertical mark at the point on the VAS continuum corresponding to their current sleep with those at the time before they joined the study. The investigator hand-measured the distance from the left end of the scale to the mark and recorded the result to the nearest whole millimeter (range of scores: 0-100 mm). Measurements were hand-checked by a study monitor.

In the 5GS patients indicated that they had either "very bad", "bad", "fair", "good" or "very good" sleep (rated 1-5 respectively). Patients were not allowed to review their previous visual analog scale or serial scale measurements. The values corresponding to the nights of LSEQ recordings were averaged to yield 2 contrasts for each patient (i.e., when changing from placebo to drug and when changing from drug to placebo washout).

Statistical Analyses

The analysis evaluating the relationship between the 5GS and the VAS was based on an all-patients-treated approach that included all patients who recorded sleep quality after dosing on both the 5GS and VAS (N=394). The comparison of 5GS and VAS assessments was not the main purpose of the study and no power statements were made in this regard. This however does not invalidate the comparison of the 5GS and VAS presented here.

To assess the agreement between the amounts of change in sleep quality as measured by 5GS and the VAS, two approaches have been used. (1) Changes from baseline values on the 100 mm VAS are calculated on a 10-grade scale. Spearman rank correlations were computed for each patient's pair of 5GS and the VAS measures of change at treatment; (2) Changes from baseline values on the 100 mm VAS were calculated in mm. The pairs of 5GS and VAS measures of changes at treatment were compared by correlation analyses, with Pearson correlation coefficients.

To evaluate the sensitivity to change of the 2 measures, we used the concept of efficiency as defined by Anderson and Chernoff.[13] By this approach, efficiency was defined by the equation $E=d/SDd$, in which E is efficiency, d is the mean change in the measure for the group, and SDd is the SD of the change measures. Efficiency is not dependent on the sample size. Higher efficiency of a measure means more power to detect a change and thus better sensitivity to change.

To assess how well the 2 measures of change correlate, contingency tables were constructed of change scores for sleep quality. Pearson Chi-Square and Somers'd tests were employed to determine whether there is concordance between the two measures.[14] By t test analysis, we determined the probability that the number of individuals whose change assessments differed by 2 or more positions on the contingency table could have arisen by chance.

*Table 1. Changes from baseline in quality of sleep assessed by 5GS and VAS scales**

Group	Change from Baseline in 5GS Measurement†	Change from Baseline in VAS Measurement‡	P (t-Test)*
Total cohort	-0.24(-11.8)	-1.26(-18)	0.01
Baseline QON < 3 cohort	-0.75(-34)	-1.66(-24)	0.41
Baseline QON = 3 cohort	-0.2(-6.6)	-1.17(-16)	0.002
Baseline QON > 3 cohort	0.2(4.5)	-0.98(-14)	< 0.001

*Values c in parentheses are the percentage changes. Negative values indicate improvement. P values are derived from the percent difference data. †Changes are calculated from baseline values on a 5-grade scale (0, worst, 5, best) and are multiplied by -1 to correct for differences in direction between the VAS and 5GS. ‡Changes are calculated from baseline values expressed as a 10-point VAS scale.

A difference of 2 or more positions on the contingency table axes means a clinically substantial difference in patients' evaluations of the change that has occurred in their sleep quality by the two measures (e.g., "unchanged" by one while "much better or worse," by the other or "somewhat worse" by one while "somewhat better" by the other).

We also tested whether the concordance between the VAS and 5GS scores of change in sleep quality is related to the baseline sleep quality score. On the basis of their baseline quality of sleep (QON) 5GS scores (mean values rounded to closest integer), the patients enrolled in the study were separated into 3 cohorts: those with averaged baseline QON scores > 3 (unimpaired sleep quality), those with averaged scores of 3 (minimally impaired quality of sleep), and those with averaged scores of < 3 (severely impaired quality of sleep). The 5GS and VAS were compared in each of these groups separately as done for the entire cohort. Data were analyzed using the SAS statistical software package version 6.12 and SPSS version 10.1.0.

Results

Table 1 depicts the changes from baseline in quality of sleep upon treatment, presented as a change in grades on a 5-grade scale (5GS) or on a 10-grade scale (VAS), and as percent change (calculated from baseline ratings). The means by both measures form a similar pattern by which the greatest change found was in the cohort with the most severely impaired quality of sleep at baseline. However, while the VAS measurements were of a similar magnitude in cohorts with unimpaired, minimally and severely impaired sleep quality at baseline, the 5GS change measures were markedly higher in patients with impaired sleep quality at baseline. Mean VAS percentage changes were consistently higher than those of 5GS measures for subjects with minimal or no impairment in sleep at baseline but for patients with impaired sleep at baseline mean 5GS percentage changes were higher but not significantly.

Table 2 shows the Correlation Coefficients (Spearman) between the VAS and 5GS changes when VAS is expressed as a 10 point scale.

The actual changes in 5GS and VAS scores at treatment period were also compared using Pearson correlations. The changes from baseline in 5GS and VAS scores were found to be linearly correlated, with a Pearson correlation coefficient of 0.41 (P< 0.001).

Table 3 shows the results of the sensitivity calculations; 5GS and VAS perceptions of change were equally sensitive. The 5GS was slightly but not significantly more sensitive than VAS for patients with severely impaired and unimpaired quality of sleep at baseline.

Contingency tables were created using amounts of change identified by VAS and 5GS measures. Tables 4 and 5 are examples of these comparisons in the treatment and washout periods. The axes represent the VAS and 5GS measures. The numbers on the axes show the actual instrument values transposed to a 0 to 5 sequence. The middle position on each axis represents little or no change. The on-diagonal position means agreement between the measures. As expected, the largest changes on the VAS tended to be seen for the largest changes on the 5GS. Of particular interest is the pattern of VAS score changes in patients who had no change on the 5GS at the treatment period. In this subgroup, 83 had a change in VAS score between -1 and -2 units (corresponding to improvement of 10 to 20 mm). As suggested by the approximately 50:50 split in the direction of VAS score changes for patients showing 0 change in 5GS in the washout period, it is very likely that this does not reflect any increased sensitivity of the VAS but, rather, is attributable to random variation.

Pearson Chi-Square analyses showed a significant relationship between the VAS and the 5GS measure in both the treatment

Table 2. Correlation coefficients (Spearman) of vas change scores with 5GS change scores

Group	Correlation Coefficients	P (Spearman Correlation)
Total cohort	0.50	< 0.001
Baseline QON < 3 cohort	0.44	< 0.001
Baseline QON = 3 cohort	0.46	< 0.001
Baseline QON > 3 cohort	0.26	0.002

Table 3. Sensitivity to change calculations

	Efficiency*			
Group	**5GS Change**	**VAS Change**	**5GS-VAS Difference**	**P- Value**
Total cohort	0.29	0.59	-0.30	0.98
Baseline QON < 3 cohort	0.94	0.75	0.19	0.98
Baseline QON = 3 cohort	0.27	0.56	-0.29	0.96
Baseline QON > 3 cohort	0.28	0.16	0.12	0.96

*Efficiency equals the mean change of the measurement divided by the SD of the change measurement

Table 4. Quality of sleep measurements at treatment period

	Change in QOS	**Worse**				**Better**
Change in QON		**+3 or more**	**+1 to +2**	**0**	**-1 to -2**	**- 3 or more**
Worse	-3 or more	2	1	1	2	2
	-1 to-2	6	**16**	**18**	13	4
	0	4	30	**53**	**83**	30
	+1 to +2	0	6	12	**43**	**31**
Better	+3 or more	0	1	3	10	23

($\chi 2$= 103.3; df= 16; p< 0.001, Somers'd (symmetric)= 0.367, p< 0.001) and washout ($\chi 2$= 105.9 (df= 16) p< 0.001, Somers'd (symmetric)= 0.391, p< 0.001) periods. The probability that the percentage of pairs disagreeing by 2 or more axis categories could have equaled zero by chance was rejected, p< 0.001 for the treatment (probability observed= 0.023 [9/394], expected= 0.24 [6 out of 25 cells]) as well as for the washout (probability observed= 0.021 [8/379], expected= 0.24 [6 out of 25 cells]) periods indicating that the number of individuals who differed by 2 or more positions on the axes could not have arisen by chance.

Discussion

In this analysis of data from 394 adult and elderly insomnia patients the 2 methods (5GS and VAS) for measuring sleep quality change gave concordant results. In analyses using data combined across all treatment groups in each study, VAS and 5GS scores were highly correlated. The 5GS had a slightly greater sensitivity than the VAS in patients who had impaired sleep quality at baseline. These data indicate that both methods are useful measurement tools for assessing sleep quality in insomnia clinical trials.

A similar conclusion was reached by Skovlund and Flaten[15] who used stochastic simulation, based on data from 268 migraine patients in a clinical trial, to investigate the statistical power of analyzing response as success or failure on a four grade scale (4GS) versus treatment difference score on the VAS; the VAS and 4GS appeared to have equal power. Furthermore, Lines et al[16] who compared a visual analog scale (VAS) method of assessing headache pain with a standard categorical four-grade scale (4GS) in a randomized, placebo-controlled, double-blind, clinical trial involving 792 treated migraine outpatients who received oral rizatriptan 5mg, sumatriptan 50mg, or placebo for a moderate or severe headache, made similar observations.

It is conceivable that the VAS may offer advantages over the 5GS in groups of patients who could have difficulty ascribing their sensations to a verbal category, such as children or mentally-impaired adults. However, in a study of sumatriptan in children with migraine, the VAS did not appear to be useful for determining treatment effects.[17]

Table 5. Quality of sleep measurements at washout period

	Change in QOS	**Worse**				**Better**
Change in QON		**+3 or more**	**+1 to +2**	**0**	**-1 to -2**	**- 3 or more**
Worse	-2 or more	**10**	**10**	2	1	0
	-1 to -2	18	**30**	**21**	9	0
	0	16	42	**62**	**45**	15
	+1 to +2	5	12	17	**25**	**19**
Better	+3 or more	1	1	1	**9**	**8**

It is surprising that post-treatment VAS changes in sleep quality were noted in patients who showed no change on the 5GS. One can speculate that the 5GS measure is sharply focused on the precisely defined variable at a moment in time whereas the VAS QOS measure is retrospective and captures to some extent the patient's general experience of a change in symptom over time. In the latter case, the symptom magnitude and some of its consequences mingle in the assessment. That would move the assessment away from temporal precision toward a more composite appraisal over time. It should however be noted that a similar observation was made by Lines et al study,[16] who compared the use of VAS and 4GS scales in migraine patients. In that study, the VAS was not retrospective.

Another explanation is that a change in 1 unit on the 5GS is of magnitude that represents a clinically important change in sleep quality. Minimal changes may be translated into movement of less than the 10 mm on the VAS, which is the equivalent of a movement of 1 unit on the 5GS, while clinically insignificant changes in the QOS measure are in the range of 2.4-5.6 mm) (Zisapel & Nir, unpublished data).

An important question to consider is whether the use of the VAS added anything to the 5GS in evaluating sleep quality and treatment effects. Based on results from this analysis, the answer would appear to be no. The VAS and 5GS yielded similar results in terms of determining treatment effects, and showed a high level of correlation in post-hoc analyses. The sensitivity of the two methods appeared similar with some advantage for the 5GS in cases that patients with more severe baseline pathology are involved.

References

1. Perlis ML, Giles DE, Mendelson WB et al. Psychophysiological insomnia: The behavioural model and a neurocognitive perspective. J Sleep Res 1997; 6:179-88.
2. Schneider-Helmert D, Kumar A. Sleep, its subjective perception, and daytime performance in insomniacs with a pattern of alpha sleep. Biol Psychiatry 1995; 37:99-105.
3. Costa e Silva JA, Chase M, Sartorius N et al. Special report from a symposium held by the World Health Organization and the World Federation of Sleep Research Societies: an overview of insomnias and related disorders—recognition, epidemiology, and rational management. Sleep 1996; 19:412-6.
4. Parrott AC, Hindmarch I. Factor analysis of a sleep evaluation questionnaire. Psychol Med 1978; 8:325-9.
5. Parrott AC, Hindmarch I. The Leeds Sleep Evaluation Questionnaire in psychopharmacological investigations - a review. Psychopharmacology 1980; 71:173-9.
6. Felson DT, Anderson JJ, Boers M et al. The American College of Rheumatology preliminary core set of disease activity measures for rheumatoid arthritis clinical trials. The Committee on Outcome Measures in Rheumatoid Arthritis Clinical Trials. Arthritis Rheum 1993; 36:729-40.
7. Spielman A, Yang C, Glovinsky P. Assessment techniques for insomnia. In: Kryger M, Roth T, Dement W, eds. Principles and practice of sleep medicine. Philadelphia: Saunders Co., 2000:1239-1250.
8. Roth T, Roehrs T, Vogel G. Zolpidem in the treatment of transient insomnia: a double-blind, randomized comparison with placebo. Sleep 1995; 18:246-51.
9. Walsh JK, Roth T, Randazzo A et al. Eight weeks of nonnightly use of zolpidem for primary insomnia. Sleep 2000; 23:1087-96.
10. Riedel BW, Lichstein KL. Objective sleep measures and subjective sleep satisfaction: How do older adults with insomnia define a good night's sleep? Psychol Aging 1998; 13:159-63.
11. Morin CM, Mimeault V, Gagne A. Nonpharmacological treatment of late-life insomnia. J Psychosom Res 1999; 46:103-16.
12. Zisapel N, Laudon M. Subjective Assessment of the Effects of CNS-Active Drugs on Sleep by The Leeds Sleep Evaluation Questionnaire: A Review. Human Neuropsychopharmacol Clin Exp 2002; In Press.
13. Anderson JJ, Chernoff MC. Sensitivity to change of rheumatoid arthritis clinical trial outcome measures. J Rheumatol 1993; 20:535-7.
14. Zar J. Biostatistical analysis. New Jersey: Prentice Hall, 1984;156.
15. Skovlund E, Flaten O. Response measures in the acute treatment of migraine. Cephalalgia 1995; 15:519-22:discussion 450-1.
16. Lines C, Visser W, Vandormeal K et al. Rizatripan 5mg versus sumatripan 50mg in the acute treatment of migraine (abstract). Headache 1997; 37:319-320.
17. Hamalainen ML, Hoppu K, Santavuori P. Sumatriptan for migraine attacks in children: a randomized placebo-controlled study. Do children with migraine respond to oral sumatriptan differently from adults? Neurology 1997; 48:1100-3.

CHAPTER 26

The Pharmacostrategy of Treating Sleep Disorders in Parkinsonism

Jean-Jacques Askenasy

Abstract

The present article proposes a pharmacostrategy for the treatment of sleep and arousal disturbances in Parkinson's disease (PD) patients. Among the factors that affect the quality of life in PD patients, *sleep disturbances* play a major role along with depression and inappropriate independence in daily activity. To apply this tharapeutic approach, three steps have to be achieved:

1. Correct diagnosis of sleep disorders based on detailed anamnesis of the patient and of the spouse or of the caregiver; (i) One week recording on a symptom diary (log) by the patient or the caregiver. (ii) Correct diagnosis of the comorbidities of the sleep disorders. (iii) Selection of the most appropriate sleep test: polysomnography (PSG), multiple sleep latency test (MSLT), multiple wake latency test (MWLT), Epworth Sleepiness Scale, actigraphy or video-PSG.

2. Dopaminergic adjustment to counterbalance the progression of the nigrostriatal degeneration and the hypersensitivity of the remaining neuronal terminals, which alter the normal activity of the motor centers in PD patients. Of the numerous neuronal receptors, the dopaminergic and cholinergic are pimarilly affected by the nigrostriatal degeneration. REM sleep corresponds to an harmonic interrelationship between the cholinergic and the dopaminergic activity. For this reason REM sleep deprivation induced by suppression of the cholinergic receptor activity ameliorates PD motor symptoms. In contrast L-Dopa and its agonists suppress the cholinergic receptor activity and therefore suppress REM sleep. Permanent adjustment to prevent effects of the progression of the degenerative process will diminishe aggravation of sleep and arousal disorders.

3. A specific therapy is required for the following sleep syndromes:

Sleep disturbances:
- Light fragmented sleep (LFS)
- Abnormal motor activity during sleep (AMADS)
- REM behavior disorders (RBD)
- Sleep related breathing disorders (SRBD)
- Sleepiness (EDS)
- Sleep related hallucinations (SRH)
- Sleep related psychotic behavior (SRPB)

Arousal disturbances:
- Sleep attacks (SA)
- Excessive daytime

The specific therapy includes:
- LFS: Benzodiazepines and Nondiazepines;
- AMADS: Clonazepam, Opioid, Apomorphine infusion;
- RBD: Clonazepam and dopaminergic agonists;
- SRBD: CPAP, UPPP, nasal interventions, losing weight;
- SRH: Clozapine, Risperidone;
- SRPD: Nortriptyline, Clozapine, Olanzepine;
- SA–adjustment;
- EDS-arousing drugs.

Each therapeutic approach must be tailored to the individual PD patient.

Pharmacostrategy

At the turn of the 20th century, Indian psychiatrists discovered the anti-psychotic action of the snakeroot Rauwolfia Serpentina (see Fig. 1).

Thirty years latter, the new antipsychotic chemical chlorpromazine was develepod. The study of the mechanism of action of these antipsychotic drugs, led to the discovery of their dramatic effect on cells that produce and store serotonin and dopamine. This historic landmark of neuroscience set the basis to the understanding of the "*chemical transmission*" as the main communication path between neurons.

The era of neuro-psycho-pharmacology was initiated in the early fifties of this century (reviewed in refs. 36,73). In the Nobel prize acceptance speech in December 2000, Arvid Carlsson stated: "After returning to my home university in Lund, Sweden, as associate professor of pharmacology, I discovered, together with Nils-Ake Hillarp and other collaborators, that reserpine caused depletion of the adrenal medullary hormones, as well as of noradrenaline in other tissues, including the brain. To investigate the mode of action of reserpine on the central nervous system, my colleagues and I administered DOPA to reserpine rabbits and mice. We then discovered the central stimulant action of DOPA as well as its ability to reverse the akinetic and sedative action of reserpine".[36] A new neurotransmitter was discovered. This happened in 1957, it took another 10 years for l-Dopa to become the panacea of the Parkinson's disease.

Before 1950 scientists believed that the dialog between neurons was mediated by electrical pulses. A decade later the dialog was considered to be chemical, opening numerous possibilities for manipulation of neuronal activity at the level of transmission. In parallel, a better understanding of the chemistry and biomechanics of movement was achieved.

Sleep and Sleep Disorders: A Neuropsychopharmacological Approach, edited by Malcolm Lader, Daniel P. Cardinali and S.R. Pandi-Perumal.

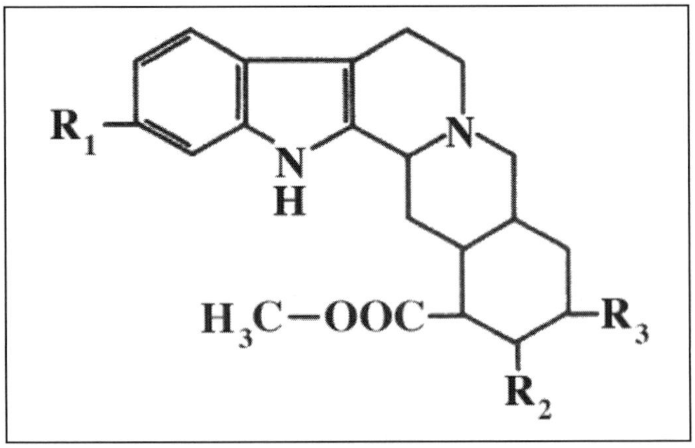

Figure 1. Chemical structure of reserpine.

At the turn of the 3rd millennium at least 100 compounds are believed to serve as neurotransmitters in the brain. The major categories include biogenic amines, peptides, aminoacids and hormones. The pharmacostrategy became an imperative concept in the therapy of brain disorders.

Pharmacological modulation of the biochemistry of sleep is a newer, fascinating, chapter in the study of sleep and treatment of its disorders. The monoamines serotonin, norepinephrine, histamine and acetylcholine modulate sleep and wakefulness. For eample, inhibition of cholinergic and histaminergic neurons is critical for sleep initiation and gamma-amino-butyric acid (GABA) a pivotal inhibitory neurotransmitter, is critical for the maintainance of sleep. The basic concept of pharmacostrategy consists of administration of precursors of neurotransmitters, according to their capacity to pass blood-brain barrier. To achieve these therapeutic goals, better understanding of several basic functions has to be achieved, including sleep, movement during normal sleep, sleep and movement in Parkinson's Disease (PD).

Normal Sleep

Two thids of our life we are awake and one third asleep. From Greek mythology until Pavlov and Sherrington, sleep was considered as a passive state of mind.[140,168] In the second half of the 20th century science proved the contrary, showing that sleep is an active state of equal physiological importance to wakefulness.[10]

However, 2000 years ago the poet and philosopher Lucretius stated an idea that became notorious at that time: "Sleep is the absence of wakefulness". If Lucretius meant to underline the dialectic unity wake/sleep, he was right and he is still right. Health is the result of an harmonic continuum of the wake state when energy is consumed, and of the restaurative sleep state. In illness this continuum is altered and sleep infitrates wakefulness and vicecersa. There is no wakefulness without sleep, and sleep without wakefulness. The physiological states, were artificially separated for didactic purpeses.

Normal sleep is a cyclic succession of two different states, the silent sleep state or nonrapid eye movement (NonREM) sleep and the active or the rapid eye movement (REM) sleep. The NonREM sleep is a cyclic succession of four stages, first two (I,II) also being named light sleep and last referred to as (III-IV) deep sleep or slow wave sleep (SWS). SWS is restorative from the perspective of protein synthesis, while REM sleep is a state of energy consumption. Although the exact impact of these sleep stages is unknown, they are of crucial importance to health. It is amazing that we can make accurate predictions about movement of stars in distant galaxies, yet many physiological processes remain shrouded in mistery.[176]

The brainstem hegemony in sleep initiation, and the spectacular "brainstem switch effect" that changes wake to sleep and sleep to wake within milliseconds was undermined by the research of Steriade.[175] The authors showed that the pontine reticular gabaergic neurons generate inhibitory postsynaptic potenials (IPSPs) that block the sinaptic transmission to thalamus, allowing *the cortex,* to generate either NonREM sleep (with oscilations ranging between 0.6-0.8.7 Hz, 1-4 Hz or 7-14 Hz), or activate REM sleep. This new concept apply to the cortex an equal role in initiating and switching to sleep as to the brainstem.

Movement during Normal Sleep

Muscle function is the basic element of the movement. The contrast between the histo-chemical similarity in fiber structure and the morphologic variety of muscles in more than one million animal species is amazing.

A man weighing 72 kg entertains cognition in an organ weighing 1.400 gr., however to achieve movement, the body recruits 36 kg. of muscle, which is approximately half of his total weight.

During wakefulness movement is triggered in the motor, premotor, and associative areas, is processed in the cerebellum, reticular nuclei of brainstem and is transmitted through the motor pthways to provoke muscle activity and movement. In the wake state the basal ganglia are involved in processing the routine automatic behavior according to cognitive, limbic and hippocampic input, while in the sleep state behavior may be the result of an independent basal ganglia activity.

Movement at will activates skeletal muscle controlled by the somatic nervous system, in association with the smooth muscle activation submitted to involuntary regulation by the autonomic nervous system.

During conscious wakefulness relaxation and contraction are in a permanent alternation. Depolarization and hyperpolarization of the cell membrane, opening and closing of the membrane channels and the positional interplay of the actin and myosin chains, are the basic frame of contraction and relaxation.

During sleep the motor activity is governed by the same states of relaxation and contraction by unconscious levels. In NonREM sleep the relaxation is characterized by a regressive gradient of the muscle tone, which parallels the deepness of sleep, and is randomlly interrupted by isolated motor unit potentials. Muscle contractions of NonREM sleep consist of **hypnic jerks** that are most abundant at the onset of sleep and **postural shifts** during transitions between the sleep stages. The hypnic jerk is an abrupt muscle flexion movement, generalized or partial and asymmetric, which may cause arousal with an illusion of falling. The electromyographic profile displays a series of complexes of 250 msec. duration in various skeletal muscles, predominantly the limbs. The postural shifts at the sleep stage transition involve activity of multiple skeletal muscles. They are more frequent in childhood at a frequency of 4.7 per hour at ages 8 to12, and decreases to 2.1 per hour at the age of 65 to 80.

In NonREM sleep the motor register displays a motionless state with reduced responsiveness at all levels. A gabaergic inhibition originating in the reticular centers, generates inhibitory postsynaptic potentials (IPSPs) that block synaptic transmission in the thalamus, with deafferentation of the motor cortex. During **slow wave sleep**, the neuronal synchronization increases.

Gabaergic blockade of the thalamus and hyperpolarisaton of the motor units through descending inhibitory postsynaptic potentials, result in low levels of neuronal output and a lesser responsiveness to afferent impulses.

In REM sleep—an unstable state—excitatory and inhibitory responses coexist. Hyperpolarisation of the alpha motor neurons induced by small and large IPSPs that originate in the nucleus pontis oralis and the nucleus reticularis gigantocellularis in the pons and medulla oblongata, results in **motor atonia of the REM sleep**. Maximal inhibitory tone of the final motor pathway allows increased activity of the motor and sensory cortices, red nucleus and cerebellum, resulting in the dream state.

The bursts of phasic REM movements arise from superimposed excitatory postsynaptic potentials (EPSPs) that originate in the brainstem and are mediated by N-methyl-D-aspartate (NMDA) excitatory synapses. The fact that IPSPs can be suppressed by strychnine, suggests that glycine plays an inhibitory role. These data explain the diminished tendon reflexes in NonREM sleep and the absent reflexes in REM sleep.

During sleep the smooth muscle tone is controlled by the autonomic nervous system, assuring continued activity of the respiration muscles and the diaphragm, bowel movements, and continued adaptation of the vascular tone and heart activity to the physical and mental needs of the body.

Sleep and Movement in Parkinson's Disease

Parkinson's Disease (PD) is caused by degeneration of the nigrostriatal dopaminergic neurons (basal ganglia disorder), provoking a movement disorder clinically characterised by gait disturbance, bradykinesia, tremor and cognitive disturbances. The classification of PD includes staging according to the severity of these symptoms as described by Hoehn and Yahr, and rating according to the Unified Parkinson's Disease Rating Scale (UPDRS).

The generally agreed concept on the etiology of PD is an environmental trigger superposed on a predisposing hereditary phenotype. Five "Parkin genes" have been proposed as possible participants in the hereditary basis of this disease.

Recently a possible involvement of RNA messangers has been suggested. The environmental factor is largely unknown, MPTP, insecticides and pesticides are suggested.

The nigrostriatal dopaminergic terminal degeneration is associated with degeneration of the cortex, brainstem and intermediolateral columns. The few remaining nigrostriatal terminals fail to metabolize a sufficient amount of L-dopa and convert it into dopamine. These remnant neurons and terminals are very sensitive to small changes in plasma levels of dopaminergic (DA) drugs. This sensitivity results in adverse reactions.

The neuro-degenerative process and the side effects of the DA drugs are the major causes of the **sleep-arousal disturbances** in PD patients.[45,103,116,132,144,210] On the one hand the neurodegenerative process imposes its direct consequences, such as bradykinesia-rigidity, psychiatric complications, disruption of the circadian rhythm and REM disturbances.[45,103,116,132,144,210] On the other hand, the side-effects of DA induce dyskinetic and dystonic movements.

The prevalence of sleep disturbances in PD patients varies according to different scores from 60-98%,[103,132] from 60% to 90%[185] and from 74% to 98%.[139]

The factors affecting the quality of life in PD patients are depression, **sleep disturbances** and dependence.[116] Determinations of the quality of life (QoL) in PD patients are emphasizing the important role of sleep.[92]

Karlson et al used the Nottingham Health Profile test on 233 PD patients to show that the most significant variables in prediction of the quality of life of PD patients are depression, **sleep disturbances (SD)** and dependence confirming the results obtained by Menza; Smith and Martinez-Martin.[92,113,116,172]

According to a self-rated methodology the prevalence of disturbed sleep in the general nonPD population, varies from a percentage of 25% in males to 41% in females.[50] Two quality of life tests related to health were elaborated for PD patients: the Parkinson's Disease Questionnaire - PDQ-39, and the Parkinson's Disease Quality of Life Questionnaire -PDQL. Both included sleep and rest measurements, despite the absence of this aspect in the UPDRS.[50] The Keyston Colorado consensus conference on establishement of an algorithm for the management of PD, held in February 1994, marked the introduction of sleep disorders in PD patients as a significant component of the therapeutic approach.[96] This approach is meant to postpone the institutionalization of PD patients, allowing the spouse or the caregiver a quiet sleep at night.

Natanel Kleitman observed that the Babinski sign is present during sleep in healthy subjects in 1963, a sign which is pathognomic of pyramidal lesions. He reasoned this phenomenon as a „rearranged organization" of the motor control during sleep. In 1971 Rucky et al dedicated a study to this phenomenon, to conclude that changes occur in the reflex responses during sleep. Chase Morales and Yanusy described the "reversal phenomenon", that originates in the reticular formation of the brainstem to ensure the switch between the wake/sleep states. The observed diminished or altered extensor reflexes, blink reflexes, somato-sensory evoked potentials, visual evoked potentials and electromyographic activity during NonREM sleep, partially reversed during REM sleep.

Studies of Sleep Disorders in PD

In 1980 a project to study the motor activity during sleep in neurodegenerative disorders was performed by Askenasy.[14] During the early studies the sleep motor activity was recorded in healthy subjects, followed by the study of sleep motor activity in PD patients. As the Mount Sinai faculty did not have at that time a sleep laboratory, I was driving the patients every evening to the sleep lab of Prof. Eliot Weitzman at Montefiori hospital, bringing them back the next morning. From the beginning the results were a function of the methodology of registration used. Certain results were obtained when recording the motor activity by surface EMG electodes, and other results when recording with implanted concentric needles electrodes on ultraviolet paper. In order to find the truth we were obliged to elaborate a very laborious methodology of recording the motor activity, concomitantly with surface and implanted needle electrodes (see Fig. 2).

The conclusions of our research referred to the different types of sleep disturbances in PD and the guidelines for a therapeutic approach. This approach intended to postpone the institutionalization of PD patinets, allowing the spouse or the caregiver a quiet sleep at night. In our view the therapeutic approach consists of 3 important steps.

First step consists of: correct diagnosis of sleep disturbances based on detailed anamnesis of the patient and of the spouse or of the caregiver; management of a symptom diary (log) by the patient or the caregiver, for one week; identification of sleep disorders caused by comorbidities (a sleep disorder not related to the PD); choice of the most appropriate test or group of tests, for the specific type of sleep disturbance among: polysomnography

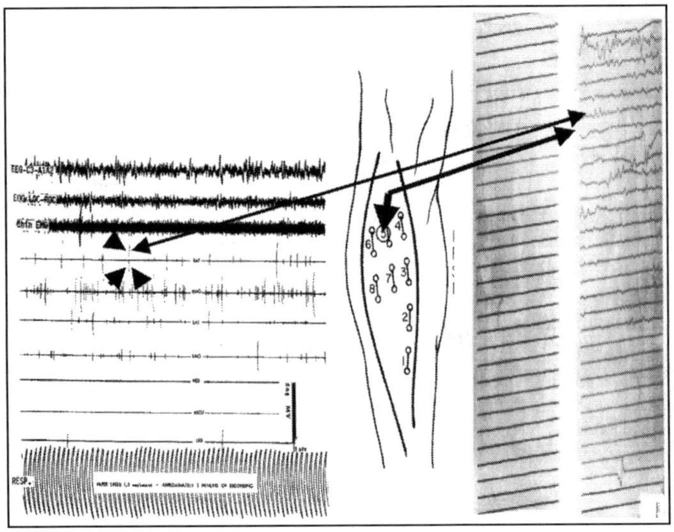

Figure 2. Concomitant polysomnographic recording with surface electrodes and electromyographic recording with needle electrodes, placed subjacent to the surface electrodes, showing the EMG aspect of a PSG signal.

Table 1. Epworth sleepiness scale (ESS)

How likely are you to doze off or fall asleep in the following situations, in contrast to feeling just tired? Even if you have not been in such situations try to work out how they would have affected you.

Choose the most appropriate number for dozing:
0 = never
1 = slight chance
2 = moderate chance
3 = high chance.

Sitting and reading	
Watching TV	_____
Sitting inactive in a public place	_____
As a passenger in a car for an hour	_____
Lying down to rest in the afternoon	_____
Sitting and talking to someone	_____
Sitting quietly after a lunch without alcohol	_____
In a car, while stopped a few minutes in the traffic	_____

(PSG), multiple sleep latency test (MSLT), multiple wake latency test (MWLT), Epworth sleepiness scale (ESS) (Table 1),[89] actigraphy (Ac)[93] or video-PSG.

PSG is useful in diagnosis of Light Fragmentd Sleep (LFS), Abnormal Motor Activity During Sleep (AMADS), REM Behavior Disorder (RBD), Sleep Related Brathing Disorder (SRBD); MSLT, MWLT and ESS is useful in Excessive Daytime Sleepiness (EDS), Sleep Attacks (SA), SRBD; Ac is useful in AMADS; and Video-PSG is useful in Sleep Related Hallucination (SRH) and Sleep Related Psychotic Behavior (SRPB).

Second step or the nonspecific therapy, is effective for all the types of sleep-arousal disturbances and includes: determine the effect of sleep on motor performance; psycho-physical assistance; and adjustment of the therapy.

Determine the Effect of Sleep on the Motor Performance in PD Patients

This information provides important hints on the interrelationship between sleep/and PD. Three possible effects of sleep on the motor performance were described in PD patients: sleep benefit (SB), sleep worse (SW) and sleep neutral (SN).

The beneficial effect of sleep on motor performance may be manifested for a duration of 30 minutes to 3 hours. SB is present in 10-55% of the PD patients.[1,28,57,83,146,181,180,198] Some studies found increased SB in PD patients of older age, lengthy disease duration, high doses of dopaminergic agonists, long duration of dopaminergic therapy and associated with hallucinations and vocalization during sleep.[43,49,117] Others did not find any difference in age stage, duration of the disease, symptomatology and duration of treatment.[55] SB may be also present in PD of early onset.[43,49,117] The incidence os sleep disorders was similar in the SB patients classified as "morning better" and SW called also "morning worse" patients. Evaluation of the effect of sleep on the motor performance to determine the subsets of SB and SW patients, may provide important insights into appearance of severe future fluctuations in the dopaminergic receptor level. The third group of SN, also termed "morning same", includes a smaller number of patients as compared to the other groups, which display less severe disease, shorter duration and low doses of DA drugs.[55] The morning motor function is probably related more to the fluctuation of the dopaminergic receptor expression and to "off" phenomena during sleep, than to an impaired sleep mechanism.[55]

Psychophysical Assistance

The obvious tendency of PD patients to live a sedentary life is directly related to the progress of the disease, which favors daily somnolence. In order to avoid the daily somnolence, 10 hygiene commandments are recommended: (1) An afternoon nap has to be allowed only in PD patients with a longstanding habit of napping; (2) PD patients have to avoid day dozing, alcohol, coffee and heavy meals before bedtime; (3) Maintenance of a regime of regular day activity with physical work; (4) Exercise may provide benefits through its effect on the cardiovascular system, on the muscle mass, on postural comples and cognition; (5) Limit fluid intake after 17:00; (6) A warm bath before bedtime; (7) Instrumental aids for getting out of and into bed; (8) Easy accessibility to water, bathroom and alarm clock at night; (9) A PD nurse has an important role in avoiding sleep disturbances of the PD patient; (10) Chronotherapy fulfills psychophysical assistance in PD patients with biological clock disorders.[35,76,105,109,173,199,206]

The Pharmacostrategy of Adjustment

Has to be performed before any specific therapeutic approach. The adjustment is implied by the evolution of the disease. Progression of the nigrostriatal degeneration and increased sensitivity of the terminals alter the normal modulator mechanisms of motor centers in PD patients. Among the numerous neurotransmitters in the nigrostriatal pathway one can distinguish two, there are transmitters with a distinct effect on REM and NonREM sleep. REM sleep corresponds to an increased cholinergic receptor activity and a decreased dopaminergic activity. This is the reason of the improvement in motor symptoms in PD patients after REM sleep deprivation, which suppresses the cholinergic receptor activity. L-dopa and its agonists suppress the cholinergic

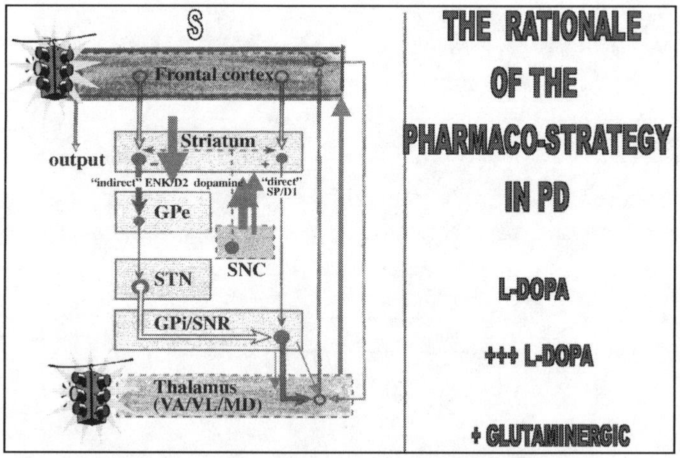

Figure 3. The nigro-straito-cortical interrelations which explain the need for permanent adjustment as a main pharmacostrategy in PD patients.

receptors and hence suppress REM sleep. L-Dopa may either potentiate wakefulness and enhance fragmentation due to involuntary movements, or potentiate NonREM sleep by deepening sleep and enhancing imobility.

Using the above mentioned sleep effects of anticholinergic drugs and dopaminergic agonists, one can improve sleep in PD patients, simply by adjustment of the dose of drugs.[13] Optimization of the dosage and of the timetable of drug adiministration, and the use of new dopaminergic agonists with fewer motor side-effects (ropinorole, pramipexole) may improve the sleep disturbance. Dopaminergic adjustment, with a beneficial effect on sleep may be achieved by use of slow release therapy, such as Sinemet CR,[178] of a selective and reversible COMT inhibitor, such as Tolcapone,[118] and agents used to improve gastric movement, such as Domperidone or Cisapride.[142] Gabapentin (a GABA analogue) may be an important adjuvant in the treatment of sleep disturbances in PD patients, which increases slow wave sleep and improves tremor, rigidity, bradykinesia and motor fluctuations.[40,147]

Recordings of the night motor activity using a wrist-worn monitor in showed significant improvement of sleep disturbances in severe forms of PD, following dopaminergic adjustment in 84 PD and 83 normal controls matched for age and sex.[194]

To achieve a succesful pharmacostrategy of the PD patient it is imperative to keep in mind the four biochemical situations (see Fig. 3):

1. Situation A: The absence of DA increases the inhibitory GABA effect of GPiSNR in the thlamo-corticothalmo circuit (TCTC).
2. Situation B: Dopaminergic drugs replace the dopaminergic effect on the striatum and remove the inhibition from the TCTC.
3. Situation C: Overcorrection with DA desinhibits the TCTC and involunatry movements intrude into sleep.
4. Situation D: The cortex activates the glutaminergic pathway to striatum to balance the overinhibition caused by DA.

Third step or the specific therapy. Is directed specifically to each sleep syndrome, and each therapeutic approach has to be tailored to each individual PD patient. Due to the complex etiology of sleep disturbance in PD, more than one type of sleep disturbance is frequently found. By improving one type of sleep disturbance, a clear improvement might be achieved for another type.

Table 2. The most frequent sleep-arousal disturbances appearing during 20 years of Parkinsonism progression, divided in 4 periods, characterized by the major symptom

Honey Moon	Motor Complications	Resistent Symptoms	Cognitive Decline
> 5 years	> 8 years	> 15 years	> 20 years
LFS	LFS	LFS	LFS
RBD	RBD	SRBD	SRBD
SRBD	SRBD	SRH	SRBD
SA/EDS	PLM-RLS	EDS	SRH
	SA/EDS		EDS

For example, successful treatment of restless legs, periodic leg movements and sleep behavior disorders, may resolve excessive daytime sleepiness.

Sleep Disturbances

Light Fragmented Sleep

Light fragmented sleep (LFS) was first described as a characteristic sleep disturbance entity of PD patients in 1981.[14] It consists of difficulties in the initiation and maintenance of sleep, synonymous with insomnia.[14] It is by far the most frequent type of sleep disturbance among PD patients,[182] with up to 80% of the PD patients complaining of LFS,[55] and present in all the stages of the disease (see Table 2).

Light and fragmented sleep is of multifactorial origin.[14,55,85,100,103,131,147,182,194,195] The implicated factors include:

1. motor origin: either due to hyper mobility expressed by resting tremor, eye blinking, dyskinesia, limb-facial dystonia, jerks, painful cramps and fragmentary myoclonus, or due to immobility with an inability to turn around in bed and stiffness.
2. disturbed breathing origin;
3. vivid dreams or nightmares;
4. nocturia which may also expose the PD patient to frequent falls, resulting in hip fractures;
5. anxiety and depression;
6. screaming interrupts sleep in 50% of the cases that display this phenomenon. The arousal due to screaming is the result of a dysfunction of the lower brainstem, the Locus Ceruleus and the pedunculo-pontine nuclei, related to REM muscle atonia. A parasympathetic deceleration of the heart rate is frequently observed in PSG.

The cause of LFS is considered to originate from disequilibrium of the sensitivity of the nigrostriatal receptors.[6,14,35,60,109,139,185,182,194] A direct consequence of LFS is daytime dozing, excessive daytime somnolence and sleep attacks.

Arousal occurs at high frequencies in PD patients that suffer of muscle hyperactivity because of disturbances in the activity of the autonomic nervous system.

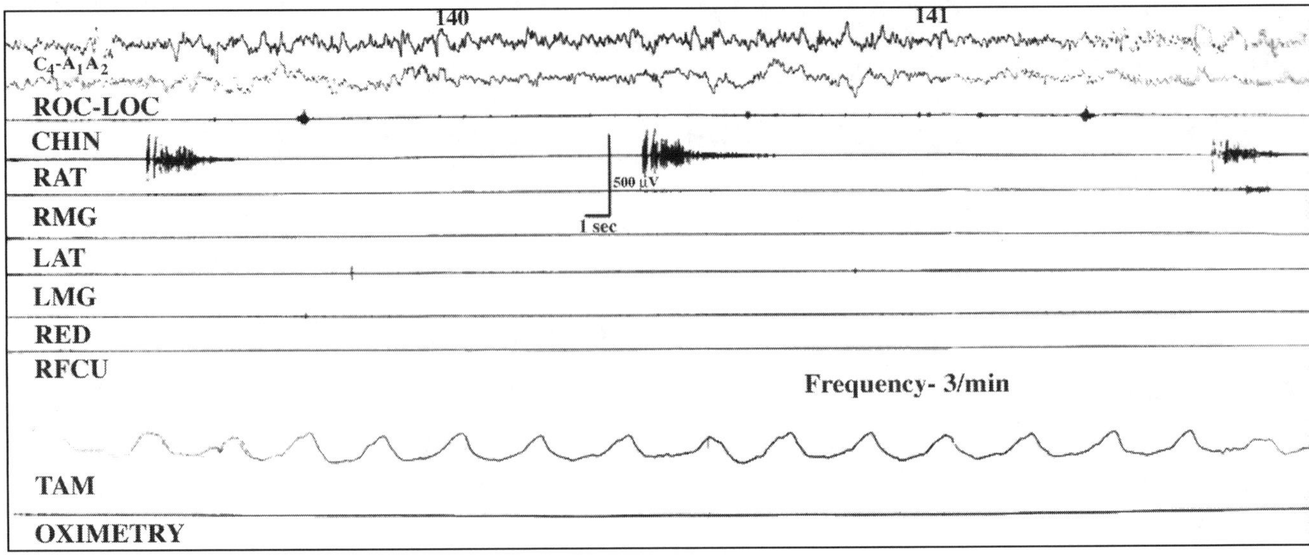

Figure 4. PSG recording. PLM bursts appearing with a rhythm of 3 per minute on right anterior tibial muscle, in a PD patient.

The Pharmacostrategy

In addition to dopaminergic adjustment and psycho-physical nonpharmacologic therapy, which is the first approach to each individual, one has to analyse several considerations. Despite of the fact that the dominant complaint of the LFS in PD patients is severe insomnia, the prescription of a sleeping pill has to be avoided because the sleep disturbance is chronic and therefore there is a risk to develop dependence. If used in some cases, the duration of the treatment has to be limited to a duration of maximum one month and several hypnotic drugs have to be alternated to avoid dependence. Short-acting nondiazepines include imidazopyridine (5-10 mg) and cyclopyrolone (7.5 mg).[58,70,71,99,122,159] The tricyclic antidepressant amitriptylina (10-40 mg), as well as the antihistaminic drug diphenhydramine (25-50 mg) have good hypnotic effects, even in nondepressive and nonallergic PD patients.

Most benzodiazepines remain attacheed to the neuroreceptor during the next morning, provoking daytime sedation, dozing, memory deficit and discoordination of the perceptual skills. This is why short acting benzodiazepines with a half-life of 3.5 hours, such as triazolam and brotizolam, are preferable rather than flurazepam which has a half-life of 60 hours (Shaw et al,1992).[82,158] The consumption of hypnotics may also cause sleep fragmentation at withdrawal.[64,108,139,151] The association of Ziclopan with Aniracetam was found to be effective.[94]

Intranasal Desmopressin was observed to be a useful tool in those PD patients that suffer of nocturia.[139]

In intractable LFS, with persistent sleep rest tremor and severe sleep muscle hyperactivity, a neurosurgical approach may be the solution. Chronic subthalamic nucleus stimulation,[128] continuous high frequency stimulation of the ventral intermediate nucleus of thalamus[8] or unilateral/bilateral pallidotomy[59] were shown to improve sleep in operated PD patients. When LFS was caused by night pain, pallidotomy was observed to be the best neurosurgical choice.[84]

Abnormal Motor Activity during Sleep (AMADS)

Periodic Leg Movements (PLM) during Sleep and Restless Legs Syndrome (RLS)

PLM wakes up the patient, while RLS occur as a post-effect of the PLM. PLM appears during sleep, while RLS appears during wakefulness and makes the patient walk due to an uncontrolled urge to move. These syndromes may appear independently, but in most cases they coexist. This is a good example of a sleep/wake continuous disturbance, a "**sleep/wake disorder**".

PLM consists of rhythmic segmental movements of the leg muscles during sleep at a frequency of 1-6 per minte, in particular the anterior tibial muscle.[42] The positive diagnosis of PLM is performed by PSG with two leads for the anterior tibial muscles of both legs (see Fig. 4).

RLS is an uncontrolled urge to move the affected (lower and upper) limbs during sleep, which oblige the patient to wake up and walk.[39,200] PD patients displaying PLM/RLS may benefit from its positive response to DA drugs[13,15,101,107,124] with a benefit of therapy up to 80% of the patients.[123]

PLM-RLS are more prevalent in some populations, as determined from population surveys[102,143] or from clinical ascertainment.[150] The segmental movements usually include extension of the big toe, ankle, knee or hip, depending on the amplitude of the neuronal discharge.[42,152] When compared with RLS, PLM are more often unilateral.[17] PLM-RLS are evident mainly in the first part of the night's sleep.[13,15] They may coexist in PD patients with sleep apnea syndrome and REM behavior disorder. Electromyographic (EMG) studies of PLM showed that movements appear in association with bursts lasting 3 to 5 seconds when the amplitude of potentials are above 130 microvolt.[152] EMG and PSG studies revealed an interesting phenomena which is more evident in late PD: the presence of burst-tremor (see Fig. 5).[16,17,152]

The etiology of PLM syndrome is unknown. Neurophysiological data and functional MRI suggested that a suprasegmental disinhibition during sleep may provoke PLM.[34,186,197,211] The

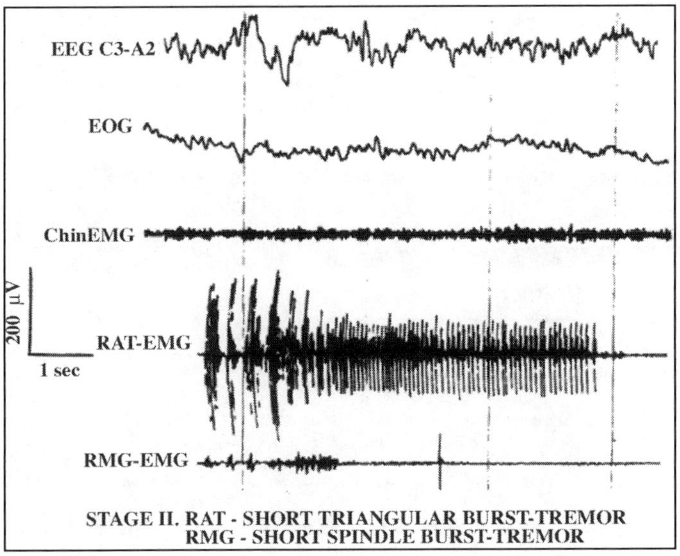

Figure 5. PSG recording. A typical burst-tremor in right anterior tibial muscle during sleep stage II in a PD patient.

efficacy of levodopa and dopamine agonists in reducing PLM favor this hypothesis.[33,202] Two-night PSG recordings of sleep in 10 PD patients, 10 multiple system atrophy patients and 10 Huntington's chorea patients, (matched for age and sex) which were free of drugs for 2 weeks, showed a significantly higher PLM index in naive PD patients. This suggests that nigrostriatal degeneration favor the PLM.[201] It was suggested that the limb movements are augmented by an "off" nocturnal phenomenon in PD patients, as a result of the more severe neuronal degeneration of the basal ganglia.[13,15,148] Others have shown that the prevalence of PLM in the general population above the age of 65 years and in PD patients is similar.[5]

PLM appear almost exclusively in (NonREM) sleep stage II in PD patients.[13,15,123] The absence of slow wave sleep in aged PD patients may explain the lack of PLM in stage III and IV. The lack of PLM in REM sleep suggests that the inhibition of the skeletal muscles is strong, and that different laws govern motor activity in sleep and in wakefulness.[5,17,18,41,127,152,170] The etiology of RLS syndrome is unknown. Studies on the PLM-RLS frequency in PD patients are still inconclusive, an issue that needs further elaboration.[125,201]

The Pharmacostrategy

Consensus conferences have elaborated the standards for treatment of PLM-RLS as an empirical approach to syndromes of unknown ethiology.[42] The Practice Parameters Standards (PPS)[42] recommend three preliminary steps. First, that prior to therapy (a) obtain an analytic anamnesis from the patient and bed partener, (b) evaluate possible comorbid conditions, (c) follow-up for side-effects and tolerance.

Second, avoid the exacerbating effect of drugs such as tricyclic agents, monoaminoxidase inhibitors (MAOI) AB (phenelzine) and MAOI B (selegiline) on PLM.[19,96]

Third, adjust the DA drug dose using slow-release agents and physical assistance such as stretching and passive movements.[42,96,185] Then the PPS recommends to initiate starting a drug therapy per se.

Several efficient drugs have been recommended for the treatment of PLM: Pergolide, Bromocriptine or Clonazepam and for the treatment of RLS: Clonidine, Gapapentin, Clonazepam and Iron. Low-dose clonazepam (0.5-2 mg), temazepam, nitrazepam, carbamazepine, clonidine or lioresal, may be helpful, and in recalcitrant cases opiates such as oxycodone or propoxyphene may be useful.[5,16-19,27,30,31,33,34,74,75,79,96,114,119-121,125,136,148,152,185,186,189,197,201,202,209,211]

In refractory cases, continuous subcutaneous 1% apomorphine infusions overnight were shown to produce a dramatic reduction of nocturnal awakenings, off periods, pain, dystonia and nocturia, which are related to PLM.[5,19,27,30,31,65,74,75,79,114,119-121,136,148,189,209] The apomorphine infusions have a short life time of approximately 34 minutes.[5,19,30,75,79,114,119-121,136,148,189]

Special attention was addressed by PPS to the treatment of PLM-RLS in pregnant women. The medication was classified according to the degree of risk from least to highest in ABC and X. Category B includes Pergolide, Category C includes L-dopa-carbidopa, Clonazepam, Propoxyphene, Codeine, Carbamazepine, Gapapentin, and Clonidine, and category X includes Temazepam.[42]

Tremor during Sleep (TDS)

The relationship between sleep and tremor has been discussed since the description of the disease by James Parkinson in 1817. In the initial report of Parkinson "Essay on the Shaking Palsy" the author stipulated that tremor disappears during sleep, and affirmed that tremor might reappear towards the end of the sleep and gradually increase its intensity.[138] At the turn of the 20th century Gowers emphazised the persistence of tremor during sleep.[72] Lord Brain considered that the alleviation of tremor during sleep is a pathognomonic sign for diagnosis of PD.[32] Our studies have elucidated this controversy.[20,21,23]

From the clinical point of view sleep attenuates tremor, hawever it can be detected by electromyography during all the stages of NonREM sleep with amplitudes up to 150 microvolt[14,17,18] (Figs. 6, 7, 8).

Its frequency of appearence decreases with the transition to deeper sleep, when the gabaergic inhibition increases.[14,17,18] The appearance of tremor during REM sleep is exceptional, and may be associated with movement preceeding or succeeding REM (Stern et al, 1968; April, 1966)[4,14,184] In rare ocasions tremor is associated with sleep spindles or K complexes.[78] The alternating pattern of tremor disappears during sleep, and both agonist and antagonist muscles are activated independently in a nonalternating pattern.[17,18]

Fragmentary Nocturnal Myoclonus (FNM)

Irregular myoclonic jerks, which appear randomly during sleep, are frequent phenomena in PD patients, in particular during light sleep.[16,78,124]

The Pharmacostrategy of Tremor and FNM

Adjustment of the DA agonists doses while avoiding administration of monoaminoxidase inhibitors (MAOI) AB (phenelzine) or MAOI B (selegiline) in the evening, and the use of low doses of clonazepam (0.5-2 mg) is the most appropriate therapeutic approach.

REM Behavior Disorder (RBD)

REM behavior disorder (RBD) is a REM violent dream associated with automatic motor behavior related to the dream content, which results in self-injury or the injury of bed partner (see Fig. 9).

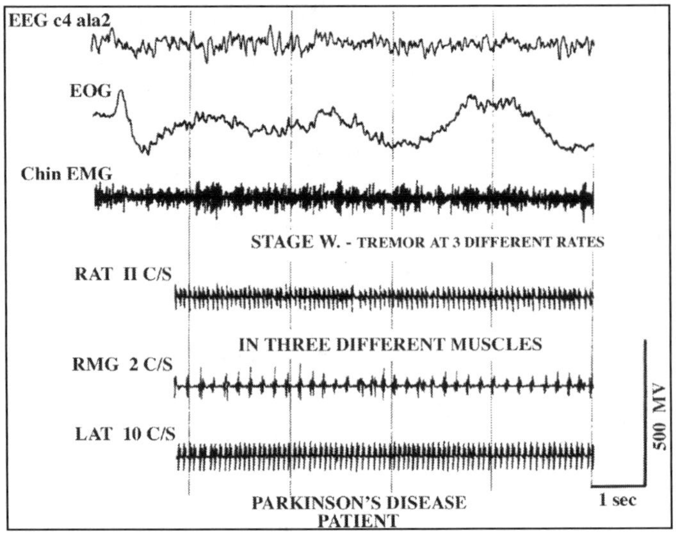

Figure 6. PSG recording. Tremor during silent wakefulness.

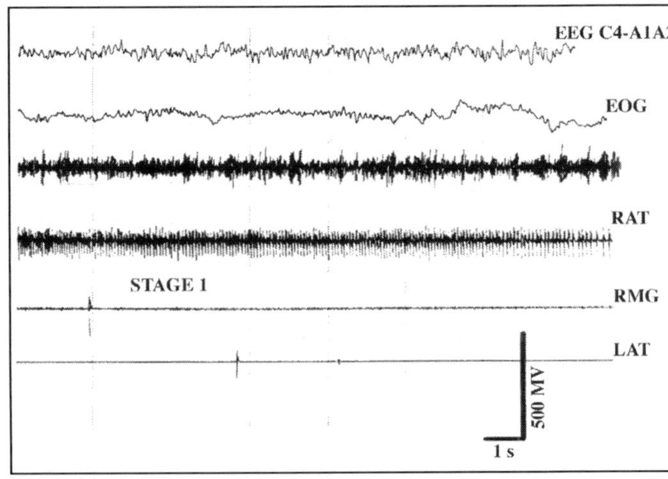

Figure 7. PSG recording. Presence of tremor in the right anterior tibial muscle, during sleep stage I, in the absence of alternating tremor in the antagonist muscle.

Historically, Hishikawa was the first to describe this type of disturbances in an alcoholic patient during a weaning period.[80] Askenasy elaborated the presence of muscle activity during REM sleep in the legs and chin of PD patients that used drugs or suffered of psychiatric conditions.[14] Schenck defined this disorder as a new type of parasomnia, and included it in the international classification of sleep disturbances in 1990 (Schenck et al, 1985;)[160,161] (Thorpy, 1990). The prevalence of RBD in PD patients is very high, between 50-75%.[46,47,162]

The positive diagnosis of RBD includes: (i) violent movements related to the dream state; (ii) tonic and phasic anomalies of the muscle tone during REM sleep; (iii) abnormal behavior during REM sleep; (iv) the absence of epileptic fits. In rare cases, RBD can be asymptomatic and may be revealed by a routine PSG.

Injuries usually evolve from the execution of defense acts, imposed by the context of the violent dreams, in the absence of muscle atonia. The absence of muscle atonia and the presence of phasic muscle activity, which characterize this disorder, has to be confirmed by the EMG applied to the legs and chin muscles of the PD patient. The relationship between self-injury and the assault of spouses is present in 32 to 64% of the patients.[134] The result of self injuries and assault attacks consists of wounds, fractures, haematomas and subdural haematomas.[134,163] The frequency of RBD episodes is variable, from a couple of times per night to one every 6 months.

The differential diagnosis includes elimination of focal epilepsy of the temporal and frontal type, confusional awakenings, sleep walking and nightmares. From the legal point of view, this aggressive behavior is an "actus reus" not associated with a "mens rea", and hence is not punishable. Upon arousal 87% of the RBD patients explain the act of defense against an attack.[134] RBD is frequntly associated with Lewy body dementia (LBD) with a significantly male predominance.[45,185] Interestingly RBD was described as a premonitory syndrome in various neurodegenrative extrapyramidal disorders such as PD, LBD, Multiple System Atrophy (MSA) Spino-cerebellar degeneration (SCD).[29,46,47,133,134,144,162,163,165,169,171,183,190,191] RBD is prevalent in PD patients prior to the onset of the clinical symptomatology.[134,163,183] In a study that monitored 100 PD patients; 15 displayed RBD,[45] and in another study 38 subjects out of 100 RBD patients developed PD at a later time.[29]

A pathophysiological approach to the understanding of RBD was presented by Jouvet in 1965, who observed the absence of muscle atonia with strange behavior during paradoxical sleep in cats with bilateral ponto-tegmental lesions.[90] The lack of muscle atonia during paradoxical sleep was proved to be caused by impaired activity of the inhibitory pathways from Locus Caeruleus (LC) to Magnocellular Nuclei.[127,145,155] Uchiama[191] published a case of RBD with severe depletion of monoaminergic neurons in LC, confirming the lack of noradrenergic inhibition in RBD.[191]

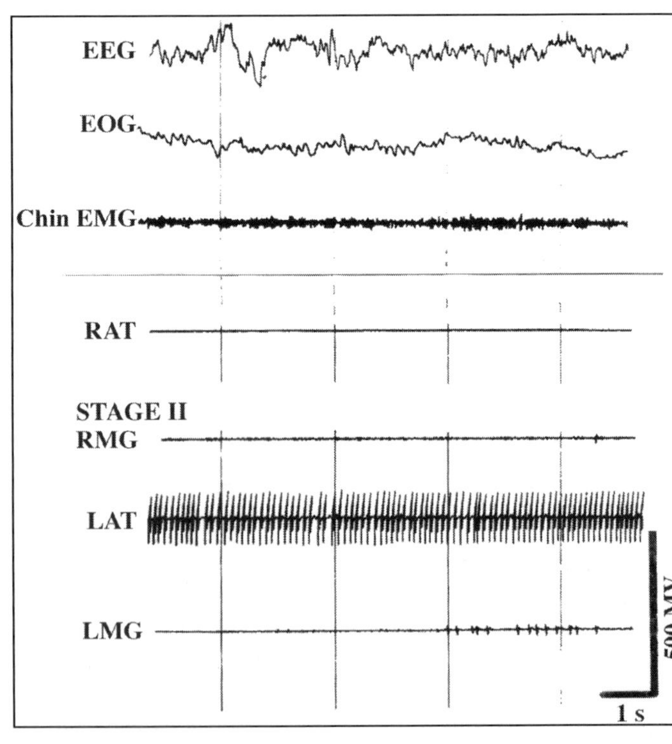

Figure 8. PSG recording. Presence of tremor in the left anterior tibial muscle during sleep stage II, in the absence of alternating tremor in the antagonist muscle.

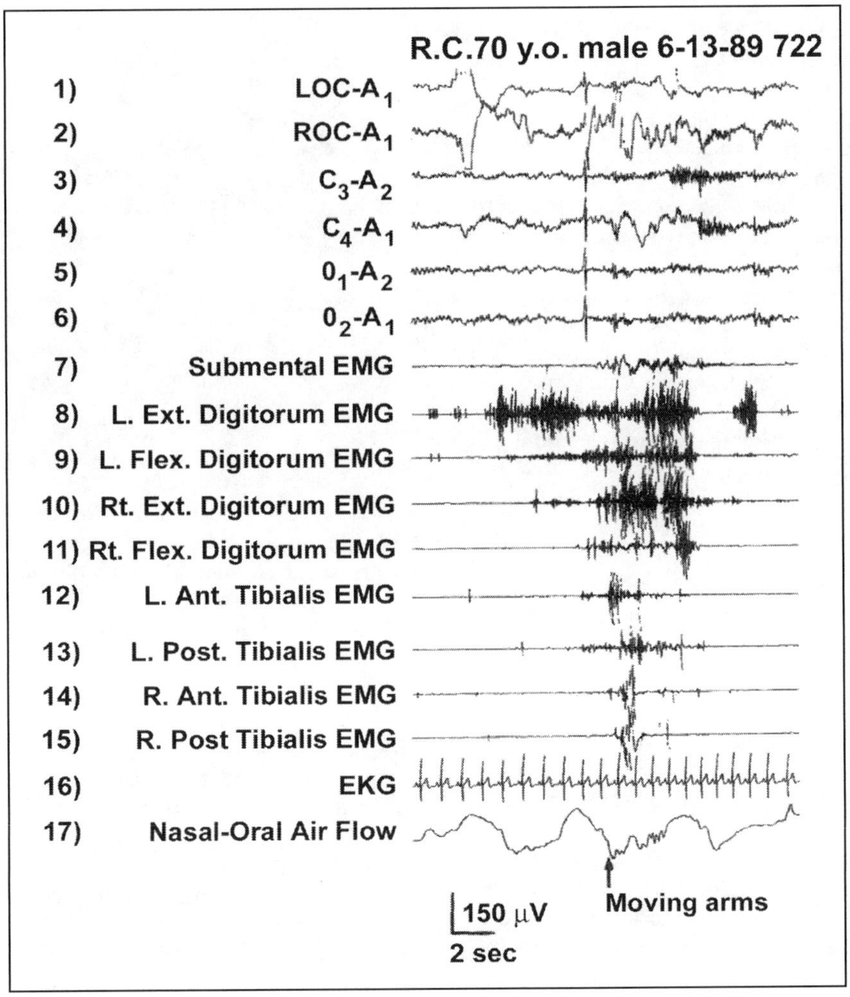

Figure 9. PSG recording. REM behavior disorder manifested by high muscular activity in various muscles during REM sleep. Reprinted with permission from Elsevier from Kryger et al, Principles and Practice of Sleep Medicine, 3rd ed. ©2000 Elsevier, Inc.

Rey[152] suggested that the excessive gabaergic inhibition from globus pallidus is the cause of the absent atonia in PD patients during REM sleep. Eisensehr et al,[53] studied 5 patients suffering of idiopathic RBD (nonPD) using monophotonic emission tomoscintigraphy. They found a significant reduction of striatal dopamine transporter as compared to normal controls, but not as severe as in PD patients. The authors concluded that RBD ia a state of impaired dopaminergic/cholinergic balance.[53] Albin[3] used positron emision tomography to confirm these findings. Interestingly Bejjany (2002) observed a PD patient treated by bilateral stimulation of the subthalamic nucleus, who developed aggressive behavior during intraoperative electrical test stimulation. This observation suggests that at the border of the limbic system, at the lateral part of the posteromedial hypothalamic region (triangle of Sano) there is a region responsible for aggressive behavior. The strong association between Levy body dementia and RBD suggests an interrelationship between the synonucleopathies and RBD.[29]

Several factors have been shown to correlate with the incidence of RBD. The peak age of appearance is 52-61 years, with a male predisposition.[134,164] RBD is associated with HLA DQB10602[162] and is frequently encountered in patients suffering of neurologic diseases such as multiple sclerosis, cranial trauma, brainstem astrocitoma, corticobasal dementia, progressive supranuclear palsy, olivo-ponto-cerebellar degeneration, Alzheimer Disease and Lewy Body Dementia. RBD may appear following the sudden arrest of the administration of REM suppressors, such as alcohol, amphetamines, cocaine, imipamine, barbiturate, anticholinesterase and tryciclic antidepressants.

The Pharmacostrategy

The specific therapeutic approach of RBD is applied along with the nonspecific approach. L-dopa and dopaminergic agonists that suppress REM may decrease the severity of RBD and may even arrest the disorder.[134,163] Clonazepam (Rivotril) is by far the most efficient drug in RBD at a dose of 0.5-2.0 mg before bedtime, with succesful results in approximately 88% of the patients.[163] Other medications that have been applied with partial benefit include tricyclics,[196] carbamazepine[2,26] and melatonine.[98]

Sleep Related Breathing Disorders (SRBD)

Are frequent among PD patients, in particular in those that suffer of PD for long periods of time.[118,181,185] These late PD patients have a tendency to a Kussmaul or Cheyne-Stokes type of breathing in the supine position during wakefulness. The upper airway resistance syndrome (UARS) and the three forms of sleep apnea syndrome (SAS) obstructive, central and mixed, may be also present in PD patients. It is interesting to note that in late PD patients, central SAS is more often present than obstructive

SAS.[6] Overweight is rare among PD patients, but when present is frequently associated with SRBD. The congenital central hypoventilation (CCHS), also termed Ondine's Curse, consists of failure of autonomic control of ventilation during NonREM sleep, which results in hypercapnia and hypoxemia. CCHS was described in PD patients that suffered of other disorders of the autonomic appearing during REM sleep. Perry et al reported a Parkinsonian syndrome with severe alveolar hypoventilation in identical twins whose autopsies showed PD with brainstem damage.[141]

A genetic predisposition was observed in PD patients with ventilation disorders.[60] These findings are supported by recent studies that describe a subset of dopaminergic neurons in the petrosal ganglia of the glossopharyngeal nerve, which convey sensory information on hypoxemia from the carotid body to the central nervous system. These dopaminergic neurons are controlled by a gene named "brain derived neurotrophic actor" (BDNF).[25,95] Dopaminergic therapy hyper-activates subset of petrosal neurons in an alternate matter, thus causing a ventilation disequilibrium. PSG is the ideal method for accurate diagnosis of SRBD, however the electronic sleep strip recently approved by FDA in conjunction with oximetry may be a simpler and more efficient method for the detection of the SRBD.

The Pharmacostrategy

Hypoxia caused by SRBD should be avoided in PD patients by the application of Continuous Positive Air Pressure (CPAP) or Bilevel Positive Airway Pressure (BiPAP) masks during the night. According to the severity of SAS and the general state of the patient, partial or total Uvulo-Palato-Pharingo Plasty and surgical nasal intervention may be considered. The use of oxygen in PD patients suffering from SRBD must be avoided because it aggravates sleep apnea, and aggravates hypoxia by elevating the response threshold to decreased O2 and increased CO2 during sleep.

Sleep Related Hallucinations (SRH)

We consider SRH as a distinct and independent entity of sleep disorders in PD patients, because the hallucinations are not necessarily an expression of psychosis. The hallucinations of the PD patients are generally visual and rarely auditory. They may appear in a minor form as a sensation of the presence of a person or an animal, and as an illusion, or in the major form as complex hallucinations with a stereotyped content of bright colors and dramatic settings.[62,86,110]

The analysis of the appearance of hallucinations at night and in reference to the sleep stages, as determined by PSG, was inclonclusive (Bliwise). The relationship between hallucinations and sleep may be present in one or more of the following four aspects: (i) hypnagogic (at sleep onset), (ii) hypnapompic (at sleep end), (iii) REM-incorporated, and in WASO (wake after sleep onset)-dependent. An association was shown between peculiar dreams and hallucinations in PD patients, suggesting a common underlying pathogenetic mechanism. The activation, at the cerebal level during REM sleep eases the formation of the imagery phenomenon.[55,62,110,129,156] An identical mechanism may cause hallucinations in narcolepsy.[8,9,86] Alternatively, the dream imagery may occur during NonREM sleep and NonREM dream.[8,9] In patients that hallucinate during the day, the hallucinations are related to a drop in the level of vigilence (Manni et al, 2002) or to a short REM sleep.[8,9] The prevalnce of SRH in PD patients with

Figure 10. A combination of watercolor painting and ink drawing. Ordinarily only one figure is seen at a given instant. Hallucinations usually take the form of a person or head, and the eyes and mouth are typically grossly distorted. Hallucinations occur in vivid color and, despite their appearance, are not threatening. Reprinted with permission from Frucht SJ, Bernsohn L. Visual Hallucination in PD, Neurology 2002: 59,1965.

altered dreams is considered to be 21 to 30%.[55,166] Another argument for the highly significant statistical relationship between hallucinations and altered dreams in PD patients is the presence of REM aberrations among PD hallucinators, as compared to PD nonhallucinators.[46,47,137]

As the hallucination imagery may extend into wakefulness (Mahowald et al, 1998; Nielsen, 2000) which is a common cause for placement in home nursing, a correct diagnosis and therapy are important.[156] Using a logistic regression analysis in 216 PD patients, three predictive factors were identified: cognitive degradation, excessive daytime sleepiness and long duration of PD.[62] One may add to these three predictive factors, the advanced age,[110] and longstanding treatment with agonistic dopaminergic drugs.[129,132]

SRH are suggested to be the consequence of brainstem lesions that affect the cholinergic and serotoninergic pathways in PD patients,[8,9,86,110] or stimulation of mesolimbic dopamine receptors by dopaminergic treatment.[46,47]

In PD patients that undergo chronic dialysis for renal insufficiency, SRH appears towards the end of the dialysis sesion.[12]

Frucht (2002) published a montage of individual images experienced and painted by an idiopathic PD patient. The 74 years old patient who was a painter, developed hallucinations during the day, following a long standing treatment with levodopa, entacapone and amantadine and he redrow them. The hallucinations were episodic, lasting seconds, without accompanying by paranoia, delusions, RBD or dementia (see Fig. 10).

Fenelon[62] in his article on hallucinations in PD patients described a typical benign hallucinations in a painter who redrow it (see Fig. 11).

The Pharmacostrategy

The therapeutic approach to SRH is initiated after exhausting the benefit of drug adjustment. The adjustment will start with the discontinuation of selegiline and amantadine, followed by anticholinergic agents and dopamine agonists. The drug dose has

Figure 11. Hypnagogic hallucination redrawn by the patient. He saw a waterfull with trickles outlining human silhouettes. Reprinted with permission from reference 62.

to be reduced gradually at bedtime, in the late evening and then in the afternoon.[46,47] It is imperative to prevent dehydration, constipation, drug interactions and dose changes of the chronic drugs, urinary or pulmonary infections.[56] Psycho-physical assistance consists of bedroom illumination at night.

When SRH persists despite adjustment of the maximally-tolerated reduction in dopaminergic agents, treatment with clozapine may be initiated. Clozapine (Leponex) is the most efficient drug, although its administration might be accompanied by side effects, such as agranulocytosis which requires frequent haematologic tests, excessive somnolence and orthostatic hypotension.[68] Clozapine is a less potent blocker of the striatal dopamine receptors and a more potent blocker of the mesolimbic dopamine receptors.[96] The starting dose of 12.5 mg at bedtime may be increased to 50 mg. Risperidone (0.5 to 1.0 mg), Olanzepine (5-15 mg), and Ondansetron (4-12 mg) can replace Clozapine if there are contraindications for administration of the drug, or because an exacerbation of the parkinsonism.[96,149,204,205,207,208]

Sleep Related Psychotic Behavior (SRPB)

Is a controversial issue because it is unclear how a psychiatric condition can appear during sleep. The expression consists of agitation, uncontrolled behavior with vocalization and screaming during NonREM and REM sleep.[66,91,179] SRPB has become increasingly common with PD due to the longevity and improved therapy.[192]

Most frequently, SRPB is caused by depression. The lower motor level of PD patients may mask some of the characteristic depressive features.[192] Assessment of depression and anxiety in 99 PD patients and 47 control subjects matched for age and sex showed significantly higher scores of disturbed sleep variables in the PD patients. The disturbed sleep variables in these depressed PD patients were directly related to on-off phenomena, the dose of L-doap and age.[116] A stepwise regression analysis of the effect of depression on sleep disorders showed a good correlation in the general population.[174] The prevalence of depression in PD patients is approximately 47%.[52] The incidence of depression in PD patients increases by 1.86% per year, as compared to age-matched nonPD subjects (0,14% per year in men and 0.19% per year in women).[52] Depressive PD patients display more often RLS. Interestingly, depressive PD patients often experience improvement of the mood, tremor, rigidity with fatigue.[51] Sleep deprivation may activate mechanisms, which are typical of suppressing cholinergic activity, which is thought to be excessive in relation to monoaminergic transmission in depressed and PD patients.[51] When comparing depressed to nondepressed PD patients, a significantly shorter REM latency was found in the depressed PD group, with a REM latency of 65 minutes (69% of 39 PD patients.[97] Anxiety is common in PD and causes insomnia, fatigue and daytime sleepiness.[77,115] Following depression and anxiety the third risk factor for PD patients to develop SRPB is dementia.[91,179,192] Taking into consideration the possibility that nocturnal wandering, panic, anxiety, depression, and paranoid delusions may appear during the WASO periods, the diagnosis of arousal in the middle of the sleep, without emergence during the day, is difficult.

Pharmacostrategy

The pharmacostrategy consists in therapy attempts to improve the QoL after an accurate diagnosis of the causes that underlie SRPB. It is warranted to start antidepressant therapy after discontinuation of selegiline.[48] Most frequent used antidepressant drugs in PD patients include: tricyclic antidepressants, MAO inhibitors and selective serotonin reuptake inhibitors with two amendments: (a) in the presence of confusion and hallucinations tricyclics should be avoided; (b) in cases of severe motor conditions serotonin reuptake blockers and MAO inhibitors should be avoided.[44,48,187,193] In very severe cases of depression, electroconvulsive therapy can be useful.[56] Except cases that suffer of depression and dementia, the drug of choice for the treatment of SRPB is low dose Clozapine (up to 50 mg/day)[130,179,188,192,193] and Olazapine (5-15 mg per day).[204] When dementia is dominant, risperidone (0.5 to 2 mg per day) may be used.[179] Ondansteron (12 to 24 mg) is effective in the treatment of paranoid delusions.[104,149] The drug of choice for treatment of anxiety is a benzodiazepine, such as alprazolam before bedtime and in more severe cases of anxiety clorazepam or prazepam may be useful. If an anxiolitic effect was not achieved by means of diazepines, additional agents may be used, including the serotoninergic anxyolitic agent Buspiron (10 to 40 mg) and imipramine (50 mg).[61]

Arousal Disturbances

Sleep Attacks

Sleep attacks (SA) have been observed in PD patients, in particular during car driving. It is expressed as a sudden and irresistible urge to sleep during the active period of the day, and occurs quite frequent in PD patients treated with antiparkinsonian drugs. The dopaminergic drugs that may cause SA include Promipexole, Ropinorole, Pergolide, Bromocriptine, Lisuride, and Piribedil.[63,69,157] Frucht reported 9 PD patients that caused accidents because they fell asleep while driving, and found that 5 of 9 patients had no warning signs before falling asleep.[69] Ferreira reported 3 PD patients that suffered of sleep attacks, while treated with ergot alkaloids and Piribedil (a DA agonist with vasodilator effect) and deduced that all dopamine agonists can in principle induce sleep attacks, despite the well-known arousal effect of L-dopa and Pergolide.[63]

Pharmacostrategy

The therapeutic approach in these cases is to gradually discontinue the therapy.

Excessive Daytime Somnolence

Excessive daytime somnolence (EDS) consists of short periods of sleep or somnolence, which are inappropriate during wakefulness, when the PD patient is in a passive state (not performing an activity), either alone or in the presence of others. EDS may appear any time during the day, even during the major arousal periods, (late morning and early evening), in a sitting or supine position. The circadian rhythm of PD patients, usually aged around 65 years, has a physiologic tendency to become biphasic, and are prone to fall asleep at siesta time (up to 20% of the nonPD elderly population has this tendency). It is considered a direct consequence of the "fragmented sleep" that characterizes the PD patients. A comparison of 90 nondepressive PD patients with 71 age-matched healthy controls, showed a significantly higher incidence of EDS in the PD patients.[195] The incidence of EDS in PD patients was also much higher (15.5%) than that of diabetic patients (4%) and healthy subjects (1%).[203] Others studies found no increased incidence of EDS in PD patients.[11,38,126] It is likely that the PD patients lose their ability to separate the wake and sleep states, and these alterante and intermingle. The PD patients that display EDS are most often in an advanced stage of the disease, are frequently disabled and show signs of cognitive decline.[181] An analysis of the frequency of the association EDS/SB (sleep benefit) on motor performance showed that PD with EDS and SB has a longest duration of L-dopa treatment, and significantly higher rate of fluctuations and dyskinesias as compared to PD patients with EDS and no SB.[181]

Although it may be stipulated that EDS is more frequent in late PD patients, as opposed to a decreased incidence of the SA syndrome, EDS has to be differentiated from excessive daytime fatigue (EDF), that causes physical tiredness and a feeling of lack of energy. EDF was reported in 43% of the PD patients.[172] The permanent feeling of excessive fatigue of these patients relates moe to the motor deficits than to the circadian factors,[195] as an expression of impaired arousal mechanism in the central nervous system.[54,153] Depression is frequently present in EDF-PD patients and antidepressant therapy may improve both conditions. The multiple sleep latency test (MSLT) is the best test to diagnose EDS. A series of 27 PD patients showed a clear association of EDS with primary disturbances in waking and REM sleep, and less relation to impaired quality and quantity of prior sleep.[181,153] Because MSLT is time-consuming the Epworth Sleepiness Scale may be used (see Table 1).[87,88]

Pharmacostrategy

The therapeutic approach to EDS includes the nonspecific measures (see the psycho-physical assistance recommendations), with emphasis on avoiding sleep periods during the day in order to prompt the night sleep. One should consider that the total sleep time in PD patients is the same as before the onset of the disease, but is more diffusely distributed throughout the 24 hours. Arousal drugs, such as caffeine, have to be used with great caution, in order to avoid hyperactivity at night. In extreme cases of very severe EDS, which is clinically difficult to differentiate from hypersomnia, Modafinil (Modiodal, Provigil 10 mg/day) may be used.

General Conclusion

We suggest the several types of sleep/arousal disturbances to be considered in PD patients, according to the following scheme:

Sleep Disturbances. Light Fragmented Sleep (LFS); Abnormal Motor Activity During Sleep (AMADS); REM Behavior Disorders (RBD); Sleep Related Breathing Disorders (SRBD); Sleep Related Hallucinations (SRH); Sleep Related Psychotic Behavior (SRPB).

Arousal Disturbances. Sleep Attacks (SA); Excessive Daytime Sleepiness (EDS).

And should be scored: 0=absent; 1=mild; 2=moderate; 3=severe. Each type of sleep-arousal disorder should be treated according to the proposed pharmacostrategy tailored for each individual patient.

References

1. Adam K, Oswald I. Protein synthesis, bodily renewal and the sleep-wake cycle. Clin Sci 1983; 65:561-567.
2. Albin RL, Koeppe RA, Chervin RD et al. Carbamazepin in REM sleep behavior disorder. Sleep 1993; 16:33-34.
3. Albin RL, Koeppe RA, Cervin RD et al. Decreased striatal dopaminergic innervation in REM sleep behavior disorder. Neurology 2000; 55:1410-1412.
4. Aldrich MS. Parkinsonism. In: Kryger MH, Roth T, Dement WC, eds. Principles and practice of sleep medicine. Philadelphia, PA: WB Saunders Co., 2000.
5. Ancoli-Israel S, Kripke D, Klauber M et al. Periodic leg movements in sleep in community dwelling elderly. Sleep 1991; 14:496-500.
6. Apps MC, Sheaff PC, Ingram DA et al. Respiration and sleep in Parkinson's disease. J Neurol Neurosurg Psychiatry 1985; 48:1240-5.
7. April RS. Observations on Parkinsonian tremor in all-night sleep. Neurology 1996; 16:720-724.
8. Arnulf I, Bejjani BP, Garma L et al. Effect of low and high frequency thalamic stimulation on sleep of patients with Parkinson's disease and essential tremor. J Sleep Res 2000; 9:55-62.
9. Arnulf I, Bonnet AM, Damier P et al. Hallucinations, REM sleep and Parkinson's disease. Neurology 2000; 55:281-288.
10. Aserinsky E, Kleitman N. Regularly occurring periods of eye motility and concomitant phenomena, during sleep. Science 1953; 118:273-274.
11. Askenasy JJM, Yahr MD. Suppression of REM rebound by pergolide. J Neural Transm 1984; 59:151-159.
12. Askenasy JJM. Unpublished observation.
13. Askenasy JJ, Yahr MD. Reversal of sleep disturbance in Parkinson's disease by antiparkinsonian therapy: A preliminary study. Neurology 1985; 35:527-32.
14. Askenasy JJM. Sleep patterns in extrapyramidal disorders. International J Neurol 1981; 15:62-76.

15. Askenasy JJM, Weitzman ED, Yahr MD. Are periodic movements in sleep a basal ganglia dysfunction? J Neural Transm 1985; 70:337-348.
16. Askenasy JJM, Yahr MD. Correlation of sleep patterns with muscle activity in patients with Parkinson's disease. In: Yahr MD, ed. Current concepts of Parkinson disease and related disorders. Amsterdam: Excerpta Medica, 1983:172-194.
17. Askenasy JJM, Yahr MD. Parkinsonian tremor loses its alternating aspect during nonREM sleep and is inhibited by REM sleep. J Neurol Neurosurg Psychiatry 1990; 53:749-753.
18. Askenasy JJM, Yahr MD. Different laws govern motor activity in sleep than in wakefulness. J Nerol Trans 1990; 79:103-11.
19. Askenasy JJM, Yahr MD. Is monoaminoxidase inhibitor induced myoclonus serotoninergically mediated? J Neral Transm 1988; 72:67-76.
20. Askenasy JJM. Sleep in Parkinson's disease. Acat Neurol Scand 1993; 87:167-170.
21. Askenasy JJM. Approaching disturbed sleepmin late Parkinson's disease: First step toward a proposal for a revised UPDRS. Parkinsonism and Related Disorders 2001; 8:123-131.
22. Askenasy JJM, Askenasy N. Inhibition of muscle sympathetic nerve activity during yawning. Clinical Autonomic Research 1996; m6:237-239.
23. Askenasy JJM. Sleep disturbances in Parkinsonism. Review J Neural Transm 2003; 110:125-150.
24. Babes V, Blocq P, Marinescu G. Atlas der patologischen histologie des nervensystems. Berlin: Hirschwald, 1906.
25. Balkowiec A, Katz DM. Brain-derived neurotrophic factor is required for normal development of the central respiratory rhythm in mice. J Physiol 1998; 510:527-33.
26. Bamford CR. Carbamazepine in REM sleep behavior disorder. Sleep 1993; 16:33-34.
27. Bastani B, Westervell FB. Effectiveness of clonidine in alleviating the symptoms of restless legs. Am J Kidney Dis 1987; 10:326.
28. Bateman DE, Levett K, Marsden CD. Sleep benefit in Parkinson's disease. J Neurol Neurosurg Psychiatry 1999; 67:384-5.
29. Boeve BF, Silber MH, Ferman TJ et al. REM sleep behavior disorder and degenerative dementia: An association likely reflecting Lewy body disease. Neurology 1998; 51:363-70.
30. Boghen D. Successful treatment of restless legs with clonazepam. Ann Neurol 1980; 8:341.
31. Bonnet MH, Arand DL. The use of triazolam in older patients with periodic leg movements, fragmented sleep and daytime sleepiness. J Gerontol 1990; 45:139-144.
32. Brain R. Clinical neurology. London: Oxford University Press, 1964:67.
33. Brodeur C, Montplaisir J, Godbout R et al. Treatment of restless legs syndrome and periodic movements during sleep with L-dopa: A double-blind, controlled study. Neurology 1988; 38:1845-1848.
34. Bucher SF, Seelos KC, Oertel WH et al. Cerebral generators involved in the pathogenesis of the restless legs syndrome. Ann Neurol 1997; 41:639-645.
35. Butler RN, Davis R, Lewis CB et al. Physical fitness: Benefits of exercise for the older patient. Geriatrics 1998; 53:46-49.
36. Carlsson A. A paradigm shift in brain research. Science 2001; 294:1021-1024.
37. Carlsson A, Lindqvist M, Riederer P. 3,4-Dihydroxyphenylalanine and 5-hydroxytryptophan as reseroine antagonists. Nature 1957; 180:1200.
38. Carskaden MA. Sleep and alertness. New York: Raven Press, 1989:53-69.
39. Chabli A, Michaud M, Montplaisir J. Peridic arm movements in patients with restless legs syndrome. Eur Neurol 2000; 4:133-138.
40. Chana P, de Marins A, Barrientos N. Gabapentin and motor fluctuations in Parkinson's Disease. Mov Disord 1997; 12:608.
41. Chase MH, Morales FR. Phasic changes in motoneuron membrane potential during REM periods of active sleep. Neurosci Lett 1982; 34:177-182.
42. Chesson Jr AL, Wise M, Davila D et al. Practice parameters for the treatment of restless legs syndrome and periodic limb movement disorder. An American Academy of Sleep Medicine Report. Standards of Practice Committee of the American Academy of Sleep Medicine. Sleep 1999; 22:961-968.
43. Clark EC, Feinstein B. The on/off effect in Parkinson's disease treated with levodopa with remarks concerning the effect of sleep. Adv Exp Med Biol 1977; 90:175-182.
44. Cohen LJ. Rational drug use in the treatment of depression. Pharmacotherapy 1997; 17:45-61.
45. Comella CL, Nardine TM, Diederich NJ et al. Sleep related violence, injury, and REM sleep behavior disorders in multiple system atrophy. Neurology 1998; 51:526-529.
46. Comella CL, Ristanovic R, Goetz CG. Parkinson's disease patients with and without REM behavior disorder (RBD): A polysomno-graphic and clinical comparison. Neurology 1993; 43(Suppl 2A):301.
47. Comella CL, Tanner CM, Ristanovic RK. Polysomnographic sleep measures in Parkinson's disease patients with treatment-induced hallucinations. Ann Neurol 1993; 34:710-714.
48. Cote L. Depression: Impact and management by the patient and family. Neurology 1999; 52(Suppl 3):S7-9.
49. Curie LJ, Bennett Jr JP, Harrison MB et al. Clinical correlates of sleep benefit in Parkinson's disease. Neurology 1997; 48:1115-7.
50. Damiano AM, Snyder C, Strausser B et al. A review of health-related quality-of-life concepts and measures for Parkinson's disease. Qual Life Res 1999; 8:235-43.
51. Demet EM, Chicz-Demet A, Fallon JH et al. Sleep deprivation therapy in depressive illness and Parkinson's disease. Prog Neuropsychopharmacol Biol Psychiatry 1999; 23:753-84.
52. Dooneief G, Mirabello E, Bell K et al. An estimate of the incidence of depression in idiopathic Parkinson's disease. Arch Neurol 1992; 49:305-7.
53. Eisensehr I, Linke R, Noachtar S et al. Reduced straiatal dopamine transportes in idiopathic rapid eye movement sleep behavior disorder. Comparison with Parkinson's disease and controls. Brain 2000; 123:1155-1160.
54. El-Ad B, Korczyn AD. Disorders of excessive daytime sleepiness-an update. J Neural Sci 1998; 153:192-202.
55. Factor SA, McAlarney T, Sanchez-Ramos JR et al. Sleep disorders and sleep effect in Parkinson's disease. Mov Disord 1990; 5:280-5.
56. Factor SA, Molho ES, Podskalny GD et al. Parkinson's disease: Drug-induced psychiatric states. In: Weiner WJ, Lang AE, eds. Advances in neurology: Behavioral neurology of movement disorders. Philadelphia, PA: Lippincot-Raven, 1995.
57. Factor SA, Weiner WJ. "Sleep benefit" in Parkinson's disease. Neurology 1998; 50:1514-5.
58. Fairweather DB, Kerr JS, Hindmarch I. The effects of acute and repeated doses of zolpidem on subjective sleep, psychomotor performance and cognitive function in elderly volunteers. Eur J Clin Pharmacol 1992; 43:597-601.
59. Favre J, Burchiel KJ, Taha JM et al. Outcome of unilateral and bilateral pallidotomy for Parkinson's disease: Patient assessment. Neurosurgery 2000; 46:344-53.
60. Feinsilver SH, Friedman JH, Rosen JM. Respiration and sleep in Parkinson'd disease. J Neurol Neurosurg Psychiatry 1986; 49:964.
61. Feldman RS, Meyer JS, Quenzer LF. Sedative hypnotic and anxyolitic drugs. Principles of neueropsychopharmacology. Sunderland, MA: Sinauer Associates, 1997.
62. Fenelon G, Mahieux F, Huon R et al. Hallucinations in Parkinson's disease: Prevalence, phenomenology and risk factors. Brain 2000; 123:733-45.
63. Ferreira JJ, Galitzky M, Montastruc JL et al. Sleep attacks and Parkinson's disease treatment. Lancet 2000; 355:1333-4.
64. Fish DR, Sawyers D, Allen PJ et al. The effect of sleep on the dyskinetic movements of Parkinson's disease, gilles de la tourette syndrome, Huntington's disease and torsion dystonia. Arch Neurol 1991; 48:210-4.
65. Frankel JP, Lees AJ, Kempster PA et al. Subcutaneous apomorphine treatment of Parkinson's disease. J Neurol Neurosurg Psychiatry 1990; 53:96-101.
66. Friedman JH. Behavioral dysfunction in Parkinson's disease. Clin Neurosci 1998; 5:87-93.
67. Friedman JH, Goetz CC, Stebbins GT. Psychotic symptoms in Parkinson's disease. J Am Geriatr Soc 1997; 45:252-3.
68. Friedman JH, Lannon MC. Clozapine in the treatment of psychosis in Parkinson's disease. Neurology 1989; 39:1219-21.

69. Frucht S, Rogers JD, Greene PE et al. Falling asleep at the wheel: Motor vehicle mishaps in persons taking pramipexole and ropinirole. Neurology 1999; 52:1908-10.
70. Ganzoni E, Santoni JP, Chevillard V et al. Zolpidem in insomnia: A 3 year post-marketing surveillance study in Switzerland. J Int Med Res 1995; 23:61-73.
71. Goa KL, Heel RC. Zopiclone: A review of its pharmacodinamic and pharmacokinetic properties and therapeutic efficacy as an hypnotic. Drugs 1986; 32:48-65.
72. Gowers WR. A manual of diseases of the nervous system. Philadelphia: Blakiston, 1901.
73. Greengard P. The neurobiology of slow sinaptic transmission. Science 2001: 294:1024-1030.
74. Guilleminault C, Flagg W. Effect of baclofen on sleep-related periodic leg movements. Ann Neurol 1984; 15:234-239.
75. Hanwerker J, Palmer RF. Clonidine in the treatment of "restless leg" syndrome. New Engl J Med 1985; 313:1228-1220.
76. Hauri P. Treating psychophysiologic insomnia with biofeedback. Arch Gen Psychiatry 1981; 38:752-758.
77. Henderson R, Kurlan R, Kersum JM et al. Preliminary examination of the comorbidity of anxiety and depression in Parkinson's disease. J Neuropsychiatry Clin Neurosci 1992; 4:257-264.
78. Hening WA, Walters AS, Chokroverty S. Motor functions and dysfunctions of sleep. In: Chokroverty Sudanshu, ed. Sleep Disorders Medicine. Butterworth – Heinemann, 1995:265-266.
79. Henning WA, Walters A, Kavey. Dyskinesias while awake and periodic leg movements in sleep in restless leg syndrome: Treatment with opioids. Neurology 1986; 36:1363-1366.
80. Hishikawa Y. Proceedings: Abnormalities of REM sleep in narcoleptics and in acute alcohol psychotics. Electroencephalography Clin Neurophysiol 1975; 39:542-543.
81. Hoehn MM. The natural history of Parkinson's disease in the prelevodopa post-levodopa eras. Neurol Clin 1992; 10(suppl 2):331-339.
82. Hoehns JD, Perry PJ. Zolpidem: A nonbenzodiazepine hypnotic for treatment of insomnia. Clin Pharmacy 1993; 12:814-828.
83. Hogl BE, Gomez-Arevalo G, Garcia S et al. A clinical, pharmacologic, and polysomnographic study of sleep benefit in Parkinson's disease. Neurology 1998; 50:1332-9.
84. Honey CR, Stoessl AJ, Tsui JK et al. Unilateral pallidotomy for reduction of parkinsonian pain. J Neurosurg 1999; 91:198-201.
85. Horiguchi J, Inami Y, Nishimatsu O et al. Sleep-wake complaints in Parkinson's disease. Rinsho Shinkeigaku 1990; 30:214-6.
86. Houeta JL, Arnulf I. Troubles psychiques et somnolence. Rev Neurol 2002; 158(HS):7S102-7S107.
87. Johns MW. Sleepiness in different situations measurd by the Epworth Sleepiness scale. Sleep 1994; 17:703-710.
88. Johns MW. Daytime sleepiness, snoring, and obstructive sleep apnea: The Epworth sleepiness scale. Chest 1993; 103:30-36.
89. Johns MW. A new method for measuring day time sleepiness: The Epworth sleepiness scale. Sleep 1991; 14:540-545.
90. Jouvet M, Delorme JF. Locus coeruleus et sommeil paradoxal. CR Soc Biol 1965; 159:859-899.
91. Juncus JL. Management of psychotic aspects of Parkinson's disease. J Clin Psychiatry 1999; 60(Suppl 8):42-53.
92. Karlson KH, Larsen JP, Tandberg E et al. Influence of clinical and demographic variables on quality of life in patients with Parkinson's disease. J Neurol Neurosurg Psychiatry 1999; 66:431-5.
93. Katayama S. Actigraphic analysis of diurnal motor fluctuations during dopamine agonist therapy. Proceedings Bangkok of the 3d ASRS congress 2000.
94. Katsunuma H, Shimizu T, Ogawa K et al. Treatment of insomnia by concomitent therapy with Zopiclon and Aniracetam in patients with cerebral infarction, cerebroatrophy, Alzheimer's disease and Parkinson's disease. Psychiatry Clin Neurosci 1998; 52:198-200.
95. Katz DM, Balkowiec A. New insights into the ontogeny of breathing from genetically engineered mice. Curr Opin Pulm Med 1997; 3:433-9.
96. Koller WC, Silver DE, Lieberman A. An algorithm for the management of Parkinson's disease. Neurology 1994; 44(Suppl 10):1-52.
97. Kostic VS, Susic V, Covickovic-Sternic N et al. Reduced rapid eye movements sleep latency in patients with Parkinson's disease. J Neurol 1989; 236:421-3.
98. Kuntz D, Bes F. Melatonin as a therapy in REM sleep behavior disorder patients open-labeled pilot study on the possible influence of melatonin REM-sleep regulation. Mov Disord 1999; 14:507-511.
99. Lader M. Zopiclone: Is there any dependence and abuse potential? J Neurol 1997; 244(Suppl):S18-S22.
100. Laihinen A, Alihanka J, Raitasuo S et al. Sleep movements and associated autonomic nervous activities in patients with Parkinson's disease. Acta Neurol Scand 1987; 76:64-8.
101. Lang AE. Restless leg syndrome and Parkinson's disease: Insights into pathophysiology. Clin Neuropharmacol 1987; 10(suppl 5):476-8.
102. Lavigne GJ, Montplaisir JY. Restless legs syndrome and sleep bruxism: Prevalence and association among Canadians. Sleep 1994; 17:739-743.
103. Lees AJ, Blackburn NA, Campbell V. The nighttime problems of Parkinson's disease. Clin Neuropharmacol 1988; 11512-519.
104. Lesser RP, Fahn S, Snider SR et al. Analysis of the clinical problems in parkinsonian and the complications of long-term levodopa therapy. Neurology 1979; 29:1253-1260.
105. Lichstein KL, Fischer SM. Insomnia. In: Hersen M, Bellak AS, eds. Handbook of clinical behavior therapy with adults. New York: Plenum Press, 1985.
106. Linazasoro G, Grandas F, Martinez Martin P et al. Controlled release levodopa in Parkinson's disease: Influence on selection criteria and conversion recommendations in the clinical outcome of 450 patients. STAR Study Group. Clin Neuropharmacol 1999; 22:74-9.
107. Linazasoro G, Masso JFM, Suarez JA. Nocturnal akathisia in Parkinson's disease: Treatment with clozapine. Mov Disord 1993; 8:171-174.
108. Linsen SM, Zitman FG, Breteler MH. Defining benzodiazepine dependence: The confusion persists. Eur Psychiatry 1995; 10:306-311.
109. MacMahon DG. Parkinson's disease nurse specialists: An important role in disease management. Neurology 1999; 52:S21-5.
110. Manford M, Andermann F. Complex visual hallucinations. Clinical and neurobiological insights. Brain 1998; 121:1819-40.
111. Marinescu G, Draganescu S, Sager O et al. Rev. Neurologique Paris. 1929:481-497.
112. Marinesco G, Rascano V. Contribution a la physiologuie pathologique du parkinsonisme. Compt Rend Soc Biol 1921; 34:1017-1020.
113. Martinez-Martin P. An introduction to the concept of "quality of life in Parkinson's disease". J Neurol 1998; 245(Suppl 1):S2-6.
114. Matthews WB. Treatment of the restless legs syndrome with clonazepam. Br Med J 1979; 281:751.
115. Mayeux R. Mental state. In: Koller WC, ed. Handbook of Parkinson's disease. New York: Marcel Dekker, 1987.
116. Menza MA, Rosen RC. Sleep in Parkinson's disease. The role of depression and anxiety. Psychosomatics 1995; 36:262-266.
117. Merello M, Hughes A, Colosimo C et al. Sleep benefit in Parkinson's disease. Mov Disord 1997; 12:506-8.
118. Micek ST, Ernst ME. Tolcapone: A novel approach to Parkinson's disease. Am J Health Syst Pharm 1999; 56:2195-205.
119. Mitler MM, Browman CP, Menn SJ. Nocturnal myoclonus: Treatment efficacy of clonazepam and temazepam. Sleep 1986; 9:385-392.
120. Moldofsky H, Tullis C, Quance G et al. Nitrazepam for periodic movements in sleep (sleep-related myoclonus). Can J Neurol Sci 1986; 13:52-54.
121. Montagna P, Lugaresi E, Plazzi G. Motor disorders in sleep. Eur Neurol 1997; 38:190-7.
122. Monti JM, Attali P, Monti D et al. Zolpidem and rebound insomnia, a double-blind, controlled polysomnographic study in chronic insomniac patients. Pharmacopsychiatry 1994; 27:166-175.
123. Montplaisir J, Boucher S, Poirier G et al. Clinical, polysomnographic, and generic characteristics of restless legs syndrome: A study of 133 patients diagnosed with new standard criteria. Mov D isord 1997; 12:61-65.

124. Montplaisir J, Godbout R, Pelletier G et al. Restless leg syndrome and periodic leg movements during sleep. In: Kryger MH, Roth T, Dement WC, eds. Principles and practice of sleep medicine. Philadelphia, PA: WB Sauders Co., 2000.
125. Montplaisir J, Godbout R, Poirier G et al. Restless legs syndrome and periodic movements in sleep: Physiopathology and treatment with L-dopa. Clin Neuropharmacol 1986; 9:456-463.
126. Morewitz JH. Evaluation of excessive daytime sleepiness in the elderly. J Am Geriatr Soc 1988; 36:324-339.
127. Morrison AR, Bowker RM. The biological significance of PGO spikes in the sleeping cat. Acta Neurobiol Exp 1975; 35:821-840.
128. Moro E, Scerrati M, Romito LM et al. Chronic subthalamic nucleus stimulation reduces medication requirements in Parkinson's disease. Neurology 1999; 53:85-90.
129. Moskovitz C, Moses K, Klawans HL. Levodopa-induced psychosis: A kindling phenomenon. Am J Psychiatry 1978; 135:669-675.
130. Musser WS, Akil M. Clozapine as a treatment for psychosis in Parkinson's disease: A review. J Neuropsychiatry Clin Neurosci 1996; 8:1-9.
131. Nakamuro T, Futamura N, Murata K et al. Screaming during sleep in Parkinson's disease. Rinsho Shinkeigaku 1998; 38:457-60.
132. Nausidea PA, Weiner WJ, Kaplan LR et al. Sleep disruption in the course of chronic levodopa therapy; an early feature of the levodopa psychosis. Clin Neuropharmacol 1982; 5:183-194.
133. Negro PJJ, Faber R. Lewy body disease in a patient with REM sleep disorder. Neurology 1996; 46:1493.
134. Olsen EJ, Boewe BF, Silber MH. Rapid eye movements sleep behavior disorder: Demographic, clinical and laboratory findings in 93 cases. Brain 2000; 123:331-9.
135. Olson WL, Gruenthal M, Mueller ME et al. Gabapentin for parkinsonism: A double blind, placebo-controlled crossover trial. Amer J Medecine 1997; 102:60-66.
136. Oshtory MA, Vijayan N. Clonazepam treatment of insomnia due to sleep myoclonus. Arch Neurol 1980; 37:119-120.
137. Pappert EJ, Goetz CG, Niedermann FG et al. Hallucinations, sleep fragmentation, and altered drem phenomena in Parkinson's disease. Mov Did 1999; 14:117-21.
138. Parkinson J. Essay on the shaking palsy. London: Sherwood Nealy & Jones, 1817.
139. Partinen M. Sleep disorder related to Parkinson's disease. J Neurol 1997; 244:S3-6.
140. Pavlov IP. Innere Hemmung der bedingten Reflexe und der schlaf ein und derselbe Prozess. Skandinavische Archive fur Pshysiologie 1923; 44:42-58.
141. Perry TL, Wright JM, Berry K et al. Dominantly inherited apathy, central hypoventilation, and Parkinson's syndrome: Clinical, biochemical, and neuropathological studies of 2 new cases. Neurology 1990; 40:1882-7.
142. Pfeiffer RF. Gastrointestinal dysfunction in Parkinson's disease. Clin Neurosci 1998; 5:136-46.
143. Phillips B, Young T, Finn L et al. Epidemiology of restless legs syndromes in adults. Arch Intern Med 2000; 160:2137.
144. Plazzi G, Corsini R, Provini F et al. REM sleep behavior disorders in multiple system atrophy. Neurology 1997; 48:1094-1097.
145. Pompeiano O. Mechanisms responsible for spinal inhibition during desynchronized sleep: Experimental study. In: Guilleminault C, Dement WC, Passuant P, eds. Advances in Sleep Research. Narcolepsy NY: Spectrum Press, 1976:3:411-421.
146. Ramm P, Smith CT. Rates of cerebral protein synthesis are linked to slow wave sleep in the rat. Physiol Behav 1990; 48:749-753.
147. Rao ML, Clarenbach P, Vahlensieck M et al. Gabapentin augments whole blood serotonin in healthy young man. J Neural Transmission 1988; 73:129-134.
148. Reuter I, Ellis CM, Ray Chaudhuri K. Nocturnal subcutaneous apomorphine infusion in Parkinson's disease and restless legs syndrome. Acta Neurol Scandinavica 1999; 100:163-7.
149. Rich SS, Friedman JH, Ott BR. Risperidone versus Clozapine in the treatment of psychosis in six patients with Parkinson's disease and other akinetic-rigid syndromes. J Clin Psych 1995; 56:556-559.
150. Rothdach AJ, Trenkwalder C, Haberstock J et al. Prevalence and risk factors of RLS in an elderly population: The MEMO study. Memory and morbidity in Augsburg Elderly. Neurology 2000; 54:1064-1068.
151. Rubio P, Burguera JA, Sobrino R et al. Sleep disorders and Parkinson's disease: Study of series. Rev Neurol 1995; 23:265-8.
152. Rye DB, Bliwise DL. Movement disorders specific to sleep and the nocturnal manifestations of waking movement disorders. In: Watts RL, Koller WC, eds. Movement Disorders: Neurologic principles and practice. New York: Mc Graw-Hill, 1997.
153. Rye DB, Bliwise DL, Dihenia B et al. Daytime sleepiness in Parkinson's disease. J Sleep Res 2000; 9:63-9.
154. Sager O, Goldhammer L, Mares A. Wissenschaft. Berlin: Annalen Akademie Verlag, 1957:338-360.
155. Sakai K, Sastre JP, Danamori N. Specific neurons in the ponto-medulary reticular formation with special references to the postural atonia during paradoxical sleep in the cat. In: Pompeiano O, Marsen CA, eds. Brain mechanism of perceptual awareness and purposal behavior. New York: Raven Press, 1981:405-410.
156. Sanchez-Ramos JR, Ortoll R, Paulson GW. Visual hallucinations associated with Parkinson's disease. Arch Neurol 1996; 53:1265-8.
157. Schapira AH. Sleep attacks (sleep episodes) with pergolide. Lancet 2000; 355:1332-3.
158. Scharf MB, Roth T, Vogel GW et al. Multicenter, placebo controlled study evaluating Zolpidem in the treatment of chronic insomnia. J Clin Psychiatry 1994; 55:192-199.
159. Schatzberg AF, Cole JO, De Battista C. Hypnotoics. Manual of clinical psychopharmacology. Washington, DC: American Psychiatric Press, 1997.
160. Schenck CH, Bundlie SR, Patterson AL. Rapid eye movement sleep behavior disorder: A treatable parasomnia affecting older adults. J Au Med Assoc 1987; 257:1786-1789.
161. Schenck CH, Bundlie SR, Ettinger MG et al. Chronic behavioral disorders of human REM sleep: A category of parasomnia. Sleep 1986; 9:203-308.
162. Schenck CH, Bundlie SR, Mahowald MW. Delayed emergence of a parkinsonian disorder in 38% of 29 older men initially diagnosed with idiopathic rapid eye movement sleep behavior disorder. Neurology 1996; 46:388-393.
163. Schenck CH, Mahowald MW. Polysomnographic, neurologic, psychiatric, and clinical outcome report on 70 consecutive cases with REM sleep behavior disorder (RBD): Sustained clonazepam efficacy in 89,5 of 57 cases. Cleve Clin J Med 1990; 57(Suppl):S9-S23.
164. Schenck CH, Hurwitz TD, Mahowald MW. Symposium: Normal and Abnormal REM sleep egulation: REM sleep behavior disorder: An update of a serie of 96 patients and a review of the world literature. J Sleep Res 1993; 2:224-231.
165. Sforza E, Krieger J, Petiau C. REM sleep behavior disorder and psychopathological findings. Sleep Med Rev 1997; 1:57-69.
166. Sharf B, Moskovitz C, Lupton MD et al. Dream phenomena induced by Chronic levodopa therapy. J Neural Transmission 1978; 43:143-151.
167. Shaw SH, Curson H, Coquelin JP. A double-blind, comparative study of zolpidem and placebo in treatment of insomnia in elderly insomnia in elderly psychiatric in- patients. J Nerv Ment Dis 1968; 147:202-210.
168. Sherrington CS. Man on his nature. New York: Doubley Day Ed, 1955.
169. Shimizu T, Sugita Y, Teshima Y et al. Sleep study in patients with spinocerebellar degeneration and related disease. In: Koella WP, ed. Sleep Basel S Karger 1981; 435-437.
170. Siegel JM. Pontomedullary interactions in the generation of REM sleep. In: Mc Ginty DJ, Drucker –Collin R, Morrison A, Parmeggiani PL, eds. Brain Mechanisms of Sleep. New York: Raven Press, 1985:157-174.
171. Silber MH, Ahlskog JE. REM sleep behavior disorder in parkinsonian syndromes. Sleep Res 1992; 21:313.
172. Smith MC, Ellgring H, Oertel WH. Sleep disturbances on Parkinson's disease patients and spouses. J Am Geriartr Soc 1997; 45:194-9.
173. Spielman AJ, Saskin P, Thorpy MJ. Treatment of chronic insomnia by restriction of time in bed. Sleep 1987; 10:45-46.

174. Starkstein SE, Preziosi TJ, Robinson RG. Sleep disorders, pain, and depression in Parkinson's disease. Eur Neurol 1991; 31:352-5.
175. Steriade M, Amza F. Coalescence of sleep rhythms and their chronology in corticothalamic networks. Sleep Research Online 1998; 1(1):1-10.
176. Stern P. Sweet dreams are made of this. Science 2001; 294:1047.
177. Stern M, Roffwarg H, Duvoisin R. The Parkinsonian tremor in sleep spouses. J Am Geriatr Soc 1997; 45:194-199.
178. Stocchi F, Barbato L, Nordera G et al. S. Sleep disoders in Parkinson's disease. J Neurol 1998; 245(Suppl 1):S15-8.
179. Stoppe G, Brandt CA, Staedt JH. Behavioural problems associated with dementia: The role of newer antipsychotics. Drug Aging 1999; 14:4154.
180. Takahashi Y, Kipnis D, Daughaday W. Growth hormon secretion during sleep. J Clin Invest 1968; 47:2079-2090.
181. Tandberg E, Larsen JP, Karlsen K. Excessive daytime sleepiness and sleep benefit in Parkinson's disease: A community-based study. Mov Disord 1999; 14:922-7.
182. Tandberg E, Larsen JP, Karlsen K. A community-based study of sleep disorders in patients with Parkinson's disease. Mov Disord 1998; 13:895-9.
183. Tan A, Salgado M, Fahn S. Rapid eye movement sleep behavior disorder preceding Parkinson's disease with therapeutic response to levodopa. Mov Disord 1996; 11:214-6.
184. Tassinari CA, Broughton R, Poire R. An electro-clinical study of nocturnal sleep in patients presenting abnormal movements. Electroencephalography Clin Neurophysiol 1965; 18:95.
185. Trenkwalder C. Sleep dysfunction in Parkinson's disease. Clin Neurosci 1998; 5:107-14.
186. Trenkwalder C, Bucher SF, Oertel WH. Electrophysiological pattern of involuntary limb movements in the restless legs syndrome. Muscle Nerve 1996; 19:155-162.
187. Treves TA, Paleacu D, Korczin AD. Treatment of depression in Parkinson's Disease. In: Koller WC, Paulson G, eds. Therapy of Parkinson's Disease. New York: Marcel Dekker, 1995.
188. Trosch RM, Friedman JH, Lannon MC et al. Clozapine use in Parkinson's disease: A retrospective analysis in a large multicentered clinical experience. Mov Disord 1998; 13:377-82.
189. Trzepacz PT, Violette EJ, Sateia MJ. Response to opioids in three patients with restless leg syndrome. Am J Psychiatry 1984; 141:993-995.
190. Turner RS, Chervin RD, Frey KA et al. Probable diffuse Lewy body disease presenting as REM sleep behavior disorder. Neurology 1997; 49:523-527.
191. Uchiyama M, Isse K, Tanaka K. Incidental Lewy body disease in a patient with REM sleep behavior disorder. Neurology 1995; 45:709-712.
192. Valldeoriola F, Molinuevo J. Therapy of behavioral disorders in Parkinson's disease. Biomed Pharmacother 1999; 53:149-53.
193. Vallderiola F, Nobbe FA, Tolosa E. Treatment of behavioural disturbances in Parkinson's disease. J Neural Transm Supplo 1997; 51:175-204.
194. van Hilten B, Hoff JI, Middelkoop HA et al. Sleep disruption in Parkinson's disease. Assessment by continuous activity monitoring. Arch Neurol 1994; 51:922-8.
195. van Hilten JJ, Weggeman M, van der Velde EA et al. Sleep, excessive daytime sleepiness and fatigue in Parkinson's disease. J Neural Trans Park Dis Demnet Sect 1993; 5:235-44.
196. Watanabe T, Sugita Y. REM sleep behavior disorder (RBD) and dissociated REM. Nippon Rinsho 1998; 56:433-438.
197. Watanabe S, Sakai K, Ono Y et al. Alternating periodic leg movements induced by spinal anesthesia in an elderly male. Anaesth Analg 1987; 66:1031-1032.
198. Weitzman ED, Hellman L. Temporal organization of the 24 h pattern of the hypothalamic-pituitary axis. In: Ferin M, Halberg F, Richart RM, eds. Biorhythmsand human reproduction. New York: Wiley, 1974.
199. Weitzman ED,Czeisler CA, Coleman RM. Delayed sleep phase syndrome. A chronobiological disorder with sleep onset insomnia. Arch Gen Psychiatry 1981; 38:737-746.
200. Wetter TC, Collado-Seidel V, Pollmacher T et al. Sleep and periodic leg movement patterns in drug-free patients with Parkinson's disease and multiple system atrophy. Sleep 2000; 23(2141):361-367.
201. Wetter TC, Pollmacher T. Restless legs and periodic leg movements in sleep syndromes. J Neurol 1997; 244(4 suppl 1):S37-45.
202. Wetter TC, Stiansky K, Winkelmann J et al. A randomized, controlled study of pergolide in patients with restless legs syndrome. Neurology 1999; 52:944-950.
203. Whitney CW, Enright PL, Newman AB et al. Correlates of daytime sleepiness in 4578 elderly persons: The cardiovascular healts study. Sleep 1998; 21:27-36.
204. Wolters EC, Jansen ENH, Tuynman-Qua HG et al. Olanzepine in the treatment of dopaminomimetic psychosis in patients with Parkinson's disease. Neurology 1996; 47:1085-1087.
205. Wong DT, Calligaro DO, Bymaster FP et al. Preclinical pharmacology of olanzepine: A novel antipsychotic agent. Brazil: Proceedings 9th World Congress of Psychiatry Rio de Janeiro, 1993.
206. Zarcone VPJ. Sleep Hygiene. In: Kryger MH, Roth T, Dement WC, eds. Principles and practice of sleep medicine. Philadelphia, PA: WB Saunders Co., 2000.
207. Zoldan J, Friedburg G, Goldenberg-Stern H et al. Ondansteron in hallucinations in advanced Parkinson's disease. Lancet 1993; 341:562-563.
208. Zoldan J, Friedburg G, Weitzman A et al. Ondansteron, a 5HT3 antagonist for visual hallucinations and paranoid delusional disorder associated with chronic l-dopa therapy in advenaced Parkinson's disease. In: Battisin L, Scarlato G, Caraceni T, Rugieri S, eds. Advances in Neurology: Parkinson's Disease. Philadelphia, PA: Lippincot-Raven, 1996.
209. Zucconi M, Cocagna G, Petronelli R. Nocturnal myoclonus in restless legs syndrome: Effect of carbamazepine treatment. Funct Neurol 1989; 4:263-271.
210. Zweig RM, Jankel WR, Hedreen JC. The pedunculopontine nucleus in Parkinson's disease. Ann Neurol 1989; 26:41-46.
211. Yokota T, Hirose K, Tanabe H et al. Sleep related periodic leg movements (nocturnal myoclonus) due to spinal cord lesion. J Neurol Sci 1991; 104:13-18.

CHAPTER 27

The Neuropharmacology of Nightmares

J.F. Pagel

Introduction

Frightening dreams are reported from all stages of sleep.[1] The night terrors associated with arousal from deep sleep (stages 3 and 4) are characterized by autonomic discharge, extreme fear, difficult arousal and easy return to sleep. Nightmares are the bad, anxiety dreams often associated with REM sleep, characterized by detailed narrative and often resulting in arousal.[2] The nightmares of posttraumatic stress disorder (PTSD) are frightening, and sometimes stereotypic dreams; often reexperiences of the individual's physical and mental trauma. PTSD nightmares occur at sleep onset (stage 1 and 2) and in REM sleep.

Much of our current knowledge relating to the neuropharmacology of dreaming comes from studies of drug effects on REM sleep (considered the electrophysiological correlate of dreaming). Much research has been limited to animal models in which no dream report is possible, with REM sleep considered as the defined correlate of dreaming. The association between REM sleep and dreaming has been shown, however, to be doubly dissociable; dreaming has been consistently shown to occur outside REM sleep and REM sleep to occur without dreaming.[3]

Current theories postulating specific neuropharmacologies for sleep and dreaming are based primarily on experimental evidence from neurotransmitter studies of single cells in animal models of dreaming in which dreaming is postulated to be equivalent to REM sleep. Because of the difficulty of this micropipette technique our current knowledge is limited to a few hundred of neurons utilizing specific neurotransmitters.[4] This data has been extrapolated to explain the functioning of the hundred billion neurons making up the CNS. Many of these neurons are likely to utilize more than one neurotransmitter and secondary signaling systems.[5-7] The ubiquitous and fast acting neurotransmitters of the CNS have a multiplicity of highly site-specific roles that may result in net changes on behavioral states and cortical arousal that directly contrasts with their effects at the microscopic level.[8] Such single cell neurotransmission data is used to support of various theories of dream phenomenology and function despite often contradictory findings.[4,9]

Sleep and dreaming are global states of consciousness. Evidence reflecting macroscopic CNS regional activity patterns may apply better to neuropsychology and the phenomenology of states such as dreaming. Such evidence includes neuropathology studies, generalized CNS drug effects and side effects, electrophysiology and CNS neuroimaging studies. Neuropathology and CNS neuroimaging have yet to be applied fully to assess CNS correlates of dreaming and nightmares. Currently our best database of knowledge of dream and nightmare neurophysiology may be the known neuropharmaological effects and side-effects of CNS active agents.[10] The rare attempts to collate this data with the single neuron animal studies have been in studies structured so as to support single neurotransmitter theories of dreaming.[11-13] The following article will present an overview of our current knowledge of the drug effects and side effects of CNS active agents as they apply to a particularly well defined type of disturbing and impactful dream – the nightmare. That meta-analytic data is integrated with data from single neuron animal studies, neuropathology and CNS neuroimaging data into theoretical constructs applied to the neuropharmacology of dreaming.

Defining the Nightmare

The DSM-IV defines a nightmare as a frightening dream, avoiding the confusing question of the general use and research definition of dream.[14] There is no generally accepted definition for dream. Based on epistemological training, the definition of dream for a sleep medicine physician (mentation reported as occurring in sleep) may actually contradict the common psychoanalytic definition of dreaming (bizarre, hallucinatory mentation occurring in either sleep or wake). Dream definitions can, however, be categorized as including three primary characteristics:

1. occurring in specific states sleep and/or waking,
2. involving a process of reporting or recall, and
3. including some form of content that may vary from an awareness of dreaming to bizarre, hallucinatory mentation.[15,16]

The terminology—anxiety dream and dream anxiety attack—in general use is interchangeable with nightmare. In other studies, however, an anxiety dream is defined to be a different area of frightening dream classification.[17] Nightmare definitions may include content variables particularly noting the "intensity" of the nightmare - characteristics of visual discontinuity, vividness, intense affect, and self participation.[18] Nightmares have been differentiated from other intense dreams by content variable including the characteristic of dream imagery to be expressed as an unmitigated perception of external reality.[18] Some definitions of nightmare include an intense fearful awakening.[19] Clinically the most commonly used definition for nightmare comes from the American Academy of Sleep Medicine (AASM), "an unpleasant or frightening dream usually occurring in REM sleep".[20] Nightmares are a associated with a variety of sleep disorders that include: nightmare syndrome, Post-traumatic stress disorder (PTSD), and REM behavior disorder; and may occur in

Sleep and Sleep Disorders: A Neuropsychopharmacological Approach, edited by Malcolm Lader, Daniel P. Cardinali and S.R. Pandi-Perumal.

association with narcolepsy, sleep paralysis, sleep starts and terrifying hypnagogic hallucinations. Psychiatric disorders associated with the complaint of nightmares include: nocturnal panic, nocturnal dissociative disorders, PTSD and Acute Stress Disorder.

Nightmare Incidence

Studies of nightmare have been limited by the high incidence and the lack of specificity of the complaint. Nightmares are common, occurring on a frequent basis in greater than 50% of the individuals in some populations.[21] 5 to 8% of the general population report a current problem with nightmares.[22,23] In a general population study of 1,049 patients with insomnia, 47% reported having at least one nightmare/week[24] In this study nightmares were more common in children, between the ages of 5 and 12, and were associated with increases in nocturnal awakenings, sleep onset insomnia and daytime recall.[25] In a two-week prospective study of college students, 47% described at least one nightmare.[26] Frequent nightmares are the most common symptom of Post-traumatic Stress Disorder (PTSD).[27] The nightmares of PTSD are often associated with disturbed sleep and altered daytime behavior that is best described as hyperarousability.[28] The incidence of PTSD following trauma varies at least in part based on the severity of trauma. Thirty percent of Vietnam War veterans were affected by PTSD, as were 68% of veterans of the 1973 Arab-Israeli conflict and 8% of veterans of the Gulf War.[29] Among the civilian population PTSD affects approximately 25% of individuals experiencing severe emotional or physical trauma or a severe medical illness[30] In some groups of patients (i.e., immigrant psychiatric patients) the incidence of PTSD exceeds 40%.[31,32]

The Association between Rapid Eye Movement Sleep (REMS) and Nightmares

Awakenings with anxiety dreams occur out of all stages of sleep.[1] Sleep terrors—an arousal disorder occurring out of stages 3 and 4—can usually be clinically differentiated from nightmares. Sleep terrors usually occur in the first half of the night, dream content is limited, autonomic and motor behavior common and sometimes associated with injuries, awakening difficult and associated with disorientation and confusion, and often followed by a return to sleep. Nightmares are classified as a REMS parasomnia.[30] Much of the physiologic and pharmacological literature has generally addressed dreaming by quantifying the physiologic correlate of REM sleep. Psychoactive medications alter the EEG associated with sleep, often affecting REMS percentage, latency, associated phasic activity, and the amplitude and frequency of sleep associated EEG rhythms.[33,34]

Researchers and theorists often consider that REMS = Dreaming. As Alan Hobson has stated referring to drug effects, "There are no dreams because there is no REM sleep".[35] Dreaming, however, dependent on definition, has been recurrently noted to occur on arousal from sleep stages other than REMS. Dream recall is similar on awakenings from sleep onset and REM sleep (> 80%).[36] Recall from other nonREM stages of sleep depends primarily on the stages distance from awake, being lowest from deep sleep (approximately 40% of awakenings associated with dream recall). The dream remembered in the morning is most likely to be from the longest, and most physiologically disturbed REMS period (most eye movements and irregularity of respiration).[37,38] Some authors have postulated that the occurrence of dreaming, defined as bizarre hallucinatory mentation, occurring outside REM sleep indicates that REM sleep must occur outside what is polysomnographically defined as REM sleep. This theory would account for the dream reports from other sleep stages and for dream like mentation reported while awake.[39] Polysomnographic evaluation of patients suffering from nightmares reveal that only a small subset have increased arousals occurring out of REM sleep (dream interruption insomnia) as would be expected from a REMS parasomnia often associated with arousal.[2,19,40] Nightmares, particularly those associated with the diagnosis of PTSD can clearly occur in Sleep Stages 1 and 2 interfering with the initiation of sleep.[41]

Neurochemical Modulation of REM Sleep—Single Neuron Data From Animal Models

A considerable body of evidence suggests that reciprocal interaction between noradrenergic "REM-off" neurons, and cholinergic and noncholinergic "REM-on" neurons in the locus coerleus acts as a control system for REM sleep.[42,43] Serotonin and norepinephrine firing and levels are markedly reduced in the locus coerleus during REMS while dopamine influences are maintained.[44] Evidence also suggests that GABA is an inhibitory intermediary neurotransmitter for this system in the locus coerleus.[45] The expression of REM sleep related physiology (ponto-geniculate-occipital (PGO) waves and atonia) also depends on a subpopulation of pedunculopontine and laterodorsal tegmental neurons that release acetylcholine to act upon muscarinic receptors. The responsiveness of this region and therefore modulation of REM sleep involves closely adjacent glutamateric neurons and alternate afferent neurotransmitters.[8,46] The systems' known neurochemical complexity has limited consistency with postulated adrenergic/cholinergic on/off switch models of REMS and dreaming.[7,8,12]

Neuroimaging of REM Sleep

PET data has consistently demonstrated decreased lateral prefrontal deactivation and limbic midline activation during REM sleep.[6,8,51,52] Portions of the ventromedial limbic related prefrontal cortices, and closely related medial subcortex and cortex reactivate in REM sleep following their deactivation relative to waking and NREM sleep.[51,52] Activation of the dorsolateral prefrontal cortex, the cingulate, precuneus and medial frontal cortex and portions of the parietal visual spatial attentional system is associated with the saccadic eye movements of REM sleep.[53] Near infrared spectroscopy indicates activation of the visual cortex during REM sleep.[54] Since the large body of molecular receptor studies (see above) have concentrated on brain stem modulators for REM sleep, it is somewhat surprising that neuroimaging data demonstrates a general pattern of cortical rather than brain stem alteration in activity associated with REM sleep.

Neuropathology and Dreaming

A cessation of dreaming has been associated with both left and right, and bilateral inferior parietal-lobe lesions as well as with deep bifrontal lesions.[13,55] Neuropsychological studies have suggested that specific cases of focal cerebral pathology are associated with the loss of dream recall, even in cases in which REM sleep continues to occur.[55] These findings suggest that both focal and diffuse cerebral pathology can induce reports of nondreaming in some patients that had previously experienced dreaming.[13,55] These studies cataloged a large number of patients who continued to report dreaming despite extensive CNS damage including those with survivable brainstem lesions.

Nightmares – Reported Pharmacologic Effects and Side Effects

The literature describing drug effects on dreaming in humans consists of retrospective human studies addressing reports of vivid dreaming/nightmares as a side effect of medication or as a symptom of medication withdrawal.[10,56] The high frequency of nightmares in both general and psychiatric populations makes interpretation of drug induced nightmare data difficult since few of these studies include blinded or placebo cohorts for comparison. Historically, several groupings of medication types have been associated with nightmare occurrence secondary either to withdrawal, toxicity or as side effects of use (Table 1). A meta-analysis of clinical trial in which disturbed dreaming and nightmares are reported is presented in Table 2A and 2B. A qualitative probability assessment is used to determine the probability that nightmares are drug induced by these agents using the following range of assessment: definite – probable – possible – doubtful.[57]

Neurotransmitter Modulating Systems Postulated to Be Involved in Dreaming and Nightmares

Acetycholine

REM sleep is affected by pharmacological alteration of cholinergic activity in the CNS.[58] Many lines of study support the hypothesis that brainstem cholinergic neurons can be excited to induce REM sleep.[4,59] Micropipette animal studies with carbacol, an acetyl-choline agonist; and neostigmine, an acetyl-cholinesterase inhibitor, have demonstrated that REMS can be induced by injections into pontine-recticular areas.[4,60] Cholinergic agents are most likely to increase percentages of REM sleep, with cholinergic antagonists tending to decrease REM sleep.[4] Cholinergic neurons in the ventral medulla may induce the suppression of motor discharge associated with the muscle atonia of REM sleep.[61] The most widely accepted neurochemical construct for REMS involves the reciprocal interaction of declining noradrenergic and serotonergic activity in the locus coeruleus and Ralphe nucleus with increased cholinergic activity at REMS onset. This disinhibition of cholinergic neurons results in phasic activation of the geniculate bodies and visual cortex and increases in PGO wave activity.[35,62] The reported side effects of agents with anticholinergic activity include, nightmares, disordered dreaming and hallucinations. This has led to theoretical constructs in which the cholinergic activity of a wide variety of pharmaceutical agents has been postulated to explain CNS side effects, including nightmares and hallucinations sometimes induced by these drugs.[11]

Pharmacological agents affecting acetylcholine metabolism have, however, rarely been reported in case reports and clinical trials to induce nightmares. The acetylcholinesterase inhibitors (donepezil, rivastigmine and tacrine) act as cholinergic agonists and are rated as possibly associated with reports of disordered dreaming based on clinical trial data (Table 2A).[10] Based on the known involvement of acetylcholine neuroreceptors with REM sleep and theoretical postulates that REMS=dreaming, it is somewhat surprising that this clinical side-effect data does not provide good support for theoretical postulates that cholinergic neurons serve as the primary neuroreceptor system involved in dreaming and nightmares.

Norepinephrine

Brainstem noradrenergic activity declines at REMS onset. An increase in noradrenergic activity, however, has been associated with PTSD.[63] Administration of an agent (Yohimbine) which activates norepinephrine neurons by blocking the alpha2-adrenergic autoreceptor can lead to an increase in PTSD symptoms (panic attacks and flashbacks).[64] Both beta-blockers and alpha agonist are used clinically in the treatment of PTSD symptoms including recurrent nightmares.[65] Antihypertensive agents in general use affect adrenergic CNS receptors. These drugs have been shown to affect both REM sleep and reports of dreaming. The reported effects of these agents on both dreams and nightmares, may be opposite to the drug's known pharmacological effects on REM sleep.[66] Decreases in dream recall occur with use of both alpha agonists (REM suppressant) and beta blockers (nonREM suppressant). An agents' effect on REM sleep may or may not be associated with an associated change in reported dreaming. The use of beta-blockers depresses REM sleep percentages yet can result in reports of increased dreaming, nightmares and hallucinations.[67] Beta-blocker and alpha-agonist agents are responsible for 34% of clinical trials in which nightmares are reported as an adverse effect.[56] Clinical trial and case report data demonstrates that beta-blockers as a class are the agents most commonly associated with patient reports of nightmares (Table 2A, 10).

Table 1. Medication types known historically to cause nightmares

Drug Classes Known to Induce Nightmares	Type of Effect	Postulated Pharmacologic Etiology for Nightmare Induction
Amphetamines and amphetamine like agents	Associated with chronic use and withdrawal	Noradreanergic/Dopaminergic
Anticholinergics	Associated with chronic use and toxicity	Cholinergic
Barbiturates and barbituate-like agents	Known side effect on withdrawal	REM Rebound
Benzodiazepines	Known side effect on withdrawal	REM Rebound/GABA
Dopamine agonists	Associated with chronic use and over-dosage	Dopaminergic
Ethanol	Known side effect on withdrawal	REM Rebound
Hallucinogenics	Toxicity and effect of use	Serotonin and norepinephrine, cognitive effects, cholinergic
Tricyclic antidepressants	Drug and withdrawal effects	Serotonin and norepinephrine

Table 2A. Medications affecting CNS neurotransmitter systems reported to induce nightmares in clinical trials and case studies

Affected Neuroreceptor Drug	Patient Reports of Nightmares – Evidence Base Clinical Trials (CT) Case Reports (CR)	Probability Assesment of Drug Effect
ACETYLCHOLINE– Cholinergic Agonists		
Donepezil	CT [3/747 report disordered dreaming]	Possible
Rivastigmine	CT [1/100-1/1000 report disordered dreaming]	Possible
Tacrine	CT [1/100-1/1000 (2076) report disordered dreaming]	Possible
NOREPINEPHRINE – Beta Blockers		
Atenolol	CT [3/20 patients]	Probable
Betaxolol and carbachol [opth.]	CR [1] – de-challenge	Possible
Bisopropol	CT [3/68 patients] : CR [1] – de-challenge	Probable
Labetalol	CT [5/175 patients]	Probable
Oxprenolol	CT [11/130 patients]	Probable
Propranolol	CT [8/107 patients]	Probable
– Norepinephrine Effecting Agents		
Atomoxetine	CT [4/269 abnormal dreams compared to 3/263 placebo group]	Possible
Deserpidine	CT – disordered dreaming listed as side effect	Possible
Guanethidine	CT [4/48 patients]	Probable
Methyl dopa	CT [infrequent reports of nightmares]	Possible
Tramadol	CR [1] – de-challenge	Possible
SEROTONIN – SSRI		
Fluoxetine	CT [1-5% - greater frequency in OCD and bulemic trials: CR [4] – de and re-challenge]	Probable
Escitalopram oxylate	CT [Abnormal dreaming – 1% 999 patients]	Probable
Nefazodone	CT [3% (372) versus 2% control]	Probable
Paroxetine	CT [4% (392) versus 1% control]	Significant
DOPAMINE - Agonists		
Amantadine	CT [5% report abnormal dreams]: CR [1]	Probable
Bupropion	CR [1] – de-challenge Possible	
Cabergoline	CT [1/188 patients]: CR [1] – de-challenge	Possible
Levodopa	CT [2/9 patients]	Probable
Pergolide	CT [2.7% (189) report abnormal dreams versus 4.5% placebo]	Doubtful
Ropinirole	CT [3% (208) report abnormal dreaming versus 2% placebo]	Probable
Selegiline	CT [2/49 reporting vivid dreams]	Probable
– Amphetamine like Agents		
Bethanidine	CT [2/44 patients]	Probable
Fenfluramine	CT [7/28 patients]: CR [1] de and re-challenge	Probable
Phenmetrazine	CT [3/81 patients]	Probable
GABA		
Flunitrazepam	CT [1/127 patients]	Possible
Gabapentin	CT [1/100-1/1000 (2074) report abnormal dreams]	Possible
Gaba hydroxy buterate	CT [nightmares >1% 473 patients]	Probable
Nitrazepam	CR [2]	Possible
Triazolam	CT [7/21 patients]	Probable
Tiagabine	CT [3/2531 patients]	Possible
Zopiclone	CT [3 – 5/83 patients]	Probable

Serotonin

Antidepressants affecting serotonin metabolism often induce REM sleep suppression. This effect is greatest for the monamine oxidase inhibitors (MAOI's) followed by the tricyclic antidepressants and the selective serotonin reuptake inhibitors (SSRIs). REM sleep suppression is not generally seen with buspiron, trimpramine and nefazodone.[68] Intense visual dreaming and nightmares are associated with clomiprimine, paroxetine and fluvoxamine withdrawal.[69] This effect could occur secondary to REM sleep rebound occurring after the withdrawal of these REM sleep suppressant agents. However, studies of reported dream recall with antidepressant use show that recall may vary independently of REMS suppression, though an initial decline in dream recall may occur during the period of initial dosing of the agents.[70] Studies of chronic steady state use and antidepressant withdrawal have shown inconsistent effects: increased dream recall with SSRI's and

Table 2B. Other drug classes reported to induce nightmares in recent case reports and clinical trials

Affected Neuroreceptor Drug	Patient Reports of Nightmares – Evidence Base Clinical Trials (CT) Case Reports (CR)	Probability Assesment of Drug Effect
ANESTHETICS		
Katamine	CR [1]	Possible*
Midazolam	CT [< 1%]	Possible*
ANTIINFECTIVES and IMMUNO-SUPRESSANTS		
Amantadine	CT [5% reporting abnormal dreams]: CR [1]	Probable+*^
Ciprofloxacin	CR [1] – de-challenge	Possible^
Erythromycin	CR [2] – de-challenge	Possible
Fleroxacin	CT [7/84 patients]	Probable^
Ganciclovir	CR [1] – de and re-challenge	Probable*^
Gusperimus	CT [13/36 patient]	Probable
ANTI-EPILEPTICS		
Ethosuximide	CT [reports of night terrors]	Possible*
Lamotrigine	CT [1/100-1/1000 report abnormal dreams]	Possible*^
Valproic acid	CR [1] – de-challenge	Possible*^
Zonisamide	CT [1/100-1/1000 report abnormal dreams]	Possible^
ANTI-PSYCHOTICS		
Chlorpromazine	CR [1] – de-challenge	Possible*^
Clozapine	CT [4%]	Probable*^
Thiothixene	CR [3] – de-challenge	Possible*^
ANTI-HISTAMINE		
Chlorpheniramine	CT [4/80 patients]	Probable*^
ACE INHIBITORS		
Captopril	CR [1]	Possible*^
Enalapril	CT [.5-1% abnormal dreaming – 2987 patients]	Probable*^
Losartin potassium	CT [> 1% dream abnormality – 858 patients]	Probable*^
Quinapril	CT Probable*^	
OTHER AGENTS – NO PROPOSED MECHANISM		
Buprenorphrine	CR [1] – de-challenge	Possible
Digoxin	CR [1] – de and re-challenge	Probable
Naproxen	CR [1] – de and re-challenge	Probable*^
Verapamil	CR [1] – de and re-challenge	Probable^

+ Agents listed in multiple classes; * Agents inducing daytime sedation as a side effect to use; ^ Agents inducing insomnia as a side effect to use

A qualitative probability assessment is used to determine the probability that nightmares are drug induced by these agents based on the Naranjo (1981) algorithm ranging from definite – probable – possible – doubtful.[57] The association between each medication and described side effect (nightmares or alterations in dreaming) in clinical trial reports is rated from significant (p< .01), to probable (reported by > 1% of population relative to controls), to possible (less than 1% difference compared to control or in studies without controls) to doubtful (minimal evidence for side-effect/drug association). This study does not include data concerning drugs for which no effects on dreaming or nightmares are reported, because of concerns as to the significance of negative reports. Older (before 1990) clinical trial data is not included since clinical trial reports for agents known to induce nightmares per case reports (examples: tricylic antidepressants, amphetamines, and benzodiazepines [Table 2] often did not include reports of nightmares or disturbed dreaming.

tricyclic antidepressants, no effect, and decreased recall.[71,72] SSRI use can be associated with patient reports of nightmares: rated as significant for paroxetine; probable for fluoxetine, nefazodone; and possible for sertraline. Other agents that have serotonin effects are also reported to induce nightmares: probable (rintaserin), and possible (venlafaxine and zonisamide) (Table 2A).

Dopamine

A cessation of dreaming has been associated with both left and right, and bilateral inferior parietal-lobe lesions.[13] Neuropsychological studies have suggested that specific cases of focal cerebral pathology are associated with the loss of dream recall, even in cases in which REM sleep continues to occur.[55] These neural tracts ablation, of which results in a global loss of dreaming, are postulated to be dopaminergic.[55] Dopamine receptor stimulation may induce nightmares. Medications efficacious in the treatment of Parkinson's disease share the tendency to cause disordered dreaming. Dopamine, bromocriptine, pergoline and other dopamine agonists can lead to vivid dreaming, nightmares and night terrors which can be the first signs of the development of drug induced psychosis.[73] While amphetamines are adrenergic agonist agents, cognitive side effects such as nightmares have been postulated to occur secondary to dopamine receptor stimulation.[56] A probable association with reports of nightmares was noted for amantadine, levodopa, ropinirole and selegiline. A possible association exists for cabergoline. The dopamine reuptake inhibitor bupropion has a possible association with nightmare reports. The association of the amphetamine-like agents fenfluramine and phenmetrazine with patient reports of nightmares is rated as probable based on both clinical trials and case report data (Table 2A).

GABA

Nightmares and intense dreaming have been associated with the REM sleep rebound associated with withdrawal from REM sleep suppressing agents. In addition to antidepressant agents, REM sleep suppressants include: ethanol, barbituates, benzodiazepines, nonbenzodiazepine hypnotics and sympathomimetic drugs.[33,34] Several authors have suggested that the nightmare inducing characteristics of these agents reflect drug effects at the GABA receptor.[74,76] Populations of inhibitory thalamic GABAnergic cells are known to be active during REMS and responsive to acetylcholine stimulation.[77]

Twenty-four percent of reports of nightmares come from benzodiazpine clinical trials.[56] The newer nonbenzodiazepine hypnotic (zopiclone, zaleplon and zolpidem) are not associated at clinical dosages with REM sleep suppression or REM sleep rebound on withdrawal yet these agents have been reported in clinical trials to induce nightmares.[78,79]

Other Agents with a Reported Association to Nightmare Induction

Much of the literature concerning pharmacological agents reported to induce nightmares is related to agents used in anesthesia. Although not clearly sleep, some induction anesthetics are reported to cause altered dreaming. An increased incidence of "pleasant" dreams are reported with profolol use.[80] The barbituate thiopental, isoflurane, and ketamine have been reported to produce disordered dreaming and nightmares.[81,82] Ketamine can induce waking hallucinations and confusion. This association has, in part, led to proposals that dreams and nightmares are hallucinatory experiences occurring during sleep.[83] Ketamine and midazolam are considered to have a possible association with patient reports of nightmares (Table 2B).

Agents Affecting Host Defence Reported to Induce Nightmares

Aristotle and Hippocrates pointed out that an association exists between infection and sleepiness. Both viral and bacterial infections can be associated with large increases in NREM sleep.[84] Such microbial-induced changes in sleep are considered part of the acute phase response, and can be induced by both muramyl peptides and endotoxins.[84,85] Some antibiotics (i.e., fluoroquinolones) are associated with patient reports of insomnia, and have been noted to induce nightmares.[85,86] The cytokines IL-1B and TNF-α, and prostaglandin E2 are known to be involved in NonREM sleep regulation.[86] Antibiotics, antivirals and immunosupressant drugs can induce in some patients the complaint of nightmares. A clear, but currently poorly defined relationship exists between host defense and infectious disease, and sleep/dreaming. Several of the agents reported to induce nightmares in clinical trials and case reports are antibiotics: fleroxacin (probable); erythromycin, and ciprofloxacin (possible association with patient reports of nightmares). Both amantadine and ganciclovir are antiviral agents considered by the authors to have a probable association with patient reports of nightmares. Gusperimus is an immunologic response suppressant that is also reported to induce nightmares (probable). These agents may induce nightmares by affecting sleep-related immunologic response to infectious disease (Table 2B).[10]

Several antiepileptics have a possible association with nightmares in both case reports and clinical trials (valproic acid, ethosuximide, zonisamide, and lamotrigine). The antipsychotic clozapine has a probable association with nightmares, with other antipsychotics (chlorpromazine and thiothixene) noted to have a possible association. Most of these agents commonly produce daytime sedation as a side effect to medication use (Table 2B [*]). The antihistamine chloropheniramine utilized for allergy symptoms and to induce sleepiness is likely the most commonly used medication addressed in this review. Chlorpheniramine has a probable association with patient reports of nightmares. Table 2B also includes agents reported to cause nightmares for which no causality is postulated.

Several of the antihypertensive agents reported as probable inducers of nightmares (captopril, enalapril, losartin potassium, and quinipril) are inhibitors of angiotensin I-converting enzyme (ACE). These agents can induce insomnia as a side effect to use. The CNS effects of these agents may be through neurotransmitter effects. In some patients ACE inhibitors lower plasma norepinephrine levels, and may inhibit presynaptic norepinephrine release and postsynaptic alpha adrenoreceptor activity.[87] The clinical significance of this finding is not known.

Summary

Neuotransmitters Systems Affecting Dreaming and Nightmares

Animal studies demonstrate that brainstem adrenergic and aminergic neuronal populations are involved with GABAminergic neuronal populations in REM sleep.[4,8] Studies of focal human cerebral pathology indicate a probable involvement of basilar frontal cortex dopaninergic neuronal tracts in dreaming.[3,55] Data from human clinical trials and case reports indicates that adrenergic, aminergic, and dopaminergic neuronal populations are likely to have prominent roles in inducing nightmares. Reports of altered dreaming and nightmares are consistently associated with agents exerting pharmacological effects on dopamine, serotonin, and norepinephrine. Beta-blockers are the agents most likely to result in patient complaints of nightmares. The strongest clinical evidence found in this meta-analysis for the association a drug with nightmare induction is for the SSRI paroxetine.[10] Most agents affecting dopaminergic neuroreceptors have been reported in clinical trials to induce nightmares in some patients. Medications altering these neurotransmitter systems are likely to induce reports of nightmares and disordered dreaming for patients taking those medications. These neurotransmitters may function in a reciprocal interaction involving a wide spectrum of neurotransmitters interacting in an intricate modulation of the cardinal sleep stages – REM and NonREM sleep.[8]

Clinical trial and case report data is less clear in its support for the association of GABA and acetylcholine receptors with dreaming and nightmare alteration with the reported nightmare/drug association rated as possible (rather than probable or significant) for the majority of drugs evaluated. The finding that different types of drugs known to affect the GABA receptor (agonists, modulators, and reuptake inhibitors) can result in patient complaints of nightmares and abnormal dreaming is suggestive that GABA may be a modulator of the neuronal populations involved in dreaming.[45,74,75] Other neurotransmitter modulators proposed to affect this system include orexin, adenosine, histamine, glycine, glutamate, nitric acid and neuropeptides.[8] The commonly used antihistamine chlorpheniramine has been reported to induce nightmares suggesting a potential role for histamine as a modulator of dreaming. The neurochemical and pharmacological basis for clinical effect for many of the agents included in (Table 1) remains poorly defined. It is possible that the induction of nightmares and altered dreaming by some of these agents is secondary to neurotransmitter effects that in the future may be better described.

Discussion

Some theorists have argued for a simplified "chemistry of conscious states' in which the adrenergic/amniergic reciprocal paradigm responsible for REM sleep induction is applied to dreaming.[4,11] Others have argued for dopamine as primary neurotransmitter of dream.[13,44] Recent single cell neurotransmission studies have served, however, to emphasize the complexity of the neurotransmission modulating system of REM sleep.[7,8] At this point in time a congruence of studies from very different experimental orientation (human case reports of nightmares and animal model studies of REM sleep modulation) indicate that this is a complex system characterized poorly by neurotransmitter models limited to the modulation one or several neurotransmitters at discrete CNS sites. The data from clinical trial and case reports of medications reported to induce nightmares indicates a complex pattern of the neurochemical systems. This is experimental data obtained from humans in which the cognitive process of dreaming is addressed. The imagery of dreaming, particularly nightmares with their externally unmitigated perception of external reality, parodies in its complex process of nonperceptual imagery the perceptually involved process of waking consciousness. Dreaming is likely to utilize the same memory processes utilized for the incorporation of waking memories.[4,58,62,77] Neurotransmitter systems utilized in nonperceptual conscious processes such as imagery are likely to be involved in dreaming as well. This conceptual view of dreaming as a state of consciousness is supported by this review of the neurochemistry of nightmares which suggests that many of the same neurotransmitters involved in the processes of waking consciousness can also alter dreaming. These findings argue that dreaming is more likely a state of consciousness rather than a simpler form of perceptual hallucination as suggested by some theorists.[35,62,83]

The neuropharmacological agents inducing nightmares in human studies do not correlate consistently with the agents reported to induce REM sleep in animal models. Pharmacological agents affecting the negative neuro-modulators of REM sleep—norepinephrine and serotonin—are those most likely to induce nightmares in human subjects. Agents that increase acetylcholine levels such as the acetylcholineserase inhibitors routinely utilized in patients with Alzheimer's disease would be expected based on animal models to increase REM sleep. The side effect of nightmares and/or altered dreaming secondary to the use of these agents is rarely reported (only 3 of 747 patients using donepizil in clinical trials reported changes in dreaming) (Table 1A). Neuropathological studies demonstrate that REM sleep is modulated in the brain stem while the presence or absence of reports of dreaming are affected by cerebral lesions.[13,55,59] These lines of evidence lend credence to the abundance of psychological studies demonstrating that dreaming occurs without REM sleep, and REM sleep without dreaming.[3,36,88]

These lines of evidence further emphasize that REM sleep is not dream sleep. REM sleep is a well defined electrophysiological process occurring regularly during sleep and present in most mammals. Neuropharmacological modulation of REM sleep occurs in the brain stem and involves the neurotransmitters: acetylcholine (+ modulator), Serotonin and norepinephrine (- modulation), GABA and Dopamine. A complex spectrum of additional enzymes, peptides, and neurotransmitters including NADPH-diaphorase, nitric oxide synthase, corticotrophin-releasing factor, substance P, and glutamate are involved in the orchestration of normal REM sleep.[7,8,46-50] Dreaming requires the ability to consciously describe a report of sleep mentation. Dreams are not restricted to REM sleep and are commonly reported from all stages of sleep.[2,36] Dreaming is no longer reported by some patients with damage to left the and right, and bilateral inferior parietal-lobe as well as with deep bifrontal lesions.[13,55] Pharmacological agents affecting adrenergic, aminergic, and dopaminergic neuronal populations induce nightmares. A spectrum of other agents including anti-infectives, anesthetics, ACE inhibitors, anti-seizure medications and antihistamines.[10]

Studies of REM sleep and studies of dreaming are studies of different yet related topics. REM sleep is a phylogenically ancient brain stem process.[8] An intact frontal and parietal cortex is required for the dream report.[13,55] Both REM and NREM sleep are physiologic states with the neurochemical and electrophysiological neurosignalling capacity for interaction between brainstem and cortex during sleep.[8,33]

Conclusion

The study of the neuropharmacology of dreaming and nightmares is an area of neuroscience in its infancy. Dreaming and nightmares are an area of study replete with grand theories and a paucity of actual data. Evidence from neuropathology studies, generalized CNS drug effects and side effects, electro physiology and CNS neuroimaging studies is beginning to be applied to this area of study. Initial findings suggest a complex neurochemical system in which neuropharmacological agents that affect REM sleep differ from those that affect dreaming.

Definitions

Sleep – A global reversible state of perceptual disengagement from the environment.

Dream – There is no generally accepted definition for dream. Based on epistemological training, the definition of dream for a sleep medicine physician (mentation reported as occurring in sleep) may actually contradict the common psychoanalytic definition of dreaming (bizarre, hallucinatory mentation occurring in either sleep or wake). Dream definitions can, however, be categorized as including three primary characteristics: (1) occurring in specific states sleep and/or waking, (2) involving a process of reporting or recall, and (3) including some form of content that may vary from an awareness of dreaming to bizarre, hallucinatory mentation.

Nightmare – Nightmares are vivid and distressing mental experiences occurring during sleep often resulting in arousal and generally associated with REM sleep.

Rapid Eye Movement Sleep (Rem Sleep) – A stage of sleep identified by the simultaneous presence of a relatively low voltage cortical electroencephalogram (EEG), an absence of activity in the antigravity muscles (atonia) and periodic bursts of rapid eye movements. Individuals awaked from REM sleep report dream mentation more than 80% of the time.

Non Rem (Nrem) Sleep – The other stages of sleep that are not REM sleep including sleep onset (Stage 1 and 2 from which dreaming may be reported more than 80% of the time) and deep sleep (stages 3 and 4) from which dreams are reported approximately 40% of the time on arousal. These stages are often described as synchronized sleep because each stage is characterized by the presence of regular sleep stage specific background cortical EEG frequencies.

Neurotransmitters – Chemical messengers active at the neural synapse fulfilling four criteria: (1) It is synthesized by the neuron, (2) It is present in the presynaptic terminal and is released in amounts sufficient to exert a defined action on the postsynaptic neuron or effector organ, (3) When administered exogenously in reasonable concentrations it mimics the action of the endogenously released transmitter exactly, and (4) A specific mechanism exists for removing it from its site of action.

Positron Emission Tomography (Pet) – A scanning device that uses low dose radioactive sugar, usually deoxyglucose-F18, to measure brain activity producing a scanning report describing variations in levels of brain area function for discrete time periods.

Posttraumatic Stress Disorder (Ptsd) - Nightmares are the most common symptom of PTSD a disorder characterized by symptoms of "hyper arousal" occurring after serious physical or mental trauma

Sleep Terrors – an arousal disorder occurring out of stages 3 and 4 – can usually be clinically differentiated from nightmares. Sleep terrors usually occur in the first half of the night, dream content is limited, autonomic and motor behavior common and sometimes associated with injuries, awakening difficult and associated with disorientation and confusion, and often followed by a return to sleep

Hallucination – A sensory perception that has the compelling sense of reality of a true perception but that occurs without external stimulation of the relevant sensory organ.

References

1. Fisher C, Byrne J, Edwards A et al. A psychophysiological study of nightmares. J Am Psychoanal Assoc 1970; 18:747-782.
2. Nielsen TA, Zadra A. Dreaming disorders. In: Kryger MH, Roth T, Dement WC, eds. Principles and practice of sleep medicine. 3rd ed. Philadelphia, PA: WB Saunders Co., 2000:753-772.
3. Solms M. Dreaming and REM sleep are controlled by different brain functions. In: Pace-Schott EF, Solms M, Blagrove M, et al, eds. Sleep and Dreaming: Scientific Advances and Reconsiderations. Cambridge: Cambridge University Press, 2003:51-58.
4. Hobson JA, Steriade M. The neuronal basis of behavioral state control: Internal regulatory systems of the brain. In: Bloom F, Mountcastle V, eds. Handbook of Psychology. Washington DC: American Physiological Society, 1986:IV(14):701-823.
5. Kandle ER. The brain and behavior. In: Kandel ER, Schwartz JH, Jessell TM, eds. Principles of Neural Science. 4th ed. New York: McGraw Hill, 2000:5-18.
6. Schwartz JH. Neurotransmitters. In: Kandel ER, Schwartz JH, Jessell TM, eds. Principles of Neural Science. 4th ed. New York: McGraw Hill, 2000:280-297.
7. Rye DB. Contributions of the pedunculopontine region to normal and altered REM sleep. Sleep 1997; 20(9):757-788.
8. Pace-Schott EF. Postscript: Recent findings on the neurobiology of sleep and dreaming. In: Pace-Schott EF, Solms M, Blagrove M, et al, eds. Sleep and Dreaming: Scientific Advances and Reconsiderations. Cambridge: Cambridge University Press, 2003:335-350.
9. Hobson JA, Pace-Schott EF, Stickgold R. Dream science 2000: A response to commentaries on dreaming and the brain. In: Pace-Schott EF, Solms M, Blagrove M, et al, eds. Sleep and Dreaming: Scientific Advances and Reconsiderations. Cambridge: Cambridge University Press, 2003:231-246.
10. Pagel JF, Helfter P. Drug induced nightmares – An etiology based review. Hum Psychopharmacol Clin Exp 2003; 18:59-67.
11. Perry E, Perry R. Acetylcholine and hallucinations: Disease-related compared to drug-induced alterations in human consciousness. Brain and Cognition 1995; 28:240-258.
12. Hobson JA. The chemistry of conscious states. Boston: Little, Brown and Company, 1994.
13. Solms M. The neuropsychology of dreams - A clinicoanatomical study. Mahwah, New Jersey: Lawrence Erlbaum Associates, Publishers, 1997.
14. Francis A. (Chair – Task Force on DSM-IV) DSM-IV – Diagnostic and Statistical Manual of Mental Disorders. 4th ed. Washington DC: American Psychiatric Association, 1994.
15. Pagel JF, (Chairman) Blagrove M, Levin R et al. Defining dreaming - A paradigm for comparing interdisciplinary studies of dream. The bulletin of the American Academy of Sleep Medicine, 1999; 6(4):34.
16. Pagel JF, Blagrove M, Levin R et al. Definions of dream - A paradigm for comparing field descriptive specific studies of dream. Dreaming 2001; 11(4):195-202.
17. Krakow B, Neidhardt J. Conquering bad dreams and nightmares: A guide to understanding, interpretation and cure. New York: Berkley Books, 1992.
18. Kuiken D, Sikora S. The impact of dreams on waking thoughts and feelings. In: Moffitt A, Kramer M, Hoffman R, eds. The Functions of Dreaming. Albany, NY: State University of New York Press, 1993:419-476.
19. Zadra A, Donderi DC. Prevalence of nightmares and bad dreams and their relation to psychological well-being. Jornal of Abnormal Psychology 2000; 109:210-219.
20. In: Thorpy M, ed. American sleep disorders association, the international classification of sleep disorders. Revised: Diagnostic and coding manuel. Rochester, MN: American Sleep Disorders Association, 1997.
21. Pagel JF. Nightmares and disorders of dreaming. American Family Physician Apr 2000; 61(7):2037-2042.
22. Bixler EO et al. Prevalence of sleep disorders in the Los Angeles metropolitan area. Am J Psychiatry 1979; 136:1257-1262.
23. Klink M, Quan SF. Prevalence of reported sleep disturbances in a general adult population and their relationship to obstructive airway diseases. Chest 1987; 91:540-546.
24. In: Moffitt A, Kramer M, Hoffman R, eds. The functions of dreaming. Albany, NY: State University of New York Press, 1993.
25. Terr LC. Nightmares in Children. In: Guilleminault C, ed. Sleep and its Disorders in Children. New York: Raven, 1987.
26. Wood JM, Bootzin RR. The prevalence of nightmares and their independence from anxiety. J Abnorm Psychol 1990; 99:64-68.
27. Fawzi MC et al. The validity of posttraumatic stress disorder among Vietnamese refugees. Journal of Trauma and Stress 1997; 10:101-108.
28. Grillon C et al. Baseline startle amplitude and prepulse inhibition in Vietnam veterans with posttraumatic stress disorder. Psychiatry Research 1996; 64:169-178.
29. Yehuda R, McFarlane AC. Conflict between current knowledge about posttraumatic stress disorder and its original conceptual basis. American Journal of Psychiatry 1995; 152:1705-1713.
30. Schenck CH, Mahowald MW. REM sleep parasomnias. Neurology Clinics 1996; 14:697-720.
31. Ekblad S, Roth G. Diagnosing posttraumatic stress disorder in multicultural patients in a Stockholm psychiatric clinic. Journal of Nervous and Mental Disease 1997; 185:102-107.
32. Ross RJ et al. Sleep disturbance as the hallmark of posttraumatic stress disorder. American Journal of Psychiatry 1989; 146:697-707.
33. Pagel JF. Modeling drug actions on electrophysiological effects produced by EEG modulated potentials. Human Psychopharmachology 1993; 8(3):211-216.

34. Pagel JF. Pharmachologic alteration of sleep and dreams - a clinical framework for using the electrophysiological and sleep stage effects of psychoactive medications. Human Psychopharmachology 1996; 11(3):217-224.
35. Hobson JA. The chemistry of conscious states. Boston: Little, Brown and Company, 1994; 258.
36. Foulkes D. Dreaming: A cognitive-psychological analysis. Hillsdale, New Jersey: Lawrence Erlbaum Associates, 1985.
37. Cohen DB. Sleep and dreaming: Origins. Nature and Function. New York: Pergamon, 1979.
38. Baekeland F, Lasky R. The morning recall of rapid eye movement period reports given earlier in the night. Journal of Nervous and Mental Disease 1968; 147:570-579.
39. Nielsen TA. A review of mentation in REM and NREM sleep: "Covert" REM sleep as a possible reconciliation of two opposing models. In: Pace-Schott EF, Solms M, Blagrove M, Harnard S, eds. Sleep and Dreaming: Scietific Advances and Reconsiderations. Cambridge: Cambridge University Press, 2003:59-74.
40. Lavie P, Hefez A, Halperin G et al. Long term effects of traumatic war-related events on sleep. Am J Psychiatry 136(2):175-178.
41. Kramer M, Kinney L. Sleep patterns in trauma victims with disturbed dreaming. Psychiar J Univ Ottawa 1988; 139(1):12-16.
42. Sakai K, Crochet S, Onoe H. Pontine structures and mechanisms involved in the generation of paradoxical (REM) sleep. Archives Italiennes de Biologie 2001; 139:93-107.
43. Siegel JM. Brainstem mechanisms generating REM sleep. In: Kryger MH, Roth T, Dement WC, eds. Principles and Practice of Sleep Medicine. 3rd ed. Philadelphia, PA: WB Saunders Co., 2000:112-133.
44. Gottesmann C. Neurophysiological support of consciousness during waking and sleep. Progress in Neurobiology 1999; 59:469-508.
45. Mallick BN, Kaur S, Saxena RN. Interactions between cholinergic and gabanergic neurotransmitters in and around the locus coeruleus for the induction and maintence of rabid eye movement sleep in rats. Neuroscience 2001; 104(2):467-485.
46. Standaert D, Saper C, Rye D et al. Colocalization of atriopeptin-like immunoreactivity with choline acetyltransferase and substance-P like immunoreactivity in the pedunculopontine and laterodorsal tegmental neculi of the rat. Brain Res 198; 382:163-8.
47. Vincent SR, Satoh K, Armstrong DM et al. Neuropeptides and NADPH-diaphorase activity in ascending cholinergic rectular system of the rat. Neuroscience 1986; 17:167-82.
48. Vincent S, Kimura H. Histochemical mappin of nitric oxide synthase in the rat brain. Neuroscience 1992; 46:755-84.
49. Austin M, Rice P, Mann J et al. Localization of corticotrophin-releasing hormone in the human locus coeruleus and pedunculopontine tegmental nucleus: An immunocytochemical an in situ hybridization study. Neuroscience 1995; 64:713-27.
50. Lavoie B, Parent A. Pedunculopontine nucleus in squirrel monkey: Distribution of cholinergic and monoaminergic neurons in the mesopontine tegmentum with evidence for the presence of glutamate in cholinergic neurons. J Comp Neurol 1994; 334:190-209.
51. Buchsbaum MS, Hazlett EA, Wu J et al. Positron emission tomography with deoxyglucose-F18 imaging in sleep. Neuropsychopharmacology 2001; 25:S50-6.
52. Nofzinger EA, Nichols TE, Meltzer CC et al. Changes in forebrain function from waking to REM sleep in depression: Preliminary analysis of [18F] FDG PET studies. Psychiatry Research: Neuroimaging 1999; 91:59-78.
53. Hong CCH, Gillin JC, Dow BC et al. Localized and lateralized cerebral glucose metabolism associated with eye movements during REM sleep and wakefulness: A positron emission tomography (PET) study. Sleep 1995; 18:570-80.
54. Igawa M, Atsumi Y, Takahashi K et al. Activation of visual cortex in REM sleep measured by 24-channel NIRS imaging. Psychiatry and Clinical Neuroscience 2001; 55:187-8.
55. Kaplan-Solms K, Solms M. Clinical studies in neuro-psychoanalysis. Introduction to a Depth Neuropsychology. London: Karnac Books, 2000.
56. Thompson D, Pierce D. Drug induced nightmares. The annals of pharmacotherapy 1999; 33:93-96.
57. Naranjo CA et al. A method for estimating the probability of adverse drug interactions. Clinical Pharmacology and Therapeutics 1981; 30:239-245.
58. McCarley R. REM sleep and depression: Common neurobiological control mechanisms, American Journal of Psychiatry 1982; 139(5):565-570.
59. Jouvet M. The paradox of sleep: The story of dreaming. MIT Press, 1999.
60. Baghdoyan HA, Monaco A, Rodrigo-Angulo M et al. Microinjection of neostigmine into the pontine recticular formation of cats enhances desynchronized sleep signs. Journal of Pharmacology and Experimental Therapy 1984; 231:173-80.
61. Chase MH, Morales FR. Control of motorneurons during sleep. In: Kryger MH, Roth T, Dement WC, eds. Principles and Practice of Sleep Medicine. 3rd ed. Philadelphia, PA: WB Saunders Co., 2000:155-168.
62. Hobson JA, Pace-Schott EF, Stickgold R. Dreaming and the brain: Toward a cognitive neuroscience of conscious states. In: Pace-Schott EF, Solms M, Blagrove M, Harnard S, eds. Sleep and Dreaming: Scientific Advances and Reconsiderations. Cambridge: Cambridge University Press, 2003:1-50.
63. Kosten TR, Mason JW, Giller EL. Sustained urinary norepinephrine and epinephrine elevation in posttraumatic stress disorder. Psychoneuroendocrinology 1987; 12:13-20.
64. Southwick SM, Krystal JH, Morgan CA et al. Abnormal Noradrenergic function in posttraumatic stress disorder. Arch Gen Psychiatry 1993; 50:266-274.
65. Kolb LC, Burris BC, Griffiths S. Propranolol and clonidine in the treatment of post traumatic stress disorders of war. In: Van der Kolk BA, ed. Posttraumatic Stress Disorder: Psychological and Biological Sequelae. Washington DC: American Psychiatric Press Inc., 1984:98-105.
66. Dimsdale J, Newton R. Cognitive effects of Beta-blockers. Journal of Psychosomatic Research 1991; 36(3):229-236.
67. Brismar K, Motgensen L, Wetterberg L. Depressed melatonin secretion in patients with nightmares due to beta-adrenoceptor blocking drugs. Acta Med Scand 1987; 221(2):155-158.
68. Gursky J, Krahn L. The effects of antidepressants on sleep: A review, Harvard. Review of Psychiatry 2000; 8(6):298-306.
69. Coupland NJ, Bell CJ, Potokar JP. Serotonin reuptake inhibitor withdrawal. Journal of Clinical Psychopharmacology1996; 16(5):356-362.
70. Pace-Schott E et al. Effects of serotonin reuptake inhibitors (SSRI) on dreaming in normal subjects. Sleep 1999; 22(Sup 1):H278D.
71. Lepkifker E et al. Nightmares related to fluoxetine treatment. Clinical Neuropharmachology 1995; 18(1):90-94.
72. Pace-Schott E et al. Enhancement of subjective intensity of dream features in normal subjects by the SSRI's paroxetine and fluvoxamine. Sleep 2000; 23(Sup. 2):A173.
73. Stacy M. Managing late complications of Parkinson's disease. Medical Clinics of North America 1999; 83(2):469-481.
74. Xi MC, Morales FR, Chase MH. Evidence that wakefulness and REM sleep are controlled by a GABAergic pontine mechanism. Journal of Neurophysiology 1999; 82:2015-2019.
75. Mallick BN, Kaur S, Saxena. Interactions between cholinergic and GABAnergic neurotransmitters in and around the locus coeruleus for the induction and maintainence of rapid eye movement sleep. Neuroscience 2001; 104:467-485
76. Oxorn D et al. The effects of midazolam on propofol-induced anesthesia: Propofol dose requirements, mood profiles, and perioperative dreams. Anesth Analg 1997; 85:553-9.
77. Steriade M. Brain electrical activity and sensory processing during waking and sleeping states. In: Kryger MH, Roth T, Dement WC, eds. Principles and Practice of Sleep Medicine. 3rd ed. Philadelphia, PA: WB Saunders Co., 2000: 93-111.
78. Ansoms A et al. Zopiclone or lometazepam in the treatment of insomnia and the effect on behavior and mood during the post-alcoholism withdrawal period. Current Therapy and Research 1991; 49:54-64.
79. Van Moffaert M et al. Comparison of zopiclone and flunitrazepam in the treatment of insomnia in depressed patients. Current Therapy and Research 1990; 48:140-153.

80. Marsh S et al. Dreaming and anesthesia: Total i.v. Anaesthesia with propofol versus balanced volitile anaeshthesia with enflurane, European Journal of Anaesthesiology 1992; 9:331-333.
81. Krissel J et al. Thiopentone, thiopentone/ketamine, and ketamine for induction of anesthesia in caesarean section. European Journal of Anaesthesiology 1994; 11:115-122.
82. Knill RL, Moote CA, Skinner MI et al. Anesthesia with abdominal surgery leads to intense. REM sleep during the first postoperative week, Anesthesiology 1990; 73(1):52-61.
83. Hobson JA. Dreaming as Delerium. MIT Press, 1999.
84. Krueger JM, Fang J. Host defense, in principles and practice of sleep medicine. In: Kryger M, Roth T, Dement W, eds. Philadelphia, WB: Saunders Co., 2000:3:255-265.
85. Krueger JM, Kubillis S, Shoham S et al. Enhancement of slow wave sleep by endotoxin and libid A. American Journal of Physiology 1986; 251:R591-597.
86. Jaffe SE. Sleep and infectious disease, in principles and practice of sleep medicine. In: Kryger M, Roth T, Dement W, eds. Philadelphia, WB: Saunders Co., 2000:3:1093-1102.
87. McEvoy GK, ed. Cardiac Drugs. AHFS Drug Information. Belthesda, Maryland: American Society of Health Systems Pharmacists, Inc., 2002:24:04.
88. Pagel JF. Nondreamers. Sleep Medicine 2003; 4:235-241.

CHAPTER 28

The Night Eating Syndrome

Grethe Støa Birketvedt and Jon R. Florholmen

Abstract

The typical behavioral characteristics of the night eating syndrome have been described as morning anorexia, evening hyperphagia and insomnia. The neuroendocrine characteristics have been described as changes in the circadian rhythm by an attenuation in the nocturnal rise of the plasma concentrations of melatonin and leptin and an increased circadian secretion of cortisol. The night eaters also have an overexpressed hypothalamic-pituitary-adrenal axis with an attenuated response to stress. The night eating syndrome appears to represent a new eating disorder, different from the established disorders of anorexia nervosa, bulimia nervosa, and binge eating disorders. It differs in the frequency and size of ingestions at night, and the elevated plasma levels of cortisol reflect increased activity of CRH, as expressed by an attenuated ACTH and cortisol response. In conclusion the mechanisms behind the increased CRH stimulation may involve alterations in the neurotransmitter systems, causing increased nocturnal appetite and disruption in the sleep pattern. This may, to some extent, explain the disturbances in the circadian secretions of melatonin and leptin and the behavioral characteristics of the night eating syndrome.

Introduction

The night eating syndrome (NES) is characterized by morning anorexia, evening hyperphagia and insomnia and was first described in 1955 by Dr. Albert Stunkard and his colleagues at the University of Pennsylvania.[1] The night eating was found to occur during periods of stress and was associated with a poor outcome of efforts at weight reduction. Depression in the evening and at night was also another characteristic included in the definition used. In the 1955 publication, clinicians observed that the syndrome only appeared among obese persons. Since 1955, NES has not been subjected to careful clinical study, but its prevalence has been estimated at 1.5% in the general population.[2] Twelve percent of the obese patients in a nutrition clinic was found to suffer from the night eating syndrome[3] and 27% and 28%, respectively, in 2 samples of severely obese persons.[2,4] The night eating syndrome has so far not been included in any of the Diagnostic and Statistical Manuals of the American Psychiatric Association.

Although night eating occurs among nonobese persons as well, the prevalence seems to increase with increasing weight, from 7.9% in a surgical sample[5] to 8.9%[3] and 15%[6] in obesity clinics.

Between 1955 and 1995 little attention was given the night eating syndrome. In 1995 Dr. Grethe Støa Birketvedt at the University of Pennsylvania started to characterize the syndrome in more detail.[7] The study reported in the American Journal og Medical Association in 1999 confirmed the elements of the night eating syndrome previously reported, including morning anorexia, evening hyperphagia, and insomnia. It quantified the evening hyperphagia, showing that food intake, which continued later than that of obese controls subjects, was 2930 kcal compared to 2334 kcal for the control subjects ($p < .055$). In Figure 1 we have shown that the cumulative caloric intake of the night eaters lagged behind that of the obese control subjects, so that at 4:00 p.m. they had consumed only 37% of their daily intake, compared to 74% by the controls ($p < .001$). The food intake of the controls then slowed while that of the night eaters continued until after midnight. During the period of 6:00 p.m. to 6:00 a.m., the night eaters consumed 56% of their caloric intake, compared to 15% for the control subjects ($p < .001$).

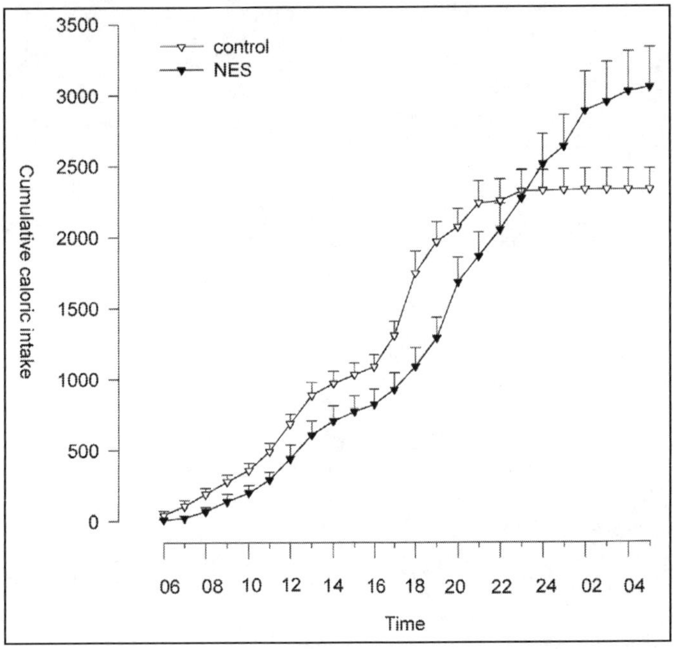

Figure 1. Twenty-four hour pattern of mean cumulative energy intake and mood for a 5-day period. The intake of the night eaters lags behind that of control subjects until 10 p.m. and then greatly exceeds it ($p < .001$). Daytime mood of the night eaters is lower than that of the controls ($p < .001$) and falls even lower during the evening and night ($p < .001$). Error bars represent SEM in all figures. NES indicates night-eating syndrome.

Sleep and Sleep Disorders: A Neuropsychopharmacological Approach, edited by Malcolm Lader, Daniel P. Cardinali and S.R. Pandi-Perumal.

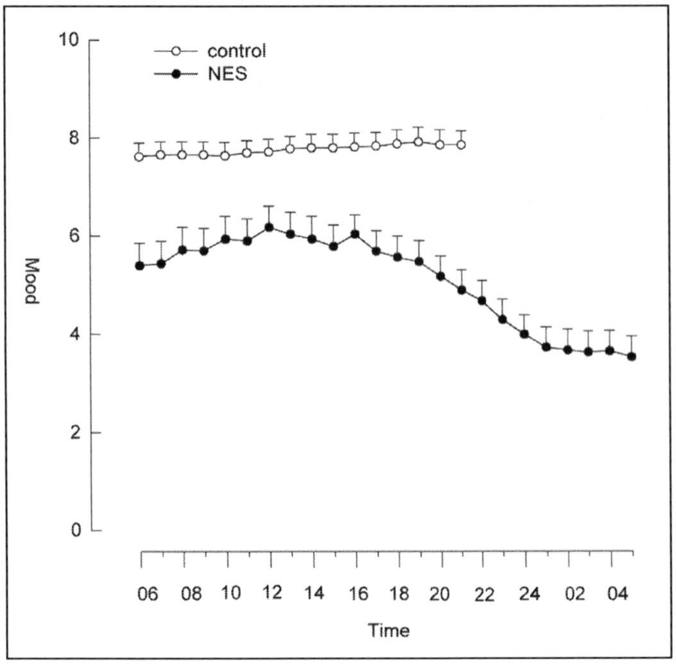

Figure 2. Daytime mood of the night eaters is lower than that of the controls (p< .001) and falls even lower during the evening and night (p< .001). Error bars represent SEs in all figures.

As also reported in the 1999 paper (Fig. 2), the mood of the night eaters was lower than that of the control subjects during the morning hours and fell significantly during the evening and night. Furthermore, after 4:00 p.m. the mood of the night eaters fell at a rate of 0.25 units per hour while that of the controls remained unchanged.

The night eaters suffered from both sleep onset and sleep maintenance insomnia, and the number of nighttime awakenings was 3.6 for the night eaters and 0.3 for the control subjects (p < .001). Half of the 178 awakenings of the night eaters were associated with food intake while none of the controls ate while they were awake (Fig. 3).

The nighttime snacks of the night eaters were not binges, but of only moderate size, averaging 271 kcal. The carbohydrate content of these nighttime snacks was 70% of the total caloric intake compared to 47% for their food intake during the rest of the day (p < .001). Furthermore, the carbohydrate to protein ratio of the nighttime snacks was 7:1. This nutrient pattern, of a carbohydrate to protein ratio, increases the availability of tryptophan for transport into the brain and conversion into serotonin with its sleep promoting properties.[8,9]

The behavioral study was complemented by a neuroendocrine study conducted in the Clinical Research Center of the University Hospital in Tromsø where subjects were admitted for 24-hour periods.[7] All subjects were women and consisted of both overweight and normal weight night eaters and overweight and normal weight control subjects, matched for body mass index and age. Unlike their usual nighttime eating, night eaters in the Clinical Research Center received four meals of 300 kcal at 8:00 a.m., 12:00 noon, 4:00 p.m. and 8:00 p.m.

Among night eaters, both overweight and normal weight, there was a marked blunting of the plasma melatonin levels at night (p < .001) (Fig. 4).

Plasma leptin levels were higher among the overweight subjects than among the normal-weight subjects (both night-eaters

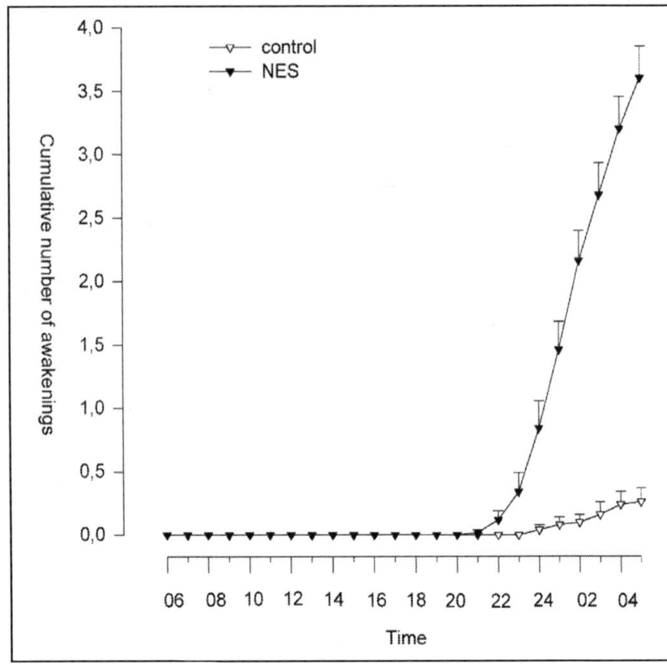

Figure 3. Nighttime snacks of the night eaters and controls

and controls) (p< .001) (see Fig. 5). When compared with their respective control groups, the rise in the nocturnal (12AM to 6 AM) plasma leptin levels was lower in both normal-weight and overweight night eaters (p< .001). The time of the highest plasma

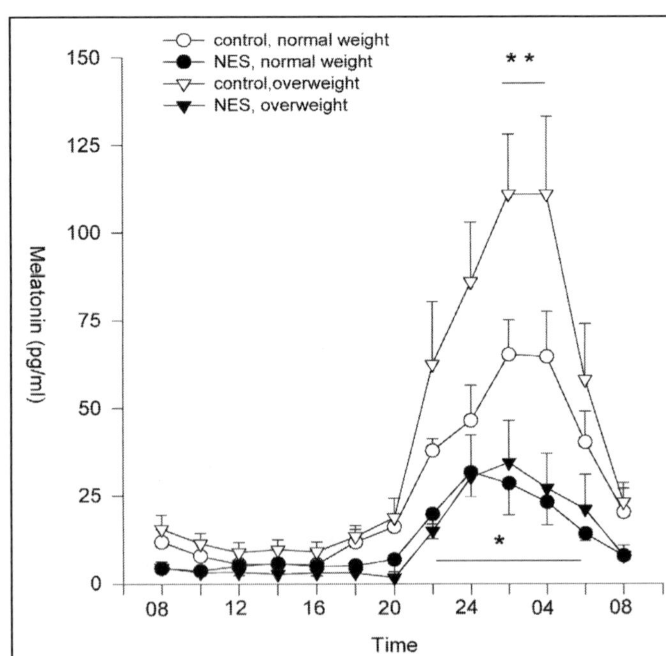

Figure 4. Twenty-four-hour mean plasma melatonin levels in overweight and normal weight night eaters and overweight and normal weight controls. The single asterisk and bracket indicate a significant difference (p< .001) in levels between the normal-weight night eaters and normal-weight control group, and between the overweight night eaters and the overweight control group. The 2 asterisks and bracket indicate significant difference (p< .001) in levels between the normal-weight and overweight control groups. NES indicates night eating syndrome.

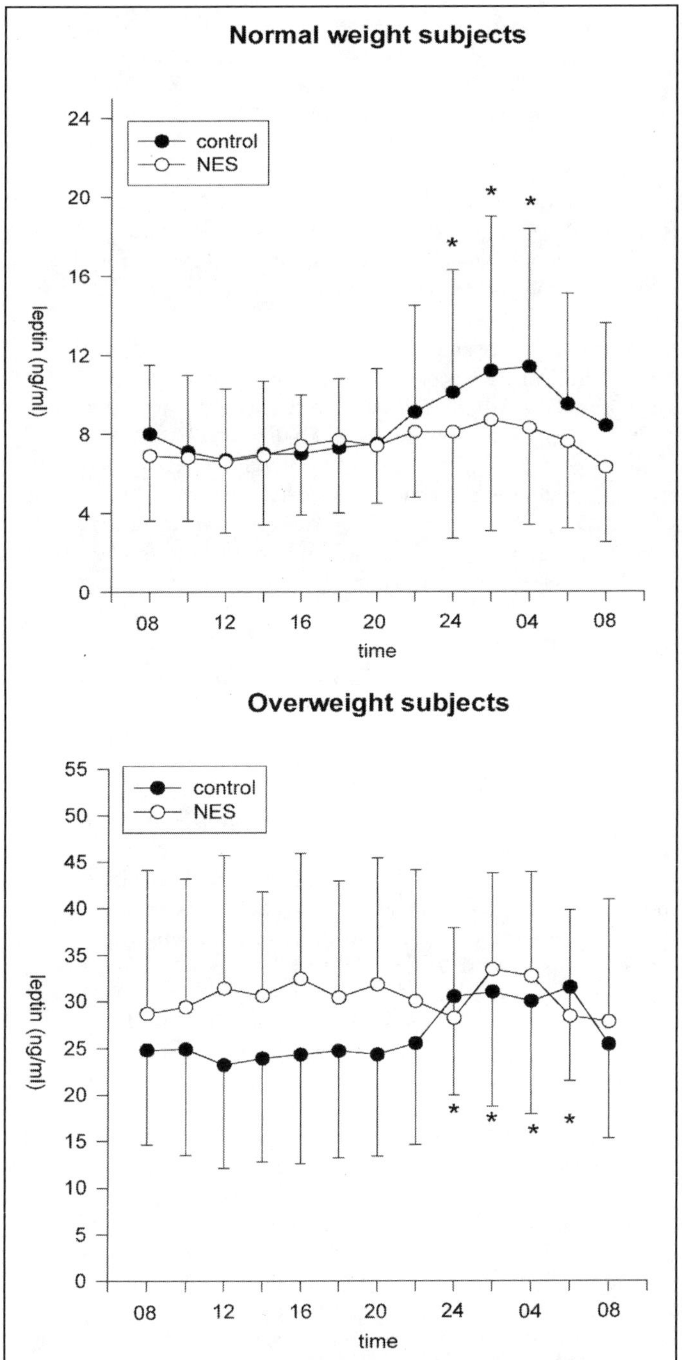

Figure 5. Twenty-four-hour mean plasma leptin levels in overweight and normal weight night eaters and overweight and normal weight controls. Asterisks and brackets indicate significant differences in levels of nocturnal concentrations (when compared with the baseline concentrations at 8 AM) between the night eaters and control subjects, both overweight and normal-weight (p< .001).

leptin concentration did not differ between the normal-weight groups (p= .28) or between the overweight groups (P= .63).

Confirming the earlier clinical impression that the night eating syndrome was associated with stress, plasma cortisol levels of the night eaters were higher than those of control subjects for most of the 24 hours (Fig. 6).

The night eating syndrome appears to be a unique combination of an eating disorder, a sleep disorder and a mood disorder.

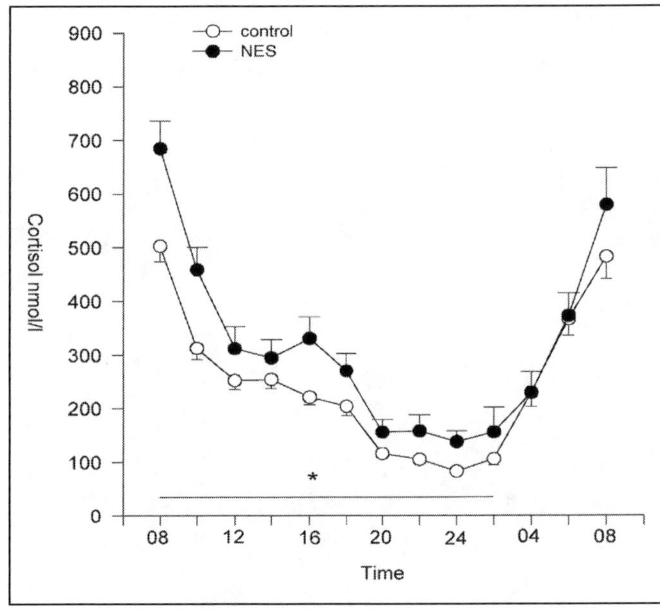

Figure 6. Twenty-four-hour mean plasma cortisol levels in subjects with and without night-eating syndrome. The asterisk and bracket indicate a significant difference between the levels of the 2 groups (p= .001).

The distinctive neuroendocrine findings are closely associated in a pattern that helps to link these findings with the behavior of the night eaters. Thus, the blunting of the nighttime rise in melatonin may contribute to the sleep maintenance insomnia[10] and to depression.[11]

The failure of leptin to rise at night must limit its usual nighttime suppression of appetite and may permit the breakthrough of hunger impulses, further disrupting sleep.

The elevated levels of cortisol reflect the clinical impression that night eating occurs during periods of life stress.

A later study[6] has confirmed circadian aspects of the Birketvedt et al study from 1999.[7] Not only did night eaters consume a greater part of their food intake during the latter part of the day than controls, but their predisposition to eat later in the day was confirmed by the greater consumption of a test meal. This study also found elevated levels of depression in night eaters as assessed by the Zung Depression Inventory.

In 2002 Birketvedt et al tested the hypothesis that night eaters have an overexpressed hypothalamic-pituitary-adrenal axis with an attenuated response to stress. They performed a 120-min corticotropin-releasing hormone (CRH) test (100 ug i.v.) in female subjects suffering from the night eating syndrome, and measured the plasma concentration of ACTH and cortisol. The results showed that, in night eaters compared with controls, the CRH-induced ACTH and cortisol response to CRH was significantly decreased (Fig. 7).

They concluded that disturbances in the hypothalamic-pituitary-adrenal axis with an attenuated ACTH and cortisol response to CRH were found in subjects with the night eating syndrome.[10]

In night eaters, there are complex neuroendocrine characteristic changes, with attenuation in nocturnal secretions of melatonin and leptin and increased secretion of cortisol. An attenuated response of ACTH and cortisol in plasma after injection of CRH has also been described. Studies performed so far may indicate that the night-eating syndrome is associated with an attenuated pituitary-adrenal response to CRH.

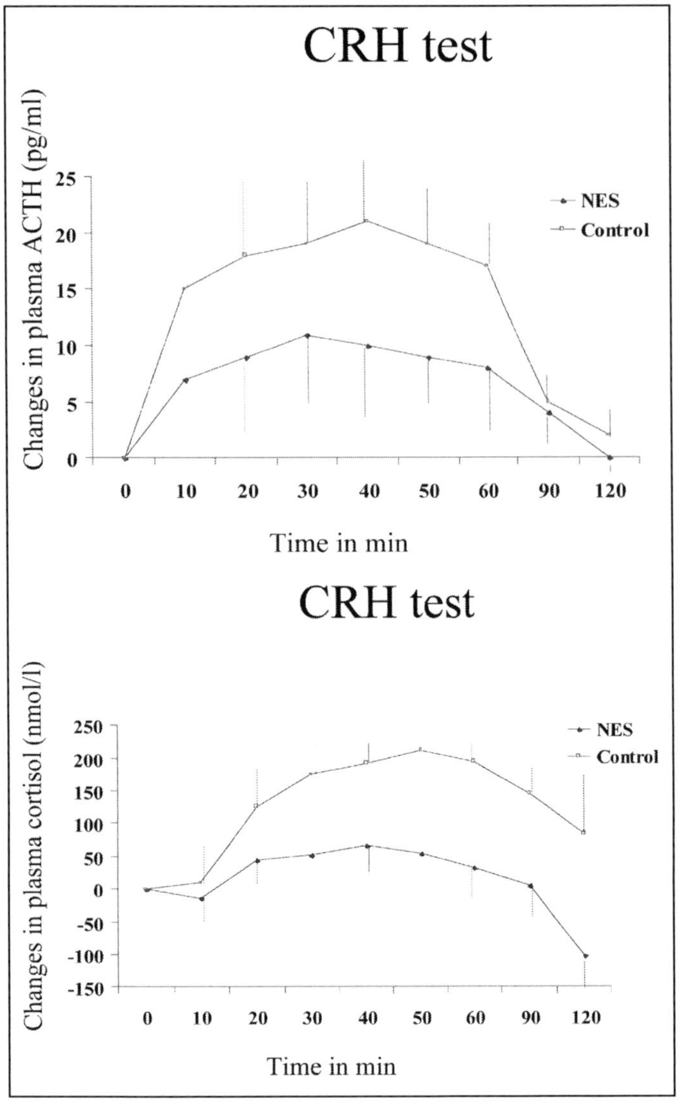

Figure 7. Effect of intravenous injection of corticotropin-releasing hormone (CRH) on plasma concentrations of ACTH in subjects with night eating syndrome and healthy controls. Effect of intravenous injection of CRH on plasma concentrations of cortisol in subjects with NES. Asterisks show difference from the control group.

A healthy response to acute stress is of great importance in our daily physical and psychological challenges. This response is mediated from the hypothalamus to the pituitary gland, which in turn mediates various neuroendocrine signals. When the stressor is gone, the neuroendocrine responses are normally terminated. Failure to terminate this response is observed in various conditions of an overexpressed HPA axis with a subsequent attenuation of the pituitary neuroendocrine response.

It is well known that night eaters experience wide variability in frequency of symptoms. They can have periods when they experience many awakenings and eating episodes, but also periods when these events are minimal.

The night eating syndrome appears to represent a new eating disorder, different from the established disorders of anorexia nervosa, bulimia nervosa, and binge eating disorders. It differs from the latter two disorders in the frequency and size of ingestions at night. The elevated plasma levels of cortisol reflect increased activity of CRH, as expressed by an attenuated ACTH and cortisol response that may well explain the disrupted sleep and appetite pattern observed in night eaters. Several other disorders, such as obesity, fatigue syndrome, anorexia nervosa, bulimia nervosa, insomnia and depression, have been linked to disturbances in the HPA axis. All of these disorders share some phenotypes with the night eaters, such as mood disruptions, eating disorders, and sleeping disorders. Whether these clinical features are the result of common pathophysiological mechanisms in the HPA axis remains to be clarified and awaits further studies.

In conclusion, subjects suffering from night-eating episodes have signs of disturbances in the HPA axis with an attenuated ACTH and cortisol response to CRH. The mechanisms behind the increased CRH stimulation may involve alterations in the neurotransmitter system, causing increased nocturnal appetite and disruption in the sleep pattern. This may, to some extent, explain the disturbances in the circadian secretion of melatonin and leptin and the behavioral characteristics of the night eating syndrome.

Differential Diagnosis

The first step in defining a disease or disorder has been taken in the case of the night eating syndrome: it is readily recognized by persons manifesting the disorder and physicians are becoming familiar with it. The differential diagnosis is with binge eating disorder and "nocturnal sleep-related eating disorders". It depends upon differences in the frequency and size of eating episodes and in the state of consciousness during them.

The night eating syndrome differs from binge eating disorder in both the far greater frequency of nighttime awakenings and in the modest size of the ingested food: 270 kcal during night-time ingestions,[7] compared to the 1300 kcal ingested during eating binges reported by others.[12] Furthermore, night eaters do not appear to suffer to the same degree from the intense food, diet, and body image disorders that characterize binge eaters.[5]

The night eating syndrome appears to differ also from the "nocturnal sleep-related eating disorders" reported from sleep research clinics, often in association with sleep walking and parasomnia states. A key distinction between the two disorders lies in the extent of consciousness during the night eating. In the sleep disorder clinic patients reported by Schenck and Mahowald,[13] 84% of night eating occurred during total or partial unconsciousness and, for the majority, during stage 3/4 sleep. By contrast, in their sleep disorders clinic,[14] Manni, Ratti, and Tartara reported that 5.8% of 120 persons referred for insomnia manifested night eating during full consciousness. Finally Aronoff et all reported in 1994 the only polysomnographic studies of a subject who ate while fully conscious found that he awoke to eat during stage II sleep. The prevalence of "nocturnal sleep- related eating disorders" appears to be far lower than that of the night eating syndrome noted above: 0.5%, or 38 cases out of approximately 8,000 polysomnographic examinations[13] as reported by Schenck and Mahowald in 1994. In a study of 221 subjects recruited for a study of binge eating disorder, Stunkard et al could report in 1996 that 54 women identified themselves as binge eaters, 15 identified themselves as night eaters, and only 12 said that they were both. In light of the inclination of many persons to consider any overeating a binge, these results suggest a difference between the two disorders.

Prognosis and Treatment

Information about prognosis in the night eating syndrome is limited, confined to retrospective accounts. Researchers suggest that it follows a chronic course, exacerbated by stressful life situations.

In contrast to the plethora of treatments that have been proposed for binge eating disorder, there have been no formal studies of the treatment of the night eating syndrome. The 1955 report on the disorder noted that long-term psychodynamic psychotherapy had been associated with improvement in some patients, attributed to a reduction in stress[1]. The blunted rise in nighttime melatonin, suggests that provision of this agent at bedtime might be a reasonable option. Several night eaters have tried to take melatonin at night with both successful and not successful results, and different food containing tryptophan has been suggested. But neither one of the treatments proposed have a successful ending.

Summary and Conclusions

The identification of two different eating disorders helps to define two subsets of obese persons who may benefit from special attention. Studies of binge eating disorder show that traditional behavioral weight reduction programs reduce binge eating and body weight and are the treatment of choice. The diagnosis of binge eating disorder may be most useful as a marker for the psychological problems that frequently affect binge eaters and that deserve treatment in their own right.

The relative ease with which the night eating syndrome can be identified suggests that, when treatments are developed, they may find ready acceptance. Anecdotal reports to the author from night eaters who have explored treatment options suggest that selective serotonin reuptake inhibitors have been helpful, as might be expected from their effects on disturbances in mood and sleep. In view of the lack of rise in nighttime melatonin, provision of this agent at bedtime would seem a rational option.

References

1. Stunkard AJ, Grace WJ, Wolff HG. The night-eating syndrome. A pattern of food intake among certain obese patients. Am J Med 1955; 19:78-86.
2. Rand CSW, Macgregor MD, Stunkard A. The night eating syndrome in the general population and among post-operative obesity surgery patients. Intl J Eat Disord 1997; 22:65-69.
3. Stunkard AJ. Eating patterns and obesity. Psychiatry Quarterly 1959; 33:284-294.
4. Rand CS, Kuldau JM. Morbid obesity: A comparison between a general population and obeseity surgery patients. Int J Obes Relat Metab Disord 1993; 17:657-661.
5. Adami GF, Meneghelli A, Scopinaro N. Night eating and binge eating disorder in obese patients. Intl J Eat Disord 1999; 25:335-338.
6. Gluck ME, Geliebter A, Satov T. Night eating syndrome is associated with depression, low self-esteem, reduced daytime hunger, and less weight loss in obese outpatients. Obes Res 2001; 9(4):264-7.
7. Birketvedt G, Florholmen J, Sundsfjord J et al. Behavioral and neuroendocrine characteristics of the night-eating syndrome. JAMA 1999; 282:657-663.
8. Berry EM, Growdon JH, Wurtman JJ et al. A balanced carbohydrate: Protein diet in the management of Parkinson's disease. Neurology 1991; 41:1295-1297.
9. Yokogoshi H, Wurtman RJ. Meal composition and plasma amino acid ratios: Effects of various proteins or carbohydrates, and of various protein concentrations. Metabolism 1986; 35:637-642.
10. Birketvedt GS, Sundsfjord J, Florholmen JR. Hypothalamic-pituitary-adrenal axis in the night eating syndrome. Am J Physiol Endocrinol Metab 2002; 282(2):E366-9.
11. Grilo CM, Shiffman S. Longitudinal investigation of the abstinence violation effect in binge eaters. J Consult Clin Psychol 1994; 62:611-619.
12. Kennedy SH, Garfinkel PE, Parienti V et al. Changes in melatonin levels but not cortisol levels are associated with depression in patients with depression and eating disorders. Arch Gen Psychiatry 1989; 46:73-78.
13. Schenk CH, Mahowald MW. Review of nocturnal sleep-related eating disorders. Intl J Eat Disord 1994; 16:343-356.
14. Manni R, Ratti MT, Tartara A. Nocturnal eating: Prevalence and features in 120 insomniac referrals. Sleep 1997; 20:734-8.

CHAPTER 29

Drug Effects on Dreaming

Mehmet Yucel Agargun and Hanefi Ozbek

Abstract

Among the proposed functions of dreaming in human being, the most research supports are mood-regulation, problem-solving, learning, and memory construction. Recent imaging techniques have provided meaningful information on functional neuroanatomy and neurophysiology of REM sleep and dreaming. In addition to serotonin, norepinephrine, and acetylcholine in terms of a reciprocal interaction between the cholinergic REM-ON and aminergic REM-OFF neurons suggested by McCarley and Hobson, dopamine has recently began to think to play a significant role in modulation of dream functions, particularly nightmares. The disinhibition of REM physiology is due primarily to dopaminergic dysfunction, specifically the removal of dopaminergic inhibition on amygdalar sites in dopamin-related syndromes including parkinson disease, REM sleep behavior disorder, and narcolepsy. The disinhibited amygdala yields the affective and personality changes, and the unpleasant dreams associated with PD, RBD, and narcolepsy as well as depression. There is limited data in the literature on drugs and dreaming or dream content even if it is well-known that many antidepressant drugs may cause nightmares or frightening dreams. In this chapter, we review and discuss the effects of medications on dreaming.

The Neuroanatomy of REM Sleep and Dreaming

Recently the functional imaging techniques, such as PET and fMRI, have been contributed to our understanding of brain functions and cognitive processing during sleep and dreaming.[1-3] These studies also provided meaningful information on functional neuroanatomy of different sleep stages. During stages 3 and 4 NREM sleep (slow wave sleep), brainstem regions, thalamic nuclei, basal ganglia, hypothalamus, prefrontal cortex, cingulate cortices, and medial regions of the temporal lobe appear to deactivate. In contrast, during REM sleep, significant increases in regional blood flow or glucose metabolism have been reported in the pontine tegmentum, thalamic nuclei, occipital cortex, mediobasal prefrontal lobes, and limbic system including the amygdala, hippocampus, anterior cingulate cortex. Dorsolateral prefrontal cortex, posterior cingulate, and pariatal cortex were found as least active regions in REM sleep. Recent data also suggest that rapid eye movements generated by mechanisms similar to PGO waves during REM sleep. The brain regions are significantly more activated in related to eye movement density during REM sleep than during wakefulness included the right geniculate body and the occipital cortex. Some of these studies also indicated that sleeping brain influenced by external stimuli, altering the normal signature profile of these quiescence REM and NREM sleep to a more stimulus-specific activity pattern. Sounds presented during NREM sleep are associated with regionally-specific responses broadly similar to those observed during wakefulness. This is probably valid for REM sleep and dream content can be potentially mapped onto a specific distribution of brain activity in REM sleep.

Although there are limited numbers of studies investigating dream process in brain-damaged patients, these studies help us to combine functional anatomy of dreams with imaging findings in clinical samples.[5] In regard with these studies, medial occipito-temporal structures, inferior parietal structures, basal forebrain pathways, frontal-limbic and temporal-limbic structures are involved in normal or disordered dream processes. The medial occipito-temporal region contributes a factor of visual representability to the overall process of dreaming. For example, the damage to this region can result in a lack of visual imagery in dreams. The inferior parietal region contributes a factor of spatial cognition to the overall process of dreaming. With damage to this region (of either hemisphere) the subjective experience of dreaming may stop completely. The connections between mediobasal frontal cortex and brainstem and diencephalic-limbic nuclei contribute a factor of appetitive interest to the overall process of dreaming. With bilateral damage to these connections, dreaming may also stop completely. The frontal-limbic region also contributes a factor of mental selectivity to the process of dreaming. Damage to the structures in this region may lead to excessive dreaming and dreamlike thinking. Excitatory activity in the limbic system contributes a factor of affective arousal to the overall process of dreaming. Seizure activity in this region, in particular the amygdala, may lead to frequent and recurring nightmares. Prefrontal cortex had been attributed for some characteristic features of dreams such as the lack of insight, distortion of the perception, and amnesia on awaking.

Neuropharmacology of REM Sleep and Dreaming

Dopamine

Dopamine (DA) was not thought to play a significant role in modulation of sleep functions until very recent years[6] However, the administration of dopaminergic agents to patients or animals tended to enhance wakefulness,[7,8] thus implying that reducing dopaminergic tone would reduce wakefulness and enhance sleep. Reduced levels of dopamine transmission appear to be associated with excessive sleepiness and conversely normal levels of DA are

Sleep and Sleep Disorders: A Neuropsychopharmacological Approach, edited by Malcolm Lader, Daniel P. Cardinali and S.R. Pandi-Perumal.

critical for maintaining wakefulness. Although there are very few studies of dreams of parkinson disease (PD) patients, a recent study[9] reported a correlation between Mini Mental State scores and length of dream report and event story structure within the dream. Contemporary clinical impression is that dreams reported by PD patients are typically vivid and unpleasant due to REM disinhibition.[6] In regard to another dopaminergic syndrome REM sleep behavior disorder (RBD), dream content typically involves the patient under some sort of threat either against himself or his wife. Most patients with RBD report that they repeatedly experienced this RBD-related "nightmare" of being attacked by animals or unfamiliar people. In a recent study,[10] the patients with narcolepsy reported positive and negative emotions (anxiety/fear) in a balanced fashion in sleep onset REM and nighttime REM sleep. The authors suggested that anxiety/fear was due to inadequate blockade of noradrenergic activity in the locus cereleus and, the propensity to experience positive emotion was increased due to cholinergic hypersensitivity and dopaminergic dysregulation. Recently, McNamara et al[6] hypothesized that the disinhibition of REM physiology is due primarily to dopaminergic dysfunction, specifically the removal of dopaminergic inhibition on amygdalar sites. The disinhibited amygdala yields the affective and personality changes, and the unpleasant dreams associated with PD, RBD, and narcolepsy as well as depression. Table 1 summarizes clinical findings in dopaminergic syndromes of sleep and dreaming.

It happens that the only monoaminergic neurotransmitter which continues to function during REM sleep is dopamine, which is known to be dysfunctional in cases of psychosis, and especially among schizophrenic patients. In a 'normal' subject, when the dopamine level increases in the extracellular environment of the brain, under the influence, for example, of amphetamines, nightmares occur (a possible prelude to psychotic decompensation), as well as psychotic-type disturbances.[11]

Serotonin

Dorsal raphe neuronal activity and serotonin release may be actively modulated during sleep and waking. These are most active during waking, less active during slow wave sleep and was lowest in activity during REM sleep.[12] REM sleep is controlled by two neuronal groups in the brainstem, the REM-on and REM-off cells. Increased activity of cholinergic REM-on cells in the pedunculopontine/laterodorsal tegmental (PPT/LDT) nuclei is believed to initiate and maintain REM sleep.[13] Activity of monoaminergic REM-off cells is reduced prior to and during REM sleep, including firing of serotonergic neurons in the dorsal raphe nucleus (DNR).[14] Stimulation of the 5-HT_{1A} heteroreceptors is hypothesised to attenuate REM sleep, while stimulation of the presynaptic 5-HT_{1A} autoreceptors is hypothesised to enhance it by suppressing serotonergic neurotransmission.[15,16] 5-HT2 receptors are related to regulation of slow wave sleep rather than REM sleep,[17] whereas 5-HT2 antagonists may increase REM sleep latency, reduce the amount of REM sleep in cats[18] and influence sleep-wake states in rats by decreasing REM sleep and mildly increasing deep slow-wave sleep.[19]

Serotonergic REM suppression causes a decrease in dream frequency while the cholinergic rebound from serotonergic suppression, i.e., during acute SSRI discontinuation, results in augmented report length and bizarreness.[20]

Norepinephrine

Noradrenergic cells of the locus coeruleus (LC) have similar discharge pattern to serotonergic cells of DNR during sleep-wake cycle; during the transition to REM sleep, discharge slows dramatically. Durig REM sleep the discharge rate is the lowest. The lowest discharge rate of LC cells during REM sleep suggests a role in the 'gating' of aspects of REM sleep. Many cells of the locus ceruleus cease firing during the transition from nonREM sleep to REM sleep (or during REM sleep). They regain their

Table 1. Clinical findings in dopaminergic syndromes of sleep and dreaming

	Personality	Vivid Unpleasant Dreams	Flat to Unpleasant Affect	Disinhibition of REM Atonia	Mesocortical Dopaminergic Dysfunction	Frontal Lobe Impairment
Depression	Ruminative, perseverative	↑ "mascochistic" dream content (dreamer attacked or damaged)	↑	?	yes	yes
Narcolepsy	?	↑ (dreamer attacked and threatened)	↑	No (rather there is inappropriate increase in sleep paralysis)	some	Frontal and temporal
PD	Somewhat rigid and perseverative	↑	↑	?	yes	yes
RBD	?	↑ (dreamer under attack or threatened)	↑	↑	yes	yes

Arrows indicate consensus findings from a majority of studies as reviewed in the text and indicate direction of change relative to healthy control subjects. ↑= increase; ↓= decrease; →= no change. Table reprinted with McNamara's permission.[6]

activity once REM sleep has terminated. Thus, the effects on norepinephrine are directly inhibitory.

Acetylcholine

Acetylcholine (Ach) is associated with REM sleep. Release of ACh in the cortex is highest during waking and REM sleep, and lowest during delta sleep. It appears that REM sleep initiation begins in the ACh neurons located in the peribrachial area of pons. This area of the brain initiates the effects of all of components of REM sleep inluding cortical desynchrony, rapid eye movement, and skeletal paralysis. This area controls cortical desynchrony directly via pathways that pass through areas of the thalamus, and, indirectly, through ACh neurons in the basal forebrain. Rapid eye movement is initiated and maintained via ACh pathways that go to the tectum at the back of the brain stem.

The Reciprocal-Interaction Model of REM Sleep Regulation

McCarley and Hobson described a reciprocal interaction between the cholinergic REM-ON and aminergic REM-OFF neurons in 1975.[21] According to this model, the REM-OFF neurons in the LC were inhibitory to the REM-ON neuronal population, while the REM-ON neurons exerted an excitatory effect on the LC REM-OFF neurons. It was proposed that the LC-neurons were active throughout, except during REM sleep, and their continuous activation inhibited the activity of cholinergic REM-ON neurons. The LC-neurons cease firing, resulting in withdrawal of the tonic inhibition from the REM-ON neurons in REM sleep. Cessation of noradrenergic neuronal activity allows an increased number of REM-ON neurons to escape from inhibition resulting in generation of REM sleep. The activation of REM-ON neurons exerts an excitatory effect on the LC-neurons resulting in inhibition of REM-ON neurons and termination of REM sleep episodes.

GABA

GABA is an inhibitory amino acid in the brain. It was found to be increased in both DRN and LC during REM sleep.[22] Microinjection of GABA agonists into DRN increases REM sleep. Thus, it may be suggested that LC neurons are tonically inhibited by GABA during REM sleep. However, there is limited literature on the effects of GABA, glutamate, and other transmitters on REM sleep.

Drugs and Dreaming

Dopaminergic Compounds, REM Sleep and Dreaming

The cluster of sleep, dream and cognitive changes associated with four disorders including PD, RBD, narcolepsy, and depression can be explained by assuming that lowered dopaminergic tone leads to a disinhibition of REM physiology and amygdalar activity and that this disinhibition of REM and amygdalar functions yields unpleasant dreams, negative affect, and frontal lobe impairment.[6] What is the evidence that the frontal lobe of the brain is responsible for dreaming? Damage to the frontal areas renders dreaming impossible but leaves the REM cycle unaffected. L-dopa leads to a massive increase in the frequency and intensity of dreaming without any effect on the frequency or intensity of REM sleep. The intensity of dreaming caused by dopamine stimulants (like L-dopa) can be stopped by drugs which block the transmission of dopamine (like anti-psychotics).[23]

Recently, Wisor et al[7] using polygraphic recordings and caudate microdialysate dopamine measurements in narcoleptic dogs, have shown that the wake-promoting antinarcoleptic compounds modafinil and amphetamine increase extracellular dopamine without affecting other putative wake promoting substances/receptors like hypocretin receptor 2. They also reported that DAT knock-out mice suffered from excessive levels of sleepiness and were unresponsive to the normally potent wake-promoting action of modafinil, methamphetamine, and the selective DAT blocker GBR12909. Another recent study[24] reported pergolide-induced sleep attacks and increased REM in PD patients.

Increased levels of dopamine resulted in more vivid, nightmarish dreams.[25] Amphetamine releases norepinephrine and inhibits its reuptake. D-amphetamine decreases the percentage of REM sleep and the length of total sleep. Long-term treatment may cause a large REM-sleep rebound.[26] After withdrawal from cocaine or amphetamines short REM-sleep latency, and increased amounts and percentage of REM sleep may occur.[27]

Cholinergic Compounds, REM Sleep and Dreaming

REM sleep is marked by cholinergic activation and aminergic suppression. Decrease of aminergic firings frees the cholinergic system from its restraints and possibly triggers REM sleep. Physostigmine and galanthamine, cholinomimetic drugs, provoked an earlier onset of REM sleep, whereas subchronic treatment with scopolamine, a cholinergic antagonist, only led to a heightening of REM density.[28] It was reported that somatostatin overcomes the REM-sleep-suppressing effect of scopolamine.[29]

Cholinergic agonists such as carbachol, bethanechol and neostigmine may induce REM sleep.[30] Cholinomimetic administration increases the occurrence of PGO waves, as well as other components of REM sleep.[31] Gillin et al[32] reported that scopolamine (a muscarinic antagonist) inhibits REM sleep and REM sleep rebound may occur when scopolamine was abruptly discontinued.

Cisdioxolane and oxotremorine-M, M2-receptor agonists, cause rapid induction of REM sleep and an increased percentage of REM sleep, while M1 agonists have no significant effect on REM sleep.[33]

Sitaram et al[34] showed that man dreams during physostigmine-induced REM sleep. They firstly pretreated seventeen normal volunteers with methscopolamine and then administered one intravenous infusion per night of either placebo or physostigmine either ten or 35 minutes after sleep onset. Their findings indicated that dreaming occurred during physostigmine-induced REM periods but that physostigmine did not alter mentation during nonREM sleep. They also noted that these dreams were similar to spontaneous REM sleep dreams in content, vividness, unusualness, and emotionality.

Serotonergic Compounds, REM Sleep and Dreaming

Lysergic acid diethylamide (LSD) shuts down serotonergic neurons, tends to increase REM sleep in humans.[35] Serotonin agonists such as fenfluramine, which releases serotonin, decrease REM sleep.[36] Drugs that decrease serotonin in the brain produce nightmares and more vivid bizarre dreams.[37]

The pineal nonapeptide hormone arginine vasotocin (AVT) dramatically increases the amount of REM sleep, decreased REM sleep latency, and induced REM periods at sleep onset in healthy prepubertal subjects.[38] Methergoline, a selective central 5-hydroxytryptamine (5-HT) receptor blocker, completely prevents AVT induction of REM sleep. This also represents that the high sensitivity of prepubertal subjects to AVT reflects an immaturity of REM triggering centers.

Noradrenergic Compounds, REM Sleep and Dreaming

Drugs that decrease norepinephrine increase the amount of REM sleep, while drugs that increase norepinephrine activity decrease REM sleep.[39] Clonidine reduces PGO-spike activity.[40] Reserpine, depletes norepinephrine as well as dopamine and serotonin by preventing their entry into the vesicles when they are released, increases REM sleep and shortens REM latency.[41]

It has been reported that alfa1-adrenergic blockers such as prazosin induces cataplexy of narcolepsy and alfa2-adrenergic agonists such as clonidine decreases cataplectic attacks.[42,43] Prazosin may also increase REM sleep.[44] On the other hand, clonidine may decrease REM sleep in humans.[45] Thymoxamine, an alfa-11 receptor antagonist, increases REM sleep.[46] Alfa-2 antagonist such as yohimbine decreases REM sleep in animals.[47] Beta blockers such as propranolol have no consistent effects on REM sleep. Drugs that decreases norepinephrine produces nightmares and more vivid bizarre dreams.[37]

Antidepressant Drugs, REM Sleep and Dreaming

In general, nonspecific monoamine reuptake inhibitors (TCAs), MAO inhibitors (MAOI), selective serotonin reuptake inhibitors (SSRIs), and selective NE reuptake inhibitors decrease or inhibit REM sleep. Mianserin and venlafaxine have also inhibitory effects on REM sleep. On the other hand, bupropion, nefazadone, tianeptine, and amineptine do not inhibate REM sleep, even may increase. Table 2 summarizes the effects of antidepressant drugs on REM sleep, slow wave sleep, and sleep continuity.

Unfortunately there is limited data in the literature on antidepressant drugs and dreaming or dream content even if it is well-known that many antidepressant drugs may cause nightmares or frightening dreams.

Roth, Kramer, and Salis concluded that some sedative-hypnotic and antidepressant agents may effect the quality of dreams, but the precise nature of the effect is yet to be determined in their review on drugs, REM sleep, and dreams.[48] In a recent study,[49] patients taking antidepressants showed lower dream recall frequency than patients without any medication. These patients reported more positively toned dreams than drug-free patients.

The first systematic study of antidepressants (TCAs) on dreaming is belonged to Riemann et al.[50] Although it does not deal directly with dream content, but with self-ratings of the pleasantness of dream reports by depressed outpatients at various stages in their treatment with trimipramine. They aimed at investigating morning and laboratory dream recall and content in patients with a major depressive disorder. They found that antidepressive treatment with trimipramine (an antidepressant which does not suppress REM sleep) led to a positive influence on patients' mood that was paralleled by a change of dream mood in a positive direction.

Armitage et al[51] compared 27 premedication dream reports from a group of patients suffering major, nonpsychotic depression with 32 dreams from the same group after they began taking antidepressant medication, either fluoxetine or nefazandone to study the effects of antidepressants on sleep physiology. They obtained the dreams after a set time of awakening each morning in the sleep laboratory.They found that a low level of dream recall both before and during medication, with only 21 of 89 patients reporting at least one dream from at least one of the two conditions. They concluded that both sets of dream reports were short, relatively bland with little emotion and when depression symptoms were improved by antidepressant treatment, dreams were less vivid with fewer scenes.

In a recent study, Kirschner[52] examined the effects of sertraline on the dream content of a young woman with generalized anxiety disorder and panic attacks. Kirschner used the major categories of Hall and Van de Castle's[53] system of content analysis to compare dream reports before and after drug treatment. In the case, prior to diagnosis and treatment, the dreamer had high levels of aggression and low levels of friendliness in her dreams and the post-medication dreams more closely approximate the female norms.

Pace-Schott et al[20] examined the dream effects of paroxetine and fluvoxamine in order to both increase clinical knowledge of these agents and to test an important potential method for probing the relationship between REM sleep neurobiology and dreaming in humans. Normal paid volunteers (4 males, 10 females) completed a 31-day home-based study consisting of: 7 days drug-free baseline; 19 days on either 100 mg fluvoxamine (7 Ss) or 20 mg paroxetine (7 Ss) in divided morning and evening doses; and 5 days acute discontinuation. The subjects were asked to write

Table 2. The effects of antidepressant drugs on sleep EEG

Drugs	REM Sleep	SWS	Sleep Continuity
Nonspecific monoamine reuptake inhibitors (TCAs)			
Amitriptyline	D	I	I
Clomipramine	D	I	I
Desipramine	D	I	I
Doxepine	D	I	I
Imipramine	D	NC	NC
Nortriptyline	D	I	NC
Trimipramine	NC	NC	I
MAOIs			
Moclobemide	D	NC	NC
Phenelzine	D	NC	D
SSRIs			
Citalopram	D	NC	NC
Fluvoxamine	D	NC	D
Fluoxetine	D	NC	D
Paroxetine	D	NC	D
Selective NE reuptake inhibitors			
Maprotiline	D	NC	I
Oxaprotiline	D	NC	NC
Viloxazine	D	D	D
Mianserine	D	I	I
Tianeptine	NC	NC	NC
Amineptine	NC	D	NC
Mirtazapine (NaSSA)	D	I	I
Venlafaxine (SNRI)	D	NC	D
Trazodone	NC-D	NC-I	I
Nefazodone	NC-I	NC-I	NC-I
Bupropion	I	D	NC

D= decreases; I= increases; NC= non-consistent

dream reports, self-scored specific emotions in their reports and rated seven general dream characteristics using 5-point Likert scales upon awakening using the Nightcap ambulatory sleep monitor. They found that mean dream recall frequency decreased during treatment compared with baseline; dream report length and judge-rated bizarreness were greater during acute discontinuation compared with both baseline and treatment and this effect was a result of the fluvoxamine-treated subjects; and the subjective intensity of dreaming increased during both treatment and acute discontinuation compared with baseline. Their findings suggest that the decrease in dream frequency during SSRI treatment may reflect serotonergic REM suppression while the augmented report length and bizarreness during acute SSRI discontinuation may reflect cholinergic rebound from serotonergic suppression. In a study[54] examining the efficacy, appropriate dose range, side-effects and clinical usefulness of citalopram in obsessive-compulsive disorder, the authors noted an increase in dreaming as an experienced adverse event.

Becker and Dufresne[55] assessed perceptual changes in 12 depressed patients treated wih bupropion, 12 patients given other antidepressants, and 12 drug-free controls. They found that bupropion was associated with vivid dreaming and changes in attention, memory, and perception. which may have contributed to its therapeutic effectiveness. The findings in this study may represent bupropion‘s REM-induced effects.

Drug-Induced Nightmares

There are several studies indicating drug-induced nightmares in the literature including nightmares secondary to drug withdrawal or drug-associated ones. In order to examine the effects of neuroleptic drugs in terms of frightening dreams were a function of differences in total dream recall, Strayhorn and Nash[56] administered a questionnaire to outpatients on bedtime-only doses and divided daily doses of tricyclic or neuroleptic drugs in a Veterans Administration Hospital Mental Hygiene Clinic, asking about frequency of dream recall, frequency of frightening dream recall, and doses and times of any medications taken. They found no significant difference between the two groups with respect to frequency of dream recall. They concluded that when a patient on bedtime doses of tricyclic or neuroleptic drugs had undesirable frightening dreams, the clinician should consider a change to divided daily doses. Lepkifker et al[57] reported four patients who experienced nightmares on fluoxetine monotherapy. Interestingly, Schenck et al[58] suggested fluoxetine as a cause of RBD.

Recently, Thompson and Pierce[59] assessed the English-language literature on drug-induced nightmares, excluding nightmares secondary to drug withdrawal or drug-associated night terrors. They considered published articles, letters, case reports, and abstracts in English were identified by MEDLINE for 33 years. They reviewed possible pharmacologic mechanisms for drug-induced nightmares, such as REM suppression and dopamine receptor stimulation. Using qualitative, quantitative, and possible pharmacologic mechanism criteria, they suggested that sedative/hypnotics, beta-blockers, and amphetamines were the therapeutic modalities most frequently associated with nightmares and dopamine agonists and dopamine receptor stimulation might considered as a possible pharmacologic mechanism.

Recently serotonin-2 antagonists/reuptake inhibitors such as trazodone and nefazodone have suggested as efficient drugs in the treatment of nightmares. Warner et al[60] reported trazodone as effective for the treatment of insomnia and nightmares associated with chronic post traumatic stress disorder (PTSD). Nefazodone used effectively in the treatment of nightmares in chronic combat-related posttraumatic stress disorder patients.[61] Similarly, mirtazapine, an alfa-2 antagonist, was found as effective for PTSD nightmares.[62] It is interesting to see these drugs as effective in the treatment of nightmares because they do not supress potently REM sleep. Thus, there is a question on whether nightmares are adaptive dreams to copy with stresful life events. Perhaps as suggested by Hartmann,[63] nightmares are most useful dreams.

Acknowledgments

Thanks are due to Lutfullah Besiroglu who participated in the final reading of this chapter.

References

1. Maquet P, Franck G. REM sleep and the amygdala. Molecular Psychiatry 1997; 2:195-196.
2. Braun AR, Balkin TJ, Wesenten NJ et al. Regional cerebral blood flow throughout the sleep-wake cycle. An H2(15)0 PET study. Brain 1997; 120:1173-1197.
3. Nofzinger EA, Mintun MA, Wiseman MB et al. Forebrain activation in REM sleep: An FDG PET study. Brain Research 1997; 770:192-201.
4. Maquet P. Functional neuroimaging of normal human sleep by positron emission tomography. J Sleep Res 2000; 9:207–231.
5. Solms M. Dreaming and REM sleep are controlled by different brain mechanisms. Behav Brain Sci 2000; 23:843-850.
6. McNamara P, Durso R, Auerbach S. Dopaminergic syndromes of sleep, mood and mentation: Evidence from Parkinson's disease and related disorders. Sleep and Hypnosis 2002; 4:119-131.
7. Wisor JP, Nishino S, Sora I et al. Dopaminergic role in stimulant-induced wakefulness. Journal of Neuroscience 2001; 21:1787-1794.
8. Agid Y, Javoy-Agid M, Ruberg M. Biochemistry of neurotransmitters in Parkinson's disease. In: Marsden CD, Fahn S, eds. Movement Disorders. New York: Buttersworth & Co., 1987:166-230
9. Cipolli C, Bolzani R, Massetani R et al. Dream structure in Parkinson's patients. Journal of Nervous Mental Disorders 1992; 180:516-523.
10. Fosse R, Stickgold R, Hobson JA. Emotional experience during rapid-eye-movement sleep in narcolepsy. Sleep 2002; 25:724-732.
11. Gottesmann C. The neurochemistry of waking and sleeping mental activity: The disinhibition-dopamine hypothesis. Psychiatry and Clinical Neurosciences 2002; 56:345-354.
12. Portas CM, Bjorvatn B, Ursin R. Serotonin and the sleep/wake cycle: Special emphasis on microdialysis studies. Prog Neurobiol 2000; 60:13-35.
13. Jones BE. Paradoxical sleep and its chemical/structural substrates in the brain. Neuroscience 1991; 40:637-656.
14. McCarley RW, Massaquoi G. Neurobiological structure of the revised limit cycle reciprocal interaction model of REM cycle control. Journal of Sleep Research 1992; 1:132-137.
15. Monti JM, Jantos H, Monti D et al. Dorsal raphe nucleus administration of 5-HT1A receptor agonist and antagonists: Effect on rapid eye movement sleep in the rat. Sleep Research Online 2000; 3:29-34.
16. Sorensen E, Gronli J, Bjorvatn B et al. Sleep and waking following microdialysis perfusion of the selective 5-HT1A receptor antagonist p-MPPI into the dorsal raphe nucleus in the freely moving rat. Brain Research 2001; 897:122-130.
17. Sharpley AL, Gregory CA, Solomon RA et al. Slow wave sleep and 5-HT2 receptor sensitivity during maintenance tricyclic antidepressant treatment. Journal of Affective Disorders 2000; 19:273-277.
18. Sommerfelt L, Ursin R. The 5-HT2 antagonist ritanserin decreases sleep in cats. Sleep 1993; 16:15-22.
19. Coenen AM, Ates N, Skarsfeldt T et al. Effects of sertindole on sleep-wake states, electroencephalogram, behavioral patterns, and epileptic activity of rats. Pharmacol Biochem Behav 1995; 51:353-357.

20. Pace-Schott EF, Gersh T, Silvestri R et al. SSRI treatment suppresses dream recall frequency but increases subjective dream intensity in normal subjects. J Sleep Res 2001; 10:129-142.
21. Hobson JA, McCarley RW, Wyzinski PW. Sleep cycle oscillation: Reciprocal discharge by two brainstem neuronal groups; Science 1975; 189:55–58.
22. Siegel JM. Brainstem mechanisms generating REM sleep. GABA release in the dorsal raphe nucleus. In: Kryger MH, Roth T, Dement WC, eds. Principles and Practice of Sleep Medicine. 3rd ed. Philadelphia: WB Saunders, 2000:724-741.
23. Kirschner N. Changes in dream content after drug treatment. Dreaming 1999; 9:195-200.
24. Ulivelli M, Rossi S, Lombardi C et al. Polysomnographic characterization of pergolide induced sleep attacks in idiopathic PD. Neurology 2002; 58:462-465.
25. Hartmann E, Russ D, Oldfield M et al. Dream content: Effects of L-Dopa. Sleep Research 1980; 9:153.
26. Rechtschaffen A, Maron L. The effect of amphetamine on the sleep cycle. Electroencephalogr Clin Neurophysiol 1964; 16:438-445.
27. Watson R, Hartmann E, Schildkraut JJ. Amphetamine withdrawal: Affective state sleep patterns and MHPG excretion. Am J Psychiatry 1972; 129:39-45.
28. Reimann D, Hohagen F, Fritsch-Montero R et al. Cholinergic and noradrenergic neurotransmission: Impact on REM sleep regulation in healthy subjects and depressed patients. Acta Psychiatrica Belgica 1992; 92:151-171.
29. Danguir J, De Saint-Hilaire-Kafi S. Scopolamine-induced suppression of paradoxical sleep is reversed by the somatostatin analogue SMS 201-995 in rats. Pharmacol Biochem Behav 1988; 30:295-297.
30. Hernandez-Peon R, Chavez-Ibarra G, Morgane PJ et al. Limbic cholinergic pathways involved in sleep and emotional behavior. Exp Neurol 1963; 8:93-111.
31. Callaway CW, Lydic R, Baghdoyan HA et al. Pontogeniculooccipital waves: Spontaneous visual system activity during rapid eye movement sleep. Cell Mol Neurobiol 1987; 7:105-149.
32. Gillin JC, Sutton L, Ruiz C et al. The effects of scopolamine on sleep and mood in depressed patients with a history of alcoholism and a normal comparison group. Biol Psychiatry 1991; 30:157-169.
33. Velazquez-Moctezuma J, Gillin JC, Shiromani PJ. Effects of specific M1, M2 muscarinic receptor agonists on REM sleep generation. Brain Res 1989; 503:128-131.
34. Sitaram N, Moore AM, Gillin JC. The effect of physostigmine on normal human sleep and dreaming. Arch Gen Psychiatry Oct 1978; 35(10):1239-43
35. Muzio JN, Roffwarg HP, Kaufman E. Alterations in the nocturnal sleep cycle resulting from LSD. Electroencephalogr Clin Neurophysiol 1966; 21:313-324.
36. Shiromani P, Gillin JC, Henriksen SJ. Acetylcholine and the regulation of REM sleep: Basic mechanisms and clinical implications for affective illness and narcolepsy. Annu Rev Pharmacol Toxicol 1987; 27:137-156.
37. Hartmann E. The nightmare: The psychology and biology of terrifying dreams. NY: Basic Books Inc., 1984.
38. Pavel S, Goldstein R, Petrescu M et al. REM sleep induction in prepubertal boys by vasotocin: Evidence for the involvement of serotonin containing neurons. Peptides 1981; 2:245-50.
39. Siegel JM, Rogawski MA. A function for REM sleep: Regulation of noradrenergic receptor sensitivity. Brain Res Rev 1988; 13:213-233.
40. Depoorte H. Adrenergic agonists and antagonists and sleep-wakefulness stages. In: Wauquier A, Gaillard JM, Monti JM et al, eds. Sleep: Neurotransmitters and Neuromodulators. New York: Raven Press, 1985:79-92.
41. Coulter JD, Lester BK. Reserpine and sleep. Psychopharmacology 1971; 19:134-147.
42. Mignot E, Guilleminault C, Bowersox S et al. Role of central a-1 adrenoceptors in canine narcolepsy. J Clin Invest 1988; 82:885-894.
43. Aldrich MS, Rogers AE. Exacerbation of human cataplexy by prazosin. Sleep 1989; 12:254-256.
44. Hilakivi I, Leppavouri A. Effects of methoxamine, an a-1 adrenoceptor agonist, and prazosin, an a-1 antagonist, on the stages of the sleep-wake cycle in the rat. Acta Physiol Scand 1984; 120:363-372.
45. Nicholson AN, Pascoe PA. Presynaptic a-2 adrenoceptor function and sleep in man: Studies with clonidine and idazoxan. Neuropharmacology 1991; 30:367-372.
46. Oswald I, Adam K, Allen S et al. A-Adrenergic blocker thymoxamine and mesoridazine both increase human REM sleep duration. Sleep Research 1974; 3:62.
47. Hilakivi I. The role of b- and a-adrenoreceptors in the regulation of the stages of the sleep-waking cycle in the cat. Brain Res 1983; 227:109-118.
48. Roth T, Kramer M, Salis PJ. Drugs, REM sleep, and dreams. In: Wolfman BB, ed. Handbook of Dreams. NY: Van Nostrand Reinhold Co., 1979:203-225.
49. Schredl M, Schafer G, Weber B et al. Dreaming and insomnia: Dream recall and dream content of patients with insomnia. J Sleep Res 1998; 7:191-198.
50. Riemann D, Low H, Schredl M et al. Investigations of morning and laboratory dream recall and content in depressive patients during baseline conditions and under antidepressive treatment with trimipramine. Psychiatric Journal of the University of Ottawa 1990; 15:93-99.
51. Armitage R, Rochlen A, Fitch T et al. Dream recall and major depression: A preliminary report. Dreaming 1995; 5:189-198.
52. Kirschner N. Changes in dream content after drug treatment. Dreaming 1999; 9:195-200.
53. Hall CS, Van de Castle R. The Content Analysis of Dreams. NY: Appleton-Century-Crofts, 1966.
54. Koponen H, Lepola U, Leinonen E et al. Citalopram in the treatment of obsessive-compulsive disorder: An open pilot study. Acta Psychiatr Scand 1997; 96:343-346.
55. Becker RE, Dufresne RL. Perceptual changes with bupropion, a novel antidepressant. Am J Psychiatry 1982; 139:1200-1201.
56. Strayhorn JM, Nash JL. Frightening dreams and dosage schedule of tricyclic and neuroleptic drugs. J Nerv Ment Dis 1978; 166:878-880.
57. Lepkifker E, Dannon PN, Iancu I et al. Nightmares related to fluoxetine treatment. Clin Neuropharmacol 1995; 18:90-94.
58. Schenck CH, Mahowald MW, Kim SW et al. Prominent eye movements during NREM sleep and REM sleep behavior disorder associated with fluoxetine treatment of depression and obsessive-compulsive disorder. Sleep 1992; 15:226-235.
59. Thompson DF, Pierce DR. Drug-induced nightmares. Ann Pharmacother 1999; 33:93-98.
60. Warner MD, Dorn MR, Peabody CA. Survey on the usefulness of trazodone in patients with PTSD with insomnia or nightmares. Pharmacopsychiatry 2001; 34:128-131.
61. Gillin JC, Smith-Vaniz A, Schnierow B et al. An open-label, 12-week clinical and sleep EEG study of nefazodone in chronic combat-related posttraumatic stress disorder. J Clin Psychiatry 2001; 62:789-796.
62. Lewis JD. Mirtazapine for PTSD Nightmares. Am J Psychiatry 2002; 159:1948-1949.
63. Hartmann E. The nightmare is the most useful dream. Sleep and Hypnosis 1999; 1:199-203.

CHAPTER 30

Sleep Problems in Primary Care

Alan G. Wade

Introduction

Sleep is a normal restorative function. What is not clear is how much sleep we need and what constitutes "normal sleep".

Complaints of sleeping difficulty are extremely common with probably about a third of the population experiencing difficulty at any one time. It is suggested that 10 percent of all patients attending the primary care physician have significant insomnia and that this is associated with marked functional impairment.[1] And yet for such a common disorder, the risks associated with the disorder, the benefit of treatment, and the optimal management are poorly defined.[2]

Specialist services tend to revolve round sleep laboratories. These are extremely resource intensive and patently inappropriate for investigating and managing large numbers of patients. And so the vast majority of patients with insomnia will inevitably be investigated and managed in primary care.

Despite this background, the average primary care physician receives little training in sleep medicine and generally acknowledges his or her ability to be at best modest.[3] Short-term training appears to improve the diagnostic and treatment rate but as with other educational initiatives, the benefit wanes with time.[4]

The vast majority of patients complaining of sleep difficulties in primary care will be complaining of too little sleep or "insomnia". The rare patient complaining of excessive sleep (as opposed to sleepiness or fatigue) or "narcolepsy" should probably be referred for specialist assessment before any treatment is instituted.

Definition

Primary Insomnia

Insomnia can be defined as a complaint by the patient of difficulty initiating or maintaining sleep or of sleep which is unsatisfactory or fails to provide restoration of normal physical or mental functioning the following day. It can be categorized as primary, secondary, related to disruption of circadian rhythm or to one of a number of special categories such as sleep apnoea and sleepwalking.

Both DSM IV and ICD,[10] the principal diagnostic instruments emphasise three important aspects of the diagnosis of primary insomnia.[5]

1. The diagnosis is based on the patient complaint of inadequate or unsatisfactory quality of sleep. The length of the patient's sleep is not important, reflecting the wide variation in the length of sleep experienced by individuals not considering themselves insomniacs.
2. Insomnia is not a diagnosis to be attached to an acute problem. Transient difficulties with sleeping in response to every day stress or disruption of normal routine are so common as to warrant neither a diagnostic label nor medical intervention. The symptoms should have been present for at least one month.
3. Perhaps most importantly of all, there should be evidence that the insomnia is interfering with the patient's ability to carry out normal tasks the following day. The most common and serious problem is associated with driving.

Secondary Insomnia

Sleep difficulties are more often attributable to another cause—physical or psychological—than they are to a primary sleep problem. In general practice therefore the diagnosis of primary insomnia tends to become one of exclusion.

Secondary insomnias can usefully be divided into those caused by medical conditions, by psychological conditions and by problems associated with medication or drugs. Although advice about sleep hygiene or the short-term use of hypnotic drugs may be helpful if the condition has become chronic, treatment of secondary insomnia is essentially that of the primary condition. As primary-care physicians, we expect to know our patient's medical history but it is surprising how frequently someone is self medicating with antacids for heartburn, analgesics for mild osteoarthritis or ephedrine based drugs for allergies or obesity without thinking to inform their physician. A careful history of OTC medication can be extremely helpful (Tables 1 and 2).

Insomnia can also be associated with the use of, or even more frequently, the attempted withdrawal from alcohol, nicotine or illicit drugs. This is an extremely important aspect of the history to elicit and to record carefully if for no other reason than that the use of hypnotics in these patients is generally contraindicated.

Circadian Rhythm Disorder

Circadian rhythm disorder is usually encountered secondary to jetlag or in shift workers where the normal sleep/wake pattern is disrupted. It can however, be encountered in otherwise normal patients who over a long period of time and for no obvious reason slip into a chaotic and irregular daily pattern. In virtually all cases, they can be managed by careful attention to sleep hygiene (Table 3).

The other group of patients who may be considered to come into the category of circadian rhythm disorder are patients with visual impairment and in particular the elderly.[6]

Sleep and Sleep Disorders: A Neuropsychopharmacological Approach, edited by Malcolm Lader, Daniel P. Cardinali and S.R. Pandi-Perumal.

Table 1. Drugs associated with insomnia

OTC	Caffeine Nicotine Amphetamines and other stimulants
Prescription drugs	Nonsteroidal agents Corticosteroids SSRIs Non-selective beta-blockers Theophyllines
Withdrawal related insomnia	Alcohol Nicotine Benzodiazepines Antidepressants

Table 2. Typical medical conditions associated with insomnia

Pain – from any cause	
Cardiovascular	Heart failure Angina
Respiratory	Dyspnoea due to COPD or Asthma Upper respiratory tract obstruction
Gastro-intestinal	GORD Diarrhoea
Neurological	Parkinson's Alzheimer's

This particular sleep disorder is associated with disruption of the normal diurnal pattern of melatonin production from the pineal gland. It has therefore been suggested that giving oral melatonin in appropriate doses to regulate the sleep pattern may be helpful. At present, however, melatonin in an appropriate dosage and formulation is not a registered drug but may be bought directly in many countries.

Presentation

Although a common problem in primary care, patients tend not to complain specifically of sleeplessness.[5] More frequently there will be a nonspecific problem of tiredness or fatigue or else the patient is being seen for an unrelated problem or routine physical examination. Occasionally the complaint will be of poor memory or concentration leading to difficulties at work or school. This can be severe enough to have threatened employment. In each of these situations it is important for the GP to consider the possibility of insomnia.

According to UK Department of Transport, 360 deaths and 24,000 accidents are directly attributable to sleep related causes.[7] And thus any road traffic accident, regardless of how apparently trivial, should trigger the taking of a thorough sleep history.

Not uncommonly, the consultation is instigated not by the patient but by the partner who complains of excessive snoring or restlessness. This should alert the practitioner to the possibility of sleep apnoea syndrome or restless legs syndrome.

Table 3. Sleep hygiene patient advice

The aim of "sleep hygiene" is to impose a pattern and establish a daily routine in which going to sleep is just one of the daily activities which constitute that pattern. This pattern will vary from person-to-person but some of the basic rules to follow and included in the suggestions below:

Try to ensure that the environment of your bedroom is as tranquil as possible
- Use shutters or blinds to exclude sunlight
- Avoid bright light and noise
- Ensure the temperature is "comfortable"

Do not lie on your bed except when you're going to sleep at bedtime
- This helps the body associate the bed with sleep

Try to develop a regular pattern during the day as well as the evening
- Take meals and exercise at similar times each day

Don't take naps
- Ideally avoid naps altogether but if this is not possible schedule them for a regular time in the middle of the day

Go to bed the same time every day
- Until a regular pattern is developed try to avoid disruption of the routine with late nights

Refrain from vigorous exercise at least 4 hours before bedtime
- A short evening walk, if it is part of your regular routine would do no harm

Develop sleep rituals
- Try to do the same things, in the same order every night starting well before bedtime

Avoid tea, coffee, other caffeine containing drinks such as cola, nicotine and alcohol for four hours before going to bed
- Although alcohol seems out of place in this list, it is important because although initially it might help, it leads to restlessness and disruption during the night.

Have a light snack before bed
- It is not good to have a heavy meal before sleeping but a light snack such as a milky drink may help

Getup at the same time every day
- This will help establish your daily routine

Examination

Examination should very much be focused on positively identifying serious sleep problems such as sleep apnoea and negatively excluding the host of physical and psychological illnesses potentially leading to secondary insomnia. The age of the patient and history will help point the way – a thirty year old is unlikely to have heart failure, but might well have atypical asthma; a fifty year old woman may be suffering menopausal symptoms.

It is important to consider the possibility of sleep apnoea in any obese patient with hypertension who complains of excessive daytime drowsiness or whose partner complains of them snoring excessively.

Investigation

In general practice, the investigation of primary insomnia depends essentially upon a careful history both from the patient and if possible from their bed partner.

The history should concentrate on:

- Length of history
- Sleep preparation
- Sleep latency
- Disturbance during the night
- Time and method of awakening
- Daytime function
- Lifestyle (e.g., shift worker)

The specific investigation of primary insomnia by general practitioners is limited. Sleep diaries can help the patient to provide a much more accurate and comprehensive history and provide the basis for a subsequent management plan. While widely advocated for the investigation of insomnia they are seriously under-used in primary care. Validated sleep diaries such as the Pittsburgh or the Karolinska are essentially research tools and in practice asking the patient to keep a simple record of the time they took to go sleep, how frequently they were disturbed each night and details of the time and process of wakening is probably sufficient for most purposes. Of vital importance is to record their sleepiness, napping and functioning during the following day.

Treatment

The treatment of secondary insomnia is relatively straight forward and successful. The basic cause of the problem will have been elucidated and treatment will hopefully resolve not only the primary problem but also the insomnia. Nevertheless occasionally the short-term use of hypnotics to assist the patient in regaining their normal sleep pattern will be a helpful and sympathetic gesture.

Primary insomnia, however, provides one of the great challenges in general practice. Some essential facts need to be established;

- Primary insomnia is a common condition
- Primary insomnia is a chronic condition
- The majority of patients prefer some form of psychological treatment to medication[8]
- There is a limited availability in most practices for psychological treatment of established efficacy
- Hypnotic medication is licensed and recommended for short-term use only in most countries
- In many countries sedative drugs are available for direct sale to patients
- Specialist sleep centres are not available and not equipped to investigate and manage large patient numbers

In terms of the strict definition, sleep difficulties should have been present for at least one month before the diagnosis is made. Very few patients with chronic insomnia present to the general practitioner in less than one month. Some patients in response to an acute life stressor may experience temporary, acute sleep difficulty. A simple explanation and advice will usually be adequate management but occasionally a short course of hypnotic medication may be helpful.

Most problems in general practice, however, gravitate round the difficulty of the patient with chronic primary insomnia of sufficient severity to warrant treatment. In general, a chronic illness will require long-term treatment. Guidelines for the use of sedatives and hypnotics in most countries suggest limiting the use to "short term" usually meaning two or four weeks. This immediately creates a potential problem. It is thus good practice that nonpharmacological treatment be used or perhaps combined with short-term hypnotic use. The prescription of hypnotics without attention to other lifestyle changes will rarely result in a satisfactory outcome.

Sleep Hygiene

"Sleep hygiene" so called, is the mainstay of management and consists of advice aimed at developing a regular pattern to the patient's life—during the day as well as at night. It should be built around resisting taking naps except at agreed times during waking hours and avoiding stimulants such as tea, coffee and nicotine for several hours prior to going to bed. Many patients will have attempted to self medicate with alcohol before presenting and it is as well to explain from the beginning that chronic use results in poorer quality asleep, tolerance and the danger of dependency. Typical suggestions for sleep hygiene are contained in Table 3.

Nonpharmacologic Treatment of Insomnia

There is evidence that, at least in the short to medium-term, many patients with insomnia would benefit from specific nonpharmacological management.[9,10] Scientific assessment, however, has been limited to specific behavioural interventions requiring specialist training and not readily available to the majority of patients. Several attempts have been made to address this problem and at least one apparently successful approach is to use trained nurses to deliver modified CBT based treatment.[11] Similarly a two-session abbreviated CBT approach has also shown benefit.[12] Nonspecific interventions such as exercise advice, while apparently attractive have very limited evidence of benefit.[13]

Pharmacotherapy

The current pattern of pharmacotherapy for insomnia in primary care is moulded by guidelines and legal considerations which are often at variance with the initial desires of the patient. Insomnia is a chronic condition: hypnotics are recommended for short term use. In general terms, patients tend to perceive long term hypnotic treatment as having greater efficacy and lesser risk than physicians.[14] This has the effect of causing strains in the doctor patient relationship and difficulties in getting a clear picture of current practice. In this situation it is vital to agree a management plan with a patient before any medication is prescribed.

Historically, based mixtures have been used to promote sleep. In the 1950s and 60s, chloral hydrate and the barbiturates were used for their sedative properties but were associated with widespread abuse and lethal overdoses. The advent of the benzodiazepines in the 1970s represented a significant step forward but again problems of abuse were identified but also more subtle problems associated with prolonged action of possible association with road traffic accidents and unsteadiness in the elderly leading to increased falls with associated fractures. In addition, withdrawal from benzodiazepine drugs may be associated with a syndrome consisting of initial insomnia followed by a prolonged period of anxiety and agitation.

The newer hypnotic agents, Zolpidem, zopiclone and zaleplon act as agonists at the benzodiazepine receptor component of the GABA complex. They are rapidly absorbed orally, have an excellent side-effect profile and have a low interaction potential.[15] Zaleplon has an ultra short half-life which makes it useful for promoting sleep even if taken during the night.[16] Zolpidem with a longer half-life will not only shorten sleep latency but also promote sleep maintenance. Both zaleplon and zolpidem are used in doses of 5 to 10 mg. Zopiclone, used in a dose of 7.5 mg for adults has a slightly longer half-life and thus greater potential for hangover sleepiness following day.

All these drugs however, are recommended for use for between two and four weeks. Despite this, in clinical practice chronic usage is not uncommon. Zolpidem has been studied for up to six months and a major review of the use of zopiclone and zolpidem has recently been undertaken and although not completely absent the abuse potential of these drugs is certainly lower than that of the benzodiazepines.[17] Alternative strategies for the long term treatment of chronic insomnia are certainly needed with noncontinuous use one alternative. Zolpidem has been used in this way[18] and a large-scale study has also been undertaken of nonnightly use extending to eight weeks without evidence of discontinuation effects or increased frequency of dosing.[19]

Benzodiazepine drugs such as temazepam are also effective in short-term use but require more careful monitoring having a greater potential for causing morning sedation, memory dysfunction, tolerance and abuse. In primary care, instituting a policy of stopping all long term users of benzodiazepine hypnotics, where necessary by substituting an alternative for a short time appears less problematical than was first thought.[20]

Primary insomnia can also be associated with symptoms of depression and there is widespread use of sedative antidepressant drugs such as the tricyclic agents for long-term use in this indication. The effectiveness and safety of the strategy has not been reported but as they are often used in relatively low doses, even elderly patients may be at negligible risk.

Possibly associated with the reluctance of many primary care physicians to prescribe for insomnia many patients have resorted to the use of ethanol, illicit drugs, OTC and herbal medications.[21] Typically OTC sleep promoters are sedative antihistamines such as diphenhydramine. They produce a poor quality of sleep, often cause drowsiness the following day and it is mainly this property which has led to them being superseded in other indications by nonsedative alternatives. Some herbal remedies have been compared to benzodiazepines in double-blind studies and apparently been found beneficial but this has to be laid against the increasing list of side-effects being documented as these agents come under increasing scrutiny.[22]

Melatonin, as a naturally occurring hormone has been widely studied. Essentially it is produced in the pineal gland in response to darkness and has widespread effects throughout the body including signaling preparation for sleep. The production of melatonin is disrupted with upset to the circadian rhythm and reduces with increasing age. It has thus been studied particularly in the management of jetlag, other circadian rhythm disorders and insomnia in the elderly.[23] At present there is no formulation available for prescription but in some countries it is widely used having been bought OTC as a herbal medication or food supplement.

Elderly

As many as 40 percent of older patients will complain of sleep difficulties at some time.

Changes in sleep pattern are part of the normal ageing process and are characterized by increased time in bed, decreased total sleep time, prolonged sleep latency and increased night-time wakening. Increasing age is also associated with physical problems and these too can contribute to insomnia.

Initial assessment should be based on a careful history obtained from the patient and, if available, from their partner. It is important first to exclude any significant psychiatric or physical disease. Typical psychiatric problems might include depression, in elderly patients there is often a large anxiety component, and early dementia where disturbed sleep can often be an early symptom. Physical problems causing sleep disturbance commonly revolve around pain, urinary problems and upper gastrointestinal pathology with acid reflux.

Having excluded the primary causes of secondary insomnia it is impractical in primary care to fully investigate every case of primary insomnia and indeed it is not necessary. A careful history with perhaps a diary card record for a week or two will usually delineate the problem. For the majority of patients a simple explanation and advice about sleep hygiene will normally suffice (Table 3). For elderly patients where this is not successful and who continue to have functional impairment, many wish some form of nonpharmacological treatment. The availability of these treatments is very patchy and current evidence of their efficacy rather limited. CBT has good evidence of efficacy but this may be short lived, other interventions such as bright light therapy have little scientific data to support their use.[10] While in general terms, exercise should be encouraged, its use in treating primary insomnia in the elderly appears of only marginal benefit.[24]

Most primary care physicians and patients will need to rely on pharmacology. This immediately raises the difficulty that for many elderly patients, insomnia is a chronic problem and may need long-term treatment.[25] This presents the dilemma of all medical interventions of balancing benefit against risk.

The Risks

It is important that we are clear about the risks involved:

- Increased falls at night
- Cognitive dysfunction or sleepiness the following day
- Dependence or addiction

The evidence supporting increased falls in elderly patients using hypnotics is not strong.[26] Where an association has been found, it has been with longer acting agents and with increasing doses.[27] It would seem a sensible precaution, therefore, to avoid all but short acting hypnotics and at the lowest doses in patients who are already unstable on their feet or have conditions such as osteoporosis where the consequences of a fall might have more serious implications. Overuse and abuse of benzodiazepines for many years has led some doctors to be reluctant to prescribe them at all but a recent large study has found little escalation of dosage, particularly in the elderly, and it would appear that for the majority of patients modern hypnotics can be used safely but with appropriate monitoring. Although not widely investigated, the successful use of "as required" medication has been reported with zolpidem and this would certainly be an attractive strategy in primary care.[28]

The Benefits

Insomnia in the elderly is associated with reduced quality of life, poor daytime functioning and a cycle which often includes increasing daytime napping, increasingly disturbed sleep patterns increasingly disrupted activity patterns and which may eventually result in major depression. Treatment of insomnia to break the cycle is a major health benefit.

The most intractable problem faced by primary-care physicians in dealing with elderly insomnia is that of the patient with dementia. Each patient must be managed individually but where support is available, it is possible to institute sleep hygiene measures with some benefit.[29] The judicious use of hypnotics, antidepressants and possibly antipsychotics will, however, be necessary in many patients.

Specific Sleep Difficulty Syndromes

These range from the relatively common and important sleep apnoea syndrome to the rarer so-called "parasomnias" such as nightmare disorder, sleep terror disorder and sleepwalking disorder.

Also common are the so-called limb movement disorders - "restless legs syndrome" where the patient complains of uncomfortable sensations and movement of their limbs and which may start before going to bed and can certainly cause delay in getting to sleep and occasionally wake the patient during the night and "Idiopathic nocturnal limb movements" which are a myoclonic jerking of the limbs and often complained of by the bed partner rather than the patient themselves. Rarely they can be of such severity as to cause awakening.

Sleep apnea is an important disorder which should always be investigated when the diagnosis is suspected and referral to a specialist for this and the other, rarer parasomnias is strongly recommended. By contrast, limb movement disorders are essentially a diagnosis based on history and at least initially management should be possible in primary care.

Sleep Apnea

Obstructive sleep apnea (OSA) is one of the commonest sleep disorders encountered in primary care.[30] It is generally accepted as being one of the most serious with significant physical, psychological and public health implications. It is also the sleep disorder par excellence which is amenable treatment and yet the current knowledge of the disorder is poor among primary-care physicians.[31] This results in many patients being put at risk and failing to benefit from treatment available. Questionnaires have been developed to assist in diagnosis but are rarely used in routine practice.

Sleep apnoea can be defined as excessive daytime sleepiness with irregular breathing at night. It is associated with reduced muscle tone in the upper laryngeal airway. When the patient falls asleep, further reduction in muscle tone results in airway narrowing and obstruction, signified by snoring, and leading to complete or partial awakening with resulting increased muscle tone. This cycle repeats itself throughout the night which results in disrupted sleep and typical tiredness the following day. Most studies suggest a prevalence for the disorder of 1-2% for middle aged men and about half that for women.

Sleep apnoea is claimed to be a significant cause of premature death but the evidence to support this is not strong. With daytime sleepiness, reduced cognitive function, impaired physical functioning, irritability and reduced quality of life it certainly has an impact on the life of the patient. There is an independent association with hypertension. Snoring, a prominent symptom of sleep apnoea, results in marital and family tension and it is this which frequently results in presentation to the general practitioner.

Sleep apnoea, however, is not a discreet entity and represents one end of the spectrum from normal quiet breathing during sleep through milder snoring to severe intermittent upper airways obstruction at relatively light sleep levels. Questionnaires have been developed for the diagnosis of sleep apnoea but none that have the sensitivity and specificity to be clinically useful and to date these have not been widely adopted.[32] The diagnosis is essentially based on sleep laboratory, EEG and oximetry studies and as such requires referral to a specialist sleep laboratory or respiratory specialist with an interest in this topic. The decision to refer a specific patient is seldom absolute but in view of the potential seriousness of the condition and the possibility of effective treatment it is important to maintain a high level of suspicion for the diagnosis.

From a practical point of view, the availability of specialist sleep laboratories to investigate every patient complaining of daytime tiredness and snoring is impractical. For patients with milder degrees of sleep apnoea simple attention to lifestyle factors such as weight reduction and smoking and alcohol cessation may be of some assistance. Sedatives can aggravate the symptoms and should be stopped. It is as well to remember that in many countries drugs such as sedating antihistamines are available without prescription and may also aggravate the symptoms. If the problem is severe, however, referral should not be delayed while these measures are undertaken.

Definite indications for referral include:

- Falling asleep in hazardous situations e.g., driving
- Excessive snoring which is severe enough to be considered for surgery
- Daytime sleepiness which is severely impairing function

Relative risk factors which might raise the level of suspicion include:

- Obesity (50 percent of patients with sleep apnoea are obese)
- Neck circumference > 17 inches (43 cm)
- Male gender
- High smoking and alcohol consumption
- Sedative drug ingestion
- Small mandible

Sleep Related Movement Disorders

One of the most common causes of insomnia seen in general practice is that of "restless legs syndrome" or RLS. It is said to occur in 10 to 15 percent of the population although not usually of sufficient severity to warrant active treatment.[33] It becomes increasingly common with advancing age and sometimes there is a family history of the disorder. RLS is a clinical diagnosis based on a history of an uncomfortable feeling often described by patients as a sensation of "crawling" or "creeping" which the patient attempts to relieve by movement. Generally the symptoms occur at rest and prevent the patient getting to sleep but occasionally are so severe that night-time awakenings also occur.

It is unusual for a cause for RLS to be found—so-called "idiopathic"—but it can be associated with the late stages of pregnancy, iron deficiency or other causes of anaemia, uraemia and neurological diseases such as parkinsonism or neuropathy. As a general practitioner, it is also worth reviewing concomitant medication as the condition can certainly be aggravated by drugs such as phenothiazines, antipsychotics and metoclopramide.

Where present, treatment of an underlying condition will usually alleviate the symptoms of RLS. More commonly however no obvious medical disorder will be found and for most patients, a simple reassurance that the condition is not indicative of serious illness combined with advice will usually suffice. What is successful in one patient unfortunately appears in some others to make matters worse and advice will be very much on a "trial and error" basis. Some patients find RLS worse if they have had a day of strenuous exercise; others need to be advised that regular modest exercise might be helpful. Likewise with heat and cold – some will benefit from heat, others from cold applications. Reducing the intake of alcohol, caffeine and tobacco when indicated by the history can be useful and instigation of a good sleep hygiene regime may also prove helpful and can certainly do no harm.

Pharmacological treatment, however, is more problematic and in general is restricted to the more severe cases. First-line agents are dopaminergic drugs such as levodopa, or carbodopa. In general, the dosage of levodopa in the 100 mg to 500 mg range will be helpful to most patients. As in Parkinson's disease, there is a tendency for augmentation to occur within a few months with recrudescence of symptoms. Dopamine antagonists such as pergolide have also been used with apparent benefit but most primary-care physicians are less conversant with the use of these agents and may well consider referral to a specialist at that stage. Other agents such as benzodiazepines, anticonvulsants (in particular gabapentin) and clonidine have all been described as being useful in small trials. In general efficacy appears to vary from time to time and frequent changing of medication may be needed to keep symptoms under control. It is worth remembering also that none of the recommended treatments are without problems of side-effects, drug interactions or addiction potential and careful consideration must be given to the benefit which will be gained by using them in a disease with a relatively benign course.

Conclusion

Insomnia is a common complaint encountered in general practice which causes considerable distress to the sufferer. These patients present both a diagnostic and management challenge for primary care physicians. Despite the frequency with which the disorder is encountered most doctors consider themselves lacking in both knowledge and expertise indicating the need for a greatly expanded educational programme. Most patients are treated with drugs and there is the need for us to resolve the discrepancy between the treatment of a debilitating and distressing chronic condition from which patients are seeking relief and the recommendation to only use short-term medication. An open informed discussion about the efficacy, benefits, dangers, side-effects and liability to abuse of longer term use of hypnotic agents in typical primary care patients is urgently required.

References

1. Simon GE, VonKorff M. Prevalence, burden, and treatment of insomnia in primary care. Am J Psychiatry 1997; 154(10):1417-23.
2. Richardson GS. Managing insomnia in the primary care setting: Raising the issues. Sleep 2000; 23(Suppl 1):S9-12; discussion S13-5.
3. Papp KK, Penrod CE, Strohl KP. Knowledge and attitudes of primary care physicians toward sleep and sleep disorders. Sleep Breath 2002; 6(3):103-9.
4. Backhaus J, Junghanns K, Mueller-Popkes K et al. Short-term training increases diagnostic and treatment rate for insomnia in general practice. Eur Arch Psychiatry Clin Neurosci 2002; 252(3):99-104.
5. Doghramji PP. Detection of insomnia in primary care. J Clin Psychiatry 2001; 62(Suppl 10):18-26.
6. Zizi F, Jean-Louis G, Magai C et al. Sleep complaints and visual impairment among older Americans: A community-based study. J Gerontol A Biol Sci Med Sci 2002; 57(10):M691-4.
7. Horne JA, Reyner LA. Sleep related vehicle accidents. BMJ 1995; 310(6979):565-567.
8. Vincent N, Lionberg C. Treatment preference and patient satisfaction in chronic insomnia. Sleep 2001; 24(4):411-7.
9. Morin CM, Hauri PJ, Espie CA et al. Nonpharmacologic treatment of chronic insomnia. An American Academy of Sleep Medicine review. Sleep 1999; 22(8):1134-56.
10. Montgomery P, Dennis J. Cognitive behavioural interventions for sleep problems in adults aged 60+. Cochrane Database Syst Rev 2003; (1):CD003161.
11. Espie CA, Inglis SJ, Tessier S et al. The clinical effectiveness of cognitive behaviour therapy for chronic insomnia: Implementation and evaluation of a sleep clinic in general medical practice. Behav Res Ther 2001; 39(1):45-60.
12. Edinger JD, Sampson WS. A primary care "friendly" cognitive behavioral insomnia therapy. Sleep 2003; 26(2):177-82.
13. Fuscaldo JM. Prescribing physical activity in primary care. W V Med J 2002; 98(6):250-3.
14. Mah L, Upshur RE. Long term benzodiazepine use for insomnia in patients over the age of 60: Discordance of patient and physician perceptions. BMC Fam Pract 2002; 3(1):9.
15. Wagner J, Wagner ML. Nonbenzodiazepines for the treatment of insomnia. Sleep Med Rev 2000; 4(6):551-581.
16. Israel AG, Kramer JA. Safety of zaleplon in the treatment of insomnia. Ann Pharmacother 2002; 36(5):852-9.
17. Hajak G, Muller WE, Wittchen HU et al. Abuse and dependence potential for the nonbenzodiazepine hypnotics zolpidem and zopiclone: A review of case reports and epidemiological data. Addiction 2003; 98(10):1371-8.
18. Hajak G, Cluydts R, Declerck A et al. Continuous versus nonnightly use of zolpidem in chronic insomnia: Results of a large-scale, double-blind, randomized, outpatient study. Int Clin Psychopharmacol 2002; 17(1):9-17.
19. Walsh JK, Roth T, Randazzo A et al. Eight weeks of nonnightly use of zolpidem for primary insomnia. Sleep 2000; 23(8):1087-96.
20. Gilhooly TC, Webster MG, Poole NW et al. What happens when doctors stop prescribing temazepam? Use of alternative therapies. Br J Gen Pract 1998; 48(434):1601-2.
21. Johnson EO, Roehrs T, Roth T et al. Epidemiology of alcohol and medication as aids to sleep in early adulthood. Sleep 1998; 21(2):178-86.
22. Houghton PJ. The scientific basis for the reputed activity of Valerian. J Pharm Pharmacol 1999; 51(5):505-12.
23. Zisapel N. The use of melatonin for the treatment of insomnia. Biol Signals Recept 1999; 8(1-2):84-9.
24. Montgomery P, Dennis J. Physical exercise for sleep problems in adults aged 60+. Cochrane Database Syst Rev 2002; (4):CD003404.

25. Morgan K, Clarke D. Longitudinal trends in late-life insomnia: Implications for prescribing. Age Ageing 1997; 26(3):179-84.
26. Wysowski DK, Baum C, Ferguson WJ et al. Sedative-hypnotic drugs and the risk of hip fracture. J Clin Epidemiol 1996; 49(1):111-3.
27. Ray WA, Griffin MR, Schaffner W et al. Psychotropic drug use and the risk of hip fracture. N Engl J Med 1987; 316(7):363-9.
28. Hajak G, Bandelow B, Zulley J et al. "As needed" pharmacotherapy combined with stimulus control treatment in chronic insomnia—assessment of a novel intervention strategy in a primary care setting. Ann Clin Psychiatry 2002; 14(1):1-7.
29. McCurry SM, Gibbons LE, Logsdon RG et al. Training caregivers to change the sleep hygiene practices of patients with dementia: The NITE-AD project. J Am Geriatr Soc 2003; 51(10):1455-60.
30. Ohayon MM, Guilleminault C, Priest RG et al. Snoring and breathing pauses during sleep: Telephone interview survey of a United Kingdom population sample. BMJ 1997; 314(7084):860.
31. Chung SA, Jairam S, Hussain MR et al. Knowledge of sleep apnea in a sample grouping of primary care physicians. Sleep Breath 2001; 5(3):115-21.
32. Netzer NC, Stoohs RA, Netzer CM et al. Using the Berlin Questionnaire to identify patients at risk for the sleep apnea syndrome. Ann Intern Med 1999; 131(7):485-91.
33. O'Keeffe ST. Restless legs syndrome. A review. Arch Intern Med 1996; 156(3):243-8.

CHAPTER 31

SSRIs and Sleep in Man

Sue J. Wilson and David J. Nutt

Introduction

The class of antidepressants known as SSRIs (selective serotonin reuptake inhibitors) is a major part of the treatment armamentarium in psychiatric practice. These drugs were licensed for use first in major depression, and subsequently various SSRIs have received licences in anxiety indications such as obsessive compulsive disorder (OCD) panic disorder (PD) generalized anxiety disorder (GAD) and post-traumatic stress disorder (PTSD). Thus a huge number of patients worldwide are taking these agents.

It is important to be aware of sleep effects of SSRIs for many reasons. The first reason is that sleep disruption is a major complaint in depression, with over 90% of patients having sleep complaints in 1 study of nearly a thousand patients with major depression in the US.[1] Insomnia is often the symptom which causes depressed patients to seek medical help, and relief of sleep disturbance is important to encourage compliance with medication and psychological treatments.

Secondly, one of the few robust biological findings in depression is alteration of sleep architecture, the structure and organisation of the various sleep stages. Depressed patients as a group have alterations in REM sleep; they enter REM sleep earlier than control subjects, and nonREM sleep appears to be reduced in the first part of sleep. These two findings suggest that there is a major disruption to both the circadian and the homeostatic drives to sleep. Antidepressants, (including ECT) nearly all alter sleep in the opposite direction to these depression-related changes, which may mean that sleep is an indirect biological marker of their therapeutic effects in the brain. Importantly, there is evidence that patients who have these sleep changes are more likely to respond to pharmacological treatments than to psychological ones.[2]

Thirdly, measurement of sleep provides a very sensitive index of serotonergic effects in the brain both in volunteers and patients, and longitudinal studies of sleep during drug administration may give important insights into the adaptation mechanisms which have been implicated in the therapeutic effects of these drugs.

SSRIs and Sleep Architecture

There is extensive literature on acute effects of SSRIs (although only 1 controlled study of sertraline and 2 of citalopram) and less on their chronic effects. Table 1 gives a summary of the effects from 37 controlled studies. The most consistent effects are on REM sleep, and these effects are similar in magnitude and direction in healthy volunteers and depressed patients.

Acute Effects on REM Sleep

The 2 effects described with all SSRIs (and also with mixed serotonin/noradrenaline reuptake inhibitors) are dose-related and consist of reduction in the overall amount of REM sleep over the night, and delay of the first entry into REM sleep (increased REM onset latency ROL). Figure 1 shows a hypnogram from a depressed patient and a normal volunteer before and after short-term treatment with paroxetine.

Some acute studies in Table 1 show less effect on REM latency than others, and this is usually due to time of dosing. SSRIs have relatively long absorption times with maximum plasma levels achieved at about 4-8 hours. If the drug is given at bedtime, plasma levels are probably not high enough early in the night to delay the first REM episode. There is a little evidence that fluoxetine brain entry may be slower than for the other SSRIs, as in one study maximum increase in REM latency after a single 40 mg dose was on the third night,[3] and other single dose studies have not shown the consistent REM effects which appeared after multiple dosing.

Chronic and Withdrawal Effects on REM Sleep

Most of the effects described above become less evident after chronic treatment, this diminution being less obvious with fluoxetine than the other SSRIs (see Table 1). Maximal effects on amount of REM sleep seem to occur after a few days, and then gradually return to baseline levels in a few weeks; studies after 8 weeks rarely show significant reductions in REM. REM amount also showed a rebound increase 4-12 days after fluoxetine withdrawal in the detailed study by Feige et al[3] and 6 days after citalopram withdrawal in van Bemmel's study.[4] REM onset latency however appears to be less susceptible to these rebound/withdrawal effects.

Effects on Sleep Continuity and Non-REM Sleep

Changes in sleep initiation and continuity after acute SSRIs are also similar in volunteers and depressed patients, and consist of increased light (stage 1) sleep, number of arousals from sleep and time spent awake at night. In general the magnitude of this arousing effect has been larger in normal volunteers, but it must be remembered that there is a large baseline difference in these 2 groups, with depressed patients starting out with very disrupted sleep. In general these sleep disturbance effects diminish over time, with most studies in depressed patients showing no worsening of sleep after a few days of treatment. The exception to this is fluoxetine, which continued to disrupt sleep continuity after 8 weeks in a large multicentre study.[5]

Sleep and Sleep Disorders: A Neuropsychopharmacological Approach, edited by Malcolm Lader, Daniel P. Cardinali and S.R. Pandi-Perumal.

Table 1. REM decrease (%) + 10-30, ++ 30-60, +++ > 60
REM latency increase (%) + 30-100 ++ 100-200, +++ > 200

Drug	Dose	Duration (d= Days, w= Weeks)	Sig. REM Decrease	Sig. REM Latency Increase	Sig. Sleep Efficiency Decrease	Sig. Stage 1 Increase
Normal volunteer studies						
Paroxetine	20/40 mg pm	single	++	++	Y	Y
	30 mg am	single	+++	+++		
	20 mg	3-7 d	++	+++	Y	Y
	20-30 mg	2-4 w	++/+	++	Y	Y
Fluvoxamine	100 mg	single	++	+	Y	Y
	150mg	4 w	++	+	Y	
Fluoxetine	20/40 mg	single	0	+	Y	
	60 mg	single	+	+	Y	
	20 mg	6-10 d	+	+	Y	
	20/40 mg	3-5 w	+	+	Y	Y
Citalopram	No single dose study					
	20 mg	3 d	++	+++	Y	Y
Sertraline	No NV studies					
Depression studies						
Paroxetine	20 mg	3 d	+++	+++	Y	
	20 mg	10 d	++	+++		
	20-40 mg	8 w	+++	++		Y
Fluvoxamine	200 mg	single	+++	+++	Y	
	200 mg	7 d	+++	+++		
	100-200 mg	3 w	++/+++	++/+++		
	100-250	12 w	+	++		
Fluoxetine	20 mg	7 d				Y
	20 mg	2 w		+		Y
	20-40 mg	8 w	+	++	Y	Y
Citalopram	20 mg	Single	++	+++		
	40 mg	5 w	++	+++		
Sertraline	~100-150	12 w		++		

Data from papers marked '*' in reference list

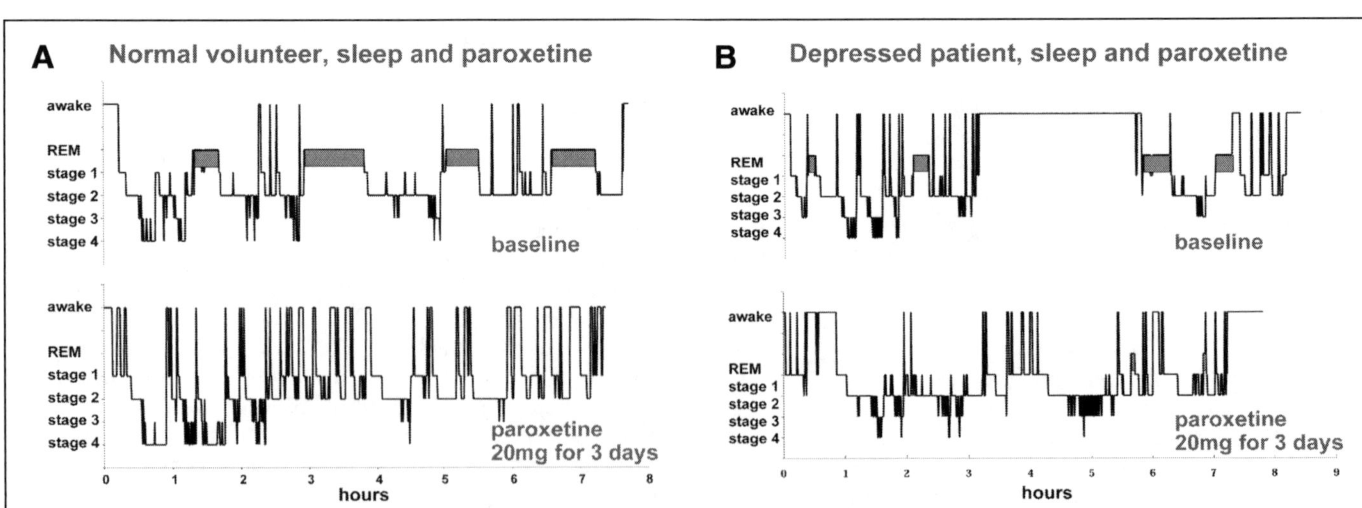

Figure 1. Hypnograms from a normal volunteer and a depressed patient before and after short-term treatment with paroxetine.

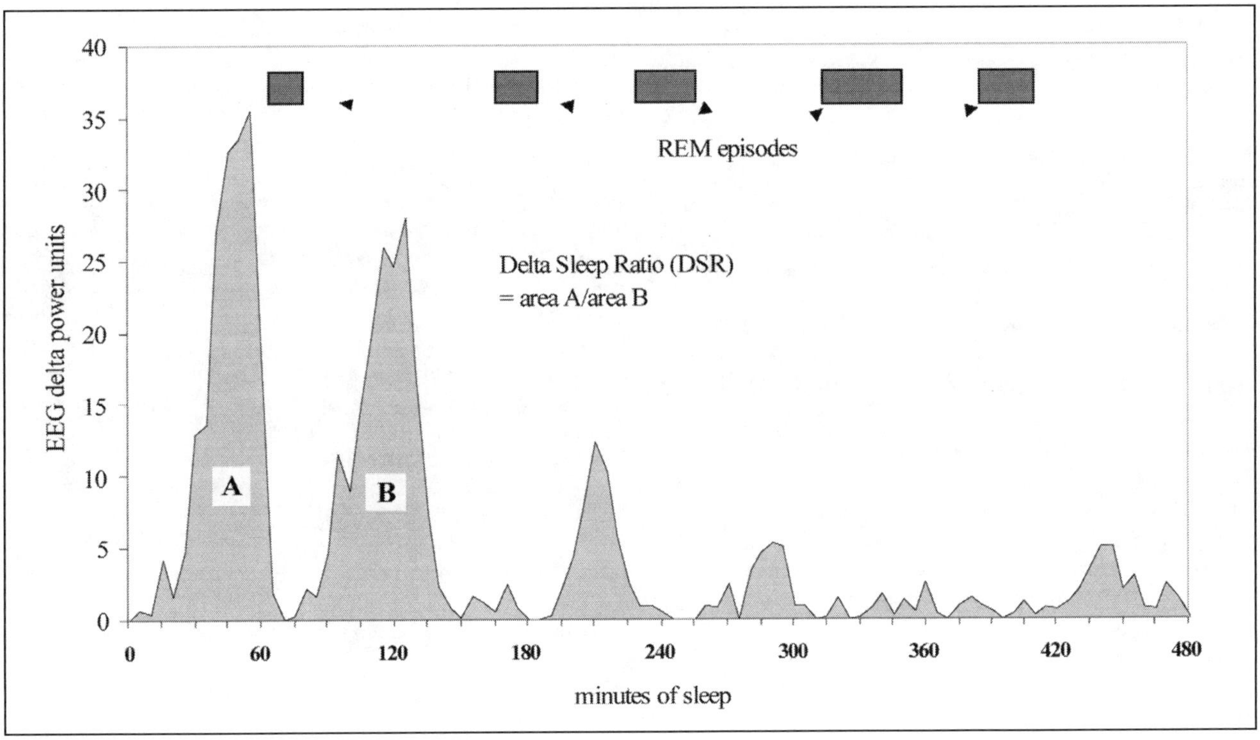

Figure 2. Evolution of delta power over the night in a normal subject.

There have been very few reports of changes in overall amount of deep (slow wave) sleep acutely after SSRIs. In normal subjects episodes of slow wave activity in non-REM sleep occur discretely, interspersed with the REM episodes (see Fig. 2) in a cyclical pattern and diminishing over the night according to the homeostatic process of sleep described first by Borbely.[6] In depressed patients the ratio of the slow EEG activity in the first sleep cycle to that in the second (delta sleep ratio) is lower than in normal subjects. In several studies this delta sleep ratio has been reported to be increased after successful antidepressant treatment.[7] However as described, SSRIs delay the first REM period, allowing the first sleep cycle to be very long, and thus the amount of slow wave activity in the first cycle is high. This increase in delta sleep ratio could therefore be an artefact caused by REM latency changes, which also occur in normal subjects. Nefazodone, one of the few antidepressants which do not suppress REM sleep, does not improve delta sleep ratio.[8]

The mechanism of REM suppression after SSRIs is probably mediated through the $5HT_{1A}$ receptor, as preclinical studies have suggested.[9,10] Fragmentation of sleep is probably not explained by this however, as potent $5HT_{1A}$ agonists do not have the marked sleep-fragmenting effects of SSRIs.[11] It may be that stimulation of post-synaptic $5HT_2$ receptors is involved, as blocking these with drugs like nefazodone improves sleep.[5,12]

SSRIs and Subjective Sleep

Insomnia is one of the adverse events occurring during SSRI treatment with reported incidence of 12-20%. In normal volunteer studies subjective worsening by SSRIs has been widely demonstrated early in treatment but after 6 nights[13] and 5 weeks[14] of fluoxetine, improvements in sleep quality were reported.

Patients' perception of sleep is extremely important in clinical practice, and in some conditions such as primary insomnia there is a poor correlation between what the patient reports about sleep, and what can be measured using polysomnography. However in depression (before treatment) this correlation is fairly high, that is the patients' sleep complaints are borne out by the objective evidence (see ref. 15). This may change during pharmacological treatment, and improvements in sleep quality have been described in remitted depressed patients whose objective sleep remains disturbed.[12,16] Thus improved mood probably results in improved perception of sleep.

Although insomnia is common in early treatment with SSRIs, somnolence is often reported as a side effect, but has not been supported by objective measures. In a study by Ridout et al[17] the visual analogue rating scale for sedation was significantly higher during 5 days' paroxetine treatment in normal volunteers but objective measures of psychomotor performance were improved. In our own study of 5 weeks' fluoxetine in normal subjects an objective measure of sleepiness, the Multiple Sleep Latency Test, showed longer average daytime sleep latency, suggesting a greater degree of alertness.

SSRIs and Movements in Sleep

There have been many case reports and small studies of movement disorders during sleep after SSRIs, but no controlled studies. Ohayon[18] in a large survey found that taking SSRIs was a risk factor for restless legs syndrome. Fluoxetine has been described as inducing large eye movements in non-REM sleep.[19] SSRIs are known to induce or exacerbate bruxism;[20] interestingly this has been ameliorated by buspirone in 2 reports, so may involve $5HT_{1A}$ mechanisms.

We have personal experience of large numbers of sleep recordings in patients and normal volunteers taking SSRIs, and especially after higher doses there is increased phasic muscle activity during sleep, although we have never quantified it. We have also seen this however in subjects taking clomipramine, dothiepin and venlafaxine.

SSRIs and Dreaming

Although nightmares are reported as an adverse event in clinical trials of SSRIs (see ref. 21) numbers were small and the presence of depression may also be a factor. Enhancement of dreaming has been noted during treatment with the SSRIs fluoxetine[22,23] and citalopram,[24] and changes in dream content after sertraline.[25] A controlled study in normal volunteers showed an increase in dream intensity but a decrease in recall with fluvoxamine and paroxetine during 3 weeks treatment, and an increase in length of reports and bizarreness during withdrawal from fluvoxamine.[26]

The mechanism for these effects is unclear. Certainly there are more awakenings during REM sleep in SSRI-treated subjects, and bouts of REM are moved towards morning wakening by the REM delaying mechanism, both of which would affect dreaming. However this does not account for increased intensity.

Use of SSRIs to Treat Sleep Disorders

As well as in depression, subjective improvement in sleep quality has been described after paroxetine treatment in PTSD,[27] with improvements in nightmares also. One small study[28] showed improvements in subjective but not objective sleep quality in primary insomnia after paroxetine. Studies in fibromyalgia[29] and chronic fatigue syndrome[30] have shown subjective sleep improvement after fluoxetine.

Paroxetine has been reported to be effective in adult night terrors[31] and sleepwalking.[32]

There are a few small case series where SSRIs have been tried in obstructive sleep apnoea syndrome, with minor improvements.[33-35]

Conclusions

SSRIs have marked dose-dependent effects on REM sleep in healthy volunteers and depressed patients, with REM onset latency being lengthened and REM amount being reduced. After weeks of treatment REM latency remains long, but the amount of REM recovers, and shows rebound after withdrawal. These changes in amount of REM sleep may reflect receptor adaptation.

Sleep continuity is worsened by SSRIs, with time awake after sleep onset and stage 1 sleep being increased early in treatment. Effects are more marked in normal volunteers than in depressed patients, probably because normal volunteers have better baseline sleep. This worsening in sleep recovers after a few days in depressed patients.

In general the deterioration in sleep continuity is not paralleled by a worsening of subjective ratings of sleep, which improve during SSRI treatment of depression.

$5HT_{1A}$ Agonists and Sleep

These drugs are partial agonists with varying intrinsic activity at $5HT_{1A}$ receptors. Buspirone is presently licensed as an anxiolytic and gepirone is under investigation for depression. Ipsapirone and flesinoxan are investigative compounds often used as pharmacological probes. Several new compounds in this class are being studied for anxiety and depression, so it is worth briefly describing the effects on sleep, particularly as the REM sleep effects of the SSRIs are thought to be mediated via the $5HT_{1A}$ receptor. There are important pharmacokinetic differences between these drugs and the SSRIs, which have an impact on their sleep effects; they act more quickly (time to maximum plasma concentration approx 15-30 min, SSRIs 4-6 hours) and are more quickly eliminated (half-life approx 2 hours, SSRIs approx 24-30 hours).

All but two of the sleep studies of $5HT_{1A}$ agonists were performed in healthy volunteers; one study looked at 6 insomniac patients[36] and one examined parallel groups of depressed patients and healthy volunteers.[37]

REM suppression is reported in nearly all the studies (single dose), particularly in the first 3 hours of sleep.[36-44] These changes appear to be modest with buspirone and more marked with ipsapirone, although no direct comparison studies are reported. If rated in the same way as the SSRIs in Table 1, the REM onset latency changes would be accorded '+' and the REM amount little changed apart from in the early part of the night. Wilson et al[11] report a comparison of two $5HT_{1A}$ agents which shows REM-suppressing effects in accordance with the differing intrinsic efficacy of the 2 drugs in preclinical studies. There are no chronic studies, although Manfredi et al[36] report 7 days dosing with buspirone and Driver et al[41] 14 days of ipsapirone in volunteers, with the REM suppression remaining but not increasing during the sub-chronic phase.

As to sleep initiation and maintenance, buspirone 5mg caused a slight increase in wake time after sleep onset[36] and ipsapirone 10mg and 20mg decreased sleep efficiency.[37,40] These latter authors also showed increased slow wave sleep in the first sleep cycle, and increased slow wave activity, after acute ipsapirone and Driver et al[41] report a decrease in SWS after 10-14 days' dosing. Interestingly the REM sleep changes and sleep fragmentation by ipsapirone were not significantly different in depressed patients and volunteers given a single dose of ipsapirone.[37]

Studies examining subjective sleep effects of buspirone 10mg and 20mg, ipsapirone 5 mg and eptapirone 2mg report no changes in sleep perception in volunteers[11,41] or insomnia.[36]

Changes in REM sleep are unsurprising but interestingly the changes in sleep continuity and slow wave sleep are unlike those of SSRIs. It may be therefore that the sleep fragmentation after SSRIs is not mediated via the $5HT_{1A}$ receptor.

References

1. Thase ME. Antidepressant treatment of the depressed patient with insomnia. J Clin Psychiatry 1999; 60(Suppl 17):28-31
2. Thase ME, Buysse DJ, Frank E et al. Which depressed patients will respond to interpersonal psychotherapy? The role of abnormal EEG sleep profiles. Am J Psychiatry 1997; 154(4):502-9.
3. *Feige B, Voderholzer U, Riemann D et al. Fluoxetine and sleep EEG: Effects of a single dose, subchronic treatment, and discontinuation in healthy subjects. Neuropsychopharmacology 2002; 26(2):246-58.
4. *van Bemmel AL, van den Hoofdakker RH, Beersma DGM et al. Changes in sleep polygraphic variables and clinical state in depressed patients during treatment with citalopram. Psychopharmacology 1993; 113:225-230.
5. *Rush AJ, Armitage R, Gillin JC et al. Comparative effects of nefazodone and fluoxetine on sleep in outpatients with major depressive disorder. Biol Psychiatry 1998; 44(1):3-14.
6. Borbely AA. A two process model of sleep regulation. Hum Neurobiol 1982; 1(3):195-204.
7. Ehlers CL, Havstad JW, Kupfer DJ. Estimation of the time course of slow-wave sleep over the night in depressed patients: Effects of clomipramine and clinical response. Biol Psychiatry 1996; 39(3):171-81.
8. Veitch W, Wilson SJ, Argyropoulos S et al. Slow waves in sleep are altered by paroxetine and nefazodone in depressed patients. J Psychopharm 2001; 15(3 suppl):A18.
9. Monaca C, Boutrel B, Hen R et al. 5-HT 1A/1B receptor-mediated effects of the selective serotonin reuptake inhibitor, citalopram, on sleep: Studies in 5-HT 1A and 5-HT 1B knockout mice. Neuropsychopharmacology 2003; 28(5):850-6.
10. Monti JM, Monti D. Role of dorsal raphe nucleus serotonin 5HT1a receptor in the regulation of REM sleep. Life Sciences 2000; 66:1999-2012.

11. *Wilson SJ, Bailey JE, Rich AS et al. Effects of 5HT1A agonists on sleep in normal volunteers: A comparison of eptapirone, buspirone, and placebo. J Psychopharm 2003; 17(3 suppl):A73.
12. *Hicks JA, Argyropoulos SV, Rich AS et al. Randomised controlled study of sleep after nefazodone or paroxetine treatment in out-patients with depression. Br J Psychiatry 2002; 180:528-35.
13. *Vasar V, Appelberg B, Rimon R et al. The effect of fluoxetine on sleep: A longitudinal double blind polysomnographic study of healthy volunteers. International Clinical Psychopharmacology 1994; 9:203-206.
14. *Wilson SJ, Bailey JE, Alford C et al. Effects of 5 weeks of administration of fluoxetine and dothiepin in normal volunteers on sleep, daytime sedation, psychomotor performance and mood. J Psychopharmacol 2002; 16(4):321-31.
15. Argyropoulos SV, Hicks JA, Nash Jr et al. Correlation of subjective and objective sleep measurements at different stages of the treatment of depression. Psychiatry Res 2003; 120(2):179-90.
16. *Wilson SJ, Bell C, Coupland NJ et al. Sleep changes during long-term treatment of depression with fluvoxamine - A home based study. Psychopharmacology 2000; 149:360-365.
17. Ridout F, Meadows R, Johnsen S et al. A placebo controlled investigation into the effects of paroxetine and mirtazapine on measures related to car driving performance. Hum Psychopharmacol 2003; 18(4):261-9.
18. Ohayon MM, Roth T. Prevalence of restless legs syndrome and periodic limb movement disorder in the general population. J Psychosom Res 2002; 53(1):547-54.
19. Schenck CH, Mahowald MW, Kim SW et al. Prominent eye movements during NREM sleep and REM sleep behavior disorder associated with fluoxetine treatment of depression and obsessive-compulsive disorder. Sleep 1992; 15(3):226-35.
20. Ellison JM, Stanziani P. SSRI-associated nocturnal bruxism in four patients. J Clin Psychiatry 1993; 54(11):432-4.
21. Pagel JF, Helfter P. Drug induced nightmares—An etiology based review. Hum Psychopharmacol 2003; 18(1):59-67.
22. Lepkifker E, Dannon PN, Iancu I et al. Nightmares related to fluoxetine treatment. Clin Neuropharmacol 1995; 18:90-94.
23. Markowitz JC. Fluoxetine and dreaming. J Clin Psychiatry 1991; 52(10):432.
24. Koponen, H, Lepola U, Leiononen E et al. Citalopram in the treatment of obsessive-compulsive disorder: An open pilot study. Acta Psychiat Scand 1997; 96:343-346.
25. Kirschner NT. Medication and dreams: Changes in dream content after drug treatment. Dreaming 1999; 9:195-200.
26. Pace-Schott EF, Gersh T, Silvestri R et al. SSRI treatment suppresses dream recall frequency but increases subjective dream intensity in normal subjects. J Sleep Res 2001; 10(2):129-42.
27. Stein DJ, Davidson J, Seedat S et al. Paroxetine in the treatment of post-traumatic stress disorder: Pooled analysis of placebo-controlled studies. Expert Opin Pharmacother 2003; 4(10):1829-38.
28. Nowell PD, Reynolds CF 3rd, Buysse DJ et al. Paroxetine in the treatment of primary insomnia: Preliminary clinical and electroencephalogram sleep data. J Clin Psychiatry 1999; 60(2):89-95.
29. Goldenberg D, Mayskiy M, Mossey C et al. A randomized, double-blind crossover trial of fluoxetine and amitriptyline in the treatment of fibromyalgia. Arthritis Rheum 1996; 39(11):1852-9.
30. Vercoulen JH, Swanink CM, Zitman FG et al. Randomised, double-blind, placebo-controlled study of fluoxetine in chronic fatigue syndrome. Lancet 1996; 347(9005):858-61.
31. Wilson SJ, Lillywhite AR, Potokar JP et al. Adult night terrors and paroxetine. Lancet 1997; 350(9072):185.
32. Bengtson H, Broman J-E, Hetta J. The effect of paroxetine on sleepwalking in 8 adults. Sleep Research Online 1999; 2(S1):131.
33. Kopelman PG, Elliott MW, Simonds A et al. Short-term use of fluoxetine in asymptomatic obese subjects with sleep-related hypoventilation. Int J Obes Relat Metab Disord 1992; 16(10):825-30.
34. Kraiczi H, Hedner J, Dahlof P et al. Effect of serotonin uptake inhibition on breathing during sleep and daytime symptoms in obstructive sleep apnea. Sleep 1999; 22(1):61-7.
35. Hanzel DA, Proia NG, Hudgel DW. Response of obstructive sleep apnea to fluoxetine and protriptyline. Chest 1991; 100(2):416-21.
36. Manfredi RL, Kales A, Vgontzas AN et al. Buspirone: Sedative or stimulant effect? Am J Psychiatry 1991; 148(9):1213-7.
37. Gillin JC, Sohn JW, Stahl SM et al. Ipsapirone, a 5-HT1A agonist, suppresses REM sleep equally in unmedicated depressed patients and normal controls. Neuropsychopharmacology 1996; 15(2):109-15.
38. Ware JC, Rose FV, McBrayer RH. The acute effects of nefazodone, trazodone and buspirone on sleep and sleep-related penile tumescence in normal subjects. Sleep 1994; 17(6):544-50.
39. Gillin JC, Jernajczyk W, Valladares-Neto DC et al. Inhibition of REM sleep by ipsapirone, a 5HT1A agonist, in normal volunteers. Psychopharmacology (Berl) 1994; 116(4):433-6.
40. Seifritz E, Moore P, Trachsel L et al. The 5-HT1A agonist ipsapirone enhances EEG slow wave activity in human sleep and produces a power spectrum similar to 5-HT2 blockade. Neurosci Lett 1996; 209(1):41-4.
41. Driver HS, Flanigan MJ, Bentley AJ et al. The influence of ipsapirone, a 5-HT1A agonist, on sleep patterns of healthy subjects. Psychopharmacology (Berl) 1995; 117(2):186-92.
42. Moore P, Seifritz E, Schlosser A et al. Rapid tryptophan depletion plus a serotonin 1A agonist: Competing effects on sleep in healthy men. Neuropsychopharmacology 2001; 25(5 Suppl):S40-4.
43. Ansseau M, Pitchot W, Gonalez-Moreno A. Flesinoxan, a $5HT_{1A}$ agonist, in major depression: Clinical efficacy and effects on REM latency and body temperature (open study). European Neuropsychopharmacology 1992; 2:313.
44. *Wilson SJ, Bailey JE, Rich AS et al. Using sleep to evaluate comparative serotonergic effects of paroxetine and citalopram. Eur Neuropsychopharmacol 2004; 14(5):367-72.
45. *Bell C, Wilson S, Rich A et al. Effects on sleep architecture of pindolol, paroxetine and their combination in healthy volunteers. Psychopharmacology Berl 2003; 166(2):102-10.
46. *Hartmann E. Sleep Research 1979; 8:98.
47. *Hendrickse WA, Roffwarg HP, Granneman BD et al. The effects of fluoxetine on the polysomnogram of depressed outpatients: a pilot study. Neuropsychopharmacology 1994; 10:85-91
48. *Jindal RD, Friedman ES, Berman SR et al. Effects of sertraline on sleep architecture in patients with depression. J Clin Psychopharmacol 2003; 23(6):540-8.
49. *Kupfer DJ, Perel JM, Pollock BG et al. Fluvoxamine versus desipramine: comparative polysomnographic effects. Biological Psychiatry 1991; 29:33-40.
50. *Nicholson AN, Pascoe PA Studies on the modulation of the sleep-wakefulness continuum in man by fluoxetine, a 5HT uptake inhibitor. Neuropharmacology 1988; 27:597-602.
51. *Oswald I, Adam K. Effects of paroxetine on human sleep. Br J Clin Pharmacol 1986; 22(1):97-9.
52. *Roschke J, Kogel P, Schlosser R et al. Analysis of sleep EEG microstructure in subchronic paroxetine treatment of healthy subjects Psychopharmacology (Berl) 1997; 132(1):44-9.
53. *Saletu B, Frey R, Krupka M et al. Sleep laboratory studies on the single-dose effects of serotonin reuptake inhibitors paroxetine and fluoxetine on human sleep and awakening qualities. Sleep 1991; 14(5):439-47.
54. *Schlosser R, Roschke J, Rossbach W et al. Conventional and spectral power analysis of all-night sleep EEG after subchronic treatment with paroxetine in healthy male volunteers. Eur Neuropsychopharmacol 1998; 8(4):273-8.
55. *Sharpley AL, Williamson DJ, Attenburrow ME et al. The effects of paroxetine and nefazodone on sleep: A placebo controlled trial. Psychopharmacology (Berl) 1996; 126(1):50-4.
56. *Staner L, Kerkhofs M, Detroux D et al. Acute, subchronic and withdrawal sleep EEG changes during treatment with paroxetine and amitriptyline: A double-blind randomized trial in major depression. Sleep 1995; 18(6):470-7.
57. *Trivedi MH, Rush AJ, Armitage R et al. Effects of fluoxetine on the polysomnogram in outpatients with major depression. Neuropsychopharmacology 1999; 20(5):447-59.
58. *Winokur A, DeMartinis NA 3rd, McNally DP et al. Comparative effects of mirtazapine and fluoxetine on sleep physiology measures in patients with major depression and insomnia. Clin Psychiatry 2003; 64(10):1224-9.

CHAPTER 32

Sleep and Antipsychotic Drugs in Schizophrenia Patients

Jaime M. Monti and Daniel Monti

Abstract

Insomnia is a common feature in schizophrenia. The sleep disturbance of either never-medicated or previously treated schizophrenia patients is characterized by an increase of stage 2 sleep latency and wake time after sleep onset, and a reduction of total sleep time and sleep efficiency. In addition, stage 4 sleep, slow wave sleep, and REM latency are consistently decreased. The limited number of studies directed at disclosing the effects of typical (haloperidol, thiothixene, flupentixol) and atypical antipsychotics (olanzapine, clozapine, risperidone) on sleep in schizophrenia patients tend to indicate that they improve sleep induction and/or sleep maintenance.

Introduction

Insomnia is a common feature of schizophrenia. To be considered as a symptom related to schizophrenia, the sleep disturbance must last for at least one month and be associated with daytime fatigue or impaired daytime functioning.[1] Although the sleep disturbances in schizophrenia could be sufficiently severe to warrant independent clinical attention, they seldom are the predominant complaint. Nevertheless, severe insomnia is often seen during exacerbations of schizophrenia, and may actually precede the appearance of other symptoms of relapse.[2] Less frequently, severe sleep disorders may complicate schizophrenia to the degree that patients can become suicidal.[3]

Polysomnographic Sleep in Schizophrenia Patients

The all-night polysomnographic sleep of never-medicated or previously treated schizophrenia patients has been compared with that of normal controls in several studies. In these studies sleep variables were grouped into sleep initiation and maintenance, Non-REM (NREM) sleep structure, and REM sleep features. Sleep continuity measures included wake time after sleep onset, number of awakenings, total sleep time, and sleep efficiency. NREM sleep structure comprised the minutes or percentage spent in each sleep stage. REM sleep was expressed in minutes or percentage of total sleep time; REM latency was the time from the first epoch of stage 2 to the first REM period.[4]

Never-Medicated Schizophrenia Patients

A total of 75 patients and 61 normal controls were included in five studies.[5-9] Their age ranged from 21.0 to 71.1 years. Two studies included only chronic patients, whereas in the other three studies acute, subacute, and/or subchronic schizophrenics were also included. In four studies sleep was assessed in a sleep laboratory during one night, which was preceded by one adaptation night, whereas in one study two consecutive nights of polysomnography were conducted, and averaged data from both nights were reported for all variables. Values corresponding to some variables were omitted in two studies. Compared with controls stage 2 sleep latency, wake time after sleep and the number of nocturnal awakenings were increased whereas total sleep time and sleep efficiency were reduced (Fig. 1). Values corresponding to stage 2 sleep were found to be reduced. Stage 4 sleep amounted to very low values in both the schizophrenia patients and the controls. The relatively low values of stage 4 sleep in the controls tends to indicate incomplete adaptation to the sleep environment. REM latency and REM sleep in minutes were decreased in three out of the five studies (Fig. 1). The values corresponding to sleep variables tended to change in the same direction irrespective of the phase of illness. Nevertheless, REM latency amounted to much lower values in studies where only chronic patients were included.

Schizophrenia Patients Previously Treated with Neuroleptics

Studies That Included a Control Group

A total of 200 young patients aged 18.0-35.0 years and 190 normal controls aged 20.0-31.3 years were included in thirteen studies.[7,10-21] As a whole patients had not been taking neuroleptics prior to the study for periods ranging from 1-2 days to 1-2 years. Concerning the phase of illness, seven studies included only chronic patients, whereas in the other six studies acute, subacute, subchronic and chronic schizophrenia patients were included. In two studies sleep was assessed in a sleep laboratory during one night, which was preceded by one adaptation night, whereas in eleven studies 2-5 consecutive nights of polysomnography were conducted, and averaged data from the recording nights were reported for all variables. Values corresponding to some variables were omitted in seven studies. Compared with controls, in schizophrenia patients stage 2 sleep latency and wake time after sleep

Sleep and Sleep Disorders: A Neuropsychopharmacological Approach, edited by Malcolm Lader, Daniel P. Cardinali and S.R. Pandi-Perumal.

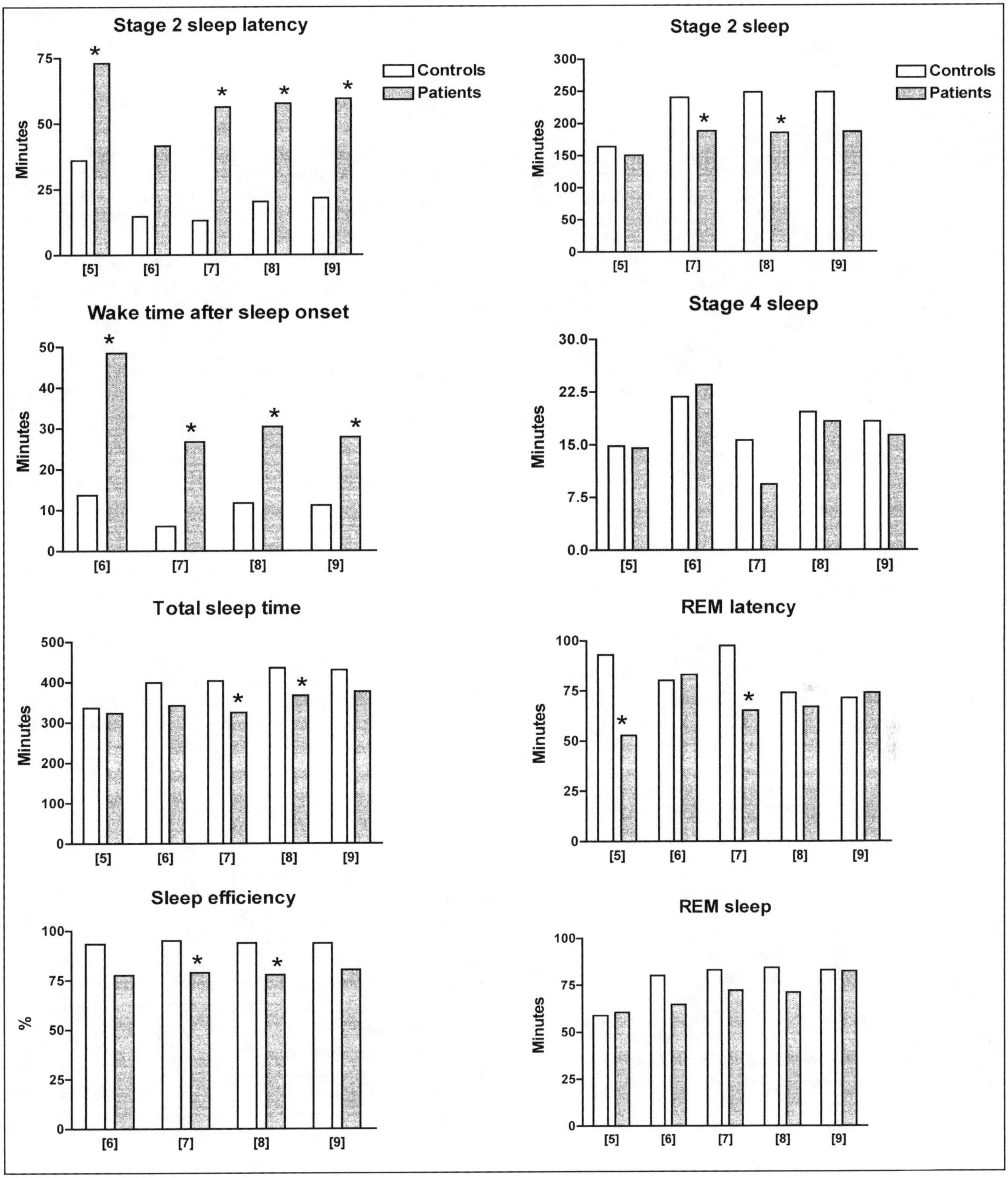

Figure 1. Left column: Sleep initiation and continuity measures of never-medicated schizophrenia patients. Numbers below bars correspond to studies cited in the text. Right column: Sleep architecture of never-medicated schizophrenia patients. Numbers below bars correspond to studies cited in the text.

onset were increased, whereas total sleep time and sleep efficiency were reduced (Fig. 2). In addition, stage 2 and stage 4 NREM sleep, and REM latency in minutes were decreased in almost all studies (Fig. 2). Thus, available data tend to indicate that sleep onset and sleep maintenance were disrupted in the schizophrenia patients irrespective of the length of the drug-free period prior to the polysomnographic study. Interestingly, the phase of illness did not substantially influence the values corresponding to sleep variables.

Studies That Did Not Include a Control Group

A total of 115 patients aged 20.2-56.0 years were included in nine studies.[22-30] As a whole patients had not been taking neuroleptics prior to the study for periods ranging from 2 to 6 weeks. In one study sleep was assessed in a sleep laboratory during one night, which was preceded by one adaptation night, whereas in eight studies 2-4 consecutive nights of polysomnography were conducted and averaged data from the recording nights were reported for all variables. Values corresponding to some variables were omitted in six studies. The mean values corresponding to stage 2 sleep latency and waking time after sleep onset were increased, whereas total sleep time, sleep efficiency, stage 4 sleep, and REM latency in minutes were reduced in most studies when compared with such measures in healthy young volunteers who slept in a sleep laboratory[31,32] (Fig. 3).

Delta Sleep and Negative Symptoms of Schizophrenia

Most polysomnographic studies have detected a reduction of visually scored stage 4 sleep and slow wave sleep (stages 3 and 4) in schizophrenia patients. In addition, the duration of visually scored stage 4 sleep or slow wave sleep has been shown to be inversely correlated with the severity of negative symptoms.[6,26] Automated delta sleep measures revealed that total and average delta wave counts were also inversely associated with negative symptoms.[20]

Consistency and Shortcomings

It can be concluded that studies of sleep in schizophrenia patients have failed to generate consistent findings. This is more evident in the early studies in which the diagnostic criteria were not specified or nonstandardized diagnostic procedures were used . In addition, some sleep studies employed scoring methods that are no longer used. Other methodological shortcomings include: small number of subjects, and such confounding variables as age, gender or phase of illness. The effects of medication merit special concern. In this respect, patients included in several studies were taking typical neuroleptics and other medications 1-2 days to 1-2 weeks before starting the sleep study. As stressed by Ganguli et al[6] and Lauer et al,[8] chronic treatment with neuroleptics, as well as their withdrawal, have profound effects on sleep maintenance, NREM sleep structure, and REM sleep features. In this regard, haloperidol continues to block dopamine D_2 receptors for weeks or months after its withdrawal, and cessation of chlorpromazine administration induces long-lasting disuse supersensitivity on forebrain dopaminergic systems mediated by changes in the receptors for the neurotransmitters.

It should also be mentioned that several sleep studies did not include a control group, or included subjects with an incomplete adaptation to the sleep environment (first night effect). In the studies by Ganguli et al,[6] Tandon et al,[7] Lauer et al,[8] and Lauer and Krieg[9] where never-medicated schizophrenia patients were included, the control groups were comprised of only young adults; yet the observed values corresponding to REM sleep in min and/or stage 4 sleep were far below those found in the normal young adult individual who is living on a conventional sleep-wake schedule and who is without sleep complaints.[32] As a result the observed differences in stage 4 sleep and REM sleep between controls and schizophrenia patients were not significant. On the other hand, in these studies where previously treated schizophrenia patients were included, control subjects seem to have been more thoroughly adapted to the recording procedure.

In spite of all these methodological shortcomings, the available evidence indicates that the sleep disturbances of either never-medicated or previously treated schizophrenia patients is characterized by a sleep onset and maintenance insomnia. In addition, stage 2 sleep and REM latency were reduced in most studies.

What Are the Mechanisms Involved in the Disruption of Sleep in Schizophrenia Patients?

The dopamine (DA) hypothesis of schizophrenia formulated in the 1960s states that the symptoms of schizophrenia depend upon the overactivity of the dopaminergic system.[33] The latter would be related to an increased density of DA D_2 receptors. Nevertheless, recent postmortem and in vivo imaging (PET) studies indicate that about 30% of patients show striatal D_2 receptor densities that do not differ significantly from matched control subjects.[34] Differences remain significant irrespective of analyzing drug-naive and drug-treated patients separately. More recent studies have refined the DA hypothesis of schizophrenia by stressing that in addition to the increased density of D_2 receptors, there is an enhanced sensitivity of dopaminergic neurotransmission. In this respect, an increased responsiveness of the nigrostriatal pathway has been shown with a pharmacological stress model and SPECT.[35,36] On the basis of these findings it could be proposed that the sleep disturbance of schizophrenia patients is related to an overactivity of the dopaminergic system. Indirect evidence derived from pharmacological studies tends to support this contention. Accordingly, the DA D_2 preferring agonists apomorphine, bromocriptine, and pergolide have been shown to enhance wakefulness and to reduce slow wave sleep and REM sleep when administered to laboratory animals at doses that interact with postsynaptic D_2 receptors.[37] In contrast, YM-09151-2, a substituted benzamide that selectively blocks D_2 receptors, increases light sleep.[38]

The possibility remains that insomnia in schizophrenia patients is not exclusively DA dependent but that other neurotransmitter systems are also involved in the disruption of this behavioral state. In this respect, recent neuroanatomical and neurochemical studies have provided evidence for a decrease of γ-aminobutyric acid (GABA)ergic neurotransmission playing a role in the pathophysiology of schizophrenia.[39,40] Neurons in the basal forebrain, preoptic area and anterior hypothalamus constitute a sleep-inducing system. Electrical stimulation of the preoptic area, including the horizontal limb of the diagonal band leads to sleep with electrocortical synchronization.[41] Moreover, recording of single neuronal activity in the preoptic/anterior hypothalamic area of several species identified populations of neurons that increased their discharge rates during slow wave sleep.[42] All these neurons contain GABA and project to the nonspecific thalamic nuclei, which in turn relate to the cerebral cortex. In addition, the GABA-containing neurons project to brain stem and hypothalamic areas where they inhibit noradrenergic serotonergic, histaminergic and cholinergic neurons involved in the promotion of wakefulness.[43] Thus, a decrease in the functional activity of the GABAergic system could add to the overactivity of the dopaminergic system in the promotion of insomnia in schizophrenia patients.

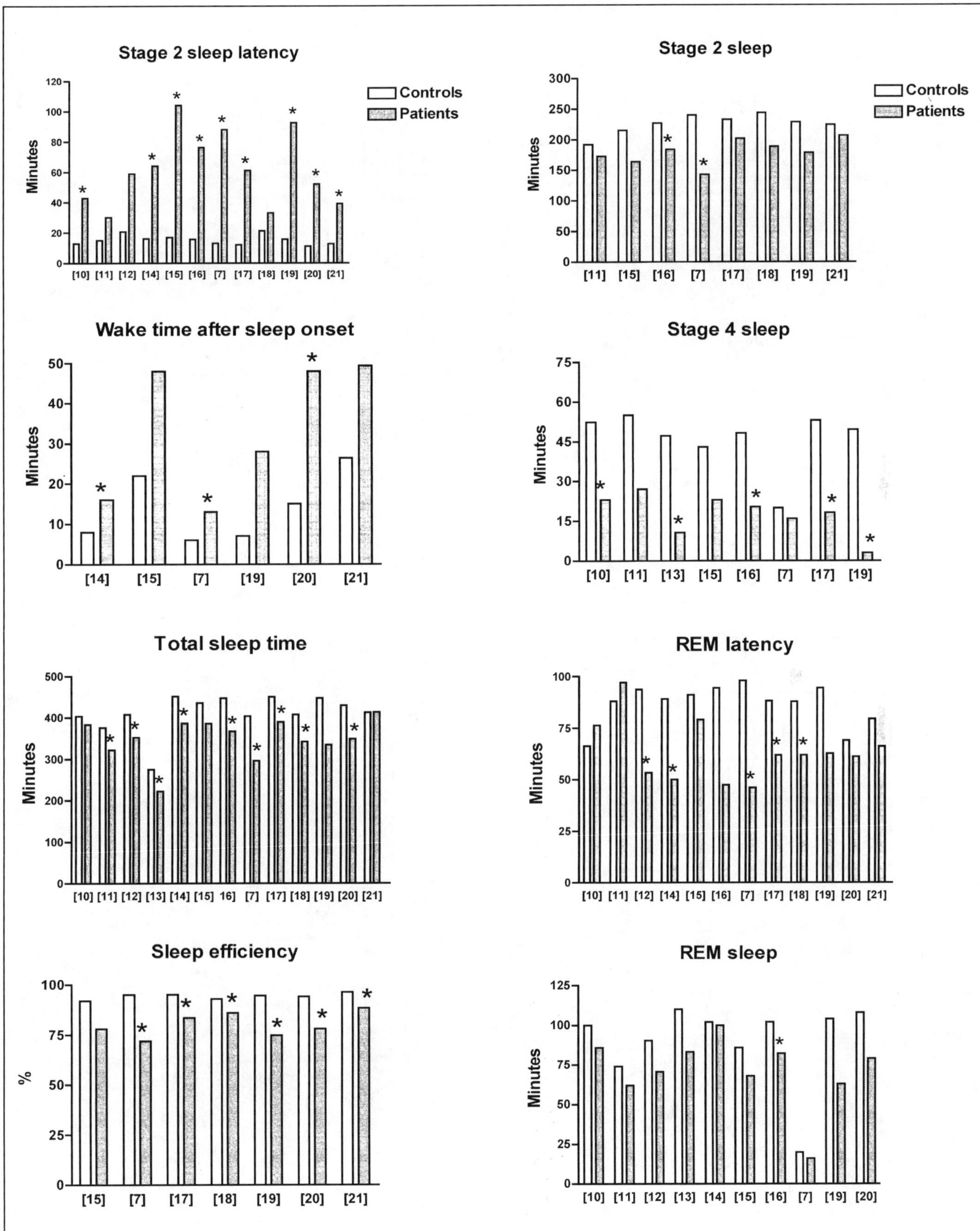

Figure 2. Left column: Sleep initiation and maintenance of schizophrenia patients previously treated with neuroleptics. The studies included a control group. Numbers below bars correspond to studies cited in the text. Right column: Sleep architecture of schizophrenia patients previously treated with neuroleptics. The studies included a control group. Numbers below bars correspond to studies cited in the text.

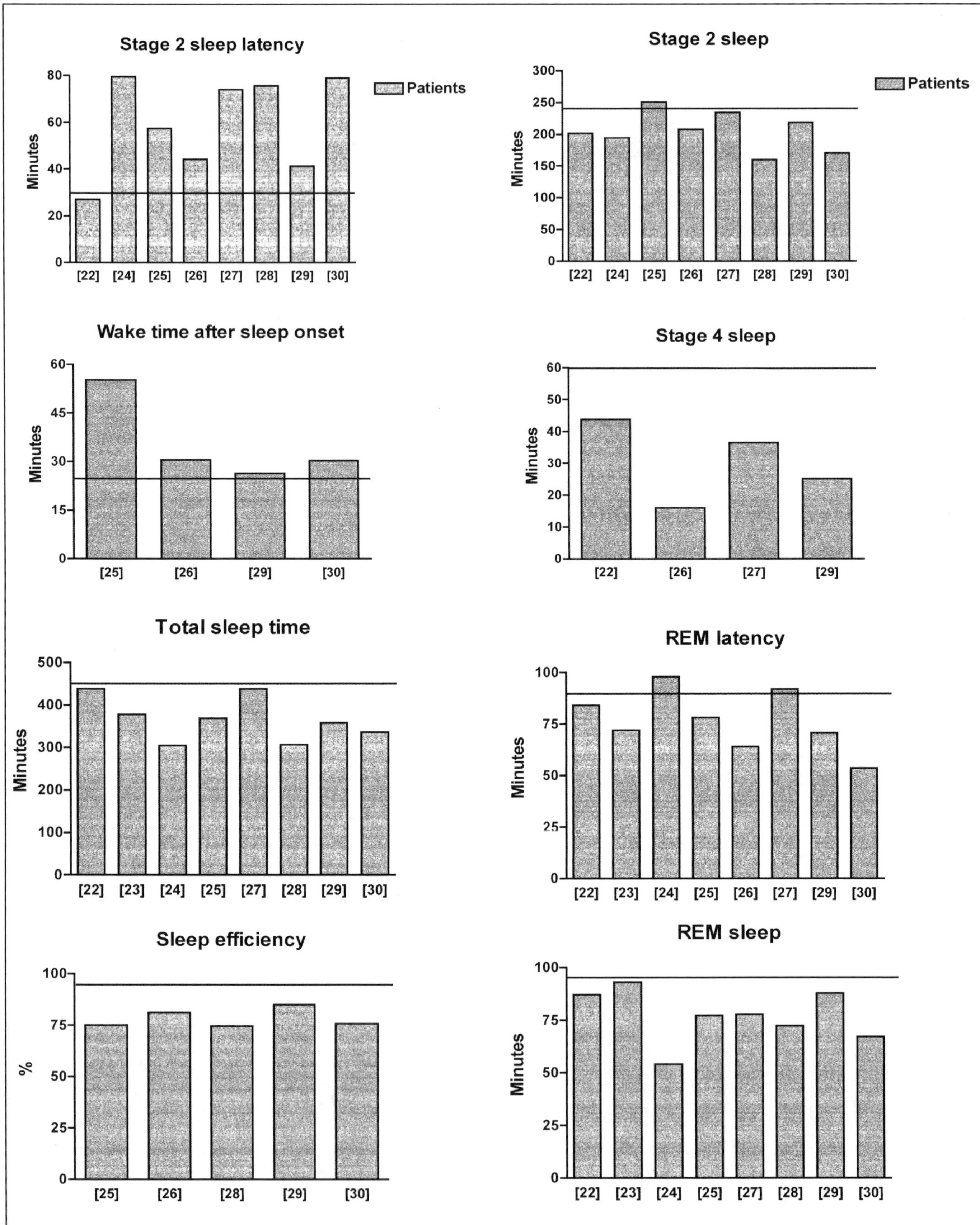

Figure 3. Left column: Sleep initiation and maintenance of schizophrenia patients previously treated with neuroleptics. The studies did not include a control group. Numbers below bars correspond to studies cited in the text. Right column: Sleep architecture of schizophrenia patients previously treated with neuroleptics. The studies did not include a control group. Numbers below bars correspond to studies cited in the text.

Table 1. Effects of antipsychotic drugs on sleep parameters in schizophrenia patients

Sleep Induction and Maintenance

Sleep Variables	Haloperidol	Olanzapine	Clozapine
Stage 2 sleep latency	decrease	no change	decrease or no change
Wake time after sleep	no change	decrease	decrease or no change
Total sleep time	increase	increase	increase
Sleep efficiency	increase	—	increase

Haloperidol: 6.6-11.4 mg/day;[29,48,49,51] Olanzapine: 10 mg/day;[50] Clozapine: 170.0-347.9 mg/day[49,64]

Effects of Atypical and Typical Antipsychotics on Polysomnographic Measures of Sleep

The difficulties inherent in sleep studies of schizophrenia patients are well known. Considering that the availability of EEG sleep data on the effects of antipsychotic drugs in schizophrenia is sparse, we also included studies in which these compounds were administered to normal controls. Our analysis was limited to the effects on sleep variables of the atypical antipsychotics olanzapine, risperidone, and clozapine, and the typical antipsychotics haloperidol, thiothixene, and flupentixol.

With one exception,[44] the investigators made use of a sleep laboratory to obtain objective evidence on the effects of antipsychotic drugs on sleep induction and continuity measures, and on sleep architecture. Although the descriptions of the characteristics of the patient population were adecuate, only in the studies by Sharpley et al[45,46] and Haffmans et al[47] was there adherence to a strict double-blind placebo-controlled design. Moreover, details of the subtypes of schizophrenia were missing in several studies. Various patient populations were included in the available studies - healthy subjects, outpatients, and hospitalised patients. The information obtained from studies involving asymptomatic volunteers has its limitations because the values corresponding to their sleep variables are within normal limits. The effects of other drugs were not adequately addressed in several studies. For example, in the studies by Dursun et al,[44] Taylor et al,[48] Wetter et al,[49] Salin-Pascual et al,[50] and the preliminary report by Haffmans et al[47] an adequate washout period between drugs was lacking. Moreover, patients included in the studies by Dursun et al,[44] Taylor et al,[48] Maixner et al,[51] and Yamashita et al[52] were also taking medication to prevent parkinsonian symptoms, including procyclidine, benztropine, or trihexyphenydil. It is well known that all these compounds have central anticholinergic actions that can modify sleep. No attempts were made to determine the short-term, intermediate-term, or long-term effects of the antipsychotic drugs on sleep variables in normal subjects or schizophrenic patients. In addition, the questions of rebound insomnia following withdrawal of the antipsychotic medication, or of tolerance to the sleep-inducing and sleep-maintaining effects of the different compounds were not addressed by the authors.

Olanzapine given to healthy subjects or schizophrenia patients reduced or had no effect on stage 2 sleep latency, respectively. On the other hand, wake time after sleep onset was reduced, whereas total sleep time was enhanced (Table 1). Concerning sleep architecture, stage 1 sleep was decreased whereas stage 2 sleep and slow wave sleep were augmented (Table 2). Olanzapine tended to disrupt REM sleep in the healthy subjects as judged by the reduction of REM sleep and the increase of REM latency.

The limited information on risperidone suggests that the compound improves sleep maintenance and increases slow wave sleep in schizophrenia patients.[47,52]

The administration of clozapine to schizophrenia patients increased total sleep time, sleep efficiency, and stage 2 sleep (Tables 1 and 2). Clozapine decreased or induced no change of stage 4 sleep, slow wave sleep or REM sleep (Table 2).

Typical antipsychotic drugs including haloperidol, thiothixene, and flupentixol reduced stage 2 sleep latency, and increased total sleep time and sleep efficiency (Table 1). Stage 4 sleep and slow wave sleep remained unchanged whereas REM latency was significantly increased (Table 2).

Thus, the limited number of studies directed at disclosing the effects of typical and atypical antipsychotics on sleep in

Table 2. Effects of antipsychotic drugs on sleep parameters in schizophrenia patients

Sleep Structure

Sleep Variables	Haloperidol	Olanzapine	Clozapine
Stage 2 sleep	no change	increase	increase
Stage 3 sleep	increase or no change	—	no change
Stage 4 sleep	no change	—	decrease, no change
Slow wave sleep	no change	increase	decrease
REM latency	increase	no change	no change
REM sleep (min)	no change	no change	no change
REM density	decrease or no change	increase	—

Table 3. Receptor binding profile of typical and atypical antipsychotics

	Haloperidol	Flupentixol	Olanzapine	Risperidone	Clozapine
D_1	-	+	+++	-	++
D_2	+++	+++	+++	+++	++
D_3	++	¿	¿	+	-
D_4	+	¿	+	+	+
$5\text{-}HT_{2A}$	+	+	+++	+++	+++
$5\text{-}HT_{2C}$	-	¿	++	+	+
α_1	+	+	+++	+++	+++
ACh (M)	-	-	+++	-	+++
H_1	-	-	+++	-	+++

From: Blin et al,[53] Horacek,[54] Stip,[55] Kapur and Remington.[56] D_1, D_2, D_3, D_4 = dopamine receptors; $5\text{-}HT_{2A}$, $5\text{-}HT_{2C}$ = serotonin receptors; α_1 = adrenergic receptor; ACh (M) = cholinergic (muscarinic) receptor; H_1 = histamine receptor; Affinity: (-) absent; (+) low; (++) intermediate; (+++) high; (¿) unknown

schizophrenia patients tend to indicate that they improve the sleep disturbance. However, methodological flaws limit the validity of the conclusions.

What Are the Mechanisms Involved in the Effects of Antipsychotic Drugs on Sleep?

Typical and atypical antipsychotic drugs bind to a wide variety of CNS receptors. The produce their effects by blocking dopamine (D_1, D_2, D_3, D_4), serotonin ($5\text{-}HT_{2A}$, $5\text{-}HT_{2C}$, $5\text{-}HT_6$, $5\text{-}HT_7$), a-adrenergic (a_1,a_2), histamine (HA) (H_1), and acetylcholine (muscarinic) receptors. Irrespective of their chemical structure antipsychotics show intermediate (clozapine) to high (haloperidol, flupentixol, thiothixene, olanzapine, risperidone) affinity for the D_2 receptor. In addition, clozapine and olanzapine have intermediate and high affinity for the D_1 receptor, respectively. In contrast to the classical antipsychotics, the newer antipsychotics show high affinity for the serotonin $5\text{-}HT_{2A}$ receptor and to a lesser extent for the $5\text{-}HT_{2C}$, $5\text{-}HT_6$ (clozapine, olanzapine), and $5\text{-}HT_7$ (clozapine, risperidone) receptor. Olanzapine, risperidone, and clozapine bind with high affinity to the a_1 adrenoceptor, whereas olanzapine and clozapine display high affinity for both the HA H_1 and the acetylcholine (muscarinic) receptor[53-56] (Table 3).

The blockade of D2 receptors has been proposed to be responsible for the improvement of positive symptoms. Moreover, the blockade of $5\text{-}HT_{2A}$ receptors contributes to the improvement of negative symptoms and of cognitive functions. On the other hand, the blockade of a_1, H_1, and acetylcholine (muscarinic) receptors induces a number of side-effects. The blockade of D_2 receptors could be partly responsible for the improvement of sleep in schizophrenia patients. However, the amelioration of sleep has been related almost exclusively to the blockade of a_1, H_1 and cholinergic (muscarinic) receptors. This is based on the premise that a_1 receptor (prazosin and related drugs), first-generation H_1 receptor (pyrilamine, diphenhydramine), or cholinergic (muscarinic) receptor (scopolamine) antagonists produce somnolence, and increased likelihood of falling asleep and reduced concentration.[57-59]

Idzikowski et al[60,61] have demonstrated that the $5\text{-}HT2_{A/C}$ receptor antagonist ritanserin selectively increases slow wave sleep in normal volunteers. A similar effect has been described in poor sleepers and insomniac patients.[62,63] Thus, the olanzapine-induced increase of slow wave sleep could partly depend on the blockade of 5-HT2A/C receptor.

References

1. American Psychiatric Association. Diagnostic and Statistical Manual of Mental Disorders, 4th ed. Washington, DC: American Psychiatric Press, 1994.
2. van Kammen DP, van Kammen WB, Peters JL et al. CSF MHPG sleep and psychosis in schizophrenia. Clin Neuropharmacol 1986; 9 (suppl 4):575-577.
3. American Sleep Disorders Association. International classification of sleep disorders, revised: Diagnostic and coding manual. Rochester, Minnesota: American Sleep Disorders Association, 1997.
4. Rechtschaffen A, Kales A. A manual of standardized terminology, techniques, and scoring system for sleep stages in human subjects. National Institute of Mental Health Publication 204. Government Printing Office. Washington D.C.: Washington Printing Office, 1968.
5. Jus K, Bouchard M, Jus AK et al. Sleep EEG studies in untreated, long-term schizophrenic patients. Arch Gen Psychiatry 1973; 29:386-390.
6. Ganguli R, Reynolds CF, Kupfer DJ. Electroencephalographic sleep in young, never-medicated schizophrenics. Arch Gen Psychiatry 1987; 44:36-44.
7. Tandon R, Shipley JE, Taylor S et al. Electroencephalographic sleep abnormalities in schizophrenia. Arch Gen Psychiatry 1992; 49:185-194.
8. Lauer CJ, Schreiber W, Pollmächer T et al. Sleep in schizophrenia: A polysomnographic study on drug-naive patients. Neuropsychopharmacology 1997; 16:51-60.
9. Lauer CJ, Krieg JC. Slow wave sleep and ventricular size: A comparative study in schizophrenia and major depression. Biol Psychiatry 1998; 44:121-128.
10. Feinberg I, Koresko RL, Gottlieb F et al. Sleep electroencephalographic and eye-movement patterns in schizophrenic patients. Compr Psychiatry 1964; 5: 44-52.
11. Caldwell DF, Domino EF. Electroencephalographic and eye movement patterns during sleep in chronic schizophrenic patients. Electroenceph. clin. Neurophysiol 1967; 22:414-420.
12. Stern M, Fram DH, Wyatt R et al. All-night sleep studies of acute schizophrenics. Arch Gen Psychiatry 1969; 20:470-477.
13. Hiatt JF, Floyd TC, Katz PH et al. Further evidence of abnormal non-rapid-eye movement sleep in schizophrenia. Arch Gen Psychiatry 1985; 42:797-802.
14. Zarcone VP, Benson KL, Berger PA. Abnormal rapid eye movement latencies in schizophrenia. Arch Gen Psychiatry 1987; 44:45-48.
15. Kempenaers C, Kerkhofs M, Linkowski P, et al. Sleep EEG variables in young schizophrenic and depressive patients. Biol Psychiatry 1988; 24:828-833.
16. Benson KL, Faull KF, Zarcone VP. Evidence for the role of serotonin in the regulation of slow wave sleep in schizophrenia. Sleep 1991; 14:133-139.
17. Benson KL, Zarcone VP. Rapid eye movement sleep eye movements in schizophrenia and depression. Arch Gen Psychiatry 1993; 50: 474-482.

18. Hudson, JI, Lipinski JF, Keck PE et al. Polysomnographic characteristics of schizophrenia in comparison with mania and depression. Biol Psychiatry 1993; 34:191-193.
19. Benson KL, Sullivan EV, Lim KO et al. Slow wave sleep and computed tomographic measures of brain morphology in schizophrenia. Psychiatry Res 1996; 60:125-134.
20. Keshavan MS, Reynolds CF, Miewald JM et al. Delta sleep deficits in schizophrenia. Arch Gen Psychiatry 1998; 55:443-448.
21. Hoffmann R, Hendrickse W, Rush AJ et al. Slow wave activity during non-REM sleep in men with schizophrenia and mejor depressive disorders. Psychiatry Res 2000; 95:215-225.
22. Brannen JO, Jewett RE. Effects of selected phenothiazines on REM sleep in schizophrenics. Arch Gen Psychiatry 1969; 21:284-290.
23. Gillin C, Buchsbaum MS, Jacobs LS et al. Partial REM sleep deprivation, schizophrenia and field articulation. Arch Gen Psychiatry 1974; 30:653-662.
24. Reich L, Weiss BL, Coble P et al. Sleep disturbance in schizophrenia. Arch Gen Psychiatry 1975; 32:51-55.
25. Brambilla F, Scarone S, Pugnetti L et al. Bromocriptine therapy in chronic schizophrenia: Effects on symptomatology, sleep patterns, and prolactin response to stimulation. Psychiatry Res 1983; 8:159-169.
26. van Kammen DP, van Kammen WB, Peters J et al. Decreased slow wave sleep and enlarged lateral ventricles in schizophrenia. Neuropsychopharmacology 1988; 1:265-271.
27. Thaker GK, Wagman AMI, Tamminga CA. Sleep polygraphy in schizophrenia: Methodological issues. Biol Psychiatry 1990; 28:240-246.
28. Taylor SF, Tandon R, Shipley JE et al. Sleep onset REM periods in schizophrenic patients. Biol Psychiatry 1991: 30:205-209.
29. Nofzinger EA, van Kammen DP, Gilbertson MW et al. Electroencephalographic sleep in clinically stable schizophrenic patients: Two-weeks versus six-weeks neuroleptic-free. Biol Psychiatry 1993; 33:829-835.
30. Zarcone VP, Benson KL. BPRS symptom factors and sleep variables in schizophrenia. Psychiatry Res 1997; 66:111-120.
31. Monti JM, Altier H. Flunitrazapam (RO 5-4200) and sleep in normal subjects. Psychopharmacologia 1973; 32:343-349.
32. Carskadon MA, Dement WC. Normal human sleep: An overview. In: Kryger MH, Roth T, Dement WC, eds. Principles and Practice of Sleep Medicine. Philadelphia: WB Saunders, 2000:15-25.
33. Carlsson A, Lindquist M. Effects of chlorpromazine or haloperidol on formation of 3-methoxytyramine and normetanephrine in mouse brain. Acta Pharmac Toxicol 1963; 20:140-144.
34. Verhoeff NP. Radiotracer imaging of dopaminergic transmission in neuropsychiatric disorders. Psychopharmacology 1999; 147:217-249.
35. Laruelle M, Abi-Dargham A, van Dyck CH et al. Single photon emission computarized tomography imaging of amphetamine-induced dopamine release in drug-free schizophrenic subjects. Proc Natl Acad Sci USA 1996; 93:9235-9240.
36. Breier A, Su TP, Saunders R et al. Schizophrenia is associated with elevated amphetamine-induced synaptic dopamine concentrations: evidence from a novel positron emission topography method. Proc Natl Acad Sci USA 1997; 94:2569-2574.
37. Monti JM, Hawkins M, Jantos H et al. Biphasic effects of dopamine D-2 receptor agonists on sleep and wakefulness in the rat. Psychopharmacology 1988; 95:395-400.
38. Monti JM, Jantos H, Fernández M. Effects of the dopamine D-2 receptor agonist, quinpirole on sleep and wakefulness in the rat. Eur J Pharmacol 1989; 169:61-66.
39. Benes FM. Emerging principles of altered neural circuitry in schizophrenia. Brain Res Brain Res Rev 2000; 31:251-269.
40. Blum BP, Mann JJ. The GABAergic system in schizophrenia. Int J Neuropsychopharmacol 2002; 5:159-179.
41. Sterman MB, Clemente CD. Forebrain inhibitory mechanisms: sleep patterns induced by basal forebrain stimulation in the behaving cat. Exp Neurol 1962; 6:103-117.
42. Szymusiak R, Steininger T, Alam N et al. Preoptic area sleep-regulating mechanisms. Arch Ital Biol 2001; 139:77-92.
43. Jones BE. Basic mechanisms of sleep-wake states. In: Kryger MH, Roth T, Dement WC, eds. Principles and Practice of Sleep Medicine. Philadelphia: WB Saunders, 2000:134-154.
44. Dursun SM, Patel JKM, Burke JG et al. Effects of typical antipsychotic drugs and risperidone on the quality of sleep in patients with schizophrenia: A pilot study. J Psychiat Neurosci 1999; 24:333-337.
45. Sharpley AL, Vassallo CM, Cowen PJ. Olanzapine increases slow wave sleep: evidence for blockade of central 5-HT_{2C} receptors in vivo. Biol Psychiatry 2000; 47:468-470.
46. Sharpley AL, Vassallo CM, Pooley EC et al. Allelic variation in the 5-HT_{2C} receptor (HT2RC) and the increase in slow wave sleep produced by olanzapine. Psychopharmacology 2001; 153:271-272.
47. Haffmans PMJ, Oolders JM, Hoencamp E et al. The effect of risperidone versus haloperidol on sleep patterns of schizophrenic patients—Results of a double-blind, randomised pilot trial. Eur Neuropsychopharmacol 2001; 11 (Suppl 3):S 260.
48. Taylor SF, Tandon R, Shipley JE et al. Effect of neuroleptic treatment on polysomnographic measures in schizophrenia. Biol Psychiatry 1991; 30:904-912.
49. Wetter TC, Lauer CJ, Gillich G et al. The electroencephalographic sleep pattern in schizophrenic patients treated with clozapine or classical antipsychotic drugs. J psychiat Res 1996; 30:411-419.
50. Salin-Pascual RJ, Herrera-Estrella M, Galicia-Polo L et al. Olanzapine acute administration in schizophrenic patients increases delta sleep and sleep efficiency. Biol Psychiatry 1999; 46:141-143.
51. Maixner S, Tandon R, Eiser A et al. Effects of antipsychotic treatment on polysomnographic measures in schizophrenia: A replication and extension. Am J Psychiatry 1998; 155:1600-1602.
52. Yamashita H, Morinobu S, Yamawaki S et al. Effect of risperidone on sleep in schizophrenia: a comparison with haloperidol. Psychiatry Res 2002; 109:137-142.
53. Blin OA. Comparative review of new antipsychotics. Can J Psychiatry 1999; 44:235-244.
54. Horácek J. Novel antipsychotics and extrapyramidal side effects. Theory and reality. Pharmacopsychiatry 2000; 33 (Suppl):34-42.
55. Stip E. Novel antipsychotics: Issues and controversies. Typicality and atypical antipsychotics. J Psychiatry Neurosci 2000; 25:137-153.
56. Kapur S, Remington G. Atypical antipsychotics: New directions and new challenges in the treatment of schizophrenia. Annu Rev Med 2001; 52:503-517.
57. Monti JM. Disturbances of sleep and wakefulness associated with the use of antihypertensive agents. Life Sci 1987; 41:1979-1988.
58. Heller-Brown J, Tylor P. Muscarinic receptor agonists and antagonists. In: Hardman JG, Limbird LE, eds. The Pharmacological Basis of Therapeutics. New York: McGraw-Hill, 1996:141-160.
59. Monti JM, Monti D. Histamine H_1 receptor antagonists in the treatment of insomnia. Is there a rational basis for use? CNS Drugs 2000; 13:87-96.
60. Idzikowski C, Mills FJ, Glennard R. 5-Hydroxytryptamine-2-antagonist increases human slow wave sleep. Brain Res. 1986; 378:164-168.
61. Idzikowski C, Cowen PJ, Nutt D et al. The effects of chronic ritanserin treatment on sleep and the neuroendocrine response to L-tryptophan. Psychopharmacology 1987; 93:416-420.
62. Adam K, Oswald I. Effects of repeated ritanserin on middle-aged poor sleepers. Psychopharmacology 1989; 99:219-221.
63. Ruiz-Primo E, Haro R, Valencia M. Polysomnographic effects of ritanserin in insomniacs: A crossed double-blind controlled study. Sleep Res 1989; 18:72.
64. Hinze-Selch D, Mullington J, Orth A et al. Effects of clozapine on sleep: A longitudinal study. Biol Psychiatry 1977; 42:260-266.

CHAPTER 33

Sleep and Epilepsy:
From Interrelationships to Influence of Antiepileptic Drugs

António Martins da Silva, Melissa Mendez, Chun Bai and S.R. Pandi-Perumal

Introduction

Patients with Epilepsy often complain of sleep difficulties: sleep disruption and excessive daytime sleepiness (EDS) are common symptoms. Less frequently patients complain about insomnia. Difficulties on the study of sleep-epilepsy interactions come out because seizures themselves disrupt sleep; sleep is in some types of epilepsy a seizurepromoting factor; and because of the pharmacological effect on sleep of Anti-Epileptic Drugs (AED). Somnolence is reported as a common AED effect.

Frucht and coworkers studied 400 institutionalised (on a tertiary center) epileptic patients and found 250 in which was possible to identify at least one seizureprecipitating factor.[30] Sleep deprivation (SD) was referred in 70 out of the 250 (28%) and sleep in 55 (22%) out of them (sleep and SD corresponded to 50% of the precipitating factors identified).

Using three key words (AED, Sleep, Epilepsy) hundreds of papers were found on a literature search. Such extensive data could be an excellent source of relevant information if presented in a practical form. Our proposal is to go through the published data looking for the established relationship between sleep and epilepsy and also the influence of AED on sleep and behaviour of epileptic patients. We try to summarise the results on a schematic approach.

Sleep and Epilepsy Interactions

Diverse types of events can be associated to sleep and epilepsy. **Ictal events**: clinical and paroxysmal; **Interictal events**: paroxysmal. All of them could occur during sleep, on sleep stage transitions or on sleep-wake transition.

Diverse authors looking at different aspects have approached the relationship between these variables:

1. Influence of waking/sleep cycle on the time of occurrence of the seizures (Declerk, 1983).[17] Within this approach epilepsies have been classified as sleep, awakening and random epilepsies;[44,45]
2. Influence of sleep cycles on epilepsy – studies on frequency, type or time of occurrence of seizures or of interictal epileptiform activity: frequency, type, location, and duration;[9,46,58,59]
3. Influence of epilepsy on sleep structure: sleep staging, sleep duration, cyclic pattern;[64,85]
4. Effects of SD and of sleep after SD on paroxysmal (ictal or inter-ictal) events;[15,80]
5. Influence of comorbidity on sleep of epileptic patients;[21,23] and
6. Influence of AED on sleep structure and performance.[68,69,73]

Influence of Waking/Sleep Cycle on the Time of Occurrence of the Seizures

The epilepsy-sleep and the circadian sleep/wake cycle interact with each other. The first studies observing the effect of the sleep/wake cycle on the expression of seizures took place in the late 1800s, when Féré[25] studied the time of seizure occurrence in hospitalized epileptic patients and two peaks become apparent: that approximately two-thirds of patients had seizures between 8 p.m. and 8 a.m., and that most seizures occurred in the early morning hours, usually between 3 a.m. and 5 a.m. Gowers also studied hospitalized epileptic patients and found that they could be grouped into those with daytime seizures (42%), those with nocturnal seizures (21%), and patients whose seizures occurred randomly across the sleep/wake cycle. Within sleep, the greatest susceptibility to seizures occurred at sleep onset and at the end of sleep.[37]

Later, Janz divided seizure disorders into 3 groups: sleep epilepsies, awakening epilepsies, and diffuse epilepsies. Each of these groups showed distinct clinical characteristics, especially regarding the age of seizure onset, seizure etiology, and prognosis.[44]

Awakening Epilepsies

These include idiopathic epilepsies. They start mainly in childhood and adolescence, with up to three-fourths beginning between the ages of 10 to 25 years. They usually have a benign course. Samples of awakening epilepsies include juvenile myoclonic epilepsy (JME), childhood and juvenile absence, and generalized tonic-clonic seizures on awakening.

We reviewed continuous data from 32 patients with JME followed up at an outpatient clinic. Notwithstanding JME's usual good clinical evolution in terms of seizure control, a group of 18 out of them was free from seizures or medication and another group of 14 was medicated and continued to have rare seizures. On this last group it was possible to find the factors considered as precipitating patients' seizures: sleep and transition to awakening or vice versa were the most common factor (71%); menses,

Sleep and Sleep Disorders: A Neuropsychopharmacological Approach, edited by Malcolm Lader, Daniel P. Cardinali and S.R. Pandi-Perumal.

alcohol abuse or photic stimulation had similar percentage (≃10% each).

Sleep Epilepsies

These are commonly idiopathic or cryptogenic forms. They have a more variable age of onset, most commonly between 10-25 years old, but one third of patients may present after the third decade. They tend to represent mostly focal or secondary generalized seizure disorders. The prognosis is variable, with conditions such as Benign Epilepsy with Centrotemporal Spikes (BECTS) having a good prognosis, while Landau-Kleffner and the syndrome of Continuous Spike-Wave during Slow-Wave sleep (CSWS), also called Electric Status Epilepticus during Sleep (ESES) have a more variable course.

Diffuse Epilepsies

These are mostly symptomatic, with over half of the patients having identifiable etiologies such as vascular disease, encephalitis, and brain tumors or head trauma. Age of onset is variable, although a high proportion of cases start in early childhood, usually secondary to congenital abnormalities or birth injuries. These patients usually have a worse prognosis for seizure control. Common examples of epileptic syndromes in this group are the West and the Lennox-Gastaut syndromes and the Progressive Myoclonic Epilepsies.[44,45,85,86]

Influence of Sleep Cycles on Epilepsy – Studies on Frequency, Type or Time of Occurrence of Seizures or of Interictal Epileptiform Activity (Frequency, Type, Location, and Duration)

Sleep cycles influence the expression or suppression of seizures. Sleep is a promoting factor of some types of epilepsy, and a large proportion of seizures occur during sleep. Bazil, 2003 reported that at least up to 20% of epileptic attacks occur during sleep,[3] and Crespel and coworkers reported that up to 61% of seizures of frontal lobe origin occur during sleep.[12] The sleep periods in which the propensities to seizures or the frequency of paroxysmal events are higher are those in which the inter-staging transitions (within sleep, between sleep and waking, falling asleep) and intrastate synchronization occur. Some epilepsy types only occur during sleep (for instance the CSWS, nocturnal frontal lobe epilepsies).

Diverse authors report the benign prognosis of sleep epilepsies.[70,97] That could be the case for some of benign forms of epilepsy of childhood or adolescence, but it is rather different for other epilepsies, such as seizures of frontal lobe origin or types occurring mainly during sleep (for example, CSWS and Landau-Kleffner Syndrome (LKS).

There have been multiple studies looking at how sleep cycles and/or sleep stages influence different seizure types—focal or generalised—by modulation of temporal and spatial (scalp location) seizure characteristics. The two main sleep cycles (NREM and REM; nonrapid eye movement and rapid eye movement in sleep) and the diverse sleep stages have different influences on seizure expression.[86] In a comparative study of seizures occurring during sleep in two patient populations, epileptic patients on routine monitoring and epileptic patients with pharmacoresistant epilepsies, Minecan et al, found several differences.[64] The authors recorded 50 seizures (42 partial and 8 secondarily generalized) in 23 epileptic patients undergoing routine monitoring, and 67 seizures (51 partial seizures and 16 secondarily generalized) on the monitoring of a group of 32 patients with pharmacoresistant epilepsies. The authors found that the occurrence of seizure during the nights was mainly in NREM Sleep stage II (by group or merging both groups of patients). The same results were found even when the authors extend the analysis to the overall nights recorded (nights in which seizures occur or not).

In general most sleep seizures have been reported to occur during NREM Sleep. Overall, seizure frequency increases during NREM sleep, with most seizures occurring during stages 1 and 2. Although seizures may occur during REM sleep, their frequency is diminished, their duration is shorter, and in some cases minimal motor behaviour has been reported. Partial seizures during sleep are also more likely to become generalized during NREM sleep, while they rarely generalize during REM sleep.

The effect of sleep on seizure expression also varies according to seizure type. In partial seizure disorders, seizures of frontal lobe are more likely to occur during sleep than those of temporal lobe origin. However, temporal lobe seizures generalize more during sleep than during wakefulness,[5] while frontal lobe seizures generalize at the same rate during wakefulness and sleep. Some studies have also raised the question of whether neocortical temporal lobe seizures are more likely to occur during REM than hippocampal seizures.[65,66]

Diverse studies have been carried out on paroxysmal activity and sleep-wake cycle. Kellaway and coworkers discussed a double oscillatory model influencing the expression/suppression of seizures.[46] Circadian (sleep/wake cycle) and ultradian sleep cycles were referred as factors influencing paroxysmal activity expression.[9,54]

Influence of Epilepsy on Sleep Structure: Sleep Staging, Sleep Duration, Cyclic Pattern

There are diverse factors determining sleep-epilepsy (ictal/interictal events) influences. For instance epileptic seizures with motor behaviour occurring during sleep disrupt sleep structure. Most relevant data concerning sleep changes in epileptic patients are related to awakenings and disruptions induced by epileptic attacks, namely those determining motor behaviour changes. This has an immediate consequence: sleep fragmentation, poor sleep compliance, excessive tendency to sleep naps, and daily hypersomnia. The patient tends to compensate for the sleep loss by taking daytime naps. In some cases patients complain of hypersomnia.

Some clinical manifestations only occur in determined sleep periods. It is well established that immediately after the occurrence of generalized tonic-clonic seizures no REM sleep periods occur. REM latency is prolonged after a generalized seizure, and there is an up to 50% decrease in REM sleep. Interestingly, no REM rebound is seen on following nights.[89]

The influence of seizure type on sleep was reported on seizures and interictal events. Dadmehr and coworkers compared the sleep parameters during 3-hour post-meal nap recordings in 28 epileptic patients versus 10 normal controls.[13] Generalized seizure patients had longer sleep onset latency (SOL). Those with partial seizures had more stage 2 sleep and greater sleep efficiency (SE) than patients with generalized seizures. The findings suggest that disturbed sleep, possibly related to aberrant arousal occasioned by generalized epilepsy or epileptogenic foci (Fig. 1), is common in seizure patients, and may be related to interictal behavioural and cognitive symptoms. Polytheraphy may have an

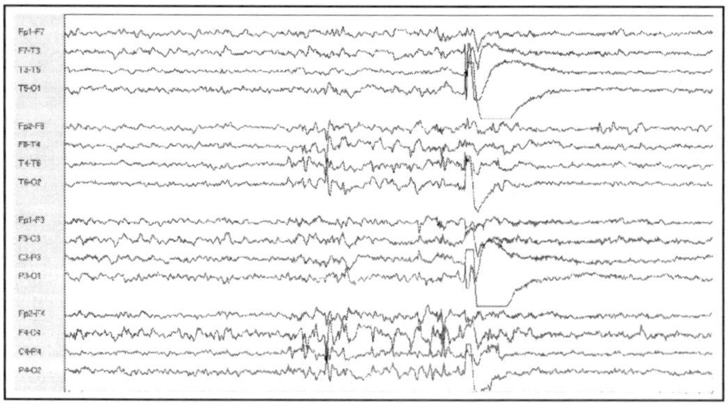

Figure 1. EEG obtained during a 24-hour ambulatory study shows repetitive spike discharges over the right at C4 resulting in an arousal from sleep (screen duration 10 seconds, 10 uV/mm).

additional effect on sleep, and may prolong or disrupt it. Schmidt et al, 1988 observed that patients who experienced an increase of several seizures in temporal relationship with poor compliance more often reported loss of sleep and more often had frequent seizures than noncompliers with a single seizure relapse or no change in seizure control. The authors claim that exacerbation of seizures through poor compliance occurs primarily in patients with frequent seizures and through loss of sleep.[84]

In general the influence of sleep on epilepsy can be summarised as:

1. The epilepsies are influenced by the sleep-waking rhythm and sleep cycles;[37,44,46]
2. About 20 to 30% of seizures occur during the night or on sleep stage transition;[3,44,45]
3. Sleep is a specially facilitating method to seizures or paroxysmal activity when other activation methods fail.[80]
4. It is also well known that sleep can enhance or suppress not only ictal or interictal events, but can also modify the temporal characteristics of the paroxysmal events. Bursts of interictal epileptiform activity are also modulated by the diverse sleep cycles,[58,59] with shorter spike duration and higher frequency rate in NREM sleep;[59]
5. During REM sleep, spikes tend to have more localized characteristics.
6. Finally, the influence of the sleep stage transition on the expression of paroxysmal events is also well known, with the synchronizing epochs characterising the cyclic alternating pattern (CAP) being the greater promoter epileptiform events or seizure expression.[88]

Two other different aspects might be considered when interactions between epilepsy, sleep, and AED were studied: the role of SD and of comorbid sleep disease in epileptic patients.

Sleep Deprivation

Because of the described influence of sleep on epilepsy another aspect that has been extensively studied is the possible role of SD or of sleep periods after SD on the activation of the ictal or interictal phenomena. Declerck and coworkers studied the sleep-wakefulness patterns recorded following 1 night of total SD in a large sample of 252 epileptic patients.[15] Data was compared to 15 normal subjects. Patients were taking various types and combinations of AED. A reduction in SOL and an increase in NREM (S3 and S4 sleep) stages were present after SD. This effect persisted in patients treated with VPA and CBZ. NREM (S1+S2) sleep increased in patients treated with phenytoin. Presence of other comedication also affected sleep parameters.

Nowadays it is established that SD may activate interictal paroxysmal activity in a proportion up to 20-50% of the studies. The higher figures come from the studies carried out on periods of sleep recorded after SD, and in young people suffering from primary generalised epilepsy. SD also improves the effect of the hyperventilation and of the intermittent photic stimulation when these two procedures are done in EEGs recorded after SD.

Comorbidity

Sleep pathologies and psychogenic events are common epilepsy comorbidities. The most frequent sleep pathology associated to epilepsy is the obstructive sleep apnoea syndrome (OSAS). OSAS is a very common disease that causes frequent sleep disruptions and sleep stage transitions. This could easily have a promoter epileptogenic effect. Another putative effect of OSAS on epilepsy enhancement could be the role of hypoxemia caused by the apnoea and the possible decrease of cerebral blood flow resulting from low blood volume from cardiac output. Both could enhance the probability of seizure occurrence. The treatment of sleep disorders improves sleep quality and may decrease seizure recurrence in epileptic patients. Multiple investigators have documented patients with refractory epilepsy who have had an over 50% seizure reduction, including several who became seizurefree, after treatment of coexisting apnea.[8,53,92]

Another frequent comorbidity is the association of periodic limb movements disorders (PLMD) in patients with epilepsy. This frequent association may be veiled by the epilepsy treatment. In fact, BZDs have been used as first line drugs to treat PLMD. Clonazepam (Fig. 2) that was also used, as AED is currently the first approach to these movement disorders. Nowadays, other AEDs (namely VPA and GBP) have been successfully used on PLMD treatment.[21-23] This double effect could mask the evidence of the association of both morbidities.

Another frequent comorbidity is the coexistence of psychogenic seizures in epileptic patients and its possible influence on sleep. Few consistent studies are available. However in a recent study no interactions between psychogenic seizures and sleep have been reported. The authors conclude that the sleep of epileptic patients with psychogenic seizures is similar to that of patients with

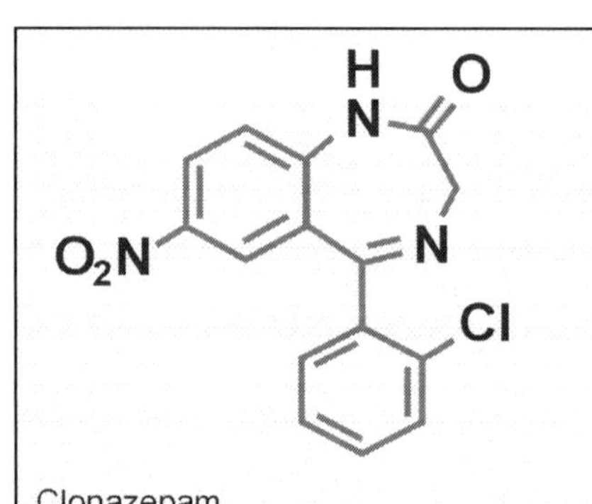

Figure 2. Clonazepam.

major depression.[3] The reasons for the differences, if any, might be on factors outside the psychogenic event.

In a overall and short comment we can conclude that the interaction between sleep and epilepsy might be summarised as: epilepsies are influenced by sleep but themselves do influence sleep; NREM sleep S2 is a relevant seizure and paroxysmal activity promoter; sleep after SD is a relevant promoting factor of paroxysmal events; treatment of concomitant sleep pathologies do improve epilepsy.

Sleep and Anti Epileptic Drug Interactions

Anti-epileptic drugs (AEDs) have various deleterious effects on sleep and on daily patient condition. Epileptic patients frequently report changes in daytime vigilance and daytime sleepiness. This could be secondary to drug treatment but it could be related also to frequent disruptions of sleep structure by night seizures, inducing changes on sleep pattern or abrupt sleep disruption. This is why in spite of old references to sleep changes induced by barbiturates or to the sleep promoter effects of BZDs; the effects of AED on sleep of epileptic patients were not well defined.

There are several factors that make it difficult to study the effect of AED sleep and to compare the results of different studies. Among these there are:

1. Diverse methodological approaches rendering the studies non comparable (also sometimes the measures used are not objective)
2. Different populations studied (as inclusion of diverse types of epilepsies grouped on the same set)
3. Differences on AED dose regimen (giving diverse distribution profiles of the AED)
4. Differences on the time and duration of treatment (frequently the information on previous treatments is scarce and incongruent and treatment durations is not always indicated)
5. Failure to control for seizures, and finally
6. Failure to control concomitant AEDs or other comedication.

In spite of these remarks is important to note that the effect of AED in mono- or in polytherapy in terms of toxicity is not well defined. In an elegant study Deckers and coworkers reported a double blind clinical trial in 130 adult.[14] Patients with untreated generalized tonic-clonic and/or partial seizures in which a combination of CBZ and VPA was compared with CBZ monotherapy with patients starting with equal drug loads. Neurotoxicity was the primary outcome measured by seizure counts, clinimetric epilepsy scales, and neuropsychological tests at up to 12 months. The authors found no statistical differences between the two treatments in the reduction of seizure frequencies, in overall neurotoxicity, or in overall systemic toxicity. The frequencies and clinimetric scores of certain adverse effects did differ (e.g., more monotherapy patients remained sedated, and more polytherapy patients gained weight). The authors concluded that no differences were found in overall neurotoxicity between monotherapy and polytherapy.

Excessive Daytime Sleepiness (EDS) has also been frequently reported, but not well documented, in epileptic patients using diverse AEDs. One of the most relevant studies carried out is the study of Palm et al in which preadolescent children with epilepsy who were seizurefree were taken off AEDs.[68,69] Multiple Sleep Latency Test (MSLT), Daily Average Sleep Tendency (DAST), EEG and questionnaires were used in order to compare seizurefree from seizures patients performances and sleep tendency during treatment and after discontinuation of therapy, to healthy controls. Patients had significantly higher daytime sleep tendency than controls, even after AED discontinuation. This implies that factors such as AED, recent seizures or complicated epilepsy cannot be evoked as the sole cause of somnolence (see Table 1).

Table 1. Median and range of sleep onset latency (SOL) (in min) (determined by MSLT)

	N	Median	Range
With AED	9	21*	10-27
6 months without AED	9	19**	14-25
Healthy Controls	18	26**	22-30

*/** P< 0.01; adapted from Palm et al, 1992

Sleep, Epilepsy and Antiepileptic Drugs Interactions

There are diverse aspects that might be considered when we study the interactions between sleep, epilepsy and AEDs. Difficulties arise from the clinical context. Because the circumstances previously described on the sleep-epilepsy interactions additional cautions need to be considered. Are the possible relationships occurring only during sleep? Are they related to seizures occurrence or are they secondary to nonspecific aspects of the type of epilepsy? And is the role of AED on sleep characteristics well defined and definitive, or is the sleep better because epilepsy improved?

The study of the relationship between AEDs and sleep has several limitations that arise from diverse factors: lack of information about previous medication, lack of data on the long term efficacy of AED, lack of treatment stabilisation, and scarce information about the influence of new AED. However epileptic patients rather frequently refer fatigue as a consistent complaint, which could partly due to the effect of AED therapy on sleep quality and SE.[32]

Partial Seizures

The information on effects of AED on sleep architecture of epileptic patients with partial or partial-onset seizures is even more confusing. In a recently published paper Legros and Bazil, 2003 investigated the sleep structure of 39 epileptic patients with partial-onset (localisation related) epilepsies recorded during 72 nights. Only patients in monotherapy and not having seizures on the recording time or on the precedent 24h were analysed. Sleep architecture was compared to a control group (adults also with localization related epilepsies not medicated).[50] The AED influences were diverse: CBZ and LTG do not significantly disrupt sleep structure, but PHT and VPA do, increasing NREM S1. Patients on GBP therapy had increased SWS.

Basic Mechanisms

Sterman, 1981 studied the spontaneous EEG activity recorded during NREM sleep by means of power spectral analysis.[87] The author identified characteristic EEG abnormalities in epileptics. In the study medicated epileptics showed differences restricted to the early morning nonREM sleep sample with significantly greater power at 4-7 Hz and lower power at 12-15 Hz. Un-medicated subjects showed these differences in both samples, also showing a significantly greater power in the delta 0-3 Hz band. The author suggests that AED exert a general normalizing influence on EEG characteristics during sleep, particularly at the beginning of the night in proximity to their administration. The significant EEG deviations were restricted to two basic rhythmic patterns during sleep. This suggests a disturbance of underlying thalamocortical mechanisms.

Methodologies Used to Study AED Influences on Sleep

The first condition to evaluate the possible role of AED on sleep structure is that sleep recordings should be obtained when epileptic seizures do not occur.

Apart from this recommendation, the effects of AEDs on sleep have been studied in various situations. The results in general are frequently contradictory, in part because of the confounding and concurrent phenomena already detailed: epileptic seizures inducing sleep disturbances, patients under polytherapy, and other sleep events occurring in the same population. Also the methodologies used to analyse the data are diverse. Series studied include different populations (different types of epilepsies), diverse doses of medications and situations: different treatment durations - effects on acute vs chronic situations, different combinations of therapy. Moreover only with the introduction of objective criteria: polysomnography (PSG), MSLT and/or the use common scales (such as Epworth Sleepiness Scale-ESS and the Awake Maintenance Task, AMT) were it possible to determine whether patients on AED therapy had impaired sleep or ability to maintain wakefulness.

Experimental data from animal models may help to interpret the relationship between AEDs and sleep. Wauquier and colleagues compared the sleep-wakefulness (S-W) patterns in 4 genetically epileptic beagles to normal beagles.[93] The only differences found reveal that epileptic dogs tend to have more and shorter S-W epochs and shortening of both REM and deep slow wave sleep (dSWS) onset latency. The diverse AED tested did not affect S-W patterns.

Classical AED

For the classical drugs, most of them first line AED, the influence on sleep has been diversely approached. Declerk studied the influence of three different antiepileptic drugs (PHT, CBZ and VPA) as chronic therapy on sleep architecture.[17] The most relevant findings seemed to be an increase of the first NREM Cycle duration (namely on NREM sleep stages 3-4). Patients were treated in monotherapy (with partial or generalised seizures) or in association to BZDs (patients studied had generalised or partial secondary generalised seizures, often with organic lesion). In this subgroup the increase on duration of Stage 3-4 of NREM was not as significant as on the others. The increase of the first REM Cycle duration was not so evident and the effect was seen mainly on partial epilepsies.

In a study conducted by Drake and associates, the sleep characteristics of 17 epileptic patients (12 with complex partial and 5 with generalised TC seizures) receiving AED in monotherapy (five treated with CBZ, five with PB, five with PHT and two receiving BZD) were assessed during the nights in which seizures did not occur.[19] TST was reduced, wakefulness increased, and SOL was prolonged in partial when compared to generalized epilepsy. REM sleep was reduced and its latency decreased in partial seizure patients. AEDs have differential effects on nighttime sleep, although both groups (more markedly in partial seizure patients) had decreased SWS. PHT increased SOL and decreased TST. CBZ increased awakening and diminished SWS and REM sleep. VPA reduced SWS. BZD decreased sleep and REM latencies. Thus, epilepsy may affect nocturnal sleep differently on partial or generalized seizures.

In a different study Salinsky and coworkers tested patients with the Awake Maintenance Task (AMT) to measure fatigue and sleepiness, to determine objectively whether patients on chronic stable AED therapy had impaired ability to maintain wakefulness.[83] The study enrolled thirty epileptic patients receiving AEDs [CBZ, PHT, PB, and VPA] that were compared to 35 healthy controls, to 12 seizure patients not taking AEDs, and to 16 patients with Multiple Sclerosis. Patients receiving AEDs had a mean total latency to sleep significantly lower when compared with the three control groups. Among patients receiving AEDs, objective EEG drowsiness did not correlate with AED levels or performance measures. Untreated seizure patients had significantly greater complaints of lack of vigor despite a near absence of objective drowsiness. The authors claim that epilepsy patients receiving chronic AED therapy have impaired ability to maintain wakefulness.

All these studies exemplify the possible influence of AED used in monotherapy.

The classical AEDs also influence other aspects of sleep characteristics. In the study reported by Drake et al on sleep spindles characteristics, frequency was significantly lower in patients with generalized epilepsy than in those with partial seizures.[20] Additionally, patients with complex partial seizures and partial seizures with secondary generalization differed significantly in spindle frequency. Spindle frequency was significantly lower in patients with polytherapy. Patients whose regimens included PB had significantly lower spindle frequencies. Spindle frequencies differed significantly between PHT and CBZ. The authors conclude that the differences in spindle frequency may be due to residual medication effects, underlying encephalopathy or physiological differences between partial and generalized epilepsy.

Phenobarbital and Sleep of Epileptic Patients

Hyperactivity and increased somnolence have been frequently reported in epileptic patients treated with phenobarbital (Figs. 3,4) and barbiturates in general.[18] During MSLT, patients on PB showed shorter mean SOL compared with VPA and control groups, though within-group variability was considerable. Patients on PB showed longer motor movement times, impaired atten-

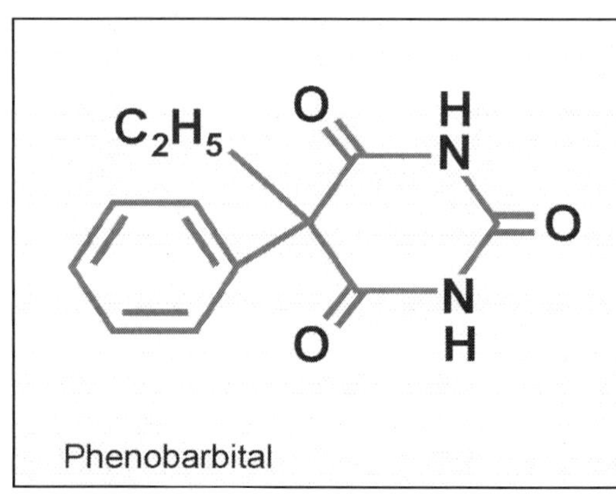

Figure 3. Phenobarbital.

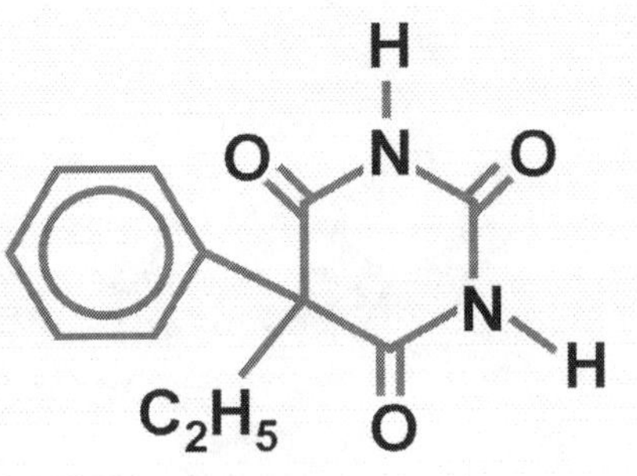

Phenobarbitone

Figure 4. Phenobarbitone.

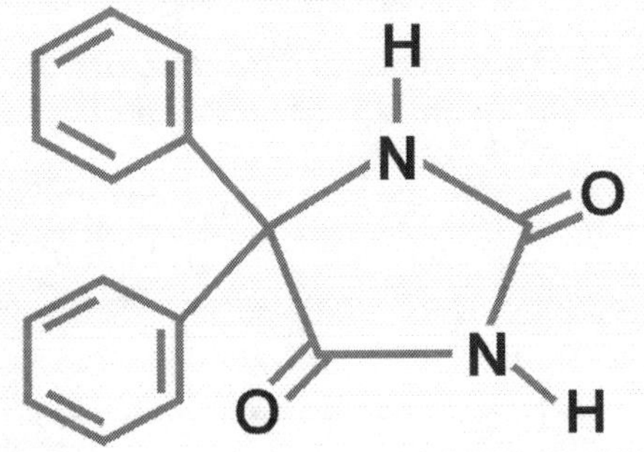

Phenytoin

Figure 5. Phenytoin.

tion (cancellation test, CT). When compared with controls, patients under PB had a reduced processing speed and a trend towards lower critical flicker fusion (CFF) threshold.[55,56] To other authors no change the overall sleep structure was observed in children with febrile seizures treated under PB.[41] In an elegant study Wolf et al, 1985 investigated 40 consecutive unmedicated newly diagnosed epilepsy patients treated with PB or PHT in a random sequence.[95] After adjusting to a steady-state, with normal AED serum levels, in all 34 patients treated with PB the short term effects of PB were the following: at the end of the night the total sleep showed shorter SOL, REM sleep and arousals decreased, the percentage of sleep stage 2 was increased, and stages 3 and 4 were unchanged. When long-term findings are compared with baseline (4 weeks after steady state reached), the persistent changes are SOL, less movements and shorter duration of awakenings. The long-term findings showed accentuation of the longer NREM sleep on the first cycles and REM sleep was postponed (the main REM changes were on the first cycle). Also influences on sleep microstructure have been reported. Spindle frequency was significantly lower with polytherapy. Patients whose regimens included PB had significantly higher spindle frequencies.[20]

A summary of influences of diverse AED effects on sleep as published in literature is displayed in Table 3.

A synopsis of how PB affects sleep:

I. Short term effects on sleep: (1) sleep latency reduced; (2) sleep stage 2 increased;*# (3) no changes on S3/4; (4) less REM sleep and awakenings;

II. Long term effects of PB on sleep: (A) increased of S2; (B) increased REM onset latency; (C) sleep spindles frequency increased.

Studies supporting it: Wolf et al, 1985 (1, 2, 3, 4, A, B); Drake et al, 1991 (C); Hirtz et al, 1993 (no changes in children with febrile convulsions).

Phenytoin and Sleep of Epileptic Patients (Fig. 5)

In the eighties, Wolf and collaborators conducted elegant studies trying to approach the sleep-AED relationship. The authors considered sleep as a modulator of seizure activity and many antiepileptic drugs as modulators of sleep. This close interaction is further complicated by another one: the possible influences of sleep organization in antiepileptic drug action, namely on differences in drug response of various epileptic syndromes. The studies were conducted studying the sleep of a set of patients recorded firstly off medications. After adjustment to therapeutic steady states of PB and PHT (as sequential sole agents in a crossover design with random sequence) diverse sleep parameters were recorded. Shorter SOL, less motor behaviour, and less number of arousal were seen with PB, especially in patients with generalized epilepsy. There were increases in S1/2, and REM sleep was decreased. The usual sleep pattern was altered; with maximal deep sleep early and maximal REM sleep late in the night. PB seemed to have maximal effect in the first REM cycle. Patients under PHT had shorter latency but also shorter sleep (S1/2), longer S3/4, with no alteration of REM sleep. The maximal alterations were observed in the third REM cycle. With both drugs, there were some differences in the response of generalized as opposed to focal epilepsies, and of awakening as opposed to sleep epilepsies. The early REM cycles seemed to be more modifiable by drugs in patients with generalized or awakening epilepsies than in patients with focal or sleep epilepsies.[94]

This study was completed by prospective longitudinal observations.[95] Patients on PHT had a shorter SOL and a decrease in sleep S1,S2 and increase of SWS mainly in the latest NREM-REM cycles. The third cycle was the most affected. TST was longer but REM did not change. Movement related arousals and body movements were increased. A decrease in REM sleep duration was observed in patients with focal epilepsy.

On the same group Roder-Wänner and coworkers reported a prospective longitudinal study of alterations in PSG correlates by PHT monotherapy for epilepsy.[79] A first dose of 100mg already caused abbreviation of SOL and an increase in SWS in the first NREM/REM cycle. After steady state adjustments, an increase in SWS was seen in the later REM cycles. The full night SWS was augmented. Sleep architecture was more affected in the third NREM/REM cycle. Effects were reversed with the continuation of therapy except for the shortening of SOL. The authors did not find significant differences in the response between generalized and partial epilepsies, seizure control and length of follow-up. The serum concentration of PHT had only a very limited influence. But a delayed further decrease of epileptic discharges during sleep was observed later in patients who were seizurefree.[79]

Other studies not always confirm the effects on sleep by PHT on chronic treatment. In 12 patients studied in a medium-long term control (longer than 6 months) only reduction on sleep latency and increased movement time during sleep remained stable. Declerck group also found an increase in S1 and 2 and a small REM reduction in patients treated with PHT. Diverse sleep

** More prominent effects on patients with generalised vs focal epilepsies; # More prominent effects on patients with photosensitivity.*

Table 2. Sleep macrostructure in epileptic patients with different drug regime, independently of type of epilepsy*

	Normal	Abnormal
Monotherapy	11	6
Polytherapy	3	9

$\chi 2$ = $p < 0.05$; n=29; *personal data

reactions were reported by Legros and Bazil, study in partial or localization related epilepsies.[50] The length of S1 was higher and REM sleep reduced. On the contrary SWS was also diminished. The sleep-PHT interactions are indicated in Table 3.

Synopsis of how PHT affects sleep (see also Table 3):

Short term effects:

1. reduced SOL;
2. decreased S2*;
3. no change or mild increase in SWS (max in first cycle);
4. decreased REM; increased body movements and awakenings.

*In patients with photosensitivity the effect is greater (signif).

Long term effects:

A. decreased sleep latency;

B. decreased sleep efficiency.

Studies: Wolf et al, 1985 (1, 2, 3, 4, *, A); Roder Wanner et al, 1987 (1, 2, 3, 4, A); Dadmeher et al (B); Declerck et al, 1985 (1, 3, 4); Drake et al, 1990 (A, B).

Carbamazepine and Sleep of Epileptic Patients

Carbamazepine (CBZ; $C_{15}H_{12}N_2O$; 5*H*-Dibenz[*b,f*]azepine-5-carboxamide) (Fig. 6 and 7) was approved by the U.S. Food and Drug Administration in 1974 for the use in psychomotor and grandmal epilepsy (generalized and partial seizures) in adults who have not responded to other agents. Efficacy of CBZ is proved not only in temporal lobe but also in frontal lobe epilepsies, with marked reduction on seizure frequency. This is why many reports address the influence of CBZ on sleep as the effect on seizures reduction, and consequently on improving of sleep stability. Many studies have been conducted on the influence of CBZ on sleep structure. Experimental data from animals give us interesting

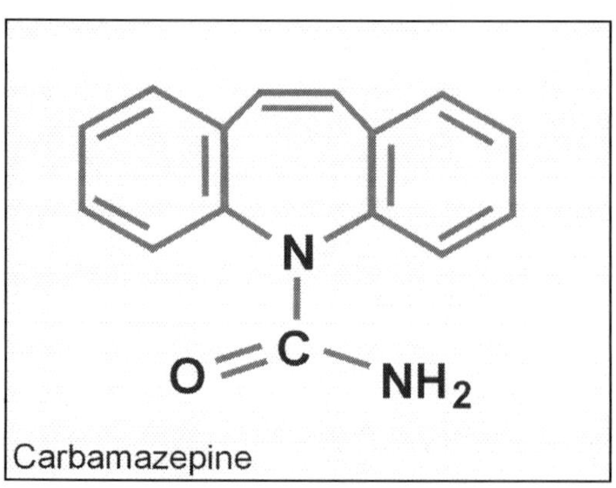

Figure 6. Carbamazepine.

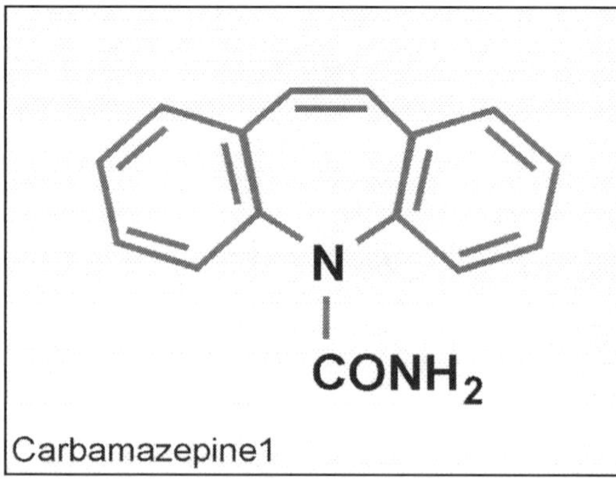

Figure 7. Carbamazepine 1.

results on REM reduction and REM fragmentation.[33,34] The same findings and also an increase of total sleep time, and reduction of stage shifts, were confirmed in patients with partial epilepsy[31,54,90,91] Similar findings have been described with the use of controlled-release (CR) formulations. The results of acute administration of CBZ-CR showed increase in stage shifts, reduction of REM sleep, and increased REM fragmentation. However these results were almost completely reversed in temporal lobe epilepsy patients in chronic treatment.[35,72] Legros and Bazil also did not find significant differences in sleep architecture in patients with localization related epilepsies and under CBZ treatment.[50] The putative action mechanism could be an increasing of 5HT levels or of adenosine receptors modulating 5HT and catecolamines effects on sleep.

In summary, reduction of SOL, increase in SWS and TST, REM reduction/fragmentation and increase of REMOL latency have been found both in immediate-release and in CR formulations. The REM modifications seemed to be reversed in chronic treatment.

Synopsis of CBZ effects on sleep (see also Table 3):

1. Increased SWS;
2. Decreased REM;
3. Increased REM fragmentation;
4. Decreased SOL;
5. Increased TST;
6. Increased REM onset latency and decreased REM density.

Studies:

Normal Releasing Formulation:

Gann et al, 1994: 12 volunteers, 1 week, (1, 4, 6);

Gigli et al, 1998,1992: Experimental Data (cats) (2, 3);

Manni et al, 1990: 14 patients, partial epilepsy, Chronic TT (2, 6, 7);

Touchon et al, 1987: 15 patients, 800mg, 1 month after TT: 5 (Gigli et al, 1997; Placidi et al, 2000).

Controlled Release Formulation.

Placidi et al, 2000 (ELT): Acute TT (400mg) + Chronic TT (400mg) – (2, 3);

Chronic TT (idem) 1 month: no changes; (ELT) 800mg – (same results as Placidi);

Mechanisms (?): Increasing of 5HT levels or of Adenosine receptors modulating 5HT and Catecolamines.

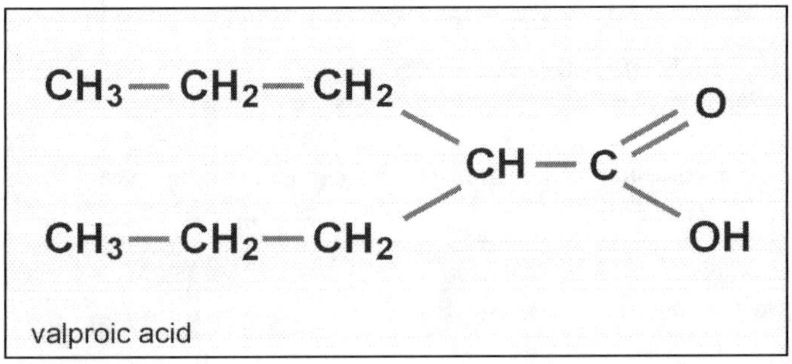

Figure 8. Valproic acid.

Valproic Acid and Sleep of Epileptic Patients

Effects of valproic acid (VPA; 2-propylvaleric acid or di-*n*-propylacetic acid) (Fig. 8) on sleep have been recently emphasised because the use of this AED on other neurological disorders. Treating patients with periodic limb movement disorders (PLMD) successfully with small VPA doses (125-600mg at bed time) Ehrenberg and colleagues noted that sleep was stabilized and daytime somnolence reduced.[23] Improvement in SE and daytime alertness, decreased sleep stage 1 and increased sleep stages 3 and 4, with REM unchanged were consistently recorded. But was the better sleep profile due to the AED or was it due to the successful treatment of the disease (PLMD)? In studies carried out in outpatient clinics, Drake and coworkers reported that patients taking VPA had reduction in SWS (findings that are not significant and also different on generalised than on partial epilepsies) that was not confirmed in other series.[19] Wolf et al, 1985 and Declerk and Wauquier, 1991 reported relative increase of Sleep Stage 1 and SWS (S3/4) without changes on REM sleep.[16,95] In a prospective study found consistent increases on sleep S1 and S3/4 after VPA therapy, without REM changes.[78] Harding and associates found, by automatic sleep analysis, a decrease in REM activity and an increase in delta activity (S3/4).[39] The manual analysis did not show differences although they followed the same trends. The authors suggest that it could be possible that the automated results may reflect EEG frequency changes rather than sleep cycle changes caused by VPA. Similar findings on Sleep S1 were reported on patients with partial (localisation related epilepsies) studied by Legros and Basil, in which the authors found that VPA disrupted sleep by increasing stage 1 sleep.[50]

Synopsis of how VPA effects sleep (see also Table 3):

1. Increased S1 during all sleep
2. Decreased S2
3. No change or an increase in SWS
4. No change in REM or arousals

Studies: Declerk and Wauquier, 1991 (1, 3, 4); Wolf et al, 1985 (1, 2, 3, 4); Harding et al 1985 (1, 2, 4); Röder, Wanner and Wolf, 1981 (1, 2, 4); Legros and Bazil, 2003.

Primidone and Sleep of Epileptic Patients

There are diverse studies addressing the influence of primidone (PRM) on sleep. Because of the comparisons made between the drug influence on the sleep of healthy volunteers and on the sleep of epileptic patients, the study of Maxion and colleagues is one of the most interesting.[62] Administration of 250mg of PRM to human volunteers at night, the sleep was affected by an increase on SWS and reduction of REM percentage and REM density found. In stable conditions at three months follow-up of epileptic patients treated with 750 mg/day of PRM in monotherapy, a shortened SOL and REM density (but not change in REM percentage) were found.[62] The sleep-PRM interactions are condensed in Table 3.

Synopsis of how PRM affects sleep:

Short term effects (Healthy volunteers single dose of 250 mg):

1. Decreased sleep latency;
2. Increased SWS;
3. Decreased REM density and percentage

Long term effects (Epileptics with total dose of 750 mg/day):

A. Decreased SOL;

B. Decreased REM density.

Studies: Maxion et al, 1975 (1, 2, 3); Wolf et al, 1985 (1, 2, 3, A, B); Röder-Wanner and Wolf, 1981 (A, B).

Ethosuximide and Sleep of Epileptic Patients

Ethosuximide (ESM; $C_7H_{11}NO_2$; 2-Ethyl-2-methylsuccinimide) (Fig. 9) is still a major antiepileptic, mostly used in paediatric cases. The effect of ESM as a drug is difficult to study, because in certain forms of epilepsy the seizure reduction has major direct effects on sleep stability, especially on stage shifts. The influence of ESM on sleep has been reported as increasing S1, in its overall percentage (in early or late night sleep).[96] This S1 increase is more robust on the immediate sleep after REM. Patients have a decrease of SWS, namely on the first NREM cycle. ESM seems to prolong the early night REM cycles, shortening the later.[95] Sleep fragmentation and EDS have also been reported, which could be a result of interaction on slow wave, restorative sleep.[78]

Synopsis of how ESM affects sleep (see also Table 3):

Short term effects (single dose 250 mg):

1. Increased S1 in all sleep cycles, especially first cycle;
2. Decreased SWS in first cycle;
3. Decreased S3;
4. Small increase in REM (first cycles);
5. Increased SOL.

Studies: Declerk and Wauquier, 1991 (1, 2, 3); Wolf, 1987 (1, 2, 4, 5); Roder Wanner et al, 1981 (1, 2, 4, 5).

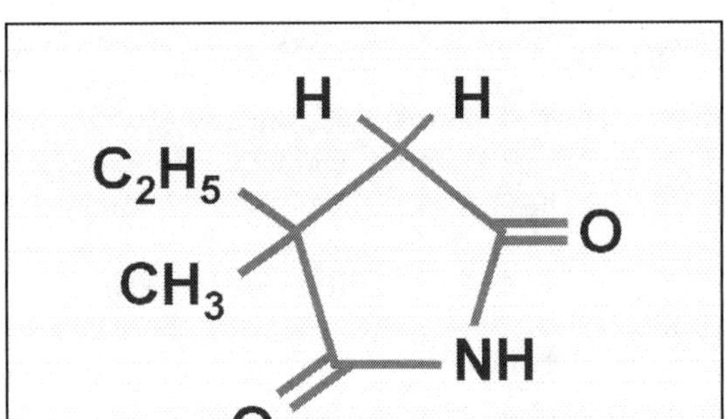

Figure 9. Ethosuximide.

Table 3. Simplified version of the summary of the effects of AEDs on sleep architecture

AEDs	Sleep Latency	TST	Sleep Efficiency	Arousals & Awakenings	% Stage 1	% Stage 2	% Stages 3 & 4	% REM
Benzodiazepines	↓	↑	↓	↓	↓	↑	↓	↓
Carbamazepine	↓	↑	↑	↓	↑	NC	↑	Mild ↓ or NC in % ↓ Density
Ethosuximide	↑	?	↓	↑	↑	NC	↓	↑
Felbamate	↑	NC (↓)	?	?	?	?	?	↓
Gabapentin	?	↑	↑	↓	↓	NC	↑	↑
Lamotrigine	↑	NC (↓) insomnia	NC	NC	NC	↑	↓	↑
Levetiracetam	NC	NC	NC	NC	NC	↑	↓	↑ latency
Phenobarbital	↓ acutely	NC	↓	↓	↑	↑	NC	↓
Phenytoin	↓	↑ acutely; NC chronically	↓	↑	↓ acutely	NC or mild ↓	NC or mild ↑	NC in %, ↓Density
Primidone	↓	NC	NC	NC	NC	NC	↑	↓
Sulthiame	Normalization of EEG				(↑) NC			
Tiagabine	NC	NC (↑)	↑	NC	↓	NC	↑	NC
Topiramate	↓	?	?	?	?	?	?	?
Valproicacid	NC	NC	NC	↑	↑	↓	NC or ↑	NC
Vigabatrin	NC	NC	?	NC	?	?	?	?

TST= Total sleep time; NC= no significant change; ? = Unknown or conflicting data; ↑ = increase; ↓ = decrease

Sulthiame and Sleep of Epileptic Patients

Sulthiame (STM) (Fig. 10) is a sulphonamide derivative. It is a *carbonic anhydrase* (CA) inhibitor with an anticonvulsant effect in the treatment of focal epilepsies, namely in children. Normalisations of the EEG in sleep, waking or in both were described under STM therapy in certain forms of epilepsy. This AED is mainly used in paediatric forms of epilepsy, namely pharmacoresistant. Experimental data in animal models (guinea pig hippocampus) showed that STM exerts its anticonvulsant effects blocking sodium channels. This seems to be different from the effects induced by blockade of the *carbonic anhydrase*.[52] Conversely in another experiment, carried out in hippocampal slices of the same animals, Leniger and coworkers studied the effects of sulthiame on pH_i and on epileptiform activity.[51] Differently, their results suggested that STM acts as a membrane-permeate *carbonic anhydrase* inhibitor whose beneficial effect on epileptiform activity results at least in part from a modest intracellular acidosis of central neurons.[51]

In spite of these controversies sleep improvement described in some series could be also a concomitant effect of seizure improvement found in the worst cases, mainly in epilepsies accompanied by encephalopathy (which is seen in many of published series).[24]

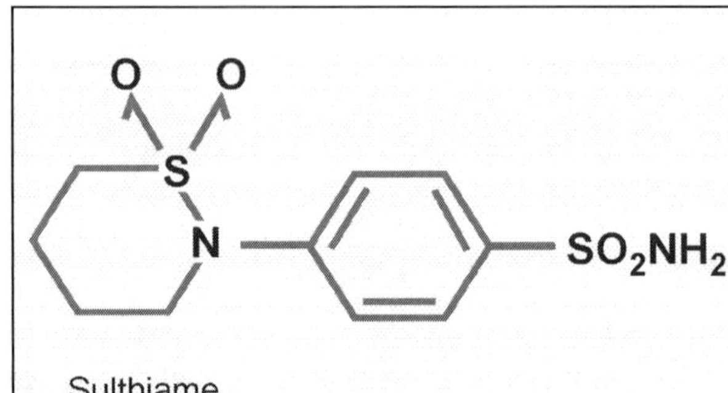

Figure 10. Sulthiame.

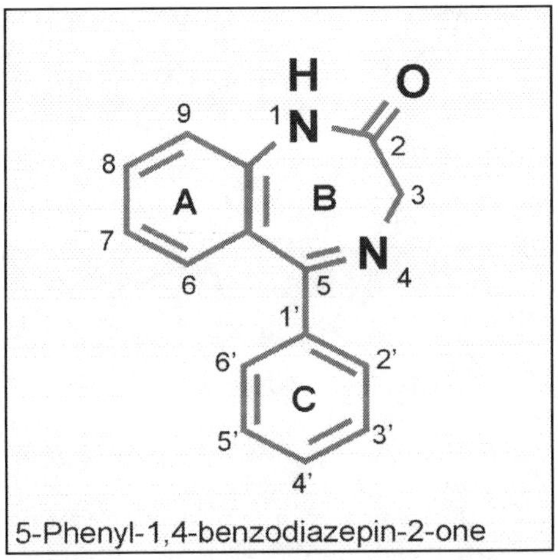

Figure 11. Benzodiazepine.

Gross-Selbeck treated groups of children with benign and encephalopathic forms of epilepsy.[38] Bast and coworkers also used this AED in benign forms of epilepsies accompanied by sleep structural changes.[1] Bast and coworkers followed-up their patients changes in sleep pattern up to six months after initiation of drug treatment. The sleep architecture and sleep-EEG normalization persisted in more than 50% of treated patients even at six months follow up interval. Increase of early sleep stages (S1 or "light sleep") was frequent.

Synopsis of STM effects on sleep: "Normalization of EEG and increase S1" (see Table 3).

Studies: Bast et al, 2003; Engler et al, 2003

Benzodiazepines and Sleep of Epileptic Patients

The uses of BZDs (Figs. 11, 12) have been widely acknowledged in the pharmacological management of epilepsy, both for acute emergencies and chronic disease management (as adjunctive therapy). In particular, BZDs play a pivotal role in the pharmacological management of status epilepticus (SE) because of the high lipophilicity, rapid brain penetration and receptor binding properties thus making them as an ideal candidate to prevent neuronal damage. BZDs are nowadays more concerned to specific forms of epilepsy (they remain relevant drugs for acute changes, repeated seizures, add-on in severe forms or for some generalised idiopathic such as JME). BZDs decrease the SOL, increase sleep stage 2, and decrease SWS, namely S3. They also cause a reduction of microarousals and an increase in REMOL and REM sleep reduction.[2,19] reported that the use of BZDs decreased SWS sleep and increases REM latencies.[19]

Synopsis of BZDs affects on sleep (see Table 3):

1. Decreased SOL;
2. Increased S2, Decreased S3;
3. Decreased "Microarousals";
4. Decreased REM;
5. Increased REMOL.

Studies: Bastien et al, 2003 (1, 2, 3, 4); Drake et al, 1990 (1, 4, 5).

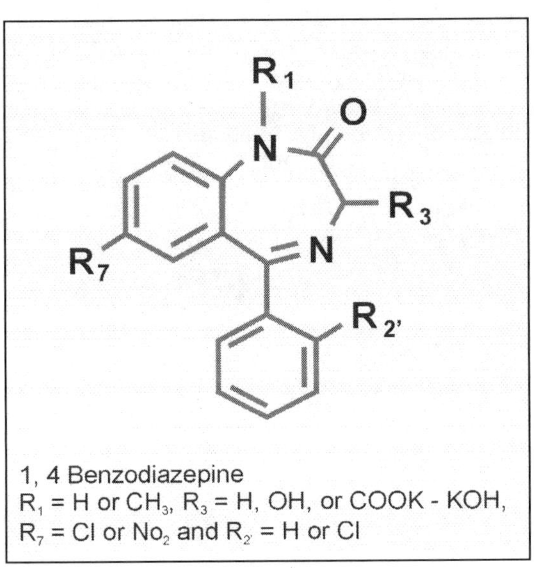

Figure 12. Benzodiazepine.

New Anti-Epileptic Drugs and Sleep

There is limited research to date and few objective neurophysiological evaluations of influences of new AEDs on epileptic patients' sleep when these were treated with such AEDs. They are frequently used as add-ons. Limitations of knowledge are higher especially because monotherapy has been less frequently used.[11]

Vigabatrin and Sleep of Epileptic Patients

Because common side effects on visual fields, Vigabatrin (VGT; $C_6H_{11}NO_2$; γ-vinyl- GABA) (Fig. 13) is only used for certain epileptic syndromes of childhood or infancy - with West Syndrome (WS) as the most common indication. VGT-induced increases in TST (by increasing S1) and REM reduction have been reported in amygdala-kindled rats.[77] Other series in humans do not report significant changes.[10]

A short synopsis of this data shows the effects of VGT on sleep architecture (Table 3):

1. No changes in SOL, TST, or awakenings in humans
2. Animal studies have shown increased TST, increased stage 1, decreased REM, and decreased awakenings.

Studies: Raol and Meti, 2000 (Exp data) (2); Bonanni et al, 1997 (add-on) (1, 3, 4).

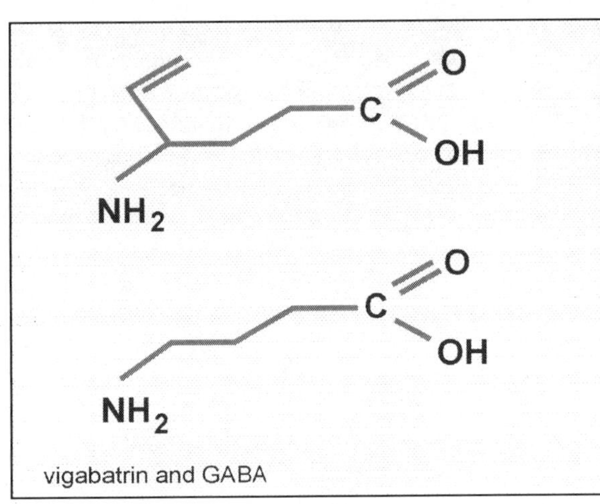

Figure 13. Vigabatrin and GABA.

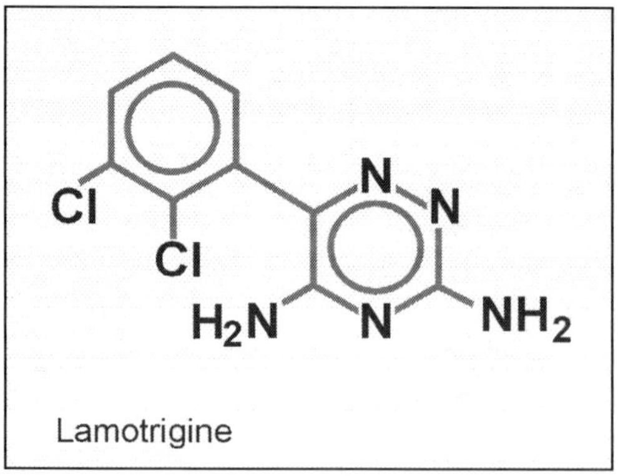

Figure 14. Lamotrigine.

Lamotrigine and Sleep of Epileptic Patients

Two different relations marked Lamotrigine (LTG; $C_9H_7Cl_2N_5$; 3,5-diamino-6-(2,3-dichlorophenyl)-1,2,4-triazine; 6-(2,3-Dichlorophenyl)-1,2,4-triazine-3,5-diyldiamine) (Fig. 14) influence on sleep: decrease of SWS percentage[72,73]or insomnia.[82] Comparing baseline with 3 months of follow up of LTG therapy, LTG increased REM sleep, reduced number of entries into REM sleep, decreased number of phase shifts, and decreased percentage of SWS.[72] Other extensively studied reports in which the changes were assessed by means of self-questionnaires complaints, using the Epworth Sleepiness scale (ESS).[26,27] Sleep changes were minimal if any. When present, difficulties on maintaining sleep are among the most relevant.[82]

Synopsis of how LTG affects sleep (Table 3):

1. No alterations in ESS;
2. Increased S2;
3. Decreased SWS;
4. Insomnia, difficult to fall and maintain asleep;
5. No significant change in "arousals" or "stage shifts";
6. Increased REM.

Studies: Placidi et al, 2000b (Add-on, 13 pts, 300mg/d) (2,3,6)*; Foldvary-Schaefer, 2002; Foldvary et al, 2001 (1,2,3,5,6);Sadler et al, 1999; * % Seizure reduction 85%; 50% Pts = Seizure free.

Gabapentin (GBP) on Sleep of Epileptic Patients

Gabapentin (GBP; $C_9H_{17}NO_2$; 1-(Aminomethyl) cyclohexaneacetic acid) (Figs. 15, 16) effects on sleep have been studied in diverse series, both at baseline and 3 months after treatment. Increase in of SWS has reported in multiple studies.[22,27,50,71,76] GBP therapy improves the sleep pattern of epileptic patients and it seems to modulate the expression of IEA with different effects in relation to the various vigilance levels.[74]

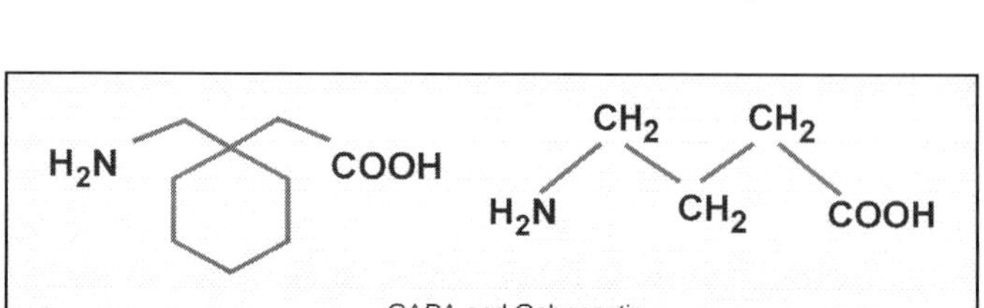

Figure 15. GABA and Gabapentin.

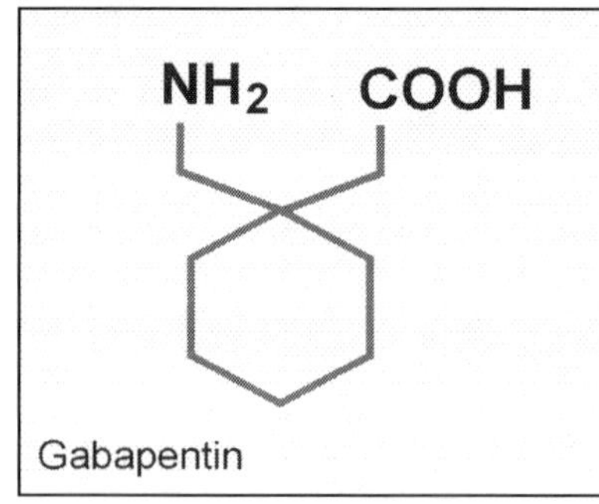

Figure 16. Gabapentin.

GBP increased REM sleep percentage, increased mean duration of REM periods, reduced number of awakenings, and reduced stage 1 sleep percentage.[72]

With increasing serum GBP levels, a study in healthy subjects showed an increase in SWS and in whole blood serotonin (5-HT; $p < 0.05$). Rao and coworkers have proposed a presumed mechanism of action of GBP on sleep and coworkers.[76] GBP inhibits monoaminergic neurotransmitters having anti-epileptogenic activity and modulates the release of 5-HT from blood platelets. To these authors, the increase in peripheral 5-HT points paradigmatically to an increase in the bioavailability of 5-HT that may account for the increase in SWS. Both drugs, LTG and GBP improve sleep stability while reducing seizures.

Synopsis of how GBP affects sleep (Table 3):

1. Increases SWS (S3/4);
2. Decrease in arousals and Stage 1;
3. Decreases PLMD;
4. Increases % REM.

Studies: Foldvary-Schaefer, 2002 (1); Rao et al, 1998 (1,2); Ehrenberg et al, 1998 (1,3); Placidi et al, 1997 (2,4) Placidi et al, 2000c.

Felbamate and Sleep of Epileptic Patients

Felbamate (FBM; also W554 and ADD 03055; $C_{11}H_{14}N_2O_4$; 2-Phenyl-1,3-propanediol dicarbamate) (Fig. 17) is used in refractory forms of epilepsy. When used the most relevant effect seems to be Insomnia. Felbamate reduced REM but did not change the total amount of sleep.

Synopsis of how Felbamate affects sleep: Insomnia (leading to cessation of treatment in some patients)

Studies: Foldvary-Schaefer, 2002; Ketter et al, 1996.

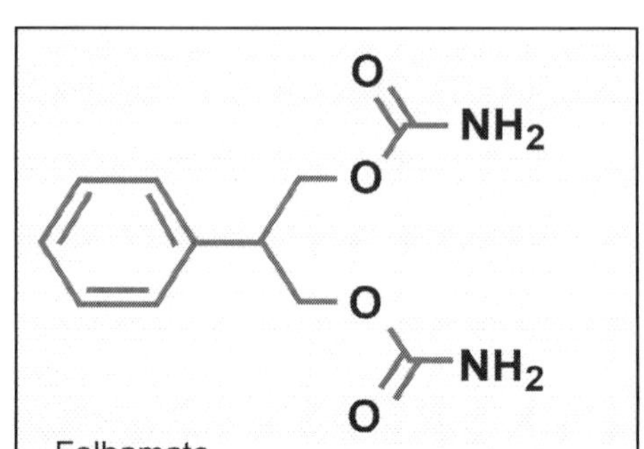

Figure 17. Felbamate.

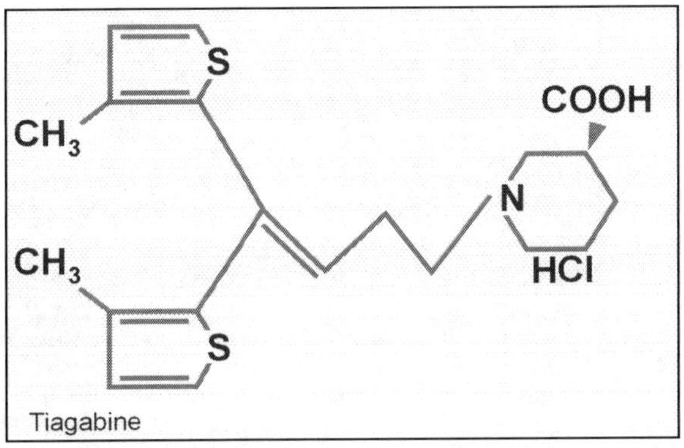

Figure 18. Tiagabine.

Tiagabine and Sleep of Epileptic Patients

There have been few studies of the effect of tiagabine hydrochloride (nipecotic acid combined to a lipophilic anchor) (Fig. 18) on sleep. Mathias and colleagues studied the acute effect of tiagabine on sleep following the administration of a single 5 mg bedtime dose to 10 healthy elderly subjects.[61] They reported a major increase in restorative sleep (SWS or S3 and S4), decreased awakenings, and an increase in SE. The effect closely resembled that of the GABA (A) agonist gaboxadol in young subjects.[60]

Synopsis of how Tiagabine affects sleep (Table 3):

1. Increased SE;
2. Decreased awakening;
3. Significant increase in SWS.

Studies: Mathias et al, 2001a,b.

Topiramate and Sleep of Epileptic Patients

Increase in somnolence with topiramate (TPM; $C_{12}H_{21}NO_8S$; 2,3:4,5-bis-*O*-(l-methylethylidene)-β-D-fructopyranose sulfamate; or 2,3:4,5-Di-*O*-isopropylidene-β-D-fructopyranose sulphamate) (Fig. 19) was reported by Marcotte and colleagues in psychiatric patients.[57] Recently Bonanni and coworkers analysed the daytime sleepiness reported in epileptic patients receiving TPM in monotherapy.[11] They conducted a study in which epileptic patients were compared at baseline and two months after a "slowly titrated" TPM monotherapy. The baseline comparisons to normal and also at 2-month intervals (to MSLT scores, Visual simples and choice reaction times, and subjective daytime sleepiness) were not significantly different.

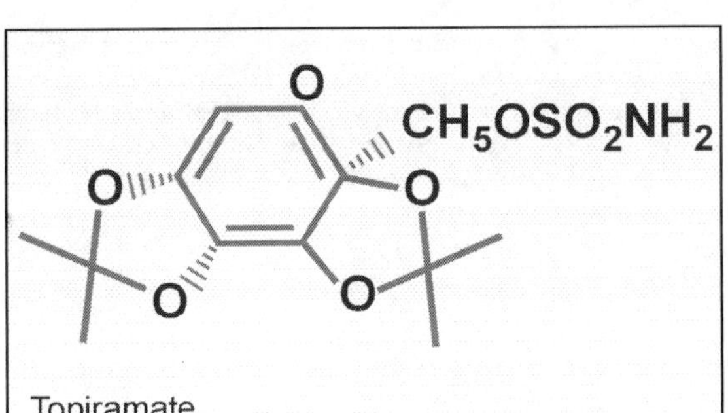

Figure 19. Topiramate.

Synopsis of how TPM affects sleep (Table 3):

1. Increase in somnolence and reduction of SOL in some studies.
2. No effect in other studies. See Marcotte D, 1998.

Levetiracetam and Sleep of Epileptic Patients

Levetiracetam ((S)-α-ethyl-2-oxo-pyrrolidine acetamide) is a new AED chemically derivative of pyrrolidine acetamide, is an ethyl analogue of piracetam. It is used as add-on therapy for partial epilepsy. Studies on the influence sleep are scarce.[6] analysed the objective and subjective effects of levetiracetam on sleep in healthy volunteers and in patients with partial epilepsy on stable CBZ monotherapy on a similar to double blind cross over placebo-controlled study design. Levetiracetam produced an increase in the time spent in stage 2 sleep, accompanied by a decrease in the time spent in S4 in the patient group and increase in REMOL in the healthy volunteers. The subjective changes noted better sleep quality (less awakenings and more restful sleep), although both healthy people and epileptic patients felt less alert on awakening. It seems that levetiracetam does affect subjective sleep perception, but does not influence objective sleep measures of sleep continuity.[6]

Synopsis of how Levetiracetam affects sleep (Table 3):

1. Increase S2 and REMOL;
2. Decrease S4;
3. Doubtful changes on sleep continuity.

Source: Bell et al, 2002.

Oxcarbazepine and other New AED and Sleep of Epileptic Patients

Other AED have been tested, namely Oxcarbazepine (OXC; a keto derivative of CBZ) and Zonizamide. The efficacy and safety of these diverse AED on a long-term basis and use criteria only recently were discussed.[28,29] Although diverse reviews do not provide sufficient information about the influence of these AED on sleep recent reports on OXC confirm previous results.[48,49,67] Somnolence is a common side effect when OXC is used in polytheray and less frequent if used as monotherapy.[7,36,40,81] But again for all these new AEDs the scarce information provided on its influence on sleep only could be suspected because in such drugs somnolence and tiredness were reported.

Summary and Future Directions

Somnolence is common in the context of side effects of AED[43] although consistent studies also report somnolence as existing in epileptic patients not treated with medications,[68] and there is no established relationship between serum levels of AED and somnolence on diverse series.[83] Some associations are more prone to give somnolence: BZD, BZD + (PB, PHT, CBZ). All AEDs seemed to have effect on sleep macro or microstructure (see Tables 2 and 3).

Diverse consequences on sleep stability have been described at short and long term, and for some AEDs the findings of different studies are contradictory.

In order to help overcome diverse difficulties to study Sleep-AED and Epilepsy interactions the following guidelines are proposed:

- Compare for seizures and randomise if no seizures.
- Test patients without recent seizures
- Study different populations and types of epilepsies at baseline

- Study diverse drugs/ intake regime/ monotherapy if possible
- Use a sequence: baseline, initial TT, control at steady state
- Use objective standardized measures (MSLT), Multiple Vigilance Test (MVT), or well-tested subjective ones, such as the Epworth Sleepiness Scale (ESS)
- Use parameters and quantitative analysis: spindles - am plitude, frequency, duration of runs; delta - (amplitude, frequency, duration of bursts); K complexes; CAP (Cy clic Alternating Pattern).

References

1. Bast T, Volp A, Wolf C et al. Sulthiame study group. The influence of sulthiame on EEG in children with benign childhood epilepsy with centrotemporal spikes (BECTS). Epilepsia 2003; 44(2):215-20.
2. Bastien CH, LeBlanc M, Carrier J et al. Sleep EEG power spectra, Insomnia, and chronic use of Benzodiazepines. Sleep 2003; 26(3):313-7.
3. Bazil CW. Epilepsy and sleep disturbance. Epilepsy & Behavior 2003; 4:S39-S45.
4. Bazil CW, Legros B, Kenny E. Sleep structure in patients with psychogenic nonepileptic seizures. Epilepsy & Behavior 2003; 4:395-98.
5. Bazil CW, Walczak TS. Effects of sleep and sleep stage on epileptic and nonepileptic seizures. Epilepsia 1997; 38(1):56-62.
6. Bell C, Vanderlinden H, Hiersemenzel R et al. The effects of levetiracetam on objective and subjective sleep parameters in healthy volunteers and patients with partial epilepsy. J Sleep Res 2002; 11(3):255-63.
7. Beydoun A, Sachdeo RC, Kutluay E et al. Sustained efficacy and long-term safety of oxcarbazepine: One-year open-label extension of a study in refractory partial epilepsy. Epilepsia 2003; 44(9):1160-5.
8. Beran RG, Holland GJ, Yan KY. The use of CPAP in patients with refractory epilepsy. Seizure 1997; 6(4):323-25.
9. Binnie CD, Aarts JHP, Houtkooper MA et al. Temporal characteristics of seizures and epileptiform discharges. Electroenceph Clin Neurophysiol 1984; 58:498-505.
10. Bonanni E, Massetani R, Galli R et al. A quantitative study of daytime sleepiness induced by carbamazepine and add-on vigabatrin in epileptic patients. Acta Neurol Scand 1997; 95:193-96.
11. Bonanni E, Galli R, Maestri M et al. Daytime sleepiness in epilepsy patients receiving topiramate monotherapy. Epilepsia 2004; 45(4):333-7.
12. Crespel A, Baldy-Moulinier M, Coubes P. The relationship between sleep and epilepsy in frontal and temporal lobe epilepsies: Practical and physiopathologic considerations. Epilepsia 1998; 39:150-57.
13. Dadmehr N, Congbalay DR, Pakalnis A et al. Sleep and waking disturbances in epilepsy. Clin Electroencephalogr 1987; 18:136-41.
14. Deckers CL, Hekster YA, Keyser A et al. Monotherapy versus polytherapy for epilepsy: A multicenter double blind randomized study. Epilepsia 2001; 42:1387-94.
15. Declerck AC, Martens WL, Wauquier A. Evaluation of the effects of antiepileptic drugs on sleep-wakefulness patterns following 1 night total sleep deprivation in epileptic patients. Neuropsychobiology 1985; 13(4):201-5.
16. Declerck AC, Wauquier A. Influence of antiepileptic drugs on sleep patterns. Epilepsy Res Suppl 1991; 2:153-63.
17. Declerck AC. Interaction epilepsy, sleep and Anti-epileptics: A clinical neurophysiological study. Lisse: Swets and Zeitlinger BV, 1983.
18. Domizio S, Verrotti A, Ramenghi LA et al. Anti-epileptic therapy and behaviour disturbances in children. Childs Nerv Syst 1993; 9:272-74.
19. Drake Jr ME, Pakalnis A, Bogner JE et al. Outpatient sleep recording during antiepileptic drug monotherapy. Clin Electroencephalogr 1990; 21(3):170-3.
20. Drake Jr ME, Pakalnis A, Padamadan H et al. Sleep spindles in epilepsy. Clin Electroencephalogr 1991; 22(3):144-9.
21. Ehrenberg B. Importance of sleep restoration in comorbid disease: Effect of anticonvulsants. Neurology 2000a; 54(Suppl 1):S33-7.
22. Ehrenberg BL, Wagner AK, Corbeit K et al. Double-blind trial of gabapentin for periodic limb movement disorder or sleep: Preliminary results (Abstract). Neurology 1998; 50:A276.
23. Ehrenberg BL, Eisensehr I, Corbeit KE et al. Valproate for sleep consolidation in periodic limb movement disorder. J Clin Psychopharmacol 2000; 20(5):574-8.
24. Engler F, Maeder-Ingvar M, Roulet E et al. Treatment with sulthiame (Ospolot) in benign partial epilepsy of childhood and related syndromes: An open clinical and EEG study. Neuropediatrics 2003; 34(2):105-9.
25. Féré C. Les Èpilepsies et les Èpileptiques. Paris: Alcan, 1890, Cited in Janz D (ref. 44).
26. Foldvary N, Perry M, Lee J et al. The effects of lamotrigine on sleep in patients with epilepsy. Epilepsia 2001; 42(12):1569-73.
27. Foldvary-Schaefer N. Sleep complaints and epilepsy: The role of seizures, antiepileptic drugs and sleep disorders. J Clin Neurophysiol 2002; 19:514-521.
28. French JA, Kanner AM, Bautista J et al. Therapeutics and technology assessment subcommittee of the American academy of neurology; quality standards subcommittee of the American Academy of Neurology; American epilepsy society. Efficacy and tolerability of the new antiepileptic drugs I: Treatment of new onset epilepsy: Report of the Therapeutics and Technology assessment subcommittee and quality standards subcommittee of the American Academy of Neurology and the American epilepsy society. Neurology 2004a; 62(8):1252-60.
29. French JA, Kanner AM, Bautista J et al. Therapeutics and technology assessment subcommittee of the American Academy of Neurology; quality standards subcommittee of the American Academy of Neurology; American epilepsy society. Efficacy and tolerability of the new antiepileptic drugs II: Treatment of refractory epilepsy: Report of the therapeutics and technology assessment subcommittee and quality standards subcommittee of the American Academy of Neurology and the American epilepsy society. Neurology 2004b; 62(8):1261-73.
30. Frucht MM, Quigg M, Schwaner C et al. Distribution of seizure precipitants among epilepsy syndromes. Epilepsia 2000; 41(12):1534-9.
31. Gann H, Riemann D, Hohagen F et al. The influence of carbamazepine on sleep-EEG and the clonidine test in healthy subjects: Results of a preliminary study. Biol Psychiatry 1994; 35(11):893-6.
32. Geurkink EA, Sheth R, Gidal BE et al. Effects of anticonvulsant medication on EEG sleep architecture. Epilepsy & Behavior 2000; 1:378-83.
33. Gigli GL, Gotman J, Thomas ST. Sleep alterations after acute administration of carbamazepine in cats. Epilepsia 1988; 29:748-52.
34. Gigli GL, Gotman J. Effects of seizures, kindling, and carbamazepine on sleep organization in cats. Epilepsia 1992; 33:14-22.
35. Gigli GL, Placidi F, Diomedi M et al. Nocturnal sleep and daytime somnolence in untreated patients with temporal lobe epilepsy: Changes after treatment with controlled-release carbamazepine. Epilepsia 1997; 38:696-701.
36. Glauser TA, Nigro M, Sachdeo R et al. Adjunctive therapy with oxcarbazepine in children with partial seizures. The oxcarbazepine pediatric study group. Neurology 2000; 27;54(12):2237-44.
37. In: Gowers WR, ed. Epilepsy and other chronic convulsive diseases. New York: William Wood Comp., 1895, Nijmegen: Reprint from Arts & Boeve, 1994.
38. Gross-Selbeck G. Treatment of "benign" partial epilepsies of childhood, including atypical forms. Neuropediatrics 1995; 26(1):45-50.
39. Harding GFA, Alford CA, Powell TE. The effect of sodium valproate on sleep, reaction times, and visual evoked potential in normal subjects. Epilepsia 1985; 26:597-601.
40. Herranz JL, Argumosa A. Caracteristicas e indicaciones de la oxcarbacepina. Rev Neurol 2002; 35(Suppl 1):S101-9.

41. Hirtz DG, Chen TC, Nelson KB et al. Does phenobarbital used for febrile seizures cause sleep disturbances? Pediatr Neurol 1993; 9(2):94-100.
42. Hoppen T, Sandrieser T, Rister M. Successful treatment of pharmacoresistent continuous spike wave activity during slow sleep with levetiracetam. Eur J Pediatr 2003; 162(1):59-61.
43. Ieiri I, Hirata K, Higuchi S et al. Pharmacoepidemiological study on adverse reactions of antiepileptic drugs. Tokyo: Chem Pharm Bull, 1992:40(5):1280-8.
44. Janz D. The grand mal epilepsies and the sleep-waking cycle. Epilepsia 1962; 3:69-109.
45. Janz D, Durner M. Juvenile myoclonic epilepsy. In: J Jr Engel, Pedley TA, eds. Epilepsy. A Comprehensive Textbook. New York, Philadelphia: Lippincott-Raven, 1997: 3:2389-2400.
46. Kellaway P, Frost Jr JD, Crawley JW. Time modulation of spike-and-wave activity in generalized epilepsy. Ann Neurol 1980; 8:491-500.
47. Ketter TA, Malow BA, Flamini R et al. Felbamate monotherapy has stimulant-like effects in patients with epilepsy. Epilepsy Res 1996; 23(2):129-37.
48. LaRoche SM, Helmers SL. The new antiepileptic drugs: Scientific review. JAMA 2004a; 291(5):605-14.
49. LaRoche SM, Helmers SL. The new antiepileptic drugs: Clinical applications. JAMA 2004b; 291(5):615-20.
50. Legros B, Bazil CW. Effects of antiepileptic drugs on sleep architecture: A pilot study. Sleep Medicine 2003; 4:51-55.
51. Leniger T, Wiemann M, Bingmann D et al. Carbonic anhydrase inhibitor sulthiame reduces intracellular pH and epileptiform activity of hippocampal CA3 neurons. Epilepsia 2002; 43(5):469-74.
52. Madeja M, Wolf C, Speckmann EJ. Reduction of voltage-operated sodium currents by the anticonvulsant drug sulthiame. Brain Res 2001; 900(1):88-94.
53. Malow BA, Fromes GA, Aldrich MS. Usefulness of polysomnography in epilepsy patients. Neurology 1997; 48(5):1389-94.
54. Manni R, Galimberti CA, Zucca C et al. Sleep patterns in patients with late onset partial epilepsy receiving chronic carbamazepine (CBZ) therapy. Epilepsy Res 1990; 7:72-6.
55. Manni R, Ratti MT, Galimberti CA et al. Daytime sleepiness in epileptic patients on long-term monotherapy: MSLT, clinical and psychometric assessment. Neurophysiol Clin 1993a; 23(1):71-76.
56. Manni R, Ratti MT, Perucca E et al. A multiparametric investigation of daytime sleepiness and psychomotor functions in epileptic patients treated with phenobarbital and sodium valproate: A comparative controlled study. Electroencephalogr Clin Neurophysiol 1993b; 86(5):322-28.
57. Marcotte D. Use of topiramate, a new anti-epileptic as a mood stabilizer. J Affect Disord 1998; 50(2-3):245-51.
58. Martins da Silva A, Aarts JHP, Binnie CD et al. The circadian distribution of interictal epileptiform EEG activity. Electroenceph Clin Neurophysiol 1984; 58:1-13.
59. Martins da Silva A, Canijo M, Cunha I et al. Biorhythmic modulation of spike and wave paroxysms by sleep naps. In: Wolf P, Dam M, Janz D et al, eds. Advances in Epileptology. New York: Raven Press, 1987:16:687-91.
60. Mathias S, Steiger A, Lancel M. The GABA(A) agonist gaboxadol improves the quality of post-nap sleep. Psychopharmacology (Berl) 2001; 157(3):299-304.
61. Mathias S, Wetter TC, Steiger A et al. The GABA uptake inhibitor tiagabine promotes slow wave sleep in normal elderly subjects. Neurobiol Aging 2001; 22(2):247-53.
62. Maxion H, Jacobi P, Schneider E et al. Effect of the anticonvulsant drugs primidone and diphenylhydantoin on night sleep in heath volunteers and epileptic patients. In: Koella WP, Levin P, eds. Sleep 1974. Basel: Karger, 1975:510-13.
63. Mendez M, Radtke RA. Interactions between sleep and epilepsy. J Clin Neurophysiol 2001; 18(2):106-27.
64. Minecan D, Natarajan A, Marzec M et al. Relationship of epileptic seizures to sleep stage and sleep depth. Sleep 2002; 25:899-904.
65. Montplaisir J, Laverdiere M, Saint-Hilaire JM. Sleep and epilepsy. In: Gottman JR, Ives JR, Gloor P, eds. Long-term Monitoring in Epilepsy (EEG Suppl no37). Amsterdam: Elsevier, 1985:215-39.
66. Montplaisir J, Saint-Hilaire JM, Laverdiere M et al. Contribution of all-night polygraphic recording to the localization of primary epileptic foci. In: Canger R, Angleri F, Penry JK, eds. Advances in Epileptology. XI Epilepsy International Symposium. New York: Raven Press, 1980:135–138.
67. Nguyen DK, Spencer SS. Recent advances in the treatment of epilepsy. Arch Neurol 2003; 60(7):929-35.
68. Palm L, Anderson H, Elmqvist D et al. Daytime sleep tendency before and after discontinuation of antiepileptic drugs in preadolescent children with epilepsy. Epilepsia 1992a; 33(4):687-91.
69. Palm L, Persson E, Bjerre I et al. Sleep and wakefulness in preadolescent children with deficits in attention, motor control and perception. Acta Paediatr 1992b; 8-1(8):618-24.
70. Park SA, Lee BI, Lee SJ et al. Clinical courses of pure sleep epilepsies. Seizure 1998; 7:369-77.
71. Placidi F, Diomedi M, Scalise A et al. Effect of long-term treatment with gabapentin on nocturnal sleep in epilepsy. Epilepsia 1997; 38(Suppl 8):179-80.
72. Placidi F, Diomedi M, Scalise A et al. Effect of anticonvulsants on nocturnal sleep in epilepsy. Neurology 2000a; 54(Suppl 1):S25-32.
73. Placidi F, Marciani MG, Diomedi M et al. Effects of lamotrigine on nocturnal sleep, daytime somnolence and cognitive functions in focal epilepsy. Acta Neurol Scand 2000b; 102(2):81-6.
74. Placidi F, Mattia D, Romigi A et al. Gabapentin-induced modulation of interictal epileptiform activity related to different vigilance levels. Clin Neurophysiol 2000c; 111:1637-42.
75. Placidi F, Scalise A, Marciani MG et al. Effect of antiepileptic drugs on sleep. Clin Neurophysiol 2000d; 111(Suppl 2):S115-9.
76. Rao ML, Clarenbach P, Vahlensieck M et al. Gabapentin augments whole blood serotonin in healthy young men. J Neural Transm 1988; 73:129-34.
77. Raol YH, Meti BL. Effects of vigabatrin on sleep-wakefulness cycle in amygdala-kindled rats. Epilepsia 2000; 41(2):128-31.
78. Röder-Wanner UU, Wolf P. Effects of treatment with dipropylacetate and ethosuximide on sleep organization in epileptic patients. In: Dam M, Gram L, Penry JK, eds. Advances in Epileptology: XII Epilepsy International Symposium. New York: Raven Press, 1981:145-53.
79. Roder-Wanner UU, Noachtar S, Wolf P. Response of polygraphic sleep to phenytoin treatment for epilepsy. A longitudinal study of immediate, short- and long-term effects. Acta Neurol Scand 1987; 76(3):157-67.
80. Rowan AJ, Veldhuisen RJ, Nagelkerke NJ. Comparative evaluation of sleep deprivation and sedated sleep EEGs as diagnostic aids in epilepsy. Electroenceph clin Neurophysiol 1982; 54:357-364.
81. Schachter SC, Vazquez B, Fisher RS et al. Oxcarbazepine: Double-blind, randomized, placebo-control, monotherapy trial for partial seizures. Neurology 1999; 52(4):732-7.
82. Sadler M. Lamotrigine associated with insomnia. Epilepsia 1999; 40(3):322-5.
83. Salinsky MC, Oken BS, Binder LM. Assessment of drowsiness in epilepsy patients receiving chronic antiepileptic drug therapy. Epilepsia 1996; 37(2):181-187.
84. Schmidt D, Reininghaus R, Winkel R. Relevance of poor compliance for seizure control. Epilepsy Res Suppl 1988; 1:141-6.
85. Shouse MN, Martins da Silva A, Sammaritano M. Circadian rhythm, sleep and epilepsy. J Clin Neurophys 1996; 13:32-50.
86. Shouse MN, Martins da Silva A, Sammaritano M. Sleep. In: Engel J, Pedley TA, eds. Epilepsy: A comprehensive textbook. Philadelphia: Lippincott-Raven Publishers, 1997:2:1929-1942.
87. Sterman MB. Power spectral analysis of EEG characteristics during sleep in epileptics. Epilepsia 1981; 22(1):95-106.
88. Terzano MG, Parrino L, Anelli S et al. Effects of generalized interictal EEG discharges on sleep stability: Assessment of by means of cyclic alternating pattern. Epilepsia 1992; 33:317-26.
89. Touchon J, Baldy Moulinier M, Billiard M et al. Sleep organization and epilepsy. Epilepsy Res Suppl 1991; 2:73-81.
90. Touchon J, Baldy-Moulinier M, Billiard M et al. Organisation du sommeil dans l'épilepsie récente du lobe temporal avant et après traitement par carbamazepine. Rev Neurol (Paris) 1987; 143:462-7.
91. Touchon J, Baldy-Moulinier M, Billiard M et al. Sleep instability in temporal lobe epilepsy. In: Wolf P, Dam M, Janz D et al, eds. Advances in epileptology. New York: Raven Press, 1987:709-711.

92. Vaughn BV, D'Cruz OF, Beach R et al. Improvement of epileptic seizure control with treatment of obstructive sleep apnea. Seizure 1996; 5:73-78.
93. Wauquier A, Van den Broeck WA, Edmonds Jr HL. Sleep in epileptic beagles and antiepileptics. Funct Neurol 1986; 1(1):53-61.
94. Wolf P, Roder-Wanner UU, Brede M. Influence of therapeutic phenobarbital and phenytoin medication on the polygraphic sleep of patients with epilepsy. Epilepsia 1984; 25(4):467-75.
95. Wolf P, Röder-Wanner UU, Brede S et al. Influence of antiepileptic drugs on sleep. In: Martins da Silva A, Binnie CD, Meinardi H, eds. Biorhythms and Epilepsy. New York: Raven Press, 1985:137-53.
96. Wolf P. Influsence of antiepileptic drugs on sleep. In: Wolf P, Dam M, Janz D, Dreifus F, eds. Advances in Epileptology. New York: Raven Press, 1987:733-7.
97. Yakub BA, Wahoed G, Kabiraj MM. Nocturnal epilepsies in adults. Seizure 1997; 6:145-49.

CHAPTER 34

Herbal Medicines and Sleep

Marcello Spinella

Living in a complex and dangerous environment as humans have for most of our evolutionary history requires one to possess effective mechanisms of arousal, both consciousness and emotional, in order to meet the demands of the environment. An organism needs to be able to arouse behaviorally in order to deal with predators and other environmental threats. While many of the physical threats of survival have arguably minimized for most individuals with the advent of civilization, some physical threats (e.g., violent crime) do remain. Moreover, a host of psychological threats to our well-being exist, including unemployment, taxes, and divorce, which can be equally distressing. In many circumstances, mechanisms of arousal serve an adaptive purpose, but when acutely or chronically over-activated, they become maladaptive, manifesting as stress, anxiety, and insomnia.

To treat these conditions of over-activation, we have developed several effective pharmacological treatments. However, there are also in existence several plant-derived medications which have also long been used. This chapter will review the known herbal medications used to promote sleep, focusing especially on those for whom pharmacological and empirical evidence exists. As can be discerned from the evidence presented here, this field is far from complete. Further research is needed in all known sleep-promoting herbal medicines, and perhaps to identify other species that have potential as sleep-aids. This chapter focuses on the psychopharmacology of herbal medications relevant to sleep.

Valerian

History and Botany

Valerian (*Valeriana officinalis*) is native to Europe and Asia, but now grows in most parts of the world.[1] It grows 50 to 100 cm in height, with an erect stem with pinnate leaves and numerous small pink-white flowers, and has a distinct, unpleasant odor. Valerian has been used for at least 1000 years, and in 16th century Europe it was used as a treatment for epilepsy.[2] It's most commonly reported uses are to treat insomnia and anxiety.[3]

Chemical Constituents

Valerian's primary chemical constituents are monoterpenes, sesquiterpenes, and alkaloids. The monoterpenes include bornyl acetate, l-borneol, valenol, valeranone, and valmane. A subgroup of the monoterpenes are the valepotriates, which include valtrate and its derivatives, baldrinal, and homobaldrinal.[4] However, valepotriates rapidly decompose in the stored herb, so their content in valerian preparations is very low.[5] The sesquiterpenes present in valerian include isovaleric acid, valerenic acid, valerenal, valeranone, and valerenol. The alkaloids found in valerian include valeranine and actinidine. The content of valepotriates and sesquiterpenes varies across species of the *Valeriana* genus. For example, *Valeriana officinalis* has relatively high content of sesquiterpenes and low content of valepotriates, while *Valeriana edulis* has a high proportion of valepotriates and low content of sesquiterpenes.[6]

Pharmacodynamic Mechanisms

The mechanisms of action of valerian are not entirely certain, but they likely involve a facilitation of GABA transmission. Low microgram concentrations of an aqueous valerian extract inhibits uptake and stimulates release of GABA from synaptosomes.[7,8] GABA release by the valerian extract is independent of Ca^{2+} and membrane depolarization, and thus is not from vesicular stores. Further, the GABA release is dependent on Na^{+} concentrations, suggesting that valerian inhibits and reverses the GABA reuptake transporter. Thus, valerian extract appears to reverse GABA uptake and release it from the cytosolic pool. However, further investigation found that it is the actual GABA content of the extract can partially account for the reversal of GABA uptake in vitro.[9] Alternately, since GABA does not pass the blood-brain barrier well, it is possible that the glutamine content of valerian extract contributes to this effect since it both crosses the blood-brain barrier and serves as a precursor for GABA.

Some binding of valerian extract occurs at $GABA_A$ receptors. Valerian extracts inhibit neuronal firing in a rat brainstem preparation in a concentration-dependent manner comparable to that of the $GABA_A$ agonist muscimol and blocked by the $GABA_A$ antagonist bicuculline.[10] While Cavadas and colleagues[11] attributed this effect to the amino acid content of the extract, Yuan and colleagues[10] showed both total extract and valerenic acid to have this effect. Valerenic acid also inhibits breakdown of GABA.[12] Low concentrations of valerian extracts enhance benzodiazepine binding ($[^3H]$flunitrazepam).[13] A flavonoid, 6-methylapigenin, has been isolated from *Valeriana wallichii* that may confer a benzodiazepine mechanism to this species.[14] The valpotriates may have central depressant effects through in vivo conversion to homobaldrinal.[15]

Collectively, valerian extracts show a variety of GABAergic mechanisms. The predominant mechanism has yet to be determined, but an additive or synergistic interaction among the several mechanisms is possible. Future research on valerian extracts needs to address any compositional differences between aqueous and nonaqueous extracts, and to determine the responsible active

Sleep and Sleep Disorders: A Neuropsychopharmacological Approach, edited by Malcolm Lader, Daniel P. Cardinali and S.R. Pandi-Perumal.

constituents. For some time it was thought to be the valepotriates, but this was later discredited due to their unstable nature.

Effects

Muscle Relaxant

Valepotriates (isovaltrate and valtrate) and valerenone were investigated show muscle relaxant effects in the guinea pig ileum, probably by a direct action on the muscle.[16] However, GABAergic drugs have muscle relaxant effects in the spinal cord. This is a likely mechanism for valerian but has yet to be explicitly investigated.

Behavioral Effects

Several behavioral effects have been reported in animals. These include suppression of the orienting response in an open-field paradigm, decreasing spontaneous and caffeine-induced motor activity, and potentiation of the behavioral actions of barbiturates.[17] Oral and intravenous doses of valtrate and acetoxyvaltrate reduce locomotor activity.[18] Valerian extracts show sedative effects in animals that are dose-dependent manner.[19,20] These effects are moderate compared to diazepam and the chlorpromazine.[20] However, valepotriates reverse the anxiogenic effects of diazepam withdrawal in rats in the elevated plus maze.

Sleep Electrophysiology

A few studies have been published which examine the effects of valerian on human sleep electrophysiology, which have produced variable results. An early study showed that a combined preparation of valerian and hops in sleep-disturbed subjects increased the amount of slow wave and REM sleep.[21] Another study using an aqueous extract (400 mg) of valerian did not show significant EEG effects, but suggested a relation between the EEG effects and subjective effects, i.e., shortened sleep latency and increased latency to first waking.[22,23] Another study using aqueous extracts showed no effects on sleep stages or EEG spectra.[24]

A controlled study of valerian in sleep was published which utilized double-blind placebo controls, and randomization with two doses of valerian (60 and 120 mg).[25] Valerian increased sleep stages 1, 2, and 3, and reduced stage 4 and REM. Dose-dependent effects were noted, where the 120 mg dose produced greater sedative effects. Peak effects occurred 2-3 hours after administration. Mood ratings did not differ, positively or negatively, between the experimental and control conditions.

The effects of valerian in poor sleepers was studied comparing it to placebo controls.[26] Valerian showed an increase in slow-wave and a decrease in stage 1 sleep. K-complex density was increased, but REM was unaltered, and no effects were reported on subjective sleep quality. The most recent polysomnographic study of Valerian used hydroalcoholic extract of *Valeriana officinalis* and *Valeriana edulis*.[27] *V. edulis* reduced the number of awaking episodes, decreased stages 1 and 2, and increased delta sleep and REM.

The physiological effects of valerian across animals and humans are consistent with its sedative effects. The human electrophysiological findings are somewhat incongruent, but the methodology must be taken into account. Earlier studies used less rigorous controls, and two used aqueous extract instead of the entire herb, likely excluding much of the hydrophobic constituents. Another difference between the two latter studies which may account for differing results is the different populations used (normal versus sleep-disturbed).[25,26] Further work using larger samples and more careful methodological controls is warranted to evaluate these inconsistencies. However, the EEG effects of valerian are consistent with a sleep-promoting effect.

Sleep Quality

Valerian improves subjective ratings of sleep particularly when taken nightly over one- to two-week periods.[28] Valerian was studied with a randomized, placebo-controlled, and double-blind crossover study using a commercially available valerian preparation.[29] The preparation used contained primarily sesquiterpenes, and very low amounts of valepotriates. The subjects were 27 adults who were seen in a medical clinic for sleep difficulties. Those receiving valerian experienced improvements were in sleep quality (89%), with a proportion (44%) rating highest quality sleep. No differences were seen between those who received either valerian or placebo first before the crossover. An absence of adverse side effects, including nightmares, was reported. This study unfortunately used a preparation which included two other herbs, which limit the conclusions that can be exclusively drawn with valerian. Another controlled study was performed that had a much larger ($N = 128$) sample.[30] Improvements were seen in sleep quality and decreases in sleep latency. Further, sleep quality was most improved in poor sleepers and tobacco smokers.

Six weeks of valerian extract (600 mg daily) reduced ratings of stress severity, producing vivid dreams in 16% of subjects in a methodologically controlled study.[31] Valerian also improved sleep in insomniacs withdrawn from benzodiazepine medication.[32] It created a decrease in wake time after sleep onset in when compared to placebo, improving this variable to the level seen in normal controls. Valerian improved sleep in a small sample of children with intellectual deficits in a randomized, placebo controlled study.[33] The greatest effectiveness was observed in children who had the most hyperactivity.

Cognitive Effects

There are few formal studies of the cognitive effects of valerian. Whereas the benzodiazepine flunitrazepam produces significant impairment the morning after administration, valerian (alone or in combination with hops) does not.[34] However, 1 to 2 hours after administration valerian produced a slight but statistically significant decrease in vigilance and processing of complex information. Although mild, this effect may contraindicate valerian use in situations where peak cognitive performance is required (e.g., driving).

Toxicity

No health hazards have yet been reported with normal use of valerian, and it has been approved by the German Commission E as a treatment for anxiety and sleep.[35] Concern has been raised over valepotriates potentially affecting the liver, based on their chemical structure. However, *Valeriana officinalis* has a low valepotriate content, they decompose rapidly in the stored herb, and they are not well-absorbed. Liver toxicity in animals or humans has never been demonstrated.[5] Thus, valeportriates, in practice, are not likely to cause toxicity.

High doses of valerian produce headache, vomiting, stupor, dizziness, and cardiac dysfunction.[3] A case was reported of a single overdose (approximately 20 times the recommended therapeutic dose), reportedly producing mild symptoms that resolved in 24 hours. Given its central depressant effects and putative GABAergic mechanisms, concurrent use with other central depressants, including ethanol, should be avoided.

Kava

History and Botany

Kava (*Piper methysticum*) is a plant native to the South Pacific islands.[36] It is also known by the names kava, awa, waka, lawena, or yaquona. It was used in those cultures for ceremonial and recreational purposes. Kava may also be given reciprocally as a gift in the resolution of a social conflict, or partaken as an after-work drink for relaxation. It is known traditionally for producing a relaxed but alert mental state.[36] The large rhizome is the part of the plant used medicinally, which is pounded, chewed, or grated and then drunk in a cold-water infusion. In the West, it is typically sold in dried, ground encapsulated form, extract, or prepared as a tea.

Chemical Constituents

The pharmacologically active chemicals most studied from the kava plant are collectively called kavalactones. The principal kavalactones are: kavain, dihydrokavain, yangonin, dimethoxyyangonin, methysticin, and dihydromethysticin.[36] Kavalactones, given in combination, produce create pharmacokinetic and pharmacodynamic synergy.[37]

Pharmacodynamic Mechanisms

Kava facilitates GABA transmission. Low micromolar concentrations enhance the binding of ligands to the $GABA_A$ receptor, potentiating binding of GABA and enhancing Cl^- influx.[38,39] Kavalactones do not alter the binding of flunitrazepam, so their effect on $GABA_A$ is not through the benzodiazepine receptor.[40] Kavalactone displacement of the $GABA_A$ agonist muscimol is highest in the hippocampus and amygdala.[39] Combination with pentobarbital caused a synergistic effect on binding.

A second mechanism by which kava produces CNS depression is by inhibition of voltage-gated ion channels. Kavain, dihydrokavain and dihydromethysticin act as noncompetitive inhibitors of the Na^+ channel.[41] Kavain blocks the [^{3}H]batrachotoxin binding, but not [^{3}H]saxitoxin binding.[42] This suggests an action at receptor site 2 of the Na^+ channel, a site common to local anesthetic drugs. Kavain inhibits L-type Ca^{2+} channels as well, and the subsequent release of glutamate.[42-44] These effects on voltage-gated ion channels are obtained in the micromolar range, and consistent with concentrations reached in the brain by peripheral administration.[45]

A third mechanism of kavalactones relevant to sleep involves monoamine effects. Kavalactones block the reuptake of norepinephrine.[46] This occurrs at concentrations of less than 10 μM, but this study only used a single kavalactone in vitro, ruling out additive and synergistic actions of combined administration. Kavalactones also inhibit MAO_B in the micromolar range.[47] Kavalactones reversibly inhibited platelet MAO_B in the low micromolar range. A dose of 120-240 mg of kavalactones for 3-4 weeks caused a 26-34% reduction of MAO_B in platelets.

Neither an acute dose of dihydromethysticin (100 mg/kg) or chronic doses of kavain altered levels of dopamine, serotonin, or their metabolites in the striatum and cortex of rats.[48] Kavalactones have complex and mixed effects on monoamine levels in the nucleus accumbens, depending both on which kavalactone is used and at what dose.[49] Short term increases in dopamine and decreases in serotonin were reported.

Effects

Neuromuscular

Kava has direct neuromuscular relaxing effects through a mechanism similar to those of local anesthetics: by a direct blocking effect on ion channels of the muscle. Given kava's central GABAergic effects, it is likely to cause muscle relaxant effects in the spinal cord.

Electrophysiology

The effects of kavain on human electrophysiology were examined in a double-blind, placebo-controlled study.[50] Dose-dependent increases were seen in delta, theta, and alpha 1 power, and decreases occurred in alpha 2 and beta power. These changes were suggestive of a sedative effect of kavain, and were maximal in frontal areas. An initial activating effect was seen at the lowest dose (200 mg) but not at the largest dose (600 mg). Event-related potentials (ERPs) were used to study the cognitive effects of kava, as compared to oxazepam during verbal memory and attention tasks.[51,52] While oxazepam impaired performance, kava actually improved performance and was associated with an enhanced task-associated potentials in frontal, parietal, and occipital areas.

Cognitive

A study of the cognitive effects of kava on psychometric tests was undertaken by Foo and Lemon.[53] Kava only produced minor effects on the digit symbol task, but no other tests of attention, reaction time, visuomotor tracking, and short-term memory. Kava and ethanol combined produced potentiated effects on subjective and objective cognitive measures. Very large doses of kavalactones (205 g, or 150 times the clinical dose) produce more pronounced cognitive and saccade impairments.[54]

Anxiolytic

Several controlled studies have been performed to assess the anti-anxiety effects of kava. The studies employed standardized doses of kavalactones ranging from 60 to 240 mg per day, and were 4 to 24 weeks in duration.[55] A meta-analysis found kava to be superior to placebo across all controlled studies. Kavain was found to be equivalent in efficacy to the benzodiazepine oxazepam.

Sleep

Despite it's potential for improving sleep, only two studies to date have formally investigated this. One study found six-week, daily treatment with kava reduced stress-induced insomnia.[56] The other study similarly found kava extract to improve anxiety-related sleep problems, ameliorating quality and restorative aspects of sleep.[57]

Toxicity

Kava has been approved by the German Commission E for treatment of anxiety and insomnia. In clinical studies of kava for anxiety, adverse effects were uncommon and did not differ across placebo and kava groups. Given kava's central depressant effects, it should not be taken with other similar drugs, including benzodiazepines, barbiturates, ethanol, or anti-seizure medications. There is one clinical report of combined administration of kava and the benzodiazepine alprazolam, causing lethargy and disorientation for several hours.[58] Concern has been raised recently over liver toxicity with kava. Several cases have been reported of liver toxicity associated with kava use, some leading to liver failure or death.[59] There are many variables to be considered in this matter, including concurrent use of other medications which could increase hepatotoxicity, amount and duration of use, the type of kava preparation used, and prior liver illness. Some short term studies report a lack of effect on liver enzymes.[60,61] Long-term aboriginal Australian users of kava showed mild elevation of certain liver enzymes which were reversible.[62] There is evidence that the extraction process using acetone is partly responsible for the hepatotoxic effect.[63] While this matter remains unresolved at present, prudent use of kava would be short-term, not in combination with other drugs affecting the liver, and in low to moderate doses. More traditional preparations might afford greater safety than many currently-used chemical extraction processes. Using combinations of herbs may also best be avoided until the their hepatic effects are known.

A scaly skin eruption called kava dermopathy has been reported to occur from kava use.[64] It is reversible, and appears to only occur with heavy chronic use, but it has not been reported in the west.

A few cases were reported of dyskinesias presenting with kava use in individuals with Parkinson's disease.[65] These consisted of dystonia, tonic head rotation, twisting of the trunk, oculogyric crises, and increased duration of "off" periods in a Parkinsonian patient. These symptoms subsided with discontinuation of kava and treatment with a cholinergic muscarinic antagonist. The authors suggest that this represents a dopamine antagonist action of kava, and raise caution about their use in the elderly.

Passionflower

History and Botany

A few members of the Passionflower family (passifloracea) have sedative and anxiolytic effects. The one most studied is *Passiflora incarnata*, although some work has been done on *P. coerulea* and *P. edulis*. The whole plant or aerial parts are used for medicinal effects. It is native to the mid- to southeastern United States. Native Americans used passionflower prepared as a tea for sedative and anxiolytic effects.[1]

Chemical Constituents

There are 3 categories of constituents in passionflower: flavonoids, maltol, and indole alkaloids. The greatest accumulation of flavonoids occurs in the leaves.[66] The indole alkaloids are small amounts (up to 0.01 percent), including harman, harmine, harmaline, and harmalol.[5]

Pharmacodynamic Mechanisms

The most studied constituent of passionflower is the flavonoid, chrysin. Chrysin was isolated from *P. coerulea*, a species closely related to *P. incarnata*. It binds to benzodiazepine receptors with micromolar affinity and competes for binding with the benzodiazepine fluniztrazepam.[67] Behavioral assays suggest that chrysin acts as a partial agonist at central benzodiazepine receptors.[68] Anxiolytic effects of chrysin are blocked by flumazenil, arguing for a benzodiazepine mechanism.[69] However, chrysin antagonized the electrophysiological effects of GABA at $GABA_A$ receptors.[70] These conflicting effects need to be reconciled with further, more careful research. A benzodiazepine partial agonist mechanism of chrysin is still possible, although other mechanisms may exist. The remaining constituents have not been well characterized for their neuropharmacolgocal action or are present in small quantities and not presumed to contribute to the psychotropic effects.

Effects

Anxiolytic

Peripherally-injected chrysin exhibits anxiolytic effects in mice coma (1 mg/kg i.p.) in the elevated-plus maze similar to diazepam, reversed by pretreatment with a benzodiazepine antagonist.[71] The anxiolytic effect was not likely due to sedation since there is no concurrent reduction in motor activity at the doses used. Unlike diazepam, chrysin does not produce muscle relaxation at higher doses. The sedative and anxiolytic effects of passionflower were examined in two other animal behavioral assays (staircase test, light/dark box choice test). Both anxiolytic and sedative effects occur, as well as potentiation of pentobarbital sedation, at 400 mg/kg of hydroalcoholic extract in mice.[71] The anxiolytic and sedative activity occur in a dose-dependent continuum.[19] *Passiflora edulis* has sedative effects as well.[72] Chronic administration of passionflower flavonoids produced anxiolytic effects and prevented the incurrence of diazepam-dependence.[73] A small, controlled trial showed passionflower extract improved anxiety in Generalized Anxiety Disorder better than placebo.[74] Passionflower performed equivalently to oxazėpam, but appeared to cause less cognitive impairment. Despite the potential of passionflower as a treatment for insomnia, particularly anxiety-related insomnia, no clinical trials have been published to date.

Cognitive

In rats, chrysin does not have any amnestic effects on either acquisition or retention in three tests of memory (inhibitory avoidance, shuttle avoidance, and habituation to an open field tests), even at higher doses than required to produce anxiolytic effects.[75] The cognitive effects of passionflower have not been examined in humans.

Anti-Seizure

Central, but not peripheral, injections of chrysin significantly reduced chemically-induced (pentylenetetrazol) seizures in mice.[76] This effect was abolished by prior injection of a benzodiazepine antagonist. This has not been tested in humans and should not be substituted for conventional treatments for seizures.

Toxicity

There have been no formal studies of the toxicity of passionflower, but adverse effects have not been reported. There is one report of a case of inflammatory vasculitis associated with a preparation of passionflower.[77] Like other herbs in this category, its putative benzodiazepine action contra indicates its combined use with other central depressants.

Chamomile

History and Botany

Chamomile refers to two similar species of plants: German chamomile (*Matricaria recutita*) and Roman Chamomile (*Chamaemelum nobile*), both of which are members of the *Asteracea* family. Chamomile has been used throughout history, including ancient Egyptian, Roman, and Greek cultures. The two species look very much like daisies, with white petals and a yellow central disc. They are both native to Europe, Africa and Asia, and have been naturalized in North America. The flowering tops are dried and often used to make a tea.

Chemical Constituents

Chamomile contains the terpenoids, (-)-alpha-bisabolol, (-)-alpha-bisabololoxides A and B, and a gauianolide lactone called matricin. Also contained are the flavonoids apigenin and apigenin-7-glucoside.[78]

Pharmacodynamic Mechanisms and Effects

Apigenin binds with micromolar affinity at benzodiazepine receptors (4 μM), but has no effect at muscarinic, α1 adrenergic receptors, or the GABA binding site of the $GABA_A$ channel.[79] However, others report an antagonistic effect at GABA channels, which are insensitive to flumazenil.[70] While another study confirmed a sedative effect of apigenin in mice, it also failed to reverse the effect with flumazenil.[69]

Apigenin showed anxiolytic effects in mice, but no anti-seizure effects.[79] At doses ten times greater than required for anxiolytic effects, apigenin showed mild sedative effects. The neuropharmacological mechanisms of chamomile have yet to be elucidated. Since doubt has been cast on a benzodiazepine mechanism of apigenin, other mechanisms by apigenin and other constituents, including flavonoids, must be evaluated.

Regardless of it's mechanisms, controlled trials of chamomile preparations have not yet been reported in humans.

Toxicity

Chamomile appears very low in toxicity. It has been listed as Generally Regarded as Safe (GRAS) by the Food and Drug Administration. Adverse reactions may include allergic reactions to the pollen in the flowers, which are uncommon.

Other Sedative Herbs

The herbal medicines in this section have far less empirical research to support their use in improving aspects of sleep. However, they is some research for each suggesting that further research is warranted.

Catnip

Catnip (*Nepeta cataria*) has a long recorded history of use and is noted for sedative properties in humans. The active agent for this effect is uncertain, but catnip has in it several terpenes, including nepetalactone. Humans have reported sedative effects of catnip, and one reported accidental ingestion by a young child reportedly produced sedative effects.[80] Controlled trials of its effects have not been reported.

Hops

Hop (*Humulus lupulus*) is a flowering vine that grows in Europe, Western Asia, and North America.[1] It has reputed anxiolytic and sedative effects, often obtained by placing the female flowers in a pillow for their fragrance. It has not been well studied, but the responsible agent is believed to be 2-methyl-3-butene-2-ol, since it produces sedation when injected intraperitoneally in mice.[81] In humans, one study found no CNS depressant effects when administered orally.[82] However, effects via inhalation have not been studied.

Skullcap

Skullcap (*Scutellaria laterifolia*) is an herb that has been used in Chinese and Western medicine for sedative and anti-seizure effects.[83] Its pharmacological and behavioral effects have not been established in animals or humans. It does contain the flavonoids baicalin and baicalein, and the amino acid glutamine, so GABAergic mechanisms are possible.[84] One methodologically-controlled study in humans showed anxiolytic effects.[85]

Lemon Balm

Lemon balm (*Melissa officianalis*) is flowering perennial plant, and a member of the mint family.[1] In mice, lemon balm has sedative effects and analgesic activity.[86] It also increased the sedative activity of a barbiturate (pentobarbital). A controlled study in humans showed improvement of memory performance in humans and increased ratings of calmness.[87] This study also demonstrated binding of lemon balm constituents to muscarinic and nicotinic receptors in human cerebral cortex tissue.

St. John's Wort

St. John's wort (*Hypericum perforatum*) has been studied extensively for the treatment of depression. Evidence supports it's use in cases of mild to moderate depression.[88,89] However, it also has potential for the treatment of sleep disorders.

Several active constituents have been identified in St. John's wort.[90] These have a variety of mechanisms including monoamine mechanisms (blocking reuptake, weak inhibition of MAO and COMT), as well as effects on adenosine, GABA, and glutamate receptors. Chronic use leads to adaptation of monoamine receptors.

Controlled studies of St. John's wort show that it increases slow wave sleep and increases REM latency.[91,92] Studies on the qualitative effects on sleep would be needed to demonstrate an improvement in sleep as well. Serious side effects from St. John's wort monotherapy have not been reported, but it's potential for interaction with antidepressants and drugs metabolized by the cytochrome P450 3A4 enzyme have been reported.[93,94]

Conclusions

There are several herbs that have central depressant effects and have been used for anxiolytic and sedative effects historically. This has been supported by neuropharmacological, animal, and human studies. Kava, valerian, and passionflower have been the best supported by research in this regard, although others show potential. Much work remains to be done in further testing the safety and efficacy of these drugs. However, with further study some of these medicines may prove favorable in the treatment of sleep-related conditions.

References

1. Kowalchik C, Hylton WH. In: Emmaus PA, ed. Rodale's Illustrated Encyclopedia of Herbs Rodale Press, 1987.
2. Temkin O. The Falling Sickness: A History of Epilepsy from the Greeks to the Beginnings of Modern Neurology. Baltimore: Johns Hopkins University Press, 1971.
3. Gruenwald J, Brendler T, Jaenicke C. PDR for Herbal Medicines. 1st ed. Montvale NJ: Medical Economics Company, 1998.
4. Houghton PJ. The biological activity of Valerian and related plants. J Ethnopharmacol 1988; 22(2):121-42.

5. Tyler V. Herbs of Choice. New York: Pharmaceutical Products Press, 1994.
6. Lindahl O, Lindwall L. Double blind study of a valerian preparation. Pharmacol Biochem Behav 1989; 32(4):1065-6.
7. Santos MS, Ferreira F, Cunha AP et al. Synaptosomal GABA release as influenced by valerian root extract—involvement of the GABA carrier. Arch Int Pharmacodyn Ther 1994; 327(2):220-31.
8. Santos MS, Ferreira F, Cunha AP et al. An aqueous extract of valerian influences the transport of GABA in synaptosomes. Planta Med 1994; 60(3):278-9.
9. Santos MS, Ferreira F, Faro C et al. The amount of GABA present in aqueous extracts of valerian is sufficient to account for [3H]GABA release in synaptosomes. Planta Med 1994; 60(5):475-6. No abstract available.
10. Yuan CS, Mehendale S, Xiao Y et al. The gamma-aminobutyric acidergic effects of valerian and valerenic acid on rat brainstem neuronal activity. Anesth Analg Feb 2004; 98(2):353-8
11. Cavadas C, Araujo I, Cotrim MD et al. In vitro study on the interaction of Valeriana officinalis L. Extracts and their amino acids on GABAA receptor in rat brain. Arzneimittelforschung 1995; 45(7):753-5.
12. Riedel et al. 1982.
13. Ortiz JG, Nieves-Natal J, Chavez P. Effects of Valeriana officinalis extracts on [3H]flunitrazepam binding, synaptosomal [3H]GABA uptake, and hippocampal [3H]GABA release. Neurochem Res 1999; 24(11):1373-8.
14. Wasowski C, Marder M, Viola H et al. Isolation and identification of 6-methylapigenin, a competitive ligand for the brain GABA(A) receptors, from Valeriana wallichii. Planta Med 2002; 68(10):934-6.
15. Houghton PJ. The scientific basis for the reputed activity of Valerian. J Pharm Pharmacol 1999; 51(5):505-12.
16. Hazelhoff B, Malingre TM, Meijer DK. Antispasmodic effects of valeriana compounds: An in-vivo and in-vitro study on the guinea-pig ileum. Arch Int Pharmacodyn Ther 1982; 257(2):274-87.
17. Dunaev VV, Trzhetsinskii SD, Tishkin VS et al. Biological activity of the sum of the valepotriates isolated from Valeriana alliariifolia. Farmakol Toksikol 1987; 50(6):33-7.
18. Holzl J, Fink C. Effect of valeprotriate on spontaneous motor activity in mice. Arzneimittelforschung 1984; 34(1):44-7.
19. Della Logia R, Tubaro A, Redaelli C. Evaluation of he activity on the mouse CNS of several plant extracts and a combination of them. Riv Neurool 1981; 51(5):297-310.
20. Leuschner J, Müller J, Rudmann M. Characterisation of the central nervous depressant activity of a commercially available valerian root extract. Arzneimittelforschung 1993; 43(6):638-41.
21. Muller-Limmroth W, Ehrenstein W. Experimental studies of the effects of Seda-Kneipp on the sleep of sleep disturbed subjects; Implications for the treatment of different sleep disturbances (author's transl). Med Klin 1977; 72(25):1119-25.
22. Leathwood PD, Chauffard F. Quantifying the effects of mild sedatives. J Psychiatr Res 1982-3; 17(2):115-22.
23. Leathwood PD, Chauffard F. Aqueous extract of valerian reduces latency to fall asleep in man. Planta Med 1985; 2:144-8.
24. Balederer G, Borbely AA. Effect of valerian on human sleep. Psychopharmacology (Berl) 1985; 87(4):406-9.
25. Gessner B, Klasser M. Studies on the effect of harmonicum much on sleep using polygraphic EEG recordings. EEG EMG Z Elektroenzephalogr Verwandte Geb 1984; 15(1):45-51.
26. Schulz H, Stolz C, Muller J. The effect of valerian extract on sleep polygraphy in poor sleepers: A pilot study. Pharmacopsychiatry 1994; 27(4):147-51.
27. Herrera-Arellano A, Luna-Villegas G, Cuevas-Uriostegui ML et al. Polysomnographic evaluation of the hypnotic effect of Valeriana edulis standardized extract in patients suffering from insomnia. Planta Med Nov 2001; 67(8):695-9.
28. Hadley S, Petry JJ. Valerian. Am Fam Physician 2003; 67(8):1755-8.
29. Lindahl O, Lindwall L. Double blind study of a valerian preparation. Pharmacol Biochem Behav 1989; 32(4):1065-6.
30. Leathwood PD, Chauffard F, Heck E et al. Aqueous extract of valerian root (Valeriana officinalis L.) improves sleep quality in man. Pharmacol Biochem Behav 1982; 17(1):65-71.
31. Wheatley D. Kava and valerian in the treatment of stress-induced insomnia. Phytother Res 2001; 15(6):549-51.
32. Poyares DR, Guilleminault C, Ohayon MM et al. Can valerian improve the sleep of insomniacs after benzodiazepine withdrawal? Prog Neuropsychopharmacol Biol Psychiatry 2002; 26(3):539-45.
33. Francis AJ, Dempster RJ. Effect of valerian, Valeriana edulis, on sleep difficulties in children with intellectual deficits: Randomised trial. Phytomedicine 2002; 9(4):273-9.
34. Gerhard U, Hobi V, Kocher R et al. Acute sedative effect of a herbal relaxation tablet as compared to that of bromazepam. Schweiz Rundsch Med Prax 1991; 80(52):1481-6.
35. Hadley S, Petry JJ. Valerian. Am Fam Physician 2003; 67(8):1755-8.
36. Lebot V, Merlin M, Lindstrom L. Kava—the Pacific elixir: The definitive guide to its ethnobotany, history, and chemistry. Rochester, Vt: Healing Arts Press, 1997.
37. Spinella M. The importance of pharmacological synergy in psychoactive herbal medicines. Altern Med Rev 2002; 7(2):130-7.
38. Boonen G, Haberlein H. Influence of genuine kavalactone enantiomers on the GABA-A binding site. Planta Med 1998; 64(6):504-6.
39. Jussofie A, Schmiz A, Hiemke C. Kavapyrone enriched extract from Piper methysticum as modulator of the GABA binding site in different regions of rat brain. Psychopharmacology (Berl) 1994; 116(4):469-74.
40. Davies LP, Drew CA, Duffield P et al. Kava pyrones and resin: Studies on GABAA, GABAB and benzodiazepinebinding sites in rodent brain. Pharmacol Toxicol 1992; 71(2):120-6.
41. Friese J, Gleitz J. Kavain, dihydrokavain, and dihydromethysticin noncompetitively inhibit the specific binding of [3H]-batrachotoxinin-A 20-alpha-benzoate to receptor site 2 of voltage-gated Na+ channels. Planta Med 1998; 64(5):458-9.
42. Gleitz J, Friese J, Beile A et al. Anticonvulsive action of (+/-)-kavain estimated from its properties on stimulated synaptosomes and Na+ channel receptor sites. Eur J Pharmacol 1996; 315(1):89-97.
43. Ferger B, Boonen G, Häberlein H et al. In vivo microdialysis study of (+/-)-kavain on veratridine-induced glutamate release. Eur J Pharmacol 1998; 347(2-3):211-4.
44. Schirrmacher K, Busselberg D, Langosch JM et al. Effects of (+/-)-kavain on voltage-activated inward currents of dorsal root ganglion cells from neonatal rats. Eur Neuropsychopharmacol 1999; 9(1-2):171-6.
45. Keledjian J, Duffield PH, Jamieson DD et al. Uptake into mouse brain of four compounds present in the psychoactive beverage kava. J Pharm Sci 1988; 77(12):1003-6.
46. Seitz U, Schule A, Gleitz J. [3H]-monoamine uptake inhibition properties of kava pyrones. Planta Med 1997; 63(6):548-9.
47. Uebelhack R, Franke L, Schewe HJ. Inhibition of platelet MAO-B by kava pyrone-enriched extract from Piper methysticum Forster (kava-kava). Pharmacopsychiatry 1998; 31(5):187-92.
48. Boonen G, Ferger B, Kuschinsky K et al. In vivo effects of the kavalactones (+)-dihydromethysticin and (+/-)-kavain on dopamine, 3,4-dihydroxyphenylacetic acid, serotonin and 5-hydroxyindoleacetic acid levels in striatal and cortical brain regions. Planta Med 1998; 64(6):507-10.
49. Baum SS, Hill R, Rommelspacher H. Effect of kava extract and individual kavalactones on neurotransmitter levels in the nucleus accumbens of rats. Prog Neuropsychopharmacol Biol Psychiatry 1998; 22(7):1105-20.
50. Frey R. Demonstration of the central effects of D,L-kawain with EEG brain mapping. Fortschr Med 1991; 109(25):505-8.
51. Münte TF, Heinze HJ, Matzke M et al. Effects of oxazepam and an extract of kava roots (Piper methysticum) on event-related potentials in a word recognition task. Neuropsychobiology 1993; 27(1):46-53.
52. Heinze HJ, Münthe TF, Steitz J et al. Pharmacopsychological effects of oxazepam and kava-extract in a visual search paradigm assessed with event-related potentials. Pharmacopsychiatry 1994; 27(6):224-30.
53. Foo H, Lemon J. Acute effects of kava, alone or in combination with alcohol, on subjective measures of impairment and intoxication and on cognitive performance. Drug & Alcohol Review 1997; 16(2):147-155.

54. Cairney S, Maruff P, Clough AR et al. Saccade and cognitive impairment associated with kava intoxication. Hum Psychopharmacol 2003; 18(7):525-33.
55. Pittler MH, Ernst E. Efficacy of kava extract for treating anxiety: Systematic review and meta-analysis. J Clin Psychopharmacol 2000; (1):84-9.
56. Wheatley D. Kava and valerian in the treatment of stress-induced insomnia. Phytother Res 2001; 15(6):549-51.
57. Lehrl S. Clinical efficacy of kava extract WS 1490 in sleep disturbances associated with anxiety disorders. Results of a multicenter, randomized, placebo-controlled, double-blind clinical trial. J Affect Disord 2004; 78(2):101-10.
58. Almeida JC, Grimsley EW. Coma from the health food store: Interaction between kava and alprazolam. Ann Intern Med 1996; 125(11):940-1.
59. Stickel F, Baumuller HM, Seitz K et al. Hepatitis induced by Kava (Piper methysticum rhizoma). J Hepatol 2003; 39(1):62-7.
60. Gastpar M, Klimm HD. Treatment of anxiety, tension and restlessness states with Kava special extract WS 1490 in general practice: A randomized placebo-controlled double-blind multicenter trial. Phytomedicine 2003; 10(8):631-9.
61. Singh YN, Devkota AK. Aqueous kava extracts do not affect liver function tests in rats. Planta Med 2003; 69(6):496-9.
62. Clough AR, Bailie RS, Currie B. Liver function test abnormalities in users of aqueous kava extracts. J Toxicol Clin Toxicol 2003; 41(6):821-9.
63. Whitton PA, Lau A, Salisbury A et al. Kava lactones and the kava-kava controversy. Phytochemistry 2003; 64(3):673-9.
64. Norton SA, Ruze P. Kava dermopathy. J Am Acad Dermatol 1994; 31(1):89-97.
65. Schelosky L, Raffauf C, Jendroska K et al. Kava and dopamine antagonism. J Neurol Neurosurg Psychiatry 1995; 58(5):639-40.
66. Menghini A, Mancini LA. TLC determination of flavonoid accumulation in clonal populations of Passiflora incarnata L. Pharmacol Res Commun 1988; 20(Suppl 5):113-6.
67. Medina JH, Paladini AC, Wolfman C et al. Chrysin (5,7-di-OH-flavone), a naturally-occurring ligand for benzodiazepine receptors, with anticonvulsant properties. Biochem Pharmacol 1990; 40(10):2227-31.
68. Medina JH, Paladini AC, Wolfman C et al. Chrysin (5,7-di-OH-flavone), a naturally-occurring ligand for benzodiazepine receptors, with anticonvulsant properties. Biochem Pharmacol 1990; 40(10):2227-31.
69. Zanoli P, Avallone R, Baraldi M. Behavioral characterisation of the flavonoids apigenin and chrysin. Fitoterapia 2000; 71(Suppl 1):S117-23.
70. Goutman JD, Waxemberg MD, Donate-Oliver F et al. Flavonoid modulation of ionic currents mediated by GABA(A) and GABA(C) receptors. Eur J Pharmacol 2003; 461(2-3):79-87.
71. Wolfman C, Viola H, Paladini A et al. Possible anxiolytic effects of chrysin, a central benzodiazepine receptor ligand isolated from Passiflora coerulea. Pharmacol Biochem Behav 1994; 47(1):1-4.
72. Maluf E, Baros HMT, Frochtengarten ML et al. Assessment of the hypnotic/sedative effects and toxicity of Passiflora edulis aqueous extract in rodents and humans. Phytother Res 1991; 5(6):262-266.
73. Dhawan K, Dhawan S, Chhabra S. Attenuation of benzodiazepine dependence in mice by a tri-substituted benzoflavone moiety of Passiflora incarnata Linneaus: A nonhabit forming anxiolytic.
74. Akhondzadeh S, Naghavi HR, Vazirian M et al. Passionflower in the treatment of generalized anxiety: A pilot double-blind randomized controlled trial with oxazepam. J Clin Pharm Ther 2001; 26(5):363-7.
75. Salgueiro JB, Ardenghi P, Dias M et al. Anxiolytic natural and synthetic flavonoid ligands of the central benzodiazepine receptor have no effect on memory tasks in rats. Pharmacol Biochem Behav 1997; 58(4):887-91.
76. Medina JH, Viola H, Wolfman C et al. Overview—flavonoids: A new family of benzodiazepine receptor ligands. Neurochem Res 1997; 22(4):419-25.
77. Smith GW, Chalmers TM, Nuki G. Vasculitis associated with herbal preparation containing Passiflora extract [letter]. Br J Rheumatol 1993; 32(1):87-8.
78. Robbers JE, Speedie MK, Tyler VE. Pharmacognosy and Pharmacobiotechnology. Baltimore: Willams and Wilkins, 1996.
79. Viola H, Wasowski C, Levi de Stein M et al. Apigenin, a component of Matricaria recutita flowers, is a central benzodiazepine receptors-ligand with anxiolytic effects. Planta Med 1995; 61(3):213-6.
80. Osterhoudt KC, Lee SK, Callahan JM et al. Catnip and the alteration of human consciousness. Vet Hum Toxicol 1997; 39(6):373-5.
81. Hansel R, Wohlfart R, Coper H. Sedative-hypnotic compounds in the exhalation of hops, II. Z Naturforsch [C] 1980; 35(11-12):1096-7.
82. Hansel R, Wagener HH. Attempts to identify sedative-hypnotic active substances in hops. Arzneimittelforschung 1967; 17(1):79-81.
83. Wong AHC, Smith M, Boon H. Herbal remedies in psychiatrtic practice. Archives of General Psychiatry 1998; 55:1033-44.
84. Awad R, Arnason JT, Trudeau V et al. Phytochemical and biological analysis of skullcap (Scutellaria lateriflora L.): A medicinal plant with anxiolytic properties. Phytomedicine 2003; 10(8):640-9.
85. Wolfson P, Hoffmann DL. An investigation into the efficacy of Scutellaria lateriflora in healthy volunteers. Altern Ther Health Med 2003; 9(2):74-8.
86. Soulimani R, Fleurentin J, Mortier F et al. Neurotropic action of the hydroalcoholic extract of Melissa officinalis in the mouse. Planta Med 1991; 57(2):105-9
87. Kennedy DO, Wake G, Savelev S et al. Modulation of mood and cognitive performance following acute administration of single doses of Melissa officinalis (Lemon balm) with human CNS nicotinic and muscarinic receptor-binding properties. Neuropsychopharmacology 2003; 28(10):1871-81.
88. Gupta RK, Moller HJ. St. John's Wort. An option for the primary care treatment of depressive patients? Eur Arch Psychiatry Clin Neurosci 2003; 253(3):140-8.
89. Spinella M. The psychopharmacology of herbal medicines: Plant drugs that alter mind, brain, and behavior. MIT Press, 2001.
90. Butterweck V. Mechanism of action of St John's wort in depression: What is known? CNS Drugs 2003; 17(8):539-62.
91. Schellenberg R, Sauer S, Dimpfel W. Pharmacodynamic effects of two different hypericum extracts in healthy volunteers measured by quantitative EEG. Pharmacopsychiatry 1998; 31(Suppl 1):44-53.
92. Sharpley AL, McGavin CL, Whale R et al. Antidepressant-like effect of Hypericum perforatum (St John's wort) on the sleep polysomnogram. Psychopharmacology (Berl) 1998; 139(3):286-7.
93. Markowitz JS, Donovan JL, DeVane CL et al. Effect of St John's wort on drug metabolism by induction of cytochrome P450 3A4 enzyme. JAMA 2003; 290(11):1500-4.
94. Spinella M, Eaton LA. Hypomania induced by herbal and pharmaceutical psychotropic medicines following mild traumatic brain injury. Brain Inj 2002; 16(4):359-67.

INDEX

D

E

F

G

H

I

K

L

M

N

O

P

R

S

T

U

V

W

Z